W9-CCK-296

ADAPTED PHYSICAL EDUCATION AND SPORT

THIRD EDITION

Joseph P. Winnick, EdD
State University of New York, College at Brockport
Editor

Human Kinetics

Library of Congress Cataloging-in-Publication Data

Adapted physical education and sport / Joseph P. Winnick, editor.-- 3rd ed.
    p. cm.
  Includes bibliographical references and index.
  ISBN 0-7360-3324-6
    1. Physical education for handicapped persons. 2. Sports for the handicapped. I.
Winnick, Joseph P.

  GV445 .A3 2000
  371.9'04486--dc21

                                                     00-039636

ISBN: 0-7360-3324-6

Acquisitions Editor: Judy Patterson Wright, PhD; **Developmental Editor:** Katy M. Patterson; **Assistant Editor:** Amanda S. Ewing; **Copyeditor:** Arlene Miller; **Proofreader:** Erin Cler; **Indexer:** Craig Brown; **Permission Manager:** Cheri Banks; **Graphic Designer:** Fred Starbird; **Graphic Artist:** Amy J. Markstahler; **Cover Designer:** Jack W. Davis; **Photographer** (cover): Tom Roberts; **Art Manager:** Craig Newsom; **Illustrators:** Tom Janowski and Tom Roberts; **Printer:** IPC Communication Services

Printed in the United States of America   10 9 8 7 6 5 4 3 2

Human Kinetics
Web site: www.humankinetics.com

United States: Human Kinetics, P.O. Box 5076, Champaign, IL 61825-5076
800-747-4457
e-mail: humank@hkusa.com

Canada: Human Kinetics, 475 Devonshire Road, Unit 100, Windsor, ON N8Y 2L5
800-465-7301 (in Canada only)
e-mail: hkcan@mnsi.net

Europe: Human Kinetics, Units C2/C3 Wira Business Park, West Park Ring Road, Leeds LS16 6EB, United Kingdom
+44 (0) 113 278 1708
e-mail: hk@hkeurope.com

Australia: Human Kinetics, 57A Price Avenue, Lower Mitcham, South Australia 5062
08 8277 1555
e-mail: liahka@senet.com.au

New Zealand: Human Kinetics, P.O. Box 105-231, Auckland Central
09-523-3462
e-mail: hkp@ihug.co.nz

Contents

Preface

It is with a great deal of satisfaction and motivation that I write this preface for the third edition of *Adapted Physical Education and Sport*. When I wrote the preface for the first and second editions, I was confident that these books would benefit children with unique physical education needs by providing clear and concise information for teachers and others who provide quality services. In this third edition, the book is further developed and, at times, reorganized to meet today's trends in adapted physical education and sport. This preface identifies and explains major influences on the book's content and organization, briefly summarizes new and continuing features, provides an overview of the parts of the book, and closes with some comments about the value of adapted physical education and sport in the lives of youngsters today.

LEGISLATION: A MAJOR INFLUENCE ON THIS BOOK

There are many factors that shape the emphasis, approach, content, and organization of any book. This book is heavily influenced by original and current versions of two landmark laws: the Individuals with Disabilities Education Act (IDEA), which was originally signed in 1975 as PL 94-142; The Education for All Handicapped Children Act; and Section 504 of the Rehabilitation Act (originally passed in 1973 as PL 93-112).

One of the major purposes of IDEA is to ensure that all children with disabilities have available to them a free public education that emphasizes *special education* and *related services* to meet their unique needs. IDEA makes clear that the special education to be made available includes physical education, which in turn may be specified as adapted physical education. As will be clearly communicated in this book, regulations associated with IDEA define physical education as *the development of physical and motor fitness; fundamental motor skills and patterns; and skills in aquatics, dance, and individual and group games and sports (including intramural and lifetime sports)*. In this book, adapted physical education is defined in a manner consistent with key provisions of the IDEA definition. This helps schools and agencies in each state to develop and implement adapted physical education programs that are consistent with federal legislation.

The emphasis on the identification of *unique needs*, which is part of the IDEA definition of special education, is also emphasized in this book. One of the most important concepts associated with IDEA is the requirement of an *Individualized Education Program* (IEP) designed to meet the unique needs of children. This book includes numerous ways of adapting physical education to meet individual needs. 1We also provide detailed information in chapters 4 and 19 about the components of and strategies for developing individualized programs—both IEPs and Individualized Family Service Plans (IFSPs).

In IDEA, the disabilities of children ages 3 to 21 are identified and defined. Information related to physical education is presented in this book in regard to each of the specific disabilities identified in IDEA. IDEA also defines infants and toddlers (ages zero to two) with disabilities. One entire chapter on physical education services related to infants and toddlers is presented in the book.

Both IDEA and Section 504 require that education be provided in the most normal/integrated setting appropriate. In regard to physical education, this book encourages education in the regular environment to the extent appropriate. The book prepares readers for working in inclusive environments, with extensive discussions about inclusion in several chapters. However, it recognizes that education in the least restrictive environment is the "law of the land" and also prepares readers to provide services in more restrictive settings. In regard to sport, this book recommends an integration continuum that encourages participation in a variety of settings.

Therefore, this book responds to legislative requirements in many ways. These include compatibility with the definition of special education, the requirement of identifying unique educational needs of the individual, the requirement for individualized education programs, and the importance of education in the most normal, integrated setting possible. Because of the orientation used in this book, educators can be most confident that they are implementing programs that respond to societal needs that are expressed through legislation.

NEW FEATURES IN THIS EDITION

Since the second edition was written (1994–95), many changes impacting adapted physical education and sport have occurred. The authors of this book have recognized these changes and have made their chapters current. As editor, I have modified the text's orientation and organization to the extent necessary to bring it to the new millennium. One new feature in this edition is the attention given to inclusion. Our intent is to share ideas about providing services in any one of a variety of settings. The concept of inclusion is introduced in chapter 1 and general information regarding inclusion is presented in chapter 2. Each of the chapters dealing with children with disabilities now includes a section on inclusion, and several of the chapters focusing on activities also discuss and make applications relevant to inclusion.

Because individuals with disabilities are increasingly involved in sport, this book includes for the first time a separate chapter on adapted sport (chapter 3). Chapter 3 covers the most recent information available pertaining to classification, sport organizations, national governing bodies, the United States Olympic Committee, and Paralympic Games. The chapter includes information regarding school- and community-based adapted sports programs, transition services, and the role of the physical educator in adapted sport—all unique features in this text. Conceptually guiding this chapter is a sport integration continuum, which serves to stimulate thinking about sport opportunities in integrated as well as segregated sport settings.

In the past, most of the attention regarding adapted physical education has been on the 5 to 21 age range. In this third edition, separate chapters relating to infants and toddlers (ages zero to two), and preschoolers (ages three to five) have been developed. Although chapters discussing infants and toddlers were present in the second edition, those in this third edition are written by different authors who have built on previous writings, resulting in a dramatic change in the content of these chapters.

Although all chapters have been updated, major revisions are provided in several chapters. The chapter on Individualized Education Programs (chapter 4) has not only been updated in response to changes in federal legislation but includes information on the development of 504 Plans, which have been increasingly utilized in the past few years. The chapter on measurement and assessment (chapter 5) has been significantly changed. It contains updated information on standardized tests. Some of the new information added pertains to authentic assessment and the measurement of physical activity. The newly created Brockport Physical Fitness Test, a criterion-referenced health-related physical fitness test appropriate for youngsters with disabilities, is introduced in the chapter, and additional information on the test appears in appendix C. Chapter 21 on physical fitness has been dramatically changed in response to the increased attention being given to physical fitness and physical activity in society, the changes in orientation to these topics, and the new knowledge available. Finally, in response to societal interest in adventure sports and activities, this third edition gives greater attention to them.

A feature that has been expanded in this edition is the listing of resources at the end of each chapters' references. In this edition much more attention has been given to the identification of audiovisual and electronic resources. With the multitude of computer-assisted resources now available, the one concern was where to stop. Hopefully, the goal was reached of presenting the best and latest resources to expand and enrich topics covered in the body of the text.

This third edition features three new elements designed to enhance reader understanding: chapter outlines, opening chapter vignettes, and application examples. The chapter outlines consist of a brief listing of chapter headings, providing a preview of topics that will aid understanding and retention. The opening vignettes present "real-life" scenarios that introduce one or more chapter concepts to be discussed. Application examples provide readers with the opportunity to explore "real-life" situations and see how the concepts in the text can be applied to these situations to "solve" the issues at hand.

Six authors contributed to this book for the first time. They are Dr. Lauren Lieberman, writing the chapter on visual impairments and deafness with Dr. Diane Craft; Dr. Cathy Houston-Wilson, writing the chapter on infants and toddlers and coauthoring the early childhood adapted physical education chapter with Dr. Lauriece Zittel; Dr. Zittel coauthoring the early childhood adapted physical education chapter with Dr. Houston-Wilson; Dr. Monica Lepore, writing the chapter on aquatics; Dr. Michael Paciorek, writing the chapter on adapted sport; and Dr. Abu Yilla, writing the chapter on enhancing wheelchair sport performance. These six people were involved in writing five chapters, and their new perspectives result in significant changes and advances in these chapters.

CONTINUING FEATURES

In this and previous editions of this book, a good portion of the chapters focuses on general topics related to physical education and sport rather than merely covering each disability in turn. The latter approach has traditionally been associated with a medical model for the structuring of knowledge rather than a more educationally-oriented model. On the other hand, relevant information regarding disabilities is presented in the text so that relationships between disabilities and performance in physical education and sport are discussed.

This edition retains focus on physical education for youngsters within the ages of 0 to 21. No attempt is made to address the entire age span of persons with disabilities. The book is also delimited to the areas of physical education and sport and makes no claims to cover important but allied areas such as recreation or therapeutic recreation. This clear focus in regard to the scope of the text makes it manageable and easy to read.

An electronic instructor guide has again been developed to accompany this book. For each chapter, the instructor's guide provides objectives, suggestions for learning and enrichment activities, resources, and material that can be converted to transparencies. Because it is so important for college students to be aware of persons with disabilities and to teach and interact with them in a positive manner, the electronic instructor guide includes additional ideas to provide these opportunities. The guide also includes some ideas for an introductory course related to adapted physical education and sport as well as a sample course syllabus. Finally, an electronic bank of test questions has also been developed, which may be used to develop quizzes, exams, or study questions.

PARTS OF THE BOOK

The book is presented in four parts. Part I, Foundational Topics in Adapted Physical Education and Sport, encompasses chapters 1, through 7, which are designed to introduce the reader to adapted physical education and sport and cover general topics in the adapted physical education and sport subdiscipline. This includes chapters on adapted sport, individualized education programs, measurement and assessment, and behavior management.

Part II, Individuals With Unique Needs, includes nine chapters. The first eight, chapters 8 through 15, are devoted to disabilities associated with IDEA. Some of the disabilities discussed are mental retardation, learning disabilities, visual impairments, and spinal cord disabilities. These chapters provide an understanding of these disabilities, how they relate to physical education and sport, and educational implications associated with each disability covered. Particular attention is given in each chapter to inclusion and sport programs. Chapter 16 discusses youngsters with unique physical education needs without disabilities.

The third part of the book, Developmental Considerations, includes four chapters. Chapters 17 through 20 in this part cover motor development; perceptual-motor development; and adapted physical education for infants, toddlers, preschoolers, and children in early elementary programs.

Part IV, Activities for Individuals With Unique Needs, includes chapters 21 through 28. These chapters present physical education and sport activities for both school and out-of-school settings. A key aspect of this part is the presentation of specific activity modifications and variations for the various populations involved in adapted physical education and sport. This part concludes with a chapter on wheelchair sport performance. Part IV, in particular, will serve as an excellent resource for teachers, coaches, and other service providers long after they have left colleges and universities and are involved in providing quality programs for youngsters.

The appendixes consist of the latest definitions of infants, toddlers, and children with disabilities in IDEA; a list of disabled sport organizations; and information related to the Brockport Physical Fitness Test. Each of these complement information presented in the main body of the book.

CLOSING COMMENTS

As opportunities in adapted physical education and sport have increased, there has been a realization that individuals with disabilities are really individuals with abilities who are capable of much more than society has ever believed. With greater participation, the value of physical activity has been more clearly recognized and accepted. More youngsters with disabilities, parents, medical professionals, educators, and others recognize the tremendous value of physical education and sport today than ever before. This recognition and acceptance extend throughout the world, as clearly demonstrated at international symposia related to adapted physical activity and international competition in sport.

As the field of adapted physical education and sport has advanced, many recognizable subspecialties have emerged. Thus, I have assembled top people in their areas of expertise to serve as a team to write this book. The result is a book that reflects the very sharply focused thinking of a team of authorities who draw on many years of experience in adapted physical education and sport.

This book has been designed to be comprehensive, relevant, and user-friendly. It has been designed as both a resource and a text for adapted physical education and sport. As a resource, this book aids teachers, administrators, and other professionals as they plan and provide services. As a text it can be used to prepare students majoring in physical education, recreation, sport management, special education, and related disciplines. Use of this book and the study questions presented in the electronic test bank accompanying this book will help prepare professionals to pass the Adapted Physical Education National Standards (APENS) exam established by the National Consortium for Physical Education and Recreation for Individuals with Disabilities (NCPERID). Passing this exam leads to a nationally recognized certification in adapted physical education. Although the book can serve different purposes, its primary thrust is its emphasis on providing quality services. If it helps to provide quality services to individuals with unique physical education needs, my hope for it has been met.

Joseph P. Winnick

Acknowledgments

Diane H. Craft—I gratefully acknowledge the information and resources of learning disabilities contributed by C. Robin Boucher, PhD, Special Education Resource, Fairfax County, Virginia, Public Schools. My deepest thanks to Craig Smith for giving so freely of his time and experience.

David L. Gallahue—Thanks to Edwina Johansen for typing the manuscript and to my students and colleagues at Indiana University dedicated to quality physical education and sport experiences for all.

Cathy Houston-Wilson—I would like to thank Joe Winnick for giving me the opportunity to contribute to this text, to Laurie Zittel, my coauthor of the preschool chapter, for the collaborative efforts we shared, and to my little ones, Meaghan and Shannon, who continue to help me understand and appreciate early childhood development.

Luke E. Kelly—I would like to thank Joe Winnick for the opportunity to contribute to this book and for his inspiration and dedication to improving the quality of services provided in adapted physical education and sport. I would like to acknowledge my graduate students at the University of Virginia who have served as a sounding board for my ideas and as reviewers of my drafts. I would also like to thank my wife and children for their time and continuous support of my work.

Ellen M. Kowalski—I would like to thank Joe Winnick for inviting me to participate in writing this book. Special thanks to Phyllis Weikart, whose book *Teaching Movement and Dance* was the inspiration in my teaching and writing chapter 23. Finally, I would like to extend my appreciation to my students and colleagues, from whom I continue to learn.

Patricia L. Krebs—Appreciation is extended to Joseph P. Winnick for writing the section on cognitive development and part of the section on planned programs in chapter 8.

E. Michael Loovis—I would like to thank Pam Maryjanowski of the Empire State Games for the Phyiscally Challenged; Pam Abbotts of Recreation Network, Inc. in Australia; and Bethany Siegler from Sunrise Medical, Inc. in Longmont, Colorado for their assistance in acquiring photographs used in chapter 26.

Monica Lepore—I would like to thank Dr. Willie Gayle and Dr. Shawn Stevens for allowing me to use some of the material from our adapted aquatics programming textbook for chapter 24. I would also like to thank my daughter, Maria, for her support and love.

Laureen Lieberman—I would like to acknowledge Dr. Jim Mastro, Bimidji State University, for all his help in writing my chapter.

Michael Paciorek—I wish to thank Joe Winnick for inviting me to be a part of a wonderful team of professionals and to add a new and unique chapter to this book. Much appreciation is extended to my good friend and colleague Jeff Jones for his critical reviews and insight into disability sport. Special recognition and thanks is given to my wonderful wife Karen, and to my sons Clark and Clay who are always there for me.

David Porretta—Recognition goes to the following people and organizations for the preparation of photos used in chapters 12, 13, and 25: Rick Swauger and the Columbus Pioneers, Jerry McCole and the United States Cerebral Palsy Athletic Association, Flex-Foot, Inc., and Quad Rugby Today. Also, a special thanks to Joseph Winnick for inviting me to participate in this and previous editions of the book.

Sarah M. Rich—I thank Dr. Joseph P. Winnick for his leadership and dedication to this project.

Francis X. Short—Appreciation is extended to Joe Winnick for keeping me "current"; to J.J. Brewer, Kevin Head, Jim Dusen, and Sommer Tiller for helping me with the photographs; and to Judy Wright, Katy Patterson, and the rest of the folks at Human Kinetics for their hard work on the text.

Paul R. Surburg—Both chapters 15 and 16 are dedicated to Dr. Luther Schwich, who was instrumental in guiding me into the field of physical education and who, until his death from cancer, provided me with insights into the role of physical activity for persons with cancer.

Joseph P. Winnick—I wish to acknowledge the wonderful support and cooperation I have received from the outstanding authors involved in this edition. I very much appreciate the help and support I have received from many persons at the State University of New York, College at Brockport.

Abu B. Yilla—This work extends the work of Dr. Colin Higgs from the second edition of this book. Dr. Higg's permission and contributions to this chapter are greatly appreciated. The support of the Dallas Wheelchair Mavericks basketball team and the Department of Kinesiology, University of Texas at Arlington are also appreciated.

Lauriece L. Zittel—I would like to thank Joe Winnick for the opportunity to share thoughts and ideas. Chapter 20 was inspired by observing the dedication of many early childhood specialists. And to all the "little ones," your desire to grow and learn is what prompts our commitment to the area of early education.

Foundational Topics in Adapted Physical Education and Sport

Part I of this book, consisting of seven chapters, introduces adapted physical education and sport and presents topics that serve as a foundation for the book. Chapter 1 defines adapted physical education and sport and offers a brief orientation concerning its history, legal basis, and professional resources. Programmatic direction, inclusion, and qualities of service providers are also introduced. In chapter 2, the focus shifts to the organization and management of programs. Topics include programmatic and curricular direction in adapted physical education and guidelines for administrative procedures and program implementation. Program evaluation and human resources associated with adapted physical education and sport programs are also covered. Chapter 3 emphasizes information pertain-

ing to adapted sport. Following a brief introduction, the chapter covers the status and issues associated with adapted sport from the Paralympic Games to local school and community programs. Chapter 4 contains a detailed discussion of individualized education programs developed for students with unique needs, including Section 504 Accommodation Plans. Vital to the establishment of these programs are several concepts related to measurement and evaluation. Chapter 5 discusses measurement and assessment strategies and recommends specific tests for adapted physical education. Chapter 6 deals with instructional strategies related to adapted physical education. Finally, chapter 7 covers basic concepts and approaches related to methods of managing behavior.

An Introduction to Adapted Physical Education and Sport

Joseph P. Winnick

▷

▷

▷

HISTORY TELLS US THAT PEOPLE with disabilities can achieve goals many would think impossible. For example, Wilma Rudolph—despite birth defects and polio—was a triple gold medalist in the 100-meter, 200-meter, and 400-meter relays in the 1960 Olympics in Rome. Peter Gray, whose right arm was amputated, played center field for the St. Louis Browns in 1945. Other athletes with unique needs include Harry Cordellos, a sightless distance runner who ran the 1975 Boston Marathon with a sighted partner in 2 hours, 57 minutes, and 42 seconds; Tom Dempsey, born with only half of a right foot, who set a National Football League record in 1970 for the longest field goal kicked (63 yards); and Casey Martin, with a serious impairment of his leg associated with the Klippel-Trenaunay-Weber Syndrome, attained fame as a gifted golfer. Those with relatively low abilities tend not to be celebrities; however, their needs may also be unique.

Individuals who pursue a career of teaching physical education and coaching sports typically enjoy physical activity and are active participants in physical education and athletics. Often, however, they do not become knowledgeable about adapted physical education and sport until they prepare for their careers. With increased awareness, they realize that people with unique needs may exhibit abilities ranging from very low to extremely high, as the opening examples illustrate. Students begin to appreciate that there are individuals with a variety of unique needs involved with adapted physical education and sport. They learn that those with unique needs include people both with and without disabilities.

If physical education and sport opportunities are offered in educational institutions and other societal entities, it is necessary that they be made available to all students, including those with disabilities. In America it is not desirable or permissible to discriminate on the basis of disability in regard to these opportunities. Adapted physical education and sport is a field that has evolved to meet the unique physical education and sport needs of individuals. This chapter introduces the reader to adapted physical education and sport.

THE MEANING OF ADAPTED PHYSICAL EDUCATION

Because different terms and different definitions for the same term have been applied to physical education that meets unique needs, it is necessary to clarify the definition of adapted physical education. Adapted physical education is an individualized program of physical and motor fitness; fundamental motor skills and patterns; and skills in aquatics, dance, and individual and group games and sports designed to meet the unique needs of individuals. Typically, the word "adapt" means "to adjust" or "to fit." In this book, the word "adapt" is consistent with these definitions and includes the modification of objectives, activities, and methods to meet unique needs. It encompasses traditional components associated

with adapted physical education, including those designed to correct, habilitate, or remediate. Adapted physical education is a subdiscipline of physical education that allows for safe, personally satisfying, and successful participation to meet the unique needs of students.

Adapted physical education is generally designed to meet long-term (more than 30 days) unique needs. Those with long-term unique needs include persons with disabilities as specified in the Individuals with Disabilities Education Act (IDEA). According to IDEA a child with a disability means a child having mental retardation; hearing impairments including deafness, speech or language impairments, visual impairments including blindness, serious emotional disturbance, orthopedic impairments, autism, traumatic brain injury, and other health impairments; specific learning disabilities; deafblindness or multiple disabilities; and who, for this reason, needs special education and related services (Office of Special Education and Rehabilitative Services [OSE/RS], 1998). The term *child with a disability* for a child ages three to nine may, at the discretion of the state and the local educational agency, include a child experiencing developmental delays, as defined by the state and as measured by appropriate diagnostic instruments and procedures in one or more of the following areas: physical development, cognitive development, communication development, adaptive development, or social or emotional development; and who, by reason thereof, needs special education and related services (OSE/RS, 1998). Adapted physical education may also include infants and toddlers (individuals under three years of age) who need early intervention services, because they are experiencing developmental delays in one or more of the following areas: cognitive development, physical development, communication development, social or emotional development, or adaptive development; or have a diagnosed physical or mental condition that has a high probability of resulting in developmental delay; and may also include at a state's discretion, at-risk infants and toddlers. (IDEA, 1997). The term at-risk infant or toddler means an individual under three years of age who would be at risk of experiencing a substantial developmental delay if early intervention services were not provided to the individual (IDEA, 1997).

Adapted physical education may include persons with disabilities as encompassed within Section 504 of the Rehabilitation Act of 1973 and its amendments. Section 504 defines a person with a disability as any person who has a physical or mental impairment, which substantially limits one or more major life activities, has a record of such an impairment, or is regarded as having such an impairment. Although every child who is a student with a disability under IDEA is also protected under 504, all children covered under 504 are not necessarily students with a disability under IDEA. Individuals with disabilities who do not need or require services under IDEA are, nonetheless, entitled to accommodations and services that are necessary to enable them to benefit from all programs and activities available to nondisabled students.

Adapted physical education may include pupils who *are not* identified by a school district as disabled under federal legislation, but who may have unique needs that call for a specially designed program. This group might include students restricted because of injuries or other medical conditions; those of low fitness (including exceptional leanness or obesity), inadequate motor development, low skill, or those persons with poor functional posture. These individuals may require individually designed programming to meet unique goals and objectives.

According to IDEA, students ages 3 to 21 with disabilities must have an Individualized Education Program (IEP) developed by a planning committee. In developing an IEP, physical education needs must be considered, and the IEP developed may include specially designed instruction in physical education. IDEA (1997) also requires the development of an Individualized Family Service Plan (IFSP) for infants and toddlers with disabilities. Although physical education services are not mandated for this age group, they may be offered as part of an IFSP. In accordance with Section 504 of the Rehabilitation Act of 1973 and its amendments, it is recommended that an accommodation plan be developed by a school-based assessment team to provide services and needed accommodations for individuals with disabilities. Although not required by federal law, an Individualized Physical Education Program (IPEP) should also be developed by a planning committee for those who have a unique need, but who have not been identified by the school as having a disability. It is recommended that each school have policies and procedures to guide the development of all individualized programs. More specific information on the development of programs and plans is presented in chapters 2 and 4 (ages 3 to 21) and chapter 19 (ages 0 to 2).

Consistent with the least restrictive environment concept associated with IDEA, adapted physical education may take place in classes that range from integrated (i.e., regular education environments) to segregated (i.e., including only persons receiving adapted physical education). Although adapted physical education is a *program* rather than a *placement*, it is critical to realize that a program received is directly influenced by placement (the setting in which it is implemented). Whenever appropriate, students receiving an adapted physical education program should be included in regular physical education environments. Although an adapted physical education program is individualized, it can be implemented in a group setting and should be geared to each student's needs, limitations, and abilities.

Adapted physical education should emphasize an *active* program of physical activity (figure 1.1, a-b) rather than a *sedentary* alternative program. It should be planned to attain the benefits of physical activity by meeting the needs of students who might otherwise be relegated to passive experiences associated with physical education. In establishing adapted physical education programs, educators work with parents, students, teachers, administrators, and professionals in various disciplines. Adapted physical education may employ developmental (bottom-up), community-based (top-down), or other orientations and may use a variety of teaching styles. It takes place in schools and other agencies responsible for educating individuals. Although adapted physical education is *educational*, it draws on *related* services (more on related services later in this chapter), especially medically related services, to help meet instructional objectives and goals.

In this text, adapted physical education and sport are viewed as part of the emerging area of study referred to as adapted physical activity, a term that encompasses the comprehensive and interdisciplinary study of physical activity for the education, wellness, sport participation, and leisure of individuals with unique needs. Adapted physical activity encompasses the total lifespan, whereas adapted physical education focuses only on the ages of 0 to 21. Although adapted physical education may exceed the minimal time required by policies or law, it should not be supplanted by related services, intramurals, sports days, athletics, or other experiences that are not primarily instructional.

ADAPTED SPORT

Adapted sport refers to sport modified or created to meet the unique needs of individuals with disabilities. Adapted sport may be conducted in integrated settings in which individuals with disabilities interact with nondisabled participants or in segregated environments that include only those persons with disabilities (see

Figure 1.1   (a) A student with a lower-limb impairment moves briskly around a track, and (b) a student with a visual impairment confidently conquers a climbing wall.

the sport continuum presented in figure 3.1 on page 35). Based on this definition, for example, basketball is a regular sport and wheelchair basketball would be considered an adapted sport. Goal ball (a game created for persons with visual impairments in which players attempt to roll a ball that emits a sound across their opponents' goal) is adapted sport because it was created to meet unique needs.

Adapted sport encompasses "disability sport"—Paralympics, Deaf Sport, and others—which typically focuses on segregated participation in regular and/or adapted sport. The term "adapted sport" is preferred to "disability sport," in part, because it stimulates and encourages participation and excellence in a variety of settings. This orientation to adapted sport is consistent with the sport integration continuum presented in figure 3.1. It is believed that this model leads to more participation in sports by persons with disabilities and will lead to more creative offerings and grouping patterns related to sports at every level of participation. Adapted sport terminology supports the development of excellence in sport while promoting growth in sport participation within a variety of settings.

Adapted sport programs are conducted in diverse environments and organizational patterns for a variety of purposes. Educational programs are generally conducted in schools and include intramural, extramural, and interschool activities. Intramural activities are conducted within schools, involve pupils from two or more schools, and are sometimes conducted as play days or sport days at the end of instructional or intramural sessions. Interschool sports involve competition between representatives from two or more schools and offer enriched opportunities for selected and more highly skilled individuals. Adapted sport activity may also be conducted for leisure or recreational purposes within formal, open, or unstructured programs or as a part of the lifestyle of individuals and/or groups. Adapted sport activity may also be conducted for wellness, medical, or therapeutic reasons. For example, sport or adapted sport may be used as part of recreational therapy, corrective therapy, sport therapy, or wellness programs. In general, involvement in sport or adapted sport has several purposes. This book will focus on adapted sport in educational settings and in regional, national, and international competition under the governance of formalized organizations promoting sport for individuals with disabilities.

PURPOSES, AIMS, GOALS, AND OBJECTIVES

The purpose of an adapted physical education and/or sport program is to provide a sense of direction. The purpose of a specific program should be consistent with the mission or purpose of the organization within which it is associated and with the regular or general physical education or sport program available for persons without disabilities. It is important that purposes are developmentally appropriate. For this book, it will be assumed that the purpose of adapted physical education and sport is to enhance self-actualization. This enhances optimal personal development and provides benefits to society. This purpose is consistent with the humanistic philosophy interpreted by Sherrill (1998). Sherrill says that humanism is a philosophy that pertains to helping people become fully human, thereby actualizing their potential for making the world the best possible place for all forms of life.

There is no universal model or paradigm related to purposes, aims, or goal areas in adapted physical education and sport. The framework presented in figure 1.2 is offered as a skeletal reference for educational programs, which is consistent with federal legislation and the orientation used in this book. Figure 1.2 encompasses the statement of purpose presented as well as a program's aims and program and content goals.

In this orientation, the physical education and sport program aims to produce physically educated individuals who live active and healthy lifestyles, which enhance their progress toward self-actualization. The development of a physically educated individual is accomplished by maximizing the integrated cognitive, psychomotor, and affective domains of learning, which serve as goals of a physical education and sport program. Program goals are accomplished by development *of* and *through* the psychomotor domain. In figure 1.2 education of the psychomotor domain is represented by solid lines connecting content and program goals. Development through the psychomotor domain is represented by dotted lines among cognitive, affective, and psychomotor development areas.

Program goals are developed through content areas in the physical education and sport program. Content areas related to psychomotor development may be grouped in many ways. Figure 1.2 shows three general areas: physical and motor fitness; fundamental motor skills and patterns; and skills in aquatics, dance, and individual and group games and sports. These content areas are consistent with the definition of adapted physical education used in this book and the definition of physical education associated with IDEA. Each of these content areas includes specific developmental areas or sport skills. For example aerobic capacity may be a developmental area under physical fitness, and basketball is a sport within the content area of group games and sports.

The content goals shown in figure 1.2, as well as goals in affective and cognitive domains, may serve as annual goals for individualized programs. Specific skills and developmental areas associated with these goals may be used to represent short-term objectives. For example, an annual goal for a student might be to improve physical fitness. A corresponding specific short-term objective might be to improve health-related flexibility by obtaining a score of 25 centimeters on a sit-and-reach test. Objectives can be expressed on several levels to reflect the specificity desired. The emphasis given to the three developmental areas in a program for an individual should be based on the student's needs. An adapted physical education program is established to meet objectives unique for each individual.

SERVICE PROVIDERS

Individuals providing direct service are the key to ensuring quality experiences related to adapted physical education and sport. These individuals include teachers, coaches, therapists, volunteers, and others. In regard to teachers, it is important to emphasize that

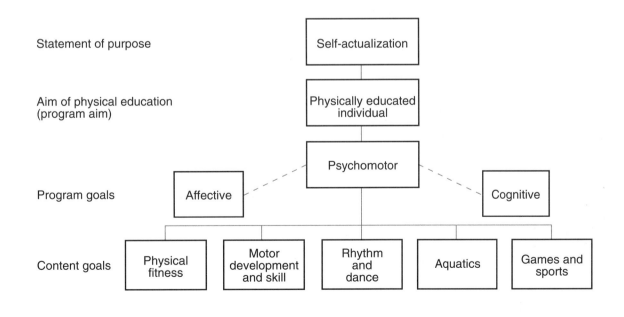

Figure 1.2   Aims and goals for an adapted physical education and sport program.

adapted physical education and sport must be provided not only by persons who specialize in this field but by regular physical educators. If services were only provided by "specialists" in adapted physical education, relatively few individuals needing services would be getting services, because there are too few "specialists" in adapted physical education.

To meet the needs of children in adapted physical education and sport, it is important that teachers of physical education assume responsibility for all children they teach. Each of these teachers must be willing to contribute to the development of *each* individual. This requires a philosophy that looks beyond "won and lost" records as the ultimate contribution in one's professional life. It requires an individual who has appropriate professional knowledge, skills, values, and a caring and helping attitude. A good teacher and/or coach of children places the development of positive self-esteem as a priority and displays an attitude of acceptance, empathy, friendship, and warmth, while ensuring a secure and controlled environment. The good teacher or coach of adapted physical education and sport selects and uses teaching approaches and styles beneficial to students, provides individualized and personalized instruction and opportunities, and creates a positive environment where students can succeed. The good teacher uses a praising and encouraging approach and creates a positive educational environment in which all students are accepted.

Frequently individuals studying to become teachers have little or no experience working with individuals with disabilities and/or those with unique physical education needs. It is important to take advantage of every opportunity to interact with persons with disabilities, to describe the value of physical activity to them, and to listen to their stories about their experiences in adapted physical education and sport. Being involved in disability awareness activities and having an opportunity to assume and function as if one has a disability provide important insights and values to prospective teachers.

BRIEF HISTORY OF ADAPTED PHYSICAL EDUCATION

Although significant progress concerning educational services for persons with disabilities is relatively recent, the use of physical activity or exercise for medical treatment and therapy is not new. Therapeutic exercise can be traced to 3000 B.C. in China. It is known that the ancient Greeks and Romans also recognized the medical and therapeutic value of exercise. However, the idea of physical education or physical activity to meet the unique needs of persons with disabilities is a recent phenomenon. Efforts to serve these populations through physical education and sport have been given significant attention only during the 20th century, although these efforts began in the United States in the 19th century.

Beginning Period

In 1838, physical activity began receiving special attention at the Perkins School for pupils with visual disabilities in Boston. According to Charles E. Buell (1983), a noted physical educator with a visual impairment, this special attention resulted from the fact that Samuel Gridley Howe, the school's director, advocated the health benefits of physical activity. For the first eight years, physical education consisted of compulsory recreation in the open air. In 1840, when the school was moved to South Boston, boys participated in gymnastic exercises and swimming. This was the first physical education program in the United States for students who are blind, and, by Buell's account, it was far ahead of most of the physical education programs in public schools. Buell, a leader in providing physical education for persons who are blind, is the author of several books and articles.

Medical Orientation

Although physical education was provided to blind people, as well as individuals with other disabilities, in the early 1800s, most students of the history of adapted physical education recognize that medically oriented gymnastics and drills began in the latter part of the century as the forerunner of modern adapted physical education in the United States. Sherrill (1998) states that physical education prior to 1900 was medically oriented and preventive, developmental, or corrective in nature. Its purpose was to prevent illness and/or to promote the health and vigor of the mind and body. Strongly influencing this orientation was a system of medical gymnastics developed in Sweden by Per Henrick Ling and introduced to the United States in 1884.

The Shift to Sports and the Whole Person

From the end of the 19th century into the 1930s, programs began to shift from medically oriented physical training to sports-centered physical education, and concern for the whole child emerged. Compulsory physical education in public schools increased dramatically, and physical education teacher training developed (rather than medical training) for the promotion of physical education (Sherrill, 1998). This transition resulted in broad mandatory programs consisting of games, sports, rhythmic activities, and calisthenics designed to meet the needs of the whole person. Individuals unable to participate in regular activities were provided corrective or remedial physical education. According to Sherrill, physical education programs between the 1930s and the 1950s consisted of regular or corrective classes for students who today would be considered normal. Sherrill has succinctly described adapted physical education during this period:

Assignment to physical education was based on a thorough medical examination by a physician who determined whether a student should participate in the regular or corrective program. Corrective classes were comprised primarily of limited, restricted, or modified activities related to health, posture, or fitness problems. In many schools students were excused from physical education. In others, the physical educator typically taught several sections of regular physical education each day. Leaders in corrective physical education continued to have strong backgrounds in medicine and/or physical therapy. Persons preparing to be physical education teachers generally completed one university course in corrective physical education. (pp. 12-13)

The Emerging Comprehensive Subdiscipline

During the 1950s, more and more pupils described as handicapped were being served in public schools, and the outlook toward them was becoming increasingly humanistic. With a greater diversity in pupils came a greater diversity in programs to meet their needs. In 1952, the American Association for Health, Physical Education and Recreation (AAHPER) formed a committee to define the subdiscipline and give direction and guidance to professionals. This committee defined adapted physical education as "a diversified program of developmental activities, games, sports, and rhythms suited to the interests, capacities, and limitations of students with disabilities who may not safely or successfully engage in unrestricted participation in the rigorous activities of the general physical education program" (Committee on Adapted Physical Education, 1952). The definition retained the evolving diversity of physical education and specifically included students with disabilities. Adapted physical education serves today as the comprehensive term for this subdiscipline.

Recent and Current Status

With the impetus provided by a more humanistic, more informed, and less discriminatory society, major advances continued in the 1960s. Many of these advances were associated with the Joseph P. Kennedy family. In 1965, the Joseph P. Kennedy, Jr. Foundation awarded a grant to the American Alliance for Health, Physical Education, Recreation and Dance (AAHPERD) to launch the Project on Recreation and Fitness for the Mentally Retarded. The project grew to encompass all special populations, and its name was changed in 1968 to the Unit on Programs for the Handicapped. Dr. Julian Stein dramatically influenced adapted physical education throughout the United States at every level as director of the unit (figure 1.3).

Figure 1.3   Julian "Buddy" Stein. Dr. Stein has provided sustained leadership in the field of adapted physical education and sport.

In 1968, the Kennedy Foundation exhibited further concern for individuals with mental retardation with the establishment of the Special Olympics. This program grew rapidly, with competition held at local, state, national, and international levels in an ever-increasing range of sports. During the mid-1960s, concern for people with emotional and/or learning disabilities had a significant effect on adapted physical education in the United States. The importance of physical activity for the well-being of those with emotional problems was explicitly recognized by the National Institute of Mental Health, U.S. Public Health Service, when it funded the Buttonwood Farms Project. The project, conducted at Buttonwood Farms, Pennsylvania, included a physical recreation component. This project was valuable for recognizing the importance of physical activity in the lives of persons with disabilities, bringing the problems of seriously disturbed youngsters to the attention of educators, and developing curricular materials to prepare professionals in physical education and recreation for work with this population.

During the same era, adapted physical education gained much attention with the use of perceptual-motor activities as a basis or modality for academic and/or intellectual development, particularly for students with learning disabilities. Newell C. Kephart, Gerald N. Getman, Raymond H. Barsch, Marianne Frostig, Phyllis Maslow, Bryant J. Cratty, and Jean A. Ayres were major proponents of motoric experiences as a basis or modality for perceptual, academic, and/or motor development. This influence, particularly strong in the 1970s, continues with much less emphasis or support today.

Contemporary direction and emphasis in adapted physical education are heavily associated with the individual's right to a free and appropriate education. Because of litigation and the passage of various federal laws and regulations, change and progress have occurred in both adapted physical education and sport. This legal impetus has improved programs in many schools and agencies, extended mandated physical education for persons ages 3 to 21, stimulated activity programs for infants and toddlers, and resulted in dramatic increases in participation in sports programs for individuals with disabilities. Legislation has also resulted in funds for professional preparation, research, and other special projects relevant to the provision of full educational opportunity for persons with unique needs. Finally, the impact of federal legislation and a strong belief in the right and value of an education in the regular educational environment have resulted in a significant movement toward *inclusion* regarding the education of children with disabilities in the United States. The elements of contemporary direction and emphasis mentioned here are covered in greater detail in several parts of this book.

THE INCLUSION MOVEMENT

Inclusion means educating students with disabilities in a regular educational setting. The movement toward inclusion was encouraged by and is compatible with the least restrictive environment (LRE) provisions associated with IDEA. Education in the LRE requires, to the maximum extent appropriate, that children with disabilities are educated with children who are nondisabled. However, a continuum of alternative environments (including segregated environments) may be used for the education of an individual if that is the *most* appropriate environment, according to LRE provisions in IDEA. A recommended continuum is presented in figure 2.1 on page 21. The inclusion movement has also been given impetus by many who believe that separate education is not an equal education and that the setting in which a program is implemented significantly influences the program provided for a child. In the 20th annual report to Congress, it was reported that 95 percent of students with disabilities, ages 6 to 21, attended schools with their nondisabled peers (U.S. Department of Education, 1998). This is the reality of inclusion and it requires appropriately prepared educators. Although this book prepares teachers to serve children in all settings, it gives special emphasis to the skills and knowledge needed to optimally educate children in regular educational environments.

LITIGATION

Much has and can be written about the impact of litigation on the guarantee of full educational opportunity in the United States. The most prominent of cases, which has served as an important precedent for civil litigation, was *Brown v. Board of Education of Topeka, Kansas* (1954). This case established that the doctrine of *separate but equal* in public education resulted in segregation that violated the constitutional rights of black persons. Two landmark cases also had a heavy impact on the provision of free, appropriate public education for all handicapped children. The first was the class action suit of the *Pennsylvania Association for Retarded Children v. Commonwealth of Pennsylvania* (1972). Equal protection and due process clauses associated with the Fifth and Fourteenth Amendments served as the constitutional basis for the court's rulings and agreements. The following were among the rulings or agreements in the case:

▸ Labeling a child as mentally retarded or denying public education or placement in a regular setting without due process or a hearing violate the rights of the individual.

▸ All mentally retarded persons are capable of benefiting from a program of education and training.

▸ Mental age may not be used to postpone or in any way deny access to a free public program of education and training.

▸ Having undertaken to provide a free, appropriate education to all its children, a state may not deny mentally retarded children the same.

A second important case was *Mills v. Board of Education of the District of Columbia* (1972). This action, brought on behalf of seven children, sought to restrain the District of Columbia from excluding children from public schools or denying them publicly supported education. The district court held that, by failing to provide the seven children with handicapping conditions and the class they represented with publicly supported specialized education, the district violated controlling statutes, its own regulations, and due process. The District of Columbia was required to provide a publicly supported education, appropriate equitable funding, and procedural due process rights to the seven children.

From 1972 to 1975, 46 right-to-education cases related to persons with disabilities were tried in 28 states. They provided the foundation for much of the legislation discussed in the next section.

LAWS IMPORTANT TO ADAPTED PHYSICAL EDUCATION AND SPORT

Laws have had a tremendous influence on education programs for individuals with disabilities. Since 1969, colleges and universities in many states have received federal funds for professional preparation, research, and other projects to enhance programs for individuals with disabilities. Although the amount of money has been relatively small, physical educators have gained a great deal from that support. The government agency that has been most responsible for administering federally funded programs related to adapted physical education is the Office of Special Education and Rehabilitative Services within the Department of Education.

Four laws or parts of laws and their amendments have impacted greatly on adapted physical education and/or adapted sport: IDEA, Section 504 of the Rehabilitation Act of 1973, the Olympic and Amateur Sports Act, and the Americans with Disabilities Act. A timeline depicting important milestones and a brief statement of the importance of these laws are presented in table 1.1.

IDEA

A continuing major impetus related to the provision of educational services for students with disabilities is IDEA (Public Law 105-17 Amendments of 1997). Definitions associated with this law can be found in appendix A. This act expanded on the previous Education for the Handicapped Act and amendments. However, IDEA reflects the composite and the most recent version and amendments of these laws (see table 1.1). This act was designed to ensure that all children with disabilities have available to them a free appropriate public education that emphasizes special education and related services designed to meet their unique needs and prepares them for employment and independent living (box 1.1). In this legislation, the term special education is defined to mean specially designed instruction at no cost to parents or guardians to meet the unique needs of a child with a disability, including instruction conducted in the classroom, in the home, in hospitals and institutions, and in other settings; and instruction in physical education (IDEA, 1997). IDEA specifies that the term related services means transportation and any developmental, corrective, and other supportive services that are required to assist a child with a disability to benefit from special education. These services include speech-language pathology and audiology services, psychological services, physical and occupational therapy, recreation including therapeutic recreation, early identification and assessment of disabilities in children, counseling services including rehabilitation counseling, orientation and mobility services, and medical services for diagnostic and evaluation pur-

poses. Related services also include school health services, social work services in schools, and parent counseling and training (OSE/RS, 1998). The act also ensures that the rights of children with disabilities and their parents or guardians are protected, and it helps states and localities provide education for all individuals with disabilities. IDEA has also established a policy to develop and implement a program of early intervention services for infants and toddlers and their families.

Definition and Requirements of Physical Education in IDEA

Regulations associated with IDEA (OSE/RS, 1998) define physical education as the "development of (a) physical and motor fitness, (b) fundamental motor skills and patterns, and (c) skills in aquatics, dance, and individual and group games and sports (including intramural and lifetime sports)." This term includes special physical education, adapted physical education, movement education, and motor development. IDEA requires that special education, including physical education, be made available to children with disabilities and that it include physical education specially designed, if necessary, to meet their unique needs. This federal legislation, together with state requirements for physical education, significantly affects adapted physical education in schools. Readers should notice that the definition of adapted physical education used in this book closely parallels the definition of physical education in IDEA.

> **Box 1.1 Highlights of the Individuals With Disabilities Education Act (IDEA)**

IDEA and its rules and regulations require:

- A right to a free and appropriate education
- Physical education be made available to children with disabilities
- Equal opportunity for nonacademic and extracurricular services and activities
- An individualized program designed to meet unique needs for children with disabilities
- Programs conducted in the least restrictive environment
- Nondiscriminatory testing and objective criteria for placement
- Due process
- Related services to assist in special education

Table 1.1 Legislative Timeline: 1973-1998

Law	Date	Importance
PL 93-112. The Rehabilitation Act of 1973	1973	Section 504 of this act was designed to prevent discrimination against and provide equal opportunity for individuals with disabilities in programs or activities receiving federal financial assistance.
PL 94-142. The Education for all Handicapped Children Act of 1975	1975	This act was designed to ensure that all children with handicapping conditions have available to them a free appropriate public education that emphasizes special education (including physical education) and related services designed to meet their unique needs.
PL 95-606. The Amateur Sports Act of 1978	1978	This act was passed to coordinate national efforts concerning amateur activity, including activity associated with the Olympic Games. This legislation led to the establishment of the Committee on Sports for the Disabled (COSD) as a standing committee of the United States Olympic Committee (USOC). The COSD coordinates American efforts for the Paralympics.
PL 98-199. Amendments to the Education for All Handicapped Children Act	1983	This act provided incentives to states to provide services to infants, toddlers, and preschoolers with handicapping conditions.
PL 99-457. Education for All Handicapped Children Amendments of 1986	1986	This act expanded educational services to preschool children ages three to five and established a discretionary program to assist states to plan, develop, and implement a comprehensive, coordinated, interdisciplinary program of early intervention services for infants and toddlers with handicapping conditions, birth to age three (or zero to two).
PL 101-476. Individuals with Disabilities Education Act (IDEA)	1990	Replaced the term "handicapped" with "disabilities"; expanded on types of services offered and disabilities covered.
PL 101-336. Americans with Disabilities Act	1990	Extended civil rights protection for individuals with disabilities to all areas of American life.
PL 105-17. Individuals with Disabilities Education Act Amendments of 1997	1997	Provided several changes in the law including provisions for free appropriate education to all children with disabilities (ages 3 to 21); extension of a "developmental delay" provision for children ages 3 to 9; emphasis on educational results; required progress reports for children with disabilities that are the same as those for nondisabled children; and changes in Individualized Education Program (IEP) requirements.
PL 105-277. Olympic and Amateur Sports Act in the Omnibus Appropriations Bill	1998	Continues services associated with the Amateur Sports Act of 1978. As a result of this legislation, the United States Olympic Committee assumed the role and responsibilities of the United States Paralympic Committee.

Free Appropriate Public Education Under IDEA

The term free appropriate public education means special education and related services that are provided at public expense, under public supervision and direction, and without charge; meet the standards of the state educational agency; include an appropriate preschool, elementary, or secondary school education in the state involved; and are provided in conformity with an IEP (OSE/RS, 1998).

Least Restrictive Environment

IDEA requires that education be conducted in the least restrictive environment. See the Student Placement application example exploring the best possible setting for a child with behavioral problems. Education in the least restrictive environment means that individuals with disabilities are educated with individuals who are not disabled, and special classes, separate schooling, or other removal of children with disabilities from the regular physical education environment occurs only when the nature or severity of the disability of a child is such that education in regular classes with the use of supplementary aids and services cannot be achieved satisfactorily (IDEA, 1997).

Relevant to education in the most appropriate setting is a continuum of instructional placements (see figure 2.1 on page 21), which ranges from a situation in which children with disabilities are integrated into a regular class to a very restrictive setting (out-of-school segregated placement).

Focus on Pupil Needs and Opportunity

IDEA implicitly, if not explicitly, encourages educators to focus on the educational needs of the student instead of on clinical or diagnostic labels. For example, as the IEP is developed, concern focuses on present functioning level, objectives, annual goals, and so on. IDEA does not require that disability labels be identified on an IEP. The associated rules and regulations also indicate that children with disabilities must be provided with an equal opportunity for participation in nonacademic and extracurricular services and activities, including athletics and recreational activities.

Section 504 of the Rehabilitation Act

The right of equal opportunity also emerges from another legislative milestone that has had an impact on adapted physical education and sport. Section 504 of the Rehabilitation Act provides that no otherwise qualified individual with a disability, solely by reason of that disability, be excluded from participation in, be denied the benefits of, or be subjected to discrimination under any program or activity receiving federal financial assistance or any program or activity conducted by an executive agency or by the United States Postal Service (Rehabilitation Act Amendments of 1992).

An important intent of Section 504 is to ensure that individuals with a disability receive intended benefits of all educational programs and extracurricular activities. Two specific conditions are prerequisite to the delivery of services that guarantee benefits to those individuals: Programs must be equally effective as those provided to nondisabled students, and they must be conducted in

Application Example: Student Placement

Setting: Individualized program planning committee meeting

Student: 10 years old with behavioral problems, inadequate physical fitness (as evidenced by failing to meet specific or general standards on the Brockport Physical Fitness Test), and below-average motor development (at or below one standard deviation below the mean on a standardized motor development test)

Issue: What is the appropriate setting(s) for instruction?

Application: On the basis of the information presented above and after meeting with the parents and other members of the program planning committee, the following was determined:

- ▶ The student will receive an adapted physical education program.
- ▶ The program will be conducted in an integrated setting with support services whenever the student's peer group receives physical education.
- ▶ The student will receive an additional class of physical education each week with two other students who also require adapted physical education.

the most normal and integrated settings possible. To be equally effective, a program must offer individuals with disabilities equal opportunity to attain the same results, gain the same benefits, or reach the same levels of achievement as peers without disabilities.

To illustrate the basic intent associated with Section 504, let us consider a totally blind student enrolled in a college course in which all other students in the class are sighted. A written test given at the end of the semester would not provide the student who is blind equal opportunity to demonstrate knowledge of the material; therefore, this approach would not be equally effective. By contrast, on a test administered orally or in Braille, the blind student would have an equal opportunity to attain the same results or benefits as the other students. In giving an oral exam, the instructor would be giving equivalent, as opposed to identical, services. (Merely identical services, in fact, would be considered discriminatory and not in accord with Section 504.) It is neither necessary nor possible to guarantee equal results; what is important is the equal opportunity to attain those results. For example, a recipient of federal funds offering basketball to the general student population must provide wheelchair basketball for students confined to wheelchairs, if the need exists.

A program is not equally effective if it results in indiscriminate isolation or separation of individuals with disabilities. To the maximum degree possible, individuals with disabilities should participate in the least restrictive environment, as represented by a continuum of alternative instructional placements (see chapter 2).

Compliance with Section 504 requires program accessibility. Its rules and regulations prohibit exclusion of individuals with disabilities from federally assisted programs because of architectural or other environmental barriers. Common barriers to accessibility include facilities, finances, and transportation. Money available for athletics within a school district cannot be spent in a way that discriminates on the basis of disability. If a school district lacks sufficient funds, then it need not offer programs; however, it cannot fund programs in a discriminatory manner.

In accordance with Section 504, children with disabilities who do not require special education or related services (not classified under IDEA) are still entitled to accommodations and services in the regular school setting that are necessary to enable them to benefit from all programs and activities available to nondisabled students. Every child who is a student with a disability under IDEA is also protected under Section 504, but all children covered under Section 504 are not necessarily students with a disability under IDEA.

Section 504 obligates school districts to identify, evaluate, and extend to all qualified students with a disability (as defined by this act) residing in the district a free appropriate public education, including modifications, accommodations, and specialized instructions or related aids as deemed necessary to meet their educational needs as adequately as the needs of nondisabled students are met. School districts across the United States are now developing 504 Accommodation Plans to provide programmatic assistance to students in order that they have full access to all activities. For example, a 504 Plan related to physical education might seek specialized instruction or equipment, auxiliary aids or services, or program modifications. A sample 504 Plan is presented in chapter 4.

The Rehabilitation Act is known as complaint-oriented legislation. Violations of Section 504 may be filed with the United States Office for Civil Rights (OCR) and, in addition, parents may request under Section 504 an impartial hearing to challenge a school district's decision regarding their children.

The Olympic and Amateur Sports Act

The Amateur Sports Act (ASA) of 1978 (PL 95-606), amended by PL 105-277, and the Olympic and Amateur Sports Act in the Omnibus Appropriations Bill, have contributed significantly to the provision of amateur athletic activity, including competition for athletes with disabilities. This legislation has led to the establishment of the Committee on Sports for the Disabled (COSD), a standing committee of the United States Olympic Committee (USOC), which includes the following sport organizations: United States Deaf Sports Federation (USDSF); Special Olympics, Inc.; Wheelchair Sports, USA; United States Cerebral Palsy Athletic Association (USCPAA); Dwarf Athletic Association of America; Disabled Sports, USA (DS/USA); and the United States Association for Blind Athletes (USABA). The COSD under this legislation has many duties regarding the continued development of sports opportunities for individuals with disabilities, including providing policy and procedural recommendations to the USOC. Also, as a result of this legislation, the USOC assumes the role and responsibilities of the United States Paralympic Committee. Additional information on the duties of these entities is presented in chapter 3.

Americans With Disabilities Act

In 1990, PL 101-336, Americans with Disabilities Act, was passed. Whereas Section 504 focused on educational rights, this legislation extended civil rights protection for individuals with disabilities to all areas of American life. Provisions include employment, public accommodation and services, public transportation, and telecommunications. Related to adapted physical education and sport, this legislation has required that community recreational facilities including health and fitness facilities be accessible and, where appropriate, that reasonable accommodations be made for persons with disabilities. It is essential that physical educators develop and offer programs for persons with disabilities that will give them the ability to participate in meaningful and competitive experiences in the community.

HISTORY OF ADAPTED SPORT

Deaf athletes were among the first Americans with disabilities to become involved in organized sports at special schools. As reported by Gannon (1981), in the 1870s the Ohio School for the Deaf became the first school for the deaf to offer baseball, and the state school in Illinois introduced football in 1885. Football became a major sport in many schools for the deaf around the turn of the century, and basketball was introduced at the Wisconsin School for the Deaf in 1906. Teams from schools for the deaf have continued to compete against each other and against athletes in regular schools.

Beyond interschool programs, formal international competition was established in 1924 when competitors from nine nations gathered in Paris for the first international silent games. In 1945, the American Athletic Association for the Deaf (AAAD) was established to provide, sanction, and promote competitive sport opportunities for Americans with hearing impairments. The earliest formal, recorded athletic competition in the United States for individuals with visual disabilities was a telegraphic track meet between the Overbrook and Baltimore schools for the blind in 1907. In a telegraphic meet, local results are mailed to a central committee, which makes comparisons to determine winners. From this beginning, athletes with visual disabilities continue to compete against each other and against their sighted peers.

Since the 1900s, wars have provided impetus for competitive sport opportunities. Sir Ludwig Guttman of Stoke Mandeville, England, is credited with introducing competitive sports as an integral part of the rehabilitation of veterans with disabilities. In the late 1940s, Stoke Mandeville Hospital sponsored the first recognized games for wheelchair athletes. In 1949, the University of Illinois organized the first national wheelchair basketball tournament, and it resulted in the formation of the National Wheelchair Basketball Association (NWBA). To expand sport opportunities, Ben Lipton founded the National Wheelchair Athletic Association (NWAA) in the mid-1950s. This organization has sponsored various competitive sports on state, regional, and national levels for individuals with spinal cord conditions and other conditions requiring wheelchair use. Another recent advancement was the creation of the National Handicapped Sports and Recreation Association (NHSRA). This organization, known today as Disabled Sports, USA (DS/USA), was formed by a small group of Vietnam veterans in the late 1960s. It has been dedicated to providing year-round sport and recreational opportunities for persons with orthopedic, spinal cord, neuromuscular, and visual disabilities.

Figure 1.4   This symbol of the Special Olympics was a gift of the former Union of Soviet Socialist Republics on the occasion of the 1979 International Special Olympic Games, hosted by the State University of New York, College at Brockport. The artist is Zurab Tsereteli.

Special Olympics was created by the Joseph P. Kennedy, Jr. Foundation to provide and promote athletic competition for persons with mental retardation. This organization held its first international games at Soldier Field in Chicago in 1968. (A symbol for the Special Olympics is shown in figure 1.4.) Special Olympics has served as the model sports organization for persons with disabilities through its leadership in direct service, research, training, advocacy, education, and organizational leadership.

During the last quarter of the 20th century, other national multisport and unisport programs have been formed. These have expanded available sport offerings to an increasing number of persons with disabilities. The latest offerings have been organized for athletes with visual impairments, cerebral palsy, closed head injury, stroke, dwarfism, and les autres (others).

The evolution of sport organizations within the United States has led to greater involvement in international competition. In fact, many American sport organizations have international counterparts (see chapter 3). Especially notable in this regard is the International Paralympics, which is discussed in greater detail in chapter 3. An individual who has provided sustained leadership to the Paralympic movement and served as president of the International Paralympic Committee is Canadian Robert D. Steadward, Rick Hansen Centre, University of Alberta (figure 1.5). In chapter 3, American organizations that participate in international games are listed. These organizations are multisport programs, that is, several sports are included as a part of these programs. In addition to multisport

Figure 1.5   Robert D. Steadward, an outstanding leader on the International Paralympic Committee.

organizations, there are several single-sport organizations, such as the American Blind Bowling Association and the National Wheelchair Basketball Association. Some of these programs are also associated with international competition. Unisport organizations provide excellent opportunities for athletes with disabilities, and several of these organizations are identified in other chapters of this book.

In the past few years, much of the impetus for sports for athletes with disabilities has been provided by out-of-school sport organizations. Although developing at a slower rate, other opportunities have begun to surface throughout the United States in connection with public school programs. An important milestone came in 1992 when Minnesota became the first state to welcome athletes with disabilities into its state school association. This made Minnesota the first state in the nation to sanction interschool sports for junior and senior high school students with disabilities. This program undoubtedly serves as a model for the rest of the country.

A few states now organize statewide competition for athletes with disabilities. Some of these are combined with community-based organized sport programs, and others are provided independently from other organized sport programs. Finally, sport programs in rehabilitation settings for individuals in communities are emerging in major cities in the United States. More detailed information on these programs is presented in chapter 3.

PERIODICALS

The increased knowledge base and greater attention to adapted physical education and sport in recent years have been accompanied by the founding and development of several periodicals devoted to the subject. Among the most relevant of these are the *Adapted Physical Activity Quarterly, Palaestra,* and *Sports 'N Spokes.* Other periodicals that publish directly relevant information from time to time include *Journal of Physical Education, Recreation and Dance; Strategies; Research Quarterly for Exercise and Sport; Journal of Visual Impairment and Blindness; Journal of Learning Disabilities; American Annals of the Deaf; Teaching Exceptional Children; American Journal of Mental Deficiency; Journal of Special Education; Therapeutic Recreation Journal; Journal of the Association for Persons with Severe Handicaps;* and *Clinical Kinesiology.*

ORGANIZATIONS

The American Alliance for Health, Physical Education, Recreation and Dance (AAHPERD) is an important national organization that makes significant contributions to programs for special populations. AAHPERD (called AAHPER before the dance discipline was added) has many members whose primary professional concern lies in adapted physical education and sport. AAHPER established a definition of adapted physical education in 1952. Over the years, its many publications, conferences, and conventions have given much attention to adapted physical education—not only on the national level, but within the organization's state, district, and local affiliates. Its professional conferences and conventions are among the best sources of information on adapted physical education and sport. At the national level, AAHPERD continues to advocate physical education, fitness, and physical activity for persons with disabilities. In the past few years, AAHPERD has reorganized to better serve persons with disabilities. An organization within AAHPERD, the Adapted Physical Activity Council, is now directly associated with adapted physical education. It is expected that this council, established in 1985, will continue to provide key professional services and leadership.

The National Consortium for Physical Education and Recreation for Individuals with Disabilities (NCPERID, or the Consortium) was established to promote, stimulate, and encourage professional preparation and research. It was started informally in the late 1960s by a small group of college and university directors of federally funded professional preparation and/or research projects seeking to share information. Its members have extensive backgrounds and interest in adapted physical education and/or therapeutic recreation. They have provided leadership and input on national issues and concerns, including the development of IDEA and its rules and regulations; federal funding for professional preparation, research, demonstration projects and other special projects; and the monitoring of legislation. The Consortium holds an annual meeting and publishes a newsletter.

The International Federation for Adapted Physical Activity (IFAPA), which originated in Quebec, has expanded to a worldwide organization with an international charter. Its primary service has been to sponsor a biennial international adapted physical activity symposium, held in Quebec (1977), Brussels (1979), New Orleans (1981), London (1983), Toronto (1985), Brisbane (1987), Berlin (1989), Miami (1991), Yokohama (1993), Oslo (1995), Quebec (1997), Barcelona (1999), and Vienna (2001). Symposia organized by IFAPA are also conducted in alternating years in regions throughout the world. The organization primarily solicits memberships from allied health therapists, therapeutic recreators, and adapted physical educators. With its international dimensions, IFAPA can disseminate valuable knowledge throughout the world.

The Office of Special Education and Rehabilitation Service, within the Department of Education, is responsible for monitoring educational services for individuals with disabilities and for providing grants to colleges and universities to fund professional preparation, research, and other special projects.

A private organization that has made a monumental contribution to both adapted physical education and sport is Special Olympics, Inc., founded by Eunice Kennedy Shriver. Although its leadership in providing sport opportunities for persons with mental disabilities is well known, this organization has provided much more to adapted physical education and sport. Specifically, it has played a key role in the attention to physical education in federal legislation and the provision of federal funding for professional preparation, research, and other projects in federal legislation through its advocacy activities. The organization has provided a worldwide model for the provision of sport opportunities; its work is acknowledged in several sections of this book.

SUMMARY

During the past few years, increasing attention has been given to adapted physical education and sport. This chapter described and presented a brief history of this field. It presented information regarding program direction and recognized the importance and characteristics of those providing services in this field. The chapter stressed the importance of litigation, legislation, and the inclusion movement on programs affecting individuals with disabilities. Finally, periodicals and organizations particularly important to adapted physical education were identified and described to serve as part of this introductory chapter.

Chapter **2**

Program Organization and Management

Joseph P. Winnick

JIMMY, AN ELEMENTARY SCHOOL STUDENT with cerebral palsy, definitely could benefit from an individualized program to meet his physical education needs. Unfortunately for Jimmy, there is a great deal of uncertainty at his school. Is he eligible for adapted physical education? In what setting should his program be implemented? How much time should he receive in physical education? Should he receive physical therapy? In his school, confusion occurs whenever a student with a unique need enrolls. Should this school have a plan to enhance the educational process for such students? The answer is yes.

This chapter includes information that is helpful in developing a plan for an adapted physical education program. For effective organization and management, schools are advised to write detailed guidelines on how to implement adapted physical education and sport programs. The guidelines can serve as a plan for operation that reflects laws, rules and regulations, policies, procedures, and practices.

CURRICULAR AND PROGRAMMATIC DIRECTION

An important early step in developing a school district plan is to develop a directional framework for physical education and sport. As discussed in chapter 1, there is no universal model for this framework so educational entities must establish or adopt their own. A framework to provide curricular and programmatic direction may include a philosophical statement (purpose), program aims and/or content goals, and objectives. The framework in figure 1.2 provides an example of a skeletal reference that may be used for school programs.

ADMINISTRATIVE PROCEDURES AND PROGRAM IMPLEMENTATION

School administrators who create and implement sound adapted physical education programs must ensure that the resources at their disposal adequately meet the needs of the pupils they serve. First, they must identify pupils who should receive adapted physical education programs, and then they must implement programs in an appropriate setting. Appropriate and innovative instructional strategies and challenging activities should be provided to all students with unique needs. Administrators must ensure that programs fit into student schedules, meet mandated time requirements, provide sport opportunities, are conducted in accessible facilities, and are appropriately funded. The next few pages will address these areas in more detail.

Identifying Pupils for Adapted Physical Education

In identifying students for an adapted physical education program, it is important at the outset to determine who is eligible and/or qualified. In some instances, the decision is obvious and an elaborate system of identification is not necessary. In other instances, determination of a unique need will be made only after detailed assessment data is analyzed and compared with established criteria for determination of a unique need.

Essentially, an adapted physical education program is for students with unique needs who require a specially designed program exceeding 30 consecutive calendar days. In selecting candidates for such a program, procedures, criteria, and standards for determining unique needs are important (they are discussed in chapters 4 and 5). The inability to attain health-related criterion-referenced physical fitness standards that are appropriate for the individual is an example of a criterion for establishing a unique need. A unique need is exhibited because individuals are expected to meet standards appropriate for them.

There are many procedures used to identify students needing adapted physical education in a school. These procedures are assocciated with **child find**, referred to in IDEA. This may include screening

- all new school entrants,
- pupils with disabilities,
- *all* pupils annually,
- referrals, or
- pupils requesting exemption from physical education.

An important child-find activity is the screening of all new entrants to the school. For transfer students, records should be checked to determine whether unique needs in physical education have been identified. In the absence of such information, the school, as part of its procedures, may decide to administer a screening test if a unique physical education need is suspected.

A second child-find source is a list of enrolled students who have been identified as having a disability in accordance with IDEA. Every student who has been so identified and whose disability is frequently associated with unique physical education needs should be routinely screened. Many children with disabilities have participated in preschool programs, and records from these programs may indicate those children with unique physical education needs.

A third activity is the annual screening of all those enrolled in the school. Such a screening might involve informal observation as well as formal testing. Conditions that may be detected through informal screening and may warrant in-depth evaluation include (but are not limited to) disabling conditions, obesity, clumsiness, aversion to physical activity, and postural deviations.

Many students are *referred* to adapted physical education. School guidelines should permit referrals from parents or guardians; professional staff members in the school district; physicians; judicial officers; representatives of public agencies with responsibility for the student's welfare, health, or education; or the student, if the student is at least 18 years of age or an emancipated minor. Referrals for adapted physical education should be received by one specifically designated person in each school.

Medical excuses or requests for exemption from physical education may lead to referrals for adapted physical education. When an excuse or request is made, immediate discussion with the family physician may be neces-

sary to determine the needed duration of the adaptation. For a period shorter than 30 consecutive days, required adjustments can be determined by the regular physical education teacher following established local policies and procedures. If the period is longer than 30 consecutive days, the procedure for identifying pupils in adapted physical education should be followed.

Instructional Placements in the Least Restrictive Environment

Individuals who are referred or are otherwise identified as possibly requiring a specially designed program should undergo a thorough assessment to determine whether a unique need exists. Chapters 4 and 5 deal with procedures for assessment. Once it is established that students have unique physical education needs and require an adapted physical education program, they must be placed in appropriate instructional settings. In this regard, it must be emphasized that adapted physical education is a *program* that may be implemented in a variety of settings. In accord with IDEA, to the maximum effort appropriate, children with disabilities must be educated with students who are not disabled and in the least restrictive environment (LRE). To comply with the LRE requirement, options on a continuum of instructional arrangements have been

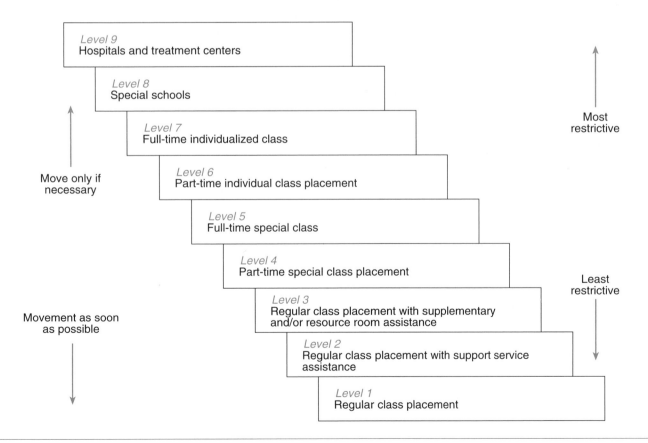

Figure 2.1   Continuum of alternative instructional placements in physical education.

advanced by various authors. Typical (although not all-inclusive) options on a continuum of instructional arrangements appear in figure 2.1. The number of options available is less important than the concept that students will be educated in the regular setting to the extent appropriate and in the environment most conducive to their advancement. The continuum presented in figure 2.1 clearly depicts more possibilities than integrated or segregated placement and thus is consistent with IDEA.

The three levels at the base of the continuum provide placement in a regular educational environment, and it is within these levels that the continuum is consistent with education in regular or inclusive environments. Level 1 placement is for students without unique needs or those whose short-term needs are met in the regular physical education program. This placement is also appropriate for individuals with unique needs requiring an adapted education program that can be appropriately implemented in the regular physical education setting.

Level 2 is for students whose adapted physical education programs can be met in a regular class environment with support service assistance. For example, some students may function well in a regular class if consultation is available to teachers and parents. In another instance, regular class placement may be warranted if a paraprofessional or an adapted physical education teacher can work with the individual with unique needs.

Level 3 is a regular class placement with supplementary and/or resource room assistance, as appropriate. Supplementary services can be provided each day or several times weekly as a part of or in addition to the time scheduled for physical education. Where indicated, the student may spend a portion of physical education time in a resource room.

Students who require part-time special class placement represent level 4. Their needs might be met at times in an integrated regular class and at other times in a segregated physical education class. The choice of setting is determined by the nature of the student's needs.

Level 5 full-time placement in a special class is appropriate for those whose unique needs cannot be appropriately met in the regular physical education setting. Levels 6 and 7 are appropriate when part- or full-time individual class placement is necessary.

Levels 8 and 9 reflect instructional placement in which needs must be met outside the regular school. In level 8, instruction is given in special schools; in level 9, instruction may be given in hospitals, treatment centers, and even at the student's home. Students in levels 8 and 9 may be placed outside the school district. In such cases, it is important to remember that the local school system is still responsible for ensuring that appropriate education is provided.

Inclusion

Within the past decade, more and more students with disabilities have been educated in regular educational environments. Inclusion has been one of the most powerful educational movements in the last decade of the 20th century. Although the movement is not *specifically* advocated as a part of IDEA, it is consistent with the requirement in the act that individuals with disabilities will be educated with nondisabled youngsters to the maximum extent appropriate. A key foundation of inclusion is the philosophy and belief that a *separate* education is not *equal*.

Although inclusion has been and continues to be a powerful force, its definitions and interpretations are varied. As defined in chapter 1, inclusion basically means educating students with disabilities in regular educational settings with nondisabled students. However, proponents of inclusion believe that it is more than integration. Craft (1996) states that "inclusion is a set of attitudes that together provide a welcoming and supportive educational environment, one that is respectful and appreciative of individual differences, and one in which all students participate regardless of gender, race, motor ability, or challenging condition (disability)."

Total inclusion differs from the LRE approach in that the LRE approach advocates education in the most normal/integrated and appropriate environment. Thus, some acceptable placements are not in a regular educational environment. Inclusion is consistent with LRE in that both approaches recognize the importance of support services for successful implementation in regular education settings. Placing students in integrated settings without needed support services is sometimes referred to as dumping, i.e., combining students with and without unique educational needs but not providing appropriate support services.

One of the most common subject areas for inclusion to take place has been physical education. Although students with disabilities requiring an adapted physical education program are commonly assigned to regular classes, teachers of physical education are often not prepared to provide optimal service delivery in the classroom. With this need in mind, Block (1994) has suggested a systematic approach to including students with disabilities in regular physical education settings. His model includes determining what to teach the student with disabilities, analyzing the regular physical education curriculum, determining the modifications needed from the regular physical education program, determining the support services needed, preparing the regular physical education teachers for implementing the program, preparing regular education students for inclusion, and preparing support personnel for their roles. Block (1994) discusses this approach in detail in his book written to help prepare teachers to include students with disabilities in regular physical education environments.

This chapter is written with the assumption that both adapted physical education and regular physical education programs for students with disabilities may be implemented in an inclusive setting, that students with disabilities will pursue a regular program in a regular setting if needs do not warrant an adapted physical education program, and that an adapted physical education program may be conducted in a noninclusionary setting.

Advocates of the inclusion movement point to many benefits for students. They believe that inclusion should be implemented, because it provides students with a more stimulating and motivating environment; it provides enhanced opportunities for students with disabilities to develop social skills and age-appropriate play skills; it facilitates the development of friendships between students with and without disabilities; and it provides well-skilled role models, which are helpful for the development of skills in all developmental domains. In summary, advocates feel that an inclusive education is the best education. Those less supportive of the inclusion movement state that students with disabilities in an inclusive setting may receive less attention and time on a task than other classmates; some teachers are not adequately prepared for successful inclusion and do not possess the interest and motivation to teach in inclusive settings; regular education students will be held back in their educational development; inclusion is too expensive if it means providing support services and decreasing class sizes; and school districts use the inclusion movement as a way of saving money by combining students with and without disabilities but not providing support services for a successful educational experience. Although these differences in opinion are held, it must be remembered that education in the most normal/integrated setting possible is a constitutional right in the United States and must be supported.

Although this section of the chapter focuses on inclusion for physical education instruction, it must be emphasized that integration and inclusion have relevance for sports also. Schools have a responsibility to provide both curricular and extracurricular experiences in the most normal/integrated settings possible. This is reflected by the use of the term adapted sport rather than disability sport in this book when referring to sport opportunities for individuals with disabilities. There is a need to examine and provide opportunities in sport that include both individuals with and without disabilities in regular and/or inclusionary settings. The term "disability sport" tends to be descriptive of only one sport setting. Use of this term as an all-encompassing term for sport related to persons with disabilities inherently impedes creativity and participation in unified sport and in unified sport settings. Chapter 3 provides a sport integration continuum that reflects an inclusionary view that is less confining than an orientation that emphasizes disability. Movement toward this more inclusionary orientation toward sport in no way

needs to impede or restrict the aim for excellence that is characteristic of many segregated sport opportunities encompassed under the disability sport umbrella. Thus, athletes with disabilities should be encouraged to develop themselves optimally to participate and compete at the highest level of athletic excellence, even though the setting may be segregated.

Guidelines for Instruction

Regardless of the setting in which adapted physical education programs are implemented, it is helpful to formulate guidelines related to instruction and to include them in an adapted physical education plan. This section presents information that is relevant to the development of instructional guidelines pertaining to curriculum, teaching styles, and organizational options.

The curriculum that is employed for students receiving adapted physical education should be based upon the scope and sequence of the school's physical education plan. It should provide activities that are developmentally appropriate and contribute to the self-actualization of each individual. Various options are available when providing curriculum experiences for students receiving adapted physical education. Writing in regard to the inclusion of students with severe disabilities into regular classes, Craft (1996) suggests four curricular options that may be applied in a variety of settings to accommodate all students: (1) same curriculum, in which students with disabilities participate in activities all other students are doing (figure 2.2); (2) multilevel curriculum, in which students are all involved in a lesson with the same curriculum areas but are pursuing different objectives at multiple levels based on their needs; (3) curriculum overlap, in which

Figure 2.2 Tandem cycling is an inclusive activity in which all students participate in the same activity.

a group of students work in the same lesson, but perform different or modified activities; and (4) alternate curriculum, in which alternate activities are provided (figure 2.3). Although Craft suggests these options for inclusive settings, they may also be applied to more restrictive settings. This book presents information in several chapters to help teachers implement various options.

When implementing physical education programs in regular/inclusive as well as more restrictive settings, attention needs to be given to analyzing and selecting teaching styles that provide optimal educational benefits to all students in the class. Certainly no one style can be selected to meet the divergent learning styles of all students. In chapter 6 of this book, various styles used for instruction are identified and contrasted considering the type and severity of disabilities, learning styles and support needs of the students, and instructional settings.

As with teaching styles, there are also a variety of organizational strategies that need to be analyzed and selected when providing instruction. These include team teaching, supportive teaching, peer and cross-age teaching, and independent learning. In chapter 6, these styles are reviewed from the perspective of their effectiveness for individualizing instruction and enhancing instruction in various settings. In regard to providing services in regular/inclusionary settings, it is important to employ organizational strategies that enhance belonging, interplay, and interinvolvement to the extent possible and appropriate. In addition to general organizational strategies, teachers can apply specific instructional strategies to enhance quality integration, provide opportunity for physical activity, and enhance the self-esteem of all students. Specific instructional strategies involving inclusion are presented in several chapters throughout this book. An example of inclusion strategies can be found in the Inclusion application example. Many strategies are implicitly presented as modifications and variations in physical activity and are discussed in chapters focusing on specific activities.

Figure 2.3 Inclusive settings allow students to perform modified or different activities while participating in class.

Class Size and Placement

Class size is an important variable that must be considered in placement. Unfortunately, class sizes for physical education have all too often been excessive. If quality instruction is expected, class sizes should not exceed 30 students. When students with unique needs in physical education are included in the regular educational environment, the number of students in the class must be adjusted according to the nature and severity of the disability and supplementary aides and services available. Special classes should not exceed 12 students; this number should be reduced to 6 students when extraordinary needs are exhibited. In certain instances, individualized instruction may be warranted. The number of students in special or regular classes should be adjusted in consideration of the number of professionals, paraprofessionals, and aides that are available in each class to provide assistance. *Chronological age affects placement as well. Age differences within a class should never exceed 3 years unless students are 16 or older.* School officials should know and comply with their state laws and regulations governing class size and composition. Each school district should specify policies regarding class sizes and support services.

Scheduling

Scheduling becomes an important matter as decisions regarding the setting for instruction are made. There are many approaches to scheduling that can accommodate various instructional arrangements. One effective method is to schedule supplementary and resource services and adapted physical education classes at the same time as regular physical education. A large school might have four physical education teachers assigned to four regular settings during a single period with a fifth teacher assigned to provide adapted physical education services. Other instructional arrangements might provide extra class time or alternate periods to supplement participation in regular physical education classes. In one scheduling technique used in elementary schools, a youngster placed in a special academic class joins an appropriate regular physical education class. This arrangement meets the student's need to be integrated in physical education while receiving special support in academic areas. Too often, students who could have been integrated in a regular physical education class are placed in a special class primarily on the basis of academic or other performance unrelated to physical education. Schools in which students are permitted to elect courses or units often find scheduling problems reduced because students may choose activities that fit their schedules in which they can participate with little or no adjustment.

Application Example: Inclusion

Setting: 7th grade physical education class

Student: 13-year-old student with mental retardation and limitations in motor coordination

Unit: Basketball

Task: Dribbling and weaving around cones, alternating hands with each dribble.

"Dribble the ball around five cones in a weaving manner, return, give the ball to the next person in line, then sit at the end of the line."

Application: The physical educator might include the following modifications of the task:

▶ Permit the use of either the same or alternating hands.

▶ Permit the skipping of alternate cones.

▶ Use a different ball size.

▶ Dribble for a shorter distance around fewer cones.

Time Requirements

It is important to clearly specify in school plans the time requirements for a program of adapted physical education. The frequency and duration of the required instructional program should at least equal that of students receiving regular physical education programs. If state time requirements for regular physical education instruction are specified for various grade levels, and if adapted physical education students are placed in ungraded programs, the school's guidelines should express equivalent time requirements, using chronological age as the common reference point. A local school district plan should communicate the federal requirements for physical education.

Physical education should be required of all students and should be adapted to meet unique needs. In cases of temporary disability, it is important to ascertain how long the student will require an adapted physical education program, and a standard should be set to distinguish temporary and long-term conditions. For this book, a short-term condition ends within 29 consecutive calendar days and can be accommodated by the regular classroom teacher. To the extent possible and reasonable, participation in physical activity rather than alternative, sedentary experiences should be required. For evaluation purposes, schools should set standards for the proportion of their students who participate in physical education but not physical activity. When many students meet physical education requirements through inactivity, it is time to reassess the local program.

School districts also need to deal with the issue of permitting participation in athletic activities as a substitute for physical education. Although coordination of instruction and sport participation (regular or adapted) is necessary, substitution should not be made unless it is approved in the student's IEP and the practice fits in with the school district's overall physical education plan. Ordinarily, substitution should not be permitted.

School districts must also clarify and coordinate instructional time requirements with related services. For instance, time spent in physical therapy must not supplant time in the physical education program. If appropriate guidelines are developed, few, if any, students should be exempt. Physical education will mean physical activity and will not be replaced by related services or extracurricular activities.

Sports

An adapted physical education plan should include general guidelines on sport participation and its relationship to the physical education program. In view of the detail involved in implementing a comprehensive extra-class sport program, a specific operating code should also be developed for local use. It is recommended that the extra-class sport program be established on the assumption that the sport program and the adapted physical education program are interrelated and interdependent. The extra-class program should build on the basic instructional program in adapted physical education and should be educational in nature.

A sport program should emphasize the well-being of the participants in the context of games and sports. It is also important to ensure participation to the extent possible and reasonable. Health examinations before participation and periodically throughout the season, if necessary, will also enhance safe participation. Athletes with disabilities should receive, at minimum, the same medical safeguards as other athletes.

For an interscholastic program with several schools participating, it is important to have a written statement of the principal educational goals as agreed to by the board of education, the administration, and any other relevant individuals or groups. The statement should reflect a concern for student welfare, an interest in the educational aspects of athletic competition, and a commitment to the development of skills that will yield health and leisure benefits during as well as after school years.

In the past few years, increasing attention has been given to providing sport opportunities for individuals with disabilities. In response to the intent of Section 504 of the Rehabilitation Act of 1973, educational and extracurricular opportunities must be provided in the least restrictive (most normal/integrated) setting possible on the basis of a continuum of settings ranging from the most restrictive (segregated) to the least restrictive (integrated). Chapter 3 presents a framework for a sport continuum to help enhance integration to the maximum extent possible, help guide decisions on sport participation, and help stimulate the provision of innovative opportunities.

Facilities

The facilities available to conduct programs in adapted physical education and sport may significantly affect program quality. A school's overall athletic facilities should be operated in a way that makes them readily accessible to individuals with disabilities. In fact, Section 504 rules and regulations prohibit exclusion of individuals with disabilities from federally assisted programs because of architectural, program, or other environmental barriers. Provision of access may dictate structural changes in existing facilities. All new facilities should be constructed to ensure accessibility and usability.

In planning facilities to facilitate adapted physical education and sport, attention must be given to indoor and outdoor areas, including teaching stations, lockers, and rest rooms. Indoor facilities should have adequate activity space that is clear of hazards or impediments and is otherwise safe. The environment must have proper lighting, acoustics, and ventilation. Ceiling clearance should permit appropriate play. Floors should have a finish that enables ambulation in a variety of ways for individuals with disabilities. When necessary, protective padding should be placed on walls. Space should be allotted for wheelchairs to pass and turn.

Like indoor areas, outdoor areas should be accessible and properly surfaced. Facilities should be available and marked for various activities including special sports. Walkways leading to and from outdoor facilities should be smooth, firm, free of cracks, and at least 48 inches wide. Doorways leading to the facilities should have at least a 36-inch clearance and be light enough to be opened without undue effort; if possible, they should be automatically activated. Water fountains with both hand and foot controls should be located conveniently for use by persons with disabilities. Colorful signs and tactual orientation maps of facilities should be posted to assist individuals with visual disabilities.

Both participants in adapted physical education and athletes need adequate space for dressing, showering, and drying. Space must be sufficient for peak periods. The design of locker rooms should facilitate ambulation and the maintenance of safe and clean conditions. Adequate ventilation, lighting, and heating are necessary. The shower room should be readily accessible and should provide a sufficient number of shower heads. The facilities should be equipped with grab rails. Locker rooms should include adequate benches, mirrors, and toilets. Persons with disabilities frequently prefer horizontal lockers and locks that are easy to manipulate. Planning must ensure that lockers are not obstructed by benches and other obstacles. All facilities should, of course, be in operable condition. Well-designed rest rooms should have adequate space for manipulation of wheelchairs, easily activated foot and/or hand flush mechanisms, grab rails, and toilet and urinal levels to meet the needs of the entire school population.

Swimming pools are among the most important facilities. Pool design must provide for safe and quick entry and exit (figure 2.4). Water depth and temperature should be adjustable to meet learning, recreational, therapeutic, and competitive needs. Careful coordination of pool use is usually necessary to accommodate varying needs. Dressing, showering, and toilet facilities must be close by, with easy access to the pool.

Students in adapted physical education and sport programs must have equal opportunity to use normal/integrated facilities. Too often, physical education for individuals with disabilities is conducted in boiler rooms or hallways. Small class size should not be used as a reason to exclude persons with disabilities from equitable use of facilities. Such a practice is discriminating and demeaning to both pupils and school personnel.

Budget

An equitable education for a student with unique needs is more costly than that of a student without unique needs. To supplement local and state funds, the federal government, through a variety of programs, provides money for the education of people with unique needs. Funds associated with IDEA are specifically earmarked. To enhance receipt of federal funds for physical education, physical educators must be sure that they are involved in IEP development.

Funds associated with IDEA are available to help provide for the excess costs of special education (i.e., costs that exceed student expenditure in regular education). These funds "flow" through state education departments (which are permitted to keep a certain percentage) and on to local education agencies. The flow-through money can be used to cover excess costs already met by states. Because adapted physical education deals with students who are not disabled as well as with those who are, it is advantageous for schools to fund teachers in physical edu-

Photo courtesy of the New York State School for the Blind. Printed with permission.

Figure 2.4   Swimming pools with easy access ramps and adjustable bottoms are more convenient for students in adapted physical education than regular pools.

cation, whether regular or adapted, from the same local funding source rather than to rely on federal money. This is justifiable, because states are responsible for the education of all their students.

In addition to meeting needs identified in IEPs, funding must support the preparation of teachers to provide quality services in adapted physical education and sport. For example, funds are needed for in-service education, workshops, clinics, local meetings, professional conferences and conventions, program visitations, and so on. Schools also need funds to maintain up-to-date libraries and other reference materials.

Interscholastic teams made up of students with disabilities must receive equitable equipment, supplies, travel expenses, officials, and so on. While the funding level for curricular and extracurricular activities in a local community is not externally dictated, available funds cannot be used in a discriminatory fashion (e.g., available to males but not females, or available to students without disabilities but not to students with disabilities).

HUMAN RESOURCES

A quality program in adapted physical education and sport depends to a great extent on the availability of quality human resources and the ability of involved personnel to perform effectively as members of a group. People are needed to coordinate and administer services, fulfill technical and advocacy functions, and provide instruction. Many of these functions are carried out in important committees. To provide high-quality services for adapted physical education and sport, the teacher must work with various school and IEP committees. In doing so, it is helpful to understand roles and responsibilities and to realize that the concern for students with

unique needs is shared by many. This section identifies key personnel and discusses their primary roles and responsibilities. Many perform their responsibilities by serving on committees identified in chapter 4.

Director of Physical Education and Athletics

Although not a universal practice, it is desirable for all aspects of physical education and sport programs to be under the direction of an administrator certified in physical education. Such centralization enhances coordination and efficiency in regard to personnel, facilities and equipment, budgeting, professional development, and curriculum. The director of physical education and athletics should oversee all aspects of the program, including the work of the coordinator of adapted physical education if that position exists.

Because adapted physical education and sport are often in the developmental stage and not a well-advocated part of the total program, the physical education director needs to demonstrate genuine concern and commitment to this part of the program. A positive attitude serves as a model for others to emulate. With the help of other administrative personnel, the director can help the program in adapted physical education and sport by ensuring adequate funding, employing qualified teachers, and providing support services. The director must also be knowledgeable about adapted physical education and sport to work effectively with individuals and groups outside the department. The director must work with other directors, coordinators, principals, superintendents, and school boards and must have positive professional relationships with medical personnel including physicians, nurses, and therapists. Other important relationships are those with parents, teachers, students with disabilities, and advocacy groups. For this reason, it is important that the director of physical education and sport be kept informed about all students who are identified as having unique needs.

Adapted Physical Education and the Sport Coordinator

To provide a quality comprehensive school program in adapted physical education and sport, schools are advised to name a coordinator. In a small school, this might be a part-time position; in larger schools, a full-time adapted physical education coordinator may be needed. Although most states do not require a special endorsement, credential, or certification to teach adapted physical education, it is best to select an individual who has considerable professional experience. If possible, the coordinator should have completed a recognized specialization or concentration in adapted physical education and, where applicable, should meet the state competency requirements for certification. The National Consortium for Physical Education and Recreation for Individuals with Disabilities (NCPERID) provides certification for the teaching of adapted physical education. A person demonstrating these competencies is likely to be knowledgeable and genuinely interested in serving in that particular role. If a school cannot employ a person who has preparation in adapted physical education, the coordinator's duties should be entrusted to someone who demonstrates genuine interest.

The particular role and functions of the coordinator depend on the size of the school, the number and types of students with disabilities within the school population, and the number and types of students involved with adapted physical education and sport. Generally, however, the coordinator needs to assume a leadership role in various functions associated with adapted physical education and sport. The specific functions often differ more in degree than in kind from those performed by regular physical educators. Table 2.1 identifies typical functions associated with adapted education and sport and indicates who is responsible for those functions. Functions may overlap or be shared; specific lines of demarcation should be drawn to suit local conditions.

Many schools employ adapted physical education teachers in positions other than that of adapted physical education and sport coordinator. Although these teachers will certainly support functions of the coordinator, they are primarily involved in carrying out instructional programs in a variety of settings. In addition, they may help to implement adapted sport programs, for example, by developing and implementing sport days, arranging competition in out-of-school programs, preparing participants, and coaching teams. Most adapted physical education teachers are very much involved in working with IEP committees.

Regular Physical Educator

Although adapted physical educators may be employed by a school, the regular physical educator plays an extremely important role in implementing quality programs in adapted physical education and sport. Table 2.1 presents several functions that are shared by or are the primary responsibility of regular physical educators. For example, they play an important role in screening. They may also implement instructional programs in integrated environments and help implement sport programs. One of the most important tasks is referral. In the area of management/leadership, the regular physical educator generally plays a secondary role. With the present-day trend of including more and more students with disabilities (with or without unique needs) in regular classes, it is often the responsibility of regular physical educators to implement such programs.

Nurse

The school nurse is an allied health professional with an important part in the successful development and implementation of adapted physical education and sport programs. The nurse must be knowledgeable about the adapted physical education and sport program and, ideally, should serve on its committee on adapted physical education. The nurse can be a valuable resource by helping to interpret information that is required for individual education planning. If time permits, the school nurse can assist the physical education staff in testing students, particularly in the case of postural screening. The nurse can also provide a valuable service in keeping medical records, communicating with physicians, and helping parents and students to understand the importance of exercise and physical activity.

Physicians

Physicians are among the allied health professionals who have an important relationship with the school's adapted physical education and sport program. The physician's role is so important that it is often addressed in federal, state, or local laws, rules, and regulations. In some instances, states look to the school physician for the final decision on participation in athletic opportunities. Also very important, physicians should provide and interpret medical information on which school programs are based. Using this information, the IEP planning groups plan appropriate programs. The school physician also has an important responsibility for interpreting the adapted physical education and sport program to family physicians and other medical personnel.

Table 2.1 Primary Responsibility for Functions Relevant to Adapted Physical Education and Sport

Function	Responsibility	
	Regular physical educator	Adapted physical educator/ coordinator
Measurement, assessment, evaluation		
• Student screening	X	X
• In-depth testing		X
• Student assessment and evaluation		X
• Adapted physical education and sport program evaluation		X
Teaching/coaching		
• Implement instructional programs for students with short-term unique needs	X	
• Implement instructional program to meet long-term unique needs in integrated environments	X	X
• Implement instructional and sport programs with guidance of adapted physical educator	X	
• Implement adapted sport program		X
Management/leadership		
• Consultation		X
• In-service education		X
• Advocacy/interpretation		X
• Recruitment of aides/volunteers		X
• Chair adapted physical education committee		X
• Liaison with health professionals	X	X
• Referral and placement	X	X
• Organization of adapted sport program		X

In states where physical education is required of all students, physicians must know and support laws and regulations. They must be confident that if a student is unable to participate without restriction in a regular class, adaptation will be made. Physicians should be aware that quality physical education and adapted physical education have changed considerably since many of them attended school, and they need to understand the nature of these changes. In other words, they need to be aware of their roles and responsibilities in well-established modern programs.

One of a physician's important functions is to administer periodic physical examinations. Examination results are used as a basis for individualized student evaluation, program planning, placement, and determination of eligibility and qualification for athletic participation. It is desirable for students with unique physical education needs to receive an exam every three years beginning in the first grade. Examinations should be annual for those assigned to adapted physical education because of medical referrals. School districts that do not provide physical examinations should require adequate examinations by the family physician. For students encompassed under IDEA, medical examinations must be given in accordance with state and local policies and procedures. For athletic participation, exams should be administered at least annually.

Coaches

Adapted sport programs should be operated under the direction of qualified school personnel. Where an adapted program includes interschool athletic teams, standards for coaches must be consistent with those for the regular interschool athletic program. Teachers certified in physical education may be permitted to coach any sport, including those whose participants have disabilities. Ideally, coaches of teams composed primarily of individuals with unique needs should have additional expertise in adapted physical education.

Coaches must follow acceptable professional practices, including maintaining a positive attitude and insisting on good sportsmanship, respect, personal control, and willingness to improve professionally through in-service programs, workshops, and clinics.

Related Service Personnel

Under IDEA, related services means transportation and such developmental, corrective, and other supportive services that are required to assist children with disabilities to benefit from special education. Related services include speech-language pathology and audiology services; psychological services; physical and occupational therapy; recreation, including therapeutic recreation; social work services; counseling services, including rehabilitation counseling; orientation and mobility services; and medical services for diagnostic or evaluation purposes only. Related service providers who often impact physical education include occupational and physical therapists. According to the rules and regulations for the implementation of IDEA, occupational therapy includes improving, developing, or restoring functions impaired or lost through illness, injury, or deprivation; improving ability to perform tasks for independent functioning when functions are impaired or lost; and preventing, through early intervention, initial or further impairment or loss of functioning. The same rules and regulations define physical therapy as services provided by a qualified physical therapist. These services have traditionally included physical activities and other physical means of rehabilitation as prescribed by a physician. The rules and regulations specify that recreation includes assessment of leisure function, therapeutic recreation services, recreation programs in schools and community agencies, and leisure education.

Much has been written about the relationship of adapted physical education and the related services of physical and occupational therapy. Often the lines of responsibility among these areas are blurred. What is clear is the fact that related services must be provided if a student requires them to benefit from direct services. For example, both physical and occupational therapy must be provided to the extent the student needs them to benefit from physical education or other direct services in the school program. IDEA specifies that physical education must be made available to children with disabilities. Also, states have their own requirements concerning the provision of physical education. Clearly, physical therapy and adapted physical education are not identical, and related services should not supplant physical education or adapted physical education (which are direct services under IDEA).

Several assumptions about the role of physical education may underlie the decision of who will design programs to improve the physical fitness of students with disabilities. First, it is clearly the physical educator's responsibility to design these programs. Thus, the physical educator is involved with the development of strength, endurance, cardiorespiratory endurance, and flexibility (range of motion). His or her responsibility concerns *both* affected and unaffected parts of the body. For example, individuals with cerebral palsy should be helped to maintain and develop their physical fitness. When dealing with the affected parts of the body, the physical educator should consult and coordinate with medical and/or related service personnel in program planning and implementation.

Sometimes improvements in physical development cannot be attained by a physical educator using the usual time allotments, methods, or activities associated with physical education. In such cases, physical or occupational therapy can enhance physical fitness development. Activities included in the physical education programs of youngsters with disabilities should be those that are typically within the scope of physical education. These are the kinds of activities subsumed under the definition of physical education in the rules and regulations of IDEA and included in the scope of physical education described in chapter 1. Although the physical educator involves children in exercise, it is important *not* to limit physical education to an exercise prescription program. Instead, it is important for the physical educator to offer a broad spectrum of fun and well-liked physical education activities. A youngster who requires a specific exercise to the extent that it would encompass an entire physical education period should meet this need in class time added to the regularly scheduled physical education period or it should be a provided related service. This approach would permit involvement in a broad spectrum of activities within the regularly scheduled physical education class. Physical educators should help students appropriately use wheelchairs and supportive devices in physical education activities. They must, therefore, be knowledgeable about wheelchairs and other assistive devices. However, it is not their responsibility to provide functional training in the use of those aids for basic movement or ambulation.

It is vital that physical educators consult physicians and other medical personnel as they plan and implement programs. Such consultations should be consistent with each school's adapted physical education program.

Although much can be written and discussed concerning roles and responsibilities, very often the quality of services provided depends upon the interpersonal relationships of service providers. Successful situations are those in which professionals have discussed their roles and responsibilities and work hard to deliver supportive services to benefit individuals with unique needs.

PROGRAM EVALUATION

At the beginning of this chapter, the importance of an adapted physical education and sport plan to guide program organization and management was stressed. Once in place, the plan can serve as a basis for program evaluation. The implementation aspects of adapted physical education should be evaluated annually, and the total plan should be evaluated at five-year intervals. Program evaluation should draw on data collected from a variety of relevant sources. Helpful to evaluation is the use of a rating scale or checklist. An instrument for evaluation is least threatening if used for self-appraisal. Evaluation provides a point of departure for identifying and discussing strengths and weaknesses and developing a schedule to remedy weaknesses.

SUMMARY

Well-organized and managed programs in adapted physical education and sport are enhanced and characterized by written plans, which serve as guides on how to implement programs based on laws, rules, policies, procedures, and practices. This chapter presents information that may be used for the development of guidelines for plans. Information is presented in four categories: curricular program and direction, administrative procedures, human resources, and program evaluation. Criteria statements in the rating scale presented in appendix B, which is recommended as part of program evaluation, implicitly reflect many recommendations in this chapter.

Adapted Sport

Michael J. Paciorek

▷

▷

▷

AS THE INDIVIDUALIZED EDUCATION PROGRAM (IEP) meeting was coming to a close, Mr. and Mrs. Watts had just finished listening to their son Max's physical education teacher describe the program of instruction she had proposed for him. Max is nine and has cerebral palsy. He was to be included in the regular physical education program two days per week and enrolled in an adapted physical education class with a smaller number of students for another day. The Watts were glad to see such quality programming but inquired about Max participating in a sport program. The physical education teacher Ms. Clark stared at them with a blank look on her face until she stammered, "Max has spastic cerebral palsy and uses a power wheelchair. There are no sport programs for him."

This chapter will provide information on the responsibility of the physical educator in providing meaningful sport participation and training opportunities for students with disabilities. It will describe a continuum of sport participation and will identify many multi and unisport organizations available for people with disabilities. While participation in sport programs should not be used as a substitute for adapted physical education, after reading this chapter teachers should be able to provide parents, like the Watts, with information on the many sports resources available for their child. Reading throughout this chapter, the physical educator should realize that the concept of least restrictive environment also applies to sports and athletic opportunities.

Although medical personnel and educators have seen for many years the potential that sport participation can have for students with disabilities, opportunities for such participation have been slow to develop. Although limited sport opportunities have been available for older individuals for some time, sport opportunities for children with disabilities have been minimal. As noted in chapter 1, federal legislation affirms the rights of students with disabilities to have equal access and opportunities to physical education, intramural, and school sport programs.

As defined in chapter 1, adapted sport refers to sport modified or created to meet the unique needs of individuals. For many people the definition of adapted sport refers only to competitive athletic opportunities. While this chapter will focus on those opportunities to a great degree, the physical educator should refer to adapted sport in the broadest sense possible and use it to encompass not only those competitive athletic experiences but also those that include leisure-time recreational pursuits that enable the person with a disability to practice healthy living outside of the school setting. Adapted sport should be viewed as legitimate sport and competition of a high quality and not just as a social experience.

INTEGRATION CONTINUUM

In the past few years, increasing attention has been given to providing sport opportunities for people with disabilities. In response to the intent of Section 504 of the Rehabilitation Act of 1973, educational and extracurricular opportunities must be provided in the least restrictive (most normal/integrated) setting possible on the basis of a continuum of settings ranging from the most restricted (segregated) to the least restrictive (integrated). Figure 3.1 presents a framework for a sport continuum to enhance integration to the maximum extent possible, guide decisions on sport participation, and stimulate the provision of innovative opportunities. Winnick (1987) originally published the continuum and related material in this section.

The continuum, which relates to the provision of programs in the least restrictive environment, encompasses opportunities in intramural and interschool programs as well as out-of-school sport and leisure programs. (For purposes of this book, all these activities are encompassed under the designation of sport.) The athlete has various options available to participate in regular sport or adapted sport through a five-level continuum. Regular sport implies that the athlete with a disability can participate in a sport without modifications. Adapted sport implies that the athlete with a disability competes in a regular sport with the aid of some assistive device (sledge hockey) or rule modification (smaller field, smaller goal, and fewer players for cerebral palsy soccer), or that the athlete participates in a sport specially designed for a particular disability, such as quad rugby, a sport designed especially for quadriplegics who use wheelchairs; or goal ball, a game designed specifically for individuals with visual impairments. More about these activities can be found in chapters 23 through 27.

Levels 1 and 2 of the continuum are essentially regular sport settings, distinguished only by a need for accommodation. In regular sport, the setting is integrated. Individuals with disabilities should be given equal opportunities to qualify for participation at these levels. An example of level 1 participation would be an athlete with mental retardation running the 800-meter dash for a high school track team or an athlete with an amputation participating in regular sport competitions with the use of a prosthesis (figure 3.2).

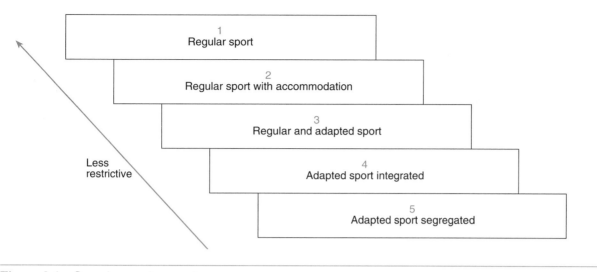

Figure 3.1   Sport integration continuum.

From "An Integration Continuum for Sport Participation," by J.P. Winnick, 1987, *Adapted Physical Activity Quarterly*, 4, p. 158. Copyright 1987 by Human Kinetics. Reprinted by permission.

In accordance with Section 504, schools and agencies should provide modified or special activities only if they operate programs and activities in the most normal and appropriate setting; qualified students with disabilities are given the opportunity to participate in programs and activities that are not separate and different; qualified students with disabilities are able to participate in one or more regular programs and ac-

Photo courtesy of Bob Radocy, Therapeutic Recreation Systems.

Figure 3.2   A child with an amputation should be able to participate in regular sport competition with the use of a prosthesis.

tivities; and students with disabilities are appropriately placed in full-time special facilities (Department of Health, Education, & Welfare, 1977; Winnick et al., 1980).

A bowler who is blind competing in regular sport competition with only the accommodation of a guide rail exemplifies level 2. Other examples may include an athlete with a physical disability participating on a regular high school track team using a field or throwing chair, or a blind swimmer competing with sighted swimmers using tap sticks to know when to begin flip turns (figure 3.3). In making accommodations at level 2, Section 504 rules and regulations require that any accommodation provided be reasonable and allow individuals with disabilities equal opportunity to gain the same benefits or results as other participants in a particular activity. At the same time, accommodations should not confer an unfair advantage on an individual with a disability. In the case of the bowler who is blind, the guide rail serves as a substitute for vision for the purpose of orientation to the target, while the throwing chair provides the stability needed for the field athlete during competition. Because the activity remains essentially unchanged for all participants and no undue advantage is given, this accommodation constitutes regular sport participation.

Perhaps the most well-publicized level 2 situation occurred when professional golfer Casey Martin was denied participation on the Professional Golf Association (PGA) tour in the spring of 1998. Martin, who has a rare circulatory disorder in his leg that makes it painful to walk moderate distances without severe pain, uses a cart when he golfs. The PGA alleged that the use of a cart constituted an unfair advantage. Martin's suit, based on the Americans with Disabilities Act, was successful. He was allowed to compete using a cart on the PGA tour.

Photo by Michael Paciorek.

Figure 3.3   A partner signals a blind swimmer when to begin flip turns when competing with sighted individuals.

Level 3 includes both regular and adapted sport conducted in settings that are partly or fully integrated. Those with a disability may compete against or coact with all participants in a contest, including competitors with and without disabilities. For instance, an athlete participating in a wheelchair (adapted sport) may compete against all runners in a marathon, including athletes with and without disabilities; nondisabled athletes run on foot (regular sport). In another level 3 example, a nondisabled athlete and an athlete with a disability may cooperate as doubles partners in wheelchair tennis. The nondisabled partner, playing ambulatory, is permitted one bounce before returning the volley (regular sport), whereas the athlete with a disability, using a wheelchair, is permitted two bounces (adapted sport).

The Special Olympics Unified Sports program is another example of athletes with and without disabilities participating together. Unified Sports, begun in 1989, is a program that places athletes with mental retardation and their peers without metal retardation on the same team for training and competition. Athletes benefit from physical and mental challenges by participating in a variety of competitions organized by Special Olympics or by community sport organizations. The use of Unified Sports rules and guidelines on age and ability grouping help ensure that all athletes play a meaningful and valued role on the team. Special Olympics now offers Unified Sports in all its summer and winter sports.

Level 3 also includes situations in which an athlete participates part-time in regular sport and part-time in adapted sport. For example, a person who is blind may participate in regular competition for powerlifting but in adapted sport competition for goal ball. Level 3 activities either include athletes with and without disabilities integrated and participating in regular sport with adaptation made for the athletes with disabilities, or athletes with disabilities participating part-time in adapted sport and part-time in regular sport.

At level 4, athletes both with and without disabilities participate in a modified version of the sport. At this level, competition or coaction must include both an individual with a disability and a nondisabled participant. One example is a game of tennis in which athletes with and without disabilities use wheelchairs in their competition against opponents who are likewise in wheelchairs.

At level 5, athletes with disabilities participate in adapted or regular sport in a totally segregated setting. For example, athletes with mental retardation compete against each other in the Special Olympics program, athletes with physical disabilities compete in wheelchair fencing, or two teams of youngsters who are blind compete in a local goal ball tournament. The Challenger Division created by Little League Baseball for athletes with physical disabilities is another level 5 example.

The conceptual framework for the sport continuum is based primarily on the degree of integration (with cooperator/coactor and/or competitor) and sport type (traditional/adapted). The continuum stresses association or interaction between athletes with and without disabilities, the key ingredient in *integration*. To some extent, but not all, the continuum reflects the severity of the disability. This is because the *nature* of the disability and the *ability* to perform, as related to a specific sport, are greater factors than *severity* of condition.

SPORT ORGANIZATIONS

As legislation on behalf of individuals with disabilities has led to more inclusion in all aspects of daily living, there has been an explosion in the number of organizations providing sports programming for persons with disabilities. These can be categorized into multisport and unisport organizations.

Multisport Organizations

Multisport organizations are those that provide training and athletic competition in a variety of sports for individuals with a particular disability. For instance, the United States Cerebral Palsy Athletic Association provides competition in 10 sports for individuals who have cerebral palsy, stroke, or traumatic brain injury. These organizations serve in much the same capacity as an able-bodied sport national governing body—to oversee the development and conduct of their sports and to promote athletic involvement for their members. In addition to the athletic competition involved, athletes have used these organizations as support groups and to discuss relevant issues. There are seven multisport disabled sports organizations affiliated with the United States Olympic Committee including the Dwarf Athletic Association of America (DAAA), Disabled Sports USA (DS/USA), Special Olympics, Inc. (SOI), United States Association of Blind Athletes (USABA), United States Cerebral Palsy Athletic Association (USCPAA), USA Deaf Sports Federation (USADSF), and Wheelchair Sports USA (WSUSA). Each organization provides sports opportunities in different ways. Some organizations, such as USCPAA, rely on sport technical officers to oversee programs, while others such as Wheelchair Sports USA and the USA Deaf Sports Federation are divided into specific sport federations. The USA Deaf Sports Federation is a Non-Paralympic sports organization. Their athletes compete internationally in the World Games for the Deaf (WGD), not the Paralympic Games. More detailed information on the multisport organizations can be found within the specific disability chapters in this text.

Unisport Organizations

Unisport organizations promote sport participation in a single sport, either for a specific disability or for multiple disabilities. For instance, the North American Riding for the Handicapped Association (NARHA), through its affiliate programs, offers therapeutic and competitive horseback riding for individuals regardless of disability. The Handicapped Scuba Association International (HSAI) offers individual and instructor-training programs in a similar manner, while the Achilles Track Club affiliates offer road racing opportunities for athletes with disabilities. Examples of unisport organizations that promote one sport for individuals with a specific disability include the United States Quad Rugby Association (USQRA), for individuals with spinal cord injuries (figure 3.4); the American Amputee Soccer Association (AASA); and the National Beep Baseball Association (NBBA), for individuals who are blind or visually impaired, to name a few.

THE OLYMPIC AND AMATEUR SPORTS ACT AND THE ROLE OF THE UNITED STATES OLYMPIC COMMITTEE

The Amateur Sports Act (ASA) of 1978, amended in 1998 as the Olympic and Amateur Sports Act PL 105-277 (within the Omnibus Appropriations Bill), is perhaps the one piece of legislation that provided the catalyst for the explosion in adapted sport. This legislation reorganized the United States Olympic Committee (USOC) and administration of amateur sports in the United States. The amended act strengthened the linkage between the USOC and athletes with disabilities by including the Paralympic Games and amateur athletes with disabilities within the scope of the act and the USOC. With the available resources and prestige attached to the USOC, sport organizations related to persons with disabilities have a new sense of hope in building their programs. The USOC Constitution was rewritten to reflect the new commitment to adapted sport:

To encourage and provide assistance to amateur athletic programs and competition for amateur athletes with disabilities, including, where feasible, the expansion of opportunities for meaningful participation by such amateur athletes in programs of athletic competition for able-bodied amateur athletes. (USOC Constitution, 1998)

Photo courtesy of Specialized Sports Unlimited.

Figure 3.4   The United States Quad Rugby Association is an example of a unisport organization.

The USOC mission now includes duties as the United States Paralympic Committee (USPC), with responsibilities for the development and training of elite athletes with disabilities. Other duties include the organization of the Paralympic Games when held in the United States and the selection of the United States Paralympic team. Because opportunities for elite athletes are becoming available, children with disabilities now have role models to emulate as they begin their training programs in local communities. Services provided to Paralympic athletes, while improving, are still far from reaching parity with services provided to Olympic athletes.

The Committee on Sports for the Disabled (COSD) was established to assume these duties and to promote the following efforts:

- Develop interest and participation throughout the United States in sports for individuals with disabilities
- Work with USOC national governing bodies to carry out the mandate of involving amateur athletes with disabilities in sports programs for able-bodied athletes
- Disseminate information on physical training, equipment design, coaching, and performance analysis for athletes with disabilities
- Encourage and support more research in areas such as sports medicine and sports safety pertaining to athletes with disabilities
- Guarantee that Olympic training facilities are fully accessible to athletes with disabilities
- Seek appropriate allocation of all funds available to the USOC for sports for athletes with disabilities

The visibility and impact of the COSD grew significantly during the 1980s and '90s, advancing the role of athletes with disabilities and their programs within the amateur sports movement. Due to COSD efforts, inclusion of athletes with disabilities in events held primarily for athletes without disabilities has been seen in demonstration events at the summer and winter Olympic Games since 1984, and in multidisability and multievent demonstration events at the United States Olympic Festivals. The COSD continues to encourage and promote a cohesive working relationship among the 41 sport national governing bodies (NGBs), the USOC/USPC, and the Paralympic sports organizations.

The COSD remains a bylaw committee of the USOC and continues with its primary responsibility of providing policy and procedural recommendations to the USOC Board of Directors regarding amateur sports programming in the United States for people with disabilities (Paciorek & Jones, 1994).

The USOC/USPC (which are one and the same) works with both the sport national governing bodies and the Paralympic sports organizations to provide support services to elite athletes with disabilities in training and competing in the Paralympic and other quadrennial games for athletes with disabilities. The USOC Disabled Sports Services department manages the internal activities necessary to facilitate this process. Disabled Sports Services also serves as the contact representative for all issues related to National Paralympic Committee concerns.

RELATIONSHIP OF DISABLED SPORTS ORGANIZATIONS TO NATIONAL GOVERNING BODIES

National Governing Bodies (NGBs) and International Federations (IFs) are organizations dedicated to the development and promotion of specific sports. They generally sanction competitions and certify officials. By law, each of the national governing bodies that are affiliated with the United States Olympic Committee must allow participation by people with disabilities. NGBs and IFs now routinely disseminate information on adapted sport participation through the use of subcommittees, information on specific programming options, or the inclusion of adapted sports rules. For example, the United States Racquetball Association includes wheelchair racquetball rules in their official rule book. Some NGBs, such as United States Swimming (USS) and the United States Tennis Association (USTA), offer extensive information on adapted sport programming in their literature. Some disabled sport organizations have merged with regular sport organizations. The International Wheelchair Tennis Federation has merged with the International Tennis Federation and now offers expanded programming for tennis players with disabilities. The trend toward including athletes with disabilities in regular sport organizations is called vertical integration. NGBs and IFs are excellent resources for the physical educator looking for information on adapted sport or regular sport programming.

PARALYMPIC GAMES

The Paralympic Games are the equivalent of the Olympic Games and are primarily for athletes with physical disabilities or visual impairments. Individuals who are deaf compete in the World Games for the Deaf (WGD). The Paralympics began in 1948 when the Stoke Mandeville Games were held in Aylesbury, England. This was the same year the games of the XIV (14th) Olympiad were being held in London. The idea of holding these games is attributed to a doctor, Sir Ludwig Guttman, who included sport in his rehabilitation for people with spinal cord injuries.

The first Paralympic Games were held in Rome in 1960, the same year that the Italians hosted the games of the XVII (17th) Olympiad. These first Paralympic Games drew 400 athletes representing 23 countries. Since then the Olympic and Paralympic Games have led a parallel existence, and whenever possible held in the same city (Tokyo, 1964; Seoul, 1988; Barcelona, 1992; Atlanta, 1996; and Sydney, 2000) or in the same country (Heidelberg, 1972; Toronto, 1976; and New York, 1984).

The 1992 Barcelona Paralympics marked the first time that a joint Olympic/Paralympic Committee was used. International Olympic Committee President Juan Antonio Samaranch was so deeply moved and impressed by the caliber of the Barcelona Paralympic Games that he decreed that after 1996 all bids for the Olympic Games must be submitted by joint Olympic/Paralympic Committees. "The Paralympic Games have been as successful as the Olympic Games," said Samaranch, "that it is an indication that we must take another look at, and think seriously about, the subject of disabled people. It was the same organizing committee and the same volunteers that worked out very well." Due in part to the success of these games, the Olympic Games, starting with the 1996 games in Atlanta, Georgia, now include two events for athletes with physical disabilities—the 1,500-meter wheelchair race for men and 800-meter wheelchair race for women—as demonstration events.

Although full inclusion and merger with the Paralympic Games may not be possible or feasible, the recognition of events for athletes with disabilities in the Olympic Games demonstrates the strides made on behalf of athletes with disabilities in being recognized as true athletes.

In this way, the Paralympic Games, which began life as an event with strong social implications and with therapeutic ends, have become the most important sporting event for people with disabilities. Every four years the participation of elite athletes at the Paralympic Games provides proof of the progress being made in terms of competitiveness and athleticism. The records broken and the marks achieved, as well as the increase in international attention, further demonstrate the great success of these games.

The Paralympic Games are recognized by the International Olympic Committee (IOC) and sanctioned by the International Paralympic Committee (IPC), a member organization of the IOC. The five international disabled sport federations, each representing various disability groups under IPC jurisdiction, provide technical guidelines for classification criteria to the Paralympics. Every country that participates in the Paralympic Games has identified a national counterpart to these federations (table 3.1). For example, the United States Association for Blind Athletes (USABA) works with the International Blind Sports Association (IBSA) for all international and Paralympic competitions.

Table 3.1 Relationship Among U. S. Olympic Committee Disabled Sport Organizations and International Sport Federations

American sport organization	International counterparts	Disability areas
Disabled Sports USA (DS/USA)*	International Sports Organization for the Disabled (ISOD)	Amputations, winter sports for several disability areas
Dwarf Athletic Association of America (DAAA)*	International Sports Organization for the Disabled (ISOD)	Dwarfs
Special Olympics*	Special Olympics, Inc. (SOI)	Mental retardation
USA Deaf Sports Federation (USADSF)**	International Committee on Silent Sports (CISS)	Deafness
United States Association for Blind Athletes (USABA)*	International Blind Sports Association (IBSA)	Blindness
United States Cerebral Palsy Athletic Association (USCPAA)*	Cerebral Palsy-International Sports and Recreation Association (CP-ISRA)	Cerebral palsy, head injury, stroke
Wheelchair Sports USA (WSUSA)*	Stoke Mandeville Wheelchair Sports Federation (SMWSF)	Spinal cord injured amputee, other wheelchair users

* Paralympic affiliated organization

** Non-Paralympic organization (participates in World Games for the Deaf-[WGD])

Note: Special Olympics, Inc., in conjunction with the USOC, has assumed the responsibility for athletes participating in the Paralympic Games for the mentally handicapped in the United States.

CLASSIFICATION

Classification systems have been widely used in sports to allow for a fair and equitable starting point for competition (Richter, Adams-Mushett, Ferrara, & McCann, 1992). Youth football has minimum and maximum weight restrictions for players, while youth soccer and baseball may have age and gender restrictions; the express purpose of these restrictions is to provide for maximum enjoyment and fairness and to aid in the prevention of injuries. Within a specific type of disability there may be a wide continuum of ability or physical characteristics. For instance, the level of acuity in people with visual impairments may vary, as do the levels of physical involvement in a person with cerebral palsy. This continuum of severity of disability is present in all disabilities. It is accepted that some form of classification must be used for athletes with disabilities, but the most equitable type of classification remains a topic of debate and complexity. Classification systems are generally of two types: medical and functional.

Medical classification verifies minimum disability and is not concerned with the functional ability of the athlete (Davis & Ferrara, 1996). Examples include the level of visual acuity for a blind athlete (figure 3.5), the level of spinal cord injury, or the location of an amputation. This evaluation provides a medically related equal starting point for competition. Success or failure in competition now depends on the physical skill and level of training of athletes.

The functional classification systems identify how an athlete performs specific sport skills (Davis & Ferrara, 1996). Functional systems combine medical information with performance information to evaluate an athlete's sport-specific skills needed in an athletic event in addition to the medical condition. For instance, classifiers may observe athletes with cerebral palsy performing their sport to determine range of motion and physical capabilities prior to classification. This classification system can be used for single- and cross-disability competitions. In other words, function is primary and medical is secondary. Whichever classification system is used, it should ensure that the training and skill level of the athlete becomes the deciding factor in success, not the type or level of disability (Paciorek & Jones, 1994). Sport and/or health care professionals who have completed a certification course perform classification for national and international competition. Classifiers usually consist of physicians, physical therapists, occupational therapists, and others knowledgeable in kinesiology and disability. Classifications related to specific disabilities can be found in the chapters in this text that discuss these disabilities.

A challenge facing leaders in cross-disability competitions, such as the Paralympics, is the great number of classification categories. Having a high number of classes produces an organizational nightmare because of the lack of numbers in each classification for competition (Paciorek & Jones, 1994). This is especially true in Europe and rural areas of the United States where sufficient numbers of individuals with similar disability conditions are not available for competition. A low number of entrants may diminish the quality of the competition or force the cancellation of events; this may confuse spectators and devastate youth competitors. To meet this challenge, the functional system of classification has been used as a means of allowing individuals with *different disabilities* to compete against each other.

Photo by Michael Paciorek.

Figure 3.5   These B-1 classified blind runners compete against others of similar visual acuity.

For instance, competitors in a powerlifting event may include representatives from cross-disability groups, such as cerebral palsy, spinal cord injured, amputations, and dwarfism. Theoretically, if the classification system is valid, each of these competitors has an equal chance of winning based on their training and skill. Unfortunately, while this system of classification may be well suited to certain sports such as target shooting and archery, it may not be appropriate for such events as swimming and track. Richter et al. (1992) note that the functional classification system, when used in certain cross-disability events, may discriminate against certain disabilities. The IPC continues to analyze and refine the classification systems now in use.

SCHOOL AND LOCAL COMMUNITY-BASED ADAPTED SPORT PROGRAMS

While international competitions such as the Paralympic Games, Special Olympics World Games, and World Games for the Deaf tend to receive much of the media's attention, similar to regular sport, very few athletes with disabilities will have the necessary skills or opportunity to participate in such events for elite athletes. The majority of athletes with disabilities will participate in grassroots and youth sport programming in their local communities. Grassroots programming may include local park and recreation programs, parent-sponsored sports programs, cross-disability sports programs affiliated with rehabilitation centers, and sports sponsored through public school athletic programs.

Section 504 of the Rehabilitation Act of 1973 and the Americans with Disabilities Act (ADA) have had great implications for the participation of youths with disabilities in sports. Together, these complementary antidiscrimination pieces of legislation require agencies that provide sport programs, whether or not they receive federal funds, to provide comparable opportunities to individuals with disabilities. No sports program may discriminate against players or coaches with disabilities, including school-based programs. All players have the right to play on recreational sports teams and must be afforded the opportunity to try out for elite level travel teams, even if special, segregated programs such as Challenger baseball leagues are available. Schools, recreation programs, and communities must make reasonable accommodations to ensure that individuals with disabilities have equal access to programs. Reasonable accommodations may require the purchasing of an adapted piece of equipment, such as a special, wider bench for a powerlifter with a spinal cord injury, or the use of facilities that are readily accessible to individuals with disabilities, such as the use of a community fitness center.

In the past few years, much of the impetus for sports for athletes with disabilities has been provided by out-of-school sport organizations. Although developing at a slower rate, other opportunities have begun to surface throughout the United States in connection with public school programs. An important milestone came on November 11, 1992, when Minnesota became the first state to welcome athletes with disabilities into its state high school association. This made Minnesota the first state in the nation to sanction interschool sports for middle and high school students with disabilities. Prior to this time (since 1981) athletics for students with disabilities in Minnesota were established and conducted by the Minnesota Association for Adapted Athletes. MAAA welcomed the development of incorporation into the Minnesota High School League.

The 1999 High School Athletics Survey conducted by the National Federation of High Schools (NFHS) indicates that only four states offer any type of adapted sports programming: Minnesota, Maryland, Connecticut, and Iowa. Minnesota continues to lead the nation in programming for students with disabilities with more than 50 middle and high schools offering coed adapted sports competition in soccer, floor hockey, and softball to more than 1,000 students with physical and mental impairments. Future plans in Minnesota include exploring the feasibility of having teleconferenced competition. The sport of bowling or powerlifting, for instance, can have a televised link provided at sites many miles apart. Athletes can bowl or lift during their turns and watch the monitor for the performance of their competitors. This has tremendous potential for individual sports such as archery and target shooting, to name a few. Teleconferenced competition will help provide increased competitive opportunities for athletes in rural locations or in places where numbers of athletes with disabilities may be low. This program continues to serve as a model for the rest of the country.

The Connecticut Interscholastic Athletic Conference (CIAC) has formed a partnership with Special Olympics to offer Unified Sports programming. Participation is open to all public schools in Connecticut. At the elementary level, students engage in noncompetitive athletic activities designed to develop skills in a variety of sports, while middle and high school students compete in statewide Unified Sports tournaments. The CIAC currently holds tournaments in the sports of basketball, soccer, softball, and volleyball. The program boasts a participation of 1,300 athletes and 120 schools throughout the state.

A few other states now provide statewide competition for athletes with disabilities. For example, Iowa offers track-and-field competitions at their state high school championships, and New York offers regional and statewide competition for individuals with disabilities (other than mental retardation) in connection with its Empire State Games for the Physically Challenged. These games, conducted with government financing, serve as an alternative to New York's Empire State Games designed for athletes without disabilities.

Sport programs in rehabilitation settings are also increasing due to the tremendous health benefits (Paciorek & Jones, 1994). Most of the larger programs are centered in major cities. Examples include the Wirtz Sports Program at the Rehabilitation Institute of Chicago's Center for Health and Fitness (figure 3.6); the Shepard Center in Atlanta, Georgia; and the Craig Rehabilitation Hospital in Denver, Colorado. These centers generally offer wide-based multisport, cross-disability programming for individuals in the community.

TRANSITION SERVICES

An important goal of physical education and adapted physical education programs is to provide students with the functional motor skills, knowledge, and opportunities necessary for lifelong, healthy independent living. Adapted sport programs can play a vital role in assisting with independent living and should be seen and used as logical extensions of the school-based program. These programs should provide for a clear transition between the school and community living. PL 105-17 The Individuals with Disabilities Education Act (IDEA) is very clear on this issue. It requires that beginning no later than age 14 each student's IEP must include specific transition-related content. Beginning no later than age 16, a statement of needed transition services that includes interagency responsibilities or any needed linkages must also be a part of each student's IEP.

IDEA section 602 (30) defines transition services in the following:

The term transition services means a coordinated set of activities for a student with a disability, designed within the outcome oriented process, which promotes movement from school to post-school activities, including post-secondary education, vocational training, integrated employment, continuing and adult education, adult services, independent living, or community participation. The coordinated set of activities shall be based on the individual student's needs, taking into account the student's preferences and interests, and includes instruction, related services, community experiences, the development of employment and other post-school adult living objectives and, when appropriate, acquisition of daily living skills and functional vocational evaluation.

Photo courtesy of Specialized Sports Unlimited.

Figure 3.6 The Rehabilitation Institute of Chicago-Center for Health and Fitness offers multisport programming for individuals with disabilities. World-class racer Linda Mastandrea works out on a wheelchair roller.

Although the law specifies that a transition plan must be provided by age 16 at the latest, transition services in physical education and sport should begin at a much earlier age to foster independence and inclusion in community living. Taking into account the specific needs and desires of each student, the physical educator should use the IEP to identify and recommend extracurricular programs including intramurals and sports. These activities will not only facilitate the goals of adapted physical education but may also facilitate goals of occupational and physical therapy. By linking these interdisciplinary services, there is a greater chance that these recommendations will be implemented. Often, once parents are aware of programs available for their child, they can have more influence than teachers with school officials in encouraging the development of such opportunities. Adapted physical education and adapted sport can provide a vital link in the transition plan to assist the person with a disability in becoming fully integrated into community life. The Transition Services application example shows how one student's IEP lists various transition options.

Although people with disabilities participate in sports for the same reasons (fun, socialization, fitness) as those without disabilities, it can be argued that the benefits are greater for those with disabilities due to existing social and cultural barriers that must be overcome. Although great strides have been made in public perceptions, inclusion, transportation, and accessibility of programs, challenges must be overcome in each of these areas before people with disabilities are able to gain full access to opportunities. Sherrill (1998) describes self-actualization as the ultimate outcome or benefit of adapted physical activity programs. It is implied that self-actualization is a lifelong process that leads to independence and is closely associated with empowerment, the process by which individuals gain control over their lives. Participation in sports may assist a person with a disability to develop greater independence through active, healthy living. Although Sir Ludwig Guttman realized the potential of sport in his rehabilitation program at Stoke Mandeville Hospital in England as early as 1946, it has taken many years for the full impact to be realized.

ROLE OF THE PHYSICAL EDUCATOR IN ADAPTED SPORT

Opportunities for students without disabilities to participate in sport and leisure programs within and outside of school are well documented. Most middle and high schools provide a full range of sports for the interested student. Opportunities for students with disabilities, however, are not always apparent or available. Although federal legislation supporting these programs

has been available since 1973, many schools have been slow to respond to the needs and desires of their students with disabilities. Numerous court cases have documented the rights of students with disabilities to have access to the same quality of programs available to students without disabilities.

As long as physical educators do not understand the important role they play in facilitating sport participation for students with disabilities, students such as Max, who was introduced in the opening scenario, will sit idly by the sidelines watching others participate and wondering what it feels like to experience the thrill of participation and competition. To help students like Max, the physical educator can facilitate participation in sports and leisure programs in the following ways:

- ▶ Realize that community recreation programs and extracurricular activities provide an excellent means to extend physical education programs into the community.
- ▶ Speak with students and parents concerning their specific interests in sports and leisure pursuits. The goals of adapted physical education should be tied to the functional needs of the students. If the student has an interest in in-line skating or bicycling, those skills should be incorporated into the school's physical education curriculum for that student.

Application Example: Transition Services

Setting:	IEP committee meeting
Student:	A 17-year-old student who is mentally retarded with spina bifida and a wheelchair user
Issue:	The student's parents ask about the educational program that will teach their child about community recreation opportunities.
Application:	Because federal law requires that transition services be included in a student's IEP by age 16, the physical educator has already considered the available options. The IEP contains the following information:

- ▶ Each week the student will be exposed to various community recreation opportunities. Every Thursday the student and other members of his class will leave school for community recreation programs. The student will attend swimming and fitness classes at the local fitness facility and receive instruction on the use of the equipment.
- ▶ Other facilities, such as the local bowling alley and the town's recreation department, will be visited on alternate weeks.
- ▶ The physical educator has planned trips to the disability sports program at the local rehabilitation hospital. The student will participate in the center's competitive sport program of sitting volleyball and sailing.
- ▶ The physical educator has collaborated with the special education resource teacher to include instruction on the use of public transportation to access the recreational programs.

▶ Through physical education classes, ensure that students with disabilities have developed the appropriate functional motor skills and knowledge they will need to participate in a wide variety of extracurricular activities. Providing sports programming in the community will be useless if the participant lacks the skill or knowledge to participate. Focus primarily on those activities that are lifelong in nature.

▶ If the student cannot perform a skill due to a disability, modify the activity to allow participation. The chapters related to specific sports in this text describe some excellent ways of modifying activities.

▶ Students without disabilities should be exposed to the capabilities of athletes with disabilities. Promote adapted sport by obtaining videos of adapted sporting events and showing them to all students. Decorate your gym with posters of athletes with disabilities and focus on the capabilities and accomplishments of athletes with disabilities. Sponsor a wheelchair basketball game or goal ball game in your gym.

▶ Develop a resource file of adapted sport programs within and outside of your community, and communicate these resources to parents and students. Contact the state or national governing bodies related to sports to obtain information on available programs for athletes with disabilities. Many families of children with disabilities are not aware of the sport and leisure programs that are now available. Community park and recreation programs may offer programming for individuals with disabilities. National disabled sport organizations have local chapters with established programs and teams at the local and regional level that provide opportunities for competitive sport.

▶ Work within the school district to assist in developing intramurals, competitive sports activities, and recreation programs for students with disabilities. Approach area schools and combine students if necessary to develop teams. Some state high school athletic associations are developing competitive sport opportunities for students with disabilities at the middle and high school levels. Contact state high school athletic federations to see what opportunities are available.

▶ Use the IEP to recommend appropriate extracurricular sport programming, such as intramurals.

▶ Appreciate the skill levels that individuals with disabilities can attain with training and encouragement.

▶ Become familiar with the sports that are available to individuals with disabilities. Consult the resources at the end of this chapter and information in other chapters in this book.

The unqualified success of national and international sports events for people with disabilities, such as the Paralympics, World Games for the Deaf, and Special Olympics, has documented the ability of people with disabilities to become elite athletes who train and compete just as hard as their counterparts without disabilities. Television and media exposure of adapted sport continues to improve. Highly publicized court cases, such as disabled golfer Casey Martin's successful attempt to compete in the PGA while using a golf cart, emphasize that people with disabilities can compete in regular competitions if reasonable accommodations are provided that do not change the nature of the sport or alter the skill required. No longer should one equate *disability* with *lack of ability*. Elite athletes with disabilities complete the wheelchair marathon in less than 1.5 hours, athletes who are blind complete bicycle races of 80 miles or more, athletes with cerebral palsy play soccer at a high level, and athletes with amputations run the 100-meter dash in times less than 2 seconds off the Olympic record. Competitions such as goal ball, wheelchair soccer, and Special Olympic gymnastics now draw large numbers of spectators. People with disabilities are more visible in our communities, such as wheelchair users playing tennis, or children with cerebral palsy playing soccer and baseball, or people who are blind riding tandem bikes or in-line skating with sighted friends. It is clear that people with disabilities have the same desires and needs to participate in sports as do people without disabilities. The physical educator plays a vital role in facilitating these experiences.

AND WHAT ABOUT MAX?

Shortly before this book went to press, Max's physical education teacher and physical therapist contacted the United States Cerebral Palsy Athletic Association and were told of an affiliated program in their area. Max and his parents discussed the various activities that were available. Max made the final decision on what activities he wanted to pursue. Participation criteria were added to Max's IEP, and, even though he has severe physical limitations, he now participates in boccie, slalom, and various field events modified for his classification. As part of his physical therapy, Max will receive sport-specific exercises that will help him become proficient in his chosen activities. With help from his physical education teacher, Max learned about world-class cerebral palsy boccie players Jim Thompson and Kenny Johnson, who he now views as role models (figure 3.7). Max hopes to equal the success of his role models one day by competing on the USCPAA boccie team in the Paralympic Games.

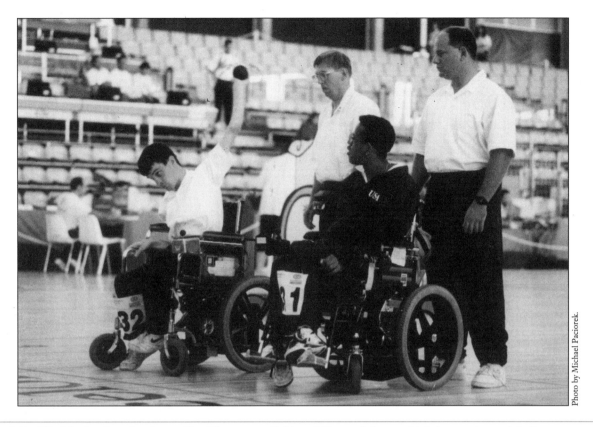

Figure 3.7   World-class boccie players Jim Thompson (left) and Kenny Johnson (right) are role models for youngsters throughout the country.

SUMMARY

In the past 25 years, many advances have occurred in adapted sport. Federal legislation, the success of some disabled sports organizations, increased awareness of health benefits by physical educators and medical personnel, and involvement with the United States Olympic Committee are some reasons for the growth of adapted sport. Many opportunities exist today for people with disabilities to participate in sport and leisure activities throughout communities. Opportunities exist at regional, national, and international levels for the individual interested in high levels of competition through multi and unisport organizations. A continuum of involvement exists within the least restrictive environment in regular or adapted sport. Physical educators of today must be resource persons. They need to be aware of the programs in their communities for their students with disabilities who can take the skills and knowledge they have acquired in adapted physical education and use them. Adapted sport should be an important part of the transition plan specified in the IEP for students who have disabilities.

Chapter 4

Individualized Education Programs

Francis X. Short

▷ **M**RS. SHOCKLEY, A PHYSICAL EDUCATION teacher at the local elementary school, was doing some paperwork at the dining room table when her husband looked over her shoulder.

"What's all this?" he asked.

▷ "I'm working on a new IEP for one of my students," she said. "I have a committee meeting tomorrow."

"All of this stuff is for one kid?"

▷ "Yeah, but it's not that much—mostly test results, a copy of his current IEP, and some notes I've made."

"Is an IEP like a lesson plan?"

"No. I still have to write lesson plans, too. Mostly the IEP lists the student's goals and objectives and the resources that will be provided to reach them. Parents also have some input, and we can use the IEP to monitor progress during the year."

"Sounds like a lot of work for one student."

"Well, it does take a little time, but it really is helpful in developing an individualized program for a student with a disability. Besides, the IEP is required by law."

▷ "Required by law? Well, then, what happens if he doesn't make the goals and objectives you're writing? Can you get in trouble?"

"Relax, Perry Mason, I won't be going to the Big House if Stevie does only one curl-up!"

Federal legislation has mandated that students with disabilities receive individualized programs. Depending on their age and other circumstances, the students' individualized programs may be described in Individualized Education Programs (IEPs), Individualized Family Service Programs (IFSPs), or Section 504 Accommodation Plans. Furthermore, students who are not disabled but who have unique needs in physical education also may have individualized programs developed for them. Such programs, while not required by federal legislation, are recommended and are called Individualized Physical Education Programs (IPEPs). This chapter provides an overview of these programs and discusses in more detail the requirements of and procedures for developing IEPs, 504 Plans, and IPEPs.

OVERVIEW OF INDIVIDUALIZED PROGRAMS

When President Ford signed PL 94-142, the Education for All Handicapped Children Act of 1975, the provision of special education in the United States was significantly changed in a number of ways. One significant change was the provision that students classified as having a disability should have IEPs. More recent legislation, culminating with the Individuals with Disabilities Education Act (IDEA), has reaffirmed the importance of IEPs in developing appropriate educational plans for students with disabilities. An IEP is a written document that essentially describes the student's current level of educational achievement, identifies goals and objectives for the near future, and lists the educational services to be provided to meet those goals. IDEA requires that IEPs be developed for all students with disabilities between the ages of 3 and 21.

IDEA also has provisions for addressing the developmental needs of infants and toddlers with disabilities. Local agencies may provide early intervention services for infants and toddlers in accord with state discretion. These services are detailed in an IFSP, a document written for all eligible participants. (The IFSP is described in chapter 20.) IDEA therefore addresses the needs of students (ages 0 to 21) who meet the criteria of infants, toddlers, and children with disabilities as defined by the act (see chapter 1 for a description of IDEA).

Some students with disabilities, however, may not meet the criteria to qualify for special education services provided by IDEA. Youngsters with the following conditions might be in this group: HIV/AIDS, alcohol abuse, substance abuse, asthma, diabetes, attention deficit disorder, or mild learning disabilities not requiring special education (French, Henderson, Kinnison, & Sherrill, 1998). These students would not have IEPs, but they may be entitled to appropriate accommodations and services tailored to meet their individual needs as provided in a Section 504 Accommodation Plan.

In physical education there may be yet a third group of students (in addition to those covered by IDEA and others covered by 504) who may require individualized programs. This would be a group of students who are not considered disabled (by either piece of legislation) but who have unique needs in physical education. Students recuperating from injuries or accidents, those convalescing from noncommunicable diseases, and those who are overweight or have low skill levels or low levels of physical fitness may fall into this category. While not required by legislation, it is recommended that school districts develop IPEPs to document a program modified to meet students' unique physical education needs.

THE STUDENT WITH A DISABILITY

The specially designed program for any child who has been identified as disabled by the school district in accordance with IDEA is dictated by the IEP. Local districts may determine and design their own IEP format. Consequently, it is not unusual for neighboring school districts to use different IEP forms. While formats may vary, however, each IEP must contain certain information.

COMPONENTS OF THE IEP

Although local IEP forms may include additional information, IDEA requires eight components. Each of these required components is discussed below, and sample physical education information that might be included in an IEP is shown in figure 4.1.

Present Level of Performance

Every IEP must include a statement of the child's present levels of educational performance, including how the child's disability affects the child's involvement and progress in the general curriculum or, for preschool children, how the disability affects the child's participation in appropriate activities (IDEA, 1997). The present level of performance (PLP) component is the cornerstone of the IEP. The information presented in all subsequent components is related to the information set forth in the PLP. If the PLP is not adequately and properly determined, chances are that the student's specially designed instructional program will not be the most appropriate. PLP statements should be objective, observable, and measurable; they should accurately reflect the child's current educational abilities. Ordinarily, the PLP component consists primarily of test results. These results can come from standardized tests with performance norms or criteria or from less formal teacher-constructed tests (including authentic assessments). Although both types of results are appropriately included in the IEP, it is recommended that, when possible, the PLP contain some standardized test results. Standardized results can help determine a unique need in a particular area, can help to illustrate the effects of a disability on the child's involvement in the general curriculum, and can provide stronger justification for an educational placement.

PLP information should be presented in a way that places the student on a continuum of achievement—that is, the test results shown should discriminate among levels of ability. For this reason, tests on which students attain either minimum (0 out of 10) or maximum (10 out of 10) scores are not very helpful in determining PLP (although some criterion-referenced tests may simply have pass/fail standards). Also, in situating a student on this continuum, the PLP component should note, to the extent possible, what the individual *can* do, not what the individual cannot do. Finally, PLP information should be presented in a way that is immediately interpretable; it should not require additional detailed explanation from the teacher. When standardized test results are included, it is helpful when percentiles, criterion-referenced standards, or other references are presented as well as the raw scores; teacher-constructed tests should be adequately described so the conditions can be replicated at a later date. (See chapter 5 for more information on measurement in adapted physical education.)

Annual Goals and Short-Term Objectives

Also required on every IEP is a statement of measurable annual goals, including benchmarks or short-term objectives. These goals and objectives relate to meeting the child's needs that result from the disability so that the child can be involved in and progress in the general curriculum, and meeting each of the child's other educational needs that result from the child's disability (IDEA, 1997).

An annual goal is a broad or generic statement designed to give direction to the instructional program. Once the PLP information has been obtained and studied, the teacher should identify one or more content areas to be emphasized in the student's program. An annual goal focuses on the student's unique needs as identified in the PLP. In fact, this linkage is a key element in writing an annual goal statement. The annual goal must clearly relate to information presented in the PLP component. For instance, if the PLP contains only information on ball-handling skills, it would be inappropriate to write an annual goal for swimming, physical fitness, or any other content area unrelated to ball-handling skills. The need for emphasis on a particular content area must be demonstrated in the PLP.

While the annual goal is broad or generic, a short-term objective (STO) is narrow and specific. An STO is a statement that describes a skill in terms of action, condition, and criterion. Action refers to the type of skill to be performed (e.g., running). Condition indicates the way the skill is to be performed (e.g., running 50 yards). Criterion refers to how well the skill is to be performed (e.g., running 50 yards in 8.5 seconds). Conditions and criteria used in physical education usually relate to such concepts as "how fast," "how long," "how far," or "how many," although it is also appropriate to describe "how mature." For instance, a 10-year-old student may throw a ball into a wall target 9 out of 10 times from a certain distance. The teacher may be pleased with the accuracy score ("how many"), but if the student does not step with the opposite foot when throwing, the teacher may

Individualized Education Program

Student's name: Thomas Hernandez

Age: 10

Present Level of Performance

1. Completed 5 laps in the 16-m PACER (Brockport Physical Fitness Test [BPFT] specific standard is 9 laps).
2. Sum of triceps and subscapular skinfold measures is 24 mm (BPFT preferred general standard is 11-22 mm).
3. Performs 10 modified curl-ups (exceeds the BPFT specific standard but does not meet the minimal general standard of 12).
4. Dominant grip strength is 8 kg (BPFT specific standard is 12 kg).
5. Scores 15 sec. on the extended arm hang (BPFT specific standard is 23 sec.)
6. Successfully performs the trunk lift at 12 in. (BPFT general standard is 9-12 in.).
7. Back-saver sit-and-reach score for right and left sides is 6 in. (BPFT general standard is 8 in.).
8. Scores 18 out of 26 for locomotor skills on the Test of Gross Motor Development (TGMD) (2nd percentile).
9. Scores 18 out of 19 for object control skills on the TGMD (50th percentile).

Effect of disability: Thomas's low fitness levels and poor locomotor skills reduce the potential benefit from the general physical education program; some individualized attention is recommended.

Annual Goals and Short-Term Objectives

1.0 Thomas will improve his aerobic functioning.

1.1 Thomas will complete 9 laps on the 16-m PACER (meets BPFT specific standard).

2.0 Thomas will improve his muscular strength and endurance.

2.1 Thomas will do 12 modified curl-ups (meets BPFT minimal general standard).

2.2 Thomas will score 12 kg on dominant grip strength (meets BPFT specific standard).

2.3 Thomas will score 23 sec. on the extended arm hang (meets BPFT specific standard).

3.0 Thomas will improve his flexibility.

3.1 Thomas will score 8 in for each leg on the back-saver sit and reach (meets BPFT general standard).

4.0 Thomas will improve his locomotor skills.

4.1 Thomas will score 25 on the locomotor skills section of the TGMD (50th percentile).

Statement of Services and Supplementary Aids

Thomas will participate in physical education classes conducted in the regular setting three days per week and receive individual instruction two days per week. No special equipment or aids are required.

Statement of Participation in Regular Settings

Approximately 60 percent of Thomas's physical education will be with nondisabled students and 40 percent will be individualized. Individualized instruction will emphasize fitness and locomotor skills.

Assessment Modifications

The BPFT includes appropriate modifications for fitness assessment, and Thomas can take the TGMD without modification.

Schedule of Services

Thomas will have physical education in the regular setting on Mondays, Tuesdays, and Thursdays from 9:30-10:15. His individual instruction will take place on Wednesdays and Fridays from 1:15-2:00.

Transition Services

None at this time. (Thomas is 10 years old.)

Procedures for Evaluation and Parental Report

Criteria for evaluation are contained in the short-term objectives. Progress on the objectives will be monitored weekly. A final assessment will be conducted during the week of June 1. The Brockport Physical Fitness Test and the Test of Gross Motor Development will be used for this evaluation. A written report describing Thomas's progress on his goals and objectives will be sent to the parents at least quarterly.

Figure 4.1 Sample physical education information for an IEP.

---

not be pleased with the quality of the movement pattern ("how mature"). In this case the teacher might write an STO that describes a movement pattern, rather than an accuracy score, to be attained.

Just as an annual goal must relate to PLP information, STOs must relate to an annual goal. If an annual goal stresses the content area of "eye-hand coordina-tion," the STOs should include skills such as throwing, catching, and striking. Furthermore, it is important that the student's baseline (pretest) ability appear in the IEP, usually in the PLP component. For example, a short-term objective might specify that a student will be expected to do 15 curl-ups in 60 seconds at some future date; this statement has little meaning unless it is known how many curl-ups the student can do now. In fact, the easiest way to write an STO is to take a well-written PLP statement, copy the action and condition elements verbatim, and make a reasonable change in the criterion. (The teacher must use professional judgment to determine what constitutes a "reasonable" expectation for improvement.) It should be noted that although STOs are helpful in identifying activities to be conducted in class, they are not meant to supplant daily, weekly, or monthly lesson plans.

Statement of Services and Supplementary Aids

A third required component of the IEP is a statement of the special education and related services and supplementary aids and services to be provided to the child (or on behalf of the child) as well as a statement of the program modifications or supports for school personnel that will be provided for the child. These services, aids, and supports are provided to help the child progress toward the annual goals, to be involved and progress in the general curriculum as well as in extracurricular and nonacademic activities, and to participate with children with disabilities as well as nondisabled children (IDEA, 1997).

Once the present level of performance is determined and annual goals and short-term instructional objectives are written, decisions must be made regarding the student's educational placement, additional services (if any) to be provided, and the use of special instructional media and materials as necessary. The placement agreed on should be considered the least restrictive environment for the student.

In addition to appropriate placement, other special education and related services may be prescribed. A special education service is one that makes a direct impact on educational objectives, for example, physical education. Provisions for this service should be specified in this component of the IEP. A related service makes an indirect impact on educational objectives and therefore should be presented only to the extent that it will help the student benefit from a special education service. Examples include physical therapy; therapeutic recreation; occupational therapy; psychological services; and speech, language, and/or hearing therapy.

In some cases special instructional materials may be required for the education of children with disabilities. These materials should also be listed in this IEP component. Modified pieces of equipment such as a beep baseball, an audible goal locator, a snap-handle bowling ball, or a bowling ramp are examples.

Statement of Participation in Regular Settings

The IEP must contain an explanation of the extent, if any, to which the child will not participate with nondisabled children in the regular class and in other extracurricular and nonacademic activities (IDEA, 1997). If a child is removed from his or her regular physical education setting to implement the adapted physical education program, for instance, it would be noted in this component of the IEP. Usually this explanation includes a percentage of time the student is excluded from (and/or included in) the regular educational setting and for what kinds of activities.

Assessment Modifications

Another required IEP component is a statement of any individual modifications in the administration of state or districtwide assessments of student achievement that are needed in order for the child to participate in such assessment. Furthermore, if it is determined that the child will not participate in a particular assessment, the IEP must include a statement as to why that assessment is not appropriate for the child and how the child will be assessed (IDEA, 1997). So, if a school district routinely administers physical education tests (physical fitness, fundamental motor skills, sports skills, aquatics, etc.) to its students, the appropriateness of those tests for a student with a disability must be considered. In this case, test item modifications or test item substitutions must be provided as necessary and noted on the IEP.

Schedule of Services

A sixth required component is the projected date for the beginning of the services and modifications listed earlier in the IEP and the anticipated frequency, location, and duration of those services and modifications (IDEA, 1997).

Transition Services

The IEP also requires that beginning at age 16 (or younger if determined appropriate), a statement of needed transition services for the child must be added (IDEA, 1997). This component, therefore, includes goals and specific actions to help the student make a successful transition from the school-based educational program to another community-based option that will occur no later than age 22. Many students, for instance, may eventually be enrolled in vocational training programs, some may have the opportunity to go on to college, and others may enter alternative adult service programs (e.g., group homes, sheltered workshops). School personnel attempt to prepare the student for entrance into the most appropriate option when he or she "ages out" of school.

Procedures for Evaluation and Parental Report

The final required component of the IEP is a statement of how the child's progress toward the annual goals will be measured and how the child's parents will be regularly informed of that progress (IDEA, 1997). This component is used to specify how and when the student's progress will be evaluated. In most cases progress is determined by testing the objectives written earlier. The evaluation should indicate the extent to which the progress is sufficient to enable the child to achieve the goals by the end of the year. Evaluation can be scheduled to occur at any time within 12 months from the time the IEP takes effect; the IEP *must* be reviewed at least annually. Parental notification must take place at least as often as parents are informed of nondisabled students' progress (e.g., the frequency of regular report cards).

DEVELOPMENT OF THE IEP

Procedures for developing an IEP vary slightly from state to state, but essentially the process involves two steps. The first is to determine if the student is eligible for special education services; if so, the second step is to develop the most appropriate program, including the establishment of goals and objectives and the determination of an appropriate placement. The process that results in the development of an IEP usually begins with a referral. Any professional staff member at a particular school who suspects that a child might possess a disability can refer the child for an evaluation to determine eligibility for special education. A referral should outline reasons for suspecting a disability, including test results, records, or reports; attempts to remedy the student's performance; and the extent of parental contact prior to the referral (New York State Education Department, 1998). A sample referral form is shown in figure 4.2. Parents may refer their own children for evaluation when they suspect a problem. In fact, when parents enroll a youngster in a new school, the district will frequently ask if they feel their child might possess a disability.

IDEA requires that an IEP Team, consisting of one or both of the child's parents; the child's teacher; a representative of the school district who is qualified to provide or supervise the provision of special education; the child (where appropriate); and other appropriate individuals, at the discretion of either the parents or the school, be charged with the responsibility of determining special education eligibility (Department of Education, 1998). (This diagnostic team has different names in different states. In New York, for instance, it is called the Committee on Special Education. Regardless of name, however, the team's primary purpose is to determine whether a particular student has a unique educational need or needs.) In many cases the IEP Team will determine unique needs by assessing the results of standardized tests. But the team will also consider other information, such as samples of current academic work; the role of behavior, language, and communication skills on academic performance; the amount of previous instruction; and anecdotal accounts, including parental input, before reaching a final decision. On the basis of the information gathered and the ensuing discussion, the IEP Team will decide if the youngster qualifies for special education; if so, the team will recommend a program and a placement setting based on an IEP it has developed.

It should be emphasized that the IEP is a negotiated document; both the school and the parents have input into its development and must agree on its contents before it is signed and implemented. In the event the two parties cannot agree on the content of the student's IEP, IDEA provides procedures for resolving the disagreement. These due process procedures are designed to protect the rights of the child, the parents, and the school district (figure 4.3).

THE ROLE OF THE PHYSICAL EDUCATOR

Historically, many physical educators have not been actively involved in the IEP development process. It is clear, however, that if students with disabilities are to receive the free, appropriate education guaranteed by IDEA, physical education must be included in their IEPs. It is also clear that this more likely will be accomplished only if physical educators make sure that they are involved in IEP development. While physical education must be included in every IEP, the extent of IEP information required varies depending on the child's educational setting (box 4.1).

Unfortunately, many students are assigned to physical education settings more as a function of convenient scheduling than as a function of educational needs. It is not unusual for students to be evaluated by the IEP Team without a physical education assessment being performed. Consequently, students may be assigned to inappropriate physical education programs or placements, and physical education teachers may be assigned students for whom there are no assessment data, long-term goals, or short-term objectives. It is therefore recommended that school districts include the teacher of adapted physical education on the IEP Team. This educator would determine if the student qualifies for an adapted program and would help write the eligible student's initial IEP.

DEPARTMENT OF PHYSICAL EDUCATION
Referral Form

This form should be used by teachers or administrators of physical education to refer students with unique needs to chairpersons of CSE, CPSE, CAPE, or the school building administrator. Referrals should be processed through the office of the APE coordinator to the director of physical education who shall forward the referral to appropriate individuals. Referrals may be made to change the program and/or placement of the student or for any other action within the jurisdiction of the CSE, CPSE, or CAPE.

Faculty member making referral _____ Date _____

Student referred _____ Age _____ Gender _____

Present physical education class (if any) _____

Student's primary or homeroom teacher _____

A unique physical education need has been identified for the student:

By the CSE? ____ Yes ____ No

By the CPSE? ____ Yes ____ No

By the CAPE? ____ Yes ____ No

If no, give reasons for believing a unique physical education need exists _____
_____
_____

Give test results, records, or reports upon which a referral is based _____
_____
_____

Describe prior attempts to remediate student's performance _____
_____
_____

Has parental contact been made? _____ Yes _____ No **If yes, describe:** _____
_____
_____

If a recommendation for placement or other action is included as a part of this referral, indicate the recommendation: _____
_____
_____

Referral processed by: Referral initiated by:

_____   _____

(Director of physical education)   (Staff member)

Legend: **APE - Adapted Physical Education**

CSE - Committe on Special Education

CPSE - Committe on Preschool Education

CAPE - Committee on Adapted Physical Eduction

Figure 4.2   A sample referral form for adapted physical education.

IEP sequence

Due process procedures

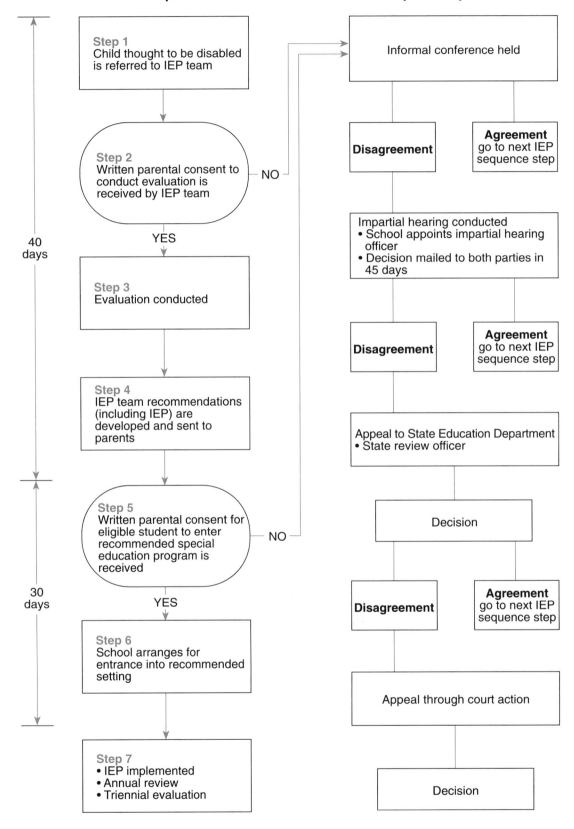

Figure 4.3   Sample IEP sequence and due process procedures.

> Box 4.1 **An Important Question**

When must physical education be described or referred to in the IEP? According to the *Federal Register* (Department of Education, 1998), the answer depends on the type of "PE program arrangement" the student is in.

▶ *Regular physical education with nondisabled students.* If a student with a disability can participate fully in the regular physical education program without any special modifications to compensate for the student's disability, it would not be necessary to describe or refer to physical education in the IEP. On the other hand, if some modifications to the regular physical education program are necessary for the student to be able to participate in that program, those modifications must be described in the IEP.

▶ *Specially designed physical education.* If a student with a disability needs a specially designed physical education program, that program must be addressed in all applicable areas of the IEP (e.g., present levels of educational performance, goals and objectives, and services to be provided).

▶ *Physical education in separate facilities.* If a student with a disability is educated in a separate facility, the physical education program for that student must be described or referred to in the IEP. However, the kind and amount of information to be included in the IEP would depend on the physical-motor needs of the student and the type of physical education program that is to be provided. Thus, if a student at a residential school for students with deafness is able to participate in that school's regular physical education program (as determined by the most recent evaluation), then the IEP need only note such participation. On the other hand, if special modifications are required for the student to participate, those modifications must be described in the IEP. Moreover, if the student needs an individually designed physical education program, that program must be addressed under all applicable parts of the IEP.

In determining eligibility for adapted physical education, it is necessary to have a strong assessment program in place. Assessment should reflect as many content areas of physical education as possible. IDEA and this book define physical education as follows: "the development of (a) physical and motor fitness; (b) fundamental motor skills and patterns; and (c) skills in aquatics, dance, and individual and group games and sports (including intramural and lifetime sports)" (Department of Education, 1998, p. 18). The physical educator, therefore, should select assessment instruments that reflect these components. Test results help the physical educator to determine unique needs and justify a program adapted to meet unique needs.

Although documentation of subaverage test performance is important in determining an adapted program, it is not the only criterion to use in making such a decision. For instance, an emotionally disturbed child might be placed in a segregated, adapted physical education class even though the child's physical and motor skills are age appropriate. A unique need in the affective domain might prevent safe and successful participation in an integrated physical education class. In this case, a placement in a segregated setting might be justified on the basis of behavioral concerns; a large class size or an emphasis on competitive activities might make a regular physical education class inappropriate for such a child.

SECTION 504 AND THE ACCOMMODATION PLAN

Although Section 504 of the Rehabilitation Act of 1973 is more than 25 years old, its implications are only now being completely realized (French, Henderson, Kinnison, & Sherrill, 1998). (See chapter 1 for a description of 504.) One such implication is that because the definition of a "qualified individual with a disability" covered under Section 504 is broader than the definition of a "child with disability" covered under IDEA, some students with disabilities will not have IEPs but nevertheless may require (and be entitled to) appropriate accommodations and services. These accommodations and services must be documented in a Section 504 Accommodation Plan, sometimes simply called a 504 Plan.

Unlike the IEP, 504 Plans do not have mandated components and, consequently, local districts usually develop their own. A sample 504 Plan is presented in figure 4.4. The plan is developed by a committee consisting of at least two school professionals (e.g., teachers, nurses, counselors, administrators) who are familiar with the student and the school district's 504 Officer (French, Henderson, Kinnison, & Sherrill, 1998). (School districts are required to have a 504 officer, a person designated to monitor the implementation of Section 504.) Technically, 504 does not require a full evaluation by a

Section 504 Accommodation Plan

Name	Date of birth	Grade
School	Date of meeting	

1. *Describe the nature of the problem:* _____

_____

_____

2. *Evaluations completed, including dates of each evaluation:* _____

_____

_____

3. *The basis for determining that the child has a disability (if any):* _____

_____

_____

4. *Describe the nature of the child's disability:* _____

_____

_____

5. *Does the disability affect a major life activity? If "Yes," explain how:* _____

_____

_____

6. *List the accommodations (i.e., specialized instruction or equipment, auxiliary aids or services, program modifications, etc.) the team recommends as necessary to ensure the child's access to all district programs:*

_____

_____

_____

_____

_____

Review/reassessment date: _____ (must be completed)

Participants (name and title)

_____   _____

_____   _____

_____   _____

_____   _____

cc: Student's cumulative file

Attachment: Information regarding Section 504 of the Rehabilitation Act of 1973
          Due Process Notice

Figure 4.4   Sample 504 Accommodation Plan.

multidisciplinary diagnostic team as does IDEA, but clearly the assessment should focus on areas of student need that may necessitate evaluations by more than one professional.

The elements of the 504 Plan found in figure 4.4 are self-explanatory, but the reader should note that item 5 (i.e., Does the disability affect a major life activity?) is particularly important. To meet the definition of a "qualified person with a disability" under Section 504, the person must have a physical or mental impairment that substantially limits one or more major life activities (New York State Education Department, 1995). Usually the accommodations listed in the plan (see item 6) are selected to help the student benefit from instruction in the regular classroom. More restrictive educational placements, while possible, would have to be justified.

THE NONDISABLED STUDENT WITH UNIQUE NEEDS

As mentioned at the beginning of this chapter, nondisabled students who have unique needs in physical education are not covered by IDEA or by Section 504. School districts, however, still must provide an appropriate education for these students.

COMMITTEE ON ADAPTED PHYSICAL EDUCATION

It is recommended that school districts establish a Committee on Adapted Physical Education (CAPE) to address any unique needs of the nondisabled student in physical education. This committee should consist of at least three members: the director of physical education or designee, the school nurse, and the teacher of adapted physical education. When possible, the student's regular physical education teacher should also be a member of the committee, and a school administrator should be available for consultation. The function of CAPE is to determine the student's eligibility for an adapted program; define the nature of that program, including placement; and monitor the student's progress. A procedure that is recommended is outlined in figure 4.5 and discussed in the seven steps below. These may be modified as necessary to meet the needs of local school districts.

Step 1

Referrals to CAPE ordinarily are made to the chairperson of the committee by a physical educator, a family physician, a parent, or even the student, when it is felt that the student has a unique need in physical education. CAPE should consider only those referrals where the needs are thought to be long term (more than 30 days).

Step 2

When CAPE receives a referral form (from a source other than the parents), it should notify the parents of the referral and indicate that the committee will be considering an adapted program. The notification should point out that physical education is a required subject area under state law (where applicable) and that "blanket" excuses, waivers, or substitutions are not appropriate options; that development of an "adapted" program would not mean that the district considers the student to be "disabled" under IDEA or 504; and that any change in program will be reviewed periodically (at least annually). Parents should also be invited to submit their own concerns or aspirations for their child's physical education program.

Step 3

In the case of a medical excuse or referral, CAPE should contact the family physician to determine the nature of the condition or disease and the impact on physical education. A sample form to be used for this purpose is shown in figure 4.6. CAPE should also consult the student's regular physical education teacher (to determine the student's performance level and any difficulties the student experiences in his or her current program). It also may be necessary to conduct additional testing to better understand the student's strengths and weaknesses.

Step 4

After considering all the information collected in step 3, CAPE must decide if an adapted program is appropriate for this student. If the student does not have a unique need, the parents are informed and the process is over. If, however, the student is eligible for an adapted program, CAPE must develop an IPEP for the student. The IPEP is similar to the IEP and should include program goals, present level of performance (including any medical limitations), short-term objectives, placement and schedule of services, and a schedule for review.

Step 5

Parents are notified of CAPE's decision. If the student is eligible, parents should receive an explanation of the adapted program and a copy of the IPEP. It is also recommended that the regular physical education teacher and, in the case of a medically initiated referral, the family physician receive copies of the IPEP as well. (The district should have due process procedures comparable to those depicted in figure 4.3 in place, if the parents do not agree with CAPE's decision or with the program outlined in the IPEP.)

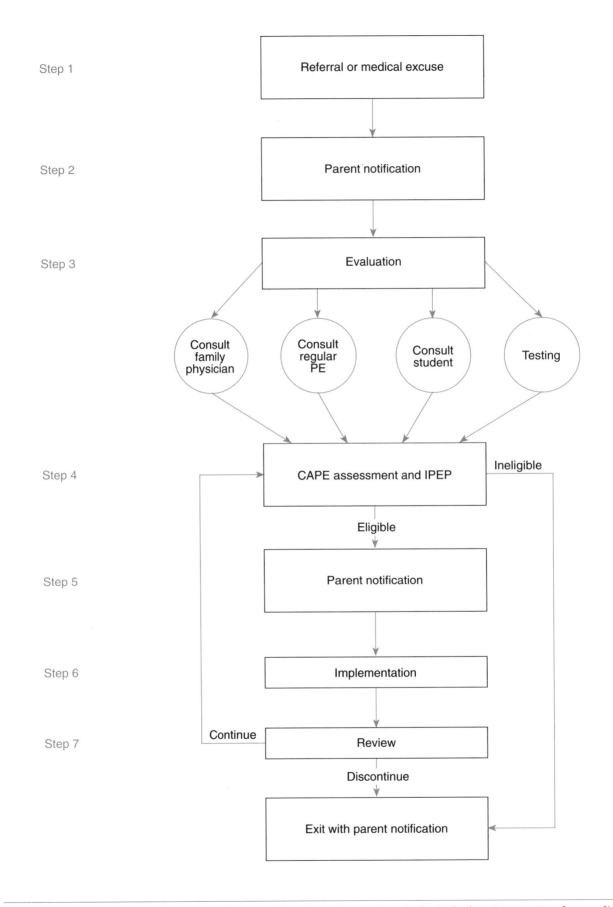

Step 1	Referral or medical excuse
Step 2	Parent notification
Step 3	Evaluation
	Consult family physician — Consult regular PE — Consult student — Testing
Step 4	CAPE assessment and IPEP — Ineligible
	Eligible
Step 5	Parent notification
Step 6	Implementation
Step 7	Review — Continue — Discontinue
	Exit with parent notification

Figure 4.5 A recommended procedure for evaluating the need for adapted physical education services for nondisabled students.

Ocean Bay Park Central School District
Physical Activity Form

To: _____, MD

From: _____ Chair, Committee on Adapted Physical Education

_____
Address

Re: _____   _____
          Name of student   Grade in school

The Committee on Adapted Physical Education (CAPE) has received a medical excuse (or referral) regarding this student. All students registered in the schools of New York are required by IDEA to attend courses of instruction in physical education. As necessary, it is required that courses are modified to meet individual student needs. The physical education classes are approximately _____ minutes in length and are held _____ times a week.

The final responsibility for the determination of a student's individualized program rests with the Committee on Adapted Physical Education. Medical information from you will assist the committee in making a decision. If further clarification is needed, the school physician will arrange a conference with you.

Diagnosis: _____

Specific physical limitations: _____

_____

Within the physical limitations listed, this student can otherwise engage in: (check one)

_____ Vigorous physical activity   _____ Mild physical activity

_____ Moderate physical activity   _____ Minimal physical activity

Other recommendations for program modification: _____

_____

Do you wish the student to return for reevaluation? Yes_____ No_____

If so, when? _____

This is to certify that I have examined_____

and recommend that he/she participate in a physical education program within the guidelines specified above for a period of _____ weeks/months.

Physician's
signature: _____

Date: _____

Figure 4.6   Sample physician form.

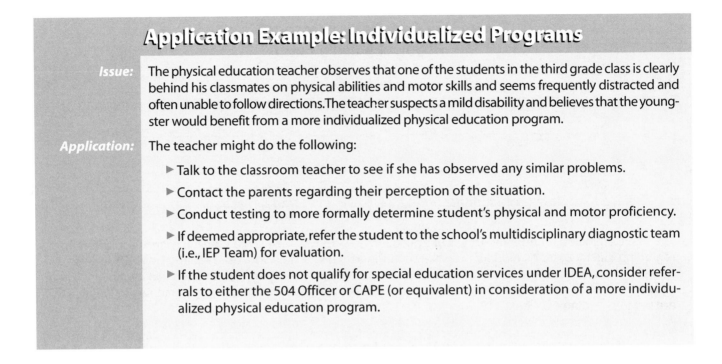

Application Example: Individualized Programs

Issue: The physical education teacher observes that one of the students in the third grade class is clearly behind his classmates on physical abilities and motor skills and seems frequently distracted and often unable to follow directions. The teacher suspects a mild disability and believes that the youngster would benefit from a more individualized physical education program.

Application: The teacher might do the following:

▶ Talk to the classroom teacher to see if she has observed any similar problems.

▶ Contact the parents regarding their perception of the situation.

▶ Conduct testing to more formally determine student's physical and motor proficiency.

▶ If deemed appropriate, refer the student to the school's multidisciplinary diagnostic team (i.e., IEP Team) for evaluation.

▶ If the student does not qualify for special education services under IDEA, consider referrals to either the 504 Officer or CAPE (or equivalent) in consideration of a more individualized physical education program.

Step 6

The adapted program described in the IPEP is implemented. Most IPEPs can probably be implemented in an integrated placement. In cases where a segregated placement is recommended, however, districts must obtain parental permission prior to changing the placement, unless the board of education has different and appropriate procedures.

Step 7

The IPEP will be in effect for the time specified under "schedule for review." At the conclusion of this time, CAPE evaluates the student's progress and decides whether to continue or discontinue the program.

SUMMARY

Students may be eligible for adapted physical education, regardless of disability, if they exhibit a unique need. IDEA and Section 504 of the Rehabilitation Act of 1973 provide procedures for the modification of educational programs for students with disabilities.

These individualized programs are spelled out in documents required by the legislation. Infants and toddlers (ages zero to two) who meet the criteria established by IDEA may receive early intervention services under the act. These services must be documented in an IFSP. Similarly, individualized programs for qualified children (ages 3 to 21) also are governed by IDEA. In this case, however, the individualized program is documented in an IEP. Not all students with disabilities qualify for services under IDEA, but they may still be entitled to modified programs under Section 504. These modified programs must be described in a Section 504 Accommodation Plan. Finally, there may be nondisabled students in a school district who have unique needs in physical education. These youngsters would not qualify for services under either IDEA or 504 but might still require a modified physical education program. Although not required by law, it is suggested that districts develop written IPEPs to document the adapted physical education program. Consider the scenario presented in the Individualized Programs application example.

Chapter 5

Measurement and Assessment

Francis X. Short

▷ **A**T A RECENT PHYSICAL EDUCATION staff meeting, teachers were discussing some of their frustrations about how students are assigned to physical education classes. "Something's definitely wrong," said Mr. Webb. "I seem to have three or four kids in every regular class who can't keep up with their classmates and should probably be learning more elementary skills than the others. Then my special education class comes into the gym, and two or three of those kids could handle the regular class, no sweat."

"That's because kids are assigned to physical education without any prior consideration of their ability," said Mrs. Nelson. "What we need to do is establish some kind of a testing program to help get these students into the most appropriate program."

"Okay, but what should we test and which tests should we use?" asked Ms. Mestre.

"Not only that," said Mr. Ferruggia, "but what cutoff scores would we use to decide on the best program for each youngster?"

Measurement and assessment serve a number of purposes in physical education. One of the more critical purposes in adapted physical education, to assess program eligibility, is characterized in the above scenario. The goal of this chapter is to familiarize the reader with many of the important concepts of measurement and assessment as they relate to adapted physical education. Measurement and assessment strategies, measurement and assessment in adapted physical education, and tests and measures for use in adapted physical education are the major topics covered.

MEASUREMENT AND ASSESSMENT STRATEGIES

The purpose of this section is to provide a brief review (or overview) of terminology and approaches generally used in physical education. The first subsection attempts to distinguish between norm-referenced and criterion-referenced standards as they are commonly used in physical education. The second subsection briefly discusses what will be referred to as standardized approaches to measurement and assessment. The third describes what will be called alternative (or authentic) types of measurement and assessment. Alternative approaches include the use of rubrics, task analyses, and portfolios. The final subsection, called Synthesis, discusses the relationship between standardized and alternative approaches.

Standards for Assessment

In the title of this chapter, "measurement" refers to the process of administering tests to students and obtaining scores that represent students' abilities (or traits). The term "assessment" (sometimes called "evaluation") pertains to a subsequent step where some kind of judgment is made about those test scores. These judgments generally are based on some standard of performance. The types of standards most commonly used to assess test scores in physical education are norm referenced and criterion referenced.

Norm-Referenced Standards

Examples of norm-referenced standards include percentiles, age norms, T scores, z scores, and other kinds of "standard scores." Norm-referenced standards generally are established by testing large numbers of subjects from specifically defined groups (usually gender and age groups and sometimes disability groups) and analyzing and summarizing the scores (often in one or more tables). These standards allow teachers to compare one student's performance against the performance of others from a particular peer group (e.g., 10-year-old girls). Norm-referenced standards, therefore, provide evaluations relative to other students and lead to statements such as "Tucker's setting ability is above average for boys his age," "Tamiko is two years behind her age group on motor-based developmental milestones," or "Tim's body mass index scores are high for his age." Norm-referenced standards are usually associated with the standardized strategies discussed later in this chapter.

Criterion-Referenced Standards

Where norm-referenced standards typically provide scores for a theoretical "average" youngster, criterion-referenced standards provide levels of "mastery" for the skill or ability being evaluated. This mastery score represents an acceptable level of performance for the test in question as determined by expert opinion, research data, logic, experience, or other means. The American Red Cross, for instance, uses criterion-referenced tests to certify lifeguards. Candidates for certification must demonstrate a level of aquatic performance deemed necessary to save lives.

Where norm-referenced assessment is a function of the scores of other students, criterion-referenced assessment is not. If the Red Cross standards for lifeguard certification were norm-referenced, perhaps the top 25 percent of the class would qualify for certification, but that top 25

Table 5.1   FITNESSGRAM and Brockport Physical Fitness Test (BPFT) Test Items Arranged by Components of Health-Related Physical Fitness

FITNESSGRAM	BPFT	
Aerobic capacity	**Aerobic functioning**	
PACER	PACER (20 m and 16 m)	
Mile run	Mile run	
Walk test	Target aerobic movement test	
Muscular strength, endurance, and flexibility	**Musculoskeletal functioning**	
Curl-ups	Reverse curl	Flexed arm hang
Trunk lift	Seated push-up	Extended arm hang
Push-up	40-m push/walk	Trunk lift
Modified push-up	Wheelchair ramp test	Curl-up
Pull-up	Push-up	Modified curl-up
Flexed arm hang	Isometric push-up	Target stretch test
Back-saver sit-and-reach	Pull-up	Shoulder stretch
Shoulder stretch	Modified pull-up	Modified Apley test
	Dumbell press	Modified Thomas test
	Bench press	Back-saver sit-and-reach
	Dominant grip strength	
Body composition	**Body composition**	
Percent fat (triceps and calf skinfolds)	Skinfolds (triceps and calf; triceps and subscapular; triceps only)	
Body mass index	Body mass index	

percent may or may not have the "mastery" necessary to save lives. With the criterion-referenced approach, theoretically all of the students in the class could qualify or, on the other hand, none could qualify. Certification is a function of the ability to meet mastery standards, not a function of how one student compares to others. Criterion-referenced assessment, therefore, makes judgments about competency and leads to statements such as "Grace's skinfold measures are in the healthy fitness zone," "Joe demonstrates a mature catching pattern," or "Louise is unable to push her wheelchair up a standard ramp."

Recently greater attention has been placed on the development of tests with criterion-referenced standards in physical education. The FITNESSGRAM (Cooper Institute for Aerobics Research, 1999a) is a good example.

The standards associated with FITNESSGRAM represent a level of performance that is thought to be necessary to attain objectives related to good health and improved function. For each test item (table 5.1), standards for a "healthy fitness zone" are provided by gender and age (5 to 17+). The healthy fitness zone is defined by a lower-end score and an upper-end score. All students are encouraged to achieve at least the lower-end criterion. There is little emphasis placed on going beyond the upper-end criterion, however, since health-related objectives can be attained simply by staying within the healthy fitness zone. As FITNESSGRAM demonstrates, criterion-referenced standards can be associated with standardized tests, but they can also be found with some of the alternative strategies discussed later in this chapter.

Standardized Strategies

Standardized strategies generally involve testing students in controlled environments. Students with like characteristics are tested in exactly the same way under exactly the same conditions. By controlling, or standardizing, the environment, teachers hope to improve the validity and reliability (including the objectivity) of the test. This kind of testing, however, is sometimes called "obtrusive," because it requires the teacher to establish a somewhat artificial testing environment, one that may be different from the actual environment where the skill or ability will be used.

In a standardized volleyball test, for instance, students may be asked to continually set or pass (i.e., bump) a ball above a line on a wall as many times as possible in 30 seconds. All students can be tested under exactly the same conditions (and their scores can be easily compared to some standard of performance), but the conditions are artificial because the game of volleyball ordinarily does not require players to handle a ball in "rapid fire" sequence nor, for that matter, to volley it against a wall. Many commercially available tests are "standardized tests."

Alternative Strategies

In recent years standardized testing strategies have come under increasing criticism. Critics, among other things, argue that many of the skills tested on standardized batteries have little "functional relevance" for the student (e.g., a test may require a student to touch her nose with her index finger as an indication of kinesthetic awareness, but this is not a skill that youngsters use a great deal either at home or at school); standardized tests are less likely to provide information important for instruction (e.g., when a standardized assessment reveals that a student is "below average" on a test, it may not directly identify *why* the performance was "below average"); and, because the testing environment is artificial (or obtrusive), it is unknown if the test results will generalize to other situations (e.g., if a student can do a good job of passing a volleyball against a wall, will that skill transfer to a game situation?) (Block, Lieberman, & Connor-Kuntz, 1998).

Proponents for alternative strategies suggest that authentic assessment is an approach that closely links assessment to instruction (so that it has day-to-day applicability for the teacher) and where assessment takes place in "natural" (unobtrusive) settings. It is designed to directly measure the skills that students need for successful participation in physical activity, and it places a premium on subjective evaluation done through observation. Three techniques associated with authentic assessment—rubrics, task analysis, and portfolios—are briefly discussed.

Rubrics

Rubrics, or scoring rubrics, are gaining in popularity as a way of measuring student performance. Closely related to checklists and rating scales, a rubric provides a mechanism for a teacher to match a student's performance to one of multiple levels of achievement through a specific set of criteria (a form of criterion-referenced standards). A sample rubric for the volleyball serve is provided in box 5.1.

One of the distinct advantages of testing and evaluating through rubrics is that students know exactly what they have to do to get the best possible score. For instance, in the volleyball serve example an overhand serve is clearly valued in the scoring scheme. Because even a poor overhand serve scores better than a good underhand serve, students know what they need to practice and can analyze their own performances (as well as that of their classmates) accordingly. The volleyball serve can be evaluated with a scoring rubric while students are playing an actual game; hence, it is "unobtrusive" and done in a "natural" environment. Rubrics also are easily modifiable. For higher-performing students, for instance, the volleyball rubric could be rewritten to include location and ball movement variables, and/or it could include a level for jump serving. For lower-performing students, the rubric might include the use of a lighter ball, a lower net, a closer serving line, or it might place a higher value on the underhand serve.

In adapted physical education settings, assessment through rubrics translates nicely to Individualized Educa-

---

> **Box 5.1 Sample Rubric for the Volleyball Serve**

5   Serves overhand. Toss brings ball consistently above the attacking shoulder. Wrist is firm at contact. Ball is rarely served out-of-bounds and can be served deep into the opponent's court with a low trajectory (ball velocity is good).

4   Serves overhand. Toss brings ball consistently over the attacking shoulder. Wrist is firm at contact. Ball is usually inbounds, although depth is inconsistent and trajectory tends to be high (ball velocity is moderate).

3   Serves overhand. Toss is inconsistent so that the ball is not always above the attacking shoulder. Wrist tends to "collapse" on contact. Ball is frequently served out-of-bounds (often in, or short of, the net; ball velocity is slow).

2   Serves underhand. Ball is usually inbounds, although the depth of the serve is often short and the trajectory tends to be high (ball velocity is moderate).

1   Serves underhand. Ball is usually out-of-bounds. Serves that are inbounds tend to be short and/or weak (ball velocity is slow).

tion Program (IEP) development and personalized instruction. For example, if a student's present level of performance on the volleyball serve is "3," a reasonable objective (or objectives) would be provided by the criteria associated with level "4" (i.e., a majority of serves inbounds, a more consistent ball toss above the attacking shoulder, a firm wrist at contact). One of the keys to successfully incorporating rubrics into the measurement and assessment strategy is to clearly define the criteria associated with each level of performance. Explicit definitions will help to reduce the subjectivity associated with alternative assessment.

Task Analysis

While there are a number of different approaches to task analysis, in essence, this technique "involves breaking skills down into smaller, sequentially ordered steps" (Houston-Wilson, 1995). So, a teacher might list the steps (or subtasks) necessary to complete a task and use that list to evaluate performance. When students are tested on that task or skill, the teacher checks the steps that the student demonstrates and leaves blank those that were not demonstrated. Consequently, if a student is unable to complete the task, the teacher will have some information as to why the student was unsuccessful (i.e., failure to perform the unchecked steps in the task analysis) and, therefore, a strategy for improvement. (See chapter 6 for more information on task analysis, including an example of a task-subtask analysis for rope jumping.)

A sample task analysis for performing the isometric push-up is provided in box 5.2. The isometric push-up is broken down into six sequential steps. In this example, however, the scoring system is more sophisticated in that the tester evaluates the student's performance relative to the amount of assistance the student requires to complete each step (i.e., independent, partial assistance, total assistance). In this manner the teacher is provided with even more information on the

---

> **Box 5.2 Task Analysis for an Isometric Push-Up**

Objective: To execute an isometric push-up correctly for three seconds.
Directions: Circle the minimal level of assistance an individual requires when correctly performing a task. Total each column. Total the column scores, and enter the total score achieved in the summary section. Determine the percentage of independence score using the chart in the summary section. Record the amount of time the position is held for the product score.

Isometric push-up	IND	PPA	TPA
1. Lie facedown.	3	(2)	1
2. Place hands under shoulders.	(3)	2	1
3. Place legs straight, slightly apart, and parallel to the floor.	(3)	2	1
4. Tuck toes under feet.	3	(2)	1
5. Extend arms while body is in a straight line.	3	2	(1)
6. Hold position for three seconds.	3	2	(1)
Sum of column scores:	6	4	2

Key to levels of assistance:

IND = Independent; the individual is able to perform the task without assistance.
PPA = Partial physical assistance; the individual needs some assistance to perform the task.
TPA = Total physical assistance; the individual needs assistance to perform the entire task.

Summary		Percentage of independence		
Total score achieved	12	6/18 = 33%	11/18 = 61%	16/18 = 88%
Total score possible	18	7/18 = 38%	12/18 = 66%	17/18 = 94%
% of independence score	66%	8/18 = 44%	13/18 = 72%	18/18 = 100%
		9/18 = 50%	14/18 = 77%	
Product score	1 sec.	10/18 = 55%	15/18 = 83%	

Reprinted, by permission, from K. Houston-Wilson, 1995, Alternate assessment procedures. In *Physical best and individuals with disabilities*, edited by J. Seaman (Reston, VA: AAHERD).

student's performance, and a "percentage of independence score" can be calculated (in this case 66 percent) as a summary score. This kind of task analysis is especially useful with students who have more severe disabilities and frequently require assistance during physical activity.

The previous example might be called a rational task analysis because it describes a model of performance based on the requirements of the task, but it does not take into account the limitations of the performer (Davis & Burton, 1991). A type of task analysis that does consider the performer is ecological task analysis, in which the teacher analyzes the relationship among the task, the environment, and the performer. For instance, the task goal might be to get from the bottom of a ramp to the top. Students may choose to achieve the goal in any number of ways (e.g., walking, running, skipping, etc.). Students in wheelchairs probably would choose to propel their chairs up the ramp, but they might also choose, for instance, to crawl up the ramp.

Based on the movement strategy selected, the teacher must consider environmental variables and performer variables that could be manipulated to aid student success (or, possibly, to make the task goal more challenging). For students who choose to propel a wheelchair up the ramp, environmental variables would include the slope, texture, and length of the ramp as well as wheelchair variables such as condition of the chair, seat position, rim position, and camber (tilt angle of the back wheels). Performer variables might include hand function, upper-body strength, and propulsive technique. When the task is analyzed in this manner, the teacher is more likely to understand more of the variables that will influence completion of the task goal and be able to more effectively design interventions that will lead to student success. For a more complete examination of ecological task analysis, see Davis and Burton, 1991 and Burton and Miller, 1998.

Portfolios

An important aspect of alternative or authentic assessment is the collection of evaluation data on multiple occasions in multiple settings. Another important aspect is documenting student learning through exhibits and work samples (rather than simply through some kind of grade). Consequently, the use of a student portfolio system is often associated with the alternative approach. Teachers can identify any number of items that students can choose to include in their physical education portfolio. Examples include, but are not limited to, videotapes, test results (including standardized tests), teacher observations, peer evaluations, rating scales, checklists, journals, and logs.

Materials, however, are not just "dumped" into a portfolio. Teachers must establish criteria for what goes in the portfolio and how it will be evaluated. In addition to material that specifically addresses progress on learning objectives, Melagrano (1996) suggests other criteria, including items that represent "something that was hard for you to do," "something that makes you feel really good," "something that you would like to work on again," "something that represents your 'best' skill," and/or "something that represents a work-in-progress."

Synthesis

As adapted physical education enters the new millennium, the push for alternative or authentic types of assessment has gained momentum. Clearly, authentic approaches hold the most promise for directly linking testing and learning. When conducted properly, authentic approaches constantly inform students of their progress (progress which, to a large extent, they monitor themselves) and what they need to work on next. Testing and learning, in a sense, become seamless because students learn, in part, from the testing program.

Alternative approaches, however, are not without weaknesses. As mentioned earlier, alternative approaches put a premium on subjective observation. As Hensley (1997) has noted, such assessment practices "have frequently been criticized on the basis of questionable validity and reliability, being susceptible to personal bias, generosity error (the tendency to overrate), lack of objective scoring, as well as the belief that they are conducted in a haphazard manner with little rigor." Furthermore, the Individuals with Disabilities Education Act (IDEA) requires that the unique needs of youngsters be determined using valid, reliable, objective, and nondiscriminatory instruments. Authentic assessment strategies often do not meet these criteria.

Consequently, it is the recommendation here that measurement and assessment strategies include both standardized and alternative elements. Instructors should recognize that there are two continua that impact the selection of tests in adapted physical education. Tests vary among their psychometric properties (including validity and reliability) and their authentic properties, and, as depicted in figure 5.1, there tends to be a relationship between these variables. Because the establishment of appropriate levels of validity and reliability requires controlled circumstances, tests with stronger psychometric qualities tend to have lower authentic qualities. Conversely, those with stronger authentic qualities tend to have weaker psychometric properties. Most tests, therefore, tend to fall close to the diagonal line drawn in figure 5.1. As a general rule, teachers should select tests on or above the diagonal line and avoid those below it. Test selection, however,

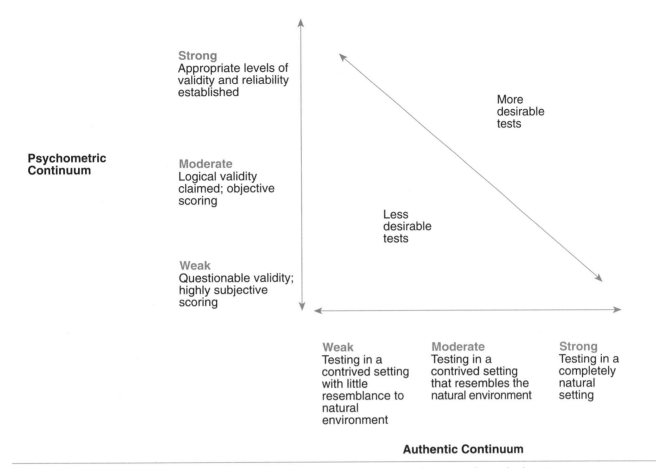

Figure 5.1   Relationship between psychometric and authentic properties of tests in physical education.

also will be influenced by the purpose of the testing. When important educational decisions pertaining to a student (e.g., eligibility for adapted physical education services) will be based, at least in part, on testing, the teacher should give preference to tests with stronger psychometric properties. When a student is learning a skill to be used in a particular context (e.g., dribbling a basketball for eventual use in a game situation), the teacher should give preference to tests with stronger authentic properties to monitor student progress. The application of both standardized and alternative assessment strategies in adapted physical education is explored further in the next section.

MEASUREMENT AND ASSESSMENT IN ADAPTED PHYSICAL EDUCATION

While there are numerous reasons for assessment in physical education, there are two primary purposes in adapted physical education: determination of unique need and providing a basis for instruction. The relationship of measurement and assessment to each of these functions is discussed in the following subsections.

Determination of Unique Need

As discussed in the preceding chapters, students who are suspected of having unique physical and motor needs should be referred to the special education diagnostic team or a Committee on Adapted Physical Education (CAPE). Referrals must document the reasons why the student should be considered for an adapted physical education program. Measurement and assessment at the referral level is usually called screening. The primary purpose of screening is to document the need for an in-depth evaluation to determine if the student has a unique need in physical education.

Determining unique need is critical for two reasons. First, a student must have a unique need to be eligible for adapted physical education services. This is true both for students who are considered to be disabled under IDEA and for nondisabled students who are low in physical-education-related abilities. Second, once a unique need is determined, that need serves as the basis for establishing appropriate IEP goals and, possibly, objectives for the student. These aspects of unique need—eligibility and goals and objectives—are discussed in the following paragraphs.

In dealing with the question of eligibility, a distinction must be made between an adapted physical education *program* and the *placement* to which a student is assigned. A student might qualify for an adapted program but receive that program in a regular class placement. Placement, therefore, is established after the appropriate program has been determined. When a student is referred to the special education diagnostic team (or CAPE) on the basis of preliminary screening, the committee must first determine whether the student is eligible for the adapted physical education program. It will probably be necessary for the adapted physical educator to conduct more assessments to determine whether or not the student has a unique need. A long-term unique need must be established for a student to qualify for adapted physical education. In the absence of a medically based referral, the criteria for entry into the adapted program generally should be based on psychomotor performance. Usually measurement and assessment for the purpose of determining program eligibility should focus, to the extent possible, on standardized testing (i.e., tests with strong psychometric properties and standards for evaluation).

A number of states have developed specific criteria for admission into adapted physical education. In states without such criteria, it is recommended that school districts adopt local criteria for admission into adapted physical education. It is also recommended that districts consider one or more of the following (or similar) standards for admission based on test results that measure developmental aspects of physical education:

▶ The student scores below the 15th percentile.

▶ The student scores more than one standard deviation below the mean (e.g., a T score less than 40).

▶ The student exhibits a developmental delay of at least two years when age norms are used (although the use of age norms is not recommended for students beyond seven years of age).

▶ The student fails to meet criterion-referenced standards on one or more areas evaluated.

▶ The student fails to meet 70 percent or more of the competencies in the physical education curriculum.

Since formal testing often takes place under "artificial" conditions, districts might also consider additional criteria. For example, corroboration of standardized test results through observational techniques, authentic test results, or a temporary trial placement might also be required.

Once eligibility has been established, based on a documented unique need, appropriate goals and objectives must be written. Typically, annual goals and short-term objectives are selected to improve the area(s) of unique need. Students in adapted physical education often have the same goals as the regular program (e.g., to improve physical fitness, ball-handling skills), but the specific objectives will usually be different. Objectives often will be different for a student with a unique need because different activities may have to

be substituted for those in the regular program or, perhaps, because different performance criteria will have to be adopted. While teachers have to rely on their professional judgment in setting appropriate performance criteria, these standards always should consider the student's current level of performance. In some cases the teacher may wish to use standardized data to help establish reasonable objectives. For instance, if a student has mental retardation, the teacher might consult the criterion-referenced standards in the Brockport Physical Fitness Test (Winnick & Short, 1999a) to determine an appropriate grip strength score for an 11-year-old girl. The teacher would see that the specific standard (one adjusted for the influence of mental retardation) is 12 kilograms and that the minimal general standard (one unadjusted for disability) is 19 kilograms. Armed with this information and knowledge of the student's present level of performance, the teacher is in a good position to establish a performance criterion that constitutes a reasonable expectation. In other cases, a teacher may wish to use authentic forms of assessment (e.g., rubrics and task analyses) to provide the criteria for an instructional objective.

After the goals and objectives have been determined, the most appropriate placement for obtaining them must be selected. It should be kept in mind, however, that every effort should be made to keep the student in a regular class placement. Teachers should attempt to modify activities and methodologies so that the student's objectives can be met in the regular class. While there was one primary criterion for admission into the program (i.e., performance), there are a number of considerations in the selection of the appropriate placement. Students, for instance, may have to be assigned to a more restrictive placement (see continuum of alternative placements in chapter 2) if they are unable to understand concepts or safety considerations being taught in the regular class placement or if they are unable to maintain appropriate peer relations or engage in other forms of acceptable social behavior.

Placement also may depend, at least in part, on what is being taught in the regular class. A student who uses a wheelchair, for example, could probably meet appropriate goals and objectives for individual sports such as swimming, weightlifting, and track and field in a regular placement. Conversely, the same student might be assigned to a more restrictive setting when team sports such as volleyball, soccer, or football are being played in the regular class (although alternative activities also could be offered within the same placement). A third important consideration in placement is the input received from the student or the student's parents.

Providing a Basis for Instruction

The role of measurement and assessment does not stop after the student has been assigned to an appropriate physical education program. Progress on the goals and objectives should be monitored closely. As suggested earlier, alternative tests can be designed and used for this purpose. Students, for instance, can work (individually, in pairs, or with the teacher) from task sheets or cards that include rubrics or task analyses for a particular activity. Teachers help students to devise practice regimens that will promote learning, as evidenced by scoring at a higher level on the rubric or by demonstrating previously missing techniques on the task analysis. Skills learned in practice situations also need to be transferred to more "natural" environments (e.g., activities of daily living, games, sports).

At the conclusion of the instructional program or unit, the teacher should conduct final testing to determine the student's "exit abilities." In some cases, grades are awarded based on this final assessment. Whether the program is graded or nongraded, progress should be evaluated in terms of the written goals and objectives. For nongraded situations, Melagrano (1996) suggests a progress report form that lists the student's goals as a checklist where teachers can check "achieved," "needs improvement," or "working to achieve" for each of the goals listed. Summary sheets from portfolios can also be used as a way to evaluate exit abilities. Occasionally, teachers may wish to give awards to students on the basis of their final test performances.

TESTS AND MEASURES FOR USE IN ADAPTED PHYSICAL EDUCATION

There are numerous published tests available to physical educators. Most of these tests would be considered more "standardized" than "alternative" as these terms have been used in this chapter. That is to say that most published tests tend to have established levels of validity and reliability, provide norm-referenced or criterion-referenced standards, and require controlled testing environments. Some of these tests, however, do contain alternative elements such as rubric-like scoring systems (e.g., Test of Gross Motor Development) or task analytic sequences (e.g., Special Olympics Sports Skills Program Guides).

Available tests in physical education measure a wide range of traits and abilities. Most, however, seem to fall within five traditional areas of physical and motor development/ability: reflexes and reactions, rudimentary movements, fundamental movements, specialized movements (including sports skills, aquatics, dance, and activities of daily living), and physical fitness. (Readers should note that these categories are somewhat arbitrary and do not encompass all possibilities. In some situations, for instance, teachers may routinely test and assess the posture and/or the perceptual-motor abilities of their students.) More recently

a sixth area, physical activity, has gained more attention. The remainder of this section of the chapter is devoted to a discussion of tests or measures from each of these six areas. One instrument from each area is highlighted. The highlighted instruments are meant to be representative of a particular content area and are recommended or used by many adapted physical educators. As you read the following material, remember that other tests are available within each area and that teachers always have the option of designing alternative measures to either augment or replace "store bought" instruments as appropriate. In adapted physical education, there are always circumstances when published instruments prove to be inappropriate for a particular student, and teachers must modify or design instruments in accordance with the student's abilities. (Additional tests are listed in this chapter's "Resources" section.)

Testing Reflexes and Reactions

The assessment of primitive reflexes and postural reactions is becoming increasingly common in adapted physical education. (See chapter 17 for more information on reflexes and reactions.) As educational services are extended to infants and toddlers, as well as to those with more severe disabilities (especially those that are neurologically based, such as cerebral palsy), physical educators will need to understand the role of reflexes and reactions on movement.

Since primitive reflexes normally follow a regular sequence for appearing, maturing, and eventually disappearing, they are particularly helpful in providing information on the level of central nervous system maturation. If a primitive reflex persists beyond schedule, presents an unequal bilateral response (e.g., present on one side, but absent or not as strong on the other), is too strong or too weak, or is completely absent, neurological problems may be suspected. When primitive reflexes are not inhibited, they will undoubtedly interfere with voluntary movement, because muscle tone involuntarily changes when reflexes are elicited. The Milani-Comparetti Motor Development Screening Test is a good example of an instrument that assesses reflexes and reactions.

Milani-Comparetti Motor Development Screening Test for Infants and Young Children

▶ Purpose: The Milani-Comparetti Test (Meyer Rehabilitation Institute, 1992) is designed to assess motor development in young children, birth to 24 months. Although the instrument has obvious use for infant and toddler programs, the inclusion of a number of reflexes and reactions in the battery makes it appropriate for older developmentally delayed individuals as well, especially those with cerebral palsy.

▶ Description: In all there are 27 items associated with this test. Nine of the items are classified as "spontaneous behaviors," which test for head control in four postures (vertical, prone, supine, and pulled from supine); body control in three postures (sitting, all fours, and standing); and two "active movements" (standing from supine and locomotion). The remaining 18 items are called "evoked responses" and test 5 primitive reflexes (hand grasp, asymmetrical tonic neck, moro, symmetrical tonic neck, and foot grasp) and 13 righting, parachute, or tilting reactions. In evaluating most of the spontaneous behaviors, testers are required to evaluate the progression of development. For instance, for "all fours" the child will progress from a prone position propped up by forearms and hands, to hands and knees, and finally to hands and feet (i.e., "plantigrade"). For evoked responses, the tester need only note whether the reflex or reaction was absent or present. Age norms are associated with each of the test items.

▶ Reliability and validity: Interobserver and test-retest data provide evidence of acceptable levels of reliability. Age norms were established based on the performance of 312 subjects. Content validity is claimed based on general acceptance of test items by physicians and therapists.

▶ Comment: The primary advantage of the Milani-Comparetti is its relative ease of administration, which is due, in part, to the limited number of test items. Physical and occupational therapists are likely to have experience with this test and may serve as resources to the adapted physical educator.

▶ Availability: Meyer Rehabilitation Institute, University of Nebraska Medical Center, 600 South 42nd Street, Omaha, NE 68198-5450. Web site: **www.unmc.edu/mrimedia/catalog.html#cc**

Testing Rudimentary Movements

Rudimentary movements are the first voluntary movements (see chapter 17). Reaching, grasping, sitting, crawling, and creeping are examples of rudimentary movements. Most instruments that assess rudimentary movements do so in some kind of developmental milestone format; that is, a series of motor behaviors that are associated with specific ages are arranged chronologically and tested individually. By determining which behaviors the child can do, the teacher can estimate the child's developmental age (since each milestone has its own "age norm") and can provide future learning activities (i.e., the behaviors in the sequence that the child cannot currently do). The Peabody Developmental Motor Scales (PDMS-2) is an example of this approach with some additional enhancements (other instruments are discussed in chapters 20 and 21).

Peabody Developmental Motor Scales

▶ Purpose: The Peabody Developmental Motor Sclaes (Folio & Fewell, 2000) assesses the motor development of children ages birth to five years in both fine and gross motor areas. Items are subcategorized into the following six areas: reflexes, stationary (balance), locomotion, object manipulation, grasping, and visual-motor integration.

▶ Description: The PDMS-2 includes 249 test items (mostly developmental milestones) arranged across six areas. The items are arranged chronologically within age levels (e.g., 0 to 1 month, 6 to 7 months, 18 to 23 months), and each is identified as belonging to one of the six categories being assessed (e.g., reflexes, locomotion). It is recommended that testers begin administering items one level below the child's expected motor age. Items are scored on a 0, 1, 2 basis according to specified criteria. Testing continues until the "ceiling age level" is reached (a level for which a score of 2 is obtained for no more than 1 of the 10 items in that level). Composite scores for gross motor quotient (reflexes, stationary [balance], locomotion, and object manipulation), fine motor quotient (grasping and visual-motor integration), and total motor quotient (combination of gross and fine motor subtests) are possible.

▶ Reliability and validity: Empirical research has established adequate levels of reliability and validity. Information is provided for a variety of subgroups as well as for the general population.

▶ Comment: The PDMS-2 appears to have certain advantages over other rudimentary movement tests. First, the large number of test items represent a larger sample of behaviors that exist in many other tests. Second, the six categories help teachers to pinpoint exactly which areas of gross motor development are particularly problematic. Finally, the scoring system and availability of normative data provide the teacher with more information on student performance than many other tests. Supplementary materials, including a software scoring and reporting system and a motor activity program, also are available in conjunction with PDMS-2.

▶ Availability: Pro-Ed, 8700 Shoal Creek Boulevard, Austin, TX 78757. Web site: **www.proedinc.com/store/9280.html**

Testing Fundamental Movements

The fundamental movements are those typically associated with early childhood (see chapter 17). Examples include throwing, catching, skipping, and hopping. Some fundamental movement test instruments use the developmental milestone approach discussed under rudimentary movements, but most utilize some type of point system to evaluate either the process of the fundamental movement or its product. Process-oriented approaches generally attempt to break down (or task analyze) a movement into its component parts and then evaluate each component individually. This approach assesses the quality of the movement, not its result. Product-oriented approaches, on the other hand, are concerned primarily with outcome. Product-oriented assessment is more concerned with the quantity of the movement (e.g., how far? how fast? how many?) than with its execution. The Test of Gross Motor Development (TGMD-2) emphasizes a process-oriented approach to the assessment of fundamental movements.

Test of Gross Motor Development

▶ Purpose: The Test of Gross Motor Development (Ulrich, 2000) has four purposes: to design a test representing gross motor content frequently taught in preschool and early elementary grades including special education; to develop a test that could be used by various professionals with a minimum amount of training; to design a test with both norm-referenced and criterion-referenced standards; and to place a priority on the gross motor skill sequence rather than the product of performance.

▶ Description: The test measures ability in 12 gross motor areas divided into 2 subtests. The locomotor subtest is comprised of the run, gallop, hop, leap, horizontal jump, and slide. The object control subtest consists of striking a stationary ball, stationary dribble,

Figure 5.2 A student attempts the kicking item from the object control subtest of the Test of Gross Motor Development.

catch, kick (figure 5.2), underhand roll, and overhand throw. For each skill the tester is provided with an illustration, equipment/condition requirements, directions, and performance criteria. Children receive one point for meeting each of the performance criteria given for each of two trials. These criterion-based scores can be added and compared to norm-referenced standards. Age norms are provided in half-year increments for ages three to eight for both subtests.

▶ Reliability and validity: Reliability coefficients are quite high (generally .84 to .96). Evidence of content-related, criterion-related, and construct-related validity is provided.

▶ Comment: The sound process of test construction should provide the user with a good deal of confidence that scores obtained by children accurately reflect their fundamental movement abilities. The fact that both criterion-referenced and norm-referenced standards are provided increases the test's utility.

▶ Availability: Pro-Ed, 8700 Shoal Creek Boulevard., Austin, TX 78757. Web site: **www.proedinc.com/store/9260.html**

Testing Specialized Movements

Of the five content areas described in this section, specialized movements (see chapter 17) present the greatest challenge for making summary statements relative to assessment and for choosing one test instrument as an example. This is due to the wide variety of possible activities that could be tested under this heading. Safrit (1990), for instance, lists 185 different skills tests across 25 different sports. Sports skills tests can take many forms, but frequently they are criterion referenced and often teacher constructed. Teachers who work with students with disabilities who compete in special sport programs, such as those sponsored by Wheelchair Sports, USA; the United States Association of Blind Athletes; or the United States Cerebral Palsy Athletic Association, are encouraged to develop their own tests specific to the event in which the athlete competes. One example of a sports skills test that can be used with athletes with disabilities comes from the Special Olympics Sports Skills Program Guides.

Sports Skills Program Guides

▶ Purpose: The Sports Skills Program Guides (Special Olympics, Inc. 1995-99) complement or supplement existing physical education and recreation programs for people with disabilities (aged eight and beyond) in sports skills instruction.

▶ Description: The Sports Skills series is a curriculum that includes 25 sports divided into 3 categories: Special Olympic Summer Sports, Special Olympic Winter Sports, and Nationally Popular

Sports. Although not a test instrument per se, assessment is a critical aspect of the program. The assessment is criterion referenced, and testers check off those skills that the student is able to perform. For instance, in Athletics there are 11 different test items corresponding to a variety of track-and-field events. Within each test, testers check the skills an athlete can demonstrate (e.g., "performs a single leg takeoff for a running long jump") and total the checks. Scoring levels for Athletics range from "Beginner" (1 to 13 points) to "Superstar" (66 to 80 points).

▶ Reliability and validity: No information has been reported, but content validity probably could be claimed, since the tests reflect task analyses of sports skills developed by "experts" in the field.

▶ Comment: A primary advantage of the program guides is convenience; a teacher or coach can adopt the existing task analytic curricula for a wide variety of sport activities. The program has been in use with participants with mental retardation for some time and has been shown to have good utility for that group. A disadvantage is that neither reliability nor validity of the various test instruments has been formally established.

▶ Availability: Special Olympics, Inc., 1325 G Street, Suite 500, Washington, DC 20005. Web site: **www.specialolympics.org**

Testing Physical Fitness

Over the years there have been numerous standardized tests of physical fitness available to teachers. The American Alliance for Health, Physical Education, Recreation and Dance (AAHPERD) alone has published the Youth Fitness Test (AAHPERD, 1976b) and the Health Related Test (AAHPERD, 1980), as well as fitness test modifications for students with either mild mental retardation (AAHPERD, 1976a) or moderate mental retardation (Johnson & Londeree, 1976). Over the past 20 years there have been two prominent trends in fitness testing that have impacted adapted physical education. The first has been a move more toward measuring and assessing health-related physical fitness rather than skill-related fitness. The second trend has been to emphasize the development of criterion-referenced standards rather than norm-referenced standards.

Today, AAHPERD has joined with the Cooper Institute for Aerobics Research and Human Kinetics to create the American Fitness Alliance (AFA), a national resource center for fitness- and physical activity-related products and services. The AFA's objective is to promote physical activity and fitness throughout a lifetime. The AFA recommends two tests of physical fitness, both of which are health related and criterion referenced.

The first is FITNESSGRAM, which was introduced earlier in this chapter. The second is the Brockport Physical Fitness Test (BPFT) (Winnick & Short, 1999), which extends the health-related, criterion-referenced approach to youngsters with disabilities. (A list of items associated with both of these tests is provided in table 5.1.)

Brockport Physical Fitness Test

▶ Purpose: The Brockport Physical Fitness Test (Winnick & Short, 1999a) provides a health-related, criterion-referenced physical fitness test appropriate for youngsters (ages 10 to 17) with and without disabilities.

▶ Description: The test battery includes 27 different test items from which teachers may choose. (A brief description of all test items can be found in appendix C.) Typically, students would be tested on between four and six test items from three components of fitness: body composition, aerobic functioning, and musculoskeletal functioning (muscular strength, endurance, and flexibility) (figure 5.3). Although specific test items are recommended for youngsters with mental retardation, cerebral palsy, visual impairments, spinal cord injuries, and congenital anomalies and amputations (as well as for those in the general population), teachers are encouraged to "personalize" testing. Personalization involves the identification of health-related concerns pertaining to the student, establishment of a desired fitness profile for the student, selection of components and subcomponents of fitness to be assessed, selection of test items to measure those components, and selection of health-related, criterion-referenced standards to evaluate fitness. Teachers, therefore, have the option to modify any of the elements of the testing program as outlined in the test manual. Both "general" and "specific" standards are available, as appropriate. A general standard is one that is appropriate for the general popu-

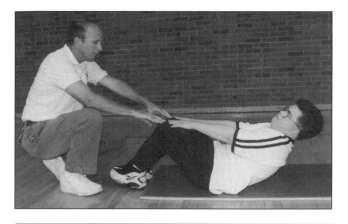

Figure 5.3   A student performs the modified curl-up, a test of musculoskeletal functioning from the Brockport Physical Fitness Test.

lation and has not been adjusted in any way for the effects of a disability. A specific standard is one that has been adjusted for the effects of an impairment or disability. Specific standards are available only for selected test items for particular groups of youngsters.

▶ Reliability and validity. The test items in the BPFT are believed to be valid and reliable. Evidence for validity and reliability is provided in a lengthy technical report (Short & Winnick, 1999), which readers can access through the Fitness Challenge Software, the computer program developed to support the BPFT.

▶ Comment: The BPFT was patterned after the FITNESSGRAM and many of the standards, especially for the general population, were adopted from that test. Teachers in "inclusive" settings, therefore, should find it relatively easy to go back and forth between the tests as necessary. In addition to the test manual, the BPFT Kit includes a training guide (Winnick & Short, 1999b), the computer software, an instructional videotape, skinfold calipers, curl-up strips, and the PACER audio CD/cassette.

▶ Availability: Human Kinetics, P.O. Box 5076, Champaign, IL 61820. Phone 800-747-4457. Web site: **www.humankinetics.com**

Measuring Physical Activity

Recent research has illuminated the positive relationship between regular physical activity and health, and many physical education programs are promoting physically active lifestyles to their students as a primary goal of the program. Consequently, it is becoming increasingly important for physical educators to find ways of measuring physical activity that describe the status of a student's activity levels and to document changes in those levels. Currently there appears to be four general types of activity measures available to teachers: heart rate monitors, activity monitors (e.g., pedometers, accelerometers, motion sensors), direct observation, and self-report instruments (Welk & Wood, 2000). The use of heart rate and activity monitors, despite their accuracy, have limited applicability in many school situations because of the cost of such devices and because only a small number of children can be measured at one time. Coding student activity through direct observation is not expensive, but it can be time consuming because only a few youngsters can be monitored at one time by a trained observer. (These approaches—heart rate monitors, activity monitors, direct observation—might actually have more utility in adapted settings than regular settings because of potentially smaller class sizes.)

Almost by default, therefore, self-report instruments probably represent the most appropriate means of measuring physical activity in most school settings. Self-report instruments require students to recall and record their participation in physical activity over a specified period of time (usually from one to seven days). Although there are a variety of self-report instruments available (see Welk & Wood, 2000 for examples), in general all seek to quantify the frequency, intensity, and duration of students' physical activity. A computer software program called ACTIVITYGRAM provides teachers with a relatively easy method for measuring student physical activity.

ACTIVITYGRAM

▶ Purpose: ACTIVITYGRAM (Cooper Institute for Aerobics Research, 1999b) records, analyzes, and saves student physical activity data and produces reports based on that data.

▶ Description: ACTIVITYGRAM is part of the FITNESSGRAM 6.0 software package. The program prompts youngsters to recall their physical activities over the previous day in 30-minute time blocks. Students select activities from within six categories: lifestyle activity, active aerobics, active sports, muscle fitness activities, flexibility exercises, and rest and inactivity. Students are also asked to rate the intensity of the activity (light, moderate, vigorous). The printed ACTIVITYGRAM report summarizes the entered data in three ways. It reports the total number of minutes of at least moderate-level activity for up to three days; it provides a time profile for each day showing levels of activity (rest, light, moderate, hard) across the hours of the day; and it gives an activity profile that describes which of the six activity categories the student pursued.

▶ Reliability and validity: Due to the subjective nature of recall and reporting, there are measurement limitations associated with all self-report instruments. Nevertheless, the Previous Day Physical Activity Recall instrument, upon which ACTIVITYGRAM is based, has been shown to provide valid and reliable estimates of physical activity and also accurately identifies periods of moderate to vigorous activity (Weston, Petosa, & Pate, 1997).

▶ Comment: Although designed primarily with nondisabled students in mind, ACTIVITYGRAM has good utility in adapted physical education settings. While specific activities may vary (e.g., running vs. pushing a wheelchair), the six categories of physical activity are appropriate for most students both with and without disabilities. Younger children and those with intellectual disabilities, however, may have trouble recalling and entering activity data. Peer tutors, teacher aides, or parents could be trained to make direct observations and to enter the data on behalf of the student who might be experiencing difficulties using the system.

▶ Availability: American Fitness Alliance, Human Kinetics, P.O. Box 5076, Champaign, IL 61820. Web site: **www.americanfitness.net**

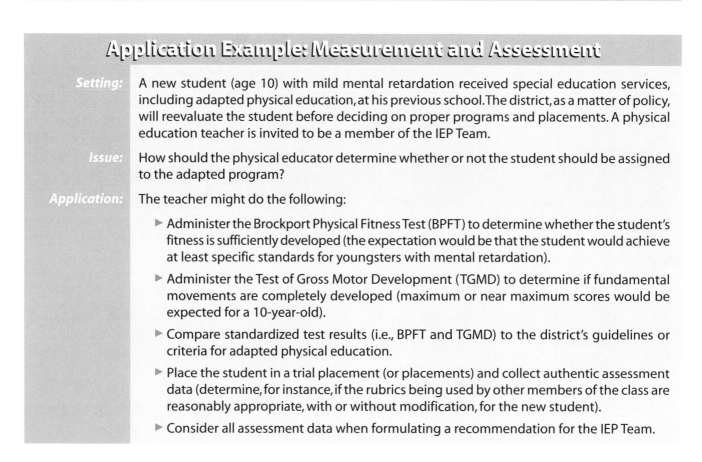

Application Example: Measurement and Assessment

Setting: A new student (age 10) with mild mental retardation received special education services, including adapted physical education, at his previous school. The district, as a matter of policy, will reevaluate the student before deciding on proper programs and placements. A physical education teacher is invited to be a member of the IEP Team.

Issue: How should the physical educator determine whether or not the student should be assigned to the adapted program?

Application: The teacher might do the following:

▶ Administer the Brockport Physical Fitness Test (BPFT) to determine whether the student's fitness is sufficiently developed (the expectation would be that the student would achieve at least specific standards for youngsters with mental retardation).

▶ Administer the Test of Gross Motor Development (TGMD) to determine if fundamental movements are completely developed (maximum or near maximum scores would be expected for a 10-year-old).

▶ Compare standardized test results (i.e., BPFT and TGMD) to the district's guidelines or criteria for adapted physical education.

▶ Place the student in a trial placement (or placements) and collect authentic assessment data (determine, for instance, if the rubrics being used by other members of the class are reasonably appropriate, with or without modification, for the new student).

▶ Consider all assessment data when formulating a recommendation for the IEP Team.

SUMMARY

Measurement and assessment serve a number of functions in adapted physical education. Most importantly, they provide a means for determining if a student has a unique need in physical education, and they provide a foundation for structuring learning experiences and monitoring progress. This is illustrated in the Measurement and Assessment application example. Measurement and assessment strategies can range on a continuum from techniques with stronger psychometric properties conducted in less natural environments (often called standardized assessment) to those with weaker psychometric properties conducted in more natural environments (often called authentic assessment). Each of these approaches has strengths and weaknesses and should

be selected in accordance with the purposes of the assessment. It is recommended that these approaches be combined to yield a more complete picture of the student. Ordinarily, assessment in adapted physical education should focus on physical fitness and motor development/ability (including reflexes, rudimentary movements, fundamental movements, activities of daily living, sports skills, aquatics, and dance, as appropriate). Standardized tests are commercially available for each of these areas and are especially useful for summative purposes such as determining unique need. Standardized tests, however, may not be appropriate for every youngster with a disability. Authentic techniques are useful when standardized tests are inappropriate and/or are especially useful for formative or instructional purposes such as monitoring student progress on a daily basis.

Chapter 6

Instructional Strategies for Adapted Physical Education

Sarah M. Rich

▶ **T**HOMAS D. HOPKINS RECENTLY GRADUATED cum laude from Central College with a bachelor of science degree in physical education. While a student he earned the nickname TD because he played varsity football for 4 years, scored 32 touchdowns, and was an All-American Division III running back his senior year. He also helped coach a Special Olympic track team each spring and volunteered as a basketball official for the city youth league for three years.

During the summer after graduation, TD was offered a job from the Coolwater School District in west Texas. Coolwater is a small district with two elementary schools and a junior-senior high school. He would be teaching physical education classes for grades 7 to 12, as well as being the head football coach.

TD was excited to learn from the principal that the Coolwater football team had won in their league last year and that many players would be returning, including star player Jim Brooks who played middle line backer on defense, fullback on offense, and was deaf.

TD also learned that he would be teaching a class of students ranging in ages from 16 to 21, all of whom were moderately mentally retarded. In addition to Jim Brooks, TD would have a student with a visual impairment, one who was a wheelchair user due to a diving accident, and a student with an above the elbow amputation due to a farm machinery accident included in his regular physical education classes.

Teaching and coaching at Coolwater would be a big challenge for TD. He was glad that his educational experience had included methods of working with children of all abilities. He was confident he could teach and coach students of various abilities to reach their highest levels of development.

Through the rest of the summer, TD planned modifications that would allow Jim Brooks to be an even more effective football player, designed activities and methods of instruction that would challenge his class of students with mental retardation, and devised strategies to allow other students with special needs to successfully participate in the regular physical education curriculum.

In August, just before the start of school, he came across this poem by an unknown author. It found a place on his bulletin board to serve as a reminder for him to try his best to meet the unique needs of all his students, especially those with disabilities.

The children are there

Beyond the hurt and the handicap

Beyond the defect and the difference

Beyond the problem and the probing

How can we help them?

How can we set them free?

The purpose of this chapter is to provide information about developing an appropriate professional philosophy, providing challenging and appropriate learning environments, and using effective instructional approaches and models to facilitate learning. This information will assist the teacher in structuring a learning environment that optimizes the acquisition of motor skills and sport activities. It will also assist teachers in interacting effectively and efficiently with their students.

As this information is used, it may be necessary to modify some of the principles presented based on the unique abilities of a teacher and the specific needs of students. By understanding the variables that impact on learning, the teacher can enhance the learning process, allowing students to reach their maximum potential (i.e., "set them free").

GENERAL PHILOSOPHICAL APPROACHES TO ADAPTED PHYSICAL EDUCATION AND SPORT

Modern physical education programs, including those adapted to meet unique needs, follow educational approaches that recognize students' individual needs. The philosophical approaches that have most influenced adapted physical education are the humanistic and the behaviorist approaches.

Some people find it difficult to accept all the tenets of one of these approaches and are more comfortable combining some of the philosophical concepts of each to form a set of compatible beliefs that provide them with a sound approach. This combination of philosophies is known as the eclectic approach.

Humanism

Humanism is a philosophical approach that emphasizes the development of self-concept, positive interpersonal relationships, intrinsic motivation, and personal responsibility (Sherrill, 1998). Humanistic physical education uses physical activity to help individuals develop self-esteem, self-understanding, and interpersonal relationships. It attempts to identify and meet unique needs, abilities, and interests through individualized instruction that incorporates student choice. Sherrill (1998) believes that adapted physical educators who employ a humanistic approach will assist their students to develop positive self-concepts so that they will become intrinsically motivated and achieve their highest potential in physical education and sport.

Humanistic theory became popular in the 1950s. The humanistic movement resulted mainly from the works of Abraham Maslow (self-actualization theory), Carl Rogers (fully functioning self-theory), and their followers. Maslow, Rogers, and their disciples espoused the theory that development occurs naturally and results in a healthy, self-actualized person.

Since students and athletes with disabilities often begin physical activity with low levels of self-esteem, skill, fitness, and motivation, many physical education teachers and coaches employ humanistic principles to assist in the development of these attributes. They develop programs that value individual differences and allow individuals to build confidence in their abilities, experience success, enjoy physical activities, and develop healthier lifestyles (figure 6.1).

Sherrill (1998) suggests that a humanistic philosophy will assist individuals to reach their fullest potential by offering learning, living, and work opportunities that are congruent with those of society in general. At the same time, individual needs are met through a continuum of services ranging from highly restrictive to fully inclusive settings.

Figure 6.1   Developing a positive self-image about physical activities is important to all individuals regardless of age. Often older athletes can give younger ones a "sense of self," which allows them to achieve at advanced levels.

Sherrill (1998) says that the characteristics of a self-actualized person, particularly self-concept and body image, are important ones for physical educators to develop in their students through movement experiences. She says that nondisabled students tend to develop positive affective behaviors without planned intervention by their teachers, but children with disabilities need special help to develop these attributes. It is, therefore, important that adapted physical education teachers help students develop positive body images and self-concepts so that they will be intrinsically motivated to become all that they can be in physical education and sport activities.

In order to reach affective domain goals in physical education, students need to feel good about themselves. Teachers play an important role here. A success-oriented program is instrumental in improving self-concept. Teachers must create a climate of success by adapting approaches and methods to meet the needs, interests, and abilities of all students. Successful experiences in physical education will promote positive social interaction.

Self-worth can further be increased by providing positive feedback and praise, using appropriate task and activity analyses, assisting students to perceive their skill development as important, and assisting students through leisure counseling to identify activities they will continue to enjoy after their school years are completed.

Hellison (1985) has also worked on a framework to demonstrate the essence of humanistic goals for physical education. These goals focus on human needs rather than on such things as fitness or sport skills development. These goals are developmental and hierarchical in nature. They include self-control, involvement, self-responsibility, caring, and student sharing.

One recurring theme of humanistic philosophy is to meet the unique needs and interests of each student through individualized instruction. Its emphasis on sensitivity to individual differences and the development of the total person makes humanism a viable philosophy on which to base physical education programs.

Behaviorism

The behavioral approach advocates a systematic planned organization of the environment to achieve a desired behavioral response from an individual. It can be used to facilitate learning and enhance social behavior. This approach uses procedures derived from the principles developed by psychologists such as B.F. Skinner and Bandura to systematically change performance. It is based on the premise that the teacher can best ensure that learning occurs by structuring the environment.

Many principles associated with behavior modification are routinely and informally applied daily by teachers employing this approach. The behavior modification approach can be developed into "formal behavior programs." In these programs, students learn control and appropriate behaviors through the following steps:

- ▶Defining the specific behavior to be developed or changed
- ▶Determining a baseline or present level of performance
- ▶Establishing one or more terminal goals
- ▶Implementing a behavioral intervention program

Auxter, Pyfer, and Huettig (1997) are among the leaders in adapted physical education who espouse the behavior modification approach. They suggest that behavior can be changed through the application of behavioral principles. They postulate that if physical educators are going to apply behavioral principles they must have precise behaviorally stated objectives, and the learner must have a reason for wanting to meet those objectives. If these prerequisites are present, then it is possible to promote effective learning by applying systematic behavioral techniques.

One example of the behavioral approach being used to facilitate skill acquisition and develop appropriate social behaviors is the Data Based Gymnasium Model (Dunn, Morehouse, & Fredericks, 1986). This approach requires teachers to task-analyze motor skills and social behaviors, develop specific objectives, and enhance their development using instructional procedures based on behavioral principles (this is discussed later in the chapter).

The main focus of behaviorism is to promote the attainment of skills that will make an individual self-sufficient. In the physical education setting, a behaviorist philosophy results in a program that is very goal oriented and structured. It strives to minimize incorrect responses by providing a highly structured learning environment that is oriented toward functional skill development and success. This orientation makes behaviorism a popular and appropriate approach in the adapted physical education setting.

Eclectic Approach

Both the humanistic and behavioristic approaches, along with other educational philosophies such as naturalism and realism, are relevant to skill facilitation and behavioral management in adapted physical education. However, the greatest value of these philosophies for the physical educator may be best realized by taking the best elements of each. This is known as the eclectic approach.

There are times when, because of the variety of individual needs, a humanistic approach is preferable. At other times, such as when there is a severe develop-

mental delay, the behavioristic approach may be more expedient and effective. Sometimes it is beneficial to integrate several philosophical approaches; when this occurs, an eclectic approach results. This allows the physical educator to enhance student success through facilitating optimal achievement.

TEACHING STYLES

Another choice crucial to successful teaching is to match the teaching styles to the learning styles and characteristics of the learners. A teaching style is a method of presenting material. Mosston and Ashworth (1990) have described several learning strategies and classified them on a continuum, from direct teacher-centered styles to indirect student-centered styles. The continuum reflects the amount of decision-making responsibility allocated to the teacher and to the learner. In a teacher-centered approach, the teacher makes most of the decisions on such matters as the structure of the learning environment, lesson content, entry level of the students, and starting and stopping times. As one moves along the continuum, decision-making responsibility is shifted increasingly to the student. Decisions assumed by students include solving movement problems, finding a beginning level for a task, and determining the amount of time to be spent on each problem. While many teaching styles were identified by Mosston and Ashworth, only four are discussed in this chapter: command, task, guided discovery, and problem solving. These styles appear to have the greatest utility for providing adapted physical education students with a challenging learning experience.

Direct Versus Indirect Styles

Teaching styles can also be described as direct or indirect. Direct teaching styles, are the more traditional teacher-centered styles, where the teacher makes most of the decisions about performance. These styles are recommended for individuals functioning at the severe/profound level, students who benefit from structure (e.g., those with autism), and students with behavioral disabilities. Individuals who are learning at an advanced skill level also may profit from these methods. Teachers who follow a behaviorist philosophy generally prefer to use direct teaching styles.

Indirect styles are more child-centered methods of teaching. They permit the learner to take an active role in the learning process through problem solving, experimentation, and self-discovery. The indirect styles are most beneficial for students in adapted physical education who are high functioning, preschool infants and toddlers, learning basic motor skills, or learning skills not requiring one correct response.

Command Style

The command style is probably the most commonly used teaching style in adapted physical education. Decisions are made predominately by the teacher concerning lesson content, organization of the learning environment, and acceptable standards of performance.

Use of the command style can be illustrated in a separate adapted physical education setting by a class of students, all of whom are in wheelchairs. The group is gathered around the teacher, who explains how to execute the overhand throw. A demonstration follows, with the teacher using a chair or wheelchair to make the demonstration more relevant for the class. The students are then sent to practice the skill using tethered balls or rebounders with ball returners attached, so that the individual can retrieve the ball easily and have maximum continuous practice. The teacher moves from student to student, assisting each one with ball control, skill improvement, and motivation. At the end of the lesson, the teacher asks questions to review the major aspects of throwing and to measure the students' comprehension of the skill.

Advantages associated with the command style are as follows:

▶Teacher control

▶Minimal investment of time in group organization

▶Knowledge of the expected outcomes

This is an effective style for use with large groups, in one-on-one instruction, or when the teacher wants all of the students to practice the same task at the same time. It is particularly effective for children with behavioral problems, those with severe/profound mental retardation, elementary-aged children, and those who require external control.

Disadvantages of the command style are as follows:

▶Little thought about how to accomplish the skill is required on the part of the learner

▶Little creativity allowed in terms of motoric response

▶Little variation of response is permitted

Also, because this method stresses the attainment of one correct response, it is generally insensitive to individual differences. This is a severe limitation when working with individuals whose motor abilities may vary greatly, as in an inclusive setting, or whose impairment inhibits production of the one correct response. For example, when working with a class that contains children with orthopedic impairments, it may be impossible for all children in the class to execute an overhand tennis serve on command by the teacher.

Task Style

The task style of instruction requires the teacher to develop a series of tasks that progressively lead to the achievement of an instructional objective. The teacher develops task cards, which are given to the students. For learners who are unable to read, task cards can be made with pictures or Braille writing, or even recorded on audiocassettes to meet individual needs. Additional resources, such as filmstrips, posters, videotapes, books, and three-dimensional models, assist learners in mastering tasks. The tasks must be presented in a manner that allows the learner to determine when the assignment has been successfully completed. For example, in a lesson using a task approach to teach dribbling, a task card may instruct the student to dribble through a five-cone obstacle course without losing control of the ball. After a successful evaluation is completed by the teacher, the learner, or a peer, the learner moves on to the next task in the sequence (e.g., negotiating the same obstacle course while dribbling with the other hand). This method allows the teacher to organize the learning setting so that all students can be working on tasks concurrently in a safe environment. Examples of the task method specifically designed for learners with disabling conditions include the I CAN curriculum (Wessel, 1979), the Data Based Gymnasium (Dunn et al., 1986), and the Special Olympics Sports Skills Program (1985).

Advantages of the task method are numerous and include the following:

▶Encourages individuals to work at their own pace

▶Allows individuals to practice tasks appropriate to their abilities

▶Diminishes competition with others

▶Allows the teacher to control the difficulty of the task

▶Enhances success for each student

▶Enables the teacher to work with students on an individual basis

▶Encourages the maximum use of equipment, facilities, and aides or volunteers

The task method is particularly effective for use in inclusive settings. It can be used effectively with children who can work independently, who may need extra time to master tasks, and who can be assisted in tasks by peers, aides, or paraprofessionals.

Disadvantages of the task style are as follows:

▶Less structure in the learning setting in comparison to command style

▶Possible increase in safety concerns

▶Possible increase in distractions

▶Appearance of disorganization because many activities are going on simultaneously

Guided Discovery Style

Guided discovery uses teacher-designed movement challenges to help students attain a specific movement goal. Students are encouraged to discover movement solutions that meet the criteria stated by the teacher. Using questions or short statements, the teacher guides the student in a progressive series of steps or subchallenges toward the desired outcome or challenge goal (Nichols, 1996).

Following is an example of the use of guided discovery in teaching a class of students with emotional disturbances to throw a softball for distance. The teacher breaks the skill down into subchallenges.

> *Challenge:* "How far can you throw the softball using the overhand throw?"
>
> *Subchallenge 1:* "How should you stand to get the most distance?" If students do not respond motorically, the teacher can cue the response by giving them ideas, such as feet together behind the line, feet parallel but wide apart, or stride position. "Which foot should be forward?"
>
> *Subchallenge 2:* "How high should the ball go to travel the farthest?" Possible cues for this subchallenge would be "Will it travel farther close to the ground?" "If you throw the ball as high as you can, will it travel a long distance?"

These are examples of subchallenges that would help a student find the answer to the teacher's challenge "How do you throw a softball the greatest distance?"

Advantages of the guided discovery style are as follows:

► Encourages creativity
► Allows students to discover how various parts of the body contribute to movement patterns
► Enhances self-concept as students receive positive feedback while shaping their responses to obtain the desired outcome

This method is appropriate with mature students and students whose cognitive abilities permit the required thought processes. According to Gallahue (1993), it is appropriate for toddlers and preschoolers who are experimenting or not yet ready to give one correct answer. It can also be used in inclusive classes, because it allows students to achieve success while discovering how to perform skills in an appropriate manner.

Disadvantages of the guided discovery style are as follows:

► May require more time to achieve movement goals
► May require more preparation on the part of the teacher
► May require more patience and a great deal of feedback by the teacher to bring about the same level of performance as the command style

Problem-Solving Style

This method resembles the guided discovery method in presenting a series of movement challenges to the learner. However, in contrast to guided discovery where one specific movement is the goal, the problem-solving style emphasizes the development of multiple solutions to a given problem posed by the teacher. This style encourages students to develop as many solutions as possible, provided that the solutions meet the criteria stated by the teacher.

In an instructional episode, in which students are encouraged to display different ways of rolling effectively, the teacher might pose the following challenges:

► "Staying within the mat areas and not getting into another child's self-space, roll in as many different ways as you can."
► "Roll with your body as long as possible."
► "Can you find a way to roll with your arms out to the side?"
► "Can you combine three different rolls and roll in a circle?"

After each challenge, the teacher lets students experiment with a variety of movement solutions. Often, additional questions are needed to elicit a variety of responses, especially with students who learn primarily by imitation and those who have had limited motor experience. Children who have spent most of their time in wheelchairs, toddlers, preschool children, and children in inclusive settings may benefit from being taught by this method.

Advantages of the problem-solving style are as follows:

► Wide allowances made for individual differences
► Emphasis placed on cognitive process and creativity
► Acceptance of skill execution that varies from the expected norm

The problem-solving style treats as correct any response to the movement challenge that meets the criteria; this is important in cases where disabling conditions inhibit a particular response. Problem solving boosts learners' self-esteem by allowing them to experience success.

Disadvantages of the problem-solving style are as follows:

► The great amount of time required for students to fully explore possible movement solutions
► Lack of structure
► Absence of an absolute outcome

The range of instructional styles allows educators to teach the same content (especially physical education) using a variety of methods. In selecting a style, the teacher should consider its appropriateness for desired objectives. The choice of teaching style should also reflect such factors as the teacher's preferences and personality; the students' ages, experience, learning style, and disabling conditions; the stage of learning; and the skills to be taught. The learning environment may also influence the choice of style. Practical considerations such as equipment, space, available time, and number of students may make one teaching style preferable to the others.

Some techniques that are useful for integrating students with and without disabilities into physical education are

- ▶modifying the activity to equalize competition,
- ▶permitting substitution or interchange of roles,
- ▶including activities that require a partner, small group, or individual-object interaction,
- ▶modifying activities so that the group assumes a disability while participating,
- ▶limiting play space if movement capabilities are restricted,
- ▶using activities that accentuate the individual's abilities rather than disabilities, and
- ▶avoiding or modifying elimination-type activities.

MOTOR LEARNING

Research supports the assertion that, when students are taught by methods that complement their learning characteristics, they are motivated and learn more easily (Webster, 1993). By understanding how a student learns best, the educator can facilitate the learning process considerably. Material may be presented in one of three approaches: the whole, the part-whole, and the progressive part.

Whole Method

The whole method of learning should be used when the skill to be learned is relatively simple or made up of few parts. An example using this method to teach an elementary school student with learning disabilities how to skip would involve the teacher demonstrating the skill and then asking the child to perform the whole skill, or perhaps taking the child by the hand and skipping together. For this child the whole approach might be more successful than practicing a step-hop with one foot and a step-hop with the other foot. Alternate ways to use the whole method to teach skipping include having the student view a videotape of a child skipping before trying it or to play Follow the Leader, having the leader include skipping as part of the game.

This method may be preferable for students who have difficulties in conceptual learning and are unable to relate the several parts of a skill to its whole. It is also the method of choice for students with short attention spans and students who learn best by imitation. However, one requirement is that the learner be able to remember the skill to be learned, its specific movements, and its sequence. Examples of motor skills that commonly can be presented through the whole approach include running, catching, striking, and jumping.

Part-Whole Method

The part-whole method requires individuals to learn skills by practicing one part at a time and then combining them to perform the whole skill. This method works best for individuals who can concentrate on and accomplish small tasks. It may not be desirable for those who have difficulty integrating various parts into a whole, even with the guidance of the instructor. In using this method, the physical educator breaks down the complete task into meaningful parts; this process is known as task analysis. The teacher should make each part an end in itself, thus providing a sense of accomplishment, even though the learner may not be able to master the entire task or terminal objective.

The individual parts of the developmental sequence or the task analysis can be taught using a concept known as chaining. **Chaining** refers to the successful learning of a task or part that is then combined with other parts to master a skill. Chaining can be completed in a forward or backward direction. **Forward chaining**, the traditional method, involves starting with the first, or most basic, step or level and building toward the end product. **Backward chaining** involves starting with the end product and working backward toward the basic parts.

An example of a skill that can be analyzed into meaningful tasks and thus is amenable to the part-whole method is the beginner's backstroke in swimming. The skill can be divided into the back float, back glide, back glide with flutter kick, arm action, and leg action. Each of these tasks can give the learner a feeling of accomplishment and lead to successful movement in the water. Many individuals with disabilities may not master the elementary backstroke in its entirety, yet they can succeed in mastering its various elements. This is an example of forward chaining. Examples of skills that are more effectively taught using backward chaining include kicking an object or throwing.

Progressive-Part Method

The progressive-part method involves teaching the most fundamental part of the skill first and then building on this base to present the next part. When the first two parts are learned, they are combined and succeeding parts are added until the whole skill is mastered.

Special Olympians might use this method in learning the triple jump. The coach would first work to develop the athlete's hopping ability and then combine the hop with a step. When the athlete masters these two skills, the final skill, jump, would be added.

With the progressive-part method, the teacher or coach should be alert to problems that can occur if an individual fails to learn one part of the progression and should take care to provide the opportunity for success at each part level. This method lets individuals master a skill at their own pace, practicing the most difficult parts while still progressing toward the overall objective. Programs that incorporate the principles of task analysis and employ the part-whole and the progressive-part methods include the Data Based Gymnasium (Dunn et al., 1986) and I CAN (Wessel & Zittel, 1998).

Principles of Motor Learning

Application of the principles that form the scientific foundations of physical education can result in better learning experiences for children with special needs. To ensure success, the physical educator or coach must understand the learning process, know how to adapt it to meet individual needs, and promote the learner's active involvement in the process.

Learning can be enhanced through the application of the principles of motor learning. The area of motor learning focuses on the process by which people acquire motor skills. By understanding and applying the motor learning guidelines, the teacher or coach may help students with unique needs to achieve their maximum potential (box 6.1).

FACILITATING SKILL DEVELOPMENT

Three ways of facilitating skill development will be discussed: task analysis, activity analysis, and activity modification. Task analysis enables the teacher to identify and sequentially develop the components inherent in various skills. Activity analysis assists the teacher in selecting appropriate activities by identifying the physical, cognitive, affective, and social components that are needed for optimal student success. Once an appropriate activity has been selected, activity modifications may be necessary to meet the special needs of the learner.

Task Analysis

Task analysis can facilitate the acquisition of a skill by separating the skill into meaningful parts. Task analysis is the process of identifying the components of a skill and then ordering them from simple to complex (Sherrill, 1998). There are several types of task analysis, including traditional, information processing, rational, anatomical, and ecological. Two common types of task analysis will be discussed here: traditional and ecological.

A traditional task analysis (TTA) provides a systematic hierarchical method for analyzing skill mastery from fundamental levels to advanced levels. It offers a way of breaking a skill down into its components or related subtasks. These components can then be arranged in sequence from easy to difficult, and short-term goals can be determined.

Before beginning instruction the teacher must decide where on the developmental continuum of

---

> **Box 6.1 Guidelines for Teaching Based on Principles of Motor Learning**

1. To accommodate differences in learning styles, use a variety of teaching techniques.

2. To facilitate learning, present skills to individuals when they have reached the appropriate developmental level.

3. To increase the likelihood of continued participation, provide enjoyable learning experiences in which the individuals succeed.

4. To motivate individuals to continue their participation, reinforce their efforts.

5. To increase learning, present the principal focus of the lesson early.

6. To enhance performance, provide appropriate practice opportunities.

7. To accommodate differences in individuals' rates and amounts of learning, individualize instruction.

8. To encourage individuals to persist in their skill acquisition efforts, provide them with appropriate feedback.

9. To assist individuals in transferring previously learned skills to new learning situations, identify common aspects of situations.

10. To help individuals make more rapid progress in learning, set meaningful and realistic goals.

11. To increase the retention of skills, provide opportunities for mastery and select skills that have personal meaning to the individual.

> Box 6.2 **Example of a Task Analysis for Rope Jumping**

Main task: For Mary to jump rope for a full (overhead) turn four times without verbal cue

Prerequisite skills: Ability to jump, ability to stand erect

I. Jump over painted line once over and back without verbal cue.

a. Walk line down and back heel-to-toe with verbal cue.

b. Face line with toes, jump over once with verbal cue.

c. Stand parallel to line, jump over and back without verbal cue.

II. Jump over still rope over and back twice without verbal cue.

a. Face rope with toes, jump over once with verbal cue.

b. Stand parallel to rope, jump over and back with verbal cue.

c. Stand parallel to rope, jump over and back twice without verbal cue.

III. Jump over wiggly (snake) rope 2 in. off ground without verbal cue.

a. Jump wiggly rope on ground once with verbal cue.

b. Jump wiggly rope 1in. off ground over and back with verbal cue.

c. Jump wiggly rope 2in. off ground over and back without verbal cue.

IV. Jump over half-turned rope without verbal cue four times.

a. Stand and jump on the back swing twice with verbal cue.

b. Stand and jump on the forward and back swing twice with verbal cue.

c. Stand and jump on the forward and back swing four times without verbal cue.

V. Jump rope a full turn overhead four times without verbal cue.

a. Stand and jump once as the rope makes a full turn with verbal cue.

b. Stand and jump twice as the rope makes a full turn with verbal cue.

c. Stand and jump four times as the rope makes a full turn without verbal cue.

---

components the student is functioning and whether the learner has met the relevant prerequisite skills. An example of a TTA is presented in box 6.2.

In writing a TTA, it is important to include all steps, describing each component in observable behavioral terms. This thoroughness will facilitate the identification of short-term behavioral objectives and thus simplify the process of IEP development.

This traditional type of task analysis is popular and has many advantages. Many curricular resources in adapted physical education and special education use this type of task analysis. It provides teachers with an easily understood, practical way of sequencing an activity so that students can move from their present level of function toward appropriate short-term objectives. This type of task analysis can be applied to the development level of skill as well as the higher level of sports skills and can accommodate a wide range of teaching methodologies.

Some disadvantages of TTA are that it tends to deemphasize differences among learners in regard to optimal sequencing of the subtasks. It places more emphasis on the task than on the learner. It is also difficult to complete TTA on some skills that do not fit into the whole-to-part format easily, for example a standing dive in swimming.

An alternative to TTA is ecological task analysis (ETA). ETA is a "process of changing relevant dimensions of a functional movement task to gain insight into the dynamics of the movement behavior of the students, and to provide teachers with clues for developing strategies" (Davis & Burton, 1991, p. 160). It was developed to encourage creative thinking by providing teachers and students with choices, which allow teachers to use indirect teaching methods to reach specific outcomes. Balan and Davis (1993) assert that the ETA provides "processes of assessment and instruction, which differs from the traditional teacher-directed methods" (p. 54). They suggest that a movement form is the result of the interaction of three major elements: the task goal, the environment, and the student's learning characteristics.

Table 6.1   Example of Ecological Task Analysis

Steps	Application
Select and present task goal	Student will propel an object from third to first base before the batter can run from home to first base.
Choice	Object propulsion by throw, kick, carry, push, roll, or other. *Sample qualitative measure*: Does the object reach base before the runner? *Sample quantitative measure*: How long does it take for the object to travel from third to first base? (use timer)
Manipulation	*Task variables:* trajectory and speed of propulsion method; texture of object, size or weight of the object. *Performance variables*: base of support, body rotation, type of propulsion, weight transfer.
Instruction	Direct instruction with demonstrations by teacher for selected skill/movement pattern, manipulate variables, compare results.

ETA is designed both as a method of instruction and as an assessment tool. Davis and Burton (1991) outline four basic steps in setting up an ETA. Step one is to identify the functional movement task or what is to be accomplished, which is often written in the form of a behavioral objective. Step two allows for student choice. These choices include choice of skill, movement pattern, environment, and implement. During this step the student practices the skill/movement to be learned and determines which methods work best for the conditions. Step three is called manipulation. During this time the teacher/coach guides the student in trying different strategies to achieve success by using environmental information. As the student works toward her goal, the teacher needs to provide meaningful feedback and motivation. Teachers may also change task dimensions to challenge students and allow discovery of new, more effective motor responses. This direct instruction by the teacher and the observation of results comprise step four of the process. Table 6.1 is an example of an ETA.

In utilizing the ETA approach, it is important that the environment allows for maximum participation by providing sufficient space, equipment, and movement options. Student choices must be encouraged and affirmed. Before direct instruction by the teacher is provided, students should be given ample time to explore choices and achieve a level of success.

ETA has some distinct advantages over the TTA format. It places more emphasis on the specific needs and abilities of the student in regard to selection of task goals, movement/skill choices, and ways of determining success. It encourages modification and manipulation of variables by establishing direct links between the task and the constraints of the performer and the environment (Burton & Miller, 1998). Finally it provides students with opportunities to explore and discover their abilities.

Some problems exist for the advocates of the ETA. Presently this approach is more theoretical. It is not yet widely utilized in adapted physical education curricular materials, but it is gaining popularity because the skill needed to accomplish a task is the basis of authentic assessment. More research needs to be done to refine and further develop ETA.

It is important for adapted physical educators to utilize task analysis in its various forms to individualize instruction and assure that each student achieves to his maximum. Task analysis is an important tool in designing IEPs.

Activity Analysis

Activity analysis is a technique for determining the basic requirements of an activity that are needed for optimal student success in performing the activity. By breaking an activity into components, the teacher can better understand the specific value of a certain activity and modify it to fit an individual learner's needs, if necessary. Thus, once a student's needs have been determined, the teacher can use activity analysis to assess whether a specific activity can meet those needs.

Activity analysis facilitates the selection of program content based on the teacher's stated objectives. In making the analysis, the teacher must determine the physical, cognitive, social, and administrative requirements for performing the activity. When analyzing an activity from a physiological perspective, the teacher should examine such factors as basic body positions required, body parts utilized, body actions performed, fundamental movement patterns incorporated, coordination needed, fitness level required, and sensory systems used. Cognitive factors that should be examined include the number and complexity of rules, as well as the need for memorization, concentration, strategies, and perceptual and academic skills. Social factors to consider in the activity analysis include the amount and type of interaction and communication required and whether the activity is cooperative or competitive. Such administrative demands as time, equipment, facility needs, and safety factors must also be determined. Table 6.2 provides an example of an activity analysis for table tennis.

Table 6.2   An Activity Analysis for Table Tennis

Activity demands	Activity	Table tennis
Physical demands	1. Primary body position required?	Standing
	2. Movement skills required?	Bending, grasping, catching, hitting
	3. Amount of fitness required?	Minimal fitness level
	a. Strength	Low ability to hold paddle
	b. Endurance	Low cardiovascular requirements
	c. Speed	Quickness is an asset
	d. Flexibility	Moderate
	e. Agility	High level desirable
	4. Amount of coordination required?	Important, especially eye-hand coordination
	5. Amount of energy required?	Little
Social demands	1. Number of participants required?	Two for singles, four for doubles
	2. Types of interaction?	Little; one to one
	3. Type of communication?	Little verbal, opportunity for nonverbal
	4. Type of leadership?	None
	5. Competitive or cooperative activity?	Competitive mainly; cooperative if doubles
	6. Amount of physical contact required?	None in singles; much in doubles
	7. Noise level?	Minimal
Cognitive demands	1. Complexity of rules?	Moderate
	2. Level of strategy?	Moderate
	3. Concentration level?	Moderate
	4. Academic skills (reading, etc.) needed?	Ability to count to 21 and to add
	5. Verbal skills needed?	None
	6. Directional concepts needed?	Yes
	7. Complexity of scoring system?	Simple
	8. Memory required?	Little
Administrative demands	1. Time required?	Can be controlled by score or time
	2. Equipment needed?	Paddles, balls, tables with nets
	3. Special facilities required?	Area large enough to accommodate tables
	4. Type of leadership required?	Ability to instruct small group
	5. Safety factors to be considered?	Space between tables

Application Example: Activity Modification

Setting:	High school physical education class
Student:	17-year-old student who is legally blind with a visual acuity of 2/200
Unit:	Indoor archery
Issue:	Modifications for safe and successful inclusion of student with visual impairment
Application:	After discussion among the physical education teacher, the student with the visual impairment, and class members, the following policies were instituted:

- ▶ Student with visual impairment's arrows will be identified with Braille tape by the arrow notch.
- ▶ A special target face will be used on which each scoring area will have a different texture.
- ▶ A raised rope will be used by all students as the shooting line.
- ▶ The archer with visual impairments will use an L-shaped wood template to align his feet and the target.
- ▶ A tape recorder will be attached to the back of the target to give auditory direction for the archer.
- ▶ A rope will be attached to the target and foot templates will be placed on the ground to allow independent movement between the target and the shooting line.
- ▶ No one will go past the shooting line until an audible signal to retrieve arrows is given.
- ▶ Classmates will volunteer to assist with scoring, finding stray arrows, and so on when asked by the student with visual impairments.

Activity Modifications

Task analysis and activity analysis provide the basis for activity or game selection. After choosing the most appropriate activities to meet the individual's learning goals, the educator must then modify them to the extent necessary. An example of an activity modification for a student with a visual impairment in an archery class is described in the Activity Modification application example. Modifications appropriate for a softball game provide an example. Activity modification for students with below-average mental abilities might include simplifying the rules or practicing lead-up activities. For those with visual impairments, the playing field might be made smaller, with base paths having a different texture than the field. Equipment might be adapted to include sound devices in the ball and in the bases to aid players with visual disabilities and lighter balls and batting tees for those with physical disabilities.

Modified rules for making "outs" or a change in the number of participants may be required for players with limited mobility. Modifications enable players to compete at their highest level of ability. For example, boccie can be played using a ramp by individuals with little or no use of their upper extremities (figure 6.2a) or by individuals who use a wheelchair (figure 6.2b). When modifying an activity, the educator should try to maintain its inherent nature so that it is as close to the traditional activity as possible.

Modifications need not affect every component of an activity; they should be limited to those that are necessary to meet individual needs. It may be necessary to modify the rules to make them more easily understandable, facilities to make them accessible, scoring to equalize competition, or the number of participants to ensure maximum participation or modifications.

ORGANIZATIONAL AND METHODOLOGICAL TECHNIQUES

Various organizational and methodological techniques may be employed to individualize instruction. They include team teaching, supportive teaching, peer and cross-age tutoring, and independent learning.

Team Teaching

If there are individuals with unique needs in an inclusive physical education class, it is often best for two or more teachers to instruct the class together so that individual differences can be accommodated. This ap-

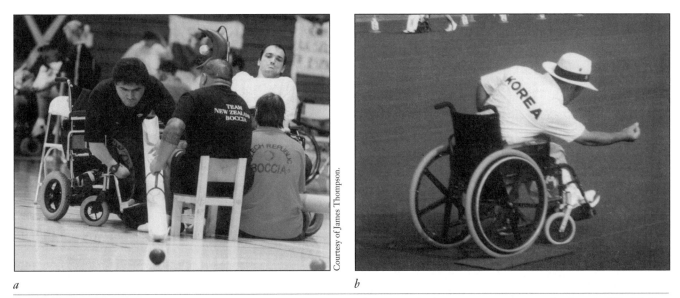

Courtesy of James Thompson.

a   *b*

Figure 6.2   (*a*) Providing Class I athletes with cerebral palsy with a ramp facilitates competition at the Boccie World Cup in 1998. (*b*) No modification is needed to allow athletes with mobility impairments to compete.

proach, called team teaching, is especially important in settings where educators are not well prepared to work with students who have unique needs. This technique is often used in inclusive settings where an adapted physical educator and perhaps an aide work with a child with special needs to allow her to participate with her peers.

An example of successful team teaching in an inclusive setting is described in an article by Block, Zeman, and Henning (1997), where 11-year-old Jimmy, who is nonverbal, severely mentally retarded, moderately cerebral palsied, and behaviorally disordered, participates safely and successfully with 25 to 30 peers in a 6th grade basketball unit. In class Jimmy had a personal aide and shared an adapted physical education teacher with two other students with disabilities. The adapted physical educator provided consultative services to the class's physical educator as well as assisting Jimmy in class. This allowed the regular physical educator to successfully include Jimmy and the other students with disabilities in the class without devoting more than an average amount of time to Jimmy. Students in the class did not appear to resent his presence and actually offered to assist Jimmy with his skills. As a result Jimmy was able to achieve his IEP physical education goals.

Supportive Teaching

Often when students with disabilities are included in regular physical education classes, it is necessary for an aide or a volunteer assistant to help the student. This assistance supports the physical education teacher's efforts to include the child fully in the activity of the regular class and promotes successful participation by the student. The supportive teaching approach is extremely valuable to a student who is just beginning the integration process. It eases the adjustment process and allows the student to concentrate on the material being taught.

Peer and Cross-Age Tutoring

Peer tutoring involves same-age students helping with instruction. Peer tutors can be used to reduce the student-teacher ratio. Cross-age tutoring programs enlist students of different ages (such as high school juniors or seniors) to work with young children receiving adapted physical education. Cross-age tutoring can provide satisfaction to the tutors while increasing the level of learning for students with disabilities. The use of peer-age tutors has become common in inclusive physical education classes. Houston-Wilson, Dunn, Van der Marss, and McCubbin (1997) reported that the use of trained peer tutors assisting students with developmental disabilities resulted in improved motor performance when this technique was used in an integrated setting over a 36-day period. Figure 6.3 illustrates the peer-tutoring concept.

Independent Learning

Independent learning gives students the opportunity to work and progress at their own rates. This technique is particularly useful when it is necessary to exclude a student from certain activities. Independent learning materials may take several forms. For example, the teacher may prepare a series of task cards for the student specifying skills to be accomplished.

Sometimes the student and the teacher might make a formal agreement designating the tasks the student must master to earn a certain grade. This is known as a **learning contract**. Such contracts allow teachers to

Figure 6.3   Peer tutor assists a student with visual impairment improve muscle strength.

personalize grading to reflect the unique conditions of each learner.

Computer-assisted instructional modules also allow students with disabilities to learn at their own pace. They progress through the use of interactive episodes and programmed learning.

PRESCRIPTIVE PLANNING AND INSTRUCTIONAL MODELS

Several curricular models, developed since the 1970s, have proved to be successful in providing a quality physical education experience for individuals with disabilities and have promoted the development of skills needed for inclusion. When used as guides, these models can help enhance the teaching process at various levels and ensure that students with unique needs are taught in an efficient and effective manner. They include I CAN Primary Skills K-3 (Wessel & Zittel, 1998), Data Based Gymnasium (Dunn et al., 1986), Project ACTIVE (Vodola, 1976), M.O.V.E. (Bidabe, 1995), Motor Development Curriculum for Children (Werder & Bruininks, 1988), Physical Best (American Alliance for Health, Physical Education, Recreation and Dance [AAHPERD], 1999a; 1996, Dunn, Vandermass, and McCubbin), Brockport Physical Fitness Test and training guide (Winnick & Short, 1999a; 1999b), and Moving to Inclusion (Active Living Alliance for Canadians with a Disability, 1994). Planned curricular programs for infants, toddlers, and preschoolers are incorporated in chapters 19 and 20.

I CAN

I CAN (Individualized instruction, Create social leisure competence, Associate all learnings, Narrow the gap between theory and practice) is a comprehensive physical education and leisure skills program appropriate for children with unique needs. It is developmental in nature and provides a continuum of skills from preprimary motor and play skills to sport, leisure, and recreation skills. The I CAN program offers a balance between psychomotor skills and affective and social development. It provides for individualized instruction at each individual's level of ability and lets learners progress at a rate and in a manner appropriate to their learning styles and motivation/interest levels.

I CAN is useful for children whose developmental growth is slower than average, including those with physical and mental disabilities as well as those with specific learning disabilities and social-emotional behavioral problems. In its boxed version (Wessell, 1979), I CAN consists of three major components: preprimary motor and play skills; primary skills; and sport, leisure, and recreation skills.

The Achievement Based Curriculum (ABC) model (Wessel & Kelly, 1985) is an enhancement to the I CAN curriculum. It includes a systematic sequential process that enables teachers to plan, implement, and evaluate instructional programs for individuals based on selected goals and objectives (Kelly, 1989).

A revised edition of *I CAN: Primary Skill K-3* was developed by Wessel and Zittel in 1998. In spiral book form, it is less costly than its predecessor while providing many user-friendly teaching resources. It provides a physical education curriculum based on a performance-based instructional model with feedback methods to improve and modify instruction based on student performance. Figure 6.4 provides a graphic view of the instructional model associated with the program. It illustrates key components that identify feedback procedures and continuous progress strategies based on student program goals.

Loop Teaching Materials provide the major components of the K-3 physical education program. They consist of Locomotor and rhythm skills, Orientation skills, Object control skills, and Personal-social participation skills. Also included are criterion-referenced performance standards, checklists, class performance sheets, progress reports, and assessment procedures.

The I CAN Primary Skills curriculum model, when combined with the SMART START Preschool Movement Curriculum (Wessel & Zittel, 1995), provides children with a sound foundation in motor skills. Figure 6.5 demonstrates a child with Down syndrome practicing kicking a stationary ball. SMART START provides teachers and caregivers a developmentally appropriate movement curriculum for preschool aged

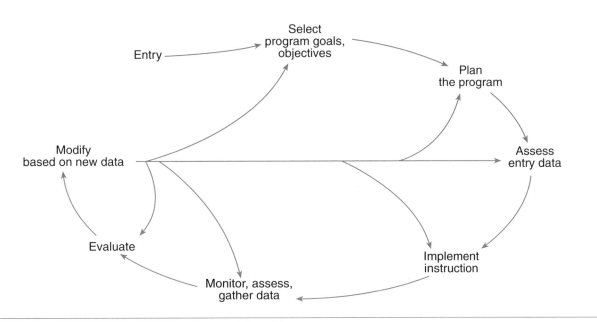

Figure 6.4   Performance-based instructional model used by I CAN.

Reprinted, by permission, from J. Wessel and L. Zittel, 1998, *I CAN Primary Skills: K-3* (2nd ed). (Austin, TX: Pro-Ed Publishing).

children of all abilities. SMART START is discussed in greater detail in chapters 19 and 20. Plans are in progress to provide additional I CAN curriculum materials in sports, dance, gymnastics, and game skills for older elementary-aged children. These curriculum materials will be a valuable tool for physical educators, adapted physical educators, and classroom teachers who wish to provide a sound physical education program to their students.

The Data Based Gymnasium

The Data Based Gymnasium (DBG) is a prescriptive instructional model that provides ways to initiate and analyze behavior, emphasizes provision of feedback, and recommends ways to manage the learning environment (Dunn et al., 1986). The DBG also provides an exercise, sport, and leisure curriculum. Specific sport skills are task analyzed, and the tasks are sequenced as phases that represent shaping behaviors. Although the DBG is similar in design to many other instructional models, it is unique in its specific delineation of behavior modification techniques as a means of accomplishing task and terminal objectives.

Originally, the DBG model was designed to provide a physical education program for individuals with severe disabilities. However, its principles are applicable in some form to many instructional situations involving students with a variety of unique learning needs who would benefit from behavior management.

The DBG instructional model emphasizes three essential elements:

▶ Cue. The cue is a condition, signal, or request to the learner designed to influence the occurrence of a behavior. Cues can be verbal, like the command "stand up," or nonverbal, like printed directions or a demonstration of the desired activity.

▶ Behavior. Behavior is what a person does in response to a cue. In the DBG model it is a component of the task that the student is to learn or the skill that the student is being asked to master. The teacher presents cues that will result in the performance of the targeted behavior. If the desired terminal behavior is too complex for the student to achieve, it must be broken down into simpler, enabling behaviors. When the enabling behavior is accomplished (forward chaining), the student may then be capable of the more complex terminal behavior. With moderately and severely disabled individuals, a reverse chaining sequence may be used. In the DBG model, this means providing assistance with the initial parts of the task but letting the student perform the last component independently. Progressing, the student is allowed to perform additional components of the task unaided. The sequence thus builds systematically from the final component to the initial one using a reverse chaining process.

▶ Consequence. After a behavior is performed, the student must receive information about its success or failure. Feedback, or consequences, can be positive or negative in nature. Positive feedback is called a reinforcer, while negative feedback is a punisher. The model stresses the delivery of appropriate positive consequences to increase the probability of a behavior's being repeated.

Figure 6.5 Utilizing the I CAN K-3 protocol helps a child with Down syndrome learn to kick a stationary ball.

DBG includes a task-analyzed game, exercise, and leisure sport curriculum. In addition, the DBG model provides examples of systematic data collection techniques that can help the instructor to successfully implement the program. More detailed information related to the DBG can be found in a book by Dunn et al. (1986) and in a chapter in a text by Dunn (1997).

Project ACTIVE

The ACTIVE program (All Children Totally InVolved Exercising) was designed by Dr. Thomas Vodola (1976) to ensure that every child, regardless of disability, would have a chance to participate in a quality physical education program. Project ACTIVE incorporates a test-assess-prescribe-evaluate planning process and includes normative as well as criterion-referenced tests in the areas of motor ability, nutrition, physical fitness, and posture. These are specially designed instructional programs for individuals with mental retardation, learning disabilities, orthopedic impairments, sensory impairments, eating disorders, and breathing problems. They also include activities for postoperative convalescent children and normal and gifted individuals. Karp and Adler (1992) have revised and updated the ACTIVE materials. Project ACTIVE is an innovative, cost-effective program that has been widely used throughout the United States.

M.O.V.E. Project

The Mobility Opportunities Via Education (M.O.V.E.) (Bidabe, 1995) model was first developed in Bakersfield, California in 1985. It is a top-down activity-based curriculum developed to assist students with profound multiple disabilities to learn the basic motor skills needed for everyday activities in the home and community. M.O.V.E. endeavors to improve an individual's ability to sit, stand, and walk. It provides a sequence of age-appropriate motor activities that are valuable to the individual's quest for independent movement in the home and community. The M.O.V.E. curriculum provides a "Top Down Motor Milestone Test" for determining present level of function and many detailed task analyses for the skills to be learned. Completion of a two- to four-day training course is recommended for those wishing to implement this program.

Body Skills: A Motor Development Curriculum for Children

This model is a motor development curriculum for children ages 2 to 12 years (Werder & Bruininks, 1988). It can be used in a variety of settings. It is particularly useful in inclusion settings to determine the present level of function of each class member. The curriculum provides a systematic procedure for the assessing, planning, and teaching of gross motor skills. It provides units in body management, locomotion, body fitness, object movement, and fine motor development.

This program consists of a series of folders, one for each of the 31 skill areas covered. Each folder has a pictorial developmental sequence of the skill for criterion-referenced assessment purposes. It also provides a series of activities that utilize the skill and suggestions for adaptations and record-keeping forms.

The strongest asset of this program is the formal assessment, the Motor Skills Inventory (MSI). It provides a criterion-referenced test for measuring a student's prestart developmental level. Using record forms, the teacher relates the scores from the MSI to a standardized measure, the Bruininks-Oseretsky Test of Motor Proficiency (Bruininks, 1978). This correlation can be very helpful when developing IEPs.

Physical Best

The new Physical Best (AAHPERD, 1999a; 1999b) is an expanded comprehensive health-related physical fitness education curriculum designed to meet the abilities of all children. It provides students with the knowledge and skills needed to participate actively in society. All activities included in the program are designed to meet the National Association for Sport and Physical Education (NASPE) standards and enhance the ability of students to meet recommended physical activity guidelines.

Two books, *Physical Best Activity Guide—Elementary Level* (AAHPERD, 1999a) and *Physical Best Activity Guide—Secondary Level* (AAHPERD, 1999b) provide instructional activities appropriate for use by all children. Both books include competitive and noncompetitive activities that range from very active to less strenuous, and encourage students to develop aerobic fitness, muscular strength and endurance, and flexibility. Each activity includes suggestions for the inclusion of students with unique abilities. The guides also feature sections on principles of effective teaching, motivation, and nutrition.

Another program that is complementary to the Physical Best program is the You Stay Active Program (AAHPERD and Cooper Institute for Aerobics Research, 1995). It is designed to supplement AAHPERD's Physical Best/FITNESSGRAM materials and to promote and reward participation in lifelong physical activity. It also promotes the inclusion of students with disabilities in all its activity suggestions.

The Brockport Physical Fitness Test and Training Program

As a part of Project Target, funded by the U.S. Department of Education, materials have been developed to evaluate and develop the health-related physical fitness of youngsters with disabilities, ages 10 to 17. The first of these is the Brockport Physical Fitness Test (Winnick & Short, 1999a), developed to assess health-related physical fitness. The second is a training program (Winnick & Short, 1999b) designed to develop the health-related physical fitness of youngsters with disabilities. The training guide includes information to enhance the development of cardiorespiratory endurance, body composition, muscular strength and endurance, and flexibility and range of motion of individuals with a variety of disabilities. The testing and training materials were designed to be as compatible as possible with the FITNESSGRAM and the Physical Best training materials.

Moving to Inclusion

The Active Alliance for Canadians with Disability (1994) developed the Moving to Inclusion curriculum. It consists of nine books, available in English and French, which address a variety of disability groups such as visual impairment, cerebral palsy, and amputation. Each book provides ideas for individualizing and modifying activities. They all espouse the philosophy that persons with disabilities should have the right to choose activities that interest them, assume any risks inherent in the activities, and utilize assistance in performing activities on an individualized basis.

Moving to Inclusion provides several useful assessment tools including Basic Movement Skills Observation Profile, Transport Skills Checklist, and Object Control Skills Checklist. These are particularly valuable when planning to include students with disabilities in regular physical education classes.

SUMMARY

Many factors affect teaching effectiveness in the adapted physical education and sport setting. Educators who are capable of employing a variety of approaches and teaching methods and matching them to the learning needs of individuals can better structure the learning environment to ensure student satisfaction and success. Flexibility permits educators to teach the same content in different ways, thus allowing them to more fully adapt their teaching to meet individual needs. It also allows teachers to optimize student skill development by successfully implementing the principles of motor learning.

Teachers can facilitate learning by employing task analysis, activity analysis, and activity modification. These techniques are valuable in helping educators to meet learners' goals and accommodate their needs. Organizational and methodological techniques allow teachers to maximize learning opportunities by decreasing the teacher-student ratio and structuring independent learning experiences.

Several planning and instructional resources have been developed to meet the needs of unique individuals. They include I CAN, the Data Based Gymnasium instructional model, Project ACTIVE, M.O.V.E., the Body Skills Motor Development Curriculum, Physical Best, Brockport Physical Fitness Test and training program, and Moving to Inclusion. In addition to these behaviorally oriented models, teachers may draw on the humanistic orientation to physical education. These resources offer valuable assistance to physical educators in designing quality physical education programs to optimize the learning and inclusion of individuals with unique needs.

Behavior Management

E. Michael Loovis

> **M**R. SMITH IS THE PHYSICAL education teacher at Middlefield Junior High School. He has been assigned an eighth grade class into which several students with behavioral disorders have been integrated. Although the class is reasonably well behaved, one student Robert has a difficult time staying on task, and he is constantly uncooperative, often getting into arguments with other students in class. Mr. Smith has instituted a token economy system with Robert such that time on-task and reduced frequency of uncooperative behavior earn points that will be displayed prominently on a chart in the gymnasium. During the first day that the token economy was in effect, Mr. Smith noticed that Robert was paying attention to his instructions and participating effectively in drills. Mr. Smith walked up to Robert and said, "Good work, Robert, you've earned one point." Then Mr. Smith placed one point on the chart. A little later Robert told a teammate that he had done a "good job." Mr. Smith approached Robert and said, "That was a very nice thing to say to Sam; you've just earned another point." At this rate Robert has a good chance to earn three points, which is the number required to exchange points earned for backup reinforcers. Robert had previously determined what those reinforcers would be with help from Mr. Smith. If Robert earns the prescribed number of points, he can redeem them for the opportunity to choose from a menu of reinforcers that include his choice for culminating activity of the day, free time in the gym on Friday, or serving as captain for his group during the next class period.

For years, educators confronting problems of inappropriate behavior have employed several remedial practices, including corporal punishment, behavior management techniques, suspension, and expulsion. In most cases use of these practices has been and continues to be *reactive*: A particular method is applied after some misbehavior has occurred. However, many of the problems that educators face on a daily basis could be prevented if they took a more *proactive* stance toward managing student behavior. Implicit in this statement is the discontinuation of ineffective practices (Mendler, 1992). For example, it will be difficult for a physical educator to establish an appropriate instructional climate in a self-contained class for students with behavior disabilities when several students persist in being verbally and physically abusive toward the teacher and other class members.

From another perspective, behavior management approaches have been used successfully to facilitate skill acquisition—either directly, through systematic manipulation of content, or indirectly, through arrangement of the consequences of performance to produce more motivation. For example, a behavior management approach can help a physical educator to determine the appropriate level at which to begin instruction in golf for students with mental retardation and to maintain the students' enthusiasm for learning this activity over time. Behavior management has also been used to teach those appropriate social behaviors considered essential to performance in school, at home, and in other significant environments. This chapter offers several proactive approaches to help teachers and coaches achieve the goals and objectives of their programs in a positive learning environment. The following approaches will be highlighted: behavior modifi-

cation, psychoeducational, psychodynamic, ecological, psychoneurological, and humanistic. Initially, behavior modification will be discussed in detail.

BEHAVIOR MODIFICATION

Behavior modification is a systematic process in which the environment is arranged to facilitate skill acquisition and/or shape social behavior. More specifically, it is the application of reinforcement learning theory derived from operant psychology. Behavior modification includes such procedures as respondent conditioning (the automatic control of behavior by antecedent stimuli), operant conditioning (the control of behavior by regulating the consequences that follow a behavior), contingency management (the relationship between a behavior and the events that follow behavior), and behavioral modeling, also called observational learning (learning by observing another individual engaged in a behavior). All have one thing in common— The planned systematic arrangement of consequences to alter an individual's response or at least the frequency of that response. As it relates to an IEP designed to improve physical fitness, this arrangement could involve the use of rewards to cause students with mental retardation to engage in sustained exercise behavior when riding stationary bicycles. It could also mean establishing a contract with a student who has cerebral palsy to define a number of tasks to be completed in a unit on throwing and catching skills.

To understand behavior modification, it is necessary to know some basic terminology. On the assumption that behavior is controlled by its effect on the environment, the first step toward understanding the management of human behavior is to define the stimuli that influence people's behavior. A measurable event that may have an influence on behavior is referred to as a *stimulus*. Reinforcement is a stimulus event that increases or maintains the frequency of a response. In physical education, reinforcement may be thought of as feedback provided directly or indirectly by the teacher or coach. Reinforcers can be physical, verbal, visual, edible, or active in nature. Examples of reinforcers include a pat on the back (physical), an approving comment like "Good job!" (verbal), a smile (visual), a piece of candy (edible), and a chance to bounce on a trampoline (active). All of these examples are usually considered positive reinforcers. Positive reinforcers, or rewards, are stimuli that individuals perceive as good, that is, as something they want. If a response occurs and it is positively reinforced, the likelihood of its recurring under similar circumstances is maintained or increased. For example, a teacher might praise a student who demonstrates appropriate attending behavior during instruction in the gymnasium. If praise is positively reinforcing to that student, the chances of the student's attending to instruction in the future are strengthened. Positive reinforcement is one of the basic principles of operant conditioning described in this section. Figure 7.1 illustrates this and the other principles that this chapter examines. The presence of aversive or "bad" stimuli—something that individuals wish to avoid—is commonly called negative reinforcement. If a response occurs and if it successfully averts a negative stimulus, the likelihood of the desired response recurring under similar circumstances is maintained or increased.

Because positive and negative reinforcement can produce similar results, the distinction between them may not be readily apparent. An example may clarify the difference. Suppose that the student mentioned in the previous paragraph had been talking to a friend while the teacher was explaining the lesson, and that this talking was distracting to the teacher and the rest of the class. If the teacher had warned that continual talking would result in after-school detention or a low grade for the day (possible aversive stimuli), and if the student perceived the stimuli as something to avoid, then the likelihood that the student would attend to instructions would increase. By listening in class the student would have avoided staying after school or having the grade reduced. This is an example of negative reinforcement because the stimulus increased the likelihood of a desired behavior through the avoidance of an aversive consequence rather than the presentation of a positive one.

Just as teachers and coaches seek to maintain or increase the frequency of some behaviors, they may wish to decrease the occurrence of others. When the consequence of a certain behavior has the effect of decreasing its frequency, the consequence is called punishment. Punishment can be either the presentation of an aversive stimulus or the removal of a positive stimulus. The intention of punishment is to weaken or eliminate a behavior. The following scenario illustrates the effect of punishment on the student in our previous example. The student has been talking during the instructional time. The teacher has warned the student that continual talking will result in detention or a grade reduction. The

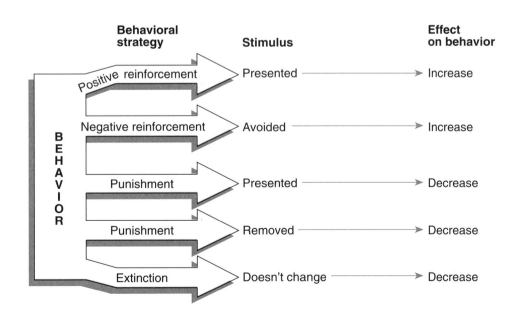

Figure 7.1   Principles of operant conditioning.

Table 7.1   Behavior, Consequence, Classification, and Probable Effect

Behavior exhibited	Consequence	Classification	Effect
Susan engages effectively in drills.	Teacher praises Susan.	Positive reinforcement	Susan will continue to do well in drills.
Juan forgets to wear his gym uniform.	Teacher suggests that next time he will lose five points.	Negative reinforcement	Juan wears his gym uniform.
Tenora forgets her tennis shoes for the second day in a row.	Teacher deducts five points from her grade.	Punishment	Tenora will not forget her shoes next class.
Bill makes aggravating noises during class.	Teacher ignores Bill's noises.	Extinction	Bill will stop making noises during class.
Chris is overly aggressive during game play in class.	Teacher withdraws opportunity to participate in free time at end of class.	Punishment	Chris will tone down his aggressiveness during game play.

student ignores the warning and continues to talk. The consequence for talking when one is supposed to be listening, at least in this situation, will be the presentation of one of the two aversive stimuli.

A slightly different scenario illustrates the notion of punishment as the removal of a positive stimulus. Our student, still talking after the teacher's warning, is punished by being barred from a five-minute free-time activity at the end of class—an activity perceived as a positive stimulus. The removal of this highly desirable activity fits the definition of punishment and will weaken or eliminate the disruptive behavior in future instructional episodes.

In contrast to punishment, withholding of reinforcement after a response that has previously been reinforced results in the extinction or cessation of a behavior. Extinction differs from punishment in that no consequence follows the response; a stimulus (aversive or positive event) is neither presented nor taken away. For example, teachers or coaches who pay attention to students when they clown around may be reinforcing the very behavior they would like to see extinguished. If they ignore (i.e., stop reinforcing) undesirable behavior, it will probably decrease in frequency.

Table 7.1 summarizes the basic principles of behavior modification. Each principle has a specific purpose; application of the principles requires that teachers or coaches analyze behaviors carefully before attempting to change them.

The examples provided in table 7.1 illustrate the range of learning principles available to teachers and coaches who wish to change the behavior of students. No one principle is necessarily the best choice all the time; the principle applied will depend on the specifics of the situation. Later in this chapter and again in chapter 10, recommendations for the use of learning principles are provided that originate from the work of Dunn, Morehouse,

and Fredericks (1986). After reading the section on their rules of thumb (chapter 10), readers are encouraged to review table 7.1 to better understand the place of selected learning principles in behavior modification strategies.

TYPES OF REINFORCERS

Several types of reinforcers can be used in behavior management. These include primary (unconditioned), secondary (conditioned), and vicarious reinforcers. The use of highly preferred activities to control the occurrence of less preferred responses, known as the Premack Principle, is another reinforcer.

Primary Reinforcers

Primary, or unconditioned, reinforcers are stimuli that are necessary for survival. Examples include food, water, and other phenomena that satisfy biological needs such as the needs for sleep, warmth, and sexual stimulation.

Secondary Reinforcers

Secondary, or conditioned, reinforcers acquire their reinforcing properties through learning. A few examples are praise, grades, money, and completion of a task. Because secondary reinforcers must be learned, stimuli or events often must be paired repeatedly with other primary or secondary events before they will become reinforcers in their own right.

Vicarious Reinforcers

The essence of vicarious reinforcement consists in observing the reinforcing or punishing consequences of another person's behavior. As a result of vicarious reinforcement, the observer will either engage in the behavior to receive the same positive reinforcement or avoid the behavior to avert punishment.

Premack Principle

According to the **Premack Principle** (Premack, 1965), activities that have a high probability of occurrence can be used to elicit low-probability behaviors. To state it another way, Premack implies that activities in which an individual or group prefers to engage can be used as positive consequences or reinforcers for activities that are not especially favored.

SCHEDULES OF REINFORCEMENT

When using the behavior modification approach, the teacher or coach must understand when to deliver a reinforcer to attain an optimal response. During the early stages of skill acquisition or behavior change, it is best to provide reinforcement after every occurrence of an appropriate response. This is called **continuous reinforcement**. After a behavior has been acquired, continuous reinforcement is no longer desirable or necessary. In fact, behavior is best maintained not through a process of continuous reinforcement but through a schedule of intermittent or partial reinforcement. Several types of intermittent schedules exist; however, the two most commonly used are ratio and interval schedules.

In **ratio schedules**, reinforcement is applied after a specified number of "defined" responses occur. **Interval schedules**, on the other hand, provide reinforcement when a specified time has elapsed since the previous reinforcement. A ratio schedule, therefore, is based on some preestablished number of responses, while an interval schedule is based on time between reinforcements. Associated with each of these major schedule types are two subtypes, **fixed** and **variable**. When these types and subtypes are combined, four alternatives for dispensing reinforcement are available. Both fixed and variable ratio schedules of reinforcement produce high response rates (e.g., providing praise every second successful throw or every third successful throw on average). In the case of interval reinforcement, a fixed schedule produces a high response rate just prior to the time for the next reinforcement. Conversely, there is a cessation of responding after reinforcement (e.g., praising first successful throw at end of each one-minute interval). Variable intervals of reinforcement, on the other hand, produce consistent response rates because the individual cannot predict when the reinforcer will be dispensed (e.g., praise first successful throw at end of an average three-minute interval).

PROCEDURES FOR INCREASING BEHAVIOR

Once it is determined that either a behavior not currently in a student's repertoire is desired or that the frequency of a given behavior needs to be altered, it is necessary to define the targeted response in measurable and observable terms. That response will be an observable instance of performance having an effect on the environment. After clear identification has been made, behavioral intervention can begin. The following discussion highlights several of the more commonly used strategies for increasing desirable behavior. These include shaping, prompting, chaining, modeling, token economy, fading, and contingency management.

Shaping

Shaping involves administering reinforcement contingent on the learning and performance of sequential steps leading to development of the desired behavior. Shaping is most often employed in the teaching of a new skill. Once the terminal or desired behavior has been learned, it is no longer necessary to perform all steps in the progression. For example, the use of shaping to teach a dive from a 1-meter board could include the following progression: kneeling dive from a 12-inch elevation, squat dive from a 12-inch elevation, standing modified dive from a 12-inch elevation, squat dive from a diving board, and standing modified forward dive from the diving board. Once the dive from the one-meter board has been learned, there is no longer any reason to practice the steps in the progression.

Chaining

Unlike shaping, which consists of reinforcing approximations of a new terminal behavior, **chaining** develops a series of discrete portions or links that, when tied together, lead to enhanced performance of a behavior. Chaining is distinct from shaping in that the steps necessary to achieve the desired or terminal behavior are performed each time that the response is emitted. In shaping, the steps in the progression are not considered essential once the terminal behavior has been mastered.

There are two types of chaining: forward and backward. In forward chaining, the initial step in the behavioral sequence occurs first, followed by the next step, and so forth until the entire sequence has been mastered (see the Behavior Management application example). A student learning to execute a layup from three steps away from the basket would take a step with the left foot; take a step with the left foot while dribbling once with the right hand; repeat the previous step and add a step with the right foot; and repeat the previous step with an additional step and jump off the left foot up to the basket for the layup attempt. In some cases a student's repertoire is either limited and/or the last step in the sequence is associated with a potent reinforcer. Under these circumstances, it may be necessary to teach the last step in a behavioral sequence first, followed by the next-to-last step, and so on until the entire sequence is learned. This is called backward chaining.

Application Example: Behavior Management

Setting:	An elementary physical education class
Student:	9-year-old student with high functioning autism with delays in throwing and catching
Unit:	Fundamental motor skills and patterns
Task:	Throwing with a mature, functional throw—at the very least stepping with the leg opposite the throwing arm
Application:	The physical educator might do any of the following:

- ► Use forward chaining of the throwing mechanics.
- ► Use a visual prompt such as a footprint to aid in stepping action.
- ► Use continuous reinforcement until correct throwing is well established.
- ► Fade visual prompt (e.g., footprints) as student becomes more consistent in performance.

Prompting

Events that help initiate a response are called prompts. These are cues, instructions, gestures, directions, examples, and models that act as antecedent events and trigger the desired response. In this way the frequency of responses and, subsequently, the chances of receiving reinforcement are increased. Prompting is very important in shaping and chaining procedures. A prompt that is commonly used during instruction with individuals with mental retardation and autism is "redirection." The purpose of a "redirect" is to communicate an alternative means to engage the learner's attention on the task at hand. It can be in the form of physical or gestural cues including pointing, touching materials, or touching the person's hand if touching can, in fact, be tolerated (Jones, 1998).

Fading

The ultimate goal of any procedure to increase the frequency of a response is for the response to occur without the need for a prompt or reinforcer. The best way to reach this goal is with a procedure that removes or fades the prompts and reinforcers gradually over time. Fading reinforcers means stretching the schedule of reinforcement so that the individual has to perform more trials or demonstrate significantly better response quality in order to receive reinforcement. For example, a person who has been receiving positive reinforcement for each successful basket must now make two baskets, then three baskets, and so forth before reinforcement is provided.

Modeling

Modeling is a visual demonstration of a behavior that students are expected to perform. From a behavioral

perspective, modeling (compare it to vicarious reinforcement) is the process in which an individual watches someone else respond to a situation in a way that produces reinforcement or punishment. The observer thus learns vicariously.

Token Economy

Tokens are secondary reinforcers that are earned, collected, and subsequently redeemed for any of a variety of backup reinforcers. Tokens, which could be poker chips or checkmarks on a response tally sheet, are earned and exchanged for consumables, privileges, or activities—the backup reinforcers. A reinforcement system based on tokens is called a token economy. Establishment of a token economy includes a concise description of the targeted behavior or behaviors along with a detailed accounting of the numbers of tokens administered for performance of targeted behaviors.

Contingency Management

Contingency management means changing behavior by controlling and altering the relationship among the occasion when a response occurs, the response itself, and the reinforcing consequences (Walker & Shea, 1988). The most sophisticated form of contingency management is the behavioral contract. Basically, the contract (which is an extension of the token economy) specifies the relationship between behaviors and their consequences. The well-developed contract contains five elements: a detailed statement of what each party expects to happen; a targeted behavior that is readily observable; a statement of sanctions for failure to meet the terms of the contract; a bonus clause, if desirable, to reinforce consistent compliance with the contract; and a monitoring system to keep track of the rate of positive reinforcement given (Kazdin, 1989).

PROCEDURES TO DECREASE BEHAVIOR

There will be occasions when some behavior of an individual or a group should be decreased. Traditionally, decreasing the frequency of behavior has been accomplished using extinction, punishment, reinforcement of alternative responses, and time-out from reinforcement. This section will stress the management techniques that are positive in nature, because they have been successful in reducing or eliminating a wide range of undesirable behaviors. Moreover, they model more socially appropriate ways of dealing with troublesome behaviors and are free of the undesirable side effects of punishment. Reinforcement is ordinarily thought of as a process to increase, rather than decrease, behavior. Consequently, extinction and punishment are most often mentioned as methods for decreasing behaviors. This section highlights the use of reinforcement techniques to decrease behavior (Parrish, 1997).

Reinforcement of Other Behavior

Reinforcing an individual for engaging in *any* behavior other than the targeted behavior is known as differential reinforcement of other behavior. The reinforcer is delivered as long as the targeted behavior (e.g., inappropriate running during the gym class) is not performed. Therefore, the student receives reinforcement for sitting on the floor and listening to instructions, standing quietly and listening to instructions, sitting on the bleachers and listening to instructions—anything other than inappropriate running during class. This reinforcement has the effect of decreasing the targeted response.

Reinforcement of Incompatible Behavior

This technique reinforces behaviors that are directly incompatible with the targeted response. For example, a student has a difficult time engaging cooperatively in games during the physical education class. The opposite behavior is playing cooperatively. The effect of reinforcing cooperation during game playing is the elimination of the uncooperative response. Unlike reinforcement of other behavior, discussed previously, this strategy defines diametrically opposed behaviors—playing uncooperatively versus playing cooperatively—and reinforces instances of the positive behavior only.

Reinforcement of Low Response Rates

With a technique known as differential reinforcement of low rates of responding, a student is reinforced for gradually reducing the frequency of an undesirable behavior or for increasing the amount of time during which the behavior does not occur. For instance, a student who swears on the average of five times per day would be reinforced for swearing only four times. This schedule would be followed until swearing was eliminated completely.

The three techniques just discussed use positive reinforcement to decrease the frequency of undesirable behavior. More traditional methods of decreasing such behaviors are described below, in recognition of the breadth and diversity of behavior management techniques.

Punishment

Normally, punishment is thought of as the presentation of an aversive consequence contingent on the occurrence of an undesirable behavior. In the Skinnerian (or operant psychology) tradition, punishment also includes the removal of a positively reinforcing stimulus or event, which is referred to as response cost. In either case, the individual is presented with a consequence that is not pleasing or is deprived of something that is very pleasing. The student who is kept after school for being disobedient is most likely experiencing punishment. Likewise, the student who has failed to fulfill a part of the contingency contract in the class and thus has lost some hard-earned tokens that "buy" free time in the gymnasium is experiencing punishment. In each case the effect is to reduce the frequency of the undesirable behavior. Walker and Shea (1988) detail the advantages and disadvantages of using punishment. One advantage is the immediacy of its effect; usually there is an immediate reduction in the response rate. In addition, punishment can be effective when a disruptive behavior occurs with such frequency that reinforcement of an incompatible behavior is not possible. Punishment can also be effective for temporarily suppressing a behavior while another behavior is reinforced. The disadvantages are several; they include undesirable emotional reactions, avoidance of the environment or person producing the punishment, aggression toward the punishing individual, modeling of punishing techniques by the individual who is punished, and reinforcement for the person who is delivering the punishment. Additionally, physical punishment may result in physical abuse to the person, although that may not have been the intent.

Time-Out

Time-out is an extension of the punishment concept. We have mentioned that punishment often involves the removal of a positive event. The time-out procedure is based on the assumption that some positive reinforcer in the immediate environment is maintaining the undesirable behavior. In an effort to control the situation, the individual is physically removed from the environment and consequently deprived of all positive reinforcement for a specified time. The three basic types of time-out procedures are observational, exclusion, and seclusion (Lavay, French, & Henderson, 1997). In observational time-out, the student is removed from an activity but is permitted to watch as classmates engage in the lesson. Exclusion time-out, on the other hand, isolates the student within the physical education setting without opportunity to observe what is going on in the lesson. Finally, seclusion time-out completely isolates the student by taking him away from the physical education setting.

IMPLEMENTING A BEHAVIOR MODIFICATION PROGRAM

On a daily basis, behavior modification is used in some form by most people—parents, teachers, coworkers, and students. In ordinary situations, however, its use may not be thorough and regular. On the other hand, the deliberate application of reinforcement learning principles in an attempt to change behavior is a systematic, step-by-step procedure. Minimally, there are four steps that, if implemented correctly, provide a strong basis for either increasing or decreasing the frequency of a particular behavior. These steps include identifying the behavior, establishing a baseline, choosing the reinforcer, and scheduling the reinforcer.

Identifying the Behavior

The first step is to identify the behavior in question. This is not as easy as it may sound, because it entails fulfillment of two criteria. The first criterion is that the behavior must be observable; specifications that distinguish one behavior from another are clearly established. Measurability is the second criterion; it assumes that the frequency, intensity, and duration of a behavior can be quantified. Sportsmanship, for example, could be defined as the number of times students compliment their opponents for good performance during a game.

Establishing a Baseline

With the targeted behavior identified, it is necessary to determine the frequency, intensity, and duration of its occurrence. This process, known as establishing a baseline, consists of observing the individual or group in a natural setting with no behavioral intervention taking place. Baseline determination should occur across a minimum of three sessions, days, periods, classes, or trials. Baseline is important as the comparison against which any programmatic gains are measured; it requires precise and accurate recording based on the criteria described in the preceding step. There are several recording systems; the choice depends on the nature of the behavior being observed. The most frequently used recording systems note event, duration, and interval. Event recording entails counting the exact number of times a clearly defined behavior occurs during a given period (e.g., the number of acts of good sportsmanship during a game or class period). Duration recording, on the other hand, measures the amount of time a student spends engaged in a particular behavior (e.g., cumulative time demonstrating good sportsmanship during a game). When reliable estimates of behavior are desired and when these observations are made during specific time periods, interval recording may be used. An example of interval recording is counting the number of 10-second intervals during which students demonstrate good sportsmanship, as defined in the first (identification) step of the behavior management program.

Choosing the Reinforcer

Once it has been determined that a behavior requires modification and the baseline data confirms this suspicion, it is essential to the success of the behavior modification program to choose the most effective reinforcer. Two important factors influence this choice. First, the type of reinforcer that will be effective in a given situation depends on the individual. Not all reinforcers work with all people; therefore, it is necessary to ascertain which is best. A second factor is quantity. Within limits, more reinforcement is probably better. However, when teachers and coaches reinforce in excessive amounts, *satiation* results and the reinforcer loses its value and effectiveness.

Scheduling the Reinforcer

With the reinforcer chosen, the next step is to schedule its use. The previous discussion of schedules is applicable here, with one very important reminder. When initiating a behavior change strategy, it is advisable to reinforce continuously. Once the behavior has shown desired change, reinforcement should be reduced gradually. It is this shift that maintains the new behavior at a desirable rate. One last word on scheduling: The longer reinforcement is delayed, the less effective it becomes.

USES OF BEHAVIOR MODIFICATION IN PHYSICAL EDUCATION AND SPORT

Dunn and Fredericks (1985) suggest that evidence supports the use of behavior modification in both segregated and inclusive programs for students with special needs. Dunn, et al. (1986) have developed a Data Based Gymnasium (DBG) for teaching students with severe and profound disabilities in physical education. Successful implementation of the DBG is dependent on systematic use of the behavioral principles discussed in this section. Additionally, Dunn et al. have provided rules of thumb that guide the use of behavioral techniques in teaching skills and/or changing social behaviors. These include the use of naturally occurring reinforcers such as social praise or extinction (i.e., ignoring a behavior). Tangible reinforcers such as food, toys, or desirable activities, which are earned as part of a token economy system, are not instituted until it has been demonstrated that the consistent use of social reinforcement or extinction is ineffective.

In skill acquisition programs, task-analytic phases and steps are individually determined, and students move through the sequence at a rate commensurate with their abilities. For example, a phase for kicking with the toe of the preferred foot consists of having students "perform a kick by swinging the preferred leg backwards and then forwards, striking the ball with the toes of the foot, causing the ball to roll in the direction of the target" (Dunn et al., 1986). Steps represent distances, times, and/or number of repetitions that may further subdivide a particular phase (e.g., kicking the ball with the toe of the preferred foot a distance of 10, 15, or 20 feet). Decisions about program modifications or changes in the use of behavioral strategies (rules of thumb) are made on an individual basis after each student's progress is reviewed. Further discussion of the DBG and the specific rules of thumb for managing inappropriate behavior in physical education with students with severe disabilities are presented in chapter 10.

Advantages of using behavior modification are that it considers only behaviors that are precisely defined and capable of being seen; it assumes that knowing the cause of a particular behavior is not a prerequisite for changing it; it encourages a thorough analysis of the environmental conditions and factors that may influence the behavior(s) in question; it facilitates functional independence by employing a system of least prompts, that is, a prompt hierarchy is used that is ordered from least to most intrusive; and it requires precise measurement to demonstrate a cause-and-effect relationship between the behavioral intervention and the behavior that is changed.

These disadvantages should be considered before a behavior management program is implemented: actual use of behavioral principles in a consistent and systematic manner is not as simple as it might seem; behavioral techniques may fail, because what is thought to be the controlling stimulus may not be so in reality; and behavioral techniques may not work initially, requiring more thorough analysis by the teacher to determine if additional techniques would be useful and to implement a new approach immediately, if necessary.

EXAMPLE OF BEHAVIOR ANALYSIS

The process for implementing a behavioral system, which is commonly referred to as behavior analysis, requires reasonably strict adherence to a number of well-defined steps. The following example illustrates the teaching of a skill using a limited number of concepts.

▶ Skill—Standing long jump
▶ Baseline—In a pretest condition, the student is observed on three different occasions performing the long jump with faulty mechanics, most notably in the takeoff and landing portions of the jump.
▶ Objective—When requested to perform a standing long jump, the student will jump a minimum of three feet, demonstrating appropriate form on takeoff, in the air, and on landing.
▶ Choose reinforcer—Teacher determines that social reinforcement, namely, verbal praise, is effective.
▶ Scheduling reinforcer—Teacher decides to use continuous reinforcement initially and then switch to a variable ratio as learning and performance increase.
▶ Prompt—Using the system of least prompts, the instructor would employ prompts in the order presented from least to most intrusive depending on the ability of the student to perform the task: (a) "Please stand behind this line and do a standing long jump" (verbal prompt); (b) "Please stand behind this line, bend your knees, swing your arms backward and forward like this, and jump as far as possible" (verbal plus visual prompt); and (c) "Please stand behind this line, bend your knees, feel how I'm moving your arms so they swing back and forth like this, and jump as far as possible" (physical guidance).
▶ Behavior—Student acknowledges prompts, correctly assumes long jump position, and executes long jump as intended.
▶ Reinforcement—Teacher says, "Good job!" (verbal reinforcement).
▶ Subsequent behavior—Student is likely to maintain or improve on the performance.

OTHER APPROACHES

The management of behavior has been the concern of individuals and groups with various theoretical and philosophical views. No fewer than five major approaches have been postulated to remediate problems associated with maladaptive behavior. Of these five models, two guide most educational programs today. They are the behavioral and psychoeducational approaches. One of these approaches, behavior modification, has already been discussed. The second, psychoeducational, will be discussed along with the psychodynamic, ecological, psychoneurological, and humanistic approaches. These interventions will be discussed only briefly here. Resources will be suggested for those who wish to further explore a particular intervention and its primary proponents.

Psychoeducational

The psychoeducational approach assumes that academic failure and misbehavior can be remediated directly if students are taught how to achieve and behave effectively. It balances the educational and psychological perspectives. This approach focuses on the affective and cognitive factors associated with the development of appropriate social and academic readiness skills useful in home, school, and community. Its proponents recognize that some students do not understand why they behave as they do when their basic instincts, drives, and needs are not satisfied. The cause of inappropriate behavior, however, is of minimal importance in the psychoeducational approach; it is more important to identify students' potentials for education and to emphasize their learning abilities. Diagnostic procedures include observational data, measures of achievement, performance in specific situations requiring particular skills, case histories, and consideration of measures of general abilities.

The psychoeducational approach focuses on strengthening the student's ego and teacher self-knowledge. This is accomplished through compensatory educational programs that encourage students to acknowledge that what they are doing is a problem, to understand their motivations for behaving in a certain way, to observe the consequences of their behavior, and to plan alternate responses or ways of behaving in similar circumstances.

When a behavioral crisis occurs (or shortly thereafter), a teacher prepared in the psychoeducational approach conducts a life-space interview (LSI), a term first used by Redl (1952). The purpose of LSI is to help the student either overcome momentary difficulties or work through long-range goals.

The psychoeducational approach assumes that making students aware of their feelings and having them talk about the nature of their responses will give them insight into their behavior and help them develop control. This approach emphasizes the realistic demands of everyday functioning in school and at home as they relate to the amelioration of inappropriate behavior.

There are several strategies that teachers can use to implement the psychoeducational approach. These include self-instruction, modeling and rehearsal, self-determination of goals and reinforcement standards, and self-reward.

Teachers are in an advantageous position to encourage students to use self-instructional strategies. This means helping students to reflect on the steps in good decision making when it is time to learn something new, solve a problem, or retain a concept. Fundamentally, the process involves teaching students to listen to their private speech, that is, those times when people talk either aloud or subvocally to themselves. For example, a person makes a faulty ceiling shot in racquetball and says, "Come on, reach out and hit the ball ahead of the body!" Self-instruction can be as simple as a checklist of questions for students to ask themselves when a decision is required. The following questions represent an example of the self-instructional process: What is my problem? How can I do it? Am I using my plan? How did I do?

In the modeling strategy, students who have a difficult time controlling their behaviors watch others who have learned to deal with problems similar to their own. Beyond merely observing the behavior, the students can see how the models respond in a constructive manner to a problematic situation. Modeling could include the use of relaxation techniques and self-instruction. Students can also learn appropriate ways of responding when time is provided to mentally rehearse or practice successful management techniques.

Another strategy that has proved effective in helping students control their behavior works by including them in the establishment of goals, reinforcement contingencies, or standards. An example of this process occurs when a group of adolescents with behavior disabilities determines which prosocial behaviors each member needs to concentrate on while on an overnight camping trip. Likewise, the group establishes the limits of inappropriate behavior and decides what the consequences will be if anyone exceeds the goal.

A final strategy used in the psychoeducational approach is self-reward. It involves preparing students to reward themselves with some preestablished reinforcer. For example, a student who completes the prescribed tasks at a practice station immediately places a check on a recording sheet posted at that station. When the checkmarks total a specified number, the student is thus instrumental not only in seeing that the goal of the lesson is achieved, but also in implementing the reinforcement process in an efficient manner.

Psychodynamic

Most closely associated with the work of Freud, the psychodynamic approach has evolved as a collection of many subtheories, each with its own discrete intervention. The focus of this approach, in any case, is psychological dysfunction. Specifically, the psychodynamic approach strives to improve emotional functioning by helping students understand *why* they are functioning inappropriately. This approach encourages teachers to accept students but not their undesirable behaviors. It emphasizes helping students to develop self-knowledge through close and positive relationships with teachers. From the psychodynamic perspective, the development of a healthy self-concept, including the ability to trust others and to have confidence in one's feelings, abilities, and emotions, is basic and integral to normal development. If the environment and the significant others in it are not supportive, anxiety and depression may result. Self-perceptions as well as perceptions of others can become distorted, and the result can be impaired personal relationships, conflicting social values, inadequate self-image, ability deficits, and maladaptive habits and attitudes.

In an attempt to identify the probable cause(s) of inappropriate behavior, the psychodynamic approach uses various diagnostic procedures such as projective techniques, case histories, interviews, observational measures of achievement, and measures of general and specific abilities. Through interpretation of diagnostic results and an analysis of prevailing symptoms, the cause of the psychic conflict is, ideally, identified. Once the primary locus of the emotional disturbance is known, an appropriate treatment can be determined.

Conventional pyschodynamic treatment modalities include psychoanalysis, counseling interviews, and psychotherapeutic techniques (such as play therapy and group therapy). Treatment sessions involve the student alone, although some therapists see only the parents. Recently, family therapy has become popular, with students and parents attending sessions together. However the session is configured, it is designed to help students develop self-knowledge.

The psychodynamic approach, including psychoanalysis and psychotherapy, is not regarded as an exemplary approach to intervention. Its inadequacies include the following: diagnostic study is time consuming and expensive; the results of diagnostic study yield only possible causes for emotional conflict; and therapeutic outcomes are similar regardless of the nature of the intervention—whether the students are seen alone or with their parents, or are seen in play therapy or in group counseling.

Teachers should be aware that implementation of psychodynamic theory as a means to manage behavior does not preclude working with groups. It need not have as its primary goal increasing students' personal awareness. It also does not imply that teachers should be permissively accepting, deal with the subconscious, or focus on problems other than those presenting real concerns in the present situation. The preceding are prevalent misconceptions about the psychodynamic approach.

Ecological

The ecological approach has as its basic assumption a disturbance in the student's environment or ecosystem. In effect, the student and the environment affect each other in a reciprocal and negative manner (i.e., some characteristic of the student disturbs the ecosystem, and the ecosystem responds in a way that causes the student to further agitate it). The problem has been and continues to be that the student is typically the one who is blamed for disturbed behavior with little or no consideration given to the context in which the disturbed behavior occurs.

Evaluative procedures for assessing the cause(s) of disturbed behavior are difficult at best. Educators have purported to use a five-phase process for collecting ecological data: describing the environment, identifying expectations, organizing behavioral data, summarizing the data, and establishing goals.

Additionally, the Behavior Rating Profile (BRP-2) examines behaviors in several settings from several different points of view. It consists of three student rating subscales where students rate themselves, one teacher rating scale, one parent rating scale, and a sociogram (a diagram or chart that uses connecting lines to indicate choices made in groups). Its purpose is to define deviant behavior specific to one setting or to one individual's expectations (Coleman, 1992).

The goal of the ecological approach is not just to stop some disturbed or unwanted behavior, but to change the environment in substantive ways so that it will continue to support desirable behavior once the intervention is withdrawn (Hallahan & Kauffman, 1997). Generally speaking, the focus of intervention within the ecological approach is on either a single ecosystem or on a combination of ecosystems. Interventions that focus on single ecosystems, however, are frequently unsuccessful, given the interrelatedness and interdependence of each unit or subsystem as it operates within the larger ecosystem (Cullinan, Epstein, & Lloyd, 1991).

Educational applications of the ecological approach are designed to make environments accommodate individuals rather than having individuals fit environments. In a classroom this could require physical and psychological adaptations such as individual or small-group work areas, time-out areas, or reinforcement centers. It means having teachers create environments where students succeed rather than anticipate failure. Changes in the home ecosystem might require parental involvement, respite care, or family therapy. At times, changes in several ecosystems are required (e.g., home, school, and community).

Present-day attempts to establish ecologically based programs include the creation and implementation of behavioral interventions as part of the IEP. Schools may no longer expel students with special needs regardless of whether or not the action was related to their disability; however, services may be provided in alternative settings. To develop an effective behavioral intervention, educators must engage in a three-step process: perform a functional assessment of the student's behavior, determine and implement intervention strategies, and evaluate the results. Assessing the student's behavior includes observation and recording of behavior patterns in a variety of settings (e.g., classroom, playground, cafeteria, gymnasium), with the intention of profiling the student's conduct. Establishing a baseline helps teachers understand either why a student is behaving in a certain way or what/who is controlling or reinforcing a student's behavior.

After completion of the functional assessment, teachers develop and implement an intervention. The intervention should set realistic goals, including the acceptance of behaviors that would ordinarily not be deemed appropriate. For example, a student who fights with a teammate each time he misses a shot during a basketball drill will realistically be permitted to shout or curse at that same teammate, as long as he does not physically assault the student. As many people as possible that work with the student should develop the intervention plan. Behavioral interventions must be team based so that the student experiences consistency. The student should also participate in development of the behavioral intervention.

Evaluation of the intervention plan is the final step. Teachers must assess how effective the plan is and if changes or modifications are necessary. Most experts agree that this should be done monthly, if not weekly.

Psychoneurological

The central focus of the psychoneurological approach is neurophysiological dysfunction. Closely associated with the medical model, this approach relies on diagnostic techniques that explore specific signs and symptoms. Physicians attempt to localize problems using neurological soft signs such as the results of gait and postural assessments as well as timed, repetitive movements and visual motor sequencing tasks, to name a few. Lack of definitive results may prompt the use of electroencephalography (EEG), computed tomography (CT), or magnetic resonance imaging (MRI) if lesser diagnostics prove uneventful. Identification of students with disabilities in this realm is made on the basis of general behavioral characteristics and specific functional deficits. General behavioral characteristics include hyperactivity, distractibility, impulsiveness, and emotional lability. Specific functional deficits, on the other hand, include disorders in perception, language, motor ability, and concept formation and reasoning. The manifestation of these characteristics and deficits is attributable to injury or damage to the central nervous system.

The psychoneurological approach places considerable importance on etiological factors. The integrity of the central nervous system is assessed on the basis of performance in selected activities or tests such as walking a line with eyes closed, touching a finger to the nose, and reacting to various stimuli such as pain, cold, and light. Additionally, a neurological examination including an EEG is often part of the diagnosis. Following diagnosis, treatment can include drug therapy, surgical procedures, physical therapy, sensory integrative therapy, and developmental training.

A major trend associated with the psychoneurological approach is drug therapy, albeit the use of drugs as a means of controlling or modifying behavior cuts across several behavioral approaches. Students are medicated for the management of such behaviors as short attention span, distractibility, impulsiveness, hyperactivity, visual motor impairments, and large motor coordination problems. Psychotropic drugs represent the major category of substances used to control emotional, behavioral, and cognitive changes of individuals with disabilities (Rinck, 1998). According to Kalachnik, et al. (1998), the psychotropic drugs include medications typically classified as antipsychotic, antianxiety, antidepressant, antimania, stimulant, or sedative-hypnotic. The most common categories of psychotropic drugs for the purpose of this discussion are stimulants, neuroleptics, and antidepressants.

Stimulants

Stimulants are administered primarily for the management of attention deficit and hyperactive disorder (ADHD). Ritalin (methylphenidate hydrochloride), the most commonly prescribed stimulant, and Dexedrine (dextroamphetamine sulfate) are adjunctive therapy and are used with students who experience moderate to severe hyperactivity, short attention spans, distractibility, emotional lability, and impulsiveness. Cylert (pemoline) is another stimulant that is frequently prescribed; however, its use is contraindicated with children under six years of age. Possible side effects include loss of appetite, weight loss, and insomnia. These effects are noted to be more profound in individuals with mental retardation (Arnold, Gadow, Pearson, & Varley, 1998).

Neuroleptics

These substances remain the most widely prescribed class of psychotropic medications for individuals with mental retardation. These drugs are also referred to as tranquilizers. They are used to control bizarre behavior in psychotic adults. In children they are used to control hyperactivity, aggression, self-injury, and stereotypic behavior. Tranquilizers are classified as either major or minor. Major tranquilizers, also referred to as antipsychotics, include such drugs as Mellaril (thioridazine hydrochloride), Thorazine (chlorpromazine hydrochloride), and Haldol (haloperidol), all of which are prescribed for the management of psychotic disorders, including severe behavior disorders marked by aggressiveness and combativeness (Baumeister, Sevin, & King, 1998). Valium (diazepam), Miltown (meprobamate), and Librium (chlordiazepoxide hydrochloride) are considered minor tranquilizers and are typically classified as anxiolytic or antianxiety drugs; they are used to relieve anxiety in normal life situations (Werry, 1998). Possible side effects of the major and minor tranquilizers include dizziness, drowsiness, vertigo, fatigue, diminished mental alertness that could impair performance in physical activities, and movement disorders, referred to as the extrapyramidal effect.

Antidepressants

Antidepressants are prescribed for adults to alleviate depression. In children they have a more diverse function, including not only the management of obsessive-compulsive disorder (OCD), affective disorders, and ADHD, but also the treatment of nocturnal enuresis and, to a lesser extent, ritualistic behavior, self-injurious behavior (SIB), and aggression. Perhaps because of the documented side effects of the antidepressant drugs—ataxia, muscle weakness, drowsiness, anxiety, sleep disruption, decreased appetite, increased blood pressure, and mental dullness (Sovner, et al. 1998)—physical educators and coaches should know when students are receiving such medication. Commonly prescribed antidepressants include Anafrani (clomipramine) and Tofranil (imipramine), which is used for ADHD where patients show excessive side effects or fail to respond favorably to stimulants. Other commonly prescribed drugs include Luvox (fluvoxamine) and Prozac (fluoxetine), although their efficacy with children has not yet been determined.

Humanistic

Based on the work of Maslow (1970), the humanistic approach has as its basis the self-actualization theory—Maslow's hierarchy of human needs. In the hierarchy, five primary human needs are identified and arranged in ascending order, from the most basic to the most prepotent. According to this theory of motivation, the human, who seeks to meet unsatisfied needs, will satisfy lower-level needs first and then satisfy needs at progressively higher levels as lower ones are met. A person lacking food, safety, love, and esteem would probably hunger for food more strongly than for anything else. But when the need for food was satisfied, the other needs would become stronger. In the need hierarchy, self-actualization is the fulfillment of one's highest potential. Maslow considered the following to be some of the attributes of self-actualized people: accepting, spontaneous, realistic, autonomous, appreciating, ethical, sympathetic, affectionate, helpful, intimate, democratic, sure about right and wrong, and creative.

Self-actualization is the process of becoming all that one is fully and humanly capable of becoming. In large measure this is developed naturally by persons without disabilities. Persons with disabilities, on the other hand, may not achieve the same relative status as their peers who are not disabled, without the intrinsic motivation to become all they are capable of becoming. The desire to move persons with disabilities toward self-actualization is supplied, at least initially, by persons who care about these individuals as persons first and only then as persons with disabilities.

In the fifth revision of her text on adapted physical education, Sherrill (1998) has continued to draw extensively and build upon self-actualization theory, as well as Roger's concept of the fully functioning self, to guide a humanistic orientation to adapted physical education in general and to affective development in particular. She is not only the spokesperson for the humanistic philosophy but also its instrumentalist in that she applies its concepts in the gymnasium and on the sports field. In an effort to demonstrate how the humanistic philosophy can be translated into action, Sherrill suggests that teachers and coaches of students with disabilities do the following:

►To the degree possible, use a teaching style that encourages learners to make some of the major decisions during the learning process. This implies that students should be taught with the least restrictive teaching style or the one that most closely matches their learning styles.

►Use assessment and instruction that are success oriented. No matter where students score in terms of the normal curve, they have worth as human beings and must be accepted as such.

►Listen to and communicate with students in an effort to encourage them to take control of their lives and make personal decisions affecting their physical well-being. Counseling students to become healthy, fit, and self-actualized requires the skills of active listening, acceptance, empathy, and cooperative goal setting. Such interaction helps persons with disabilities reinforce their internal locus of control.

►Use teaching practices that enhance self-concept. The following practices are recommended: (a) show students that someone genuinely cares for them as human beings, (b) teach students to care about each other by modeling caring behavior in daily student and teacher interactions, (c) emphasize social interaction by using cooperative rather than competitive activities, and (d) build success into the instructional plan through the careful use of task and activity analysis.

Hellison is another physical educator who has been instrumental in disseminating the humanistic viewpoint (Hellison, 1995; Hellison & Templin, 1991). He has developed a set of alternative goals for physical education that he believes focus on human needs and values rather than on fitness and sports skills development per se. Hellison has applied the model in a basketball unit that is described in chapter 10. The goals are developmental in nature and reflect a level-by-level progression of attitudes and behaviors. Specifically, they include self-control and respect for the rights and feelings of others, participation and effort, self-direction, and caring and helping.

Level 0: Irresponsibility. This new level is intended to define students who fail to take responsibility either for their actions or inactions; these students blame others for their behaviors and typically make excuses.

Level I: Self-control and respect for the rights and feelings of others. The first level deals with the need for control of one's own behavior. Self-control should be the first goal, according to Hellison, because learning cannot take place effectively if one cannot control impulses to harm other students physically and verbally.

Level II: Participation and effort. Level II focuses on the need for physical activity and offers students one medium for personal stability through experiences in which they can engage on a daily basis. Participation involves getting uninterested students to at least "go through the motions," experiencing different degrees of effort expenditure to determine if effort leads to improvement, and redefining success as a subjective accomplishment.

Level III: Self-direction. Level III emphasizes the need for students to take more responsibility for their choices and to link these choices with their own identities. Students at this level can work without direct supervision and can take responsibility for their intentions and actions. At this level, students begin to assume responsibility for the direction of their lives and to explore options in developing strong and integrated personal identities. This level includes developing a knowledge base that will enhance achievement of their goals; developing plans to accomplish their goals, especially more difficult ones; and evaluating their plans to determine their success.

Level IV: Caring and helping. Level IV is the most difficult for students; it is also not a requirement for successful engagement in the responsibility model. At level IV students reach out beyond themselves to others, to commit themselves genuinely to caring about other people. Students are motivated to give support, cooperate, show concern, and help. Generally speaking, the goal of level IV is the betterment of the entire group's welfare.

Hellison recognized that the goals only provide a framework and that strategies must be employed to cause students to interact with self-control and respect for the rights and feelings of others, to participate and show effort, to be self-directed, and to demonstrate caring and helping behavior on a regular basis. He suggests six general interaction strategies to help reach the goals. These include awareness talks (post levels on gym wall and refer to them frequently); levels in action (students can be invited to play in a cooperative game); individual decision making (students are permitted to redefine success or negotiate an alternative to a planned activity); group meetings (students discuss issues of low motivation or difficulty in being self-directed); reflection time (students record in a journal or discuss how they did during class in relation to the goals they had established); and counseling time (students discuss their patterns of abusive behavior and possibly their underlying motives for such behavior). This last strategy gives students the opportunity to talk with the teacher about specific problems that may be preventing them from achieving their goals within specified levels of the responsibility model.

SUMMARY

Lack of discipline has been identified as one of the most significant problems confronting public school teachers. An equally significant, albeit insidious problem, is the replacement of old, worn out thinking about how discipline should be delivered in favor of effective, proactive practices. A number of behavior management systems are available that can significantly reduce the need to discipline students. As students with behavior disabilities are integrated into a regular class and demonstrate persistent disruptive behavior, a behavior management program designed to alleviate the problem might include any of the following: behavior modification using a token economy system; a psychoeducational approach using the life-space interview; a psychodynamic approach incorporating play therapy; a psychoneurological approach using medication; a humanistic approach using Hellison's Responsibility Model; an ecological approach using a behavior intervention plan that includes not only the school but also the student's home and community environments; or a combination of approaches.

Moreover, physical educators and coaches will be called on to contribute as members of schoolwide behavioral management systems that are the latest attempts to provide not only sound and consistent discipline policies but also to address the need for positive behavioral instruction. These models (ERIC/

OSEP Special Project, 1997) share several common features:

- ▶ Total staff commitment to managing behavior, whatever approach is taken
- ▶ Clearly defined and communicated expectations and rules
- ▶ Consequences and clearly stated procedures for correcting rule-breaking behaviors
- ▶ An instructional component for teaching students self-control and/or social skill strategies
- ▶ A support plan to address the needs of students with chronic, challenging behaviors

Part II

Individuals With Unique Needs

The first section of this part includes eight chapters that relate to individuals with disabilities who are specially categorized in accordance with the Individuals with Disabilities Education Act (IDEA). These chapters are followed by a chapter discussing children with unique physical education needs who have not been classified as disabled by IDEA.

The disabilities discussed in the first eight chapters include mental retardation (chapter 8); learning disabilities and attentional deficits (chapter 9); behavioral disorders (chapter 10); visual impairments and deafness (chapter 11); cerebral palsy, stroke, and traumatic brain injury (chapter 12); amputations, dwarfism, and les autres (chapter 13); spinal cord disabilities (chapter 14); and other health-impaired conditions (chapter 15). These chapters examine not only the causes of these conditions but the various types of conditions and the characteristics of affected groups. Particular attention is given to inclusion and teaching methods and activities that meet a variety of unique needs. Sport programs associated with each disability are presented. These chapters are critical for giving service providers the background and understanding they need about the individuals with unique needs they are preparing to serve.

Chapter 8

Mental Retardation

Patricia Krebs

▷ ▷ ▷ ▷ ▷ ▷ ▷

LORETTA CLAIBORNE WAS BORN BLIND, crippled, and with mental retardation in the projects of York, Pennsylvania. After numerous surgeries to enable vision and correct her clubbed feet, she finally walked at the age of four and talked at the age of seven. Forbidden to participate in school sports because she was in special education, Loretta ran to get away from the bullies. In 1971, at the age of 18, she became a Special Olympics athlete. Twenty-five years later, in 1996, Loretta Claiborne received the prestigious Arthur Ashe Award for Courage at ESPN's Espy Awards. In 1999, Disney aired a made-for-TV movie about her life, and Loretta appeared on *The Oprah Winfrey Show*. Along the way, Loretta completed three Boston Marathons with a best time of three hours and three minutes, placing among the top 100 of all women finishers each time. In 1981, she received the Spirit of Special Olympics Award. In 1988 she finished in the top 25 women in the Pittsburgh, Pennsylvania Marathon and was named Special Olympics Female Athlete of the Year. In 1991, Loretta was named to Special Olympics Inc.'s Board of Directors and was selected by *Runner's World* magazine as the "Special Olympics Athlete of the Quarter Century." The following year she was inducted into the York, Pennsylvania Sports Hall of Fame and the William Penn High School Alumni Hall of Fame—the same high school that barred her from the track team because she had mental retardation. Loretta introduced U.S. President Bill Clinton at the 1995 Special Olympics World Summer Games opening ceremonies in New Haven, Connecticut and received an Honorary Doctorate Degree of Humane Letters from Quinnipiac College in Hamden, Connecticut, the only known person with mental retardation to receive an Honorary Doctorate. One of Loretta's most memorable races was a marathon in Harrisburg, Pennsylvania. Running strong, Loretta noticed another runner beginning to falter. He told her he didn't think he could make it to the finish line. Loretta slowed her pace and stayed with the man throughout the race encouraging him on. They crossed the finish line together. It was truly a case of one champion helping another. The other runner? *Former world heavyweight boxing champion Larry Holmes!*

The purpose of this chapter is to familiarize readers with the intangible concept of mental retardation and how it is manifested in children and adults. In addition to defining mental retardation, this chapter explores the many causes of mental retardation and the characteristics of individuals with mental retardation who have mild to severe limitations. Physical activity planning, with an emphasis on including students with mental retardation into regular physical education and sport programs, is covered, and the Special Olympics movement and the Paralympic Games for those with an intellectual disability are contrasted.

DEFINITION OF MENTAL RETARDATION

Mental retardation is a heterogeneous group of disorders with myriad causes. It is characterized by cognitive limitations as well as functional limitations in areas such as daily living skills, social skills, and communication. Over the years, the definition of mental retardation and the criteria for its classification have changed, dramatically affecting the incidence of mental retardation.

The current American Association on Mental Retardation (AAMR) definition of mental retardation,

adopted in May 1992, states that "mental retardation refers to substantial limitations in present functioning. It is characterized by significantly subaverage intellectual functioning, existing concurrently with related limitations in two or more of the following applicable adaptive skill areas: communication, self-care, home living, social skills, community use, self-direction, health and safety, functional academics, leisure, and work. Mental retardation manifests before age 18."

Thus, three criteria must be met for an individual to be diagnosed as having mental retardation: *Significant subaverage intellectual functioning* refers to a person scoring below 70 to 75 on an intelligence test. There are two intelligence tests used extensively throughout the world: the Stanford-Binet Intelligence Scale and the Wechsler Intelligence Scale for Children—Revised (WISC-R).

Existing concurrently with related limitations in two or more of the following applicable adaptive skill areas is the second criterion. In addition to scoring below 70 to 75 on an intelligence test, significant limitations must exist in 2 or more of the 10 adaptive skill areas listed. Adaptive skills refer to the individual's ability to mature personally and socially with age. Maturity is measured according to the individual's development in each of the 10 skill areas listed. The third criterion is that *mental retardation manifests before age 18.*

Table 8.1   Pre-1992 Classifications of Severity of Mental Retardation Based on IQ Scores

Intelligence test score	Mental retardation level
50-55 to 70-75	Mild
35-40 to 50-55	Moderate
20-25 to 35-40	Severe
Below 20-25	Profound

CLASSIFICATIONS OF MENTAL RETARDATION

There are many systems for classifying mental retardation: behavioral, etiological, and educational. Until 1992, intelligence test scores also determined the level of severity of mental retardation (table 8.1) (Burack, Hodapp, & Zigler, 1998; Accardo & Capute, 1998).

In 1992, AAMR changed its classification from four levels based on IQ scores to two levels based on functioning levels and intensity of needed supports within the adaptive skill areas. Under the new classification system, one either does or does not have mental retardation. There are no levels of mental retardation based on IQ scores. This represents a diminished role for the IQ score in definition of and classification for mental retardation. Rather, there are only two levels—mild and severe—classifying the degree of limitation. These levels are based on functioning in the 10 adaptive skill areas listed in the definition and on the amount of support the individual needs in a particular environment (e.g., school, home, community, etc.). There are four levels of support defined in the new classification system:

Intermittent—short-term supports needed during lifespan transitions (e.g., job loss)

Limited—support on a regular basis for a short period of time (e.g., job training)

Extensive—ongoing support with regular involvement; not time limited (e.g., long-term work or home living support)

Pervasive—constant and highly intense; potentially life-sustaining support

A problem with classification systems is that they assign labels to people. These labels tend to trigger absolute behavioral expectations and negative emotional reactions by society. Labels also provoke preconceived ideas about individuals' abilities, disabilities, and potential. Many people urge the elimination of the term mental retardation because it is stigmatizing and erroneously used as a global summary about people with mental retardation. Individuals with mental retardation present a diversity of abilities and potential,

and the educator must be prepared to accept this diversity.

INCIDENCE OF MENTAL RETARDATION IN THE POPULATION

According to normal probability theory, it is estimated that 2.28 percent of the total population (with no known organic dysfunction) of any society has mental retardation (2.15 percent + .13 percent = 2.28 percent). It is also estimated that .76 percent of the total population has a known organic dysfunction that causes mental retardation. These figures (2.28 percent plus .76 percent rounded to 3.0 percent) are used to estimate the incidence of individuals with mental retardation in a particular geographic area (figure 8.1).

CAUSES OF MENTAL RETARDATION

There are more than 500 disorders in which mental retardation may occur as a specific manifestation (Murphy et al., 1998). These disorders are categorized according to when in the gestational period they occur—prenatally, perinatally, or postnatally. Depending on the population studied and the methods used, it has been reported that causes of mild and severe mental retardation are as explained in table 8.2 (Hagberg & Kyllerman, 1983; McLaren & Bryson, 1987; Yeargin-Allsopp et al., 1997).

While the most prevalent known cause of mental retardation is fetal alcohol syndrome, in some studies accounting for up to half of all individuals with mental retardation, sophisticated genetic mapping research has recently determined that X-linked disorders are the most prevalent inherited disorders manifesting mental retardation (1 in 450 births). X-linked disorders are caused by a recessive sex gene defect expressed in twice as many males as females. This is because males inherit one X chromosome and one Y chromosome. Females inherit two X chromosomes. Therefore, a male who receives the defective X gene will almost always display the disorder, whereas the female with a normal dominant X gene to oppose the abnormal recessive X gene will be clinically normal. However, she risks transmitting the defective gene to half her offspring, including the sons, all of whom will be affected. An affected male with an X-linked disorder will transmit the defective gene to all his daughters, who will be carriers. His sons will be both clinically and genetically normal for the disorder.

About half of the population with mental retardation has more than one possible causal factor, and often it reflects the cumulative or interactive effects of these factors (McLaren & Bryson, 1987; Scott, 1988).

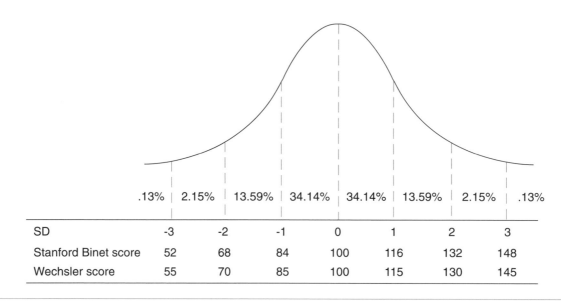

Figure 8.1   The normal curve and intelligence quotient scores.

For example, low birth weight is often considered to be an organic etiology, but other psychosocial factors such as maternal youth, poverty, lack of education, and inadequate prenatal care may also contribute to low birth weight. A multifactorial approach to etiology accounts for whether the causal factors affect the parents of the person with mental retardation, the person with mental retardation, or both, and extends the types of etiologic factors into the following four groupings:

Biomedical: relate to biologic processes, such as genetic disorders or nutrition

Social: relate to social and family interaction, such as stimulation and adult responsiveness

Behavioral: relate to potentially causal behaviors, such as dangerous (injurious) activities or maternal substance abuse

Educational: relate to the availability of educational supports that promote mental development and the development of adaptive skills (AAMR, 1992)

Similarly, prevention efforts are directed toward the parents and/or the person at risk for developing mental retardation. Primary prevention efforts are directed toward the parents of the person with mental retardation and aimed at preventing the problem from occurring (such as programs that prevent maternal alcohol abuse). Secondary prevention efforts are directed toward the person who is born with a condition that might result in mental retardation and aimed at limiting or reversing the effects of existing problems (such as dietary programs to treat persons born with phenylketonuria). Tertiary prevention efforts (such as programs for physical, educational, or vocational habilitation) are directed toward the person who has mental retardation and aimed at improving the individual's level of functioning.

COGNITIVE DEVELOPMENT

The 1992 AAMR definition of mental retardation is functionally oriented. While this orientation is useful in determining individual limitations in present functioning, it is limited for understanding the dynamic nature of intellectual functioning and how it changes from age to age in the developmental process. A developmental orientation toward intelligence facilitates effective instruction and programming.

In order to establish a developmental orientation, it is necessary to draw on the extensive work of Piaget (1952). In his vast writing, Piaget proposed that children move through four stages of cognitive development: the sensorimotor stage (ages 0 to 2), the preoperational stage (ages 2 to 7), the stage of concrete operations (ages 7 to 11), and the stage of formal operations (ages 11 or 12 to adulthood). These stages are depicted in table 8.3.

Sensorimotor Stage

During the sensorimotor stage, children develop, use, and modify their first schemata. For Piaget, schema or schemata (plural) are forms of knowing that develop, change,

Table 8.2   **Categorization of Mental Retardation (MR)**

Gestation period	Mild MR	Severe MR
Prenatal	7-23%	25-55%
Perinatal	4-18%	10-15%
Postnatal	2-4%	7-10%

Table 8.3   **Stages of Cognitive and Play Development**

Age (years)	Piagetian cognitive developmental stage	Type of play	Play group
0	Sensorimotor	Practice play and ritualization	Individual
1			
2	Preoperational	Symbolic	Egocentrism and parallel play
3			Reciprocal play (progressive reciprocity in
4			dyads, triads, etc)
5			
6			
7	Concrete operations	Games with rules	Large-group play
8			
9			
10			
11	Formal operations		
12			
13			
14			
15			

expand, and adapt. A schema may be a simple response to a stimulus, an overt action, a means to an end, an end in itself, an internalized thought process, or a combination of overt actions and internalized thought processes. Examples of schemata include tossing a ball, grasping, and sucking. During the sensorimotor stage, for example, the child develops the schema of grasping, which is internally controlled and can be utilized in grasping the mother's finger, picking up different objects, or picking up an object from various angles. A schema often will function in combination or in sequence with other schemata, as when the child throws the ball. In this activity the child combines the schemata of grasping and releasing.

To a great extent, the sensorimotor period is one in which the child learns about the self and the environment through the sense modalities. A great deal of attention is given to physically manipulating and acting on objects and on observing the effects of such actions. The child is stimulated by objects in the environment and observes how objects react to actions applied to them. Through exploration, manipulation, and problem solving, children gain information about the properties of objects such as texture, size, weight, and resiliency as they drop, thrust, pull, push, bend, twist,

punch, squeeze, or lift objects that have various properties. At this stage, the functioning of the individual is largely sensorimotor in nature, with only rudimentary ability to manipulate reality through symbolic thinking.

Preoperational Stage

The preoperational stage includes the preconceptual substage, which lasts to about the age of four, and the substage of intuitive thought, which spans the ages of four to seven. Toward the end of the sensorimotor period, the child begins to develop the ability to symbolically represent actions before acting them out. However, the symbolic representation is primitive and limited to schemata associated with one's own actions. During the preoperational stage, progress occurs as the child is able to represent objects through language and use language in thinking. The child now can think about objects and activities and manipulate them verbally and symbolically.

Concrete Operations Stage

During the stage of concrete operations, children achieve operational thought, which enables them to develop mental representations of the physical world and manipulate these representations in their minds (operations). The fact that operations are limited to those of action, to the "concrete," or to those that depend on perception distinguishes this stage from the stage of formal operations. The fact that the child is able to develop and manipulate mental representations of the physical world distinguishes this stage from earlier stages. In the stage of concrete operations, the child is able to mentally carry through a logical idea. The physical actions that predominated in earlier stages can now be internalized and manipulated as mental actions.

During the period of concrete operations, there is increased sophistication in the use of language and other signs. In the preoperational phase, the child developed word definitions without full understanding of what the words meant. In the stage of concrete operations, language becomes a vehicle for the thinking process as well as a tool for verbal exchange. In this stage children are able to analyze situations from perspectives other than their own. This decentering permits thinking to become more logical and the conception of the environment to be more coherently organized. During the period of concrete operations, children's thinking becomes more consistent, stabilized, and organized. At the same time, although children are able to perform the more complex operations just described, they are generally incapable of sustaining them when they cease to manipulate objects or when the operations are not tied to physical actions.

Formal Operations Stage

Children functioning at the stage of formal operations are not confined to concrete objects and events in their operations. They are able to think in terms of the hypothetical and to use abstractions to solve problems. They enter the world of ideas and can rely on pure symbolism instead of operating solely from physical reality. They are able to consider all of the possible ways a particular problem can be solved and to understand the effects of a particular variable on a problem. Individuals at this stage have the ability to isolate the elements of a problem and systematically explore possible solutions. Whereas children at the stage of concrete operations tend to deal largely with the present, those functioning at the formal operations stage are able to be concerned with the future, the remote, and the hypothetical. They can establish assumptions and hypotheses; test hypotheses; and formulate principles, theories, and laws. They are able not only to think but to think about what they are thinking and why they are thinking it. During this stage, individuals are able to use systems of formal logic in their thinking.

Piaget and Play

Piaget calls behaviors related to play **ludic** behaviors, which are engaged in to amuse or excite the individual.

He holds that behaviors become play when they are repeated for functional pleasure. Activities pursued for functional pleasure appear early in the sensorimotor period. According to Piaget, the most primitive type of play is practice play or exercise play. The child repeats clearly acquired skills (schemata) for the pleasure and joy of it. The infant repeats movements such as shaking a rattle over and over and exhibits pleasure in doing so. This type of play does not include symbolism or make believe.

Later in the sensorimotor period, the play in which the child engages is called ritualization. More and more schemata are developed and are used in new situations. Play becomes a happy display of mastered activities; gestures are repeated and combined as a ritual, and the child makes a motor game of them. As progress is made, the child forms still newer combinations from modified schemata. For example, a child may follow the ritual of sleeping after being exposed to the stimuli associated with sleeping (pillow, blanket, thumb sucking, etc.). Toward the end of the sensorimotor period, the child develops symbolic schemata and mental associations. These symbolic schemata enable a child to begin to pretend or make believe in play. Some authors refer to symbolic play as make-believe play. Throughout the sensorimotor stage, play is individual or egocentric, and rules are not a part of it.

The preconceptual period within the preoperational stage marks the transition between practice play and symbolic play. At the preconceptual stage, the child's play extends beyond the child's own actions. Also, new ludic symbols appear that enable children to pretend. At ages four through seven (the intuitive thought stage), there is an advance in symbolic play. The child relates a story in correct order, is capable of a more exact and accurate imitation of reality, and uses collective symbolism (other people are considered in play). The child begins to play with one or more companions but also continues to display parallel play. In play the child can think in terms of others, and social rules begin to replace individual ludic symbols. For example, games of tag and games related to the hiding of a moving object appear. Although play is egocentric, opportunities for free, unstructured, and spontaneous play are important. Children at this level are not positively responsive to intuitive thought. There is an advancement from egocentricity to reciprocity in play. Therefore, opportunities for cooperative play become appropriate. It should be remembered that the collective symbolism associated with cooperative play is at its beginning in this period. Guessing games, games based on looking for missing objects, games of make believe, and spontaneous games are stimulating for children during this stage. The fact that children are responsive to tag games, for example, indicates that they are beginning to play with others and to think of others in their play.

At the stage of concrete operations, children's play exhibits an increase in games with rules. Such rules may be "handed down," as in cultural games, or developed spontaneously. In addition, there is an expansion of socialization and a consolidation of social rules. Thus, the playing and construction of group games with rules becomes very attractive to children. As the child enters and moves through this period, play becomes less concerned with make believe and pretending and becomes more concerned with "reality." At this stage, play may be structured, social, and bound by rules. Although some children may be ready for such games by the seventh birthday, children with retarded cognitive development may not be ready for them until after adolescence, if then.

Application of Cognitive Development to Teaching

There are many teaching implications associated with cognitive theory. Due to space considerations, only a few examples are presented. First, since language is more abstract than concrete, teachers need to reduce verbalization of instructions and emphasize tactile, kinesthetic, visual, and other more concrete forms of instruction. Children who can't readily transfer learning or apply past experiences to new situations need more gradual task progressions in smaller sequential steps and need to learn and practice skills in the environments in which they will be used. It is also important to consider the level of cognitive development in teaching rules and game strategies. As cognition develops, more complex rules and strategies can be introduced. Finally, language development should be considered in verbalization. For example, it is often helpful to emphasize action words and simple sentences when communicating instructions rather than to use multiple complex sentences. Feedback on quality of performance should be short and specific. Cognitive theory serves as a basis for some of the organizational and instructional methods suggested later in the chapter.

LEARNING CHARACTERISTICS

The area in which individuals with mental retardation differ most from other individuals is in cognitive behavior. The greater the degree of retardation, the lower the cognitive level at which the individual functions. Most adults with severe limitations function at the sensorimotor cognitive stage. Adults with the mildest limitations may not be able to progress beyond the level of concrete operations. Others may be limited to simpler forms of formal operations or may be incapable of surpassing the preoperational substage.

Although the learning process and stages of learning are the same for both, children with mental retardation learn at a slower rate than nonretarded children and hence achieve less academically. The learning rate of children with mild limitations is usually 40 percent to 70 percent of the rate of nonretarded children. Children

with severe limitations are often incapable of traditional schooling. Although self-contained classes and separate schools for children with severe limitations exist in most school systems, the primary educational objectives for them involve mastery of basic life skills and communication skills needed for their care. Whereas adults with severe limitations who function at a higher level can learn to dress, feed, and toilet themselves properly and can even benefit from work activities, they will most likely need close supervision and care throughout their lives.

SOCIAL AND EMOTIONAL CHARACTERISTICS

Although children with mental retardation exhibit the same ranges of social behavior and emotion as other children, they more frequently demonstrate inappropriate responses to social and emotional situations. Because they have difficulty generalizing information or learning from past experiences at the same rate or capacity as nonretarded children, they are more often exposed to situations they are ill prepared to handle. Children with mental retardation often do not fully comprehend what is expected of them, and they may respond inappropriately because they have misinterpreted the situation rather than because they lack appropriate responses.

Educational programs for children with mental retardation should always include experiences to help them determine social behaviors and emotional responses for everyday situations. Personal acceptance and development of proper social relationships are critical to independence. As with nonretarded individuals, the reason most individuals with mild limitations lose jobs is inadequacy of social skills, such as poor work habits and the inability to get along with fellow workers.

PHYSICAL AND MOTOR CHARACTERISTICS

Children with mental retardation differ least from nonretarded children in their physical and motor characteristics. Although most children with mental retardation evidence developmental motor delays, these seem to be related more to the cognitive factors of attention and comprehension rather than to physiological or motoric deficits.

Generally, the greater the retardation, the more lag in attaining major developmental milestones. As a group, children with mental retardation walk and talk later, are slightly shorter, and usually are more susceptible to physical problems and illnesses than other children. In comparative studies, children with mental retardation consistently score lower than nonretarded children on measures of strength, endurance, agility, balance, running speed, flexibility, and reaction time. Although many youngsters

Photo courtesy of Special Olympics Maryland.

Figure 8.2   Many children with mental retardation have other conditions as well.

with mild limitations can successfully compete with their nonretarded peers, students with severe limitations tend to fall four or more years behind their nonretarded peers on tests of physical fitness and motor performance (health and safety adapted skill area).

In general, the fitness and motor performance of nonretarded children exceed that of children with mild limitations, who in turn perform better than children with severe limitations. The performance of boys generally exceeds that of girls, with the differences between the sexes increasing as the degree of limitation increases (Eichstaedt, Wang, Polacek, & Dohrmann, 1991; Londeree & Johnson, 1974). Flexibility and balance seem to be the exceptions to the generalizations just stated. Whereas nonretarded girls show greater flexibility and balance than nonretarded boys, boys with mental retardation show greater flexibility and balance than girls with mental retardation. Winnick and Short (1999a) recommend that children ages 10 to 17 with mental retardation and mild limitation in physical fitness should achieve levels of aerobic capacity, body composition, flexibility, abdominal and upper-body strength, and endurance necessary for positive health, independent living, participation in physical activities, and performance levels approaching those of their nondisabled peers. Winnick and Short (1999b) also offer activity guidelines for developing these functional and physiological fitness levels in children with mental retardation and other disabilities.

Also, children with Down syndrome exhibit more flexibility than other children with mental retardation (Eichstaedt, et al. 1991; Rarick, Dobbins, & Broadhead, 1976; Rarick & McQuillan, 1977). Down syndrome children tend to have hypotonic musculature and hypermobility of the joints, which permits them greater than normal body flexibility and, because of weak ligaments and muscles, places them at greater risk of injury.

Many children with mental retardation are hypotonic and overweight. Nutritional guidance and fitness activities may be necessary to enable a student to perform at a higher skill level. Disproportionate bodies pose many problems with body mechanics and balance. Activities conducted on uneven surfaces or requiring rapid change of direction can be frightening to the student and pose greater risk of injury and of failure. The "institutional" walk of the nonathletic child with mental retardation is characterized by a shuffling gait with legs wide apart and externally rotated for balance. Body alignment is in a total body slump. Club hands and feet, postural deviations, and cerebral palsy are all prevalent among youngsters with mental retardation; the physical educator must take them into consideration when planning the program for each child (figure 8.2).

Mental retardation often coexists with other disabilities in children (table 8.4). The occurrence and number of coexisting conditions increase with increasing level of severity of mental retardation (Murphy, et al. 1998).

DOWN SYNDROME

Down syndrome is the most recognizable genetic condition associated with mental retardation. One in 700 children is born with Down syndrome. In the United States, approximately 5,000 such children are born each year. Although fathers are genetically responsible for the abnormality in about 25 percent of all cases, women over the age of 35 present the highest risk (1 in 290) of having a child with Down syndrome. At age 40 the risk increases to 1 in 150 births, and at age 45 the risk is 1 in 20 births (Cunningham, 1987).

Cause

Down syndrome results from one of three chromosomal abnormalities. The most common cause is trisomy 21, so named because of the presence of an extra #21 chromosome. This results in a total of 47 chromosomes instead of the normal 46 (23 chromosomes received from each parent). A second cause of Down syndrome is nondisjunction. This occurs when one pair of chromosomes fails to divide during meiotic cell division, resulting in 24 chromosomes in one haploid cell and 22 in the other. A third and rare cause of Down syndrome is translocation, which occurs when two chromosomes grow together so that, while appearing to be one chromosome, they actually contain the genetic material of two chromosomes.

Characteristics

Although there are more than 80 clinical characteristics associated with Down syndrome, the following are the most common physical characteristics, some of which are seen in figure 8.3:

Figure 8.3   Special Olympics athlete with Down syndrome.

Broad hands and feet with stubby fingers and toes

Eyes slanted upward and outward with exaggerated folds of skin

Hypermobility of the joints

Mild to moderate obesity

Perceptual difficulties

Poor balance

Poor muscle tone

Poor vision and audition

Protruding, fissured tongue

Short legs and arms in relation to torso

Short neck and small low-set ears

Short stature

Small head; flat face and back of head

Small mouth; thin lips

Small nose with a flat bridge

Sparse, fine hair

Transverse crease on the palms

Underdeveloped respiratory and cardiovascular systems

White spots in the iris of the eyes

Children with Down syndrome also tend to have many medical problems. Approximately 40 percent of these individuals develop congenital heart disease. They also have a greater risk of developing leukemia. Bowel defects requiring surgery and respiratory infections are common. Individuals with Down syndrome age more rapidly, and almost all who live beyond age 40 develop Alzheimer's disease (Zigman, Silverman, & Wisniewski, 1996). All individuals with Down syndrome have mental retardation. Individuals with Down syndrome are

becoming increasingly integrated into our society and institutions—schools, health care systems, community living, and the workforce. While there is always a degree of slow development and learning difficulties associated with Down syndrome, attainment and functional ability are much higher than previously thought possible when individuals with Down syndrome attend institutions and segregated schools.

Physical Education Programming

The many medical problems of individuals with Down syndrome require medical clearance for activity participation and careful planning of the physical education program. Aerobic activities and activities requiring maximal muscular contraction must be adapted and carefully monitored. Muscle hypotonia (low muscle tone) and hypermobility (above-normal mobility) of the joints often cause postural and orthopedic impairments such as lordosis, ptosis, dislocated hips, kyphosis, atlantoaxial instability, flat pronated feet, and forward head. Exercises and activities that cause hyperflexion are contraindicated because they put undue stress on the body that may result in hernias, dislocations, strains, or sprains. Rather, exercises and activities that strengthen muscles around the joints, thereby stabilizing them, should be encouraged. Poor eyesight and hearing will require teachers to employ adapted equipment and teaching strategies typical for those with sensory impairments.

TESTS

Assessment is necessary to determine the status and needs of a student with mental retardation. Children with mild limitations who need only intermittent or limited supports in the health and safety or in the leisure adapted skill areas will, in many cases, be able to take the same tests as children without retardation. In some instances, standardized tests may require modification, or standardized tests specifically designed for students with mental retardation may need to be selected. Children with mild limitations should also be

Table 8.4   **Coexistence of Mental Retardation (MR) With Other Disabilities**

Coexisting disability	With mild MR	With severe MR
Autism	9%	20%
Epilepsy	4-7%	20-32%
Cerebral palsy	6-8%	25-30%
Sensory deficits	2%	11%

able to successfully complete analytic and holistic rubrics (a list of characteristics describing performance for each point on a fixed scale) (Hensley, 1997) as well as portfolios (collection of a student's work gathered over time) (Lund, 1997). However, children with severe limitations who require extensive or pervasive supports in the health and safety or in the leisure adapted skill area often lack the levels of physical fitness, motor ability, motivation, and understanding required to perform standardized test items. Winnick and Short (1999a) recommend using task analysis or measuring physical activity energy expenditures as an alternative to standardized tests designed to measure physical fitness for those with severe limitations. Assessment incorporating teacher-developed rubrics and task analysis may be the most appropriate methods for measuring the physical education abilities of this population. Three tests that have been specifically developed for use with students with mental retardation and are recommended are Brockport Physical Fitness Test (Winnick & Short, 1999a), designed to measure health-related physical fitness; the Ohio State University Scale of Intra-Gross Motor Assessment (Ohio State SIGMA) (Loovis & Ersing, 1979), designed to measure motor development; and the Special Olympics Sports Skills Guides (1995-1999), designed to measure gross motor development and/or motor skills and sports skills. Additional information regarding testing is presented in chapter 5.

ORGANIZATIONAL AND INSTRUCTIONAL METHODS

Although many of the organizational and instructional methods used in teaching nondisabled children can be applied to those with mental retardation, certain methods are often stressed in the teaching of students with mental retardation. When employed, these methods ensure successful positive experiences for students with mental retardation in a physical education class where maximum participation takes place in a controlled environment.

Organizational Methods

Organizational methods used in teaching children with mental retardation include learning stations, peer instruction and cross-age tutoring, community-based instruction, and partial participation.

Learning Stations

Learning stations divide the gymnasium/playing field into smaller units where each unit is designed for students to learn or practice a specific skill or sport. Students can be assigned to a specific learning station for the entire activity period or can rotate from station to station after a specific amount of time has passed or a learning goal has been achieved. Learning stations permit flexibility and provide safe and successful learning experiences for students with mental retardation as well as nondisabled students in an inclusive class. They allow students to progress at their own pace. Stations can focus on a theme such as physical fitness, tennis, or motor skills. Stations can enhance full integration while accommodating large numbers of students.

Peer Instruction and Cross-Age Tutoring

One of the most exciting developments in special education programs is the use of peers (other students) to help children with unique needs. It is common for young children to rely on slightly older peers as role models. Cross-age tutoring is an excellent way of providing children with mental retardation role models whom they can imitate. Peer instruction and cross-age tutoring increase personalized instruction time for students with mental retardation.

Community-Based Instruction

Teaching skills in the "real" environment, where the skills ultimately will be used, is preferable to artificial environments such as the classroom or gymnasium. Students with severe limitations do not generalize well from one environment to another. Thus, teaching skills in artificial environments forces a teacher to reteach these skills in community environments. It is, therefore, more efficient to teach skills in environments where they will be used. One of the critical steps in the process of teaching skills in natural environments is identifying and prioritizing environments in which these skills will actually be used. For example, teaching students to access and use community health club facilities, bowling facilities, or community pools is preferable to teaching these activities in school gymnasiums or pools. The availability of certain facilities in a particular community will determine whether a skill is truly functional.

Partial Participation

If a student with mental retardation can acquire some of the skills needed to participate in an activity, the parts of the skills that cannot be performed can be compensated through physical assistance or adaptations of equipment. Figure 8.4 illustrates the use of partial participation and physical assistance in heading a soccer ball. Often, peer tutors can provide the physical assistance while modified equipment and rule changes are adaptations that allow the student to participate in an inclusive setting. For example, a student with cerebral palsy who uses a motorized wheelchair can be assigned a specially lined area of the soccer field. If the soccer ball enters this lined area, the peer tutor stops the ball.

Courtesy of Special Olympics Maryland. David Trozzo, photographer.

Figure 8.4   Partial participation with physical assistance.

The student then has five seconds to maneuver her wheelchair to touch the ball. If the student is successful, the peer tutor then kicks the ball to a member of the student's team. If the student is unsuccessful, the peer tutor then kicks the ball to a member of the opposing team.

Instructional Methods

There are a number of instructional methods for teaching children with mental retardation: concrete and multisensory experiences, data-based teaching, task analysis, behavior management, move from familiar to unfamiliar, consistency and predictability, choice making, and activity modification.

Concrete and Multisensory Experiences

Because children with mental retardation are slower in cognitive development, their mental operations may be confined to concrete objects and events. Therefore, concrete tasks and information are more easily learned and utilized than their abstract counterparts. Instruction must, therefore, be concrete, emphasizing only the most important task cues. Because verbalization is more abstract, demonstration or modeling, physical prompting, and/or manipulation of body parts should accompany verbal instructions. Verbal instructions and cueing should be shortened and simplified and use specific action words (e.g., "run, walk, hop" as opposed to "go").

Data-Based Teaching

Data-based instruction involves carefully monitoring a student's progress and how such factors as environmental arrangement, equipment, task analysis, time of day, levels of reinforcement, and cueing techniques affect progress. By charting student progress, both teacher and student can often determine when an objective will be accomplished (e.g., complete two laps around the track). Cooperative charting by teacher and student can provide the motivation and direction to help a student estimate how long it will take to accomplish any new task, set personal objectives to accomplish a task within a reasonable margin of error, and identify practice techniques for reaching the objectives set.

Task Analysis

Since children with mental retardation are generally unable to attend to as many task cues or pieces of information as children of normal or above-normal intelligence, breaking down skills into sequential tasks is an important instructional approach in working with children with mental retardation. Many planned programs discussed later in the chapter employ task analysis.

Behavior Management

Applying the appropriate behavior management principles of cueing, reinforcing, and punishing is critical to the success of task analyzing skills and teaching all the smaller behaviors that enable the student to learn and perform the skill. Behavioral principles must be systematically employed and coordinated. Behaviors to be influenced must be pinpointed, and systems must be designed to promote change in the identified behaviors. There is substantial evidence that the shorter the time lapse between student performance and feedback, the more learning is facilitated. This is especially true for persons with severe limitations. An excellent example of a program that systematically employs behavioral principles in the teaching of children with severe limitations is the Data Based Gymnasium developed by Dunn, Morehouse, and Fredericks (1986).

Move From Familiar to Unfamiliar

Because students with mental retardation have difficulty applying past experiences and previously learned information to new, though similar, tasks, they are more likely to view each new task as a novel one. Therefore, the progression from familiar to unfamiliar must take place gradually and be strongly reinforced. A teacher should begin to teach well within the range of student skill and comprehension. Tasks to be learned should be divided into small meaningful steps, presented and learned sequentially, and then rehearsed in total, with as little change in order as possible. A word of caution: Often, children with mental retardation have short attention spans, and, although progression to new tasks should be gradual, the teacher should plan many activities to sustain the student's attention. For example, if the lesson is practicing the fundamental motor skill of hopping, the teacher may need to plan 10 different hopping activities in a 20-minute lesson!

Consistency and Predictability

Consistency of teacher behavior helps establish and maintain a sound working relationship between teacher and students. When students know what to expect, they can plan their behaviors with the certainty of what the consequences will be. Children with mental retardation are often less flexible in accepting or adapting to new routines. Therefore, day-to-day consistency of class structure, teacher behavior, and expectations of students is important in optimizing their learning.

Choice Making

Activities for persons with mental retardation are often provided without considering individual preferences. Choice making allows students who often have very little control of their bodies and environments to have some control of their activity programs. Choice making can consist of allowing students to choose which activity they want to play, which ball they prefer, how they would like to be positioned, who they would like to assist them, when they need to stop and rest, and so forth.

Activity Modifications

Often, students with mental retardation can successfully participate in physical education and sports alongside their nondisabled peers if challenging skills are modified to enable successful participation. This is particularly true for children with mental retardation who have associated health or physical impairments or who have more severe limitations. Activities can be modified by substituting fundamental motor skills and patterns for more highly developed sport skills, by reducing the speed of skill execution or the force required for successfully executing a skill, and/or by reducing the distance required for successful skill execution.

EDUCATION OF STUDENTS IN INCLUSIVE SETTINGS

Approximately 39 percent of all students with mental retardation are receiving some support services while in an inclusive class, and 54 percent are receiving specially designed instruction in separate classes within the school (U.S. Department of Education, 1998). Physical education teachers face the task of providing successful, enjoyable, and challenging learning experiences for all students in inclusive classes. Their teaching strategies must ensure that students with mental retardation will comprehend instructions and achieve success in the inclusive gymnasium. In particular, teachers can enhance effective integration by using these teaching methods presented earlier:

▶ Employing the principle of partial participation

▶ Being consistent and reliable, thereby reducing the amount of uncertainty facing students who are integrated

▶ Shortening and simplifying instructions

▶ Using well-defined and distinctive cues

▶ Employing multisensory teaching strategies

▶ Modifying instruction and activities by reducing the speed of skill execution and reducing the force required for successfully executing a skill

ACTIVITIES

In selecting activities for students with mental retardation, the physical educator should be aware of the games, activities, and sports enjoyed by children within the neighborhood and offered by local recreation agencies. These activities are appropriate choices for the physical education class. Communication with local recreation agencies can result in cooperative programming and facilitate successful inclusion of students with mental retardation into structured community-based play groups.

Selection of Activities According to Chronological Age

Activities selected and skills taught should be based on a student's chronological age and activities that the student's same-age peers enjoy. However, students' functional abilities and mental ages must be considered when determining *how* to present skills and activities. Teaching chronologically age-appropriate skills as well as teaching functional skills frequently used by all persons in natural, domestic, vocational, community, and recreational environments minimizes the stigmatizing discrepancies between students with and without disabilities. Conversely, selecting activities based on mental age often involves keeping students with more severe limitations at the lower end of the developmental continuum working on "prerequisite skills" that often are nonfunctional and have an extremely low chance of being used in daily living or community recreation and sports programs. A particular need of individuals with mental retardation is to develop the motor skills and physical fitness levels they will require for optimal vocational training and successful use of leisure time.

Activities for Students With Mild Limitations

Often students with mild limitations excel in sports, and sports may be their primary avenue for success and self-esteem. Since students with mild limitations generally need only intermittent or limited support during physical education, they are more likely to be included in physical education classes than in any other subject. Their physical and motor needs are generally like those of students without mental retardation; therefore, physical education activities for students with mild limitations often will be the same as or similar to those for their nondisabled peers.

While students with very mild limitations often excel in physical education and sports, most students with mild limitations generally do not achieve high skill levels. Still, basketball, soccer, hockey, baseball, and dancing are activities often popular among adolescents with mild limitations even though concepts of team play, strategy, and rules are sometimes difficult for them to learn. Highly skilled people with mild limitations can learn strategy and rules through concrete teaching experiences. Skill and sport activities like those fostered by Special Olympics yield success and enjoyment for students with mental retardation.

Activities for Students With Severe Limitations

Individuals with severe limitations have not traditionally been placed into inclusive public school classes but rather into special classes, schools, or institutions, be-cause they generally require extensive or pervasive supports. However, more and more students with severe limitations are functioning successfully in inclusive classroom settings when the necessary and appropriate support systems are in place (see the Inclusion of Person with Severe Limitations application example). Their level of mental and motor functioning is very basic. Their activity is generally characterized by little student interaction (i.e., parallel play), with most interactions occurring between teacher and student. School-age individuals with severe limitations generally need an educational program that utilizes sensorimotor skills, fundamental skills, movement patterns, and physical and motor fitness development.

Sensorimotor programs involve the stimulation of a child's senses so that those sensory channels are developed enough to receive information from the environment. Functional senses then permit the child to respond to the environment through movement and manipulation. In these programs, children are taught the normal infant motor progression of head control, crawling, grasping, releasing, sitting, creeping, and standing. Many students with severe limitations do not walk before the

Application Example: Inclusion of Person With Severe Limitations

Setting: Secondary physical education class

Student: 14-year-old student with mild mental retardation has severe limitations and uses a motorized wheelchair

Unit: Basketball

Issue: How to include the student in the basketball game in a meaningful way for both the student and the other basketball players

Application: The physical educator, after consultation with an adapted physical education specialist, uses partial participation with peer assistance as follows:

▶ A special area of the basketball court is marked off for use by the student and peer tutor.

▶ If the basketball enters this area, the peer tutor stops the ball. The student has five seconds to maneuver his wheelchair to touch the ball (any challenging task can be used).

▶ If the student is successful, the peer tutor then throws the ball to a member of the student's team. If the student is unsuccessful, the peer tutor throws the ball to a member of the opposing team.

The following suggestions would also work:

▶ Within the special area, set up a target (smaller basket, box, trash can, etc.) and a throwing line a challenging distance from the target.

▶ If the basketball enters the special area, the peer tutor stops the ball. The student has 10 seconds to move his wheelchair to the spot where the ball entered the area and then to the throwing line, where he propels either the basketball or a more suitable smaller/lighter ball at the target.

▶ If the student hits the target, his team scores a basket. If the student misses the target, the peer tutor throws the ball to a member of the opposing team who is standing underneath the basket.

age of nine, and many never become ambulatory. Individuals with the most severe limitations may not readily respond to their environments and may exhibit little or none of the curiosity that would motivate them to investigate the environment and learn. Even the most rudimentary skills must be taught. Through partial participation and activity modification, most students with severe limitations can participate in physical education classes alongside their nondisabled peers.

Realistically, most students with severe disabilities are unable to independently perform most age-appropriate functional skills. However, the addition of physical assistance and technologies in the form of adapted equipment, switches, and computers enable many such students to participate in chronologically age-appropriate functional activities in natural environments.

Planned Programs

Several established programs are good resources for instructional processes and methods, assessment procedures, and activities relative to physical education and sport for individuals with mental retardation. Several of these, in fact, were originally designed for use with populations with mental retardation. In one of the first contributions, Thomas M. Vodola (1978) developed Project ACTIVE (All Children Totally InVolved Exercising). Project ACTIVE includes a systematic instructional process (test-assess-prescribe-evaluate), norm-referenced tests for measurement of physical and motor ability, and a variety of other tests to help assess abilities in physical education. The project also provides many activity ideas for learning. Karp and Adler (1992) revised project ACTIVE in 1992.

An important program with particular relevance for individuals with severe limitations is the Data Based Gymnasium (Dunn, 1997). The Data Based Gymnasium (DBG) offers a behaviorally oriented instructional model for teaching students with severe limitations and a system for analyzing behavioral principles for the socialization of behaviors. Finally, DBG includes a game, exercise, and leisure sport curriculum. Specific skills within the curriculum are broken down into tasks and steps that are sequenced as phases representing shaping behaviors. Students are reinforced for successfully completing tasks that approximate the terminal or targeted behavior. DBG includes a clipboard instructional and management system that helps to identify present status, objectives, and progress on skill development. Although the original text on DBG is no longer in print, many of the original materials and ideas are incorporated in a text by Dunn (1997).

Mobility Opportunities Via Education, or M.O.V.E. (Kern County Superintendent of Schools Office, 1995), is a top-down, activity-based curriculum for those with severe limitations; it combines natural body mechanics with an instructional process that helps students acquire increased amounts of independence necessary to sit, stand, and walk. M.O.V.E. uses education as a means to acquire motor skills by having participants naturally practice their motor skills while engaging in other educational or leisure activities. The motor skills sequence is age appropriate and based on a top-down model of needs rather than a developmental sequence of skill acquisition. The motor skills are usable to the participant into adulthood and range from levels of zero self-management to independent self-management.

The *Special Olympics Sports Skills Program Guides*, created by Special Olympics, Inc., are also helpful in the development of sports skills. Initially designed for children and adults with mental retardation, the guides are presented in a series of manuals that present long-term goals, short-term objectives, task-analyzed activities, sport-specific assessments, and teaching suggestions for various sports. Sports skills program guides are available in all official summer and winter sports.

The many planned programs discussed in this section are particularly relevant to people with mental retardation. However, because several have application for other populations in adapted physical education, they are discussed in more detail in other chapters throughout this book.

SPECIAL OLYMPICS

Special Olympics was created in 1968 by Eunice Kennedy Shriver and the Joseph P. Kennedy, Jr. Foundation. It is an international sports training and competition program open to individuals with mental retardation eight years of age and older, regardless of their abilities. Children ages five to seven with mental retardation may participate in Special Olympics training programs but not in Special Olympics competitions. The mission of Special Olympics is to provide year-round sports training and athletic competition in a variety of Olympic-type sports for children and adults with mental retardation.

Summer and Winter Special Olympics Games are held annually as national, program (state or province), sectional, area or county, and local competitions. World Summer Special Olympics Games, which take place every four years, began in 1975. World Winter Special Olympics Games, also held every four years, began in 1977. Additional Special Olympics competitions that include two or more sports are defined as tournaments. To advance to higher levels of competition in a particular year (i.e., from local through area and sectional to program competition), an athlete must have trained in an organized training program for higher level competition in the sport(s) in which he is entered. To advance, an athlete must have placed first, second, or third at the lower level of competition in the sport(s). Instruction for eight-week training programs is provided in the *Special Olympics Sports Skills Program Guides*.

The showcase for acquired sports skills of Special Olympians in training is the many Special Olympics competitions held throughout the year. These competitions have the excitement and pageantry associated with the Olympic Games—including a parade of athletes, lighting of the torch, an opening declaration, and reciting of the Special Olympics oath. In addition to showcasing their skills, Special Olympics athletes often have the opportunity to meet celebrities and local community leaders, experience new sport and recreational activities through a variety of clinics, enjoy an overnight experience away from home with friends, and develop the physical and social skills necessary to enter school and community sport programs.

Official Special Olympics summer sports include the following:

Aquatics
Athletics
Basketball
Bowling
Cycling
Equestrian sports
Football (soccer)
Golf
Gymnastics
Powerlifting
Roller skating
Softball
Tennis
Volleyball

Official winter sports include the following:

Alpine and cross-country skiing
Figure and speed skating
Floor hockey

To provide consistency in training, Special Olympics uses the sports rules of the International Sports Federation (given the responsibility by the International Olympic Committee for handling the technical aspects of Olympic Games) to regulate a sport, except when those rules conflict with *Official Special Olympics Sports Rules: 1996-1999 Revised Edition* (1997).

Because of the wide range of athletic abilities among people with mental retardation, Special Olympics training and competition programs offer motor activities training for athletes with the severest limitations, team and individual sports skills, modified competition, and regulation competition in most sports.

Special Olympics has developed three pioneer programs to help integrate Special Olympics athletes into existing community and after-school sports programs. In the first program, Sports Partnerships, students with mental retardation train and compete alongside interscholastic or club athletes. Varsity and junior varsity athletes serve as peer coaches, scrimmage teammates, and boosters during competition. Athletes with mental retardation compete in existing interscholastic or club league competitions. For example, in a track-and-field meet, the varsity 100-meter race is followed by a Special Olympics 100-meter race. In distance races, all athletes start together. At the end of the meet, individual and school scores are tabulated for varsity and partnership teams. In team sports (soccer, softball, basketball, volleyball), partnership teams compete just prior to and at the same site as the varsity or junior varsity games.

The second program, Unified Sports, creates teams with approximately equal numbers of athletes with and without mental retardation of similar age and ability. Unified Sports leagues can be part of a school's interscholastic, intramural, or community recreation sports program. Currently these leagues are established in six sports: bowling, basketball, golf, softball, volleyball, soccer, and distance running.

The third program, Partners Club, brings together high school and college students with Special Olympics athletes to perform regular sports skills training and competition and to spend time enjoying other social and recreational activities in the school and community. The Partners Club should be a sanctioned school club with all the accompanying benefits.

All athletes with mental retardation should be able to earn school athletic letters and certificates, wear school uniforms, ride team buses to competitions, participate and be recognized in school award ceremonies, and represent their schools in Special Olympics local, area or county, and state competitions.

Athlete Leadership Programs (ALPS) encourage athlete self-determination, include athletes in policy and program discussions, help athletes discover new roles in Special Olympics, create opportunities for athletes to maximize their potentials, and encourage an attitude of service "with" as well as "for" athletes. ALPS include Shriver Global Messengers (public speakers), Athlete Congress, Athletes as Board Members, Coaches, Officials, Volunteers, Employees, and Media Reporters.

PARALYMPIC GAMES FOR THOSE WITH AN INTELLECTUAL DISABILITY

The Paralympic Games are the Olympic equivalent for the world's top athletes with disabilities. They include athletes with spinal cord injuries, amputations, blindness, deafness, cerebral palsy, intellectual disabilities, and les autres. Unlike Special Olympics, which provides competition for all trained athletes with mental retardation regardless of ability, the Paralympic Games provide international competition for elite athletes, 15 years and older, with intellectual disabilities who can meet minimum qualifying sport standards. They are conducted every four years just after the Olympic Games and at the same venues.

The Association Nacional Prestura de Servicio (ANDE) and the International Association of Sport for the Mentally Handicapped (INAS-FMH) held the first Paralympic Games for the mentally handicapped in September 1992 in Madrid, Spain. Fifty-six elite athletes with mental retardation competed in swimming and athletics at the 1996 Paralympic Games in Atlanta. The 2000 Paralympic Games were held in Sydney, Australia.

SAFE PARTICIPATION

If the physical educator plans activities appropriate to the academic, physical, motor, social, and emotional levels of children with mental retardation, there will be few contraindications for activity. Special Olympics has prohibited training and competition in certain sports that hold unnecessarily high risk of injury, especially injury that could have lifelong deleterious effects. Prohibited sports are the javelin, discus, and hammer throw; pole vaulting; boxing; platform diving; all martial arts; fencing; shooting; contact football and rugby; wrestling; judo; karate; nordic jumping; and trampolining.

Most individuals with Down syndrome have some increased flexibility of joints, called ligamentous laxity, which can affect any of their joints. Atlantoaxial instability describes an increased flexibility between the first and second cervical vertebrae of the neck. The instability of this joint could place the spinal cord at risk for injury if affected individuals participate in activities that hyperextend or radically flex the neck or upper spine. About 13 to14 percent of individuals with Down syndrome show evidence of instability by x-ray only and are asymptomatic. The condition can be detected by a physician's examination that includes x-ray views of full flexion and extension of the neck. Only 1 to 2 percent have symptoms that may require treatment. Symptoms may include neck pain or persistent head tilt, intermittent or progressive weakness, changes in gait pattern or loss of motor skill, loss of bowel or bladder control, increased muscle tone in the legs, or changes in sensation in the hands and feet. Physical education teachers are encouraged to follow the lead of Special Olympics in restricting individuals who have atlantoaxial instability from participating in activities that, by their nature, result in hyperextension, radical flexion, or direct pressure on the neck and upper spine. Such activities include the following:

Alpine skiing

Any warm-up exercises placing undue stress on the head and neck

Butterfly stroke

Certain gymnastics activities

Diving

Heading the soccer ball

High jump

Because many children with mental retardation, particularly those with Down syndrome, are cardiopathic, students should receive activity clearance from a physician. Appropriate activities within the limitations specified by the physician should then be individually planned.

Another common condition of individuals with mental retardation is muscular hypotonia or flabbiness. Infants with this condition are often called floppy babies. Although hypotonia decreases with age, it never disappears, and hernias, postural deviations, and poor body mechanics are prevalent because of insufficient musculature. The physical educator again must be careful not to plan exercises and activities that are beyond the capabilities of individuals with muscular hypotonia, because they can lead to severe injury. Abdominal and lower-back exercises must be selected with care, and daily foot strengthening exercises are recommended.

SUMMARY

Mental retardation is one of the most prevalent disabilities. It is a condition that may be viewed from both functional and developmental perspectives. Mental retardation has numerous causes that result in varied characteristics influencing success and participation in physical education and sport. This chapter has suggested teaching methods, tests, and activities appropriate for this population and has briefly reviewed selected planned programs relevant to students with mental retardation. One of the major programs associated with physical education and sport is Special Olympics, which has made a significant impact on both instructional and competitive opportunities for children and adults with mental retardation. The 2000 Paralympic Games for elite athletes with intellectual disabilities were held in Sydney, Australia, just after the Olympics. Most people with mental retardation have been and continue to be successfully involved in physical education and sport experiences.

Chapter 9

Learning Disabilities and Attentional Deficits

Diane H. Craft

BILLY IS A NINE-YEAR-OLD BOY who is intellectually bright and artistically gifted. He can dash off a sketch that shows unusual artistic talent, but he has great difficulty writing legibly, his letters seemingly strewn all over the page. When he can give an oral report instead of a written one, though, his excellent grasp of the subject matter becomes apparent. Billy also has difficulty skipping, walking a balance beam, or changing directions quickly in a tag game.

Dashiki uses her creativity and well-developed imagination to tell delightful children's stories. She has a wonderful sense of humor, full of perceptive and witty insights. But most of the time she keeps these insights to herself, preferring to remain inconspicuous among her outgoing eighth grade classmates. She struggles to hide the fact that she cannot yet read.

Todd, a fifth grader, excels in athletic endeavors, with seemingly endless stamina and strength. Sitting still, though, for more than a few minutes is really difficult for him. Also, Todd is completely disorganized in his schoolwork. He can rarely find his homework and struggles so hard today with math concepts that he had seemed to have finally grasped just yesterday. And when expected to do a task involving multiple steps, Todd frequently gives up on it almost immediately.

José shows a mechanical aptitude well beyond his 10 years. He is always taking apart and reassembling household appliances to learn how they work. Yet José also shows inconsistencies in his academic performance. Socially, he has difficulty making and keeping friends. In conversation, he frequently makes inappropriate remarks, or his comments come at the wrong time. When other children are trying to be serious, José will often crack a joke. When teased playfully, José may cry or yell at his playmates. Such inappropriate reactions hurt his relationships and make other children want to exclude him from their play.

These four children each have wonderful talents to share but also experience difficult challenges, often accompanied by frustration and failure in aspects of their academic work. Each of these children has a learning disability or attention deficit or both. The purpose of this chapter is to introduce physical educators to the area of learning disabilities and attention deficits to create an awareness of the realities faced by students with these disabilities. Topics include definitions of learning disabilities (LD) and attention-deficit/hyperactivity disorder (ADHD), descriptions of common characteristics along with descriptions of general approaches to working with children with LD and/or ADHD, and practical teaching suggestions.

OVERLAP OF LD AND ADHD

A learning disability (LD) is characterized by a discrepancy between academic potential and achievement that is not due to mental retardation, emotional disturbance, or environmental disadvantage; it is due, rather, to a disorder in one or more of the basic psychological processes involved in understanding or in using language, and it affects academic performance. Although a learning disability can be characterized and identified, it does not exist in isolation. It is only a part of all the strengths and challenges that characterize that individual who has learning disabilities.

To understand the impact a physical education teacher can have on a student with LD, read the following quote from a successful physical education college student with LD and ADHD.

The educator who helped me deal with my learning disability and hyperactivity the most was my third grade teacher. She never gave up trying to find ways that would help me learn. And over the year she did find ways to help me learn. She truly believed in me and my abilities. She gave me hope!

Attention-deficit/hyperactivity disorder (ADHD) is characterized by inattention, impulsivity, and overactivity. ADHD is a term that applies to many, but certainly not all, students with learning disabilities (Cantwell & Baker, 1991). Delong (1995) estimates 44 to 80 percent of the individuals with LD also have ADHD. Deficits in attention coupled with hyperactivity can result in significant underachievement in school and resulting feelings of inadequacy. The overlap between ADHD and LD is enough that this chapter will address them together. The term LD and/or ADHD will be used to denote that the information presented may apply to children who have one or both of these conditions.

It is possible for someone with LD and/or ADHD to overcome challenges or the limitations imposed by others to excel in an area of his strength. Albert Einstein, Thomas Edison, Woodrow Wilson, and Greg Louganis are notable examples of people with learning disabilities who, as adults, have contributed successfully in their specialties. Thomas Edison, the inventor, was called abnormal, addled, and mentally defective. Woodrow Wilson, the scholarly 28th president of the United States, did not learn his letters until he was 9 years old and, like Einstein, did not learn to read until about age 11. It is

---

LD Definition in the Individuals With Disabilities Education Act (IDEA)

Section 300.541 criteria for determining the existence of a specific learning disability:

A team may determine that a child has a specific learning disability if

1. The child does not achieve commensurate with his or her age and ability levels in one or more of the areas listed in (2), when provided with learning experiences appropriate for the child's age and ability levels; and

2. The team finds that a child has a severe discrepancy between achievement and intellectual ability in one or more of the following areas:

▶ Oral expression
▶ Listening comprehension

▶ Written expression
▶ Basic reading skill
▶ Reading comprehension
▶ Mathematics calculation
▶ Mathematics reasoning

The team may not identify a child as having a specific learning disability if the severe discrepancy between ability and achievement is primarily the result of

1. a visual, hearing, or motor impairment
2. mental retardation,
3. emotional disturbance, or
4. environmental, cultural, or economic disadvantage.

Reprinted with permission from the Diagnostic and Statistical Manual of Mental Disorders, 4th ed. Copyright 1994 American Psychiatric Association.

important to underscore the notion that learning disabilities are found in an enormous variety and in many combinations of personal strengths and challenges. This means that physical education teachers need to work with each child to identify areas of learning disability and help the child develop accommodation and coping strategies.

Whether destined for fame or not, students with learning disabilities or attention deficits have endured others' criticism and their own frustrations for their inability to master many things that come easily to most. They are the students most likely to be at risk to quit school, lose jobs, and experience difficult relationships throughout their lives. Further, they are more apt to experience low self-esteem, have painful memories of their childhood and schooling experiences, and experience more serious depression than students without learning disabilities or attention deficits.

There are no "cures" for LD or ADHD. Nor are these disabilities outgrown, although the characteristics may change over time. Fortunately, though, teachers can play a very important role in helping students like Billy, Dashiki, Todd, and José identify successful strategies to compensate and cope with their learning disabilities. Also, teachers can help them appreciate their strengths—and sometimes their giftedness—and value their uniqueness (Rief, 1993). "Getting the right kind of help from the professional field, especially understanding, flexibility, and determination from the child's teachers, is invaluable" (Rief, 1993, p. 148).

Distinguishing between children with LD and those with ADHD can be confusing. It is not important what label the child wears, be it LD or ADHD. What is important is to identify the child's strengths, needs, and behaviors of concern, and then work with the child to develop successful learning strategies.

WHAT IS LD?

The following definition of a specific learning disability is found in the regulations of the Individuals with Disabilities Education Act Amendments of 1997 (IDEA):

The term specific learning disability means a disorder in one or more of the basic psychological processes involved in understanding or in using language, spoken or written, which may manifest itself in an imperfect ability to listen, think, speak, read, write, spell, or do mathematical calculations.

The specific learning disability includes such conditions as perceptual disabilities, brain injury, minimal brain dysfunction, dyslexia, and developmental aphasia.

The specific learning disability does not include a learning problem that is primarily the result of visual, hearing, or motor disabilities; of mental retardation; of emotional disturbance; or of environmental, cultural, or economic disadvantage (PL 105-17, Individuals with Disabilities Education Act). The criteria for determining the existence of a specific learning disability is found in box 9.1.

All children, to some degree, and at one time or another, may have difficulty in some of these areas. But only when these behaviors occur in more than one setting, persist over an extended period of time, and interfere with learning do they need special attention. The term "specific" was added to learning disabilities to underscore that these children have learning difficulties only in specific areas (e.g., reading, speaking, calculating), and that there are other areas of learning where they are at least of average ability and are sometimes gifted in their learning.

In practice the most prevalent basis for classification appears to be academic achievement that is at least two years below age and ability level in one or more of the following areas: oral expression, listening comprehension, written expression, basic reading skills, reading comprehension, mathematics calculation, or mathematics reasoning. While common, this classification is not consistent with the approaches recommended in the learning disability professional literature (Tomasi & Weinberg, 1999). Some of the frustration and concern shared by professionals working in this field are reflected in the following statement by Moats and Lyon (1993, p. 287): "Much of our research-based thinking about learning disabilities in the United States is predicated on information that is obtained from ambiguously defined school-identified samples of children who have been administered technically inadequate measurement instruments and tests."

Increasingly, if children are achieving at grade level, they are not considered to have learning disabilities, and services are not provided no matter how great the ability-achievement discrepancy. This denial of services to children with learning disabilities can be very damaging.

Children with learning disabilities frequently also have *difficulty with social skills*. These difficulties may appear in three areas: initiating and maintaining relationships, getting and using the right social feedback, and speaking and understanding social language (Levine, 1990).

Learning disabilities are lifelong conditions. Although specific manifestations of the learning disability may change over time, the learning disability does not usually disappear "despite the best efforts of teachers, therapists, and parents to remediate them" (Raskind, Goldberg, Higgins, & Herman, 1999). As children with learning disabilities mature, researchers seek to understand what factors predict success as adults. Raskind and others (1999) have identified *the success attributes of self-awareness, perseverance, proactivity, emotional stability, goal setting, and social support systems.* It may be helpful to note that IQ and academic achievement are not powerful predictors of success. Based on this study, teachers may wish to place a high priority on helping children with learning disabilities develop these success attributes. Physical education becomes very important, because it is a subject area that lends itself to working on the personal and social skills that are identified as predictors of success among adults with learning disabilities. If these personal and social skills can be learned in physical education, it may become extremely important to the future success of individuals with learning disabilities.

WHAT IS ADHD?

The defining feature of ADHD is behavior that seems "inattentive, hyperactive, and impulsive to an extent that is unwarranted for [the person's] developmental age and is a significant hindrance to their social and educational success" (Reason, 1999, p. 85). Currently, ADHD is divided into three types: ADHD, combined type; ADHD, predominantly inattentive type; and ADHD, predominantly hyperactivity-impulsive type. Please review the Diagnostic and Statistical Manual of Mental Disorders, fourth editon (DSM-IV) criteria for diagnosing ADHD, presented in box 9.2, to gain an understanding of behaviors commonly observed.

Barkley (1997) challenges the DSM-IV view that ADHD is a disorder of attention and instead proposes it may be a disorder of inhibition. He suggests that children with ADHD are unable to attend because they are unable to inhibit their responses to attend to most everything in the environment.

The challenges to the child with ADHD are considerable, as summarized by Barkley (1997). Children with ADHD carry a substantial risk for school failure (90 percent), retention in grade (35 to 50 percent), failure to graduate from high school (36 percent), and underachievement in employment (50 percent). More than one-half of all children with ADHD will progress into conduct disorder, delinquent activities, or violations of the rights of others, and as many as one-third will progress into early substance experimentation and abuse. Antisocial personality disorder will develop by adulthood in approximately one child in six with ADHD.

Any of the behaviors described in box 9.2 are normal in childhood. Only when these behaviors occur in more than one setting and are developmentally inappropriate (because agemates have outgrown them) are the behaviors of concern. A child will display a particular combination of the above behaviors because of her unique combination of strengths, weaknesses, skills, and interests. There are also documented positive educational outcomes associated with ADHD. For example, the stories told by children with ADHD tend to be more creative than those of their classmates. Individuals with ADHD can have the potential for energy, leadership, and spontaneity (Zentall, 1993).

ADHD is conceptualized as an intrinsic disorder that is presumed to be caused by a central nervous system dysfunction. It is often studied from three perspectives: neuroanatomical, neurochemical, and neurophysiological. The neuroanatomical approach studies brain areas thought to control attention and inhibit motor activity. The neurochemical approach studies specific neurotransmitters that aid communication among the neural pathways thought to be involved in ADHD. The neurophysiological approach studies the dynamic interaction between the neurochemical and anatomical components of the brain (Riccio, Hynd, Cohen, & Gonzalez, 1993).

ADHD is not recognized as a disability category in IDEA, but some students with ADHD are served under IDEA's "other health impairment" category. To qualify, it must be demonstrated that "a child's heightened alertness to environmental stimuli results in limited alertness with respect to the educational environment" (IDEA Regulations 1997,

Diagnostic Criteria for Attention-Deficit/Hyperactivity Disorder As Listed in DSM-IV

A. Either (1) or (2)

1. Six (or more) of the following symptoms of inattention have persisted for at least six months to a degree that is maladaptive and inconsistent with developmental level:

   Inattention

   a. Often fails to give close attention to details, or makes careless mistakes in schoolwork, work, or other activities

   b. Often has difficulty sustaining attention in tasks or play activities

   c. Often does not seem to listen when spoken to directly

   d. Often does not follow through on instructions and fails to finish school work, chores, or duties in the work place (not due to oppositional behavior or failure to understand instructions)

   e. Often has difficulty organizing tasks and activities

   f. Often avoids, dislikes, or is reluctant to engage in tasks that require sustained mental effort (such as schoolwork or homework)

   g. Often loses things necessary for tasks or activities (e.g., toys, school assignments, pencils, books, or tools)

   h. Often is easily distracted by extraneous stimuli

   i. Often is forgetful in daily activities

2. Six (or more) of the following symptoms of hyperactivity-impulsivity have persisted for at least six months to a degree that is maladaptive and inconsistent with developmental level:

   Hyperactivity

   a. Often fidgets with hands or feet or squirms in seat

   b. Often leaves seat in classroom or in other situations in which remaining seated is expected

   c. Often runs about or climbs excessively in situations in which it is inappropriat (in adolescents or adults, may be limited to subjective feelings of restlessness)

   d. Often has difficulty playing or engaging in leisure activities quietly

   e. Often on the go or often acts as if driven by a motor

   f. Often talks excessively

   Impulsivity

   g. Often blurts out answers before questions have been completed

   h. Often has difficulty waiting turn

   i. Often interrupts or intrudes on others (e.g., butts into conversations or games)

B. Some hyperactive-impulsive or inattentive symptoms that caused impairment were present before age seven.

C. Some impairment from the symptoms is present in two or more settings (e.g., at school, work, and home).

D. There must be clear evidence of clinically significant impairment in social, academic, or occupational functioning.

E. Symptoms do not occur exclusively during the course of a pervasive developmental disorder, schizophrenia, or other psychotic disorder and are not better accounted for by another mental disorder (e.g., mood disorder, anxiety disorder, dissociative disorder, or a personality disorder).

Code Based on Type

Attention-Deficit/Hyperactivity Disorder, Combined Type

If both criteria A1 and A2 are met for the past six months

Attention-Deficit/Hyperactivity Disorder, Predominantly Inattentive Type

If criterion A1 is met but criterion A2 is not met for the past six months

Attention-Deficit/Hyperactivity Disorder, Predominantly Hyperactivity-Impulsive Type

If criterion A2 is met but criterion A1 is not met for the past six months

20 U.S.C. Section 300. 7 [c] [9]). Others with ADHD may be served under IDEA's learning disabilities category.

INCIDENCE OF LEARNING DISABILITIES

Children with specific learning disabilities form the largest group of learners with special needs in the United States. In addition, the learning disabilities category continues to outpace other exceptionalities with its rate of increase. Over three quarters of the students with learning disabilities in the United States spend most of their time in regular education classes, as opposed to special education classes. Thus, it is very likely that most physical education teachers will have the opportunity to teach many students with learning disabilities in their regular physical education classes.

The U.S. Department of Education indicates that as of 1997 there were more than 2.5 million children classified as having learning disabilities. This is nearly 4.5 percent of all school-age children and represents 51 percent of all children receiving special education services. The ratio of learning disabilities among boys to girls is at least 3:1. Among the states in the United States, the bases for classification vary greatly.

WHAT ARE THE SUSPECTED CAUSES OF DISABILITIES THAT ARE SO PREVALENT?

The causes of learning disabilities are largely unknown because the causes are probably multiple, cumulative, and interrelated, making them especially difficult to isolate and identify. Practitioners can only focus on the observed behaviors while researchers seek further knowledge about the underlying causes of learning disabilities.

The brain is responsible for perceiving, integrating, and/or acting on information. LD and/or ADHD may result when one or more parts of the brain are not functioning optimally. The causes of these disabilities are largely unknown. Suspected causes include neurological, genetic, and environmental factors.

Neurological factors could include known or suspected brain damage due to infections, head injuries, anoxia, or fetal alcohol exposure. There may or may not be signs of brain irregularities in neurological tests. Scientific research has not proved that prenatal exposure to drugs causes learning disabilities, but, based on the behavior of these children, such a link appears likely.

Genetic factors are also suspected of causing learning disabilities. The incidence is higher among children whose parents and grandparents also appeared to have learning disabilities.

Finally, the role of environmental factors such as toxins (e.g., lead poisoning) and inadequate nutrition (e.g., food additives, preservatives, refined sugar) in learning remains controversial. Increasingly physicians practicing in the area of environmental medicine are appreciating the possibility that some allergic responses may present symptoms that are characteristic of learning disabilities and/or hyperactive behaviors (Rapp, 1992).

CHALLENGES FACED BY STUDENTS WITH LD AND/OR ADHD

Students with attention deficits often face challenges in the areas of attention, impulsivity, and behavioral overarousal. Students with learning disabilities may also face challenges in the areas of perception, organization, and generating appropriate learning strategies. Each of these areas are defined and illustrated in the following section. Further information on teaching methods that may be helpful to students facing these challenges is provided in the final section of this chapter.

Attention

Attention is a complex cognitive ability that has many components that go far beyond the simple admonishment "pay attention." Consider the following aspects of attention, modified from Boucher (1999):

▶ A child who is unable to listen to the teacher describe the tennis serve while simultaneously imitating the movement of the serve shows difficulty with divided attention.

▶ A child who often daydreams, looking at children across the court rather than attending to the task at hand, shows difficulty with focused attention.

▶ The child who, when given the task of shooting foul shots, may shoot once and start to retrieve the ball but then becomes distracted by the chart on the wall or a person passing by the gymnasium door shows difficulty with selective attention.

▶ The child who is unable to remain on task when assigned a repetitive chore, such as shooting many foul shots, shows difficulty with sustained attention or persistence.

▶ The child who is unable to remain vigilant and poised while waiting for the pass of a baton during a relay race shows difficulty with readiness to respond.

Impulsivity

ADHD is characterized by a lack of impulse control—a person's seeming inability to not respond to the many, many distractions that vie for one's attention at any given moment. While not a cure-all, teachers can seek to reduce impulsivity, and thereby increase attention, through modifying the environment and task to help the learner.

Teachers can help otherwise inattentive children attend for longer periods of time through reducing distractions in the environment while increasing the interest and novelty of the task at hand. **Attentional underarousal** refers to not thinking before acting. Children who act before thinking about the consequences of the action, showing a lack of control or restraint on motor behavior or thought process, are acting impulsively. Impulsive children display little inhibition when they wish to speak or act. Upon arriving at the gymnasium door with the class, impulsive children may see a ball on the floor across the room, run over, pick it up, and throw wildly in the direction of the basket before reflecting on the fact that they were to do warm-ups when first entering the gymnasium. There is little consideration of the consequences of the action. They act first, then think.

Behavioral Overarousal

Behavioral overarousal also characterizes students with ADHD, as described by Boucher (1999).

Students with ADHD tend to be excessively restless and overactive, behaviorally and emotionally. Difficulty in controlling bodily movements is especially trying in situations in which they are expected to sit still for long periods of time. The speed and intensity with which they go to the extreme of their emotion is much greater than that of their same-age peers (Goldstein & Goldstein, 1990). Thus, while the student with ADHD may become quickly upset about something, he or she will also forget it just as quickly and move on to something else. This is frustrating and confusing to many adults. Often, adults say that these students lack a sense of guilt. In truth, these students wear their emotions on their sleeves and are able to dust off their feelings and carry on while the rest of us are focusing on our feelings and their behaviors. (p. 90)

Perception

Perception refers to the recognition and interpretation of stimuli received by the brain from the sense organs. Many, but not all, children with learning disabilities have difficulty with visual perception, auditory perception, and/or proprioception. **Visual perception** involves figure-ground discrimination, spatial relationships, and visual-motor coordination. **Auditory perception** involves figure-ground perception, auditory discrimination, sound localization, temporal auditory perception, and auditory-motor coordination. **Proprioception** involves kinesthetic perception, including body awareness, laterality, and directionality.

Organization

Organization refers to a systematic approach to learning. Individuals with learning disabilities may show a haphazard approach to learning a task. According to Sutaria (1985), students who are disorganized have difficulty with one or more of the following central processing tasks:

- Organizing thoughts and materials logically
- Dealing with quantitative and spatial concepts
- Seeing beyond the most superficial meanings or relationships, thus thinking only in concrete or stimulus-bound ways
- Applying rules to problem solving
- Arriving at logical conclusions or predicting outcomes
- Making generalizations
- Remaining flexible in their thinking
- Benefiting from experiences that do not mesh with their existing language system

This disorganized behavior significantly reduces their ability to deal with novel situations.

Generating Appropriate Learning Strategies

Creating learning strategies that are helpful in learning may be particularly difficult for many students with learning disabilities. Torgesen (1980) attributes the low task performance of children with learning disabilities to the use of inappropriate or inefficient learning strategies rather than to structural or capacity limitations. Capable learners may spontaneously generate appropriate learning strategies. Upon observing a new step, a folk dancer may develop verbal cues for each move that the dancer then repeats silently to guide learning the step. Many individuals with learning disabilities may not spontaneously generate or use strategies such as **verbal mediation** (saying cues aloud to oneself) to guide movement. Teachers can help provide these strategies.

BEHAVIORAL CHARACTERISTICS AMONG CHILDREN WITH LD AND ADHD

There are a number of characteristics that can be seen in children with LD or ADHD. Motor performance and physical fitness are two areas to consider.

Motor Performance

Widely varying criteria for classifying children as LD and/or ADHD have made conducting research in this field quite difficult. The problem is exacerbated by the enormous variability of performance within each child with LD and/or ADHD. Consequently, the research on motor performance is less than conclusive.

Some students with learning disabilities are average or even gifted athletes. Others are below average in motor behavior. Researchers studying the motor performance of 369 students with learning disabilities found 12 percent were normal or above average; 75 percent had moderate difficulties; and 13 percent had severe difficulties, averaging 2 to 3 years below the norm (Sherrill & Pyfer, 1985). Motor skills that included motor planning using the hands, perceptual-motor development, hand control and speed, and balance were especially challenging. Other researchers (Bruininks & Bruininks, 1977; Haubenstricker, 1983) have determined that motor skills involving balance, and visual-motor and bilateral coordination were areas of deficit for students with learning disabilities.

Of the children with learning disabilities who do show difficulties with motor behavior, there are some areas that seem especially challenging. A descriptive summary of these motor behaviors follows.

A frequently documented characteristic of children with learning disabilities is difficulty with dynamic balance (Cinelli & DePaepe, 1984). Physical educators may observe children falling frequently and lacking coordinated movements as a reflection of difficulty in dynamic balance. There may be delays in motor development and basic motor skills such as the gallop, skip, or hop (Brunt, Magill, & Eason, 1983). There may be delays in *fine motor skills* such as cutting with scissors, tying shoelaces, writing, and drawing. *Extraneous movements* such as flailing arms while skipping may be present (Haubenstricker, 1983). Perseveration may be exhibited by some youngsters. Perseveration refers to continuing to perform a movement long after it is required. It is thought to be somewhat involuntary rather than a voluntary action that can be unlearned once it is brought to the child's attention. For example, perseveration can compound the difficulty of jumping rope, because the child may find it neurologically difficult to stop jumping or slow the pace of jumping. Arrhythmical patterns during motor performance may be exhibited. These refer to the inability to maintain a constant rhythm during a repetitive task such as finger tapping or jumping rope. A child may begin jumping rope in a smooth manner but within the first six jumps has accelerated and added extraneous movements until the rope tangles in the feet. This is typical of many children as they first learn to jump rope, but it may be more difficult for some children with learning disabilities to develop the coordination to move past this stage. Motor planning contributes to smoothly executing skills involving balance and fine and gross motor movements. Many children with learning disabilities find developing a plan before starting to execute the movement to be very challenging. Misapplied force (kicking a soccer ball too hard), premature or delayed responses (kicking too soon or too late), and inappropriate responses to complex sequences of stimuli (kicking the ball into one's own goal) all reflect problems in motor planning (Haubenstricker, 1983). The *variability of performance* is at once the most frustrating and challenging aspect of learning disabilities. The motor behaviors of many children with learning disabilities are consistently inconsistent. A child who has worked very hard to learn the gallop one week may be unable to gallop the next week.

Physical Fitness

Harvey and Reid (1997) studied 19 children with ADHD (but no LD), nearly all of whom were on stimulant medication. They found that "as a group, performance of the children in fitness and fundamental gross motor skills was below average when compared to the norms of children of similar age and gender" (p. 195). These children also showed low cardiovascular endurance and poor overall fitness coupled with a high percentage of body fat. Variability is a hallmark of ADHD, yet nearly all of these children had consistently low scores. These children appeared to be poorly skilled, unfit, and overweight, even though they were regarded as hyperactive. It was not possible to determine the extent to which this uniformly poor performance may be a result of the medication.

Other Behaviors

Physical fitness, motor skill, and athletic competence are key factors in determining the self-concept of children with disabilities (Kahn, 1982). Thus, athletic incompetence and academic failure can contribute to the low self-concept of uncoordinated children with learning disabilities. Typically, the uncoordinated child is selected last when teams are chosen, an embarrassing experience that does not make any child feel wanted or valued.

Some students with learning disabilities face substantial social difficulties. Physical education provides many opportunities for cooperating together with teammates. As such, it can serve as a laboratory in which to teach social skills to students with learning disabilities.

Difficulty interpreting nonverbal cues with accuracy might underlie the social imperception that characterizes many children with learning disabilities. A child with difficulty perceiving nonverbal cues may have difficulty interpreting facial expressions. Hence, that child can misread social cues and behave inappropriately. The imperceptive person may not note the cues of a furrowed forehead, sidelong glance, clenched jaw, or incline of the body away from the speaker that may communicate dissatisfaction and a desire to get away from the speaker. Frustrated because the speaker seems to be ignoring the listener's signals of displeasure with the conversation, the listener may finally walk away, leaving the imperceptive speaker at a loss to explain what happened.

Other perception problems may also lead to social difficulties. For example, one first grade boy with learning disabilities was observed to score a point in a playground game (an infrequent event for this uncoordinated boy), after which he ran over to his cheering teammates and whacked two of them on the back with his hand. The teammates immediately stopped cheering the boy's play and instead yelled at him. It seems that the boy had seen ball players on television give each other congratulatory pats on the back after a good play. He was attempting to do the same but did not perceive the cue that the pats should be soft and not hard blows. Such social imperception does little to contribute to an individual's popularity.

GENERAL APPROACHES TO LD AND/OR ADHD

This section describes various approaches for working with children with LD and/or ADHD. Physical educators may be involved in some approaches, but others, such as the decision to prescribe medication, are beyond their responsibility. Yet it can be useful to understand these general approaches so that physical educators can support them as appropriate. Specific physical education teaching methods will be discussed at the end of the chapter.

There are many different causes of learning disabilities. It follows, therefore, that there would be many different ways of teaching children with learning disabilities. Currently there is no one approach that is universally supported but rather several, each of which has been successful with some students with learning disabilities but unsuccessful with others. Three of these approaches, behavior management, a multisensory approach, and a multifaceted approach, are described briefly here. For the sake of consistency, the physical educator may wish to determine which approach is used successfully in the child's classroom and/or home and follow the same approach where practical.

Behavior Management

Behavior management refers to strategies that use reinforcement or punishment to increase desirable behaviors and decrease undesirable behaviors (Fiore, Becker, & Nero, 1993). This approach helps the teacher analyze the child in his environment and understand how specific behaviors function for the child in that environment. For example, Todd persists in talking while the teacher is instructing the group, and the teacher persists in telling Todd to be quiet each time he is disruptive. In effect, Todd's disruptive behavior is reinforced by the teacher's attention. Understanding this, the teacher can modify her behavior so that attention is paid to Todd only when he is quiet, thus rewarding him for the appropriate behavior instead of the inappropriate one. Use of the behavior management approach, then, helps the teacher analyze the situation and make modifications within the situation. For an extended discussion of behavior management, please refer to chapter 7.

Multisensory Approaches

Multisensory approaches emphasize teaching through areas of learning strengths. Multisensory approaches, then, focus on the use of three or more of the sensory channels in the teaching-learning process.

The sensory modalities that are typically used in this fusion are visual, auditory, kinesthetic, and tactile. A teaching method that integrates three of these modalities might have the child watch a demonstration of a movement (visual), listen to the teacher describe the specific movement (auditory), and be physically manipulated through the movement (kinesthetic). For example, a student might learn letters by looking at a printed letter, hearing its name, feeling its shape by tracing the letter, and moving the entire body to form the letter's shape. The multisensory approach, in turn, influenced the "learning style" movement. Many special educators have promoted the idea that teaching children with learning and attentional difficulties is best accomplished by matching the teaching method to the child's preferred learning style. Skillful teaching generally has been affected by this movement; good teaching practice appeals to varied learning styles simultaneously. The multisensory approach has not only been utilized extensively to assist children with learning and attentional difficulties but has also become standard teaching practice.

Multifaceted Approach

Multifaceted approach is an eclectic approach to working with children with LD and/or ADHD. It is often more effective in schools than any single approach. This means a variety of approaches are used in conjunction. A list of suggestions follows (modified from suggestions in Rief, 1993):

▶Behavior management can be used both at school and at home. Avoid abandoning the behavior management technique if the first few reinforcers tried do not influence behavior. Continue to seek a strong reinforcer for this particular child.

▶Family counseling is recommended with ADHD, because the entire family is affected. It can be very reassuring to parents/caregivers to learn that they are not the cause of the difficulties faced by their child. It can be helpful to view the problem as outside of the child and to tackle the problem together as a family, learning strategies to help cope with the daily challenges of inattentive or hyperactive behavior. Mary Sheedy Kurcinka's book *Raising Your Spirited Child* is written from this perspective.

▶Individual counseling helps the child develop coping strategies, stress reduction techniques, learning strategies, and a positive self-concept. Counseling, an important source of training in social skills, may help the child learn to attend to social cues and thus decrease social imperception.

▶Cognitive therapy teaches children with ADHD the skills to regulate their own behaviors as well as to use "stop and think" techniques to counteract impulsivity.

▶Numerous school interventions, including behavior management/therapy, cognitive-behavioral therapy, and applied behavioral strategies, can be used.

▶Medical intervention may be in the form of drug therapy. Please refer to box 9.3 for further information on medications.

Parent education helps parents/caregivers learn all they can about LD and/or ADHD so they can better assist their children. Welcome and inform caregivers. Seek their assistance. Caregivers can be extremely important advocates for getting the support systems needed to teach children with special needs in physical education.

Physical activity is an important intervention in that it can help to reduce stress and focus attention. It also provides a socially appropriate outlet for energy in a child with ADHD. Noncompetitive activities that do not demand quick reaction and accuracy may work best.

Teachers are encouraged to observe the effect of physical activity on each individual child. Some children may, in fact, become overstimulated when participating in high energy, unstructured physical activity.

TEACHING CHILDREN WITH LD AND/OR ADHD

In teaching any child, it may be helpful to carefully observe the interaction of the individual with the task and environment. Each child with a learning disability has a unique and preferred way of learning. The teacher has the challenge of discovering ways to structure the environment and task to best help the child with LD and/or ADHD. Perhaps one of the most important things for a teacher to do is to act on the conviction that this child *can* learn. *The children with disabilities in our classes may be the ones from whom we can learn the most about how to teach.* The rest of this chapter provides ideas for organizing the environment and task to discover ways to help students with LD and/or ADHD learn.

Specific Teaching Strategies to Focus and Maintain Attention

Recall that 90 percent of children with ADHD experience failure in school. School failure is also common among children with learning disabilities. Thus, it is important that physical education teachers make the extra effort to reach and teach children with LD and/

▶ **Box 9.3 Medication**

For ADHD, medication should be viewed as part of a total treatment program. Other treatments include educational, social, psychological (for example individual or group behavioral therapy), parent education (to address discipline and limit setting), and tutoring. Most experts agree that treatment should address multiple aspects of the child's functioning. When prescribed, hopefully as a last resort, drug treatment for ADHD commonly involves the central nervous system stimulants methylphenidate (Ritalin), amphetamine and dextroamphetamine (Dexedrine), or pemoline (Cylert), which decrease impulsivity and hyperactivity, increase attention, and have a stabilizing effect in children due to a presumed activation of the brain stem arousal system and cortex to produce the stimulant effect.

Ideally, parents, teachers, school nurses, and physicians should work collaboratively to monitor medication effects, as adverse reactions can occur. Particularly, Ritalin, while a mild stimulant, may produce nervousness, dizziness, dyskinesia, and drowsiness. There have also been reports of growth suppression with this drug. Ritalin should not be used in children under age six. Amphetamines are associated with possible dizziness, restlessness, and tremors. Cylert should not ordinarily be considered as a first-line drug because of an association with life-threatening liver failure.

The use of these drugs does not necessarily result in improved academic or social performance, because other endogenous and exogenous factors interact (such as the child's own attitude toward learning and the parent's adherence to the prescribed medication schedule). These drugs are fast acting, taking effect within one to eight hours (one to four for Ritalin, four to eight for Dexadrine), and many parents have taken their children off the drugs during weekends and holidays. When the medication wears off, mood swings or irritability may be observed during the "rebound," and teachers may observe behavioral changes during transition periods when optimal medication dosages are being determined. Parents should consult with their physicians about the pros and cons of time-release medications, as these involve less disruption of the child's school day.

or ADHD. Many of the following suggestions are also sound teaching techniques for all children, whether in segregated or inclusive settings. Keeping students' attention is crucial for teaching. Obviously, for students with LD and/or ADHD the task is more difficult than usual. The following paragraphs offer principles for maintaining students' attention.

Use a highly structured, consistent approach to teaching. Establish a routine and repeat it day after day. Students with ADHD have greater difficulty tolerating instructional delays. Prior organization and smooth transitions during instruction are essential teaching skills. For example, every day students enter the gym and go to their spots marked on the floor. Instruction begins with an established warm-up routine, followed by the introduction and practice of new skills, participation in a game or dance, and a return to the same floor spots for the cool-down and review. The same warm-up may be used for several sessions to enable the child to learn it well and anticipate the structure of the class. The teacher may wish to create an audiocassette tape of the warm-up routine with music and verbal instructions and appoint a skilled student to lead and demonstrate each exercise. Thus freed from calling out instructions and demonstrating, the teacher may move among the students as they warm up, providing individual assistance to students with learning disabilities. Perceptual concepts, such as laterality, can be incorporated in the warm-up; for example, "Turn to the left as you run in place, twist to the right as you do sit-ups, and stretch and reach to the ceiling with your left hand."

Establish class rules. There are few class rules, but these are understood and equally applied to all students. Posting these rules provides visual cues for children who do not process auditory information well or who are impulsive. Silently pointing toward the poster enables the teacher to prompt the child effectively. Sample rules for young children could include these:

Listen when others are speaking.

Keep your hands to yourself.

Wear sneakers.

Try every activity.

Use a behavior management program to teach children the sequence "attend-think-act" to decrease their hyperactivity. Three common behavioral interventions appear especially useful: positive reinforcement, short verbal reprimands, and response cost. Continuous positive reinforcement can reduce activity level, increase time on task, and improve academic performance of students with ADHD. Short reprimands are often effective when they immediately follow the behavior. ("Billy, hands to yourself!") Avoid long or escalating reprimands. Short, quick reprimands work better. Response cost, a program in which a child can earn points/tokens for appropriate behavior (but can also lose them for inappropriate behavior), can help improve on-task behavior and task completion.

Clarify all expectations. Teach what is acceptable behavior in physical education through practicing, modeling, and reviewing behavioral expectations and rules. Follow through to consistently apply clear, fair consequences. Use proximity control to redirect students with attentional or behavioral problems. Stay close and use eye contact, a hand on the shoulder, or silent pointing to prompt students about appropriate behavior. Provide ample notice of transitions. For example, use the last minutes of class to practice relaxation techniques and calm the students before they leave.

Select activities that emphasize moving slowly and with control to decrease hyperactivity and impulsivity. Use "slow" races with such challenges as "How slowly can you do a push-up? A forward roll?" Include instruction in relaxation. For example, dim the lights and ask all students to go to their floor spots and lie on their backs with their eyes closed. Present relaxation activities through guided imagery ("Imagine that you are relaxing in a warm place under the hot sun.") or progressive relaxation ("Tense your right arm. Hold it. Now relax. Feel the tension flow from your arm. Feel how limp your arm has become."). Check that students are, in fact, relaxing by gently lifting a limb off the floor to see if it feels loose and heavy. If it does, then the person is probably relaxed. These activities may help children who seem to know only the feeling of "fast" and "tense" learn the feeling of "slow" and "relaxed." Within each lesson, alternate active games with relaxation training or passive games, using a format of work, rest, work, rest. The periods of rest are designed to give hyperactive students a chance to slow down before they become so excited that they are out of control.

Teach in a quiet, less stimulating environment to decrease distractibility. Reduce background noise. Keep the gymnasium neat, clean, and well ordered, with unnecessary equipment stored out of sight. Facing distractible children toward the corner so that their backs are to the other activities in the gym also helps focus attention.

Use verbal mediation to teach especially disorganized, distractible children. Verbal mediation is a strategy in which children are encouraged to plan aloud what they will do to focus attention on the task at hand. Distractible children may be coached to repeat directions aloud, for example, "First, I will get the ball from the box, then I will shoot foul shots until I score 10 points." If the children are off task, the teacher asks, "What is the task you are to do?" and prompts the children to refocus attention. On-task behavior is also reinforced. Praise children when they are attending to the task as expected. Catch them being good.

Provide objectives to help disorganized learners focus. Plan the instructional objectives for each lesson and share these objectives with the students. Explaining these global objectives at the beginning of the lesson lets everyone know the behavior that is to be learned. For example, if the objective of the lesson is to throw a ball at one of several targets using a mature overhand pattern, tell this to the class. Together with the students, identify the critical elements of the mature throwing pattern. An example of a critical element may be "step with the other foot as you throw."

Highlight relevant cues. Many students with learning disabilities have difficulty selectively attending to the relevant aspects of the skill to be learned. For example, place footprints on the floor so that children standing on the footprints will have their sides to the target. Provide a visual cue by placing a third footprint where the children will step as they transfer their weight during the throw. Provide verbal cues as well. Teach children to say aloud "step and throw" as a cue to transfer weight while releasing the ball. Also, teach children how to use feedback. For example, after the ball is thrown, ask children to identify where the ball landed in relation to the target ("The ball hit below the target.") and to identify what they need to change in the next throw ("I need to release the ball sooner."). The intent is to help the children use feedback in perfecting future performances.

Give instruction using more than one sense. Some students' preferred mode of learning is visual, others is auditory, and still others is kinesthetic. To accommodate the variety of learners found in any class, and especially those classes with children who have learning disabilities, teach using all three modalities. Use clear, short directions that present the global information without using elaborate descriptions. Use demonstrations liberally to add visual cues to verbal directions. Physical educators may wish to show a picture of the skill to be learned in addition to explaining the skill. A photograph of the child putting away equipment may be placed next to the storage space to visually illustrate what is expected at cleanup time. Physically assist the child through the skill to provide kinesthetic cues. Children with ADHD have difficulty attending to subtle cues, so highlight cues and teach children to look for subtle social cues.

If children are not following directions, it may be that they do not understand them. Some instruction in basic concepts may be necessary. Children may not follow the directions "skip around the outside of the circle" because they do not know how to skip, or they do not understand that the classmates holding hands form a circle, or they're not sure where the "outside" of the circle is located.

Encourage motor planning. Ask children to explain what they will do before they begin to move. This explanation will ensure that children give some advanced thought to how and where they will move. "Students with ADHD are aware they fail to plan ahead or anticipate final steps. Their attempts to plan are disorga-

nized in nature" (Zentall, 1993, p. 147). Teachers play a vital role in helping these students organize themselves. Teachers can provide a framework for organization and cue children to follow it. Physical educators may wish to use pictures, color codes, and short verbal cues to help these children organize themselves. And, as always, follow a routine that gives organization to the lesson.

Change the task to add novelty. Children with ADHD may have an attentional preference for novelty and active learning (Zentall, 1993). For example, if a child is practicing striking objects, substitute a novel object before the child's attention wanes. If a child's attention wanes after five repetitions of striking a Wiffle ball, then make the task novel every three or four repetitions by substituting a beanbag, then a Nerf ball, then a balloon, and then a koosh ball. In this manner the child may practice striking as many as 20 times.

Inclusion: Teaching Children With LD and/or ADHD in Regular Physical Education

Many of the children with LD and/or ADHD can be included in regular physical education with their typical peers. The suggestions offered below are applicable to typical children as well as those with LD and/or ADHD.

Select appropriate curricula. There are four key points in the selection of curricula appropriate for children with LD and/or ADHD.

▶Teach to mastery to enhance students' self-efficacy and self-concept. Avoid conducting a curriculum that only samples activities rather than provides practice until skills are mastered. Task analyze and use progressions to guarantee success at the early stages of learning. A data-based approach to instruction may be helpful. Daily or weekly progress is recorded so that, even when progress is slow and inconsistent, improvement over time can be demonstrated to the students.

▶Review previously acquired basic skills before teaching more advanced skills. Anticipate that children with learning disabilities may be inconsistent performers, particularly of skills taught during previous lessons. Daily review is important, as are opportunities for continued practice of newly acquired skills.

▶Minimize highly competitive team games requiring precise, skilled responses if these skills are beyond the capability of the child. Include team games only when the child has the skill to compete successfully. Forget hitting a softball pitched from the mound if the child with learning disabilities, or any child for that matter, does not have the prerequisite skills to be successful in this task. Hitting the ball off of a tee may be a better activity for the student's present skill level.

Noncompetitive activities that do not have high demands for quick reactions and accuracy may work better. Also, avoid elimination activities in which the skilled players get the most practice and unskilled players get the least practice. Reconsider teaching a competition-based curriculum that provides the most success for only the motor elite. Similarly, avoid team games that require some to be "losers." Many of these children already see themselves as losers; teachers have a responsibility to change, not reinforce, this concept. To decrease anxiety, individualize instruction so that a child does not have to perform in front of classmates a skill that he has not yet mastered. Look to a curriculum that involves cooperative learning, group initiatives, or swimming, running, gymnastics, or perhaps yoga and martial arts such as aikido or tai chi. Focus on doing one's personal best.

▶ Include activities in perceptual-motor skills, gross and fine motor skills, and balance and body awareness in the curriculum. Be alert to the role that perception plays in motor skill performance. If children have difficulty catching, it may be due to perceptual problems such as visual tracking or figure-ground discrimination, in addition to inadequate bilateral coordination or slow reaction times.

Teach socialization and cooperation. Teachers can use a number of strategies to foster students' cooperation and bolster the self-esteem of children with LD and/or ADHD.

▶ Use cooperative learning to increase students' social interactions and self-concepts. In this approach, children succeed through working together. Cooperative games and competition against oneself may be more appropriate alternatives to highly competitive team sports and games, especially in the elementary grades. Games such as Slowest Races, Cooperative Beach Ball and Volleyball, and Project Adventure Challenges may be more appropriate than the traditional team sports. Cooperative activities have been shown to increase social interaction. Further information on using cooperative learning in teaching may be found in Rief (1993).

▶ Eliminate embarrassing teaching practices that force comparison among students. Examples of such practices include posting the fitness scores of everyone in the class or "choosing up teams" in front of the class. No one enjoys knowing that he was the last choice. Instead, recognize the "most improved" and good performances of students with learning disabilities.

▶ Group students to maximize appropriate behavior. Capitalize on the principle of imitation learning by putting students who show poor social skills with students who are good models. One student with poor social skills per group of 6 to 10 is a reasonable ratio.

Get support systems. Support systems are excellent for both teachers and students with special needs. But support systems don't happen by themselves. Teachers have to make them happen. Here's how:

▶ Use peer tutors to further individualize instruction. A mature classmate, an upper-grade student, or an adult volunteer may work with the one or two challenging children. The peer tutor helps children who are distractible remain focused on the task through verbal prompts such as asking, "And what do you do now?" With peer tutors, the teacher can divide a large class into small groups that work in separate areas of the gym. This may help distractible children. Balance the number of hyperactive and hypoactive children in the group. Experiment with various combinations of students to find the mix of personalities that works best. Also, seek opportunities for children with LD and/or ADHD to serve as peer tutors to their classmates or younger children. It builds self-esteem to be helpful to others.

▶ Enlist parent/caregiver assistance. Caregivers can be important advocates in helping to get the support systems teachers need to successfully teach all children. Through the IEP mechanism, caregivers can request support systems and make themselves available to their child's physical education teacher so that the child's needs and those of other classmates may be met in the regular physical education class (Craft, 1996). Examples of support systems include adapted physical education consultation services, assistance of a paraprofessional, or reduction in class size.

▶ Work collaboratively with others. Collaborate with students' other teachers, resource specialists, administrators, parents, physicians, school nurses, psychologists, and others who might share ideas they have found successful and unsuccessful in helping students.

Teachers Who Work Well With Students With LD and/or ADHD

Understanding students with learning disabilities and/or attentional difficulties is further helped by descriptions from adults of what teaching styles and approaches worked best for them as children. Adults have recounted the frustration and failure they experienced as children with learning disabilities. In addition to stories about teachers who accused them of laziness, there are stories about the teachers who really helped them or who served as mentors. These teachers shared the following characteristics (modified from Rief, 1993).

▶ Teachers were flexible, committed, and willing to work with the student on a personal level. These teachers put a great deal of effort into teaching these children with learning disabilities and, in turn, reaped the rewards of knowing they had made a significant difference in the lives of these children.

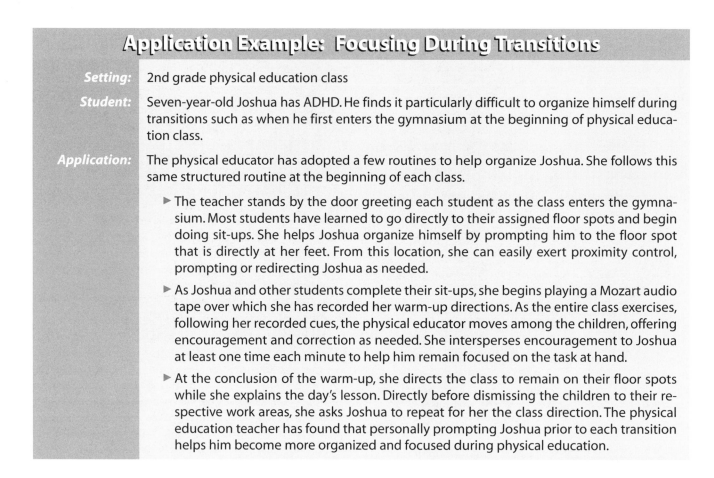

Application Example: Focusing During Transitions

Setting: 2nd grade physical education class

Student: Seven-year-old Joshua has ADHD. He finds it particularly difficult to organize himself during transitions such as when he first enters the gymnasium at the beginning of physical education class.

Application: The physical educator has adopted a few routines to help organize Joshua. She follows this same structured routine at the beginning of each class.

► The teacher stands by the door greeting each student as the class enters the gymnasium. Most students have learned to go directly to their assigned floor spots and begin doing sit-ups. She helps Joshua organize himself by prompting him to the floor spot that is directly at her feet. From this location, she can easily exert proximity control, prompting or redirecting Joshua as needed.

► As Joshua and other students complete their sit-ups, she begins playing a Mozart audio tape over which she has recorded her warm-up directions. As the entire class exercises, following her recorded cues, the physical educator moves among the children, offering encouragement and correction as needed. She intersperses encouragement to Joshua at least one time each minute to help him remain focused on the task at hand.

► At the conclusion of the warm-up, she directs the class to remain on their floor spots while she explains the day's lesson. Directly before dismissing the children to their respective work areas, she asks Joshua to repeat for her the class direction. The physical education teacher has found that personally prompting Joshua prior to each transition helps him become more organized and focused during physical education.

►Teachers sought training and knowledge about learning disabilities. Recognizing that the problems were more likely physiological and biological in nature, they avoided the interpretation that "these children are out to get us deliberately" (Rief, 1993, p. 5). Instead, these teachers saw the talents of these people as children and worked with them in making changes and accommodations. They were also able to identify coping strategies so that the person's strengths could emerge, despite the learning disabilities.

►These teachers cultivated administrative support. Administrators need to be aware of the behaviors and teaching strategies for working with children with LD and/or ADHD. Their help is essential in implementing behavior programs that might include removing a child from the class when the behavior is disrupting the teacher's ability to teach or other students' ability to learn. Educate administrators as to the importance of distributing children with learning challenges across classes and not scheduling large groups of students with LD and/or ADHD in the same physical education class. Such "dumping" is unfair to all involved.

►Teachers respected students and avoided embarrassing or humiliating them in front of others. The overriding characteristic of these teachers who really made a difference in the lives of people with learning dis-

abilities is their belief in the student. When one approach did not work, the teachers tried a second, and a third, and a fourth, and a fifth, and so forth. "These children are worth the extra effort and time" (Rief, 1993, p. 10).

SUMMARY

Teachers can play a key role in helping children with learning and/or attentional difficulties develop compensatory strategies and coping skills. Also, teachers can promote a positive self-concept, enabling these children to develop confidence and satisfaction about their accomplishments and contributions. Many children with learning disabilities or attentional difficulties perform very well in physical education. For them it may be the one area where they can experience success during the school day, because there is seldom the requirement to sit still, read, or write in physical education. Physical educators can use this opportunity to showcase the motor skills of these children. For others with learning disabilities whose motor performances are not refined, physical education can still be an enjoyable experience. But it will only be enjoyable if conducted in a manner that does not dwell on the child's motor deficits.

In preparing to teach children with LD and/or ADHD, teachers may ask the following questions. What are the strengths and needs of this child? How does this child learn? How can I as the teacher change the environment and the task to help this child learn? What approaches and programs are already in place for this child at school and at home? What support systems will I need and how will I work to get them? And, finally, how can I collaborate with others to assure this child's success? Refer to the Focusing During Transitions application example for a sample solution to organizing a child with ADHD.

Behavioral Disorders

E. Michael Loovis

▷

▷

▷

▷

▷

SEVERAL STUDENTS FROM THE SEVENTH period self-contained class for students with behavioral disorders are standing around prior to the start of physical education class. Mrs. Thomas, the physical educator, enters the gymnasium, blows her whistle, and says, "Line up and keep your mouths shut!" These students continue talking as they proceed to their predetermined spots on the gym floor where class attendance is taken. Mrs. Thomas yells that she wants the class to "keep their mouths shut and pay attention." Conversation among several students continues and Mrs. Thomas shouts at the group in a sarcastic tone, "Hey, you three, shut your mouths or you're going to the principal's office." None of the students pays any attention to Mrs. Thomas; they continue talking and mimicking her conversational tone. At this point, Mrs. Thomas sends them to the principal's office for the fifth time in two months. As they leave the gym, all three turn toward Mrs. Thomas and say, "Forget you; we'll be back!" They then slam the gymnasium door so hard that the window shatters.

Mrs. Thomas has had one of what may turn out to be many experiences with children with behavioral disorders. The key to teaching these children effectively is first understanding the types of behavioral conditions that exist and then having a grasp of instructional strategies to teach them.

This chapter encompasses two categories of behavioral conditions: autism and behavior disorders/emotional disturbance. Even though these conditions represent separate categories in Individuals with Disabilities Education Act (IDEA), the behavioral concerns that they produce relative to teaching and learning warrant their inclusion under the common label of behavioral disorders.

BEHAVIORAL DISORDERS

The fourth largest group of children and youth receiving special education are those with behavior disorders (National Center for Education Statistics, 1998). Even though the number of students with behavioral disorders has increased from 399,000 in 1992 to 438,000 in 1996, the relative percentage when compared with the number of all students with disabilities has remained constant at 7.9 percent (figure 10.1). In the past these students have been referred to as emotionally disturbed, socially maladjusted, behavior disordered, conduct disordered, and emotionally handicapped. Certain characteristics invariably associated with these students make them stand out. Not all of them exhibit the same characteristics; in fact, quite the opposite is true. Generally speaking, they can demonstrate behavior that is labeled hyperactive, distractive, and/or impulsive. Some of these students may exhibit aggression beyond what is considered normal or socially acceptable. Some may lie, set fires, steal, or abuse alcohol and/or drugs. Some may behave in a manner that is considered withdrawn; they may act immature or behave in ways that tend to highlight feelings of inadequacy. Another segment of this population may demonstrate behavior directed in a very negative way against society; these individuals are known as juvenile delinquents. Increasingly, others fit the category referred to as "at risk." These individuals are mired in an incompatibility between themselves and school, resulting in low academic achievement and high dropout rates (Rossi, 1994).

According to IDEA emotional disturbance is defined in this way:

The term means a condition exhibiting one or more of the following characteristics over a long period of time and to a marked degree that adversely affects a child's educational performance:

1. *An inability to learn that cannot be explained by intellectual, sensory, or health factors*
2. *An inability to build or maintain satisfactory interpersonal relationships with peers and teachers*
3. *Inappropriate types of behavior or feelings under normal circumstances*
4. *A general pervasive mood of unhappiness or depression*
5. *A tendency to develop physical symptoms or fears associated with personal or school problems (Federal Register, October 12, 1997, p. 55069)*

The terms emotional disturbance and behavioral disorder (BD) are used synonymously in this chapter. Identification of individuals with emotional disturbance is perhaps the most perplexing problem facing school and mental health professionals. In addition, consideration is given to the ever-expanding number of children and youth who are at risk. For example, students who are suspended or expelled for disobedience or opposition to authority figures are being diagnosed with Oppositional Defiance Disorder (American Psychiatric Association [APA], 1994) and are eligible for services under IDEA. Therefore, it is beneficial to understand the three qualifiers that appear in the first paragraph of the federal definition, namely, duration, degree, and adverse effects on educational performance.

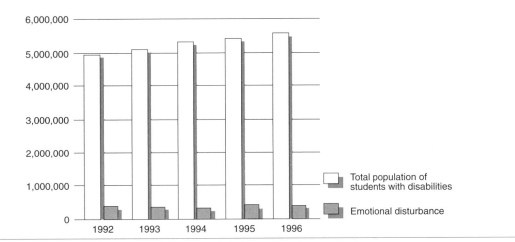

Figure 10.1   Relative percentage of students with behavior disorders compared to all students with disabilities.

Long Period of Time

This qualifier includes behavioral patterns that are chronic in nature, for example, a persistent pattern of physical and/or verbal attacks on a classmate. It excludes behaviors that conceivably could be construed as emotional disturbance but that are situational in nature and thus understandable or expected. For example, a death in the family, a divorce, or another crisis situation could alter a student's behavior in a way that makes it appear aberrant.

Marked Degree

Under consideration here are the magnitude and duration of a behavior. Intensity of behavioral displays, such as intensity of an altercation with a classmate, is considered. For example, a violent physical and verbal attack on a fellow student that requires extensive crisis intervention from teachers and counselors in contrast to a "pushing and shoving" match would qualify under this criterion. Also noted is the amount of time a student engages in a particular behavior, for example, if these attacks occur frequently.

Adversely Affects Educational Performance

There must be a demonstrable cause-and-effect relationship between a student's behavior and decreased academic performance. This requires, at the very least, determining if students are performing at or near the level they would be expected to attain without a behavioral disorder.

Behavioral Disorders in Public School Settings

When endeavoring to understand students with mild and moderate BD, which is the group most likely to be found in an integrated classroom setting, a behavioral classification appears most serviceable. Quay (1986) has done the best work in dimensional classification. Stud-

ies have shown that several dimensions (i.e., conduct disorder, anxiety-withdrawal, immaturity, and socialized aggression) are consistently found in special education classes for students who are emotionally disturbed. In 1987, Quay and Peterson, using the Revised Behavior Problem Checklist, expanded the dimensions that had been identified previously. The six new dimensions (some of which are essentially the same as those listed previously) include the following:

▶ Conduct disorder involves attention-seeking behavior, temper tantrums, fighting, disruptiveness, and a tendency to annoy others.

▶ Socialized aggression typically involves cooperative stealing, truancy, loyalty to delinquent friends, association with "bad" companions, and freely admitted disrespect for moral values and laws.

▶ Attention problems—immaturity characteristically involves short attention span, sluggishness, poor concentration, distractibility, lethargy, and a tendency to answer without thinking.

▶ Anxiety-withdrawal stands in considerable contrast to conduct disorders, involving, as it does, self-consciousness, hypersensitivity, general fearfulness, anxiety, depression, and perpetual sadness.

▶ Psychotic behavior insinuates saying things over and over and expressing strange, farfetched ideas.

▶ Motor excess suggests restlessness and an inability to relax.

An extension of Quay's work included the identification of two primary dimensions of disordered behavior, namely, externalizing and internalizing (Achenback, Howell, Quay, & Conners, 1991). Externalizing behavior involves attacks against others, which parallels Quay and Peterson's original conduct disorder and socialized aggression, while internalizing behavior involves internal mental or emotional conflict such as depression and anxiety, which approximates Quay and Peterson's anxiety-withdrawal and immaturity dimensions.

Causes of Behavioral Disorders

Several factors conceivably having a causal relationship to BD have been identified: biological, family, school, and cultural factors. In addition, society is slowly becoming aware of children and students who are at risk. Although space does not permit a detailed discussion of the forces or factors that place students at risk, it is sufficient to recognize that broad societal factors have been shown to correlate with poor educational performance. These include poverty, minority racial/ethnic group identity, non-English or limited-English language background, and specific family configurations (e.g., living in a single-parent household, limited education) (Rossi, 1994).

Biological Factors

According to Kauffman (1997), several biological aberrations may contribute to the etiology of BD. These include genetic anomalies, difficult temperament, brain damage or dysfunction, nutritional deficiencies, physical illness or disability, and psychophysiological disorders. With these factors identified, it is important to reiterate Hallahan and Kauffman's (1997) contention that "For most children with emotional or behavioral disorders, there simply is no real evidence that biological factors alone are at the root of their problems" (p. 200).

Family Factors

Pathological family relationships are major contributory factors in the etiology of BD. Broken homes, divorce, chaotic or hostile family relationships, absence of mother or father, and parental separation may produce situations in which youngsters are at risk to develop BD. It is also clear that there is not a one-to-one relationship between disruptive family relations and BD. Many youngsters find parental discord more injurious than separation from one or both parents. Research also points to a multiplier effect: When two or more factors are present simultaneously, there is increased probability that a behavior disorder will develop.

School Factors

It has become increasingly clear that, besides the family, school is the most significant socializing factor in the life of the child. For this reason, the school must shoulder some of the responsibility for causing BD. According to Kauffman (1997) schools contribute to the development of behavior disorders in several ways:

Insensitivity to students' individuality

Inappropriate expectations for students

Inconsistent management of behavior

Instruction in nonfunctional and irrelevant skills

Destructive contingencies of reinforcement

Undesirable models of school conduct

Cultural Factors

Frequently, there exists a discrepancy between the values and expectations that are embraced by the child, the family, and the school. Consequently, there is an increased probability that the student will violate dominant cultural norms and will be labeled as deviant (Kauffman, 1997).

Part of the problem centers on conflicted cultural values and standards that society has engendered. For example, the popular media has elevated many high-status models whose behavior is every bit as violent as the villains they are apprehending; however, students who engage in similar behaviors are told that they are incompatible with society's expectations.

Another problem area involves the multicultural perspective or, rather, a lack of it. Teachers find it extremely difficult to eliminate bias and discrimination when evaluating a student's behavior. Consequently, students are labeled as deviant when, in fact, it is only at school that their behaviors are considered inappropriate.

Other cultural factors influencing behavior include the students' peer group, neighborhood, urbanization, ethnicity, and social class. These factors are not significant predictors of disordered behavior by themselves; however, in combination and within the context of economic deprivation and family conflict, they can have an adverse affect on behavior (Kauffman, 1997).

A significant sociocultural factor that portrays the relationship between aberrant adult behavior and a spiraling incidence of behavioral disabilities is substance abuse. Children who are prenatally exposed to drugs and alcohol are affected in two ways. First, there is an increased incidence of neurological impairment, because both drugs and alcohol can cross the placenta and reach the fetus, causing chemical dependency, congenital aberrations, neurobehavioral abnormalities, and intrauterine growth retardation. Second, these children are exposed to family situations that are, at best, chaotic. Typically, these children find themselves in the social service system bouncing from one substitute care situation to another (Bauer, 1991). Sinclair (1998) reported that prenatally drug-exposed children in Head Start programs were more likely to be classified as emotionally/behaviorally disordered and placed in special education upon entrance into kindergarten. Van Dyke and Fox (1990) reported the long-range effects of fetal alcohol exposure. Their conclusions confirmed that a significant number of children who were diagnosed with fetal alcohol syndrome in the 1970s were having learning difficulties, behavioral problems, and attention deficits a decade later.

Polysubstance abuse or the use of combinations of drugs and alcohol is the typical pattern for substance abusers. Fetal exposure to cocaine is estimated at 5 percent for all births in suburban hospitals (Batshaw &

Conlon, 1997) and 15 percent in urban areas (Singer, Arendt, Farkas, Minnes, Huang, & Yamashita, 1997). Given this situation, much research has focused on prenatal and perinatal complications of cocaine use. The long-term effects of cocaine use during pregnancy are equivocal; apparently, most prenatally exposed children do not develop serious developmental disabilities (Hawley & Disney, 1992). Tarr and Pyfer (1996), on the other hand, reported that use/abuse of illicit drugs, alcohol, or both significantly affect the physical and motor development of neonates and infants exposed in utero. Generally, more research is needed, especially research that assesses the postnatal milieu in which these children are reared.

Approaches

The conceptual models that serve as the basis for understanding, treating, or educating students with behavioral disabilities are discussed more comprehensively in chapter 7. The following paragraphs highlight the psychodynamic, psychoeducational, ecological, psychoneurological, humanistic, and behavioral approaches to teaching students with BD.

Psychodynamic

The psychodynamic approach focuses on the improvement of psychological functioning. This improvement depends on helping individuals cope with deep-seated emotional problems that can result in impaired personal relationships, conflicting social values, poor self-concept, ability deficits, and antisocial habits and attitudes. Failure to alleviate the cause(s) of psychological dysfunction can lead to learning and behavioral difficulties. The psychodynamic approach uses a number of treatment modalities including psychoanalysis, counseling, play therapy, and group therapy. Because of its strong, primarily Freudian psychological orientation, this approach is less likely to be used by teachers, and, consequently, its potential benefits in the educational setting are at best speculative.

Psychoeducational

The psychoeducational approach assumes that making students aware of their feelings and having them talk about the nature of their responses will give them insight into their behavior and help them develop control. Academic failure and misbehavior are dealt with directly and therapeutically. Students are taught to acknowledge that they have a problem, understand why they are misbehaving, observe the consequences of their behavior, and plan an alternate response or way of behaving that is more appropriate.

Additionally, students are taught management procedures such as self-monitoring, self-assessment, self-instruction, and self-reinforcement as a means of promoting functional independence (Nelson, Smith, Young, & Dodd, 1991). Teachers are taught to anticipate a crisis and deal with it in an unemotional manner. In crisis intervention, they use the major tool of the psychoeducational approach, the life-space interview, which utilizes talking and experiencing to help students recognize and appreciate their feelings (Wood & Long, 1991).

Ecological

Ameliorating disturbance in a student's environment or ecosystem is the focus of the ecological model. Intervening to stop unwanted behavior is only a secondary function of this model. Changing the environment in substantive ways to reduce the likelihood of the behavior recurring once intervention is withdrawn is the primary focus. To facilitate this objective, environments (i.e., the school, home, and community) are modified to accommodate students rather than expecting that students will always make the adjustment to the environment. Classrooms are converted into environments where the likelihood of success is greatly enhanced. It means using more nontraditional methods such as cooperative learning as a mechanism for psychological adaptation to the classroom. Teachers in the ecological model are also counselors. Beyond teaching academic subjects, they also function as a supportive network when returning students to regular classrooms.

Psychoneurological

The use of the psychoneurological approach assumes the presence of neurological dysfunction, and the goal is to reduce those general behavioral characteristics such as hyperactivity, distractibility, impulsiveness, and emotional lability that make the management of behavior difficult at best. The prescribed treatment for such dysfunction within the psychoneurological model is drug therapy. The main category of medication used to manage the behavior of school-aged children is psychotropic drugs. Chapter 7 discusses drug therapy; it is sufficient here to note that the primary advantage of drug therapy is its ability to make other interventions more effective.

Humanistic

The humanistic approach is couched in humanistic psychology and incorporates many of the ideals engendered by the social movements of the 1960s and 1970s, including alternative schools, open education, and deschooling, to name a few. According to Kauffman (1997), this approach "emphasizes self-direction, self-fulfillment, self-evaluation, and free choice of educational activities and goals, but the theoretical underpinnings of humanistic models are hard to identify"(p. 116). The basic tenet of most humanistic approaches is that students will engage in appropriate problem solving and decision making if they are provided a supportive milieu. The life-impact curriculum described by Rhodes and Doone (1992) is a good example of this approach and parallels the work of Hellison and his colleagues that will be discussed later in this chapter and is also discussed in chapter 7.

Behavioral

The behavioral approach is an elaboration of learning principles such as those embodied in respondent conditioning, social learning theory, and operant conditioning. The term most commonly used to describe this approach is behavior modification. An extensive discussion of behavior modification and its principles, primarily reinforcement, is presented in chapter 7. The salient features of the behavioral approach are careful assessment of observable behavior; analysis of the effect of environmental stimuli, both those that precede a response (antecedents) and those that follow it (consequences); and systematic arrangement of the consequences to change a behavior or at least its frequency. Within the behavioral model, the area that is experiencing notoriety as it relates to students with behavioral disabilities is cognitive-behavior modification (Breen & Altepeter, 1990). With its emphasis on self-control techniques, it is a natural concomitant in programs that are attempting to move beyond the traditional Skinnerian concept of behavior management (e.g., in the psychoeducational approach described earlier). Recently, cognitive-behavioral approaches in combination with parental training programs have demonstrated a greater likelihood to produce lasting results (Southam-Gerow & Kendall, 1997). Other behavioral approaches, such as Lovaas's intervention with children with autism, are also receiving acclaim and are proving to be effective.

In this section six approaches—psychodynamic, psychoeducational, ecological, psychoneurological, humanistic, and behavioral—have been delineated. Each appears to hold some promise for intervening with students with behavioral disorders. In the hands of purists, one approach may be used exclusively; however, teachers are typically pragmatic and use a combination of approaches, even though they may have been trained to emphasize one method more than another.

AUTISM

IDEA prescribes autism as a separate category of disability and defines it as follows:

A developmental disability significantly affecting verbal and nonverbal communication and social interaction, generally evident before age 3, that adversely affects a child's educational performance. Other characteristics often associated with autism are engagement in repetitive activities and stereotyped movements, re-

sistance to environmental change or change in daily routines, and unusual responses to sensory experiences. The term does not apply if a child's educational performance is adversely affected primarily because the child has a serious emotional disturbance. (Federal Register, October 22, 1997, p. 55069)

Autism in Public School Settings

According to the fourth edition of the Diagnostic and Statistical Manual of Mental Disorders (APA, 1994), autism is the most severe form of the category Pervasive Developmental Disorders (PDD). It is distinguished by its early onset, usually during the first three years of life, occurring more often in males. Autism is likewise linked with mental retardation; 75 percent of individuals with autism also have mental retardation. Other disorders that are included in this category are Rett's Disorder, Childhood Disintegrative Disorder, Asperger's Disorder, and PDD—Not Otherwise Specified (NOS). Table 10.1 distinguishes among the various disorders subsumed under the major category of PDD.

Autism's primary diagnostic features include marked impairment in the development of socialization skills; impairment in verbal and nonverbal communication; and the development of a restricted repertoire of activity and interests, including self-stimulatory and self-injurious behaviors. Each of these diagnostic features will be examined in further detail in an attempt to develop a developmental and behavioral profile of the individual with autism.

Socialization

Impairments in the area of reciprocal social interactions are evidenced by multiple nonverbal behaviors, failure to develop peer relationships that are appropriate to one's developmental level, lack of social or emotional reciprocity, and inability to recognize others (APA, 1994). As it relates to nonverbal behaviors, individuals with autism frequently fail to demonstrate eye-to-eye gaze, their faces are frequently expressionless, and their postures and gestures are indicative of failure to regulate the social context. Individuals with autism demonstrate faulty peer relationships that are characterized by little or no interest in friendships; this includes a lack of spontaneity when it comes to sharing enjoyment, interests, or achievement. Lack of social or emotional reciprocity is demonstrated when children with autism fail to actively participate in simple social play or games as well as in symbolic play (Christopher, Boucher, & Smith, 1993). The final area of concern relative to socialization is the inability to recognize others, that is to say, individuals with autism may be oblivious to the presence of people in their environment. They may fail either to recognize that individuals have needs or to recognize another's misfortune.

Table 10.1   **Pervasive Developmental Disorder**

	Autistic Disorder	**Rett's Disorder**	**Childhood Disintegrative Disorder**	**Asperger's Disorder**	**PDD-NOS**
Onset	Prior to 3 years	Prior to 4 years	Prior to 10 years	Later than Autistic Disorder	
Gender	> Males	Females only	> Males	> Males	
Diagnostic features	Marked abnormal development in social interaction and communication; marked restricted repertoire of activity and interest	Multiple specific deficits following a period of normal functioning after birth; normal psychomotor development through first 5 months; loss of previously acquired hand skills between 5 and 30 months.	Marked regression in multiple areas of function following a period of at least 2 years of apparently normal development	Severe and sustained impairment in social interaction and development of restricted, repetitive patterns of behaving, interest, and activities	Category used when features do not meet criteria for Autistic Disorder
Associated disorders	75% are MR (moderate range)	Severe or profound MR	Severe MR	General medical conditions; nonspecific neurological symptoms; motor clumsiness	

MR = mental retardation

Communication

Marked impairment in the development and use of verbal and nonverbal communication is another characteristic of individuals with autism (APA, 1994). Difficulties with the development and use of expressive language are manifested in several ways, including a delay in or total failure to develop spoken language; a use of language that is idiosyncratic, including language that is stereotyped and repetitive; and an inability to initiate or sustain conversations with others. A contributory factor in the failure to develop developmentally appropriate language is the inability to engage in spontaneous make-believe play or social imitative play.

Ritualistic and Compulsive Behavior

The final diagnostic feature is a pattern of behavior, interests, and activities that is characterized as repetitive, stereotypic, and restrictive (APA, 1994). Individuals with autism are preoccupied with one or more ste-

reotyped and restricted patterns that are abnormal either in intensity or focus. They demonstrate unyielding adherence to specific, nonfunctional routines or rituals, and they frequently engage in bizarre motor mannerisms (e.g., self-stimulating behaviors). Individuals with autism also frequently engage in self-injurious behaviors (SIB), including punching, scratching, biting, eye gouging, and severe head banging (Rosenberg, Wilson, Maheady, & Sindelar, 1997).

Causes of Autism

The exact etiology of autism is unknown. It is generally agreed, however, that structural or neurochemical alterations of the central nervous system are the primary causes. Additionally, authorities contend that certain organic and metabolic conditions are associated with autism. Examples include tuberous sclerosis, congenital rubella syndrome, phenylketonuria (PKU), Fragile X Syndrome, and viral infections such as herpes simplex.

Autism has also been associated with conditions such as Down syndrome and infantile spasms. In the case of PKU and tuberous sclerosis, there is an obvious genetic link. Epidemiological studies clearly support a familial tendency toward autism (Mauk, Reber, & Batshaw, 1997).

Approaches to Autism

Certainly the conceptual models described in chapter 7 and summarized briefly in this chapter will have relevance to any discussion of treatment interventions with students with autism. It is the intention of this section to discuss three specific approaches that have proven effective with students with autism. This is not meant to be an all-inclusive examination of existing programs. The programs include Treatment and Evaluation of Autistic and Related Communication Handicapped Children (TEACCH), Cognitive-Developmental Systems Theory (Miller Method), and the behavioral approach espoused by Lovaas.

TEACCH

Developed in the early 1970s by Eric Schopler, TEACCH is housed on the campus of the University of North Carolina. TEACCH promotes an individual through programming based on a person's skills, interests, and needs. Individual assessment highlights personal understanding and acknowledges the "culture of autism," which suggests that, as a group, individuals with autism are distinct with different characteristics that do not necessarily make them inferior. TEACCH focuses on balanced expectations. They are neither high nor low when compared with what would normally be expected. Rather they begin where the person is and proceed as far as the person can go. TEACCH does not espouse full inclusion for all individuals with autism; it is more important that the individual feel comfortable with where he is and not attempt to fit into some predetermined mode of behavior that appears "normal." At the core of the TEACCH approach is structured teaching (Mesibov & Shea, 1997). Designed to promote the use of skills independently, structured teaching focuses on organizing the physical environment, developing daily work schedules and systems, providing for clear and explicit expectations, and using visual cues for presenting materials and prompting instructional sequences. A more detailed description of structured teaching appears later in this chapter.

Miller Method

Arnold Miller and Eileen Eller-Miller are cofounders of the Language and Cognitive Development Center, which is an accredited nonprofit school, clinic, and training facility for the treatment of PDD in Boston, Massachusetts. Their work (Miller & Eller-Miller, 1989) is based on their interpretation of Cognitive-Developmental Systems Theory, which basically formulates that aberrant or disordered behavior, the kind often associated with individuals with autism, is the result of systems that fail to include people or to permit normal exploration. Philosophically, the Miller Method maintains that each child, regardless of how withdrawn or disorganized, is attempting to cope with his environment. Therefore, this approach is grounded in trying to assist each person to assess and respond to one's world, to understand others and express oneself either through spoken or signed language, and to generalize newly learned capacities to the home environment.

Initially the Miller Method assesses each individual's "unique reality" using what is referred to as the Umwelt Assessment. For example, the Swinging Ball Task examines children's awareness of and ability to cope with a large object that approaches and moves away from them. The Miller Umwelt Assessment Scale has 16 such items. Based on the assessment data, a program is planned that expands and transfers each child's reality system, limited though it may be. The program attempts to impact the child's reality system by using what is referred to as a "repetitive sphere of activity." For example, a child who cannot climb stairs to go down a slide is guided through the activity repeatedly; the pace of the activity may be rapid so that the child connects one aspect of the task with the other aspects. Another interesting component of this approach is the use of elevated equipment. Elevated boards that are positioned two to four feet off the ground are used to help children focus better, to be more aware of their bodies, and to generally follow directions better. This latter aspect has obvious implications for physical education.

Lovaas

O. Ivor Lovaas is associated with the classic behavioral management approach as discussed in chapter 7. The essence of this approach is to reward instances of positive behavior and to minimize cases of negative behavior. Effective behavioral training of individuals with autism includes unambiguous instructions, prompting, positive reinforcement, shaping, and discrimination training (i.e., learning when and when not to respond). When a behavior is selected for acquisition, it is repeated five or six times a minute for up to two hours. This frequent repetition increases the rate of learning and reduces the tendency of individuals with autism to withdraw. Over time physical prompting becomes less necessary, and the learning situation requires less structure. This approach has been in existence since the 1970s. Initial efforts produced positive results, but individuals reverted back to previous behavior when intervention ceased. However, a longitudinal study (McEachin, Smith, & Lovaas, 1993) confirmed that gains made during 40 hours per week of one-to-one treatment for a period of 2 years or more with 19 children under the age of 4 years were preserved when reevaluation was conducted at 11.5 years of age.

Three prominent approaches in the field of autism have been presented. Although different in their foci and methods, they share an interesting characteristic—they all staunchly advocate structured intervention for individuals with autism, regardless of the level of severity.

PHYSICAL EDUCATION AND SPORT ACTIVITIES

According to Steinberg and Knitzer (1992), effective physical education programs for students with emotional and behavioral disturbances are the exception rather than the rule. They suggest that this is especially perplexing in light of increased academic performance and decreased student absenteeism when students are involved in vigorous and systematic exercise programs. Poor motor performance in BD students is often attributed to various indirect factors, such as attention deficits, poor work habits, impulsivity, hyperactivity, feelings of inadequacy, and demonstration of aggressive behavior, rather than some innate inability to move well.

Because some students with behavioral disabilities may demonstrate a lag in physical and motor abilities, the physical educator must provide them with appropriate developmental activities. The emphasis should be on physical conditioning, balance, and basic movement. The development of fundamental locomotor and nonlocomotor movements will also require attention. In addition, it may be necessary to emphasize perceptual-motor activities, because students with behavioral disabilities often demonstrate inadequacies in this area. Students with autism generally demonstrate poor motor skills (Morin & Reid, 1985). Consequently, programs should emphasize fundamental motor skills and patterns, individual games and sports, and developmental activities that increase physical proficiency. One such program based on a model developed by Dr. Kiyo Kitahara is referred to as Daily Life Therapy (Quill, Gurry, & Larkin, 1989).

The Daily Life Therapy approach, as employed by the Boston Higashi School, uses physical exercise as one of its central features. The program is dominated by running. During the first months in the program, children run outside 2 to 3 times per day for 20 minutes each session. An additional daily period includes gymnastics, aerobic exercise, and martial arts. One additional hour of daily outdoor play includes soccer, basketball, biking, and climbing on play equipment. The exercise regime is modified for preschool children and incorporates long walks (to replace running), roller skating, biking, and sprinting as well as movement games requiring motor imitation.

Exercise programs have been shown to exert a positive influence on disruptive behavior. As little as 10 or 15 minutes of jogging daily has produced a significant reduction in the disruptive behavior of children (Yell, 1988). Levinson and Reid (1993) confirmed similar results with 11-year-old autistic children in their study. Elliot, Dobbin, Rose, and Soper (1994) reported a reduction in maladaptive and stereotypic behaviors in adults with autism and mental retardation subsequent to vigorous, aerobic exercise. Rosenthal-Malek and Mitchell (1997) confirmed the results of Elliot et al. when they demonstrated a reduction of self-stimulatory behaviors in adolescents with autism. As it relates to Asperger's Syndrome, children do have definite motor problems and perform at lower levels on the Test of Motor Impairment-Henderson Revision than do peers of the same age (Manjiviona & Prior, 1995). Miyahara, Tujii, Hori, Nakanishi, Kageyama, and Sugiyama (1997) reported a high incidence of motor delay as it pertained to dexterity, ball skills, and balance skills in children with Asperger's Syndrome when tested with the Movement Awareness Battery for Children.

In the physical education and sport setting, teachers and coaches who are going to work with students with autism should solicit as much information as they can about the student from teachers, counselors, psychologists, and parents before starting the program. To establish the most effective instructional situation, teachers and coaches should be provided with the most up-to-date information possible relative to changes in the student's schedule, medication, or behavior management program (Mangus & Henderson, 1988). Additionally, teachers and coaches should understand that students with autism require a structured teaching environment; therefore, activities such as entrance into the gymnasium, method of starting the class, and the conduct of the class itself should be consistent and predictable.

Many physical education programs use games to accomplish goals and objectives established for individuals and classes. Because students with behavioral disorders often lack fundamental skills, they frequently are incapable of demonstrating even minimal levels of competence in these games. As a result, they have an increased tendency to act out—perhaps with verbal and/or physical aggression—or to withdraw, which further excludes them from an opportunity to develop skills.

In an effort to promote the most positive learning environment, Hellison (1995) and Hellison and Georgiadis (1992) outlined a nontraditional approach to working with inner-city, at-risk youngsters using basketball as the primary vehicle for empowering students to learn personal and social values. Employing Hellison's Responsibility Model (discussed in chapter 7) as the philosophical underpinning, the "coaching club" is a before-school program in Chicago's inner city. It offers students the opportunity to explore movement through a progression of five levels: self-control (control of one's body and temper); teamwork (full participation by all team members); self-coaching; coaching another team member; and applying skills learned in the program outside the gym to school, home, and neighborhood. Playing ability is not a prerequisite. This program exists to promote social responsibility. Likewise, extrinsic rewards are unnecessary, because students are motivated to reach level IV (Coach) on the program's evaluation system (Hellison & Georgiadis, 1992). Level IV consists of the following characteristics: has good attendance; is coachable and on task at practice; does not abuse others or interrupt practice; is able to set personal goals and work independently on these goals; possesses good helping skills—can give cues, observe, and give specific positive feedback as well as general praise; encourages teamwork and passing the ball; listens to her players and is sensitive to their feelings and needs; puts the welfare of his players above her own needs (such as the need to win or "look good"); and understands that a coach's basketball ability is not the key to being a good coach—the above characteristics are.

Relaxation is another program component that deserves a special place in the normal movement routine of many students with BD. Making the transition from gymnasium to classroom can be difficult for students with hyperactive behavior. This difficulty is not a reason to eliminate vigorous activity from these students' programs; rather, it is a reason to provide additional time, a buffer, during which the students can use relaxation techniques they have been taught. The ability to play effectively is crucial to success in physical education. Because games are a part of the physical education program for most students with BD, it is essential for teachers to be aware of the direct relationship between the type of activity chosen and the degree to which inappropriate behavior is likely to occur. The type of programming chosen directly relates to the amount of aggression demonstrated by students during activity. Body contact, simpler rules, and fewer skill requirements reduce some of the specific variables that seem to control aggression. Not to be overlooked is the New Games approach, which has a cooperative rather than a competitive orientation. In light of the problems surrounding self-concept and the antisocial behavior exhibited by some students with BD, the least desirable situation is one that prescribes winners and losers or that rewards overly aggressive behavior.

MODIFICATIONS FOR TEACHING AND PROGRAMMING IN PHYSICAL EDUCATION AND SPORT

There is not one consistently correct way of instructing children with behavioral disabilities. Effective teachers will tailor their instructional techniques according to their students' types of BD.

General Instructional Considerations

Teachers and coaches are typically not prepared to handle the nature and magnitude of problems confronted when working with students with behavioral disorders. Therefore, they are easily frustrated. If not managed effectively, this frustration can lead to inappropriate decision making and interaction with students; this behavior can exacerbate a potentially volatile situation. At the very core of effective instruction with students who exhibit externalizing behaviors, such as fighting and temper tantruming, is an examination of the ways people communicate. To communicate effectively individuals must be able to give and receive information clearly and use the information to achieve a desired result. Giving and receiving information clearly is a goal of effective interpersonal communication that results from active listening.

Active Listening

Three basic skills are essential to the technique of active listening. The first is **attending**, a physical act that requires the listener to face the person speaking, maintain eye contact, and lean forward, if seated. These actions communicate to the speaker that the listener is, in fact, interested in what is being said. **Listening**, the second component skill, means more than just hearing what is being said. It involves the process of "decoding," an attempt by the listener to interpret what has been said. Consider the following example:

Sender code: "Why must we do these exercises?"

Possible receiver decoding:

a. "He doesn't know how to perform them and is embarrassed to admit it."

b. "He's bored with the lesson and is anxious to get to the next class."

c. "He's unclear about why the exercises are necessary and is seeking some clarification."

Suppose that the most accurate decoding was *a* or *c*, but the listener decoded the message as *b*. A misunderstanding would result, and the communication process would start to break down. Situations like the one just portrayed occur frequently, with neither the speaker nor the listener aware that a misunderstanding exists. The question them becomes "What can be done to assure that the correct message is being communicated?" The answer is contained in the third and final step of the active listening process. The third component of active listening is **responding**. The listener sends back the results of her decoding in an attempt to ascertain if there are any misunderstandings. In effect, the listener merely restates the interpretation of the sender's message.

The communication process called active listening can help prevent misunderstandings, facilitate problem solving, and demonstrate warmth and understanding. As with any new skill, it requires practice for maximum effectiveness.

Verbal Mediation

A second communication technique is **verbal mediation**, which involves having individuals verbalize the association between their behaviors and the consequences of that behavior. Of particular importance in verbal mediation is having students take an active role in the process rather than passively hearing teachers make the association for them. The following situation illustrates verbal mediation: A student has just earned 10 minutes of free time by successfully completing the assigned drill at a circuit training station.

Teacher: "What did you do to earn free time?"

Student: "I followed directions and completed my work."

Teacher: "Do you like free time?"

Student: "Yes, it's fun."

Teacher: "So when you do your work, then you can have fun."

Student: "Right."

Teacher: "Good for you! Keep up the good work."

In this example, the teacher has facilitated the student's verbal mediation of the positive association between the appropriate behavior and the positive consequence.

Conflict Resolution

Teaching students with BD entails a greater than average risk of confrontation as a way of resolving conflict. This does not imply that interpersonal confrontation need be punitive or destructive. On the contrary, a healthy use of confrontation provides the opportunity to examine a set of behaviors in relation to expectations and perceptions of others as well as to establish rules.

The goal of confrontation is resolution of conflicts through constructive behavior change. Several steps are necessary in reaching this desired goal through confrontation: making an assertive, confrontive statement (one that expresses honestly and directly how the speaker feels about another's behavior); being aware of common reactions to confrontation; and knowing how to deal effectively with these reactions.

Although assertive confrontation can be an effective means of resolving conflicts, it requires skillful use of each step in the process. Without question the most crucial component of the conflict resolution process is the formulation of an effective confrontive statement. There are three main components, namely a nonjudgmental description of the behavior causing the problem, a concrete effect that the behavior is having on the person sending the message, and an expression of the feelings produced from the concrete effect of the behavior. Together these components form what is referred to as an **I-message**. When combined with active listening, which further reduces defensiveness, the conflict resolution process can be an excellent means to avoid major conflicts, as exemplified in the following conversation:

Teacher: "Robert, your disruptive behavior during class is causing me a problem. When you argue and fight with the other students in class, I have to stop teaching. It's distracting to me and I'm frustrated." (I-message)

Student: "I get that stuff at home. I don't need it here."

Teacher: "I see. Lately, you're having some problems at home with your parents." (active listening)

Student: "My dad and I have been fighting all week."

Teacher: "You're really upset about the problem you're having with your father." (active listening)

Student: "Yeah! I don't know how much longer I can put up with his bulls—t."

Teacher: "So you're angry because of the situation at home and it's carrying over into school." (active listening)

Student: "Yeah! I know you're upset about me fighting and not getting along in class. You know I've tried to get along."

Teacher: "You're a little surprised that it's such a problem for me even though the incidents are not always all that extreme." (active listening)

Student: "Well, not really. I see what you're saying. You have to stop teaching and stuff. Mostly I'm taking my anger out on the guys in class. I'll just have to remind myself that it's not their fault that my old man and me are not getting along, and I'll just try harder not to get angry and fight with the guys, OK?"

Teacher: "That would sure help me. Thanks, Robert."

Student: "No sweat!"

Application Example: Behavioral Disorders

Setting: High school physical education class

Student: Tenth grade student with a conduct disorder who is frequently abusive both verbally and physically to his peers and constantly challenging the authority of the physical educator

Issue: What is the best approach to handling this situation?

Application: The physical educator should

▶ employ the conflict resolution process,

▶ examine those events or conditions that seem to spark these episodes of abusive behavior,

▶ praise instances of appropriate behavior, and

▶ implement Hellison's Responsibility Model.

In this example, the teacher has blended effectively the use of an I-message and the skill of active listening to diffuse a situation that could have erupted into a major confrontation between teacher and student.

Beyond effective communication there are more specific instructional strategies that teachers must learn and practice in order to be competent with students with behavioral disorders. In the final part of the chapter, the discussion will center on the specific strategies and techniques that, research suggests, make it more likely that physical educators and coaches will be successful teaching students with behavioral disorders.

Mild to Moderate Behavioral Disorders

There is no single, consistently correct method of teaching children with behavioral disorders. Effective teachers will tailor their instructional techniques to the children they are teaching according to the students' types of BD.

Specific Instructional Strategies

In order to succeed with students who are overtly aggressive and acting out as well as with higher functioning students with autism, teachers will find it necessary to employ a number of intervention techniques (Walker, 1995). Many of these techniques are found in chapter 7, and the reader is encouraged to review them as necessary. In addition to those found in chapter 7, teachers desiring to control students with conduct disorders should establish a few, clear rules that govern expectations for gymnasium conduct. Students can be included in the process of developing these rules so they might be internalized better: praise students for appropriate behavior including desirable, nonaggressive contact with peers; examine antecedent events or conditions (i.e., events that set the stage for certain behaviors to

occur); and in the case of students who are aggressive and who act out, it may require separation from a particular peer who causes trouble. Teachers can also treat antisocial behaviors as errors in learning and determine whether students lack clear knowledge of the skill or whether they know what the skill is, but will not use it (e.g., a student who can describe intentionally fouling a peer during a basketball game as inappropriate, but who does it anyway). Such a student is demonstrating a performance deficit, while a student who does not understand his inappropriate behavior, upon questioning, is demonstrating a skill deficit. Involve students in self-monitoring and self-control training. Have students record and evaluate target behaviors, prompt themselves to employ remedial strategies, and apply consequences according to prearranged contracts (see the Behavioral Disorders application example).

In physical education, students with behavior disabilities who function at higher levels can be taught with a humanistic approach. Generally speaking, some techniques suggested by Sherrill (1998) for improving self-concept are singularly applicable with this population; for example, teachers should strive to

▶ conceptualize individual and small-group counseling as an integral part of physical education instruction,

▶ teach students to care about each other and to show that they care,

▶ emphasize cooperation and social interaction rather than individual performance,

▶ stress the importance of genuineness and honesty in praise,

▶ increase perceived competence in relation to motor skill and fitness, and

▶ convey that they like and respect students as human beings, for themselves as whole persons, not just for their motor skills and fitness.

Hellison's Responsibility Model

More specifically, the approach outlined by Hellison (1995) has immediate relevance for practitioners confronted with students who lack self-control and consequently present management problems. Hellison's four primary goals or developmental levels—self-control and respect for the rights and feelings of others, participation and effort, self-direction, and caring and helping—are instrumental in overcoming many current discipline and motivation problems in schools. (Hellison's levels are discussed in chapter 7.) Hellison has likewise identified six basic strategies that foster attainment across all developmental levels: awareness talks, levels of action, individual decision making, group meetings, reflection time, and counseling time. These strategies are "processes for helping students to become aware of, experience, make decisions about, and reflect on the model's goals" (Hellison & Templin, 1991, p. 108).

Because self-control is basic to Hellison's approach, awareness talks that aid in understanding the program's purpose are very important. Examples of this strategy include reminding students about the levels, illustrating attitudes and behaviors associated with levels as they occur, and conducting student-sharing sessions regarding the importance of levels.

Levels of action are the second strategy that can be employed at each level of the model. For example, at level II students can be encouraged to play a cooperative game. At level IV students who know how to engage in a certain activity can help those who do not.

Individual decision making is built into the model and confronts students at each level. At level I, abusive students can exercise choice by sitting out of an activity or modifying their unacceptable behavior. Level III choice could involve students analyzing the disparity between their program goals and the selection of activities they choose. For example, a student who wants to increase aerobic capacity but who chooses to participate in bowling with friends needs to be confronted with deciding what is more important.

Group meetings are the fourth strategy; it is used primarily with levels I and II. Student sharing invites discussion of what constitutes self-control, and it encourages students to think about the consequences of violating the established self-control rules. It also provides students with the opportunity to evaluate aspects of the program, including the instructor.

Another strategy is reflection time. At the conclusion of every class, students are given time to reflect on what occurred during the session, specifically as it related to their plans and the degree to which they were successful in accomplishing their goals. A variety of techniques are used for this purpose, including journals, checklists, and student discussions.

Counseling time is the final strategy. Time is allotted for discussions about specific problems, things that the teacher has observed, and/or general impressions of the program. Counseling is a part of each level; for example, at level I it could involve a discussion of a student's pattern of aggressive behavior. At level II it might involve a discussion of the degree of effort students are expending in relationship to the goals they have established for themselves.

Severe Behavioral Disorders

Students with severe behavior disorders require more intense programming efforts. This group, which has been characterized by Dunn, Morehouse, and Fredericks (1986), includes students who are self-indulgent, aggressive, noncompliant, self-stimulatory, or self-injurious.

Specific Instructional Strategies

Teachers who work with students who demonstrate any of the previously mentioned behaviors will find it necessary to employ a structured teaching approach. The following suggestions highlight the structured approach used by the TEACCH program and include organizing the physical setting, scheduling activities, and structuring one's teaching. Physical educators may wish to consider modifying their teaching to include arranging the teaching space, including the placement of equipment, to aid independent functioning and to prevent distraction; making materials easily accessible and clearly marked; designating work areas with well-defined boundaries; and providing not only class, but also individual schedules as part of the structure in the gymnasium, especially for students with autism. For example, students can be forewarned about their schedules either verbally or visually by using pictures or drawings representing activities. They can likewise be alerted of an impending transitioning to another activity by using pictures or by hearing a timer that signals the change. Teachers can also systematize and organize teaching methods to include appropriate directions (i.e., directions can be provided verbally or nonverbally with the use of pictures and drawings) and use prompts effectively (e.g., if a striking task were planned, one would not introduce an oversized bat while at the same time using a batting tee for the first time).

In terms of actually teaching a student with autism, a characteristic of major concern is stimulus overselectivity. Overselectivity implies that during skill acquisition an individual attends to vague aspects of relevant cues or fixates on completely irrelevant cues. This has implications for how teachers or coaches will provide sensory stimuli, how they will prompt an action, how they will fade a prompt once it is no longer necessary or appropriate, and how they will promote skill generalization. Prompts are cues, instructions, demonstrations, or other events that signal the initiation of a response. **Extrastimulus prompts** extend from full physical prompting (sometimes referred to as forced responding) to visual prompting to verbal prompting. Extrastimulus prompting has been shown to be effective for students with autism (Collier & Reid, 1987), and physical prompting has been shown to be slightly more effective than visual prompting (Reid, Collier, & Cauchon, 1991). When using physical prompting, teachers and coaches should use light touches and apply them in several locations (Connor, 1990). One also needs to be aware that some students with autism react negatively to being touched; consequently, physical prompting would be contraindicated with these students. When using prompts, one of the goals of instruction is to use the least intrusive prompt. For example, one would not use physical prompting if the individual will initiate a correct response after seeing a demonstration of the desired behavior.

Generalization, or the ability to demonstrate a skill in an environment different from the one in which the skill was learned, is difficult for students with autism (Mangus & Henderson, 1988). Teaching for generalization involves providing instruction in the most gamelike or natural environment possible (e.g., learning to bowl at the local bowling alley versus learning to bowl in a gymnasium). It likewise implies using equipment that resembles or closely approximates that which will be used in the culminating game or activity (e.g., learning to use an aluminum or wooden bat versus learning to play softball with a plastic bat). Additionally, adapted equipment should be faded as soon as possible (Connor, 1990).

The manner in which practice is organized is also an important consideration for students with autism. Practice that employs task variation, or what is commonly referred to as distributed practice, seems most beneficial when teaching skills (Weber & Thorpe, 1992). Task variation, or practicing a variety of tasks interspersed with periods of rest, serves two important functions for students with autism: It provides an opportunity to administer a higher frequency of reinforcement, and it accommodates a short attention span that is characteristic of this population of students. In terms of student-teacher ratios, the ideal is 1:1. Mangus and Henderson (1988) recommend the use of volunteers, teacher aides from special education, parents, and students from regular education classrooms in order to achieve as closely as possible the ideal ratio.

Data Based Gymnasium

Using the basic steps of behavioral programming discussed in chapter 7, Dunn and associates developed the Data Based Gymnasium (DBG). This program incorporates behavioral principles in a systematic effort to produce consistent behavior on the part of adults who work with students with BD and eventually to bring students' behaviors under the control of naturally occurring reinforcers. To this latter end, instructors use natural reinforcers available in the environment, for example, praising a desirable behavior to strengthen it or ignoring an undesirable one to bring about its extinction. Tangible reinforcers such as token economies are introduced only after it has been demonstrated that the consistent use of social reinforcement or extinction will not achieve the desired behavioral outcome.

In an effort to equip teachers with a consistent treatment procedure, Dunn and associates have prescribed a set of consequences, or rules of thumb, that are applied to inappropriate behavior. There exists a specific rule of thumb for self-indulgent, noncompliant, aggressive, and self-stimulatory behaviors. **Self-indulgent** behaviors include crying, screaming, tantruming, and repetitive, irritating activities and noises. **Noncompliant** behaviors consist of instances when students say no when asked to do something. They also include forgetting or failing to do something because students choose not to do what is asked. In addition, noncompliance includes doing what is requested, but in a less than presentable way. Verbal or physical abuse directed toward an object or a person is considered **aggressive** behavior. More specifically, hitting, fighting, pinching, biting, pushing, and deliberately destroying someone's property are examples of aggressive acts. The **self-stimulation** category includes behaviors that interfere with learning because students become engrossed in the persevering nature of the activities. Examples include head banging, hand flapping, body rocking, and eye gouging.

The rules are as follows:

▶ Ignore students who engage in self-indulgent behaviors until the behavior is discontinued, and then socially reinforce the first occurrence of an appropriate behavior.

▶ Ignore noncompliant verbalizations; either lead students physically through the task or prevent students from participating in an activity until they follow through on the initial request. Compliance with any requests is immediately reinforced socially.

▶ Punish aggressive behavior immediately with a verbal reprimand, and remove the offending student from the activity. Social reinforcement is given when students demonstrate appropriate interaction with other persons or objects.

▶ The rule of thumb for self-stimulatory behavior is a formal behavioral program. An in-depth discussion of formal behavior modification principles and programs is presented in chapter 7.

INCLUSION

The inclusion of students with behavioral disorders and autism in the "regular" class should be based primarily on the concept of the least restrictive environment; a plan for inclusion should take into account the frequency and intensity of behavioral episodes. These students will demonstrate behavioral characteristics ranging from mild to severe. For those with mild behavioral profiles, the regular class is easily the placement of choice. The decision becomes more difficult if students' behaviors are unmanageable, or worse if they endanger either themselves or others in the class.

In terms of students with behavior disorders, inclusion is facilitated much of the time by development and implementation of a behavioral management plan (see discussion in chapter 7). These plans describe in detail student expectations and consequences if behavioral expectations are not achieved. When the student with BD is included in the regular classroom, the teacher can maintain appropriate behavior by moving around the classroom or gymnasium continuously while monitoring potential troublesome situations, remaining emotionally detached (not taking unruly behavior personally) (Henley, 1997), using cooperative learning and peer tutoring, holding discussions during class time to troubleshoot problem areas, and providing for planned opportunities to learn and practice desired behaviors (Carpenter & McKee-Higgins, 1996). Students who have severe behavior disorders such that they are either disruptive or interfere with the operation of the regular class and/or are harmful to other students in the class may require a segregated approach.

Much like BD, autism can range from mild to severe. At the mild end, students can benefit from placement in the regular class. Several strategies can make this a profitable placement, including having a trained aide accompany the student to the gymnasium; using peer tutors or developing a "buddy system"; establishing well-defined and executed transitions between and among activities; posting the daily schedule for the class and providing a schedule for the student with autism; using visual prompts and cues to indicate the activities for that day and the schedule that the student will follow; and having a structured teaching plan. Students who present severe behavioral profiles, such as self-injurious behavior (SIB), should not be included in the regular class unless an aide accompanies them and then only if the student is likely to be involved in the learning process.

SUMMARY

Behavioral conditions as used in this chapter correspond to the categories of behavioral disorders/emotional disturbance and autism as defined in IDEA. The psychodynamic, psychoeducational, ecological, psychoneurological, humanistic, and behavioral approaches were examined for their effectiveness in habilitating students with behavioral disabilities. Autism was considered, with special attention given to its major diagnostic features. Three major approaches—TEACCH, the Miller Method, and the behavioral method of Lovaas—were discussed. Suggestions were provided for teaching students with identified behavioral disorders in integrated physical education classes, including specific instructional strategies gleaned from the special education literature. Effective interpersonal communication was discussed, specifically, active listening, verbal mediation, and conflict resolution. The work of Hellison and the contributions of Dunn and his colleagues were cited as effective approaches to physical education for students with behavioral disabilities and students considered "at risk."

11

Visual Impairments and Deafness

Diane H. Craft ◆ Lauren Lieberman

▷ **M**R. JOSEPH HAS TWO STUDENTS with visual impairments included in his regular middle school physical education classes. Peter has partial vision due to retinitis pigmentosa. Sarah is totally blind. Both of her eyes were removed due to retinal blastoma. She now wears prosthetic eyes. Throughout the year, Mr. Joseph strives to teach units in which Peter and Sarah can participate with

▷ minimal adaptations and then include the students in regular physical education on a flexible schedule. Mr. Joseph gives both Peter and Sarah the option of deciding in which of the two physical education units offered they each would like to participate. Both students just finished the

▷ self-defense unit with Mr. Joseph. Now he is beginning a basketball unit. Across the hall, the other physical educator is beginning a swimming unit.

Sarah opts for swimming. She has no sight, making it difficult for her to play basketball and similar sports that depend heavily on seeing the ball. But Sarah can participate in swimming with

▷ few, if any, adaptations. Peter opts for basketball. His neighbors enjoy shooting hoops and, Peter hopes to improve his game enough that he can join them.

On this particular afternoon, Peter stood ready to receive the basketball but was not aggressively moving toward the ball. The student teacher made the teammates pass Peter the ball twice.

▷ In each case the ball was immediately stolen. Mr. Joseph, who thought all students were involved in his physical education class, saw for the first time today what the actual involvement was for Peter—15 seconds of contact with the ball over a 35-minute class period. Such a low level of participation went against Mr. Joseph's philosophy of teaching.

The next day he decided to take action to increase Peter's involvement in his physical education

▷ class. Mr. Joseph reread the chapter on blindness in his college text in adapted physical education, browsed the Internet for additional appropriate adaptations, and made a call to his district's adapted physical education specialist. One week later the scene in Peter's physical education class was quite

▷ different. First, two-thirds of the class time was now devoted to skill practice with the final third for game play. Rules were changed. To teach passing skills and cooperation, each team was now required to pass the ball to each team member before scoring. Peter's team always wore yellow jerseys to increase their visibility. There was a five-second rule when Peter had the ball—no defense for five seconds. On that day Peter scored two baskets for his team and passed the ball 10 times!

Just as those with sight, persons with visual impairments want to be accepted and respected as individuals, with the visual impairment as just one of many personal characteristics, not the defining characteristic. Educators should keep in mind this quote inspired by athlete Tim Willis (figure 11.1): "Lack of sight does not mean lack of vision." The challenge for physical educators, as it is for Mr. Joseph, is planning and teaching so that the individuals with visual impairments can actively participate in physical education, recreation, and sport in school and throughout their lifetimes. But sometimes the adaptations to minimize visual impairments are the least of the teachers' concerns. Often it is the emotional and social issues resulting from how others treat persons who are blind that present the greater challenge. It can be that learned helplessness—the countless acts of overprotection by family, teachers, and well-intended strangers and the eagerness of others to "do for" instead of expecting persons with visual impairments to do for themselves—can teach persons with visual impairments that they are helpless and very limited in what they can do for themselves. Much of the information in this chapter seeks to provide ideas for overcoming the "learned helplessness" as much as accom-

modating the actual visual impairment. So, the task for the physical educator is to make the activity adaptations as needed but also to expect students with visual impairments to do much on their own and to be active, contributing members of the class and community.

DEFINITION OF VISUAL IMPAIRMENT

What is a visual impairment? Who is blind? The educational definition from the regulations of the Individuals with Disabilities Education Act (IDEA) is as follows:

Visual impairment, including blindness, means an impairment in vision that, even when corrected, adversely affects a child's educational performance. The term includes both partial sight and blindness. (PL 105-17, Individuals with Disabilities Education Act, 1997)

These classifications of visual impairments are presented in box 11.1.

Most people who are blind still have some useful sight. Perhaps one in four individuals with visual im-

Figure 11.1   When at age 10 Tim Willis lost his sight, he decided that was all he would lose—and nothing more. Years later, as a Class B1 totally blind athlete, Tim has broken 13 national records and 2 world records, making him one of the top track athletes in the world (USABA, 1998).

pairments is totally blind. Among students with visual losses, about half of the students became blind before or at birth. The incidence of visual impairments among school-age children is low—so low that most physical educators will teach only a few students with visual impairments during their careers. Yet, these teachers will teach and influence those few for years.

CAUSES OF VISION LOSS

There are multiple causes of vision loss. While most causes are associated with aging, occasionally loss of vision occurs before or at birth (congenital) or in childhood or later (adventitious). Some of the causes of blindness are presented in the following list:

▶ Macular degeneration: Causes a loss of acuity and a blind spot in the middle of the vision field (figure 11.2a).

▶ Retinal blastoma: A form of cancer that often leads to the removal of the eye. A prosthetic eyeball is often worn.

▶ Rubella: During the third trimester of pregnancy, complications from rubella may cause limited usable vision.

▶ Albinism: Characterized by a lack of pigment in the iris and throughout the body. Eyes are sensitive to

light, and tinted glasses may need to be worn inside and outside to help reduce glare. Students with albinism may prefer to work away from windows and inside the gymnasium rather than in the glare of sunlight. Students may also have nystagmus in which there is a small, involuntary, rapid movement of the eyeball from side to side.

▶ Retinitis pigmentosa: An inherited, progressive disease in which first night blindness occurs and then peripheral vision may be lost (figure 11.2b)

▶ Ushers: A cause of deafblindness, this is an inherited condition resulting in a hearing loss (usually profound deafness) and is present at or soon after birth. There is also a progressive loss of vision caused by retinitis pigmentosa.

▶ Glaucoma: Caused by the inability of intraocular fluid to drain continuously. The resulting increase in pressure can eventually lead to total blindness. Visual loss may be gradual, sudden, or present at birth.

▶ Cataracts: Characterized by an opaque instead of a clear lens, possible sensitivity to light and glare, and squinting to compensate for impaired vision (figure 11.2c). Cataracts can occur as a consequence of rubella.

▶ Retinitis of prematurity: Occurs in some infants who are born prematurely, resulting in reduced acuity or total blindness.

▶ **Box 11.1 Classifications of Visual Impairments**

Visual impairment: Umbrella term encompassing total blindness and partial sight

Partial sight: Can read print through use of large print and/or magnification

Blind: Inability to read large print even with magnification

Legal blindness: Visual acuity of 20/200 or less in the better eye even with correction, or a field of vision so narrowed that the widest diameter of the visual field subtends an angular distance no greater than 20 degrees (20/200)

Travel vision: Ability to see at 5 to 10 feet what the normal eye can see at 200 feet (5/200 to 10/200)

Motion perception: Ability to see at 3 to 5 feet what the normal eye can see at 200 feet; this ability is limited almost entirely to the perception of motion

Light perception: Ability to distinguish a strong light at a distance of three feet from the eye, but the inability to detect a hand movement at three feet from the eye (<3/200)

Total blindness: Inability to recognize a strong light that is shone directly into the eyes

Figure 11.2 Three photos of what a person with a visual impairment might see. *(a)* In age-related macular degeneration, central vision is impaired, making it difficult to read or do close work. *(b)* In retinitis pigmentosa, night blindness develops, and tunnel vision is another frequent result. *(c)* In cataracts, a clouding of the lens causes reduced ability to see detail.

Reprinted, by permission, from I. Bailey and A. Hall, 1990, *Visual Impairment: An Overview* (NY: American Foundation for the Blind). Courtesy AFB.

CHARACTERISTICS OF PERSONS WITH VISUAL IMPAIRMENTS

There is tremendous diversity among individuals with visual impairments, but there are some characteristics that seem to occur with greater frequency than among individuals with sight. The following sections discuss some of these characteristics. The reader is cautioned to remember that these are only generalities and are not true of all persons with visual impairments.

Affective and Social Characteristics

Habits such as rocking, hand waving or finger flicking, or digging the fingers into the eyes are examples of repetitive movements some persons with visual loss or multiple disabilities develop. These repetitive movements are also known as blindisms or self-stimulation. While there is much debate as to the reason for these movements, adults can simply accept these movements as variations of nervous habits most people show. Or, parents and teachers may together decide that it is in the child's best interest to stop these movements in certain situations. If these movements are not affecting the child's education or those individuals around the child, then some self-stimulating behaviors may be best overlooked.

Fearfulness and *dependence* may characterize some individuals with visual losses, whether the loss is due to congenital or adventitious blindness. These characteristics may develop, not as a result of the lack of vision, but rather as a result of the *overprotection* experienced by many persons who are blind. This overprotection usually leads to reduced opportunities for students to freely explore their environments, thus creating possible delays in perceptual, motor, and cognitive development.

Motor, Physical, and Fitness Characteristics

The lack of sight does not directly cause any unique motor or physical characteristics. But the reduced opportunity to move, which often accompanies blindness, may result in many distinct characteristics. The following are possible motor, physical, and fitness characteristics of visual impairments and the corresponding implications.

Characteristic: The reduced opportunity for "rough and tumble" play with parents, heightened protective instincts of parents and caregivers, the infant's own fear of being moved suddenly, the lack of vision that motivates movement, and the lack of opportunity to observe others moving all may contribute to motor delays among students who are blind.

Implication: Students with visual losses *must* be provided with the opportunity and motivation to move in a safe environment to minimize the development of motor delays. Enthusiastically encourage and positively reinforce movements by physically demonstrating and verbally encouraging students that it is safe to move. Motor delays are not usually observed among individuals who are now blind but who were sighted for years. These individuals need corrective feedback from observers, as a substitute for monitoring their own movements, so that skills once mastered are retained.

Characteristic: Postural deviations may be prevalent among individuals with partial sight who may hold their heads in unique positions to maximize vision. Postural deviations may be more pronounced among individuals who have been blind since birth and never have had the opportunity to see how others sit, stand, and move. Examples include a shuffling gait, forward tilt, or rounded shoulders.

Implication: Corrective postural exercises may help improve posture and thereby reduce stress on the body. Verbal or physical prompts can remind the person to "sit up tall" and "head up."

Characteristic: Body image and balance may also be less developed among students who are blind. This may be due to decreased opportunities for regular physical activity through which balance and body image are refined. Also, sight enhances balance by "seeing" a reference point for aiding stability.

Implications: Participation in activities such as dance, yoga, and movement education can be excellent means of developing body image and balance.

Characteristic: A slow shuffling gait characterizes many persons who are blind. Try closing your eyes and walking across the room, down the hall, and outside. (Ask a friend to come along to watch for safety.) It is likely that when unaided by sight, your gait showed shorter strides, a pronounced shuffle, slower pace, more time spent in the support phase, and a tendency, over a distance, to veer away from the stronger leg.

Implication: Physical educators can increase stride length by generating greater hip extension during the drive phase and greater hip flexion during the recovery stage of the sprint run (Arnhold & McGrain, 1985; Wyatt & Ng, 1997).

Characteristic: Health-related fitness levels of individuals with visual impairments are generally below those of their sighted peers (Lieberman & Carron, 1998). This may be of particular concern because Buell (1966) suggests that individuals with visual impairments need higher levels of physical fitness than their sighted counterparts. He notes that there are many instances in which individuals without sight need to spend more energy to reach the same goals as those with sight.

Implication: Health-related fitness levels of persons who are blind have the potential to match those of their sighted peers, with the possible exception of running due to the inefficiency of running with a partner. Lack of opportunity to train and difficulty arranging for sighted guides for running can be obstacles to developing fitness. It is very important for physical educators to develop fitness programs for students with visual impairments and to work with families in finding fitness activities students can pursue at home.

In summary, the missing component in the development of normal patterns of movement and fitness among students with visual impairments is *experience*, not ability. If it is lack of experience rather than lack of ability that delays motor development for individuals with visual impairments, then it is the physical educator's responsibility to provide movement experiences and to stimulate and motivate students to move. The physical educator must also encourage the students to *feel safe and good about their movements*. This can be accomplished both through instruction in physical education and through collaboration with parents/caregivers and other influential persons in the child's life. The following section offers suggestions for helping students with visual impairments enjoy physical activity.

INCLUSION: TEACHING STUDENTS WITH VISUAL IMPAIRMENTS IN REGULAR PHYSICAL EDUCATION CLASSES

With the current emphasis on inclusive education, many students with disabilities are taught in regular physical education classes. This inclusive approach can work very well if support systems are provided as needed. When teaching a student with a visual impairment in physical education, keep in mind that this student *can* do many of the same physical activities that sighted peers can do. Simple modifications, such as changing the ball color to one that contrasts more sharply with the background, are sometimes all that are needed to enable the student with partial sight to participate fully in class. Other times, more extensive support systems are needed, such as team teaching with an adapted physical education specialist, to help a student with visual impairments and severe behavioral disorder to remain focused and on task throughout the class. All too often students with visual impairments are placed in regular physical education classes without the necessary support systems. This "dumping" under the guise of inclusion is unfair to the student, the classmates, and the teacher. A result of this dumping can be the teenager who is blind and has never bicycled (tandem), roller skated, or jogged with a friend; has never danced, bowled, or done a host of other physical activities that are well within the person's capabilities but beyond her experience. What can we as teachers do to assure that the inconvenience of a visual impairment does not lead to a person who is physically inactive and inexperienced? How can we best provide physical education to students who are blind in inclusive settings? Here are some examples of support systems:

▶ Trained peer tutors or volunteers to assist in providing kinesthetic and auditory cues to students with visual impairments

▶ Specialized equipment, such as audible balls and buzzers

▶ Services of a teaching assistant, mobility instructor, physical therapist, or low-vision specialist, if such a person is needed to include a student with multiple challenges, including a visual impairment, in physical education

Enlist the assistance of parents and other caregivers in advocating for appropriate support systems. Also, collaborate with others. Problem solve with a resource specialist in adapted physical education or with physical education teachers at residential schools for students with visual impairments. Seek to have any needed support systems recorded on the student's Individualized Education Program (IEP).

Fortunately, the teacher is not the only source of ideas for adaptations. The student with the visual limitation and all students in the class are also excellent sources of ideas for adaptations. Teachers have had much success in developing accommodations by making it a class effort—How can we modify this game so that (student with visual impairment) can participate in a meaningful way?

Generally, the same developmentally appropriate, individualized curriculum used for all physical education students can be used to meet the needs and allow the inclusion of a student with a visual impairment. Developmentally inappropriate curricula, such as those that emphasize competitive sports and focus only on "playing the game," showing preference to the "motor elite," cannot be readily adapted to include students with visual impairments. When students with visual impairments repeatedly serve as human cones, scorekeepers, and spectators or are excused from physical education to go to the library, they are not participating in meaningful, active ways. Physical educators may unwittingly do disservice to students who have visual impairments when segregating them from participation in class activities. Individuals with visual impairments have the same need to move and can gain the same delight from movement as others enjoy. The skill in and love of movement can be developed through active participation in physical education. Students with visual impairments should also be introduced to sports, games, and activities. Lifetime activities such as tandem biking, running, goal ball, wrestling, judo, and bowling should be emphasized.

The following three sections will present ideas for adapting instruction for students with visual impairments based on learning about the student's abilities, fostering the student's independence, and exploring options for instructional modifications. Please refer to the Inclusion of a Student With Impaired Vision application example.

Learning About the Students' Abilities

Assess to determine each student's present level of performance. Generally, students with visual impairments are capable of taking the same formal assessments and attaining the same standards as students with sight. In case of mobility limitations due to blindness, teachers can *modify* mobility test item standards, use assistive devices, or use sighted guides based on suggestions throughout this chapter. For example, auditory cues may be added so students know where to throw and how far to skip or run. Caution should be used when interpreting test norms that do not include students with visual impairments or are based on test items that have been modified. The Brockport Physical Fitness Test (Winnick & Short, 1999) assesses the health-related fitness of youngsters who are blind. Refer to chapter 5 for additional information on measurement and evaluation. In addition to standardized assessment tests, educators may also utilize authentic assessment techniques. Authentic assessment instruments are developed based on the curriculum and particular unit of instruction (Block, Lieberman, & Conner-Kuntz, 1998).

The teacher may seek to *learn the answers to the following five critical questions* in addition to assessing current performance:

▶What can the student see? Ask the student, "*What can you see?*" rather than "How much can you see?" Ask the same question of others familiar with the student's vision including the student's previous teachers, parent/caregiver, and low-vision specialist. Read the student's educational file for further information regarding *what* the student can see.

▶When (at what age) was the loss of vision experienced, and over what period of time did it progress? Is it still progressing? Ask the student when the vision loss was experienced. If a student gradually lost all vision due to disease from the ages of 8 to 10, the student may require less time preparing for specific activities, such as hitting a sound-emitting softball and running to a sound-emitting base, than a student who is congenitally blind and has never had the chance to see a game of baseball. A student with congenital blindness may need more detailed explanations that do not depend on analogies for which the student has no basis of understanding. For example, the teaching cue "jump like a bunny" may not be useful to a child who does not have any visual memory of rabbits moving. The teacher may wish to teach the child how to "jump like a bunny."

▶How can I maximize use of existent sight? Learn from the students what helps them see best. For most visual impairments bright lighting maximizes vision. For some conditions, such as glaucoma and albinism, however, glare is a problem. Teach these students in lighting free of glare to maximize vision. When outside, these students may need tinted glasses or a hat to reduce glare.

Application Example: Inclusion of a Student With Impaired Vision

Setting: Middle school physical education class

Student: Andrea is a very active sixth grade student who enjoys biking, swimming, in-line skating, and studying judo after school and on weekends. She has retinitis pigmentosa resulting in a progressive visual impairment. She has some residual vision and is adjusting to her current visual ability.

Issue: Andrea's IEP states that she must be fully included in her two 45-minute physical education classes each week, but her teacher believes that her blindness limits her abilities. He has her exercising on a stationary bike and walking the perimeter of the gym while her classmates are involved in a variety of physical activities and games. Andrea is becoming more and more frustrated and it is affecting her performance in other subject areas.

Application: The following was determined as a result of a meeting with the school administrator, the teacher of the visually impaired, and Andrea's physical educator:

  ▶ Andrea will participate in all physical education activities with her peers.

  ▶ To assist him in adapting games, sports, and activities, the physical educator will purchase books such as *Games for People with Sensory Impairments*, *Adapted Physical Education and Sport*, and *Adapted Physical Activity, Recreation and Sport*.

  ▶ The teacher will search the Internet for information and adaptations to assist in full inclusion of Andrea.

  ▶ The school will buy brighter balls, auditory balls, and set up a guide wire system in the gymnasium and outside on the track.

  ▶ The curriculum will include a variety of games and activities conducive to full participation by Andrea, such as weight training, aerobics, goal ball, and cross-country skiing.

  ▶ Three of Andrea's friends will be trained as peer tutors to assist her in class.

  ▶ An additional class with supplemental physical education instruction will be offered should Andrea need more time to grasp a skill or activity.

▶Are there any contraindicated physical activities? To determine if there are any contraindicated (not recommended) physical activities, seek medical consultation. Note the cause of blindness. Consult reference texts on the etiology of visual impairments found in libraries, and read about the specific visual impairment the student has. There are few contraindications for individuals with total blindness, as there is no sight to seek to preserve. For persons with partial sight there might be activity restrictions imposed in an effort to preserve the remaining sight. Discuss the condition with the student's eye specialist to determine what, if any, restrictions are necessary, especially following recent eye surgery. For example, jarring movements that could cause further detachment are usually contraindicated with a detached retina. Contact sports, as well as diving and swimming underwater, may need to be avoided. Glaucoma occurs when the fluid within the eye is unable to drain and pressure increases within the eye, causing blindness. Inverted positions and swimming underwater are often contraindicated with glaucoma because of increased pressure on the eye.

▶What are the student's favorite scholastic, social, and physical activities? Also ask what adaptations the student prefers, for example, How do you like to run? With a sighted guide? Using a guide wire? Independently on a clearly lined track? Learn about the opportunities that exist for the child to participate in physical activities with family, friends, and in the community. Seek to incorporate these activities and adaptations into the child's physical education program.

Fostering the Students' Independence

Consider the following suggestions for teaching.

Develop positive attitudes toward students with visual impairments. The *attitude* of the teacher is the determining factor in the classroom. If the teacher is truly interested in teaching all students, including a student with a visual impairment, and if this teacher makes accommodations to teach to the diverse needs of all students without a fuss, then the students in the class will pick up on this attitude and will be more likely to also accept students with visual impairments or other differences.

Encourage participation of students with visual impairments in physical activities. To encourage participation, discuss with students how they feel about various activities. Respect any fears they might express, and work with the students to create an environment in which they feel safe and are more willing to participate in physical activity. The following incident may help illustrate the importance of talking with one student to understand her concerns and encouraging her to participate. A young girl who had been totally blind since birth performed very well in swimming. But she seemed unable to learn to swim underwater. When asked why she had such difficulty learning to swim underwater, she replied, "No one will see me if I go underwater. That scares me." After it was explained that water is clear and can be seen through, the girl was reassured and soon learned to swim underwater.

Help parents see their children's abilities. Parents are most children's biggest advocates. Based on their fears for the child who does not have use of a major sense organ, some caregivers may dwell on what the child can't do rather than on what the child can do. In these cases the child will rarely have the opportunity at home to try new things, play games that sighted children play, or take risks. Physical educators can *educate parents* about their children's capabilities by sending home data used in assessment, personal notes or newsletters describing students' accomplishments, photographs of the student doing activities, ribbons and certificates for accomplishments, and lists of the student's favorite physical activities. By learning about their children's current performance and abilities, parents may have a better understanding of their children's potential and may allow or even encourage more opportunity for physical activity at home and in the community.

Challenge students with visual impairments so they might feel successful. Reward and recognize all students' accomplishments to help *improve self-esteem* and *motivate* them to continue to work on their next goals. Teachers help as they work with the student to set realistic goals and implement a reward system to recognize achievement, such as through bulletin boards, student newspapers, and announcements.

Expect the student to move as independently as possible during physical education. At the beginning of the school year, thoroughly orient the person with a visual impairment to playing fields, gymnasiums, and equipment to increase independence and feelings of security. Together, identify *landmarks* that can help students orient themselves. For example, a landmark might be the mats along the walls at either end of the gymnasium in contrast to the paneling along the side walls. Enable individuals to walk around and touch everything as often as needed to create the visual map in their minds that will enable them to negotiate the area with confidence. Always keep equipment in the same position within the gymnasium. If a change is necessary, forewarn and reorient students. In the locker room, provide a lock that opens with a key rather than a combination for a student with a visual impairment. A trained peer tutor may be most helpful with mobility, skill acquisition, and feedback for children who are totally blind. If scheduling permits, meet with the student and peer tutor a few minutes before class to introduce the student to the concepts and movements to be taught that day. Box 11.2 offers more suggestions of things to keep in mind when teaching students with visual impairments.

Young children with visual impairments need additional incentives to move. Fundamental phylogenetic

When Teaching Students Who Are Blind, Remember...

► What seems like ordinary, everyday happenings may need to be explained. Narrate during a game so the students can understand what everyone is doing.

► Some experiences are outside of your child's direct experience. Tell the child about the birds that just flew overhead and allow the child to touch a bird.

► Your child may need help in putting parts together to form a whole. Allow the child to feel the entire playground set, the entire gymnasium space, and the entire trampoline.

► Imitation is harder. Physically guide children through movements rather than show them.

► Feedback is needed because your children cannot always tell how they are doing. Tell the children where they threw the ball. Even better, tie bells on the target so children can hear when they hit it (modified from Ferrell, 1984).

skills (those skills that humans typically develop such as crawling, sitting, walking, etc.) are more difficult to learn due to the deprivation of the most motivating sense—sight. Parents, specialists, teachers, and therapists must work together to give the child a reason to move and develop basic motor skills such as sitting, scooting, crawling, rolling, and walking. The following are some suggestions to instill a yearning for movement in the young child with a visual impairment:

► Bell Balloon Bash: The child chases a bell balloon around the room while crawling, walking, or running. Guide the child, if necessary, and encourage the child to kick the balloon when possible (Lieberman & Cowart, 1996).

► Parachute Swing: Two people swing a parachute while the child is inside.

► Incline Roll: Place the child on top of a low incline and allow the child to roll or crawl to a motivating sound source at the bottom.

► Scooter Pull: The child sits on a scooter and holds one part of a hula hoop. The teacher holds the opposite part of the hula hoop and pulls the child around the gymnasium.

► Therapy Ball Push: In an assisted sitting position (an adult sits directly behind the child, helping to support the child's back), the child pushes a large, heavy therapy ball to a sound source or a person. This builds upper-body strength to assist in crawling and creeping.

► Rebounder Heaven: The child jumps on a rebounder (minitrampoline) while holding the wall, the teacher, or a bar for support.

► Movement Exploration: Place the child on a mat and ask the child to move forward, backward, high, low, fast, slow, and so on.

Exploring Options for Instructional Modification

So much of what a person learns is not directly taught, but rather learned through watching and listening to what is happening around them. Information that is not directly taught but rather observed from the surroundings is called incidental learning. When there is a visual impairment, the opportunity for incidental learning is dramatically reduced. Teachers need to directly teach much more information to a child with a visual impairment than other children. Use every opportunity to explain what is happening around the child and why.

The following system of least prompts (Dunn, Morehouse, & Fredericks, 1986) provides examples specific to teaching children with visual impairments. Note that the instructor may use one or more of the following teaching strategies to help the student perform the skill.

Verbal explanations by the instructor

► Explain what the child is to do in simple terms.
► Use the child's preferred mode of communication.
► If the child does not understand the first time, repeat in a different way.
► If the child has any usable vision, demonstrate to increase understanding.
► Give feedback using precise, unambiguous descriptions. A statement such as "Hold the racket three to four inches above your left shoulder" provides more feedback than "Hold the racket like this." Use of precise language benefits both students who are visually impaired and sighted.
► Include students with visual impairments during spectator events by assigning a student "announcer" to describe the action for spectators, much as a radio announcer describes a ball game. Select an announcer with a lively sense of humor to make the event more fun for everyone.

Demonstrations by the instructor

► Show the child the desired skill or movement.
► Demonstrate within the child's field of vision.
► Ask someone close to the student's size and ability to model.
► Use whole-part-whole teaching when possible. Demonstrate the whole skill, then the parts (based on task analysis), and then the whole task again.

Physical assistance or guidance from the instructor

▶ Assist the student physically through the movement.

▶ Record which skills require physical assistance, including how much and where on the student's body assistance was needed. If asked for legal purposes, the teacher can explain when, where, and why the teacher touched a student.

▶ Forewarn the student before giving physical assistance, to avoid startling the student.

▶ Fade assistance to minimal physical prompts as soon as possible.

Brailling

▶ Allow the student to feel a peer or the instructor execute a skill or movement that was previously difficult to learn using the three previous approaches (Lieberman & Cowart, 1996).

▶ Tell the student where and when to feel you or a peer executing a skill.

▶ Document how much assistance was given, when and where the student felt you or a peer, and why, for legal purposes.

▶ Repeat brailling as many times as necessary to ensure understanding.

▶ Combine brailling with the other teaching methods to increase understanding.

Add sound devices

▶ For softball, use a large playground ball that is hit off a bounce, or use a beeper ball (a ball with an audible buzzer inside, available for purchase through Telephone Pioneers of America). The base coach calls to direct the batter to the base.

▶ Make playground balls audible by cutting the ball, inserting bells, and then resealing the ball with a bicycle tire patch.

▶ Make scoring a goal audible by tying bells onto net goals. Everyone can hear the jingling sound when a goal is scored.

Enhance visual cues

▶ Most individuals with visual impairments do have some *residual vision*. Evaluate each activity to decide what types of visual cues are needed and how to highlight these cues. Color, contrast, and lighting are important. Be sure to ask students what enhances their vision.

▶ Use colored tape to increase the contrast of equipment with the background, such as high jump standards and poles or the edges of a balance beam.

▶ Use brightly colored balls, mats, field markers, and goals that contrast with the background for most students with visual impairments. Individuals with albinism and glaucoma, however, need solid colored objects under nonglare lights.

▶ Make the gymnasium lighting brighter (or darker for students who have difficulty with glare or who tend to self-stimulate on the bright lights).

SPORTS FOR ATHLETES WITH VISUAL IMPAIRMENTS

There are two major organizations that work with athletes who have visual impairments. The goal of the United States Association for Blind Athletes (USABA) and the International Blind Sport Organization (IBSA) is to promote competitive sport opportunities for athletes with visual impairments throughout the world. These organizations also seek to change attitudes toward individuals with visual impairments. The motto of the USABA is "If I can do this, I can do anything."

USABA

USABA is the major sport organization in the United States for athletes 14 years of age and older who have a visual impairment. Organized in 1976, USABA provides competitive sport opportunities at the local, state, regional, national, and international level. Persons with visual impairments may choose to participate in integrated sports with their sighted peers or in sports exclusively for individuals with visual impairments, or both. Many individuals with visual impairments enjoy participation in sports for the blind because of the opportunity to meet and compete with others who are also blind. Physical educators are encouraged to contact USABA for the location of the nearest local sport organization for athletes who are blind, and then share this information with students who are blind.

While there is increased recognition and opportunities for elite athletes who are blind, there remain few programs designed to bring youth with visual impairments into sport. Physical educators can help by reviewing their physical education curriculum, noting those sports offered that are also USABA sports. While teaching these sports, physical educators are encouraged to make an extra effort to promote the sport among any students with visual impairments and then assist them in learning how to pursue this sport through USABA.

USABA offers competition in nine sports:

Alpine and Nordic skiing

Athletics (track and field)

Goal ball

Judo

Figure 11.3   Running using *(a)* a sighted guide, *(b)* a tether, and *(c)* a guide wire.

Powerlifting

Swimming

Tandem cycling

Wrestling

USABA athletes who reach an elite athletic level compete in the World Blind Championships and in the summer and winter Paralympics.

USABA classification for competition is based on residual vision, as shown in table 11.1. Rules for each sport are modified slightly from those established by the national sport organizations. For example, track and field follows most National Collegiate Athletic Association (NCAA) rules, except that *guide wires* are used in sprints; *sighted guides* may be used on distance runs; hurdles are eliminated; and jumpers who are totally blind touch the high bar and then back off and use a one- or two-step approach. The use of guides depends entirely on the athlete's visual classification and the particular event. Guides facilitate the activity by running alongside the athlete, with both runners holding on to a tether. Alternatively, *stationary guides* are positioned around the track to call directional signals to the runner. The following list provides descriptions of various guiding techniques for running:

▶ Sighted guide: The runner grasps the guide's elbow, shoulder, or hand depending upon what is most comfortable for the runner and guide (figure 11.3a).

▶ Tether: The runner and guide grasp a tether—a short string, towel, or shoelace. This allows the runner full range of motion of the arms, while remaining in close proximity to the sighted runner (figure 11.3b).

▶ Guide wire: The runner holds onto a guide wire and runs independently for time or distance. A guide wire is a rope or wire pulled tightly across a gymnasium or track. A rope loop, metal ring, or metal handle ensures that the individual will not receive a rope burn and allows for optimal performance. The runner holds onto the sliding device and can run for as long as he wishes independently. Please note that guide wires can be set up permanently or temporarily (figure 11.3c).

▶ Sound source from a distance: The runner runs to a sound source such as a clap or a bell. This can be done as a one-time sprint or continued for a distance run.

▶ Sound source: The guide rings bells or shakes a noisemaker for the runner to hear while they run side by side. This works best in areas with limited background noise.

▶ Sighted guides shirt: The runner with partial vision runs behind a guide with a bright shirt. This must be done in areas where it is not too crowded. Be sure to pick an easily recognized color.

Table 11.1 USABA Classification for Sport Competition

Level	Classification
B1	From no light perception at all in either eye up to light perception and inability to recognize objects or contours in any direction and at any distance
B2	From ability to recognize objects or contours up to a visual acuity of 2/60 and/or limited visual field of 5 degrees
B3	2/60 to 6/60 (20/200) vision and/or field of vision between 5 and 20 degrees

▶ Independent running: A runner with travel vision runs independently on a track marked with thick white lines.

▶ Treadmill: Running on a treadmill provides a controlled and safe environment. Select a treadmill with the safety feature of an emergency stop.

▶ Wheelchair racing: An individual who is blind and in a wheelchair can use any of the above adaptations as needed. Aerobic conditioning results from pushing over long distances, whether around the track, on neighborhood sidewalks, or along a paved path.

Wrestling rules are modified slightly to require that opponents maintain physical contact throughout the match. Wrestlers with visual impairments have a long history of victories and state championships against sighted opponents (Buell, 1966).

Women gymnasts compete according to United States Gymnastics Federation rules, except that vaulters who are totally blind may start with their hands on the horse and use a two-bounce takeoff, coaches on the balance beam may warn competitors when they near the end of the beam and no jumps are used, floor exer-cise competitors may count their steps to the edge of the mat, and music may be placed anywhere near the mat to aid directionality.

Swimming also follows NCAA rules. Athletes commonly count their strokes so that they can anticipate the pool's end. Coaches may also tap a swimmer, using a long pole with soft material at the end such as a tennis ball, to signal the upcoming end of the pool. On the backstroke, flags are hung low over the pool to brush the swimmers' arms to signal the end of the pool. When necessary, a spotter may use a kickboard to protect a swimmer's head.

Goal ball is a sport specifically designed to be played by athletes with visual impairments. The object of the game is to roll a ball that contains bells past the opposing team's end line. Two teams of three players each compete on a rectangular playing area (figure 11.4). Rope covered with wide tape is placed along the gymnasium floor to create boundaries that players can feel. All players wear blindfolds so they must listen to determine the location of the ball and then stretch, dive, or lunge to stop the oncoming ball with any part of their bodies. Physical educators may wish to play goal ball in an integrated class to reverse the usual situation. With everyone wearing blindfolds, students with visual impairments may outperform students who are sighted due to increased opportunity to develop auditory perception. Additional information on goal ball is provided in chapter 25.

International Blind Sports and Recreation Association

The international counterpart of USABA is IBSA. For international competition, USABA athletes participate in the Paralympic Games and the World Blind Championships.

National Beep Baseball Association

Beep baseball is a popular modification of baseball, governed by the National Beep Baseball Association (NBBA). Competition culminates with the NBBA World Series. Further details on beep baseball are provided in chapter 25.

The next topic in this chapter is working with individuals who are deaf or hard of hearing. We begin with a scenario describing a teacher working with a student who is Deaf.

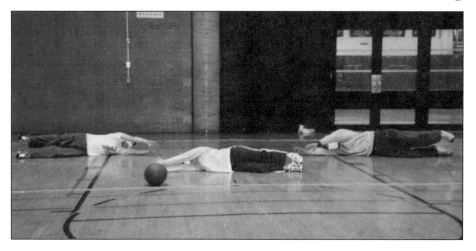

Figure 11.4   Physical educators might introduce goal ball in an integrated class as a challenge to students who are sighted.

▷ **M**RS. GOODWIN IS AN ELEMENTARY physical education teacher. She has been teaching for three years and she loves what she is doing. At the beginning of her fourth year, she had a class of second graders including a little girl named Rachel. Rachel is Deaf and uses sign language to communicate. She has a full-time educational interpreter with her named Mr. Colgan. Mrs. Goodwin had ▷ never taught a child with a disability and this was certainly a surprise. She wanted to make sure she did the right thing, so she referred back to her college text for a course in adapted physical education. She read about the minor instructional modifications she could make to ensure Rachel learned everything she was teaching. Mrs. Goodwin made sure she had picture descriptions and clear ex- ▷ planations at all the stations for her fitness warm-up. She flashed the lights on and off once to signal when she wanted the students to stop and look at her. Mrs. Goodwin made sure the class was in a semicircle when she instructed, and she used lots of demonstrations with checks for un- ▷ derstanding along the way.

▷ Mrs. Goodwin felt Rachel was learning everything she was teaching, yet she did not feel as though she knew her well, and Rachel was always the last person selected when the children chose partners. Mr. Colgan and Mrs. Goodwin decided that learning Rachel's language would solve some of these problems. Mrs. Goodwin knew that she needed to learn not only the signs to communicate ▷ to Rachel (expressive language) but also to understand what Rachel was communicating back to her (receptive language). To achieve this, Mrs. Goodwin took a few minutes out of each day and learned the signs for that day's lesson. When she was teaching kicking, she learned signs such as kick, hard, soft, far, short, stop, start, good job, more, and try again. Each class she learned a few signs, ▷ and she and Rachel taught her class the signs as well. Some of the kids were starting to pick up other signs that Mr. Colgan shared with them throughout the day, and before long, Rachel could communicate with a few of her peers. Mrs. Goodwin was becoming skilled at the physical education signs, and she noticed Rachel was trying harder than ever! She was one of the highest-skilled ▷ students in the class, and the next time Mrs. Goodwin had the students choose a partner, three students were fighting over who would be Rachel's partner!

DEFINITIONS OF HEARING LOSSES

Being Deaf in the United States today often means being a member of a subculture of American society that has its own language, its own customs, and its own way of perceiving the Deaf person's role within the hearing world. Understanding Deaf culture's perspective may be beneficial to the effective teaching of Deaf students in physical education. Awareness of this perspective might begin with the knowledge that *many Deaf people do not consider themselves disabled.* Unlike members of most populations with disabilities, most who are Deaf do not want "person first" terminology used to describe them. Many Deaf individuals prefer to be called a "Deaf person" rather than a "person who is Deaf." Dolnick (1993) notes that the use of the uppercase "D" in the word "Deaf" is a succinct proclamation by the Deaf that they share more than a medical condition; they share a culture and a language—sign language. *Hearing impairment,* a term commonly used to describe some malfunction of the auditory mechanism, *is not preferred by many Deaf individuals, because it also suggests that deafness is a disability.*

Deaf refers to a hearing loss in which hearing is insufficient for comprehension of auditory information,

with or without the use of a hearing aid. IDEA defines deaf as having a hearing loss that is so severe that the student is unable to process language through hearing, with or without the use of an amplification device. The loss must be severe enough to adversely affect the student's educational performance. (Individuals with Disabilities Education Act, 1997).

Hard of hearing refers to a hearing loss that makes understanding speech through the ear alone difficult, but not impossible. Amplification using a hearing aid and/or remedial help in communication skills are usually used by hard of hearing persons. IDEA defines hard of hearing as having a hearing loss that may be permanent or fluctuating and adversely affects the student's educational achievement or performance (Individuals with Disabilities Education Act, 1997).

Most people with hearing losses are hard of hearing, not totally Deaf. It is also important to note that two students may have the same degree and pattern of hearing loss but may utilize their residual hearing differently due to differences in age when hearing loss occurred. Motivation, intelligence, presence of disabilities, environmental stimulation, and response to a training program may also affect the degree to which residual hearing is utilized.

Amounts of hearing loss and residual hearing are both described in terms of decibel (dB) levels. The ability to

Table 11.2   Degrees of Hearing Loss in Decibels (dB)

Hearing threshold	Degrees of hearing loss	Difficulty understanding the following speech/ levels of loudness
27-40 dB	Slight—conductive loss	Faint speech
41-55 dB	Mild—use of hearing aid	Normal speech
56-70 dB	Marked	Loud speech
71-90 dB	Severe—sensory-neural loss	Shouted speech
Greater than 90 dB	Profound—use of signing	Any speech, even amplified

detect sounds in the 0 to 25 dB range is considered normal for children. Ordinary conversation occurs in the 40 to 50 dB range, while noises in the 125 to 140 dB range are painfully loud. Degrees of hearing loss are presented in table 11.2.

The age at which hearing loss occurs can greatly influence the method of communication used. Prelingual deafness refers to the condition of persons who were deaf at birth or before the development of speech and language. Postlingual deafness refers to the condition of persons whose hearing loss occurred at an age after spontaneously developing speech and language. The first three years of life are when most children learn to understand and use language in the form of oral speech. If a person is totally Deaf before learning to speak, the task of developing oral speech may be close to impossible. A prelingual Deaf person with speechreading (lipreading) training is usually no more proficient at speechreading than an untrained hearing person. (Simulate this by turning off the sound while watching people on television and try to follow what is being said. Now imagine trying to follow what is being said if you had never actually heard the words and patterns of words that would need to be seen on the lips.) Due to the difficulty most prelingual Deaf persons have in speaking and speechreading, many use sign language in place of verbal communication. In contrast, people who have already developed speech before the hearing loss occurred usually can retain intelligible speech, often with the help of additional training. It is also possible for these people to speechread with proficiency.

TYPES AND CAUSES OF HEARING LOSS

Among Deaf students, approximately two thirds have congenital deafness (present at birth) and one third have acquired deafness (develops sometime after birth).

Medical advances have enabled more severely premature babies and children with meningitis and encephalitis to survive. These survivors may have multiple disabilities, including deafness.

There are three major types of hearing loss: conductive, sensory-neural, and mixed. With a conductive loss, sound is not transmitted well to the inner ear. It is analogous to a radio with the volume on low. The words are faint but there is no distortion. Some children with conductive loss may have intelligible speech. A conductive loss can be corrected surgically or medically, because it is a mechanical problem in which nerves remain undamaged. Often hearing aids may be beneficial to increase volume. In some cases a child may be fitted with a cochlear implant to help make words clearer. This is a controversial procedure among the Deaf community.

A frequently observed condition is serous otitis media, or middle ear effusion. This condition is often treated by placing a plastic tube through the eardrum for several months to allow congealed fluid to drain away from sound-conducting inner ear bones.

A sensory-neural hearing loss is due to nerve damage. In comparison to a conductive loss, sensory-neural loss is much more severe and likely to be permanent. Children with a sensory-neural loss will have more difficulties with speech than those with a conductive loss. It is analogous to a radio that is not well-tuned. Sensory-neural losses affect fidelity as well as loudness, so there is distortion of sounds. The words may be loud, but they are distorted and garbled. While raising one's voice or using a hearing aid may help the voice be heard, the words still may not be understood. Typically, low-pitched vowel sounds are heard, but the high pitch consonants such as "t," "p," and "k" are not heard clearly. This makes it difficult to distinguish words such as "pop" from "top." Also note that only 20 to 30 percent of speech is visible on the lips. Injuries, allergies to drugs, repeated exposure to loud music, infections such as herpes viruses, and toxoplasmosis can all cause hearing loss. A mixed loss is a combination of both conductive and sensory-neural losses.

CHARACTERISTICS OF STUDENTS WITH DEAFNESS

Individuals who are classified as hard of hearing typically share the same characteristics as the general population. Their hearing impairments are mild and may not present major obstacles to speech. But most people who are postlingual profoundly deaf or prelingual severely or profoundly deaf may have unique characteristics due to the need to communicate through a means other than spoken language. These characteristics are the focus of the following section.

Language and Cultural Characteristics

American Sign Language (ASL) is the preferred means of communication within the Deaf culture. This shared language is the basis of the shared identity in the Deaf culture. Just like English, ASL is a language used to communicate, having its own grammar and structure to convey subtle nuances of abstractions, in addition to describing concrete objects. Most hearing people see only the signs that name objects and directions, such as found in figure 11.5, and are unaware of the subtleties of ASL. Currently, most prelingual Deaf children do not develop intelligible speech despite the best-known teaching methods. So, communication between hearing and Deaf people will remain a major problem until many more hearing persons learn sign. Courtesies to aid communication

> ### Box 11.3 When Communicating With a Deaf Person, Remember to

- maintain eye contact throughout the conversation,
- use paper and pencil to augment conversation,
- signal that you understand only when you really do (do not pretend to understand),
- use polite ways to gain a Deaf person's attention,
- learn to use a teletypewriter (TTY) to transmit typewritten words over the telephone lines to another TTY,
- discourage interruptions to the conversation,
- correct a Deaf person's English only if asked, and
- appreciate that many people's hearing is not improved through the use of hearing aids (Graybill & Cokley, 1993).

among hearing and Deaf persons are presented in box 11.3.

Schools for the Deaf offer a bicultural and bilingual (BI/BI) education, in which ASL is the primary language and English is taught as a second language. ASL and English are completely different languages. Deaf students who use ASL as their primary language and are learning English as a second language face challenges similar to anyone who is learning English as a second language.

Deaf students in the hearing community have decreased opportunities for incidental learning, because they cannot overhear conversations. Rarely do hearing parents, teachers, and friends sign when not addressing Deaf children. So there is little opportunity to "oversee" conversations, and continuity with life's events is often missing.

Behavioral and Affective Characteristics

As many as 1 in 16 students possesses a mild form of hearing loss that can place them at a disadvantage in school (Kottke & Lehmann, 1990). Sometimes these students are regarded as slow learners or behavior problems when their inappropriate behavior is actually due to an undetected mild hearing loss. The incidence of *impulsivity* seems to be greater among Deaf than among hearing students. Perhaps this is because Deaf students learn visually and may want to look around to check their surroundings more frequently. According to J. Marshall (personal communication, February 8, 1998), impulsivity could also be associated with deafness, because the inner ear is positioned over a conjunction of pathways to the brain, including the emotional pathway.

Motor Characteristics

If the semicircular canals of the inner ear are damaged as a part of the hearing loss, *balance problems* are likely. These balance problems may in turn cause developmental delays and motor ability delays. These balance problems occur as a result of vestibular damage, not deafness (Myklebust, 1946). Research results that show poor motor performance for Deaf students as a group could reflect the very poor performance of those students with vestibular damage along with the average performance of the rest of the Deaf students (Schmidt, 1985). Given equal opportunity to learn movements and participate in physical activity, Deaf children should equal their same-age peers in motor skills. If these children are not afforded equal opportunity, they may lag behind in motor skills. Fitness scores of hearing and Deaf students are not significantly different (Winnick & Short, 1986).

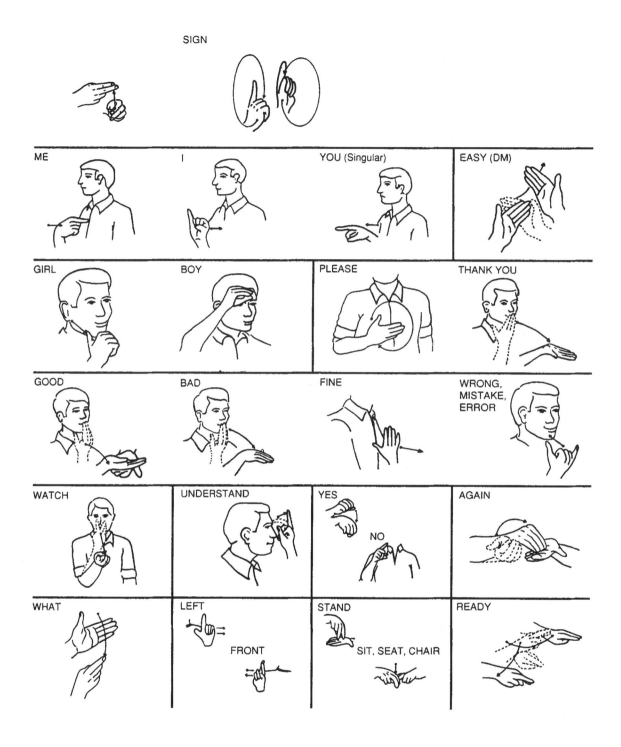

Figure 11.5 Signing for physical education.

Reprinted by permission of the American Alliance for Health, Physical Education, Recreation and Dance, 1900 Association Drive, Reston, VA 22091.

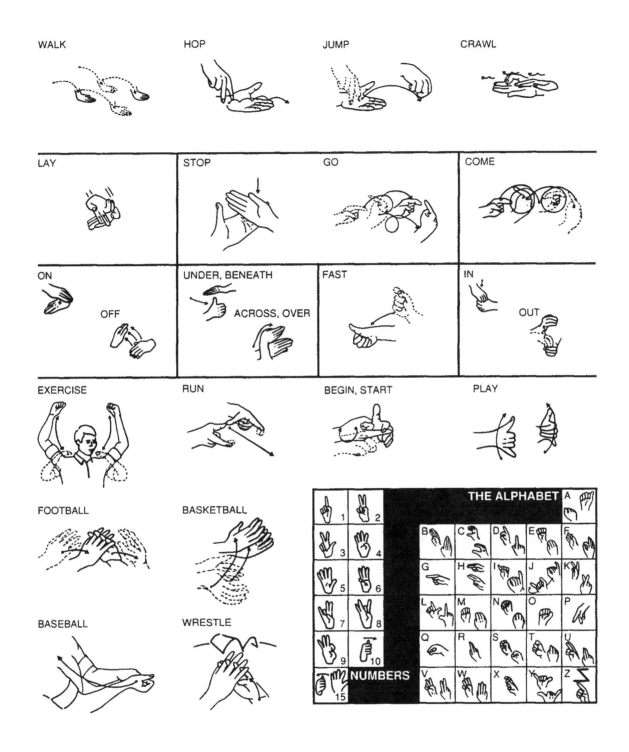

WALK HOP JUMP CRAWL

LAY STOP GO COME

ON / OFF UNDER, BENEATH / ACROSS, OVER FAST IN / OUT

EXERCISE RUN BEGIN, START PLAY

FOOTBALL BASKETBALL

BASEBALL WRESTLE

THE ALPHABET

NUMBERS

Figure 11.5 *(continued)*.

GENERAL CONSIDERATIONS FOR TEACHING PHYSICAL EDUCATION TO DEAF OR HARD-OF-HEARING STUDENTS

To the extent that speech is used for communication in the physical education class, hard-of-hearing and Deaf students will have difficulty learning. While movement is one area that is not heavily dependent on auditory cues, receiving feedback about one's movement can be problematic. Usually visual cues can be substituted for the auditory cues that are distorted or absent. Few rules, equipment, facilities, or skills require modification. According to Graziadei (1998), teaching conceptual aspects of physical education to students who are Deaf and in a hearing class with a teacher who is not fluent in sign language can be problematic.

Recall the scenario at the beginning of this section. Rachel now has a physical education teacher and friends who have basic means of communication with her. But more complex problems still face Rachel. It is unlikely that subtle sport strategy and rule concepts will be completely conveyed to her. Interpreters seldom have the sport-specific vocabulary with which to explain the subtle movement or sports concepts. Graziadei found that while the Deaf student in a hearing school is likely to perform well in physical activities, the student is likely to be limited to what can be learned through imitation and not have the opportunity to learn many of the concepts taught in physical education.

Inclusive settings may not be the appropriate placement for many Deaf students who use ASL as their primary language. Deaf students who are included in regular schools often experience isolation, social deprivation, and ridicule from teachers and peers because they lack a common language with their hearing classmates (Graziadei, 1998; Nowell & Innes 1997). If a Deaf student is placed in an inclusive classroom, it is recommended that a peer tutor program be created to alleviate these problems. Peer tutor programs have been shown to improve physical activity in inclusive physical education classes (Lieberman, Dunn, van der Mars, & McCubbin, 2000). And as tutors learn many signs, new opportunities for appropriate socialization among peers emerge. Lieberman et al. (2000) provide additional information on peer tutor programs.

SUGGESTIONS FOR TEACHING DEAF AND HARD-OF-HEARING STUDENTS

When teaching Deaf students, physical educators may find it helpful to learn the answers to the following questions. Ask the student, if possible, when seeking answers to these questions. Also, look at the student's IEP and ask questions of the parents or caregiver and teacher.

▶ What can the student hear? How can the teacher maximize any remaining hearing?

▶ What is the student's preferred mode of communication? How can the teacher maximize communication with the student?

▶ Are there any contraindications?

Most Deaf individuals have no restrictions on their participation in physical education. Children with frequent ear infections may have tubes placed in their ears. These children may need to wear ear plugs when swimming to keep water from entering their ears. Children susceptible to earaches should avoid exposing unprotected ears to cold weather. Due to balance problems and vertigo, individuals with damage to the semicircular canals of the inner ear should only climb to heights, jump on the trampoline, or dive into a pool if adequate safeguards have been taken to assure they will not be hurt if they lose their balance during these activities. Tumbling activities that require rotation, such as dive forward rolls, cartwheels, and handsprings should only be attempted with close spotting.

In inclusive physical education classes, the following suggestions will improve success, communication, and socialization. For students who have an interpreter, Graziadei (1998) has these suggestions:

▶ Encourage the interpreter to stand next to the teacher.

▶ Give lesson plans to the interpreter days in advance so both the interpreter and Deaf student can review and understand it before the lesson begins. Include a list of any specialized vocabulary.

▶ Face the Deaf student, rather than the interpreter, when addressing the student.

▶ Meet with the interpreter prior to beginning each unit to clarify sport terminology, instructional cues, and idioms that are likely to be used.

For students with residual hearing, Graziadei has these suggestions:

▶ Minimize background noise; turn off any background music and expect silence when others are speaking.

▶ Encourage the student to remove the hearing aid if there is excessive unavoidable background noise.

▶ Consider use of an audible trainer or a FM loop system (microphone worn by the teacher to amplify the voice into the hearing aid worn by the Deaf student).

These are general considerations to keep in mind when teaching Deaf or hard-of-hearing students:

▶ Use specific visual teaching cues that are easily understood.

▶ Give Deaf students copies of lesson plans modified to their reading levels.

▶ Use stations with cue cards providing written explanations and illustrations.

▶ Use clear signals for starting and stopping activities.

▶ Use demonstrations liberally.

▶ Use a scoreboard and a visual timer if playing a game.

▶ Develop highly recognizable and easily visible signals for communication at a distance.

▶ When outside, stand near the Deaf student and tap the student on the shoulder to gain attention.

▶ Enhance speechreading (lipreading) and facial cues.

▶ Face the student so that lips and facial expressions are fully visible.

▶ Avoid chewing gum, wearing a beard or mustache, or covering the mouth.

▶ Position student directly in front of the teacher.

▶ When indoors, provide bright lighting.

▶ When outdoors, take care that the student does not face into the sun.

▶ Check for understanding. Ask Deaf and hard-of-hearing students if they understand the activity before beginning. Also, help hearing students appreciate that if the Deaf students do not follow the rules, it may be because they do not understand the rules, and not because they seek to intentionally break the rules.

▶ Promote leadership skills among Deaf students.

▶ Give Deaf students choices and honor their choices.

▶ Encourage Deaf students to be team captains, group leaders, and referees.

▶ Have Deaf students help hearing students as well as the reverse.

▶ Provide clear instructions so Deaf students do not need to wait and watch others before participating.

▶ Learn as much sign language as possible if teaching students who sign. A hearing person is encouraged to practice and learn ASL, which takes as much effort as learning any other language. With this practice, a hearing person can look forward to communicating with Deaf persons beyond a superficial level.

▶ Include Deaf students in information taught during "teachable moments." These moments often occur in the middle of a game, when an interpreter is on the sidelines and the Deaf student is down the field. Two options for including Deaf students are to review the teachable moments at the end of the lesson with either the entire class or only the Deaf students, or to gather all students along with the interpreter together on the field during a teachable moment (Graziadei, 1998).

▶ Dance and rhythms can be especially helpful for Deaf students. To better feel the vibrations of music, place speakers face down on a wooden floor, turn up the base, and dance in bare feet. Butterfield (1988) suggests adding strobe lights that flash in rhythm with the music to provide visual cues.

▶ Provide extra supervision when swimming underwater for students with vestibular damage. Pair Deaf students with hearing partners, and use signals developed for communication between the teacher and the class. Flashing lights may be a quick way to gain student attention (Butterfield, 1988).

▶ Allow Deaf students the option of taking written tests with an interpreter to clarify and sign test questions. In this way, physical education knowledge is tested, not the Deaf students' knowledge of English, a second language (Graziadei, 1998).

▶ When there is only one Deaf student in the class, regularly place students in the same small groups for instruction and play. Continually teach these students signs. Keeping the composition of the groups of students constant enables students to really get to know each other and increases interaction among group members.

▶ Schedule Deaf students into the same physical education class, and encourage them to work together during class so they can help each other understand the lesson.

▶ Hold the same expectations for hearing and Deaf students with regard to motor performance, fitness, and behavior. Clearly communicate these expectations in writing to Deaf students. Educators who are unusually understanding, or offer excessive assistance when students are not prepared for class, teach these students that they can rely on others rather than accept responsibility for their own actions. Luckner (1993, p. 14) suggests that teachers ask themselves, "Am I really helping? Is the assistance I am giving helping students to cope more effectively with the world that they will be living in after graduation? Or is it better to let them experience the real consequences of their behavior?"

▶ Encourage students to become involved in Deaf sport.

Most physical fitness and motor tests can be administered to Deaf students, provided that visual cues are substituted for auditory cues in the test (e.g., dropping the arm in addition to shouting "go" to signal the start of an event). ASL is not semantically similar to English, so care must be taken when signing instructions. Be sure the person giving instructions is skilled in signing and able to give signed instructions that are semantically identical to the spoken instructions (Stewart, Dummer, & Haubenstricker, 1990). Deaf children, whose native language is ASL, perform better on motor tests when the test is administered in their native language (ASL) instead of English (Dunn & Ponticelli, 1988).

DEAF SPORT

David Stewart's (1991) view of Deaf sport follows:

There is something about being "Deaf" that is quietly comforting to those who have this identity. . . . Deaf sport can be thought of as a vehicle for understanding the dynamics of being deaf. It facilitates a social identification among Deaf people that is not easily obtained in other sociocultural contexts. . . . It relies on a Deaf perspective to define its social patterns of behaviors, and it presents an orientation to hearing loss that is distinctly different from that endorsed by hearing institutions. Essentially, Deaf sport emphasizes the honor of being Deaf, whereas society tends to focus on the adversity of hearing loss. (p. 1)

Physical educators have the extremely important role of introducing Deaf students to sport, both hearing and Deaf sport. For many Deaf students attending public schools, the majority of their exposure to Deaf culture will be through Deaf sport. Many Deaf athletes choose to compete against other Deaf athletes under the auspices of the USA Deaf Sport Federation (USADSF). The organization was established as the American Athletic Association for the Deaf (AAAD) in Ohio in 1945. Persons with moderate or severe hearing loss (55 dB or greater in the better ear) are eligible for USADSF competition.

The USADSF has national sport organizations in the following sports for men and women:

Badminton	Soccer
Baseball	Softball
Basketball	Swimming
Bowling	Table tennis
Cycling	Tennis
Flag football	Track and field
Golf	Volleyball
Handball	Water polo
Hockey	Wrestling
Skiing/snowboarding	Shooting sports

The worldwide counterpart of USADSF is the International Committee of Silent Sports (CISS). Currently, CISS is not a member of the International Paralympic Committee (IPC). Instead of participating in the Paralympic Games, Deaf sport continues to hold its own summer and winter World Games for the Deaf (WGD) every four years. Winter events include Alpine and Nordic skiing, speed skating, ice hockey, and orienteering.

Summer events include the following:

Athletics (track and field)	Men's wrestling
Badminton	Shooting
Basketball	Swimming
Bowling	Table tennis
Cycling	Tennis
Men's team handball	Volleyball
Men's soccer	Water polo

The rules followed by USADSF and CISS are nearly identical to those used in hearing national and international competitions. To equalize competition, athletes are not allowed to wear hearing aids. A few changes have been made to use visual rather than auditory cues. For example, a whistle is blown *and* a flag is waved in team sports to stop play. In track, lighting systems, placed 50 meters in front of the starting blocks and to the side of the track, are used to signal the start of a race (Bessler, 1990).

Involving youth in Deaf sport is an important objective of USADSF. The Annual Mini Deaf Sports Festival, held in Louisville, Kentucky, is specifically designed for the participation of 6- to 18-year-old Deaf students (Paciorek & Jones, 1994). This event is particularly important in fostering pride in Deaf sport and increasing involvement among Deaf youth who will form the future of Deaf sport. Physical educators are encouraged to refer to the Resource section for this chapter at the end of the book to learn more about student involvement in Deaf sport and to share this information with their Deaf students.

The sports skills of Deaf athletes span the range found in the hearing population, from unskilled to highly skilled. Deaf athletes are capable of competing as equals with hearing athletes, and some do so with much success. As far back as 1883, Deaf athletes were competing in professional sports in this country. In that year Edward Dundon became the first recorded Deaf professional baseball player and is reputed to be the reason for the development of umpire hand signals. Deaf athletes have also excelled in sports such as swimming, wrestling, bowling, and football. In fact, the huddle is claimed to have been first used by the Gallaudet University football team to prevent competing Deaf teams from eavesdropping on their plays (Strassler, 1994). Some Deaf athletes have gone on to coach hearing teams, such as Albert Berg, who became the first football coach at Purdue University in 1887.

DEAFBLINDNESS

Individuals called deafblind do not have effective use of either of the distance senses—vision or hearing. Even though the term "deafblind" suggests that these people can neither hear nor see, this is seldom accurate. The overwhelming majority of individuals who are deafblind receive both visual and auditory input, but information received through these sensory channels is usually distorted. It is more accurate to state that these people are both hard of hearing and partially sighted. Only in rare instances, such as with Helen Keller, is a person totally blind and profoundly deaf.

There are many causes of deafblindness. Understanding the cause may give an indication of the age of onset and whether there is likely to be remaining vision and hearing. **Ushers Syndrome** is a congenital disability characterized by hearing loss present at birth or shortly thereafter and the progressive loss of peripheral vision. Ushers I is congenital deafness and progressive retinitis pigmentosa, while Ushers II is adventitious deafness and progressive retinitus pigmentosa. Ushers Syndrome is the major cause of deafblindness (Sauerbuger, 1993).

Another cause of deafblindness is CHARGE Syndrome, so named because the following cluster of symptoms occur together:

C—colobma of the eye (hole in the eye)

H—heart (congenital heart defect)

A—atresia of the choanae (nasal blockage that affects eating and swallowing)

R—retardation of growth

G—genital anomalies such as undescended testes or small genitalia

E—ear malformations such as low-set, rotated, or misshapen ears

Rubella can cause deafblindness when it is contracted by the mother during the first trimester of the pregnancy. Deafblindness can also be associated with meningitis, prematurity, parental use of drugs, sexually transmitted diseases (STDs), as well as unknown causes.

Characteristics of Children With Deafblindness

While there is a tendency to focus on the medical aspects of deafblindness, physical educators can help parents begin to focus on the quality of life issues of their child with deafblindness. They can help parents learn how the deafblind person likes to communicate, what the person likes to do, and whom the person trusts. Physical educators can work to introduce the deafblind person to activities that they may come to enjoy. They can help the person experience joy in living rather than focus only on survival. Camp Abilities, a successful sports camp for youth and adolescents with visual impairments, is held for one week each summer at SUNY Brockport, New York (Lieberman & Lepore, 1998).

A major consideration with deafblindness is *isolation*. Some deafblind persons move to communities where there are several other deafblind persons. Many attend camps for deafblind persons. Physical education and sport can provide opportunities to reduce this isolation and introduce the person who is deafblind to activities such as roller skating, swimming, biking, gymnastics, and other sports.

Deafblindness presents *no* opportunity for incidental learning. This means that the student needs to be taught everything. It is becoming more common to use an **intervenor**, a person who works one on one with the deafblind person, signing exactly what is happening in the person's environment. For example, the intervenor signs that a child is on the swings over there and a woman walking her dog just passed. Signing may need to be done in the child's limited field of vision. If there is no vision, signing may be done tactually in the deafblind child's hand. Additional characteristics of deafblindness are listed in box 11.4.

Adaptations for Teaching

Many students who are deafblind will need modifications to successfully participate in regular activities. Refer to box 11.5 for considerations when teaching students with deafblindness. Modifications may include changing the rules, equipment, or environment. Examples of each modification follow.

> ## Box 11.4 Characteristics of Students With Deafblindness

- ▶ Multiplicative of deafness or blindness. The implications of both disabilities are much more severe than each disability by itself.

- ▶ Methodical. Students may be slow moving yet consistent and may do many parts of an activity slowly.

- ▶ Need for sameness. Familiar routines enable the person to move freely throughout the environment.

- ▶ Easily frustrated. The severe limitations on communication and mobility can be sources of frustration.

- ▶ Desire for communication. Students have a desire to be included, understood, and to make their own choices known.

- ▶ Enjoy coactive movement. Students may enjoy swinging, scooting, roller blading, swimming, dancing, etc.

- ▶ Isolation is a continual battle. Students have an ongoing need for socialization and true friendship.

Box 11.5 When Teaching Students With Deafblindness, Remember . . .

▶ Use multiple teaching modes such as explanation, demonstration, brailling, and physical assistance.

▶ Encourage choice making such as choice of activity and equipment.

▶ Use an ecological task analysis. There are many ways to throw. Allow the student to throw in the way that the student wants to throw. Then challenge the student to throw faster, farther, and more accurately. Next, observe if the student steps while throwing. If not, teach stepping with the throw, but do not begin by making the student step with the throw. Let the student start the way the student wants to throw. This will help with motivation to move and learn.

▶ Be flexible, patient, and creative.

▶ Provide all incidental information.

▶ Link movement to language. Teach the word for each skill learned.

▶ Learn the student's form of communication, including gestures and body language.

Rules Modification

Change the rules to fit the need of the student rather than forcing the student to accommodate to the traditional rules of the game.

Equipment

Adapt equipment as needed to provide assistance in grip as well as to increase the following: auditory awareness, limited range of motion, tactile cues, visual stimulation, and independence.

Environment

Modify the environment as needed through decreasing distractions, increasing visual cues, limiting noise, changing lighting, and increasing accessibility of the playing area.

Students with deafblindness, uncomplicated with additional disabilities, can participate in most sports, whether on a competitive or recreational level. Weightlifting, dance, roller skating, swimming, skiing, bowling, hiking, goal ball, track and field, cycling, and canoeing are some of the possibilities. Deafblind athletes wishing to compete in sports may choose to compete in sports for persons who are blind (USABA) or deaf (USADSF).

An experience with a student named Eddie may illustrate the importance of not placing ceilings on expectations for deafblind students. Eddie is 15 years old, Deaf and blind. He asked to learn to ride a unicycle. Using the same task analysis his physical education teacher had used to learn to ride, Eddie learned to ride the unicycle independently. Teaching deafblind students challenges physical educators to adapt appropriately to enable these students to learn.

SUMMARY

Children with sensory impairments are born with the same potential as their hearing and sighted peers. Early intervention and exposure to a wide variety of sports and physical activities will increase fitness and skill level and maintain a high quality of life. Physical educators are key in instilling confidence in movement among individuals with sensory impairments.

Chapter 12

Cerebral Palsy, Stroke, and Traumatic Brain Injury

David L. Porretta

> **H**E WAS ONCE ONE OF the most feared defensemen in the National Hockey League. Then, following a near-fatal automobile accident, Vladimir Constantinov acquired a brain injury severe enough that he may never skate independently again. Within a brief period of time he went from a highly skilled player to a person who needed to relearn many motor skills that he once took for granted. In this chapter, in addition to other information, you will learn that persons with traumatic brain injury, in many cases, need to relearn many of the physical education and sports skills they once took for granted before their injuries. And you will learn that physical education teachers and coaches need to be aware of other associated problems these persons will have in the learning process. No doubt Vladimir Constantinov will play hockey again. However, it may be from a wheelchair with the game adapted to his unique needs.

While the conditions of cerebral palsy (CP), stroke, and traumatic brain injury (TBI) each have their own causes, each of these conditions results in damage to the brain. As a result, people with CP, stroke, and TBI may exhibit common motor, cognitive, and behavioral characteristics. Discussing them together in this chapter highlights their commonalties. These individuals, who at one time were restricted from physical activity for fear that it would aggravate their conditions, are now encouraged to participate in a great variety of physical education and sport activities. The purpose of this chapter is to discuss CP, stroke, and TBI within a physical education and sport context. Specific information is covered relative to program and activity perspectives.

CEREBRAL PALSY

Cerebral palsy (CP) is a group of permanent disabling symptoms resulting from damage to the motor control areas of the brain. It is a nonprogressive condition that may originate before, during, or shortly after birth, and it manifests itself in a loss or impairment of control over voluntary musculature. The term cerebral refers to the brain and palsy refers to a disordered movement or posture. Depending on the location and the amount of damage to the brain, symptoms may vary widely, ranging from severe (total inability to control bodily movements) to very mild (only a slight speech impairment). Damage to the brain contributes to abnormal reflex development in most individuals; this results in difficulty coordinating and integrating basic movement patterns. It is rare for damage to be isolated to a small portion of the brain. For this reason, the person commonly exhibits a multiplicity of other impairments, which may include seizures, speech and language disorders, sensory impairments (especially those involving visual-motor control), abnormal sensation and perception, and mental retardation. CP can result from myriad prenatal, natal, or postnatal causes. Some of the more common causes are rubella, Rh incompat-

ibility, prematurity, birth trauma, anoxia, meningitis, poisoning, brain hemorrhages or tumors, and other forms of brain injury that may result from accidents or abuse. It is interesting to note that a premature infant is five times as likely as a full-term baby to have CP.

Incidence

According to the most recent figures published by the United Cerebral Palsy Associations, Inc. (1998), it is estimated that as many as 500,000 children and adults in the United States have CP. Of this number about 10 percent of the cases are considered to be acquired, that is, they occur after the 28th day of life and up to the age of 5, when the brain is 90 percent matured. Most of the acquired cases are due to head trauma. There are now fewer infants born with CP each year because of the current low birth rate and better prenatal care. It is estimated that about 5,000 babies and infants are diagnosed with CP each year in the United States. In addition, some 1,200 to 1,500 preschool-age children who were not previously identified are recognized each year as having CP. Enhanced technology and treatment in neonatal intensive care units have led to a decrease in the number of "high-risk" infants who may have otherwise acquired the condition.

Classifications

Individuals with CP typically exhibit a variety of observable symptoms, depending on the degree and location of brain damage. Over the years, classification schemes have evolved that categorize CP according to topographical (anatomical site), neuromotor (medical), and functional perspectives (of which the functional classification is the most recent).

Topographical

The topographical classification is based on the body segments afflicted. Classes include the following:

Monoplegia—any one body part involved

Diplegia—major involvement of both lower limbs and minor involvement of both upper limbs

Hemiplegia—involvement of one complete side of the body (arm and leg)

Paraplegia—involvement of both lower limbs only

Triplegia—any three limbs involved (this is a rare occurrence)

Quadriplegia—also known as total body involvement (all four limbs, head, neck, and trunk)

Neuromotor

More than 40 years ago the American Academy for Cerebral Palsy adopted a neuromotor classification system to describe cerebral palsy. This classification has undergone revisions over the years. Today, three main types are commonly described (United Cerebral Palsy Associations, 1998). It is important to understand that the characteristics described under each type may overlap the other types; they are not as distinct as one would assume.

Spasticity

Spasticity results from damage to motor areas of the cerebrum and is characterized by increased muscle tone (hypertonicity), primarily of the flexors and internal rotators, which may lead to permanent contractures and bone deformities. Strong, exaggerated muscle contractions are common, and in some cases muscles will continue to contract repetitively. Spasticity is associated with a **hyperactive stretch reflex**. The hyperactive reflex can be elicited, for example, when muscles of the anterior forearm (flexors) are quickly stretched in order to extend the wrist. When this happens, receptors that control tone in the stretched muscles overreact, causing the stretched muscles to contract. This results in inaccurate and jerky movement, with the wrist assuming a flexed position, as opposed to an extended or mid position. If muscles of the upper limb are prone to spasticity, the shoulder will be adducted, the arm will be carried toward the midline of the body, and the forearm will be flexed and pronated. The wrist will be hyperflexed, and the hand will be fisted.

Lower-limb involvement results in hip flexion, with the thigh pulling toward the midline, causing the leg to cross during ambulation. Lower-limb involvement causes flexion at the knee joint because of tight hamstring muscles. Increased tone in both the gastrocnemius and soleus muscles, along with a shortened Achilles tendon, contributes to excessive plantar flexion of the foot. A scissoring gait characterized by flexion of the hip, knee, and ankle along with rotation of the leg toward the midline is exhibited (figure 12.1). With their narrow base of support, people with a scissoring gait typically have problems with balance and locomotor activities. Because of increased muscle contraction and limited range of motion, they may have difficulty run-

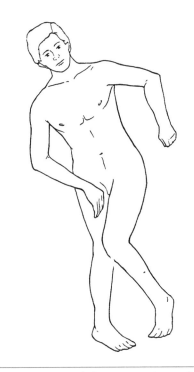

Figure 12.1 Person exhibiting spastic cerebral palsy.

ning, jumping, and throwing properly. Mental retardation, seizures, and perceptual disorders are more common in spasticity than in any other type of CP.

Athetosis

Damage to the **basal ganglia** (masses of gray matter composed of neurons located deep within the cerebral hemispheres of the brain) results in an overflow of motor impulses to the muscles, a condition known as athetosis. Slow, writhing movements that are uncoordinated and involuntary are characteristic of this type of CP. Muscle tone tends to fluctuate from hypertonicity to hypotonicity; the fluctuation typically affects muscles that control the head, neck, limbs, and trunk. Severe difficulty in head control is usually exhibited, with head drawn back and positioned to one side. Facial grimacing, a protruding tongue, and trouble controlling salivation are common. The individual has difficulty eating, drinking, and speaking. Because lack of head control affects visual pursuit, individuals may have difficulty tracking thrown balls or responding to quick movements made by others in motor activity situations. They will have difficulty performing movements that require accuracy, such as throwing a ball to a target or kicking a moving ball. A lordotic standing posture, in which the lumbar spine assumes an abnormal anterior curve, is common. In compensation, the arms and shoulders are placed in a forward position. Athetoids typically exhibit aphasia (impairment or loss of language) and articulation difficulties. It is interesting to note that in recent years very few cases of the athetoid type have been diagnosed, because it has been found that certain blood incompatibility problems can be controlled.

Ataxia

Damage to the cerebellum, which normally regulates balance and muscle coordination, results in a condition known as ataxia (figure 12.2). The cerebellum is located below and essentially behind the cerebral cortex. Muscles show abnormal degrees of hypotonicity. Ataxia is usually not diagnosed until the child attempts to walk. When trying to walk, the individual is extremely unsteady because of balance difficulties and lacks the coordination necessary for proper arm and leg movement. A wide-based gait is typically exhibited. Nystagmus, a constant involuntary movement of the eyeball, is commonly observed, and those able to ambulate frequently fall. People with mild forms of ataxia are often considered clumsy or awkward. They will experience difficulty with basic motor skills and patterns, especially locomotor activities like running, jumping, and skipping.

Functional

A functional classification scheme is commonly used today in the field of education. According to this classification system, persons are placed into one of eight ability classes, according to the severity of the disability (table 12.1). Class I denotes severe impairment, while class VIII denotes very minimal impairment. This scheme has important implications for physical education and sport, because individuals are categorized according to ability levels. For example, participants in

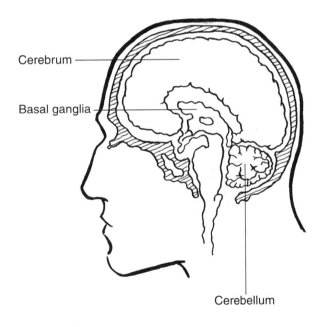

Figure 12.2 General areas of the brain involved in major neuromotor types of cerebral palsy.
Adapted, by permission, from P.D. Miller, 1995, *Fitness Programming and Physical Disability* (Champaign, IL: Human Kinetics), 16.

classes VII and VIII may be good candidates for inclusion into several regular physical education activities. Teachers and coaches can also use this system, as does the United States Cerebral Palsy Athletic Association (USCPAA) (USCPAA, 1997) and the Cerebral Palsy-International Sport and Recreation Association (CP-ISRA), to assist in equalizing competition among participants. In activities requiring competition between two individuals, players of the same classification can compete against each other. In team activities, players of the same class can be placed on separate teams so that each team is composed of players of similar functional levels. These suggestions for equalizing competition can be followed in either inclusive or noninclusive settings.

General Educational Considerations

Since it is not a disease, most medical professionals agree that CP is not treated but rather managed. Management for CP is aimed at alleviating symptoms caused by damage to the brain and helping the child achieve maximum potential in growth and development. This consists of managing both motor and other associated disabilities. Managing motor dysfunction usually entails developing voluntary muscle control, emphasizing muscle relaxation, and increasing functional motor skills. In some instances, braces and orthotic devices are used to help prevent permanent contractures or to support affected muscle groups; this is especially true for those with spasticity. Surgery can be performed to lengthen contracted tendons (especially the Achilles tendon) or to reposition an unimpaired muscle to perform the movement of an impaired one. A repositioning operation known as the Eggar's procedure relieves flexion at the knee joint and helps to extend the hip by transferring the insertion of the hamstrings from the pelvis to the femur. In rare instances, brain surgery can be performed to alleviate extreme hypertonicity. A procedure known as **chronic cerebellar stimulation (CCS)**, in which electrodes are surgically implanted on the cerebellum, has recently been developed. This procedure has demonstrated improvement; however, long-range effects on the central nervous system have not yet been documented, and it seems that this procedure results in little direct clinical application at this time.

Some research questions now being addressed by the United Cerebral Palsy Research and Educational Foundation and other such organizations are as follows: What are the factors that predispose the developing brain to injury? Why is low birth weight in full-term infants or prematurity important risk factors for CP? Which available treatments are most effective for specific disabilities of persons with CP? What are the effects of aging on persons with CP?

Table 12.1   Functional Classification Profile for Cerebral Palsy Used by Both USCPAA and CP-ISRA

Class description	Locomotion	Object control
I Severe spasticity and/or athetosis with poor functional range of motion and strength in all extremities; poor to nonexistent trunk control	Motorized wheelchair or assistance for mobility	Only thumb opposition and one finger possible; can grasp only beanbag
II Severe to moderate spastic and/or athetoid quadriplegic; poor functional strength in all extremities, and poor trunk control; classified as II lower if one or both lower extremities are functional; otherwise, classified as II upper	Propels wheelchair on level surfaces and slight inclines (lower class II with legs only); sometimes may be able to ambulate short distance with assistance	Can manipulate and throw a ball (II upper)
III Moderate quadriplegic or triplegic; severe hemiplegia; fair to normal strength in one upper extremity	Can propel wheelchair independently but may walk a short distance with assistance or assistive devices	Normal grasp of round objects but release is slow; limited extension in follow-through with dominant arm
IV Moderate to severe diplegic; good functional strength and minimal control problems in upper extremities and torso	Assistive devices used for distances; wheelchair is usually used for sport	Normal grasp is seen in all sports; normal follow-through is evident when pushing a wheelchair or throwing
V Moderate to severe diplegic or hemiplegic; moderate to severe involvement in one or both leg; good functional strength; good balance when assistive devices are used	No wheelchair; may or may not use assistive devices	Minimal control problems in upper limbs; normal opposition and grasp seen in all sports
VI Moderate to severe quadriplegic (spastic/athetoid or ataxic); fluctuating muscle tone producing involuntary movements in trunk and both sets of extremities; greater upper-limb involvement when spasticity/athetosis present	Ambulates without aids; function can vary; running gait can show better mechanics than when walking	Spastic/athetoid–grasp/release can be significantly affected when throwing
VII Moderate to minimal spastic hemiplegic; good functional ability on nonaffected side	Walks and runs without assistive devices but has marked asymmetrical action; obvious Achilles tendon shortening when standing	Minimal control problems with grasp and release in dominant hand; minimal limitation seen in dominant throwing arm
VIII Minimal hemiplegic, monoplegic, diplegic, or quadriplegic; may have minimal coordination problems; good balance	Runs and jumps freely with little to no limp; gait demonstrates minimal or no asymmetry when walking or running; perhaps slight loss of coordination in one leg or minimal Achilles tendon shortening	Minimal incoordination of hands

Adapted, by permission, from UCPAA, 1997, *United Cerebral Palsy Athletic Association Classification and Sports Manual*, 5th ed. (Newport, RI: UCPAA), 3-13.

Because of central nervous system damage, many individuals with CP exhibit abnormal reflex development, which interferes with the acquisition of voluntary movement. If abnormal reflex patterns are present, young children with CP will most likely receive some type of physical therapy designed to inhibit abnormal reflex activity in addition to enhancing flexibility and body alignment. However, recent scientific data show that passive activities and manipulations during the early years may not provide as much assistance in the remediation of abnormal reflex activity as once thought. Treatment emphasis should be directed toward having individuals perform and refine motor tasks by active self-control. Functional motor skills such as walking, running, and throwing should be developed and attained.

Attention must also be given to the psychological and social development of individuals with CP. The various disabilities associated with CP increase the possibility of adjustment problems. Because of the negative reactions that other people may have to their condition, individuals with CP may not be totally accepted. As a result, guidance from psychologists or professional counselors should be sought for both parents and their children when emotional conflicts arise.

The primary concern should be for the *total* person. Levitt (1995) advocates for a strong collaborative medical treatment approach between parents and medical professionals; this approach includes education and information, parental training skills, and emotional support. From an educational perspective, a team approach in which both medical and educational personnel work together with the parent and, when appropriate, with the student, is strongly recommended.

STROKE

Stroke (cerebral vascular accident, or CVA) refers to damage to brain tissue resulting from problems due to faulty blood circulation. Stroke can result in serious damage to areas of the brain that control vital functions. These functions may include motor ability and control, sensation and perception, communication, emotions, and consciousness, among others. In certain cases, stroke results in death. As a result, those persons who survive have varying degrees of disability, ranging from minimal loss of function to total dependency. Because of the nature of the cerebral arterial system, stroke commonly causes partial or total paralysis on either the left or right side of the body. This may be one limb (monoplegia) or body segment, or the entire side (hemiplegia). Individuals with right-sided hemiplegia are likely to have problems with speech and language and tend to be slow, cautious, and disorganized when approaching new or unfamiliar problems. On the other hand, individuals with left-sided hemiplegia are likely to have difficulty with spatial-perceptual tasks

(i.e., ability to judge distance, size, position, rate of movement, form, and how parts relate to the whole) and tend to overestimate their abilities. They often try to do things they cannot do and that may be unsafe. This has significant implications for those performing in physical education, leisure, and sport settings.

Stroke during the neonatal period may present itself in the form of seizures (Love, Orencia, & Billen, 1994). There may be no detectable neurological signs at the onset. Rather, neurological signs can appear during the first year after the stroke as motor skills develop. Children who exhibit seizures tend to have a worse prognosis for intellectual development and an increased incidence of recurrent seizures than those who do not.

There are a number of factors (risk factors) that contribute to the occurrence of stroke. These include uncontrolled hypertension (high blood pressure), smoking, diabetes mellitus, diet, drug abuse (such as heroin and cocaine), obesity, and alcohol abuse, among others. Many of these risk factors can be controlled through appropriate lifestyle changes. Over the past few years, there has been a substantial increase in the amount of knowledge regarding stroke and how it is treated, especially with regard to the promotion of healthful behaviors. As a result of better education and treatment, an increased number of people are living who otherwise might have died because of stroke.

Depending on the location of the damage, symptoms mirror those of CP and TBI. Persons may exhibit cognitive and/or perceptual deficits, motor deficits, seizure disorders, and communication problems, among others. While individuals with TBI and stroke can expect varying degrees of improvement following their injuries, individuals with CP cannot. Research indicates that children show more improvement following brain trauma (TBI and stroke) than adults.

Incidence

According to recent figures provided by the American Heart Association (1998), approximately 4 million people in the United States are living today with neurological impairments because of a stroke. Approximately 600,000 people experience a new or recurrent stroke each year. In 1995, females accounted for 61 percent of all stroke fatalities. Stroke is the third-largest cause of death in the United States following heart disease and cancer. Recent statistics also indicate that more than one-half of all individuals experiencing a stroke will survive their first stroke, although only about 10 percent of these individuals will completely recover. Males have a higher incidence rate than females, and African-Americans are more prone to strokes than Caucasians. Typically, stroke affects older segments of the population. Strokes occurring in infants, children, and adolescents are rare when compared to adults and accounts for less than 5 percent of all stroke cases (Love, Orencia, & Billen, 1994). Nonetheless, the occurrence of stroke in infants, children, and adolescents has significant implications for educators.

Classification

Although there are many types, stroke can generally be divided into two distinct categories: hemorrhage and ischemia. Hemorrhage within the brain is a result of an artery that loses its elasticity and ruptures, resulting in blood flowing into and around brain tissue. This type of hemorrhage is commonly called cerebral hemorrhage and is the most serious form of stroke. Ischemia, on the other hand, refers to the lack of an appropriate blood supply to brain tissue. The lack of blood results from a blocked artery leading to or within the brain itself. Typically, the blockage results from a progressive narrowing of the artery or from an embolism. An embolism is usually a blood clot or piece of fat deposit (plaque) that may lodge in small arteries. An insufficient or absent blood supply means that oxygen, which is vital for proper brain functioning, is absent or diminished. This interruption may be permanent or for a brief period of time. If the attack is very brief, it is termed a transient ischemic attack (TIA). About 10 percent of all strokes are preceded by TIAs and may occur days, weeks, or months before a major stroke. This type of ischemia results in full recovery; however, it may indicate a future attack that is more severe. Aside from a TIA, whether a person experiences either hemorrhage or ischemia, brain tissue (cells) dies, which results in reduced brain function or death.

General Educational Considerations

While many strokes strike without warning, some individuals will experience warning signs. Teachers and coaches should be aware of some common warning signs of stroke. These are sudden weakness or numbness of the face or arm and leg on one side of the body; sudden dimness or loss of vision in only one eye; sudden loss of speech or trouble understanding speech; sudden, severe headaches with no apparent cause; and unexplained dizziness, unsteadiness, or sudden falls, especially with any of the previous symptoms. Teachers and coaches should have students seek medical attention immediately, especially if these students have heart and/or circulatory problems or have experienced a previous stroke or brain injury.

Immediately following a stroke, persons surviving will need to be placed on a planned, systematic, and individualized rehabilitation program. However, the intensity and duration of the rehabilitation program will depend on the degree of disability. For example, a person who exhibits paresis (muscle weakness) or paralysis in one limb and retains normal voluntary movement for the remainder of the body will need little in the form of intense therapy. On the other hand, a person exhibiting complete paralysis of all four limbs will need more intense therapy. From an educational standpoint, individuals of school age will follow a school reentry program very similar to those individuals with TBI, as described in the following section.

TRAUMATIC BRAIN INJURY

Traumatic brain injury (TBI) refers to an injury to the brain that may produce a diminished or altered state of consciousness and results in impairments of physical, cognitive, social, behavioral, and emotional functioning. Physical impairments include lack of coordination, trouble planning and sequencing movements, muscle spasticity, headaches, speech disorders, paralysis, and seizures as well as a variety of sensory impairments (which include vision and hearing problems). Physical impairments often cause varying degrees of orthopedic involvement that requires the use of crutches or wheelchairs. Even when individuals exhibit no loss of coordination, motor function deficits, or sensation, apraxia may be evident. Cognitive impairments many times result in short and/or long-term memory deficits, poor attention and concentration, altered perception, communication disorders such as reading and writing skills, slowness in planning and sequencing, and poor judgment. Social, emotional, and behavioral impairments may include, among others, mood swings, lack of motivation, lowered self-esteem, self-centeredness, inability to self monitor, difficulty with impulse control, perseveration, depression, sexual dysfunction, excessive laughing or crying, and difficulty relating to others. Any or all of the above impairments vary greatly depending on the extent and location of damage to the brain and the success of the rehabilitation process. Therefore, impairments could range from mild to severe. However, with immediate and ongoing therapy, these impairments may decrease in severity. Because of the developmental nature of the central nervous system, children with TBI recover motor and verbal skills faster than adults. However, children's head injuries tend to be more diffuse than focal. A diffuse injury may affect the entire range of academic achievement and, therefore, has significant educational implications for the child.

TBI is often referred to as the "silent epidemic," because impairments continue even though no external visible signs are present on or around the face and head area. TBI can result from motor vehicle, sports, and recreation accidents; child abuse; assaults and violence; and accidental falls. In addition, TBI can be caused from anoxia, cardiac arrest, or near drowning. Motor vehicle accidents are the leading cause of TBI, falls are second, and assaults are third. As a subgroup, children are especially at risk for brain injury. According to Appleton (1998), the majority of head injuries to children are a result of

child abuse,

falls from buildings, play equipment, or trees,

injuries from objects (e.g., golf club, ball, stones, firearm),

road traffic accidents (pedestrian, passenger, cyclist),

seizures and other causes of loss of consciousness, and

sports-related injuries (e.g., horseback riding, skateboarding, rollerblading, football).

Since TBI is so common, it is now identified as a separate categorical condition in the Individuals with Disabilities Education Act (IDEA).

Incidence

According to the Brain Injury Association (1997), a national clearinghouse for brain injury, TBI is the leading killer and cause of disability in children and young adults under 45 years of age in the United States. Head injuries are 10 times more frequent than spinal cord injuries. Every year in the United States about 2 million people sustain a TBI. Of this number, hundreds of thousands will die or sustain an injury severe enough to require hospitalization. Of those that survive, approximately 70,000 to 90,000 will possess a permanent loss of function. Twice as many males are likely to sustain a TBI than females, and the highest rate of injury is among young men between the ages of 15 and 24.

Classification

Generally there are two classifications of head injury: open head injury and closed head injury (Brain Injury Association, 1997). An open head injury may result from an accident, gunshot wound, or blow to the head by an object resulting in a visible injury. On the other hand, a closed head injury may be caused by severe shaking, lack of oxygen (anoxia), and cranial hemorrhages, among others. If the head injury is closed, damage to the brain is usually diffuse; however, if the head injury is open (e.g., a bullet wound) damage is usually to a more limited area of the brain. In a closed head injury, the brain is actually shaken back and forth within the skull. This type of injury either bruises or tears nerve fibers in the brain that send messages to other parts of the central nervous system and all parts of the body. TBI can range from very mild to severe. Severe brain injury is characterized by a prolonged state in which the person is unconscious (comatose), and a number of functional limitations remain following rehabilitation. An injury to the brain can be considered minor when no formal rehabilitation program is prescribed and the person is sent directly home from the hospital. How-

ever, minor brain injury should never be treated as unimportant.

The Ranchos Los Amigos Hospital Scale describes eight levels of cognitive functioning and is typically used in the first few weeks or months following injury. These levels consist of the following:

▶ Level 1: no response—deep sleep/coma

▶ Level 2: generalized response—inconsistent and nonspecific response to stimuli

▶ Level 3: localized response—may follow simple commands in an inconsistent and delayed manner; vague awareness of self

▶ Level 4: confused/agitated—severely decreased ability to process information; poor discrimination and attention span

▶ Level 5: confused/inappropriate—consistent response to simple commands; highly distractible and needs frequent redirection

▶ Level 6: confused/appropriate—responses may be incorrect due to memory but are appropriate to the situation; exhibits retention of tasks relearned; inconsistently oriented

▶ Level 7: automatic/appropriate—appropriate and oriented behavior but lacks insight; has poor judgment and problem solving; requires minimal supervision

▶ Level 8: purposeful/appropriate—able to integrate recent and past events; requires no supervision once new activities are learned

This scale should not be used in later years as a gauge for improved function.

General Educational Considerations

Persons with TBI, depending on the severity of the injury, will need to be provided an individualized rehabilitative program. Persons suffering from severe injury will initially need to be provided with intense rehabilitation where therapy begins as soon as they are medically stable. An interdisciplinary team of medical professionals such as physicians, nurses, and speech and occupational therapists provides such a therapy program, which usually lasts for three to four months depending on the nature of the injury. For persons of school age, the rehabilitative program will take precedence over educational considerations. However, Ylvisaker, Hartwick, and Stevens (1991) suggest that medical personnel pursue the following actions shortly after admission: obtain school records to understand the student's academic, intellectual, and psychosocial baseline; initiate discussion with appropriate school officials if (when) special education programs are needed; invite school personnel to visit the medical facility and observe the student's therapy program as soon as medically feasible; and begin informal discussion of TBI, its possible outcome, and school services frequently needed.

Following the acute rehabilitation program, individuals will be provided with a long-term rehabilitation program that provides a structured environment for those who make slow improvements. As long as progress is being made, school-age individuals will remain at this level of rehabilitation and will typically receive their educational program within the rehabilitative facility.

Some persons, because of the severity of their injuries, will need extended therapy programs following long-term care. These extended and structured therapy programs may usually last from 6 to 12 months following injury, and usually emphasize cognitive skills, speech therapy, activities of daily living (ADL), the relearning of social skills, recreation therapy, and prevocational and vocational training when appropriate. Individualized educational programming continues in this environment. Only after persons have attained the maximum benefit of rehabilitation programs will they be reentered into their local educational environments.

Educators play a key role in the overall rehabilitation and educational process. Ylvisaker and Feeney (1998) stress that reentry programs need to be flexible and creative. And, because of the uniqueness of each person, no one reentry program will fit every student with TBI. Rather, educators should rely on some base principles to guide the implementation of their reentry programs. These principles include that, among others, each student presents unique cognitive, behavioral, and psychosocial challenges; assessments need to be functional, collaborative, and contextualized; supports (e.g., teacher's aide) need to be systematically reduced when appropriate; and collaborative decision making (educators, rehabilitation professionals, parents, student) needs to be fostered.

Educators need to work closely and cooperatively with the student's family to ensure the best reentry educational program possible. Walker (1997) recommends that when building effective parent-professional partnerships teachers need to remember that collaboration means sharing control with parents in educational planning; acknowledge the value of parents as the primary decision makers in determining quality of life and intervention decisions on behalf of their children; strive to establish and maintain rapport and trust in relationships with parents in order to negotiate family-centered decisions; strive for educational programs that include equal proportions of parent and professional goals; and work to resolve disagreements and interpersonal tension between them and parents. Establishing positive relationships with parents of students with TBI is important. This can be a stressful time for parents, who up to this point in their child's educational career, did not need to interact with special education teachers and therapists.

Of particular educational importance is the development and implementation of transitional plans for high school age students with TBI (Smith & Tyler, 1997). Educators are encouraged to follow a functional skills transitional approach. For some students with TBI, functional skills might be learning to plan a budget, while for others it will be acquiring independent living skills. Transitional skills related to physical education will include the learning of recreation/leisure activities such as bowling, cycling, or swimming as well as learning to access community recreation facilities. Regardless of the transitional outcomes agreed on, educators must effectively plan appropriate school experiences and establish links with community and post-school (e.g., vocational/technical schools) resources. An example dealing with transition issues in physical education can be found in the Transition application example.

The National Head Injury Foundation (now known as the Brain Injury Association) (1989) suggests a number of instructional strategies for teachers. Some of these include redirecting the student when she appears "lost," shows signs of frustration, and is not able to keep up with classmates; allowing for periods of rest in order to counteract fatigue, especially toward the end of the school day; refraining from reprimands for lack of student attention, because it may be partly caused by internal stimuli; presenting material or instructions in a slower manner (students should be encouraged to ask the teacher to repeat if needed); encouraging and teaching the student to construct flowcharts, outlines, graphs, or a diary/datebook to help organize information to facilitate memory processes; teaching the student to use special techniques to remember material; and breaking up a task into smaller and distinct parts so that they can be put together meaningfully at the end. A number of these strategies have direct implications for physical education and are discussed in detail in the section on program implications.

Since prevention is the only cure for brain injury, the Brain Injury Association's Be Headsmart campaign offers a wide variety of educational activities, from an elementary school curriculum to antiviolence initiatives to national media announcements.

PROGRAM IMPLICATIONS

All people with CP, stroke, and TBI can benefit from physical education and sport activities. The type and degree of physical disability, motor educability, interest level, and overall educational goals will determine the modifications and adaptations that are needed. With these factors taken into account, the instructional program can be individualized and personalized.

Application Example: Transition

Setting: 16-year-old male with traumatic brain injury. Student now uses a wheelchair and has lost many of the sport-related skills he learned previous to his injury.

Task: Teach functional recreation and leisure skills so the student can successfully maintain a healthy, active lifestyle during adulthood.

Application: Skills and activities once deemed important and necessary in regular secondary physical education may now need to be reevaluated relative to the functional needs of the student. In order to determine the functional recreation and leisure skills and activities to be learned, the physical education teacher needs to

▶ be a part of the student's overall individualized education planning team,

▶ consult with therapists and educators,

▶ talk with parents about their hopes and expectations for their son upon graduation,

▶ seek the student's input about his interests and desires in leisure-time activities,

▶ identify, contact, and visit community resource centers (e.g., the local YMCA, recreation, and adult fitness centers) that can provide services to the student upon graduation, and

▶ implement the student's physical education program in at least one site (when feasible) to aid the student in the transitional process.

General Guidelines

A number of general guidelines can be applied to programs for people with CP, stroke, and TBI. The guidelines that follow pertain to safety considerations, physical fitness, motor development, psychosocial development, and implications for sports.

Safety Considerations

All programs should be conducted in a safe and secure environment in which students are free to explore the capabilities of their own bodies and to interact with surroundings that will nurture their physical and motor development. Teachers and/or coaches should closely monitor games and activities, especially for those individuals who are prone to seizures or who lack good judgment (e.g., persons with TBI). In fact, about 60 percent of persons with CP exhibit seizures or have tendencies toward seizures (Ferrara & Laskin, 1997). Many of these persons take antiseizure medication, and, as a result, side effects (e.g., slowing physiologic responses to exercise, irritability, hyperactivity) may affect the person's performance in physical education and sport activities.

Students with more severe impairments will need special equipment, such as crutches, bolsters (to support the upper body while in the prone position), standing platforms (to assist in maintaining a standing posture), orthotic devices, and/or seating systems to help them perform various motor tasks (figure 12.3). However, students with mild impairments will need no spe-

cialized equipment. Because many persons with physical disabilities have difficulty maintaining an erect posture for extended periods of time, some activities are best done in a prone, supine, or sitting position. Persons with physical limitations should also be encouraged to experience as many different postures as possible not only in physical education classes but throughout the school day as well. This is especially true for persons in wheelchairs.

Because of abnormal muscle tone and reduced range of motion, many individuals with neuromotor involvement have difficulty moving voluntarily. The teacher may need to assist in the following ways:

▶Getting the person into and out of activity positions
▶Helping the person execute a specific skill or exercise
▶Physically supporting the person during activity

The teacher may also need to position students by applying various degrees of pressure with the hands to key points of the body, such as the head, neck, spine, shoulders, elbows, hips, pelvis, knees, and ankles. An example is applying both hands symmetrically to both of the individual's elbows in order to reduce flexion at the elbow joints. However, these techniques should be performed only after instruction by a therapist or physician. The ultimate aim of handling, positioning, and lifting individuals with CP is to continually encourage them to move as independently as possible. This is accomplished by gradually reducing the amount of support to key points of the body over a period of time.

Figure 12.3   Athlete with CP using crutches to assist in running.

Teachers should consult with therapists whenever possible in an effort to coordinate these procedures, especially for persons possessing severe physical disabilities. Finnie (1997) provides excellent information on appropriate ways to handle children with characteristics associated with CP. In addition, teachers should closely monitor the physical assistance that a student with a disability receives from trained peers. Peer assistance should be discouraged if it poses a safety risk.

Because the conditions described in this chapter are of medical origin, it is important that physical educators and coaches consult medical professionals when establishing programs to meet unique needs. This is especially important for students with TBI or those who are receiving physical or occupational therapy, such as students with CP and stroke.

Physical Fitness

It is generally agreed that appropriate levels of health-related physical fitness assist persons with disabilities in performing activities of daily living, recreation, and leisure activities, which, in turn, promote a healthy lifestyle. While the health-related physical fitness needs of persons with CP, stroke, and TBI are similar to nondisabled persons, some specific aspects of fitness are important to these individuals. Reduced muscular strength, flexibility, and cardiovascular endurance levels are common in persons with CP, stroke, and TBI and may lead to the inability to maintain balance, independently transfer or move one's body, perform activities of daily living, or participate in functional recreation/leisure activities. Because restricted movement

is common for individuals whose conditions are described in this chapter, it is vitally important that strength and flexibility be developed to the maximum extent possible. Weak musculature and limited range of motion, if unattended, will lead to permanent joint contractures that result in significant loss of movement capability. For example, individuals with more severe forms of spastic CP may have significant range of motion and flexibility needs. Individuals with TBI or those who have experienced a stroke may have a need to develop and sustain an adequate level of aerobic activity, especially if the trauma is recent. Whatever health-related profiles persons with CP, stroke, or TBI exhibit, a personalized approach to the enhancement of health-related fitness is recommended.

The Brockport Physical Fitness Test (BPFT) (Winnick & Short, 1999a) is the most recent fitness assessment instrument that may be used by individuals with CP, stroke, and TBI based on a personalized approach to health-related fitness. This criterion-referenced test provides test items, modifications for specific disabilities, and criterion-referenced standards for achieving fitness. The test includes components of aerobic functioning, body composition, and musculoskeletal functioning (flexibility, muscular strength and endurance) vital for achieving health-related physical fitness. Various test items may be selected within each of the three components depending on the student's individual desired profile (e.g., Target Aerobic Movement Test; upper-arm skinfold measures for body composition; Modified Apley Test for flexibility; seated push-up for muscular strength and endurance). The BPFT incorporates the eight-level functional classification system used by CP-ISRA and USCPAA (see table 12.1), which identifies the person's functional level. Based on the person's functional level, specific test items are selected. Detailed information regarding this test can be found in chapter 5.

As with others who may have low health-related fitness levels, certain precautions may need to be taken as programs are established for students with CP, stroke, or TBI. It is especially important that the teacher be sensitive to the frequency, intensity, duration, and mode of exercises and activities. Fatigue may cause the person to become frustrated, which in turn adversely affects proper performance. The instructor should permit rest periods and player substitutions when endurance-related activities like soccer and basketball are offered. It may be beneficial for those with reduced fitness levels to perform exercises and activities more frequently but with less intensity and duration. Exercises and activities that are found enjoyable by persons with CP, stroke, and TBI should be selected. Performing enjoyable exercises and activities will increase the likelihood that health-related fitness will be maintained over a lifetime.

Motor Development

CP, stroke, and TBI restrict individuals from experiencing normal, functional movement patterns that are essential to normal motor development. As a result, delays in motor control and development are common. Individuals with CP typically exhibit motor delays because they often have less opportunities to move, lack movement ability, or have difficulty in controlling movements. Individuals with varying degrees of TBI and stroke may have difficulty planning and performing movements because of damage to the motor control and related areas of the cerebrum. Children with CP and TBI are frequently unable to execute fundamental movements in an appropriate manner.

Physical education programs should encourage the sequential development of fundamental motor patterns and skills essential for participation in games, sports, and leisure activities. Authentic assessment, which emphasizes the evaluation of functional skills, should be used in physical education programs. When attempting to enhance motor development, the physical educator should be concerned primarily with the manner in which a movement is performed rather than with its outcome. The goal of every physical education program should be to encourage individuals to achieve maximum motor control and development related to functional activities (e.g., recreation/leisure and daily living activities). Standardized motor development tests recommended for use with younger students include the following:

The Denver Developmental Screening Test II

The Milani-Comparetti Developmental Chart

The Peabody Developmental Motor Scale

Psychosocial Development

Many people with CP, stroke, and TBI lack self-confidence, have low motivational levels, and exhibit problems with body image. An appropriately designed physical education program can provide successful movement experiences that not only motivate students but also help them gain the self-confidence needed to develop a positive self-image, which is vitally important for emotional well-being. A realistic body image can be developed when the physical education teacher does not expect students to perform skills and activities perfectly. Rather, it is more important that the student perform the activity as independently as possible with a specified degree of competence. The teacher should promote the attitude that it is acceptable to fail at times when attempting activities because this is a natural part of the learning process. Physical activities perceived as fun and not hard work can motivate students to perform to their maximum potentials.

Implications for Sports

Physical education teachers are encouraged to integrate many of the sport activities described in the "Adapted Sports" section of this chapter into their programs. For example, the club throw, a USCPAA field event, can be incorporated into a physical education program as a means of developing strength and can also offer an opportunity for sport competition. Other events such as bowling, archery, cycling, and boccie can be taught. Bowling, boccie, and cycling are excellent activities for persons to engage in for lifelong leisure. Team games and sports may include volleyball, basketball, soccer, and floor hockey.

Individual and dual activities may include tennis, table tennis, riflery, archery, badminton, horseback riding, billiards, and track and field. Winter activities including ice hockey, ice skating, downhill and cross-country skiing, tobogganing, and sledding are also popular in northern regions. All of the above games and sports can be offered with a view toward future competition or leisure activity.

Disability-Specific Guidelines

The previous section described a number of general program guidelines applicable to CP, stroke, and TBI. However, there are also a number of guidelines specific to each condition. These specific guidelines focus on health-related physical fitness and motor ability components, among others.

Cerebral Palsy

According to Winnick and Short (1999a), individuals with CP should possess the ability to sustain moderate physical activity (aerobic functioning), have body compositions consistent with positive health, and have musculoskeletal functions (muscular strength and endurance and flexibility) so that participation in a variety of sport and leisure activities is possible. Sustaining moderate physical activity (70 percent of maximum predicted heart rate adjusted for mode of exercise) for a period of 15 minutes represents the general aerobic standard for youngsters with CP (Winnick & Short, 1999a). The ability to perform this general standard has positive implications for sport and leisure activities. Minimum and preferred general standards are also presented for body composition and musculoskeletal functioning.

Inappropriate reflexive behaviors in persons with CP contribute to reduced aerobic activity as well as imbalances in muscle functioning and flexibility throughout various regions of the body. It can also contribute to motor coordination and equilibrium difficulties that can compromise the ability to attain acceptable health-related fitness and the ability to learn and perform various motor skills, especially those needed to perform recreation and leisure-time activities. Because of either restricted or extraneous movements, an individual with CP may exert more energy than a person without an impairment to accomplish the same task. The added energy output requires a greater degree of endurance. As a result, the duration of physical activities may need to be shortened.

When a child is receiving therapy for inappropriate reflexive behavior, it is important for the physical educator to work in conjunction with therapists to foster the suppression of certain abnormal reflexes and the facilitation of righting and equilibrium reactions. While many physical education activities help in the development of righting and equilibrium reactions, others may elicit abnormal reflexes. Some of the more common reflexes affecting the performance of physical education and sport skills include the asymmetrical tonic neck reflex (ATNR), the symmetrical tonic neck reflex (STNR), the crossed extension reflex, and the positive supporting reflex. The ATNR can prevent the effective use of implements such as bats, rackets, and hockey sticks. When present, the STNR can affect the ability to perform scooterboard activities in the prone position or other activities requiring the chin to be tucked toward the chest (e.g., looking down to control a soccer ball or catch a ground ball in baseball). Difficulty in kicking from a standing position can be affected by the crossed extension and positive supporting reflexes.

As the young student progresses in age, even with therapy, inappropriate reflexes will not be inhibited. Professionals responsible for the student's physical education program must therefore pursue attainment of functional skills, including sport skills. The attainment of functional skills such as creeping, walking, running, and throwing is important to future skill development and should be incorporated into the student's program. Asking students with CP to repetitively perform activities that elicit unwanted reflexes will not aggravate the condition of CP once they grow beyond the very early years (i.e., once they attain ages seven or eight), as once thought by professionals.

Strength

In addressing the development of strength, it is important to note that muscle tone imbalances between flexor and extensor muscle groups are common in persons with CP. For those with spastic tendencies, flexor muscles may be disproportionately stronger than the extensors. Therefore, strength development should focus on strengthening the extensor muscles. For example, even though a student may have increased tone of the forearm flexors, she may very well perform poorly on pull-ups. This being the case, one should not continue to develop forearm flexors as opposed to forearm extensors. The goal is to develop and maintain a balance between flexor and extensor muscles throughout all regions of the body. When muscular strength imbalances are present between various regions of the body, DiRocco (1999) suggests that persons with CP can use handheld weights or flexible tubing so that the appropriate resistance is applied to a particular body segment or region.

When participating in a resistance-training program, some individuals may exhibit an increase in spasticity in the involved limb or segment of the body when a contralateral nonspastic limb is involved in a resistance exercise. It is suggested that strength-building exercises be performed at a moderate speed rather than a fast speed in order to lessen the spasticity. Baxter and Lockette (1995) suggest that increased spasticity is a temporary phenomenon, and the increased spasticity should subside soon after the session. In any case, spastic muscles should not be subjected to workloads above 60 percent of maximum (DiRocco, 1999).

Individuals with CP can benefit from rigorous strength-training programs. Isokinetic resistance exercises are particularly useful for developing strength, probably because they provide constant tension through the full range of motion and aid in inhibiting jerky movements that are extraneous and uncontrolled. Moving limbs in diagonal patterns, for example, moving the entire arm across the body in a diagonal plane, encourages muscle groups to work in harmony. Such movements can be elicited by involving individuals in a variety of gross motor activities that may include throwing, striking, or kicking movements.

Flexibility

Tight muscles in both the upper and lower limbs and the hip region contribute to reduced flexibility, especially for students with spastic CP. If left unattended, restricted range of motion leads to contractures and bone deformities. Therefore, flexibility exercises and activities should be a regular part of physical education and sport programs. Individuals with spastic CP benefit from a prolonged warm-up period (15 to 20 minutes) composed of static flexibility exercises (DiRocco, 1999). The instructor may wish to begin a flexibility program session by helping students relax target muscle groups. This can be accomplished by teaching students a variety of relaxation techniques that they can then perform independently. When stretching exercises are used, they should be of a static, as opposed to a ballistic, nature, and they should be done both before and after strength and endurance activities (Surburg, 1999). If an individual is participating in a ballistic type of activity such as a club throw, ballistic stretching can be used but it should be preceded by static stretching. Surburg also recommends that stretching exercises for more severely affected body parts that are spastic should be done on a regular basis.

When flexibility exercises are done, it is recommended that fewer repetitions and longer periods of stretching be performed.

Students should be encouraged to perform stretching exercises on their own whenever possible. This type of stretching is referred to as active range of motion (performed with no assistance). Should the teacher or coach need to assist the person with spasticity in performing flexibility exercises (known as active-assistive range of motion), it is important that his hand be placed on the extensor muscle, not the flexor (spastic) muscle (Mushett, Wyeth, & Richter, 1995). For individuals who, because of severe spasticity or limited motor control, cannot voluntarily move their body parts, passive range of motion (movement performed entirely by the teacher/coach without assistance from the student/athlete) can be performed under general medical supervision.

Speed

Many students with CP have difficulty with games and sports skills that include a speed component, because quickly performed movements tend to activate the stretch reflex. However, an appropriate program can permit students with CP to increase their movement speed. Speed development activities for persons with CP will differ little from those for nonimpaired persons, except that such activities should be conducted more frequently than for students without impairments; daily activities are recommended. Students with CP should be encouraged to perform movements as quickly as possible but to perform them in a controlled, accurate, and purposeful manner. Activities having a speed component include throwing and kicking for distance, running, and jumping. Initially, the student should concentrate on the pattern of the movement while gradually increasing the speed of its execution. To develop arm and leg speed, the student can be asked to throw or kick a ball (or some other object) in a "soft" manner to a target; gradually, the throw or kick can increase in speed.

Motor coordination

Varying degrees of incoordination (dyspraxia) are common in individuals with CP and contribute to delayed motor control and development. Those who are significantly uncoordinated may have problems ambulating independently or with appliances and may need to wear protective headgear. Because they frequently fall, they should be taught, when appropriate, to fall in a protective manner. Because of abnormal movements and posture, individuals with CP have difficulty with controlling balance and body coordination. Obstacle courses, horseback riding, bicycling and tricycling, and balance beam and teeter (stability) board activities can be offered to assist in controlling movements.

Motor control difficulties notwithstanding, persons with CP (as well as those with TBI and stroke) can learn to become more accurate in their performances. Since persons with CP can have difficulty planning movements involving accuracy, they should be allowed sufficient time to plan the movement before executing it. Many times, the use of a weighted ball, bat, or other implement will assist in decreasing abnormal flailing or tremor movements. Adding weight to the implement helps in reducing exaggerated stretch reflexes, which in turn aids in controlling movements. Individuals with CP possessing motor control deficiencies resulting from athetoid, tremor, or ataxic tendencies can be expected to throw or kick for distance better and to exhibit freer running patterns than others who have limited range of motion due to spastic or rigid tendencies.

Loud noises and stressful situations increase the amount of electrical stimulation from the brain to the muscles; this tends to increase abnormal and extraneous movements, which in turn make motor activities difficult to perform. In an attempt to deal with this situation, students should be taught to concentrate on the activity to be performed. Individuals exhibiting spastic tendencies tend to relax more when encouraged to make slow, repetitive movements that have a purpose, while those with athetoid tendencies perform better when encouraged to relax before moving. Highly competitive situations that promote winning at all costs may tend to increase abnormal movements. Therefore, competitive situations may need to be introduced gradually. The teaching of relaxation techniques, which consciously reduce abnormal muscle tone and prepare the student for activity and competition, has been found to be beneficial. Another way to help individuals with CP improve general motor control and coordination is to have them construct a mental picture of the skill or activity prior to performance. This technique, called **mental imagery**, may help to integrate thoughts with actions.

In motor skill development for students with CP, the skills taught should be broken down into basic component parts and presented sequentially. This method is particularly useful for uncoordinated students seeking to learn more complex motor skills. However, because of the general lack of body coordination, activities should initially focus on simple repetitive movements rather than on complicated ones requiring many directional changes. Therefore, activities that help to develop basic fundamental motor skills and patterns, such as walking, running, jumping, throwing, catching, and so forth, should be taught.

Perceptual-motor disorders

Perceptual-motor disorders also contribute to poor motor performance. Because of these disorders, many children with CP exhibit short attention spans and are easily distracted by objects and persons in the immediate environment. Activities may therefore need to be conducted in an environment as free from distractions as possible, especially during early skill development.

Visual perceptual disorders are common among students with CP and can adversely affect activities and events that involve spatial relationships. These may include player positioning in team sports like soccer, remaining in lanes during track events, and determining distances between objects like boccia balls. Students may have difficulty with various accuracy and aiming tasks, such as throwing, tossing, or kicking an object to a specified target, as well as with activities involving various degrees of fine motor coordination, such as rifle shooting, angling, or pocket billiards.

Stroke and Traumatic Brain Injury

Previous to their brain trauma, individuals were once involved in learning and performing a host of physical education and sport skills in a normal manner. It is commonly known that the learning of motor skills requires varying amounts of cognition depending on the level of difficulty. Skills that were once thought to be quite simple to learn, now following trauma, require constant practice and planning by the person with stroke or TBI. Depending on the age of the individual at the time of trauma, some skills may have already been learned for quite some time (e.g., running, throwing, catching), while other skills have been yet to be acquired (e.g., specific sport skills). While some individuals with stroke or TBI may fully recover the motor skills lost, others with more significant and permanent injury may never regain them. In order for individuals with TBI and stroke to regain skills to their maximum potential, physical education and sport programs need to be individualized and offered on a frequent, regular basis.

Because of the nature of the disability, those with stroke and TBI may commonly exhibit weak muscles and balance difficulties. Inadequate balance may hinder the performance of many physical education and sport activities. Readers specifically interested in exercise testing and programming for persons with stroke and head injury should consult the writing of Palmer-McLean and Wilberger (1997).

Physical fitness

Acquiring and maintaining an adequate level of health-related physical fitness is important. This is especially true for persons who have been severely injured due to TBI and those who have been immobile for a long time due to stroke. The health-related fitness needs can be considered similar to persons with CP. As such, persons with stroke or TBI need to develop and sustain at least moderate levels of aerobic activity, possess body compositions consistent with good health, and possess the musculoskeletal functioning necessary to participate in a variety of sport and leisure activities. Therefore, depending on individual needs, the physical education program should allow for activities that develop and maintain these health-related components. The BPFT (Winnick & Short, 1999a) highlights items and standards that can be used in the assessment of health-related fitness for individuals with disabilities. A personalized process may be used to design a test for youngsters with stroke and TBI.

Most students may fatigue easily, especially as they begin their reentry into school. Therefore, fitness exercises and activities should be introduced on a gradual basis, and sufficient rest periods should be offered between activities, especially if physical education is offered toward the end of the school day. For persons who have been immobilized for long periods of time, Lockette and Keyes (1994) suggest that muscular strength and endurance activities performed in a gradual, progressive manner precede activities that focus on aerobic capacity.

The neurological deficits associated with head injury many times affect the person's ability to ambulate in an efficient manner. As a result, the person with head trauma uses a more significant amount of energy than would otherwise be used. According to Palmer-McLean and Wilberger (1997), appropriately planned fitness programs can improve cardiorespiratory endurance and muscle strength, thereby allowing persons to raise their energy levels to perform locomotor activities (e.g., sport, leisure, and activities of daily living) in a more efficient manner. Raising functional health-related fitness levels also increases a person's chance of living a more fulfilling and productive life.

Some persons who exhibit spasticity (similar to CP) will need to focus on relaxation and flexibility exercises and activities. Other persons who may exhibit partial paralysis will need to maintain residual functioning through muscular strength and endurance exercises and activities. Weight training and flexibility exercises and activities will not be new to the person recovering from TBI or stroke, because physical and occupational therapy rehabilitation programs typically use them.

Universal gym equipment is both convenient and safe to use to develop strength and endurance, because individuals need not be concerned with placing free weights on barbells and dumbbells. Except for the bench press exercise, persons can stay seated in their wheelchairs, although getting out of the wheelchair is important and should be encouraged. For those persons who can either independently (or with assistance) remove themselves from their chairs, isokinetic equipment (e.g., Nautilus) is also beneficial to use. However, according to DiRocco (1999), weight machines may not be the choice of equipment for persons with stroke or TBI. Muscle weakness on one side of the body may prevent the successful use of weight machines, because a number of exercises require the use of both arms or both legs simultaneously. As such, the use of free weights (e.g., dumbbells), with which limbs can be exercised individually, should be made available. For persons who do not have the availability of free weights, stretch bands are an economical way to conduct resistance training.

Since persons with more severe brain injury are more likely to be sedentary, aerobic activities that develop cardiorespiratory endurance levels should be performed on a regular basis. These may take the form of low-impact aerobics for persons who can ambulate, or aerobics done from a sitting position for those using wheelchairs. Those who can perform activities from a standing position but who have limited endurance should have the availability of a stationary object for rest or support when needed. Aquatic activities are especially good for developing physical fitness. It is recommended that individuals with TBI and stroke participate in physical fitness programs that address all areas of fitness. *The Brockport Physical Fitness Training Guide* (Winnick & Short, 1999b) includes additional exercise and activity suggestions for enhancing health-related fitness.

Motor control

Depending on the location and severity of injury, individuals with stroke or TBI have difficulty planning, initiating, and controlling gross and fine motor movements. Persons with TBI typically have difficulty performing movements sequentially. This has important implications for physical education and sport activities, especially when combinations of separate skills need to be linked together in succession. Individuals with stroke or TBI typically need to relearn movements and movement patterns that were normally performed before the injury. To assist in this process, more complex skills should be broken down into simpler subskills, and the subskills should then be practiced sequentially. Since these individuals typically have problems processing information, a sufficient amount of time should be given following directions so that movements can be planned before they are executed.

Like those persons with CP, visual perception may also be affected in persons with stroke and TBI. As such, persons may exhibit difficulty with activities requiring spatial relationships and those requiring object control (e.g., catching, kicking, striking). Of course, activities incorporating object control will need to be individualized. Adolescents with TBI and stroke should be given choices as to the type of sport and leisure activities to be learned in physical education classes. This gives the adolescent the needed sense of independence and self-control and contributes to a smooth transition from school to community.

INCLUSION

Unless otherwise decided on by individualized education program team members, students with CP, stroke, and TBI are to be included in regular physical education classes. Of course, a decision such as this must be made on an individual basis. Generally, those with mild to moderate degrees of impairment can be safely and effectively included in regular physical education settings. Most students with CP, stroke, and TBI (unless severely mentally impaired) will understand verbal and written directions, as well as rules and strategies for various games and sports. In certain cases, teachers may need to structure activities to suit the participants' abilities. For example, students with CP affecting the lower limbs could play goalie in soccer or floor hockey and could pitch or play first base in softball. Providing appropriate inclusive environments provides these students the opportunities for enhancing social and emotional development.

ADAPTED SPORTS

At almost all age levels, people with CP, stroke, and TBI now have the opportunity to become involved in competitive sports. The USCPAA offers a variety of modified sporting events (table 12.2). In addition to persons with CP, persons with stroke and TBI are included in USCPAA competition (figure 12.4). As the national governing body, USCPAA is responsible for the conduct and administration of approved sports in the United States. Athletes are able to participate in these events on the basis of their functional abilities. In addition to numerous local and regional competitions, the USCPAA conducts national championships on an annual basis.

USCPAA is a member of the Committee on Sports for the Disabled (COSD) of the United States Olympic Committee and a member of the International Sports Organization for the Disabled (ISOD). USCPAA athletes are eligible to participate in international competitions governed by ISOD as long as they meet ISOD classification standards and qualify for events. The ISOD oversees the Paralympic Games, World Championships, World Games, and World Cup competitions. The Paralympic Games are organized every fourth year with competitions in multidisabled games and sports. The World Championships are organized relative to a specific sport where single or multiple disabilities are present. More than one sport can be arranged at the same time and place. World Games, on the other hand, are organized relative to competition in one or more sports for specific disability groups (CP) or games that may deviate from existing rules. Finally, World Cup competition refers to international competition for national or club teams in team and individual sports. The international gov-

Table 12.2   Classes Eligible for USCPAA Events

Event	Class
Archery	1-8 (all classes)
Boccie (wheelchair—individual and team)	1 and 2 only
Bowling	1-8 (all classes)
Cross country (3,000 m)	5-8
Cycling	
Bicycling	5-8
Tricycling	2, 5, and 6
Equestrian	1-8 (all classes)
Powerlifting (bench press)	1-8 (all classes)
Slalom	1-4
Soccer	
Seven-a-side soccer	5-8
Indoor wheelchair soccer	1-8 (all classes)
Swimming	1-8 (all classes)
Shooting	2-8
Table tennis	3-8
Track	
60-m weave (wheelchair)	1 only
100 m, 200 m, 400 m, 800 m	2-8
1,500 m	3-4 and 6-8
4 × 100-m relay, 4 × 400-m relay	2-8
Field events	
Soft shot, precision throw, height toss, soft discus	1 only
Medicine ball thrust, distance kick	2 only
Club throw	2-6
Shot, discus	2-8
Javelin throw	3-8
Long jump	6-8

Adapted, by permission, from UCPAA, 1997, *United Cerebral Palsy Athletic Association Classification and Sports Manual*, 5th ed. (Newport, RI: UCPAA), 14-105.

erning body for USCPAA is the Cerebral Palsy-International Sports and Recreation Association (CP-ISRA).

Competition for athletes with CP, stroke, and TBI is based on the eight-level classification system described at the beginning of this chapter (see table 12.1). Athletes are placed in a specific class through two testing procedures. In the first, a functional profile is established through observation and questioning regarding the person's daily living skills. The second testing procedure involves the measurement of speed, accuracy of movement, and range of motion

for upper extremity and torso function and, for ambulant athletes, the assessment of lower-extremity function and stability. Generally, athletes compete within their designated classes in a variety of events. Table 12.2 identifies events and associated classification levels.

In addition to adult competition, USCPAA sponsors competition for junior athletes (under 18 years of age) for both wheelchair and ambulatory classes. Junior athletes from 7 to 18 years of age are placed in one of four divisions according to age levels. A Futures division is for children six years of age and below; this

Photo courtesy of Jerry McCole and USCPAA. Printed by permission.

Figure 12.4 Athlete with CP performing the distance kick field event in USCPAA competition.

division stresses participation rather than competition. Depending on age and classification, junior athletes compete in events that are the same as those for adult athletes (e.g., ambulatory soccer, swimming). The USCPAA hosts or sanctions a number of regional events across the United States each year. In addition, it hosts the annual Youth Nationals Competition, which brings together junior athletes from across the country.

Each year the USCPAA holds a number of clinics for professionals and volunteers; these clinics focus on coaching, training, and officiating techniques. The USCPAA publishes its medically approved classification and sports rules manual (USCPAA, 1997) and a separate training guide. The association also publishes a newsletter titled *Update*, which provides readers with current information regarding competition, rules, classification, and training.

SUMMARY

This chapter has described the conditions of CP, stroke, and TBI as they relate to physical education and sport. Physical and motor needs were described, and program and activity suggestions were presented. Recognizing the medical nature of these conditions, teachers and coaches are encouraged to plan activities on the basis of input from physicians and allied health professionals.

Amputations, Dwarfism, and Les Autres

David L. Porretta

▷

▷   **T**HE CROWD CHEERS AS TONY finishes first in the 100-meter dash. His time is 11.34 seconds, just 1.5 seconds off the Olympic record! No, Tony is not a member of the U.S. Olympic Team; he is a member of the U.S. Paralympic Team. Even though he was born without hands and feet and with partially formed legs and arms, Tony Volpentest is arguably one of the fastest humans today. This is one of the many accomplishments in sport today of people with disabilities, and it provides evidence that these athletes can attain elite levels of performance, given the right equipment, training, and desire for achievement.

▷

Individuals with amputations, dwarfism, and les autres (a French term meaning "the others") impairments were at one time restricted from physical activity for fear that it would aggravate their conditions. Now, these persons are encouraged to participate in a great variety of physical education and sport activities. In fact, as the chapter-opening vignette describes, persons with disabilities are approaching many of the national records set by able-bodied athletes. This chapter provides an overview of amputations, dwarfism, and les autres impairments within a physical education and sport context. Specific information is covered relative to program and activity perspectives.

AMPUTATIONS

Amputation refers to the loss of an entire limb or a specific limb segment. Amputations may be categorized as either acquired or congenital. Acquired amputations can result from disease, tumor, or trauma; congenital amputations result from failure of the fetus to properly develop during the first three months of gestation. In most cases, the cause or causes of partial or total congenital limb absence are unknown. Generally, there are two types of congenital deformities. In one type, a middle segment of a limb is absent, but the proximal and distal portions are intact; this is known as **phocomelia**. Here, the hand or foot is attached directly to the shoulder or hip without the remaining anatomical structures present. The second type of deficiency is similar to surgical amputation, where no normal structures like hands or fingers are present below the missing segment. In many cases, however, immature fingerlike buds are present; this deficiency is usually below the elbow and unilateral in nature.

Incidence

Recent estimates indicate that about 310,000 people in the United States are amputees, of whom more than two-thirds are missing a lower limb (May, 1996). Of this number, about 7 percent are below 21 years of age. Unlike the general population of amputees, those below age 21 have a greater percentage of upper- than lower-limb losses. Congenital limb losses are approximately twice as prevalent as acquired losses for those under 21 years of age.

Classification

Amputations can be classified according to the site and level of limb absence or from a functional point of view. Nine classes that are now in use by Disabled Sports, USA (DS/USA) and the International Sports Organization for the Disabled (ISOD) are identified as follows:

▷ Class A1—Double above the knee (AK)
▷ Class A2—Single AK
▷ Class A3—Double below the knee (BK)
▷ Class A4—Single BK
▷ Class A5—Double above the elbow (AE)
▷ Class A6—Single AE
▷ Class A7—Double below the elbow (BE)
▷ Class A8—Single BE
▷ Class A9—Combined lower plus upper-limb amputations

According to this system, Class A8 represents functional ability greater than Class A1. Information regarding sport competition can be found in the "Adapted Sports" section of this chapter.

General Educational Considerations

In nearly all cases, a prosthetic device is prescribed and selected for the amputee by a team of medical specialists. The prosthetic device is designed to compensate, as much as possible, for the functional loss of the limb. Devices are chosen according to the size of the individual and the area and extent of limb absence. Most authorities favor the use of a prosthetic device as early as possible following the loss of the limb, because the device tends to be more easily incorporated into the person's normal body actions than if it is introduced later. Learning to use a device takes time and effort, and some individuals with more extensive lower-limb amputations need training with canes or crutches. Special consideration within the educational environment is needed for students with prosthetic devices. For example, classroom teachers can assist therapists in helping students acquire and maintain important fine motor skills such as cutting, pasting, and drawing. Physical education teachers can assist therapists in helping students acquire and maintain important gross motor skills such as catching, throwing, and handling implements such as bats and rackets.

With recent technological advances, new types of both upper- and lower-limb prosthetic devices are now commonly seen in educational and sport settings. New types of lower-limb prosthetic devices are commonly used in sports to provide athletes with the most realistic sense of normal foot function. These devices provide an active push off in which the device propels the body in a forward and/or vertical fashion. These prostheses, made of carbon graphite, possess a type of "dynamic response" in that they can store and release energy simulating the function of a normal foot. They respond smoothly, gradually, and proportionally to pressure applied by the user. Persons are fit individually with the assistance of computer-generated designs. As such, many athletes with both below the knee and above the knee amputations use these prostheses for competitive purposes in sports such as volleyball, basketball (which involve a significant amount of jumping), and sprint and distance running (figure 13.1, a-b). There are now a number of companies that offer these dynamic response devices commercially. Recently, cosmetic foot covers have become available to amputees; these simulate the form and look of the natural foot.

These natural looking foot covers are shells that accommodate prostheses. Unless one was to closely examine the cosmetic device, one would think that the amputee had her own foot. However, the use of these cosmetic devices is of personal preference.

Individuals with limb deficiencies often have additional educational needs in the psychosocial domain. Many feel shame, inferiority, and anxiety when in social and educational settings—feelings that may result from the stares or comments from their school-age peers. The new cosmetic covers may be of help, especially if students are self-conscious about their prosthetic devices within an educational setting. Finally, individual counseling by a psychologist or professional counselor may be needed to foster healthy emotional functioning.

DWARFISM

Dwarfism is a condition in which a person is of short stature (152.4 centimeters [5 feet] or less in height). When compared to the general population, people with dwarfism (also termed little people) are shorter than 98 percent of all other people. Generally, dwarfism may result from either the failure of cartilage to form into bone as the individual grows or from a pituitary irregularity. Aside from their short stature, people with dwarfism are considered normal.

a *b*

Photo © Flex Foot, Inc.

Figure 13.1 *(a)* Earle Connor and *(b)* Kurt Collier are two athletes performing with state-of-the-art prosthetic devices.

Incidence

It is estimated that approximately 100,000 people in the United States can be labeled as possessing some type of dwarfism. Crandall and Crosson (1994) estimate that 85 to 90 percent of all infants born with dwarfism are born to nondwarf, average-size parents. Achondroplasia is considered the most common type of disproportionate dwarfism. Approximately 1 in 40,000 infants may be born with achondroplasia.

Classification

Dwarfism can be classified into two categories: proportionate and disproportionate. Those persons who possess proportionate dwarfism have all proportionate body parts, but they are very short. This type of dwarfism results from a deficiency in the pituitary gland, which regulates growth. Disporportionate dwarfism, on the other hand, is characterized by short arms and legs with a normal torso and a large head. This type of dwarfism may be caused by a faulty gene that results in failure of the bone to fully develop. Those persons with disproportionate dwarfism are the most prevalent and, of this type, achondroplasia is most common. Achondroplasia literally means the absence of normal cartilage formation and growth, and it begins in utero. This type of dwarfism may manifest itself in a waddling gait, lordosis, limited range of motion, and bowed legs. In more severe cases, where spinal involvement (such as scoliosis) and additional bone deformities are present, the individual with achondroplasia may require ambulation devices such as crutches.

Some persons with dwarfism who do not have achondroplasia may have cervical vertebrae abnormalities, similar to atlantoaxial instability in Down syndrome, which may lead to very serious neck injury. As such, the Dwarf Athletic Association of America (DAAA) requires a medical screening for all nonachondroplasia athletes before participation in running, jumping (basketball), and swimming (diving start) events.

General Educational Considerations

In general, a physical education program for persons with dwarfism can and should follow the same guidelines as one developed for individuals without impairments. On an intellectual level, students with dwarfism do not function any differently than average-size students. Therefore, these individuals are to be treated the same as other students with regard to cognitive ability and academic achievement. People with dwarfism should have the opportunity to develop a positive self-image in a psychologically safe educational environment. Many times students with obvious physical differences and limitations are held up to undue ridicule. It is the responsibility of the teacher to maintain an environment that encourages positive interactions and social contacts among all students.

Physical education teachers and coaches need to be aware of certain factors associated with physical education and sport activities for persons with dwarfism. Because of limited stature and disproportionate body segments, individuals with dwarfism may have a disadvantage in certain activities such as track and field, tennis, baseball/softball, and basketball. From a safety perspective, because of joint defects, certain activities that place undue stress on joints should be modified.

LES AUTRES

This section discusses disabilities such as muscular dystrophy, juvenile rheumatoid arthritis, arthrogryposis, multiple sclerosis, myasthenia gravis, and Guillain-Barre syndrome.

Muscular Dystrophy

Muscular dystrophy is actually considered a group of inherited diseases that are characterized by progressive, diffuse weakness of various muscle groups. Muscle cells within the belly of the muscles degenerate and are replaced by adipose and connective tissue. The dystrophy itself is not fatal, but secondary complications of muscle weakness predispose the person to respiratory disorders and heart problems. It is quite common for individuals with dystrophy in advanced stages of the disease to die from a simple respiratory infection or as a result of myocardial involvement. Symptoms of the disease may appear any time between birth and middle age; however, the majority of the cases of muscular dystrophy affect children and youth.

There are various types of muscular dystrophy, including the myotonic, facio-scapulo-humeral, limb-girdle, and Duchenne types. Myotonic muscular dystrophy, also known as Stienert's disease, manifests itself through muscle weakness and affects the central nervous system, heart, eyes, and endocrine glands. It is a slowly progressing disease generally occurring between the ages of 20 to 40 years. Congenital myotonic dystrophy is rare, occurring almost exclusively in infants of mothers with the adult form. With appropriate care, their conditions often improve; however, delayed motor development and mental retardation in late infancy and early childhood are common. The facio-scapulo-humeral type initially affects muscles of the shoulders and face and, in some instances, the hip and thigh. Life expectancy is usually normal because this type of dystrophy may arrest itself at any time. In limb-girdle muscular dystrophy, degeneration may begin in either the shoulder or the hip girdle, with eventual involvement of both. Unlike the facio-scapulo-humeral type, degeneration continues at a slow rate. Facio-scapulo-humeral

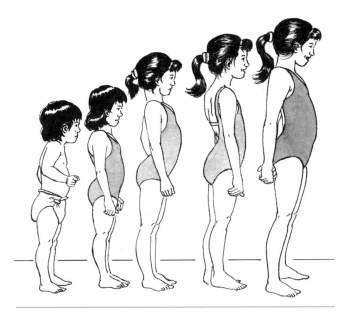

Figure 13.2 The developmental posture sequence of a child with Duchenne muscular dystrophy. Notice the increased lordosis as the child gets older.

dystrophy manifests itself during adolescence or adulthood. The limb-girdle type may be exhibited at any time from late childhood on, though it usually occurs during the teenage years. With both facio-scapulo-humeral and limb-girdle dystrophy, males and females are equally affected.

Duchenne muscular dystrophy (MD) is the most common and severe childhood form of the disease. It affects more boys than girls. Symptoms usually occur between two and six years of age. The Duchenne type is commonly referred to as **pseudohypertrophic muscular dystrophy**. A pseudohypertrophic appearance, especially of the calf and forearm muscles, is the result of an excessive accumulation of adipose and connective tissues within the interstitial spaces between degenerated muscle cells. It is yet to be determined precisely how this happens; however, the gene responsible for causing Duchenne dystrophy has been identified. Linked to this gene is a protein called dystrophin (dis-tro-pin). This protein (which is one of many) allows muscle cells to function properly; without it the muscle cells eventually die. In persons with Duchenne dystrophy, this protein is absent. Research has shown that dystrophin is attached to other proteins at the edge of muscle fibers and that it most likely helps anchor the fibers to connective tissue surrounding them.

Duchenne MD manifests itself in atrophy and weakness of the thigh, hip, back, shoulder girdle, and respiratory muscles. The anterior tibialis muscle of the lower leg becomes extremely weak, resulting in a drop foot, where the foot remains angled in a downward manner; thus, the individual is prone to falling. Steady and rapid progression of the disease usually leads to the inability

to walk within approximately 10 years after onset. The child exhibits characteristics that include

a waddling gait,

difficulty in climbing stairs,

a tendency for falling, and

difficulty in rising from a recumbent position.

An additional characteristic is a high level of creatine phosphokinase in the blood.

Lordosis frequently develops from weakness of the trunk musculature. A developmental progression of lordosis is illustrated in figure 13.2. As the disease progresses, the child will eventually need to use orthopedic devices (e.g., leg braces, walker) to continue walking. However, even with these devices and continued physical therapy, the child eventually becomes confined to a wheelchair and grows obese. Contractures may form at the ankle, knee, and hip joints, and muscle atrophy is extensive. Soon after the child begins using a wheelchair, scoliosis (a lateral curvature of the spine) also develops. Although weakness of the arms is present in the early stages of the disease, it does not cause real problems until the individual begins using a wheelchair. Then, progressive loss of strength in these muscles continues until it becomes impossible to lift objects and even lift the hands to the mouth. Death often results in about the third decade of life. With continued research, such as muscle cell transplants, a cure for this type of dystrophy may be forthcoming. However, at present, no treatment exists to stop muscle atrophy; any treatment given is basically symptomatic. A major treatment goal is to maintain ambulation as long as possible through exercise and activity.

Cognitive difficulties are present in some children with Duchenne MD; however, many of these children exhibit above-average intelligence. Sometimes slowness of movement and limitations in physical abilities are misinterpreted as cognitive difficulties. It is important to recognize that should a child have cognitive difficulties that they are stable and do not worsen as the disease progresses.

Physical education can play an important role in managing muscular dystrophy, especially when exercises and activities are performed during the initial stage of the disease. Muscular strength and endurance activities programmed on a regular basis can have a positive effect on muscular development and can serve to counteract muscular atrophy. Particular attention should be given to the development of the lower leg, hip, abdomen, and thigh, because muscles of these areas are used for locomotion. For people with weak respiratory muscles, especially those confined to wheelchairs, breathing exercises and activities should be given priority and performed on a daily basis. The Breathing Activities application example offers specific breathing activity examples for children with muscular dystrophy. Strength and endurance can be developed through aquatic activities that use water for resistance. Performed on a regular basis, flexibility activities and exercises help to develop or maintain the person's range of motion so that permanent joint contractures do not develop; flexibility activities that keep the child's attention may be chosen. Low-intensity aerobic activities are also helpful in managing obesity, which is common in persons with muscular dystrophy. Various dance movements are particularly helpful for improving flexibility and cardiorespiratory efficiency. Arm and upper-body movements for those in wheelchairs can be performed to music. Postural exercises and activities help reduce postural deviations and give the person an opportunity to perform out of the wheelchair.

Juvenile Rheumatoid Arthritis

Juvenile rheumatoid arthritis (JRA), or Still's disease, manifests itself in childhood and is one of several forms of juvenile arthritis. While the American College of Rheumatology uses the term JRA, the European League Against Rheumatism uses the term juvenile chronic arthritis (JCA). JRA is the most common type of arthritis in children and one of the more frequent chronic diseases in children today (Cassidy & Petty, 1995). As with adult rheumatoid arthritis, the cause of JRA is unknown. Depending on the degree of involvement, JRA affects joint movement. Joints become inflamed, which results in reduced range of motion. In some cases, permanent joint contractures develop and muscle atrophy is pronounced. Some authorities suggest that inflammation of the joints results from abnormal antibodies of unknown origin that circulate in the blood and destroy the body's normal structures. The disease is not inherited, nor does it seem to be a result of climate, diet, or patterns of living. It may manifest itself as early as six weeks of age. However, the highest frequency of onset is between 1 to 3 years of age and before the age of 16. According to recent figures, more than 200,000 infants, youngsters, and teenagers possess some form of arthritis, of which JRA is the most common (Arthritis Foundation, 1996). The condition afflicts twice as many girls as boys. JRA is characterized by a series of remissions and exacerbations (attacks). One cannot predict how long affected children will remain ill or the length of time they will be symptom free. Generally, the prognosis for JRA is quite encouraging: Approximately 60 to 70 percent of children will be free of the disease with no permanent joint damage 10 years after onset. Others, however, will have severe and permanent functional disability. Recent evidence indicates that adults with arthritis are at risk for depression (Cassidy & Petty, 1995). It is suggested that about 20 percent of people with arthritis exhibit depression, and this finding is similar to other populations with chronic diseases. This being the case, as children with JRA grow to adulthood, they also are at risk for depression.

At present, there is no cure for juvenile arthritis. However, research is now being conducted in the areas of genetics and immunology. Treatment for severe periods of exacerbation consists of controlling joint inflammation, which is accomplished through medicine, rest, appropriately designed exercises, and, in some cases, surgery. During acute stages, complete bedrest is strongly recommended, and excessive weight bearing by inflamed joints should be avoided. In some instances, surgery may be performed to remove damaged tissue from the joint in order to prevent greater deterioration to bone and cartilage. Total hip replacements are now performed in some cases with great success.

Application Example: Breathing Activities

Setting: Parents of a child with muscular dystrophy ask the adapted physical education teacher to suggest a couple of breathing activities that can be done at home so that their son can maintain appropriate function of his respiratory muscles.

Student: 8-year-old male with muscular dystrophy beginning to exhibit weak respiratory muscles

Application: The adapted physical education teacher recommends that the child perform the following two activities with a friend, sibling, or parent on a daily basis:

► The student and a partner face each other (about 1 meter apart) and try to keep a balloon in the air as long as possible by blowing it to each other without having it touch the ground.

► While sitting and facing each other at opposite ends of a small table (about 1 meter long), the student and a partner attempt to blow a Ping-Pong ball toward and across each other's goal (end of table).

Even during acute stages of JRA, joints should be exercised through the greatest possible range of motion at least once or twice a day, so that range of motion can be maintained. For individuals unable to exercise independently, teachers or therapists can provide partial or total assistance.

The physical education program should stress exercises and activities that help increase or maintain range of motion so that permanent contractures do not develop and normal bone density is maintained. Muscular strength and endurance activities should also be offered to decrease muscle atrophy. Isometric activities, such as hooking the fingers of both hands together and trying to pull them apart or placing the palms together and pushing, may be particularly helpful to encourage the development of hand muscles. Another hand exercise involves squeezing objects of various sizes and shapes. Hand exercises are most important in order to maintain appropriate manipulative skills. Most people with severe joint limitation or deterioration should refrain from activities that twist, jar, or place undue stress on the joints; as such, activities like basketball, volleyball, and tennis may need to be modified accordingly.

Osteogenesis Imperfecta

Osteogenesis imperfecta, also known as brittle bone disease, is an inherited condition in which bones are imperfectly formed. An unknown cause produces a defect in the protein matrix of collagen fibers. The defect reduces the amount of calcium and phosphorus (bone salts), which in turn produces a weak bone structure. Bones are very easily broken. When healed, they take on a shortened, bowed appearance. Other affected body parts that include collagen are joint ligaments, skin, and the sclera (white portion) of the eye. Joint tissues exhibit abnormal elasticity, the skin appears translucent, and the thinning sclera takes on a blue discoloration as the choroid (underlying eyeball) is exposed. There are two types of osteogenesis imperfecta (OI): congenital (present at birth) and tarda (with later onset). The congenital type is severe, while the tarda type is mild. Many students with the severe form require the use of wheelchairs. These individuals have a history of multiple fractures that result in limb deformities, and they have significant lower-limb limitations. As a result, persons with severe cases of OI are of short stature. Persons with mild degrees of OI can ambulate independently, while those with more moderate degrees of impairment use canes or crutches. There is no cure for the disease. At the present time, surgery is the most effective treatment; it consists of reinforcing the bone by inserting a steel rod lengthwise through its shaft.

Physical education activities such as swimming, bowling (with the use of a ball ramp), and the use of beach balls for striking and catching are safe to perform, because they do not place undue stress on the joints or bones. Because of abnormal joint elasticity, strength-building exercises and activities, which can increase joint stability, should be encouraged. This may be accomplished in a swimming environment. Most people with the disorder should not play power volleyball, basketball, or football unless the games are modified appropriately.

Arthrogryposis

Arthrogryposis (ar-throw-gry-po-sis), also known as multiple congenital contractures, is a nonprogressive congenital disease of unknown origin. Approximately 500 infants in the United States are born with arthrogryposis each year. The condition, which may affect some or all of the joints, is characterized by stiff joints (contractures) and weak muscles. Instead of normal muscle tissue surrounding the joints, fatty and connective tissue is present. The severity of the condition varies; an individual may be in a wheelchair or be affected only minimally. Limbs commonly exhibit deformities and can be fixed in almost any position. In addition, affected limbs are usually small in circumference, and joints appear large. Surgery, casting, and bracing are usually recommended for people with deformities. Most typically, upper-limb involvement includes turned-in shoulders, extended and straightened elbows, pronated forearms, and flexed wrists and fingers; trunk and lower-limb involvement includes flexion and outward rotation of the hip, bent or straightened knees, and feet that are turned in and down. Other conditions associated with the disease include congenital heart defects, respiratory problems, and various facial abnormalities. Individuals with arthrogryposis almost always possess normal intelligence and speech.

Because people with this disease have restricted range of motion, their physical education programs should focus on exercises and activities that increase flexibility. In addition, they should be taught games and sports that effectively make use of leisure time. In most cases, exercises and activities that are appropriate for individuals with arthritis are also acceptable for those with arthrogryposis and OI. Swimming, an excellent leisure activity, encourages the development of flexibility and serves to strengthen weak muscles surrounding joints. Other activities, modified when needed, may include cycling, miniature golf, bowling, shuffleboard, boccie, and track-and-field events.

Multiple Sclerosis

According to the most recent data provided by the National Multiple Sclerosis Society (1997), about 300,000 people in the United States have multiple sclerosis

(MS). Worldwide, it is estimated to affect between 1 and 1.5 million people (Goodkin & Rudick, 1996). In young adults, it is one of the most common central nervous system diseases today. It is a slowly progressive neurological disorder that may result in total incapacitation. Approximately two-thirds of all those afflicted with the disease experience onset between the ages of 20 and 40, and the disorder affects more women than men. The disease may manifest itself in young children or the elderly as well; however, this is rare. MS is characterized by changes in the white matter covering (myelin sheath) of nerve fibers at various locations throughout the central nervous system (brain and spinal cord); the cause is unknown. However, scientists believe that the disease may be a result of a virus attack, an immune reaction, or a combination of both. Current research studies are focusing on myelin formation and its changes, drug therapy, immunotherapy, and diagnostic tests, among others. MS is diagnosed through a variety of tests such as neurological examinations, blood tests, and magnetic resonance imaging (MRI), which can show lesions in the central nervous system.

In MS, the myelin sheath is destroyed and is replaced by scar tissue; a lesion may vary from the size of a pinpoint to about one or two centimeters in diameter. Individuals with MS may exhibit various symptoms, depending on the location of the damage. The most common symptoms are extreme fatigue, heat intolerance, hand tremors, loss of coordination, numbness, general weakness, double vision, slurred speech, staggering gait, and partial or complete paralysis. According to current estimates, about 75 percent of people with MS experience extreme fatigue. The early stage of the disease is characterized by periods of exacerbation followed by periods of remission. As scar tissue continues to replace healthy tissue, the symptoms tend to continue uninterrupted.

Because most people with MS are stricken in the most productive and enjoyable years of life, many are unable to cope emotionally with the disease. Additional stress results from the fact that there is no established treatment that can cure it. The main treatment objective is to maintain the person's functional ability as long as possible. Treatment should be directed toward preventing loss of range of motion (which would result in permanent contractures) and preserving strength and endurance. Many times the disease progresses to a point where the person needs braces or a wheelchair. Intensive therapy or physical conditioning during acute phases of MS may cause general body fatigue. Therefore, physical activities should be judiciously programmed for the individual with MS.

Recent evidence suggests that moderate physical activity on a regular basis significantly reduces nonactivity-related fatigue, which a vast majority of people with MS possess. In addition to the physiological effects, regular physical activity provides psychological benefits as well. According to the National MS Society, while current fatigue, medications are helpful, moderate, regular physical activity seems to be more effective. Mild forms of physical activity that emphasize strength and endurance should be performed for short periods of time. However, the duration and intensity of the activity should be programmed according to the individual's exercise tolerance level. Activities such as bowling, miniature golf, and table tennis are acceptable if regular rest periods are provided. In addition, a variety of stretching exercises and activities are recommended to maintain adequate range of motion. Activities incorporating balance and agility components may prove to be helpful for those exhibiting a staggering gait or varying degrees of paralysis. Many of these activities can be done in water. However, because of heat intolerance by many persons, it is strongly recommended that water temperature remain in the 80s.

Friedreich's Ataxia

An inherited neurological disease, Friedreich's ataxia usually manifests itself in childhood and early adolescence (boys and girls 7 to 13 years of age); however, it can occur as late as the age of 20. The disease was first identified in the 1860s by German neurologist Nikolaus Friedreich, who described the disease as a gradual loss of motor coordination and progressive nerve degeneration. The sensory nerves of the limbs and trunk (peripheral nerves) are affected, and there is a loss of tendon reflexes; the disease may progress in either a slow or a rapid manner. When the disease progresses quickly, many people become wheelchair bound by the late teens and early twenties. Early symptoms may be poor balance and lack of limb and trunk coordination, resulting in a clumsy, awkward, wide-based gait almost indistinguishable from the gait of ataxic cerebral palsy. This is due to the brain's inability to regulate the body's posture and the coordination of its muscle movements. Fine motor control of the upper limbs tends to be impaired, because tremors may be present. Atrophy is common in muscles of the distal limbs. Individuals typically exhibit slurred speech and are prone to seizures. Most will develop foot deformities such as clubfoot, high arches, and hammer toes, a condition in which the toes are curled because of tight flexor tendons of the second and third toes. As the disease progresses, spinal deformities such as kyphosis and scoliosis are common. The majority of individuals exhibit heart problems such as heart murmur, enlarged heart, and constriction of the aorta and pulmonary arteries. Diabetes develops in 10 to 40 percent of people with Friedreich's ataxia. Visual abnormalities include nystagmus and poor visual tracking. There is no known cure at this time; however, research is being conducted to find the gene responsible for the condition. Very recently, researchers have narrowed the gene responsible for the disease to 1 of 46 chromosomes. Therapy consists of managing foot and spinal deformities and cardiac conditions. Medication may be prescribed to control diabetes as well as cardiac, tremor, and seizure disorders. Some common diagnostic tests include electromyogram, which measures electrical activity of muscle cells; nerve conduction velocity, which measures how fast nerves are transmitting impulses; and electrocardiogram, which determines if heart abnormalities exist.

Physical education activities should be planned to promote muscle strength and endurance and body coordination. Activities that develop muscles of the distal limbs such as the wrist, forearm, foot, and lower leg are recommended. The development of grip strength is essential for activities that use implements such as rackets and bats. Individuals exhibiting poor balance and lack of coordination are in need of balance training and activities that encourage development of fundamental locomotor movements. For those with fine motor control difficulties, activities may take the form of riflery, billiards, or archery. Remedial exercises are recommended for people with foot and spinal deviations. Games, exercises, and activities should be programmed according to individual tolerance levels for those with cardiac conditions, and those prone to seizures should be closely monitored.

Myasthenia Gravis

Myasthenia gravis is a disease characterized by a reduction in muscular strength that may be minimal or severe. Worldwide, it affects about 2 out of every 1 million people each year (Watson & Lisak, 1994). It was first identified and named in 1890 by German medical professor Wilhelm Erb. Even when strength is greatly reduced, the individual still has enough strength to perform activities, but often this demands maximum or near maximum effort. In some cases, the disease is easily confused with muscular dystrophy, because muscle weakness affects the back, lower extremities, and intercostal muscles. It affects more females than males and occurs most often in the fourth decade, although some cases have shown that adolescent girls can exhibit the disease. Although the exact cause is unknown, nerve impulses are prevented from reaching muscle fibers because of the production of destructive antibodies by the immune system. The antibodies block the muscle receptors at the neuromuscular junction (points at which nerve endings meet muscle cells).

One of the main symptoms is abnormal fatigue. Muscles generally appear normal except for some disuse atrophy. Weakness of the extraocular and lid muscles of the eye occurs in about one-half of all cases; this results in drooping of the eyelid (ptosis) and double vision (strabismus). Because facial, jaw, and tongue muscles become easily fatigued, individuals may have problems chewing and speaking. Weakness of the neck

muscles may cause the person not to hold the head erect. Back musculature may also be weakened; this leads to malalignment of the spinal column, which can further restrict movement. Muscle weakness makes the execution of the activities of daily living difficult and contributes to low levels of cardiorespiratory efficiency. The disease is not progressive; it may appear gradually, or it may be sudden. It commonly goes into remission for weeks, months, or years. Affected individuals, therefore, live in fear of recurrent attacks. Among other treatments, mysathenia gravis is commonly treated with drugs that help strengthen the neuromotor transmission or suppress the immune system. However, these drugs have serious side effects when taken over a prolonged period. Most recently, a blood flow exchange process is being used; this process removes antibodies from the blood that may interfere with the transmission of nerve impulses.

Physical activities should focus on the development of physical fitness. Because people with myasthenia gravis fatigue easily, their activities should be programmed in a progressive manner and according to individual tolerance levels that take into account the duration, intensity, and frequency of the activity. When the muscles of respiration are weakened, breathing activities are strongly recommended. It is important to strengthen weak neck muscles, especially when the program includes such activities as heading a soccer ball or hitting a volleyball. Fitness levels can be maintained during acute stages through swimming activities incorporated into the program. Poor body mechanics resulting from weak musculature will ultimately affect locomotor skills; therefore, remedial posture exercises and activities should be offered.

Guillain-Barre Syndrome

Guillain-Barre syndrome (also known as infectious polyneuritis or infectious neuronitis) is a neurological disorder characterized by ascending paralysis of the peripheral nerves resulting in acute and progressive paralysis. Initially, the lower extremities become easily fatigued, and numbness, tingling, and symmetrical weakness are present. Paralysis, which usually originates in the feet and lower legs, progresses to the upper leg, continues on to the trunk and upper extremities, and finally affects the facial muscles. Symptoms usually reach their maximum within a few weeks. When initially affected, people involved in locomotor activities of an endurance nature, like distance running, typically find themselves stumbling or falling during the activity session as muscles of the feet and lower legs fatigue prematurely. The condition is frequently preceded by a respiratory or gastrointestinal infection, which suggests that a virus may be the cause. However, attempts to isolate a virus have not been successful. Some authorities believe that the syndrome is an autoimmune disease.

Although rare, Guillain-Barre syndrome occurs worldwide and affects all ages and races. It affects both males and females equally. Parry (1993) reports that worldwide, the syndrome affects 1 to 2 people per 100,000 per year. It may affect both infants and the elderly but seems to cluster in childhood and middle age. While about one-third of those stricken with the syndrome may die, the majority recover, either completely or with minimal paralysis. Acute-stage treatment includes warm, wet applications to the extremities, passive range of motion exercise, and rest.

For people who have made complete recovery, no restrictions in physical activities are needed. However, some individuals who do not completely recover exhibit weakness in limb and respiratory muscles. Their activities can focus on maintaining or improving cardiorespiratory endurance and strength and endurance of unaffected muscles. When a significant amount of weakness remains in the lower extremities, activities may need to be modified accordingly.

PROGRAM IMPLICATIONS

All people with amputations, dwarfism, or les autres conditions can benefit from physical education and sport activities. Should modifications and/or adaptations be needed, they will be determined by the type and degree of physical involvement, motor educability, interest level, and overall educational goals. Especially for persons with unique physical and motor needs, physical education instruction needs to be individualized and personalized.

General Guidelines

A number of general guidelines can be applied to programs for people with amputations, dwarfism, and les autres conditions. The guidelines that follow pertain to safety considerations, physical fitness, motor development, and implications for sports.

Safety Considerations

The physical impairments presented in this chapter are of a medical origin and, as such, the physical education teacher/coach needs to be aware of key safety considerations. Persons with physical limitations should be encouraged to experience as many different postures as possible so that contractures do not develop, leading to decubitus ulcers, which could require hospitalization. This is especially true for persons in wheelchairs such as students with advanced cases of Duchenne MD. Before getting a student into or out of a wheelchair, or positioning a person who has obvious physical limitations, it is important to consult the student's special education teacher and physical and/or occupational therapist (should the student be receiving such services).

Persons with disabilities described in this chapter may tire easily, especially when performing large-muscle activities over an extended period of time. Fatigue exhibited, for example, in MS and Duchenne MD, may cause the person to become frustrated, which in turn adversely affects proper performance. The instructor should permit rest periods and player substitutions when endurance-related activities like soccer and basketball are offered. Should teachers fail to take this fatigue factor into account and program activities without regard to students' personal intensity and duration levels, the risk of medical problems is high.

It is important that physical educators consult medical professionals when establishing physical education and sport programs to meet unique needs. This is especially important for students who are currently under the care of a physician and those currently receiving physical and/or occupational therapy. One of the medical professionals that physical education teachers should be in direct contact with is the school nurse. Typically, the school nurse understands the student's medical condition and knows whether or not the student is taking any prescribed medication. Most of the time the nurse has direct contact with the child's parents and personal physician should medical problems arise during the school day.

Physical Fitness

Reduced muscular strength, flexibility, and cardiovascular endurance levels are common in students with the physical disabilities presented in this chapter, especially those who exhibit severe forms of these conditions. For example, Scull and Athreya (1995) report that children with arthritis possess significantly lower levels of health-related fitness than chronologically age-matched nonimpaired children, and Bar-Or (1997) describes low fitness levels of children with Duchenne MD.

Today, there is a sufficient amount of evidence that appropriate levels of health-related physical fitness contribute to the overall wellness of people with amputations, dwarfism, and les autres conditions. Winnick and Short (1999) have recently developed a health-related, criterion-referenced physical fitness test— Brockport Physical Fitness Test (BPFT)— for individuals with disabilities. They recommend that youngsters with amputations and congenital anomalies be evaluated and achieve physical fitness standards that promote functioning consistent with positive health. This level of functioning includes appropriate fitness levels to adequately perform activities of daily living including physical education and sport activities. This instrument provides selected test items as well as projected standards for the evaluation of aerobic functioning, body composition, and musculoskeletal functioning (flexibility, muscular strength and endurance). Because the conditions described in this chapter are of a medical nature, certain precautions may need to be taken as health-related fitness programs are established and implemented. It is especially important that the teacher be sensitive to the frequency, intensity, duration, and mode of exercises/activities. Because of low levels of fitness, it is recommended that intermittent training be used, especially for students who fatigue easily. Activities performed with greater frequency and with less intensity and duration are recommended. Exercises/activities that are enjoyed will tend to be continued over time. Thus, there will be a greater possibility that these persons may adopt active lifestyles.

Since restricted movement is common in individuals whose conditions are described in this chapter, it is important that strength and flexibility be developed to appropriate levels of physical fitness. Range of motion is basic to overall fitness. Maximizing range of motion allows persons the opportunity to perform physical education and sport skills, as well as activities of daily living, in the most efficient and effective manner possible. Weak musculature and limited range of motion, if unattended, will lead to permanent joint contractures that result in significant loss of movement capability, which, in turn, reduce the person's level of health-related fitness. Surburg (1999) provides flexibility/range of motion guidelines and exercise/activity examples that can be applied to persons with a number of physical conditions (e.g., amputations, MS, Duchenne MD, JRA).

Motor Development

Many times amputations and les autres impairments (and to a lesser extent dwarfism) restrict individuals from experiencing normal movement patterns that are essential to normal motor development. As a result, delays in the development of motor skills and patterns are common. Lack of ability to control movements contributes to the performance of inappropriate motor skills and patterns. Children with congenital amputations, for example, are frequently unable to execute fundamental movements in an appropriate manner. And, children born with the absence of a lower limb may be delayed in acquiring locomotor patterns such as creeping, walking, and running. Muscle atrophy or weakness prevents individuals from developing the strength and endurance levels needed to perform fundamental movements vital to overall health.

Physical education programs should encourage the sequential development of fundamental motor patterns and skills essential for participation in games, sports, and leisure activities. When attempting to enhance motor development, the physical educator should be concerned primarily with the manner in which a movement is performed rather than with its outcome. For example, the teacher should be concerned with the mechanics of the movement within the physical limitations of the student (e.g., dwarfism). The goal of every physical education and sport program should be to encourage individuals to achieve maximum motor control and development within their

ability levels. The motor development of youngsters with the disabilities covered in this chapter may be assessed using standardized tests as well as less formal procedures, including task analysis and rubrics. Interested readers are encouraged to consult chapter 5 regarding information on assessment of motor development and skills.

Implications for Teaching Sports in Physical Education

Physical education teachers are encouraged to integrate many of the sport activities described in the "Adapted Sports" section of this chapter into their physical education programs. For example, sitting volleyball, a Disabled Sports, USA (DS/USA) event can be incorporated into a physical education program as a means of developing eye-hand coordination as well as offering an opportunity for sport competition. Other events included in organized competition such as DAAA, like bowling, archery, cycling, and boccie, can be taught.

Amputees

In general, a physical education program for persons with amputations can follow the same guidelines as one developed for able-bodied individuals. Aside from missing limb(s), people with amputations are considered able-bodied. However, the location and extent of the amputation(s) may require modifications in some activities.

Most persons with amputations typically use prosthetic devices in physical education activities. A person with a unilateral lower-limb amputation usually continues to use the device for participation in football, basketball, volleyball, and most leisure-time activities. As previously stated, more mechanically efficient devices are now being worn by athletes with unilateral lower-limb amputations. In some situations, a unilateral BE, AE, or shoulder amputee may consider the device a hindrance to successful performance and discard it during participation; this is common in baseball or softball. Such is the case with Jim Abbott, a professional baseball pitcher who does not wear a prosthetic device for his BE congenital amputation. Of course, in some activities the prosthetic device must be removed, as in swimming. Currently, the National Federation of State High School Athletic Association allows participating athletes to wear prosthetic devices for interscholastic sports such as football, wrestling, gymnastics, soccer, baseball, and field and ice hockey. However, a device cannot be used if it is more dangerous to other players than a corresponding human limb or if it places the user at an advantage over the opponent (Adams & McCubbin, 1991). For football, the ruling is restricted solely to BK prosthetic devices; for field hockey and soccer, upper-limb and AK prostheses are allowed, although their use is discouraged.

Table 13.1 Participation Guide for Students With Amputations in Selected Physical Education and Sport Activities

Activity	Upper extremity	Lower extremity (BK)	Lower extremity (AK)
Archery	[R]A	R	R
Baseball/softball	R	R	[R]A
Basketball	R	R	[WC]A
Bicycling	R	R	R
Bowling	R	R	R
Field hockey	R	I	[WC]A
Football	R	R	I
Golf	[R]A	R	R
Rifle shooting	[R]A	R	R
Skiing (downhill)	R	R	[R]A
Skiing (cross-country)	R	R	[R]A
Soccer	R	R	I
Swimming	R	R	R
Table tennis	R	R	R
Tennis	R	R	I
Track	R	R	[WC]A
Volleyball	R	R	[WC]A

Note. A = adapted; R = recommended; I = individualized; WC = wheelchair.

Adapted, by permission, from R. Adams and J. McCubbin, 1991, *Games, sports and exercises for the physically disabled*, 4th ed. (Philadelphia: Lea & Febiger), 181.

Participation in physical education and sport activities will require some adaptations depending on the location and extent of the amputation as well as the type of activity. Table 13.1 provides some general participation guidelines for persons with amputations.

Physical Fitness

Like people with other physical conditions, amputees may need to increase their levels of health-related physical fitness. The BPFT (Winnick & Short, 1999) can be used to assess the health-related fitness of youngsters with amputations and congenital anomalies. Depending on the site of amputation, various test items may be chosen to assess aerobic functioning, body composition, and musculoskeletal functioning. For example, a person with a unilateral AK amputation would perform the Target Aerobic Movement Test to assess aerobic functioning. In this test, a number of appropriate

physical activities can be chosen to assess aerobic functioning. An activity would be deemed appropriate if it is of sufficient intensity to reach a minimal target heart rate and to sustain the heart rate within a given target zone. Sustaining a level of moderate physical activity for a least 15 minutes would be a minimal general standard for youngsters with amputations and congenital anomalies (Winnick & Short, 1999). Other BPFT items for a youngster with a unilateral AK amputation would be a triceps and subscapular skinfold measure (body composition) and a bench press (muscular strength and endurance). Individuals with bilateral BK or AK amputations often have lower levels of aerobic functioning than upper-limb amputees, because their locomotor activities may be severely restricted. To encourage aerobic development, the physical educator should offer activities such as marathon racing or slalom events, in which a wheelchair can be used. Using an arm-propelled tricycle is also recommended, especially for those persons exhibiting at least moderate balance problems. Swimming is an excellent aerobic activity; the effect of limb absence can be minimized by the use of flippers to aid in propulsion.

Muscular strength and endurance and flexibility should be developed for all parts of the body, even at the site of the amputation/anomaly. The remaining muscles surrounding the amputation/anomaly also need to develop so that they remain in balance with the nonaffected side. For a person with only a partial limb amputation, such as an ankle or wrist disarticulation, exercises and activities should be programmed to encourage the most normal possible use of the remaining limb segment. Absence of a limb can also affect balance and leverage when performing resistance exercises/activities. Some muscle strength activities are better and more safely performed without the use of a prosthetic device. DiRocco (1999) recommends that, if a force of resistance runs through the shaft of the prosthetic device, then it would be acceptable to wear it for that particular exercise. For example, it would be acceptable for a person to wear a prosthetic device when performing the bench press, because the weight would be distributed through the shaft of the prosthesis when the person attempted to lift the weight.

Both unilateral and bilateral AK amputees have a tendency to be obese and, therefore, should be encouraged to follow a weight-reduction diet along with a program of regular, vigorous physical activity. Short, McCubbin, and Frey (1999) describe a number of desirable characteristics of a weight-loss physical activity program for persons with disabilities. Some of these characteristics include raising daily total energy expenditure, deemphasizing activity intensity and emphasizing activity duration, exercising on a daily basis, participating with partners or small groups, and using well-liked activities.

Persons who typically need to follow a weight-reduction program may have been sedentary for prolonged periods of time. Therefore, Lockette and Keyes (1994) recommend that the following five areas be assessed before a physical activity program is implemented:

Amputation type

Balance and stability

Functional range of motion

Skin integrity

Strength

Motor Skills

Limb deficiency, whether it be of the upper or lower limb, can affect the person's level of motor skills. For example, acquired BE/AE or BK/AK amputations of the dominant limb can initially result in awkward or clumsy performance of motor skills. This may be more pronounced for adolescents and adults who have already mastered motor skills with their dominant limbs (e.g., overarm throw, kicking a ball). The absence of a limb most often affects the center of gravity, to a greater degree in lower-limb amputees than in upper-limb amputees. The result is difficulty with activities requiring balance. Developing both static and dynamic balance is crucial to the performance of locomotor skills such as walking, running, hopping, or for that matter, sitting in a wheelchair. Activities that foster the development of balance and proper body alignment should be encouraged; these may include traversing an obstacle course, performing on a minitrampoline, or walking a balance beam. Speed and agility may also be adversely affected, especially in those with lower-limb deficiencies. People with unilateral AK and bilateral BK or AK amputations are most affected and may have difficulty in locomotor activities that require quick change of direction, such as basketball, football, soccer, and tennis.

While unilateral BK or BE amputees can participate most effectively in physical education and competitive sports, those with bilateral upper or lower amputations will have specific activity restrictions. Bilateral upper-limb amputees can successfully engage in activities that involve the lower extremities to a significant degree, for example, skating, soccer, and jogging.

Unilateral AK amputees can effectively participate in activities such as swimming, waterskiing, snow skiing, weightlifting, and certain field events like the shot put and javelin, which do not place emphasis on locomotion and agility. Those with bilateral BK amputations will be more limited in activities that involve jumping, hopping, or body contact like track events, football, or basketball. Bilateral AK amputees are much more restricted in their activities, usually relying part-time on a wheelchair and using crutches at other times. Activities such as archery, badminton, and riflery, which can be performed from a sitting or prone position, are appropriate.

Dwarfism

Individuals with dwarfism should be encouraged to perform in regular physical education and sport activities. The development of health-related physical fitness and motor skills are important aspects of any physical education program. As such, individuals with dwarfism should have the same opportunities to develop as other individuals. However, there are some considerations that need to be addressed in programs when developing physical fitness and motor skills. Restricted range of motion and joint defects may predispose the individual to dislocations and joint trauma. This is especially true for persons with achondroplasia. Therefore, maintaining flexibility, especially at the elbow joint, is important. Exercises and activities that place undue stress on weight-bearing joints should be avoided or modified to accommodate this limitation. For example, jogging can be replaced with walking or riding a bicycle, and volleyball can be performed with a lighter weight ball. Swimming, which promotes flexibility and cardiovascular endurance, is an excellent activity for persons with achondroplasia, because it does not place undue stress on the joints. Because of shorter limbs, the quality of movement may be affected in some activities, including ball throwing and catching, striking, and locomotor skills such as running, jumping, and hopping. Implements such as golf clubs, rackets, and hockey sticks will need to be adjusted according to the size of the individual. Failure to use appropriate-size implements will result in the execution of inappropriate and/or inefficient motor skills.

Les Autres

Atrophied or weak muscles (e.g., Duchenne MD), reduced range of motion (e.g., JRA), and balance and coordination problems (e.g., Friedreich's ataxia) many times hinder or prevent persons with les autres conditions from performing physical activities. Many of the conditions presented in this chapter are progressive (e.g., Duchenne MD, Friedreich's ataxia, MS, myasthenia gravis), that is, muscles become weaker regardless of the amount of exercise/activity. As a result, it is important to maintain current levels of muscular strength/endurance as long as possible. DiRocco (1999) suggests that persons with progressive muscular disorders do not go beyond 50 percent of their maximum resistance weight when performing muscular strength exercises/activities. According to DiRocco, the exercise intensity level is too great if functional strength does not return within 12 hours of the exercise. This should be monitored closely by the teacher/coach.

Exercises and activities that are fun increase the likelihood that they will be performed on a regular basis. For young children, it is suggested that activities be of short duration and may include rhythmic activities, active lead-up games, and obstacle courses. For adolescents, emphasis should be placed on activities with a focus on lifetime participation such as racket sports, skating, cycling, hiking, and swimming. Persons in wheelchairs can participate in activities such as sledge hockey, rugby, seated aerobics, or wheelchair tennis.

Since various limitations presented in this chapter prevent extended periods of activity, low aerobic fitness levels are common. For example, persons with MS typically exhibit general body fatigue (not related to exercise), which reduces their capacity to perform physical activity over an extended period of time. Furthermore, this inactivity makes persons with les autres conditions (e.g., Duchenne MD, amputations) prone to obesity and ultimately places them at greater risk

for coronary heart disease and other associated conditions (e.g., osteoporosis). To reduce the chance of osteoporosis, Adams (1994) recommends that persons with mild and moderate forms of orthopedic impairments (OI) participate in weight training with light resistance performed from a sitting or supine position. And, persons with mild OI can gain aerobic benefits by brisk walking, cycling, or swimming for extended periods of time. Even though Adams has recommended these activities for persons with OI, they also can be recommended for persons with other les autres conditions as well to reduce the incidence of osteoporosis and other debilitating conditions.

Severe limitations in strength and flexibility, for whatever reason, can also hinder people with les autres conditions from acquiring the motor skills needed to become successful in sport and leisure activities. The inability to perform these skills reduces the person's chances of being physically active and, in turn, places these persons at greater risk of reduced health.

Whether physical activity is performed for purposes of fitness or enhanced motor skills, proper periods of warm-up and cool-down that place emphasis on enhanced flexibility are needed. This is especially true for those with conditions that limit joint flexibility, such as JRA, OI, and other les autres conditions that may elicit spasticity. Daily range of motion exercises are recommended for those who have joint-limiting conditions such as JRA, arthrogryposis, and Guillain-Barre syndrome. Surburg (1999) offers a number of excellent recommendations in *The Brockport Physical Fitness Test Training Guide* on the development of flexibility/range of motion for persons with disabilities.

INCLUSION

As with other disabilities, students with amputations, dwarfism, and les autres conditions must be allowed to participate in regular physical education classes unless the Individualized Education Program (IEP) Team decides otherwise. Most, if not all, people with these conditions can be safely and effectively included in regular physical education settings. And, unless those with les autres conditions are severely physically impaired, most also can be safely and effectively included in regular physical education settings. Even students with severe physical impairments can succeed in regular physical education settings as long as appropriate activities are performed and support services (e.g., teacher aides) are provided. In all cases, however, decisions about inclusion must be made on an individual basis in consultation with the student's IEP Team. The impairments described in this chapter affect physical functioning rather than intellectual functioning. Students, therefore, understand verbal and written directions as well as rules and strategies for various games and sports. In certain cases, teachers may need to structure activities to suit the participants' abilities.

Students with various types of amputations or other lower-limb deficiencies could play goalie in soccer or floor hockey and could pitch or play first base in softball. These students with lower-limb amputations can be provided the option of riding a stationary bike when other students without disabilities are required to run over a long distance. Students with dwarfism can be safely and effectively included in regular physical education classes as long as the proper size equipment is used to accommodate short limbs. This would include shorter and lighter weight implements such as bats, rackets, and hockey sticks. Students with reduced muscular strength and aerobic capacity (e.g., MD, MS, Friedreich's ataxia) can also be safely and effectively included in regular physical education classes as long as activities are modified or activity options are provided. For example, when programming activities for the development of arm/shoulder strength, students with reduced strength can perform modified push-ups (performed with knees touching the floor) or chin-ups with a horizontal bar (performed from a supine position on the floor), while students without disabilities perform push-ups and chin-ups in a traditional manner. Students with joint limitations or deficiencies such as those with JRA, arthrogryposis, and osteogenesis imperfecta can also benefit from regular physical education classes. For example, when programming activities to enhance aerobic capacity, these students can swim or participate in low-impact aerobics while students without disabilities participate in more traditional forms of rope skipping, jogging/running over distance, and bench stepping.

ADAPTED SPORTS

At almost all age levels, people with amputations, dwarfism, and les autres conditions now have the opportunity to become involved in competitive sports. Organizations such as Disabled Sports, USA (DS/USA) and the Dwarf Athletic Association of America (DAAA) assist individuals in reaching their maximum potential in sport. DS/USA and DAAA offer a variety of sporting events that, in many cases, have been modified for specific disabilities. Athletes are able to participate in these events on the basis of their functional abilities. These organizations are members of the Committee on Sports for the Disabled (COSD) of the United States Olympic Committee and members of the International Sports Organization for the Disabled (ISOD).

Amputations

DS/USA sponsors organized competition for athletes with amputations (as well as competition for persons with various birth defects, visual impairments, and neurological conditions). Founded in 1967 by disabled Vietnam veterans, DS/USA is the nation's largest multisport and multidisability organization. National and international competition for athletes with amputations is based on the nine-level classification system previously described in this chapter. People with combinations of amputations not specified in the classification system are assigned to the class closest to the actual disability. For example, a combined AK and BK amputee would be placed in Class A1, whereas a combined AE and BE amputee would be in Class A5. People with single-arm paralysis are tested for muscle strength of the arms and hands. The following movements are tested and scored on a scale from zero to five (five being the greatest function):

Elbow flexion and extension

Finger flexion and extension at the metacarpophlangeal joints

Shoulder flexion, extension, abduction, and adduction

Thumb opposition and extension

Wrist flexion (dorsal and volar)

Classification for participants with single-arm paralysis is limited to A6 (AE) or A8 (BE).

Amputee competition at both the national and international levels takes place in track events such as 100-, 200-, and 400-meter dashes and 800- and 1,500-meter runs, and in field events such as shot put, discus, javelin, long jump, and high jump. National and international competition may also be offered in basketball, volleyball, lawn bowling, pistol shooting, table tennis, cycling, archery, weightlifting, and swimming (100-meter backstroke, 400-meter breaststroke, 100- and 400-meter freestyle, and 4 x 50-meter individual medley). Volleyball and basketball are offered in both sitting and ambulatory categories. In each sport, athletes of similar classifications compete with prostheses, except for those with double AK or combined upper and lower amputations.

In addition to national and regional competitions sponsored by DS/USA, amputees are eligible to compete in events sponsored by Wheelchair Sports, USA (WS/USA), the National Wheelchair Basketball Association (NWBA), and the National Federation of Wheelchair Tennis (NFWT), as long as they have an amputation of the lower extremity and require the use of a wheelchair.

Dwarfism

The DAAA was established in 1985 for the purpose of providing organized sport competition to individuals with dwarfism. While independent of the Little People of America, DAAA maintains ties with that organization. Sports include track (15, 20, 40, 60, 100 and 4 x 100-meter relay); field (shot put, tennis and softball throw, discus, soft discus, and javelin); swimming (freestyle, backstroke, and breaststroke at 25, 50, and 100 meters); basketball; boccie (individual and team); equestrian; soccer; volleyball; table tennis; and powerlifting. Separate competition is offered for men and women except for basketball, volleyball, and team boccie, in which both men and women play on the same team. Skiing is offered in the winter.

Individuals are classified for open division (ages 16 to 39) track, field, and swimming events. There are three classes for track alone, based on a ratio of standing height to sitting height, and three classes for field and swimming, based on the ratio of arm span to biacromial breadth. This system, which is now being refined, is used only for national DAAA events. For international events, the ISOD functional classification system is used.

In order to be eligible for competition, persons with disproportionate dwarfism must be equal to or less than 152.4 centimeters (5 feet) in height, while persons with proportionate dwarfism must be equal to or less than 147.3 centimeters (4 feet 10 inches) in height. Individuals participate in one of five divisions (Open, Youth, Master, Wheelchair, and Futures). Youth events (7 to 15 years) emphasize achieving one's personal best, while the Futures division is for children under 7 years of age. Here, a limited number of events are offered on a noncompetitive basis. DAAA also offers clinics and developmental events.

Les Autres

Historically, the les autres movement was associated with cerebral palsy sports. Les autres athletes performed at the National Cerebral Palsy Games in 1981 and 1983 along with CP athletes. At the National Cerebral Palsy/ Les Autres games in Michigan in 1985, les autres athletes participated in their own separate competition. In 1988 the United States Les Autres Sports Association (USLASA) held a national competition in Nashville. Until very recently, local, regional, and national competition has been held on a regular basis by the USLASA. Currently, however, les autres athletes are again being served by USCPAA. These athletes compete among themselves in events such as track and field, swimming, volleyball, archery, boccie, cycling, shooting, table tennis, wheelchair team handball, and powerlifting. Currently, the classification system for les autres athletes coincides with ISOD. Classes are divided into wheelchair and ambulatory sections. The number of eligible classes may vary with each event. Currently, for track-and-field competition, there are five wheelchair classes and five ambulatory classes, with the recent addition of three jumping classes for certain field events.

SUMMARY

This chapter has described the conditions of amputations, dwarfism, and les autres as they relate to physical education and sport. Physical and motor needs were described, and program and activity suggestions were presented. Recognizing the medical nature of these conditions, teachers and coaches are encouraged to plan activities in consultation with allied medical professionals and the person's personal physician.

14
Spinal Cord Disabilities

Luke Kelly

▶

▶

▶

▶

▶

JERRY AND RICK WERE PLAYING one-on-one basketball. Whoever reaches 15 first wins. The score is 14 all, and Jerry has the ball. Rick is on defense and knows he has to stop Jerry, or he will have to listen to him bragging the rest of the day. Rick positions himself in the center of the lane and gives Jerry an alley to his right. Rick knows Jerry has no left-hand shot and figures he can overplay his right side. Jerry starts at midcourt and drives toward Rick. Jerry makes a head fake first to the left and then to the right and then pulls up and shoots. Rick immediately yells "Travel" as the shot swishes through the net. "Travel! Who are you kidding?" responds Rick. "You overcommitted and got skunked." "Keep dreaming," says Rick, "you took four pushes on your rims without dribbling before taking that shot and that is traveling in my book." Jerry smiles, "Didn't think you could count and play defense at the same time." Rick takes the ball out and drives to Jerry's right. Jerry moves to block the lane. Rick does a 180 in his chair, dribbles the ball once on the floor, puts the ball on his lap, gives two quick pushes on his rims, picks up the ball, and shoots. "That's all she wrote!" yells Rick as the ball banks off the backboard and in. "Nice move," says Jerry. "You ready for another game?"

Rick and Jerry both suffered from spinal cord injuries that resulted in the loss of the use of their legs. They met at a rehabilitation center after their injuries and began playing wheelchair basketball. They both now play on the local wheelchair basketball team.

The focus of this chapter is to review the common spinal cord disabilities and the implications for physical education. After reading this chapter, teachers should be able to modify their programs and/or develop appropriate alternative programs to accommodate the needs of students with spinal cord disabilities. Spinal cord disabilities are conditions that result from injury or disease to the vertebrae and/or the nerves of the spinal column. These conditions almost always are associated with some degree of paralysis due to damage to the spinal cord. The degree of the paralysis is a function of the location of the injury on the spinal column and the number of neural fibers that are subsequently destroyed. Five such spinal cord disabilities will be examined in this chapter: traumatic injuries to the spine resulting in quadriplegia and paraplegia, poliomyelitis, spina bifida, spondylolysis, and spondylolisthesis. In addition, this chapter will cover orthotic devices commonly associated with spinal cord disabilities, as well as physical education and sport implications.

CLASSIFICATIONS

The physical education teacher should be aware of the different systems for categorizing spinal cord disabilities. Medical classifications are based on the segment of the spinal cord that is impaired, while sport organizations choose to classify people by their abilities in order to match similarly able athletes for competition.

Medical

As illustrated in figure 14.1, spinal cord injuries are medically labeled or classified according to the segment of the spinal column (i.e., cervical, thoracic, lumbar, or sacral) and the number of the vertebra at or below which the injury occurred. For example, an individual classified as a C6 complete has a fracture between the sixth and seventh cervical vertebrae that completely severs the spinal cord. The location of the injury is important, because it provides insight related to the functions that may be affected. The extent of the spinal cord lesion is ascertained through muscle, reflex, and sensation testing.

The actual impact of a spinal cord injury is best understood in terms of what muscles can still be used; how strong these muscles are; and what can functionally be done with these muscles in the context of self-help skills (eating, dressing, grooming, toileting), movement (wheelchair, ambulation, transfers, bed), vocational skills, and physical education skills.

Table 14.1 provides a summary of the major muscle groups innervated at several key locations along the spinal column, with implications for the movements, abilities, and physical education activities that may be possible with lesions at these locations. The functional abilities remaining are cumulative as one progresses down the spinal column. For example, an individual with a lesion at or below T1 would have all the muscles and abilities shown at and above that level.

Sport

Sport organizations that sponsor athletic events for individuals with spinal cord disabilities use different classification systems to equate athletes for competition. The most widely used system is the one used by Wheelchair Sports, USA (formerly the National Wheelchair Athletic Association—[NWAA]). This system classifies athletes by functional ability into one of several classes based on the sporting event, the degree of **muscular**

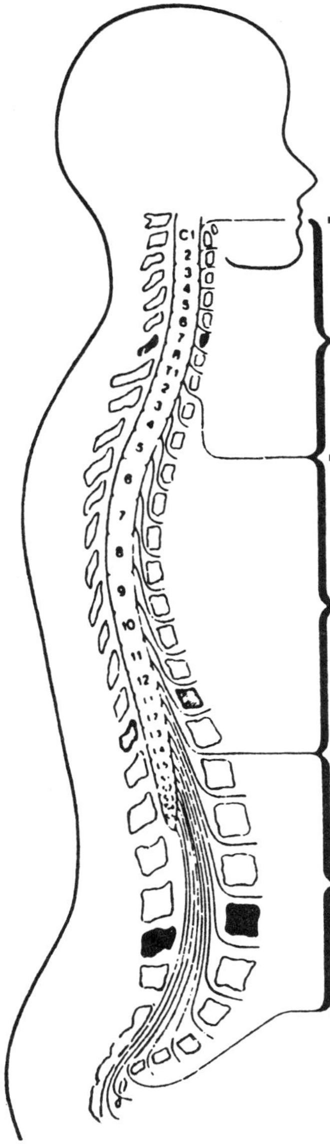

Functional Activities

QUADRIPLEGIA

PARAPLEGIA

Spinal Cord Segments		EATING	DRESSING	GROOMING	TOILETING	HOMEMAKING	DRIVING	PUBLIC TRANSPORTATION	WHEELCHAIR TRANSFERS	AMBULATION	COMMUNICATIONS	BED TRANSFER	VOCATIONAL	SEXUAL FUNCTIONING
Cervical segments C1–T1 (Neck and arm muscles and diaphragm)	C-1	*	*	*	*						*	*	*	
	C-2	*	*	*	*						*	*	*	
	C-3	*	*	*	*						*	*	*	
	C-4	*	*	*	*	*					*	*	*	
	C-5	*	*	*	*	*	*				*	*	*	
	C-6	*	*	*	*	*	*	*			✓	*	*	
	C-7	*	✓	*	✓	*	*	✓			✓	*	*	
	C-8	✓	✓	✓	✓	*	*	✓			✓	*	*	
	T-1	✓	✓	✓	✓	*	*	✓			✓	✓	*	
Thoracic segments T2–T12 (Chest and abdominal muscles)	T-2	✓	✓	✓	✓	*	*	✓			✓	✓	*	
	T-3	✓	✓	✓	✓	*	*	✓			✓	✓	*	
	T-4	✓	✓	✓	✓	*	*	✓			✓	✓	*	
	T-5	✓	✓	✓	✓	*	*	✓			✓	✓	*	
	T-6	✓	✓	✓	✓	*	*	✓			✓	✓	*	
	T-7	✓	✓	✓	✓	*	*	✓	*		✓	✓	*	
	T-8	✓	✓	✓	✓	*	*	✓	*		✓	✓	*	
	T-9	✓	✓	✓	✓	*	*	✓	*		✓	✓	*	
	T-10	✓	✓	✓	✓	*	*	✓	*		✓	✓	*	
	T-11	✓	✓	✓	✓	*	*	✓	*		✓	✓	*	
	T-12	✓	✓	✓	✓	*	*	✓	*		✓	✓	*	
Lumbar and sacral segments — Hip and knee muscles	L-1	✓	✓	✓	✓	*	*	✓	*		✓	✓	*	
	L-2	✓	✓	✓	✓	*	*	✓	*		✓	✓	*	
	L-3	✓	✓	✓	✓	*	*	✓	*		✓	✓	*	
	L-4	✓	✓	✓	✓	*	✓	✓	*	✓	✓	✓	*	
Hip, knee, ankle and foot muscles	L-5	✓	✓	✓	✓	*	✓	✓	*	✓	✓	✓	*	
	S-1	✓	✓	✓	✓	*	✓	✓	✓	✓	✓	✓	*	
Bowel, bladder, and reproduction organs	S-2	✓	✓	✓	✓	✓	✓	✓	✓	✓	✓	✓	*	
	S-3	✓	✓	✓	✓	✓	✓	✓	✓	✓	✓	✓	*	
	S-4	✓	✓	✓	✓	✓	✓	✓	✓	✓	✓	✓	*	

Legend:
- ✓ Normal or near normal function or performance.
- * Needs some type of personal and/or mechanical assistance.
- ** It can be partially available but options need to be discussed on individual basis.
- (blank) Not practical/probable.

Figure 14.1 Functional activity for spinal cord injuries.
Courtesy of Healthsouth Harmarville Rehabilitation Hospital, Pittsburgh, PA 15238.

Table 14.1 Potential Functional Abilities by Select Lesion Locations

Lesion locations	Key muscles innervated	Potential movements	Associated functional abilities	Sample physical education activities
C4	Neck Diaphragm	Head control; limited respiratory endurance	Control of an electronic wheelchair and other computer electronic devices that can be controlled by a mouth-operated joystick.	Riflery Bowling
C5	Partial shoulder Biceps	Abduction of the arms; flexion of the arms	Can propel a wheelchair with modified rims, can assist in transfers, can perform some functional arm movements using elbow flexion and gravity to extend the arm.	Swimming
C6	Major shoulder Wrist extensors	Abduction and flexion of the arms; wrist extension and possibly a weak grasp	Rolls over in bed, may be able to transfer from wheelchair to bed; improved ability to propel wheelchair independently; partial independence in eating, grooming, and dressing using special assistive devices.	Billiards Putting
C7	Triceps Finger extensions Finger flexions	Stabilization and extension of the arm at the elbow; improved grasp and release, but still weak	Independent in wheelchair locomotion, bed, sitting up, and in many cases transferring from bed to wheelchair. Increased independence in eating, grooming, and dressing.	Archery Crossbow Table tennis
T1	All upper-extremity muscles	All upper body lacks trunk stability and respiratory endurance	Independent in wheelchair, bed, transfers, eating, grooming, dressing, and toileting. Can ambulate with assistance using long leg braces, pelvic band, and crutches.	Any activities from a wheelchair
T6	Upper-trunk muscles	Trunk stability; improved respiratory endurance	Lift heavier objects because of improved stability. Can independently put their own braces on. Can ambulate with low spinal attachment, pelvic band, long leg braces, and crutches using a "swing" to gait, but still depend on wheelchair as primary means of locomotion.	Track and field Bowling Weightlifting
T12	Abdominal muscles and thoracic back muscles	Increased trunk stability; all muscles needed for respiratory endurance	Can independently ambulate with long leg braces including stairs and curbs. Uses a wheelchair only for convenience.	Competitive swimming Marathon racing
L4	Lower back Hip flexors Quadriceps	Total trunk stability; ability to flex the hip and lift the leg	Can walk independently with short leg braces and bilateral canes or crutches.	Some standing activities
S1	Hamstring and peroneal muscles	Bend knee; lift the foot up	Can walk independently without crutches. May require ankle braces and/or orthotic shoes.	Normal physical education

functioning, and actual performance during competition. Muscular functioning includes the evaluation of such actions as arm function, hand function, trunk function, trunk stability, and pelvic stability in relationship to their importance in performing a given sporting event. This classification system provides an efficient way of equating competition among a diverse group of athletes with varying types of spinal cord disabilities. The functional classification system is illustrated in table 14.2. It is important to note that, while functional abilities are the key criteria in functional classification systems, there is a relationship between these sport classifications and

the level of spinal cord damage. The approximate spinal cord lesion level associated with the National Wheelchair Basketball Association (NWBA) and several of the functional sport classifications are shown in figure 14.2.

CONDITIONS

Damage to the spinal cord can occur as a result of infectious diseases or from a variety of genetic and environmental causes. This section describes the common causes of spinal cord disabilities and the implications for planning and delivering physical education.

Table 14.2 **Functional Classifications**

Sport/class	Description
Archery	
AR1	Defined as tetraplegic archers in a wheelchair. The archer with upper cervical lesions and triceps not functional against resistance (i.e., test grades 0-3) and the archer with lower cervical lesions and good normal triceps power (i.e., test grades 4-5), wrist extensors and flexors, but without finger flexors or extensors of functional value (i.e., below grade 3 on the muscle test scale), may use a release, compound, or recurved bow, strapping, and body support. All AR1 archers are allowed to use a compound or recurved bow; a release, a finger, or any combination of the above. The equipment will be standard FITA equipment with the exception of the addition of the release and compound bow. The sighting aids must be according to current FITA shooting rules, A) outdoor target archery, article 504, archer's equipment, (a) (v). Archers who use mechanical release may receive assistance in putting arrows in their bows.
AR2	An open class for wheelchair archers. The archers use equipment according to FITA rules. For AR2 division, there must be no more than 15 cm slackness in the back upholstery of the wheelchair to be measured from the front of the main vertical support of the chair back. No strapping to the chair is allowed in the AR2 classification. The height of body support from the top of the chair to the armpit shall be no less than 11 cm.
AR3	A Standing Division for Disabled Archers will be permitted at some Wheelchair Archery, USA sanctioned events.
Field events	
F1	Events: club and discus. Have no grip with nonthrowing arm. Use resin or adhesive-like substance for grip. Discus: Have little control of the discus because finger movements are absent. Throw with a flat trajectory. Club: May club forward or may throw backward over the head. Use either thumb and index finger, index and middle finger, or middle and ring finger grip. Anatomical capability: Have functional elbow flexors and wrist dorsiflexors. May have elbow extensors (up to power 3), but usually do not have wrist palmar flexors. May have shoulder weakness. Have no sitting balance. (Old classification: 1A complete. Neurological level: C6) Note: This system applies to the spinal injured athlete. Athletes whose disability is a result of polio or other causes may show different movement and function than described here. However, the total function of the athlete in this specific event shall be similar to that of the spinal cord injury description.
F2	Events: shot, discus, and javelin. Have difficulty gripping with nonthrowing arm. Shot: Unable to form a fist and therefore do not usually have finger contact with the shot at the release point. Unable to spread fingers apart. Discus: Have no functional finger flexors (i.e., unable to form a fist). Have difficulty placing fingers over the edge of the discus, but may do so with the aid of contractures or spasticity. Javelin: Usually grip the javelin between

(continued)

Table 14.2 *(continued)*

Sport/class	Description
F2 *(continued)*	the index and middle fingers, but may use the gap between the thumb and index finger or between the middle and ring fingers. These athletes may have slight function between the digits of the hand. Anatomical capability: Have functional elbow flexors and extensors, wrist dorsiflexors, and palmar flexors. Have good shoulder-muscle function. May have some finger flexion and extension, but not functional. (Old classification: 1B complete and 1A incomplete. Neurological level: C7) Note: This system applies to the spinal-injured athlete. Athletes whose disability is a result of polio or other causes may show different movement and function than described here. However, the total function of the athlete in this specific event shall be similar to that of the spinal cord injury description.
F3	Have nearly normal grip with nonthrowing arm. Shot: Usually a good fist can be made. Can spread the fingers apart but not with normal power. Use some spreading of the fingers and can grasp the shot put when throwing. Discus: Have good finger function to hold discus. Are able to spread and close the fingers but not with normal power. Javelin: Usually grip javelin between the thumb and index finger. Have ability to hold javelin because of presence of hand muscles that spread and close the fingers. Anatomical capability: Have full power at elbow and wrist joints. Have full or almost full power of finger flexion and extension. Have functional but not normal intrinsic muscles of the hand (demonstrating wasting). (Old classification: 1C complete and 1B incomplete. Neurological level: C8)
F4	Events: shot, discus, and javelin. Have no sitting balance. Usually hold onto part of the chair while throwing. Complete class 2 and upper class 3 athletes have normal upper limbs. They can hold the throwing implement normally. They have no functional trunk movements. Incomplete 1C athletes have trunk movements with hand function like F3. (Old classification: 1C complete, 2 complete, and upper 3. Neurological level: T1-T7)
F5	Events: shot, discus, and javelin. Three trunk movements may be seen in this class: (1) off the back of a chair (in an upward direction); (2) movement in the backwards and forwards plane; and (3) some trunk rotation. They have fair to good sitting balance. They cannot have functional hip flexors (i.e., ability to lift the thigh upward in the sitting position). They may have stiffness of the spine that improves balance but reduces the ability to rotate the spine. Shot and javelin: Tend to use forward and backward movements, whereas the discus predominantly uses rotatory movements. Anatomical capability: Normal upper-limb function. Have abdominal muscles and spinal extensors (upper or more commonly upper and lower). May have nonfunctional hip flexors (grade 1). Have no adductor function. (Old classification: lower 3 and upper 4. Neurological level: T8-L1)
F6	Events: shot, discus, and javelin. Have good balance and movements in the backwards and forwards plane. Have good trunk rotation. Can lift the thighs (i.e., off the chair [hip flexion]). Can press the knees together (hip adduction). May be able to straighten the knees (knee extension). May have some ability to bend the knees (knee flexion). (Old classification: lower 4 and upper 5. Neurological level: L2-L5)
F7	Events: shot, discus, and javelin. Have very good sitting balance and movements in the backwards and forwards plane. Usually have very good balance and movements toward one side (side-to-side movements), due to presence of one functional hip abductor, on the side that movement is toward. Usually can bend one hip backward (i.e., push the thigh into the chair). Usually can bend one ankle downward (i.e., push the foot onto the foot plate). The side that is strong is important when considering how much it will help functional performance. (Old classification: lower 4 and 6. Neurological level: S1-S2)
F8	Events: shot, discus, and javelin. Stand—Standing athletes with dynamic standing balance. Able to recover in standing when balance is challenged. Not more than 70 points in the lower limbs. International F8 Class (must qualify in standing to compete internationally).

Sport/class	Description
F8 (continued)	Sit—Have normal sitting balance and trunk movements in all planes. Usually are able to stand and possibly walk with braces or by locking knees straight. Are unable to recover balance in standing when balance is challenged and will fall when attempting throws with full effort in standing. Not more than 70 points in the lower limbs, who because of poor dynamic standing balance choose to compete from a seated position. U.S. Class only.
Track	
T1	May use elbow flexors to start (back of wrist behind pushing rim). Hands stay in contact or close to the pushing rim, with the power coming from elbow flexion. The old technique is to use the palms of the hands and to push down on the top of the wheel in a forward direction. Anatomical capability: Have functional elbow flexors and wrist dorsiflexors. Have no functional elbow extensors or wrist palmar flexors. May have shoulder weakness. (Old classification: 1A complete. Neurological level: C6)
T2	Usually use elbow flexors to start but may use elbow extensors. Power from pushing comes from elbow extension, wrist dorsiflexion, and upper-chest muscles. Additional power may be gained by using the elbow flexors when the hands are in contact with the back of the wheel. The head may be forced backwards (by the use of neck muscles), producing slight upper trunk movements.
T2A	Have functional pectoral muscles, elbow flexors and extensors, wrist dorsiflexors, radial wrist movements, some palmar flexors. Have no finger flexors or extensors. (Old classification: 1B. Neurological level: C7)
T2B	Have functional pectoral muscles, elbow flexors and extensors, wrist dorsiflexors, palmar flexors, radial and ulnar wrist movements, and finger flexors and extensors. Do not have the ability to perform finger abduction and adduction (spread fingers and bring them together). (Old classification: 1C complete. Neurological level: C8)
T3	Have normal or nearly normal upper-limb function. Have no active trunk movements. When pushing, the trunk is usually lying on the legs. The trunk may rise with the pushing action. Usually use a hand flick technique for power or friction technique. May use the shoulder to steer around curves. Interrupt pushing movements to steer, and have difficulty resuming the pushing position. When braking quickly, the trunk stays close to pushing position. Anatomical Capability: Have normal or nearly normal upper-limb function. Have no abdominal function. May have weak upper spinal extension. Note: Scoliosis (curvature of the spine) usually interferes with abdominal and back muscle function. (Old classification: 1C incomplete, 2, upper 3. Neurological level: T1-T7)
T4	Have backwards movement of the trunk. Usually have rotation movements of the trunk. May use trunk movements to steer around curves. Usually do not have to interrupt pushing stroke rate around curves. When stopping quickly, the trunk moves toward an upright position. Use abdominals for power particularly when starting, but also when pushing. Anatomical capability: Have back extension, which usually includes both upper and lower extensors. Usually have trunk rotation (i.e., abdominal muscles). (Old classification: lower 3, 4, 5, 6. Neurological level: T8-S2)
Swimming	S1-S10 Freestyle, backstroke, butterfly
S1	Very severe quadriplegic with poor head and trunk control, e.g., CP1.
S2	Quadriplegic, complete below C5/6, severe muscular dystrophy; amputation of four limbs.
S3	Quadriplegic, complete below C6; a lower quadriplegic with an additional handicap; severe muscular dystrophy.

(continued)

Table 14.2   *(continued)*

Sport/class	Description
S4	Quadriplegic, complete below C7; some incomplete C5; polio with nonfunctional hands for swimming; muscular dystrophy comparable with C7.
S5	Complete quadriplegic below C8; incomplete C7 or C6 with ability to keep legs horizontal and functional hands for swimming.
S6	Complete paraplegia below T1-T8; incomplete C8 with ability to keep legs horizontal.
S7	Complete paraplegia below T9-L1; double above the knee amputee shorter than 1/2.
S8	Paraplegia L2-L3 with no leg propulsion, but ability to keep legs straight; double above the knee amputee, double below the knee amputee not longer than 1/3.
S9	Paraplegia L4-L5; polio with one nonfunctional leg; single above the knee amputee; double below the knee amputee with stumps longer than 1/3.
S10	Polio or cauda equina lesion with minimal affection of lower limbs, single below the knee amputee; double forefoot amputation.
SB1-SB10	Breaststroke
SB1	Quadriplegic complete below C6; a lower quadriplegic with an additional handicap; severe muscular dystrophy.
SB2	Quadriplegic complete below C7; muscular dystrophy comparable with C7 complete quadriplegia with no finger extension.
SB3	Complete quadriplegic below C8; complete paraplegic T1-T5; incomplete C7.
SB4	Complete paraplegia T6-T10; incomplete C8, or comparable polio.
SB5	Complete paraplegia below T10-L1; incomplete T5; double above the knee amputee shorter than 1/4.
SB6	Paraplegia and polio L2-L3 with no leg propulsion; double above the knee amputation longer than 1/4.
SB7	Paraplegia and polio L4; poor leg propulsion; below the knee amputation shorter than 1/2.
SB8	Paraplegia L5; polio with one nonfunctional leg; double below the knee amputee longer than 1/2; single above the knee amputee.
SB9	Single below the knee amputee less than 3/4.
SB10	Single below the knee amputation longer than 3/4.
SM1-SM10	Individual medley: Medley classes are calculated by (3 x S class)+(1x SB class)/by 4.

Note: Wheelchair Sports, USA also has functional classifications for table tennis and weightlifting.

Traumatic Quadriplegia and Paraplegia

Traumatic quadriplegia and paraplegia refer to spinal cord injuries that result in the loss of movement and sensation. Quadriplegia is used to describe the more severe form, in which all four limbs are affected. Paraplegia refers to the condition in which primarily the lower limbs are affected.

The amount of paralysis and/or loss of sensation associated with quadriplegia and paraplegia is related to the location of the injury (how high on the spine) and the amount of neural damage (the degree of the lesion). Figure 14.1 shows a side view of the spinal column, accompanied by a description of the functional abilities associated with various levels of injury. The functional abilities indicated for each of the levels should be viewed cautiously, because the neural damage to the spinal cord at the site of the injury may be complete or partial. If the cord is severed completely, the individual will have no motor control or sensation in the parts of the body innervated below that point. This loss will be permanent, because the spinal cord cannot regenerate itself. In many cases the damage to the spinal cord will be only partial, resulting in reten-tion of some sensation and motor control below the site of the injury. In a case involving partial lesion, the individual may experience a gradual return of some muscle control and sensation over a period of several months following the injury. This is due not to regen-eration of damaged nerves but rather to the alleviation of pressure on nerves at the injury site caused by bruis-ing and/or swelling. While damage to the spinal cord currently results in a permanent loss of function, a wealth of research is being conducted looking for ways to reverse this process. Promising research results have been found in animals using innovative drug therapies, reactivating dormant nerve cells, and using embryotic transplant therapy. For additional information on these and other advances, consult the electronic resources for this chapter at the end of the book.

Incidence

The National Spinal Cord Injury Association (1999) estimates that approximately 7,800 people suffer spi-nal cord injuries each year in the United States. Among the major causes are automobile accidents (44 percent), acts of violence (24 percent), falls (22 percent), athletic injuries (8 percent [e.g., football, gymnastics, diving]), and other accidents (2 percent). Unfortunately, a large

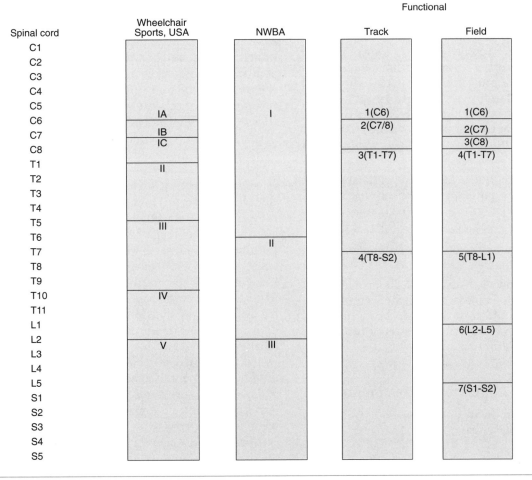

Figure 14.2   Spinal cord levels associated with functional sport classification systems.
Reprinted, by permission, from Wheelchair Sports, USA.

percentage of these injuries happen to students of high school age, with the incidence being greater among males (82 percent) than females (18 percent). It should be noted that, when spinal cord injury is suspected, proper handling of the patient immediately after the injury can play a major role in minimizing any additional damage to the spinal cord.

The American Medical Association (1990) recommends the following procedures whenever a neck injury is suspected:

> *A neck injury should be suspected if a head injury has occurred.* ***Never*** *move a victim with a suspected neck injury without trained medical assistance unless the victim is in imminent danger of death (from fire, explosion, or a collapsing building, for example).* ***WARNING:*** *Any movement of the head, either forward, backward, or side to side, can result in paralysis or death.* (pp.191-192)

Immediate Treatment If the Victim Must Be Moved:

Do not wait and hope someone else will know what to do in this situation. Do the following:

1. Immobilize the neck with a rolled towel or newspaper about four inches wide wrapped around the neck and tied loosely in place. (Do not allow the tie to interfere with the victim's breathing.) If the victim is being rescued from an automobile or from water, place a reasonably short, wide board behind the victim's head and back. The board should extend to the victim's buttocks. If possible, tie the board to the victim's body around the forehead and under the armpits. Move the victim very slowly and gently. Do *not* let the victim's body bend or twist.

2. If the victim is not breathing or is having great difficulty in breathing, tilt the head *slightly* back to provide and maintain an open airway.

3. Restore breathing and circulation if necessary.

4. Summon paramedics or trained ambulance personnel immediately.

5. Lay folded towels, blankets, clothing, sandbags, or other suitable objects around the victim's head, neck, and shoulders to keep the head and neck from moving. Place bricks or stones next to the blankets for additional support.

6. Keep the victim comfortably warm.

Treatment and Educational Considerations

The treatment of individuals with spinal cord injuries usually involves three phases:

Hospitalization
Rehabilitation
Return to the home environment

Although the three phases are presented as separate, there is considerable overlap between the treatments provided within each phase. During the hospital phase, the acute medical aspects of the injury are addressed and therapy is initiated. Depending on the severity of the injury, the hospital stay can last up to several months. Many people with spinal cord injuries are then transferred from the hospital to a rehabilitation center. As indicated by its name, the rehabilitation phase centers on adjustment to the injury and mastery of basic living skills (e.g., toileting, dressing, transfers, and wheelchair use) with the functional abilities still available. Near the end of the rehabilitation phase, a transition is begun to move the individual back into the home environment. In the case of a student, the transition involves working with parents and school personnel to make sure that they have the appropriate skills and understanding of the individual's condition and needs and that they know what environmental modifications will be required to accommodate those needs.

The abilities outlined in table 14.1 are those that can potentially be achieved by people with spinal cord injuries. Unfortunately, to achieve these abilities, individuals must accept their conditions, not be hindered by any secondary health problems, and be highly motivated to work in rehabilitation.

One of the major secondary problems associated with spinal cord injuries is psychological acceptance of the limitations imposed by the injury and the loss of former abilities. Counseling is usually a major component of the treatment plan during rehabilitation. It should be recognized that the rate of adjustment and the degree to which different individuals learn to cope with their disabilities vary tremendously.

People with spinal cord injuries are susceptible to a number of secondary health conditions. One of their most common health problems is pressure sores or decubitus ulcers. These are caused by the lack of innervation and reduced blood flow to the skin and most commonly occur at pressure points where a bony prominence is close to the skin (buttocks, pelvis, and ankles). Because of the poor blood circulation, these sores can easily become infected and are extremely slow to heal. The prevention of pressure sores involves regular inspection of the skin, the use of additional padding in troubled areas, and regular pressure releases (changes in position that alleviate the pressure). Individually designed seat cushions can be made and are used by many people to help better distribute pressure and avoid pressure sores. Keeping the skin dry is also important, because the skin is more susceptible to sores when it is wet from urine and/or perspiration.

A problem closely related to pressure sores is bruising of the skin. Because no sensation is felt in the limbs that are not innervated, it is not uncommon for them to be unconsciously bruised or irritated from hitting or rubbing against other surfaces. Injuries of this nature are quite common in wheelchair activities like basketball if appropriate precautions are not taken. Because these bruises are not felt, they can go unnoticed and eventually can become infected.

A third health problem commonly encountered in individuals with spinal cord injuries is urinary tract infections. Urination is controlled by some form of catheterization on an established schedule. Urinary infections occur when urine is retained in the bladder and backs up into the kidneys. Urinary tract infections can be very severe and usually keep the patient bedridden for a prolonged period of time, which is counterproductive for attitude, rehabilitation, and skill development. Bowel movements must also be carefully monitored to prevent constipation and incontinence. Bowel movements are usually controlled by a combination of diet and mild laxatives. In cases where bowel movements cannot be controlled by diet, a tube is surgically inserted into the intestine. The tube exits through a small opening made in the side and is connected to a bag that collects the fecal excretions.

Two other closely associated problems that frequently accompany spinal cord injuries are spasticity and contractures. Spasticity is an increase in muscle tone in muscles that are no longer innervated because of the injury. This increased muscle tone can nullify the use of other, still innervated muscles. The term "spasm" is frequently used to describe sudden spasticity in a muscle group that can be of sufficient force to launch an individual out of a wheelchair. The best treatment for spastic muscles is to stretch them regularly, particularly before and after rigorous activity. Contractures can frequently occur in the joints of the lower limbs if they are not regularly, passively moved through the full range of motion. A high degree of spasticity in various muscle groups can also limit the range of motion and contribute to contractures.

The last problem commonly associated with spinal cord injuries is a tendency toward obesity. The loss of function in the large-muscle groups in the lower limbs severely reduces the caloric burning capacity of people with spinal cord injuries. Unfortunately, a corresponding loss in appetite does not also occur. Many individuals with spinal cord injuries tend to resume their habitual caloric intake or even to increase it because of their sedentary condition. Weight and diet should be carefully monitored to prevent obesity and the secondary health hazards associated with it. Once weight is gained, it is extremely difficult to lose.

A major key to success in rehabilitation and in accepting a disability is motivation. Many individuals with spinal cord disabilities initially have great difficulty accepting the loss of previous abilities and, subsequently, are reluctant to work hard during the tedious and often painful therapy. Recreational and sport activities are commonly used in both counseling and therapy to provide reasons for working hard and as distractions. A physical educator should be sensitive to the motivational needs of a student returning to a program with a spinal cord disability. Although sport can be a motivator for many, it can also highlight the loss of previous skills and abilities.

The physical education teacher should anticipate needs in the areas of body image, upper-body strength, range of motion, endurance, and wheelchair tolerance. These needs, together with the student's functional abilities, should be analyzed to determine what lifetime sports skills and wheelchair sports are most viable for future participation. These activities then become the annual instructional goals for the physical education program.

While an individual with a spinal cord injury is still learning to deal with the injury, the physical educator can assist by anticipating the person's needs and planning ahead (see the Student Placement application example). This may involve reminding the student to perform pressure releases at regular intervals (lifting the weight off the seat of the chair, by doing an arm press on the arm supports of the chair, or just shifting the sitting position) or bringing extra towels to class to absorb extra moisture in the chair. Because spasticity and spasms are common, stretching at the beginning of class and periodically during the class is recommended. Finally, pads should be provided to prevent bruising in active wheelchair activities. As the student becomes accustomed to the condition, most of these precautions will become automatic habits. A student who has an external bag should be reminded to empty and clean it before physical education class. In contact activities, care should be taken to protect the bag from contact. For swimming, the bag should be removed and the opening in the side covered with a watertight bandage.

Poliomyelitis

Poliomyelitis, commonly called polio, is a form of paralysis caused by a viral infection that affects the motor cells in the spinal cord. The severity and degree of paralysis vary with each individual and depend on the number and location of the motor cells affected. The paralysis may be temporary, occurring only during the acute phase of the illness (in which case the motor cells are not destroyed), or permanent if the motor cells are destroyed by the virus. Bowel and bladder control, as well as sensation in the involved limbs, are not affected in this condition.

Incidence

The occurrence of polio is rare in school-age children today because of the widespread use of the Salk vaccine. The Centers for Disease Control (1995) has reported that there were no new cases of polio reported

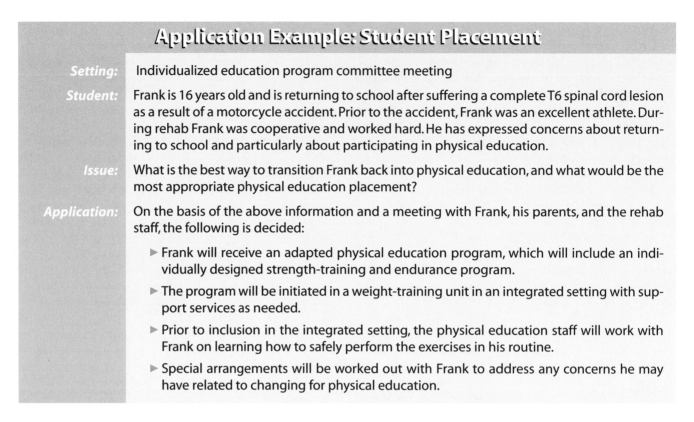

Application Example: Student Placement

Setting:	Individualized education program committee meeting
Student:	Frank is 16 years old and is returning to school after suffering a complete T6 spinal cord lesion as a result of a motorcycle accident. Prior to the accident, Frank was an excellent athlete. During rehab Frank was cooperative and worked hard. He has expressed concerns about returning to school and particularly about participating in physical education.
Issue:	What is the best way to transition Frank back into physical education, and what would be the most appropriate physical education placement?
Application:	On the basis of the above information and a meeting with Frank, his parents, and the rehab staff, the following is decided:

▶ Frank will receive an adapted physical education program, which will include an individually designed strength-training and endurance program.

▶ The program will be initiated in a weight-training unit in an integrated setting with support services as needed.

▶ Prior to inclusion in the integrated setting, the physical education staff will work with Frank on learning how to safely perform the exercises in his routine.

▶ Special arrangements will be worked out with Frank to address any concerns he may have related to changing for physical education.

in the United States from 1985-1994. This is in comparison to over 20,000 reported cases in the United States in 1952. While new cases of polio are uncommon, many individuals who have previously had polio have experienced a recurrence of many of the symptoms of polio later in life. This reoccurrence of symptoms, referred to as post-polio syndrome, affects about 25 percent of former polio victims, usually 35 to 40 years after the original onset of the disease.

Treatment and Educational Considerations

During the acute, or active, phase of the illness, the child is confined to bed. The illness is accompanied by a high fever and pain and paralysis in the affected muscles. After the acute phase, muscle testing is conducted to determine which muscles were affected and to what degree. Rehabilitation is then begun to develop functional abilities with the muscles that remain.

Depending on the severity of the paralysis, a child may require instruction in walking with crutches or long leg braces and/or using a wheelchair. When the lower limbs have been severely affected, it is not uncommon for bone deformities to occur as the child develops. These deformities can involve the hips, knees, ankles, or feet and frequently require surgery to correct.

Specific activity implications are difficult to provide for children with polio because their range of abilities can be so great. Physical educators need to accurately evaluate the abilities and limitations imposed by the condition for each student and then make appropriate placement and instructional decisions.

Many children with only one involved limb or mild involvement of two limbs will already have learned to compensate for the condition and can participate in an integrated physical education setting. Others with more extensive or severe involvement may require a more restrictive setting for the provision of adapted physical education services. Care should be taken not to totally remove these children from inclusive physical education. Whatever the degree of involvement, these children have typical play interests and the desire to be with their classmates.

Regardless of the physical education placement, the emphasis should be on optimal development of the muscles the student does have. Priority should be given to lifetime sports skills and activities that can be carried over and pursued for recreation and fitness when the school years are past. Swimming is an excellent example of an activity that promotes lifetime fitness, provides recreation, and prepares one for other activities such as sailing and canoeing.

Spina Bifida

Spina bifida is a congenital birth defect in which the neural tube fails to close completely during the first four weeks of fetal development. Subsequently, the posterior arch of one or more vertebrae fails to develop properly, leaving an opening in the spinal column. There are three classifications of spina bifida, based on which structures, if any, protrude through the opening in the spine.

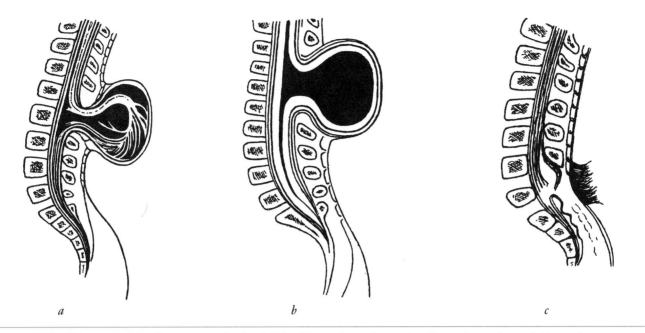

a   *b*   *c*

Figure 14.3   Diagram of the three types of spina bifida: *(a)* myelomeningocele, *(b)* meningocele, and *(c)* occulta. Reprinted, by permission, from G.G. Deaver, D. Buck, and J. McCarthy, "Spina bifida," *1951 Yearbook of Physical Medicine and Rehabilitation*, St. Louis: Mosby Inc, 1952, 10.

Myelomeningocele is the most severe and, unfortunately, the most common form of spina bifida. In this condition the covering of the spinal cord (meninges), cerebrospinal fluid, and part of the spinal cord protrude through the opening and form a visible sac on the child's back (see figure 14.3a). Some degree of neurological damage and subsequent loss of motor function are always associated with this form.

Spina bifida meningocele is similar to the myelomeningocele form, except that only the spinal cord covering and cerebrospinal fluid protrude into the sac (figure 14.3b). This form rarely has any neurological damage associated with it.

Occulta is the mildest and least common form of spina bifida. In this condition, the defect is present in the posterior arch of the vertebra, but nothing protrudes through the opening (figure 14.3c). No neurological damage is associated with this type of spina bifida.

Once detected soon after birth and surgically corrected, the meningocele and occulta forms of spina bifida have no adverse ramifications. The greatest threat in these conditions is from infection prior to surgery.

Incidence

Because some degree of neurological damage is always associated with the myelomeningocele type of spina bifida, it will be the form discussed in the remainder of this section. Approximately 1 child out of every 1,000 live births has spina bifida (Spina Bifida Association of America, 1999), and 80 percent of these children have the myelomeningocele form. The degree of neurologi-

cal damage associated with spina bifida myelomeningocele depends on the location of the deformity and the amount of damage done to the spinal cord. Fortunately, spina bifida occurs most commonly in the lumbar vertebrae, sparing motor function in the upper limbs and limiting the disability primarily to the lower limbs. Bowel and bladder control are almost always lost. The muscle functions and abilities presented in table 14.1 for spinal cord lesions in the lumbar region can also be used to ascertain what functional abilities a child with spina bifida will have.

In addition to the neurological disabilities associated with damage to the spinal cord, myelomeningocele is almost always accompanied by hydrocephalus. This is a condition in which the circulation of the cerebrospinal fluid is obstructed in one of the ventricles or cavities of the brain. If the obstruction is not removed or circumvented, the ventricle begins to enlarge, putting pressure on the brain and enlarging the head. If not treated, this condition can lead to brain damage and mental retardation, and ultimately to death. Today, hydrocephalus is suspected early in children with spina bifida and is usually treated surgically by insertion of a shunt during the first few weeks after birth (figure 14.4, a-c). The shunt, a plastic tube equipped with a pressure valve, is inserted into a ventricle and drains off the excess cerebrospinal fluid. The fluid is usually drained into either the heart (ventriculoatrial) or the abdomen (ventriculoperitoneal) to be reabsorbed by the body. Additional information regarding shunts can be found in the electronic resources section for this chapter at the end of the book.

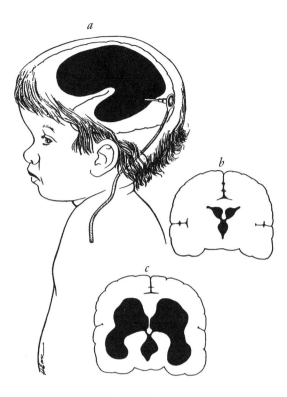

Figure 14.4   Illustration of a shunt being used to relieve hydrocephalus: *(a)* the shunt in place, *(b)* normal ventricles, and *(c)* enlarged ventricles.

Treatment and Educational Considerations

As mentioned earlier, all forms of spina bifida are diagnosed and surgically treated soon after birth. The major treatment beyond the immediate medical procedures involves physical and occupational therapy for the child and counseling and training for the parents. The therapy has two focuses: using assistive devices to position children so that they parallel the developmental positions (e.g., sitting, crawling, standing) through which a normal child progresses and maintaining full range of motion and stimulating circulation in the lower limbs. In conjunction with their therapy, children with spina bifida are fitted with braces and encouraged to ambulate. Even if functional walking skills are not developed, it is very important for the child with spina bifida to do weight-bearing activities to stimulate bone growth and circulation in the lower limbs. Parents are counseled to provide the child with as normal and as many appropriate stimuli as possible. Many parents, unfortunately, frequently overprotect and confine their children with spina bifida, which results in further delays in growth and development.

There are several important similarities and differences in the treatments of spina bifida and of the spinal cord injuries discussed earlier in this chapter. The similarities concern the muscle and sensation loss and the common problems related to these deficits:

Bone deformities

Bruising

Obesity

Postural deviations

Pressure sores

Urinary tract infections

The differences are related to the onset of the conditions. Different circumstances result in different emotional and developmental ramifications.

Generally, children with spina bifida have fewer emotional problems dealing with their condition than do children with acquired spinal cord disabilities, probably because the condition has been present since birth and they have not suffered the loss of any former abilities. However, this is not to imply that they do not become frustrated when other children can perform skills and play activities that they cannot because of their dependence on a wheelchair or crutches.

A major ramification of the early onset of spina bifida is its effect on growth and development. The lack of innervation and subsequent use and stimulation of the affected limbs retard their physical growth. The result is a greater incidence of bone deformities and contractures in the lower limbs and a greater need for orthotics (braces) to help minimize these deformities and assist in providing functional support. A concurrent problem is related to sensory deprivation during the early years of development due to restricted mobility. This deprivation is frequently compounded by overprotective parents and medical problems that confine the child to bed for long periods.

Children with spina bifida are vulnerable to infections from pressure sores and bruises. Pressure sores are most common in those who are confined to wheelchairs. Bruising and skin irritation are particular concerns for children with spina bifida who use crutches and long leg braces. These children have a tendency to fall frequently in physical activities and to be susceptible to skin irritations from their braces if the braces are not properly put on each time. Children with spina bifida have also been shown to have a greater potential to develop an allergic reaction to latex.

Bowel and bladder control present significant social problems for the child with spina bifida during the early elementary school years. Bowel movements are controlled by diet and medication, which are designed to prevent constipation. Urination is commonly controlled today by a regular schedule of catheterization, performed during the day with assistance by the school nurse or an aide. This dependence on others for help with personal functions and the inevitable occasional accident in class can have negative social implications for children with spina bifida and their classmates.

Finally, children with spina bifida have a tendency toward obesity. Several causes contribute to this tendency. The loss of the caloric expenditure typically made by the large-muscle groups in the lower limbs limits the number of calories that can be burned. Caloric expenditure is frequently further limited by the sedentary environment of these children and their limited mobility during the early years. Control of caloric intake is essential to avoid obesity. Unfortunately, food is frequently highly gratifying to these children and is overprovided by indulging parents and caregivers. Obesity should be avoided at all costs, because it further limits the children's mobility and predisposes them to a variety of other health problems.

Children with spina bifida will most likely be placed initially in some combination of adapted and regular physical education settings during the early elementary years and will be fully integrated into regular physical education settings by the end of the elementary years. It is important not to remove these children unnecessarily from inclusive physical education settings. They have the same play and social needs as the other students in their classes. On the other hand, the primary goal of physical education, to develop physical and motor skills, should not be sacrificed purely for social objectives. If the student's physical and motor needs cannot be met in an inclusive setting, the student should receive appropriate support and/or supplemental adapted physical education to meet these needs in a more restricted setting.

In summary, children with spina bifida need to pursue the same physical education goals targeted for other students. As they pursue these goals, their objectives may be different. Modifications may be needed to accommodate their modes of locomotion (crutches or wheelchair) and to emphasize their upper-body development. Emphasis should be placed on physical fitness and the development of lifetime sports skills.

Spondylolysis and Spondylolisthesis

Spondylolysis refers to a congenital malformation of the neural arches of the fourth, or more commonly, the fifth lumbar vertebra. Individuals with spondylolysis may or may not experience any back pain, but they are predisposed to acquiring spondylolisthesis. Spondylolisthesis is similar to spondylolysis, except that, in this condition, the fifth lumbar vertebra has slid forward. The displacement occurs because of the lack of the neural arch structure and the ligaments that normally hold this area in place. Spondylolisthesis can be congenital or can occur as a result of trauma to the back. It is usually associated with severe back pain and pain in the legs.

Treatment in mild cases involves training and awareness of proper posture. Individuals with spondylolisthesis frequently have an exaggerated lordotic curve in the back. In more severe cases, surgery is performed to realign the vertebrae and fuse that section of the spine.

Medical consultation should be pursued before any students with spondylolysis or spondylolisthesis participate in physical education. Children with mild cases may be able to participate in a regular physical education program with emphasis on proper posture, additional stretching, and avoidance of activities that involve severe stretching or trauma to the back. In more severe cases, an adapted physical education program may be required to provide more comprehensive posture training and exercises and to foster the development of physical and motor skills that will not aggravate the condition.

IMPLICATIONS FOR PHYSICAL EDUCATION

Assessment is the key to successfully addressing the physical education needs of students with spinal cord disabilities. Physical educators must work as part of a team to obtain the assessment data they need to provide appropriate instruction. By consulting with each other, the physical educator and the physical and/or occupational therapist can share essential information about goals and objectives for each student. The physical therapist can provide pertinent information about the muscles that are still innervated and those that have been lost, the existing muscle strength and prognosis for further development, the range of motion at the various joints, and the presence or absence of sensation in the limbs. The physical and/or occupational therapist can also provide useful information about adapted appliances (e.g., a device to hold a racket when a grip is not possible) as well as practical guidance on putting on and removing braces, adjusting wheelchairs, positioning and using restraints in wheelchairs, lifting and handling the student, and making transfers to and from the wheelchair.

Within the domain of physical education, the physical educator must be able to assess the physical fitness and motor skills of students with spinal cord disabilities. There is currently only one health-related criterion-referenced physical fitness test available, the Brockport Physical Fitness Test (BPFT) (Winnick & Short, 1999a), that is designed to accommodate individuals with spinal cord disabilities and provides appropriate standards for the evaluation of physical fitness for this population. The BPFT recommends that individuals with spinal cord injuries be evaluated in the areas of aerobic functioning, body composition, and musculoskeletal function. Criterion-referenced test items are provided for each area, with modifications for different levels of functioning based on the level of the spinal cord injury. For example, the following test items would be recommended for a youngster who is a paraplegic and uses a wheelchair: for aerobic functioning—the 15-minute target aerobic movement test; for body composition—the sum of the triceps and subscapular skinfold test; and for musculoskeletal function—the seated push-up, dominant grip strength, and a modified Apley test.

In general, individuals with spinal cord disabilities have placed significantly below students without disabilities at the same age level on physical fitness measures and in motor skill development. Winnick and Short (1985), for example, have reported that 52 to 72 percent of individuals with spinal cord disabilities in their Project UNIQUE study had skinfold measures greater than the median value for same-age subjects who were not impaired and that only about 19 percent of the girls and 36 percent of the boys with spinal cord disabilities scored above the nonimpaired median on grip strength. Winnick and Short (1984) have also reported that youngsters with paraplegic spinal neuromuscular conditions have generally lower fitness levels than normal youngsters of the same age and gender and that these youngsters do not demonstrate significant improvement with age or show significant gender differences like those found among nondisabled youngsters. These results should not be misinterpreted to mean that people with spinal cord disabilities cannot develop better levels of physical fitness. Research has shown, on the contrary, that with proper instruction and opportunity to practice these individuals can make significant improvements in physical fitness. The key is appropriate instruction and practice designed to address individual needs.

Fitness programs for individuals with spinal cord disabilities should focus on the development of all components of physical fitness. While flexibility in all joints should be a goal, particular emphasis should be placed on preventing or reducing contractures in joints where muscles are no longer innervated. These situations require a regular routine of stretching that moves the joints through the full range of motion.

Strength training should focus on restoring and/or maximizing the strength in the unaffected muscles. Care must be taken not to create muscle imbalances by overstrengthening muscle groups when the antagonist muscles are affected. Most common progressive resistance exercises are suitable for individuals with spinal cord disabilities with little or no modification. Posture and correct body mechanics should be stressed during all strength-training activities.

One of the most challenging fitness areas for individuals who have spinal cord disabilities is cardiorespiratory endurance. Work in this area is frequently complicated by the loss of the large-muscle groups of the legs, which makes cardiorespiratory training more difficult. Research has shown that persons with paraplegia typically have about only half the cardiac output compared to individuals without spinal cord injuries; and individuals with quadriplegia tend to have about only a third of the cardiac output of individuals with paraplegia (Figoni, 1997). In these cases, the principles of intensity, frequency, and duration must be applied to less traditional aerobic activities that use the smaller muscle groups of the arms and shoulders. A number of wheelchair ergometers and hand-driven bicycle ergometers have been designed specifically to address the cardiorespiratory needs of individuals with spinal cord disabilities. While it is more difficult to attain the benefits of cardiorespiratory training using the smaller muscle groups of the arms and shoulders, it is not impossible. There are a number of highly conditioned wheelchair marathoners who clearly demonstrate that high levels of aerobic fitness can be attained.

Obesity, unfortunately, is very common in individuals with spinal cord disabilities, largely because the loss of the large-muscle groups of the lower limbs diminishes their capacity to burn calories. Weight control is a function of balancing caloric intake with caloric expenditure. Because, in many cases, caloric expenditure is limited to a large degree by the extent of muscle damage and the subsequent activities that can be undertaken, the obvious solution is to control food intake.

When working on physical fitness with individuals with spinal cord injuries, safety must be a major concern. Individuals with spinal cord injuries, particularly individuals with injuries above T6, are subject to a number of unique problems such as hypotension, problems of thermoregulation, and limits on their maximal exercise heart rates. Hypotension (low blood pressure) is caused by a disruption of the sympathetic nervous system. During aerobic exercise, the body depends on the large muscles in the legs to contract and assist in pumping blood back to the heart. When the legs are not involved, blood can pool in the legs, reducing the amount of blood that is returned to the heart and, subsequently, the heart's stroke volume. Some precautions that can be taken to reduce hypotension include employing appropriate warm-up and cool-down periods as part of the workout, so the body can gradually adapt to the increased workload, and exercising in a reclined position and/or reclining an individual in the chair, if he or she experiences the symptoms of hypotension.

Thermoregulation refers to the body's ability to regulate its internal temperature in response to the outside temperature. The higher the injury on the spine, the greater the problem individuals experience with thermal regulation. When thermal regulation is an issue, care should be taken to have individuals avoid exercising in extremely cold or hot environments. Physical educators should also keep cool compresses available to help individuals cool down after aerobic workouts.

Individuals with spinal cord injuries above T6 are also subject to autonomic dysreflexia and limitations in their maximal exercise heart rates. Autonomic dysreflexia refers to a rapid increase in heart rate and blood pressure to dangerous levels. This can be triggered by a number of factors such as bowel and/or bladder distension, restrictive clothing, or skin irritation. Care should be taken to ensure that individuals empty their bowels and bladders prior to exercise and that their heart rates and blood pressures are monitored during exercise. Disruption to the normal integration of the parasympathetic and the sympathetic nervous systems also limits the maximum heart rate in individuals with injuries above T6 to approximately 120 beats per minute, which limits the aerobic training effects that can be achieved.

In recent years, several excellent resources have been published to assist physical educators in planning and implementing safe fitness programs for individuals with spinal cord injuries (American College of Sports Medicine, 1997; Goldberg, 1995; Lockette & Keyes, 1994; Miller, 1995; Rimmer, 1994; Winnick & Short, 1999b).

In addition to physical fitness and motor skill areas, physical educators should concentrate on posture and body mechanics. Individuals with spinal cord disabilities frequently have poor body mechanics as a result of muscle imbalances and contractures. Exercises and activities that contribute to body awareness and alignment should, therefore, be stressed.

In terms of sports skills, the most valuable activities are those that have the greatest carryover potential for lifetime participation. The selection of activities should provide a balance between warm- and cold-weather sports as well as indoor and outdoor sports. Preference should be given to sports that promote physical fitness and for which there are organized opportunities for participation in the community. Just about any sport (e.g., golf, tennis, swimming, skiing) can be adapted or modified so that individuals with spinal cord disabilities can participate.

The goal of assessment is to obtain the most accurate and complete data possible so that the most appropriate placement and instruction can be provided. Physical educators must be willing to devote both the time and effort required to obtain this assessment data if they wish to help their students reach their maximum potential in physical education. Due to the uniqueness of each spinal cord disability, physical educators need to use their skills in task analysis to develop their own authentic assessment tools and scoring rubrics to evaluate and teach functional motor skills. Working cooperatively with a team is the key to maximizing staff efficiency and the benefits for the students. Additional information on assessment and development of physical fitness and motor development is presented in other chapters in this text.

Once goals and objectives have been established for a student with a spinal cord disability, the next challenge is how to provide the instruction to achieve these goals in as inclusive a physical education setting as possible. Some physical education content, like physical fitness, lends itself to inclusion, because the content has to be individualized for all of the students. Since all students are at different levels of fitness in terms of their flexibility, strength, and endurance, it is common to develop individual routines based on the students' assessed needs for all students and then to use a circuit approach to have the students work on their programs. In an individualized setting like this, it is easy to accommodate the unique needs of individuals with spinal cord disabilities by defining fitness routines to meet their unique needs around the same stations as the other students. Other physical education content, such as working on sports skills like the volleyball serve, can initially appear more difficult to modify, because skills like this are traditionally taught to the class as a whole and then activities and games are used where everyone is expected to perform the same skill. For example, how do you include a wheelchair-bound student with a spinal cord injury in a volleyball game if the student cannot hit the ball hard enough when serving to get the ball over the net and cannot move quickly enough in the wheelchair to defend part of the court? What are some possible modifications? The weight of the ball could be adjusted. The student could be allowed to serve closer to the net. The student could be given a smaller zone to defend on the court after the ball is served. More students could be assigned to each side to reduce the amount of space that has to be defended.

While the physical educator is responsible for creating an inclusive setting where the needs of all students are addressed, it is also important that students with spinal cord disabilities be taught how to advocate for themselves. This form of self-advocacy involves the students being able to analyze new games and activities and then offer suggestions on how the activities could be modified so that they can successfully participate. This is a critical skill for all students with disabilities, since the physical educator will not always be there to orchestrate the modifications. A common transition step from teacher-directed accommodations to those that are student directed is to have the class help to develop modifications and accommodations. This process also sensitizes the other students to the value of making activities appropriate for all participants. The golden rules to making accommodations are that the activity must still serve its original purpose and the individuals with disabilities must be working on the same skills and/or similar skills modified to meet their unique needs. Assigning students with spinal cord impairments to roles like scorekeeper, because they are in wheelchairs, is not appropriate inclusion, because they are

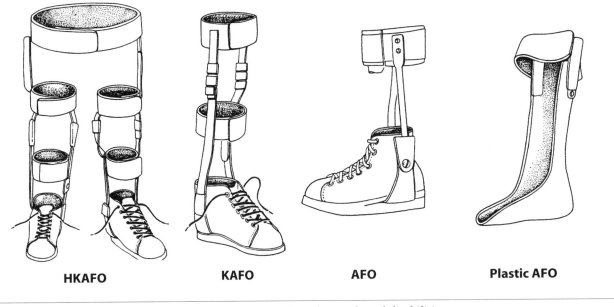

HKAFO   **KAFO**   **AFO**   **Plastic AFO**

Figure 14.5   Common orthotic devices worn by individuals with spinal cord disabilities.

not developing the skills they need to live and maintain an active healthy lifestyle. In the long term, inclusive physical education settings are beneficial for everyone. Inclusion requires that instruction be based on assessment and be individually designed. It also requires successful participation of all students in the planned drills, games, and activities, which should result in positive learning outcomes for all students.

ORTHOTIC DEVICES

Due to the neuromuscular limitations imposed by spinal cord disabilities, many individuals with these disabilities use orthotics to enhance their functional abilities. Orthotic devices are a variety of splints and braces designed to provide support, improve positioning, correct or prevent deformities, and reduce or alleviate pain. Although presented in this chapter, the use of orthotic devices is not limited to individuals with spinal cord disabilities. Orthotic devices are prescribed by physicians and fitted by occupational therapists, who also instruct users in wearing and caring for the devices. Examples of the more common orthotics are shown in figure 14.5. They are used both by individuals who are ambulatory to provide better stability and by individuals in wheelchairs to prevent deformities. These devices are commonly referred to by abbreviations that describe the joints they cover: AFO—ankle-foot orthotics, KAFO—knee-ankle-foot orthotics, and HKAFO—hip-knee-ankle-foot orthotics. Additional information on orthotics can be found in the electronic resources for the chapter at the end of the book.

Many of the newer plastic orthotics can be worn inside regular shoes and under clothing. Under normal circumstances, orthotics should be worn in all physical education activities with the exception of swimming. In vigorous activities, physical educators should periodically check that the straps are secure and that no abrasion or skin irritation is occurring where the orthotics or straps contact the skin. Orthotics can also be used to improve positioning to maximize sensory input. Figure 14.6 shows a series of assistive devices often used with children with spina bifida. The purpose of these orthotics is to allow these children with spina bifida to view and interact with the environment from the normal developmental vertical postures that they cannot attain and maintain on their own. The last device is a parapodium, or standing table. The parapodium allows individuals, who otherwise could not stand, to attain a standing position from which they can work and view the world. The parapodium frees the individual from the burden or inability to balance and bear weight and also affords complete use of the arms and hands. The parapodium can be used effectively in physical education to teach a number of skills such as table tennis. In recent years, several companies have devised ingenious modifications of the parapodium that can be easily adjusted to a number of vertical and horizontal positions.

A secondary category of orthotic devices includes canes and walkers that are used as assistive devices for ambulation. The Lofstrand or Canadian crutches are most commonly used by individuals with spina bifida and spinal cord disabilities who ambulate with leg braces and crutches. Physical educators should be aware that a person using only one cane or crutch employs it on the strong side, thus immobilizing the better arm. This should be considered and appropriate modifications (i.e., to maintain balance) should be made for skills

Normal milestones without aids

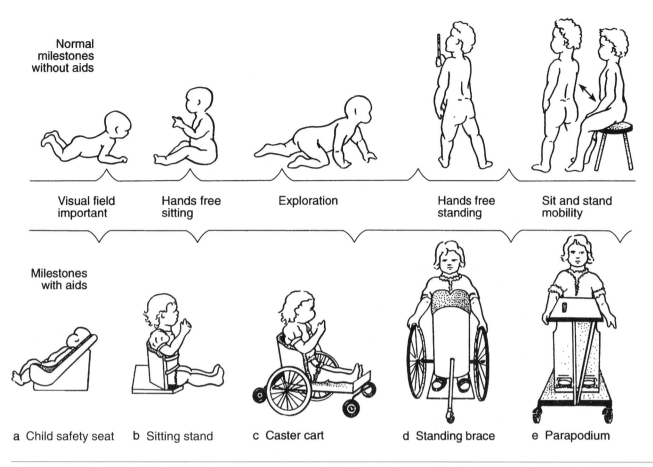

Visual field important | Hands free sitting | Exploration | Hands free standing | Sit and stand mobility

Milestones with aids

a Child safety seat | b Sitting stand | c Caster cart | d Standing brace | e Parapodium

Figure 14.6 Illustration of the use of orthotic devices to help children with spina bifida attain normal developmental postures.

in which use of the better arm is desired. For additional information on canes and crutches, consult the electronic resources for this chapter at the end of the book.

ADAPTED SPORT ACTIVITIES

Today, there are a large number of organizations that sponsor athletic programs and sporting events for individuals with spinal cord disabilities. These organizations have evolved from the need to provide athletic and recreational opportunities for individuals with spinal cord disabilities who want to participate in sport.

The NWAA, now known as Wheelchair Sports, USA, was formed in 1956 and is one of the most notable organizations sponsoring athletic events for people with neuromuscular disabilities resulting from spinal cord injuries, spina bifida, or polio. Wheelchair Sports, USA sponsors competitive events in pistol shooting, riflery, swimming, table tennis, weightlifting, archery, fencing, wheelchair slalom, and track and field. Competitors are classified into divisions by age and functional ability (see table 14.2). There are two age divisions: adult, for individuals age 16 and older and junior, for individuals 8 to 15 years old. As discussed earlier in this chapter, there is growing interest in replacing the system with a functional classification system (see figure 14.2), which is now used for international competitions.

Disabled Sports, USA plays a major role in organizing and sponsoring both competitive and noncompetitive winter sports events for individuals with disabilities. Disabled Sports, USA was formally called National Handicapped Sports (until 1994) and the National Handicapped Sport and Recreation Association (until 1976). Disabled Sports, USA is a leader in the development and dissemination of recreation/fitness and training materials related to sport and recreation for individuals with disabilities. One example is Disabled Sports USA's "Aerobics Tapes," which include a series of aerobic dance videotapes designed especially to help individuals with paraplegia, quadriplegia, amputations, and cerebral palsy develop endurance, strength, and flexibility (for more information consult the audiovisual resources for this chapter at the end of the book). Disabled Sports, USA also serves as the official governing body for sport competitions for individuals with amputations.

A number of special organizations sponsor athletic competitions in a specific sport. Although many of these organizations employ the Wheelchair Sports, USA classification system, the National Wheelchair Basketball Association (NWBA) has its own system. The NWBA was formed in 1949 and sponsors competition for men, women, and youth. In the NWBA classification system, each player is classified as a I, II, or III, depending on the location of the spinal injury and the degree of motor loss. Each class has a corresponding point value of one, two, or three. A team can comprise players in any combination of classes as long as the total point value for the 5 players does not exceed 12 and there are not more than 3 class III players playing together at the same time. The classifications are made according to the criteria described in the next section (National Wheelchair Basketball Association, 1999).

NATIONAL WHEELCHAIR BASKETBALL ASSOCIATION CLASSIFICATION SYSTEM

The criteria for the three classifications in the NWBA system are as follows:

Class I: Complete motor loss at T7 or above or comparable disability, where there is total loss of muscle function originating at or above T7.

Class II: Complete motor loss originating at T8 and descending through and including L2, where there may be motor power of hips and thighs. Also included in this class are amputees with bilateral hip disarticulation.

Class III: All other physical disabilities as related to lower-extremity paralysis or paresis originating at or below L3. All lower-extremity amputees are included in this class, except those with bilateral hip disarticulation (see class II).

The NWBA classification system provides an excellent model for equating team sports competition in physical education classes that include integrated students with spinal cord disabilities. A similar classification system based on students' skill levels could also be designed and used in physical education classes.

In the past decade, wheelchair sports have evolved from primarily recreational events into events that are highly sophisticated and competitive. Many of the initial advances in wheelchair sports were the direct result of technical advances in wheelchair design and/or research related to the postural and body mechanics of wheelchair propulsion. In recent years, attention has shifted to defining and improving training and conditioning programs for wheelchair athletes (Ferrara & Davis, 1990; Gayle & Muir, 1992; Wells & Hooker, 1990).

Specialized, organized sports programs serve as an extension of the physical education curriculum. They provide students with spinal cord disabilities with the same opportunity to gain the benefits and experiences all athletes derive from sport, and they act as an additional source of motivation. These activities also give participants an opportunity to meet and interact socially with others who have similar characteristics, interests, and needs. Finally, these sport experiences expose students with spinal cord disabilities to positive role models, who demonstrate the difference between having a disability and being handicapped. To maximize the probability that students with disabilities will both attempt and succeed in sports, physical educators must ensure that the physical education curriculum provides both instruction in the fundamental sports skills and the appropriate transition from skill development to skill application in actual sport situations.

SUMMARY

Individuals with spinal cord disabilities need to pursue the same physical education goals as other students. To successfully meet the needs of these students, physical educators must have a thorough understanding of the nature of the disabilities and the functional abilities that can be attained. They can then build on this knowledge base to determine the most appropriate placement and instructional programming for each student. Particular emphasis in the program should be placed on body awareness, physical fitness, and the development of lifetime sports skills. Adapted sports, using functional ability classification systems, provide excellent opportunities for individuals with spinal cord disabilities to apply and practice the skills learned in physical education.

15

Other Health-Impaired Students

Paul R. Surburg

▷

▷

▷

JOSH AND JANE ARE TWO students who have asthma and hate to go to physical education class. To date, their experiences in physical education class have not been rewarding or pleasant. Their former teacher's approach was that all students, regardless of any circumstance or condition, had to do all class activities without modification. This teacher felt that particularly students with asthma used their condition to get out of activities. This semester Jane and Josh will have a teacher who has national certification in adapted physical education. This teacher has already talked with them about how certain situations will be modified to accommodate their conditions. Their teacher has reminded Josh and Jane about taking their mediation an hour before physical education class, has tried to eliminate the gym of offending substances such as dusty mats, and has worked out specific warm-up and cool-down routines for these two students.

The purpose of this chapter is to provide insights into ways teachers may help students with a variety of conditions and disabilities experience safe and successful experiences in physical education. A brief explanation of each condition or disability is provided to establish some basic concepts about these conditions. Building on this background knowledge, suggestions are given regarding ways to modify and implement an individualized, condition-specific program of meaningful physical education activities.

THE OHI CLASSIFICATION

When the public laws from IDEA (PL 94-142 through PL 105-17) were discussed in earlier chapters, 1 of 11 official disabling or handicapping conditions identified was **other health impaired** (OHI). By design, this category is a broad designation that provides coverage under the law for a variety of disabling conditions and diseases. While students may be eligible for special education services under the OHI classification, there are several reasons this designation is not utilized.

First, a student with diabetes or asthma may not want to be labeled as disabled in the vernacular of certain school settings. In many states the OHI designation must be officially diagnosed by a physician. A family physician may not want to confirm the diagnosis and place a certain stigma on a student. Parents often intercede with the doctor to prevent the label of disabled being imposed on their child.

Second, in many states the development of an Individualized Education Program (IEP) that deals exclusively with physical education is a rare situation. Without certain limitations on the student in the academic learning environment, the probability of establishing an IEP only for physical education is low. Nevertheless, while OHI students may not be designated as disabled, regular and adapted physical educators must understand those students' conditions and the best way to provide meaningful physical education experiences for them.

DIABETES MELLITUS

Diabetes mellitus is a chronic disease characterized by insufficient insulin and disturbances of carbohydrate, protein, and fat metabolism. Eight million Americans know they have this disease and another 8 million have the disease but do not know it (Kriska, 1997). It is responsible for 50 percent of myocardial infarctions and 75 percent of all strokes; it is a major cause of new blindness and a causal factor for amputations (Duda, 1985). Ten percent of documented cases occur among school-age children. This statistic has direct implications for physical education programs, because there is an excellent chance of having a student with diabetes in class.

Insulin-dependent diabetes mellitus (IDDM) has, in certain cases, been linked to a viral origin; however, genetic or hereditary causes have also been associated with this type of diabetes. Also cited as a possible cause is an autoimmune reaction, where a virus may affect body cells and the immune system no longer identifies these cells as part of the body. These cells, which may be beta cells in the pancreas, are then destroyed or rejected by the body. Several factors seem to increase the likelihood of developing IDDM:

▶Emotional stress
▶Medications such as corticosteroids
▶Obesity
▶Oral contraceptives
▶Pregnancy

Types of Diabetes

There are two types of diabetes: IDDM, also known as ketosis-prone or juvenile diabetes, and non-insulin-dependent diabetes mellitus (NIDDM), also known as ketosis-resistant or maturity-onset diabetes. Insulin-dependent diabetes mellitus occurs most frequently before age 30 but may develop at any age. This type of diabetes requires daily insulin injections and dietary management. The treatment of non-insulin-dependent diabetes, by contrast, may be by diet alone or in conjunction with insulin or hypoglycemic agents. It should be noted that 80 percent of persons with NIDDM are overweight (Helmrich, et al., 1991).

Because most school-age diabetics are insulin dependent, this section will deal exclusively with this form of the disease. IDDM is an autoimmune disorder, although it can be viral or genetic, in which pancreatic cells, particularly in the Islets of Langerhans, degenerate over time; consequently, insulin production is diminished (Jimenez, 1997). Related complications are interference with the action of insulin at the cellular level, faulty storage of sugar in the liver, overproduction of sugar in the liver, and diminished utilization of sugar at the tissue level.

Insulin-dependent diabetes may become apparent dramatically with ketoacidosis or gradually with NIDDM. With a deficiency of insulin the body cannot convert glucose to glycogen for storage in the liver, and an excess of sugar accumulates in the blood. Inadequate carbohydrate utilization also affects fat metabolism. When there is improper fat usage, waste products called ketones are formed, providing a condition called acidosis.

Physical Education and the Student With Diabetes

The observation of certain procedures and protocols should ensure students with diabetes successful involvement in physical education activities.

Both the student and physical educator must be aware of the interrelationship of diet, insulin intake, and exercise (Taunton & McCargar, 1995). The site of the insulin injection may depend on the kind of exercise

planned. Insulin is a protein that cannot be taken orally, because the digestive juices would destroy it; therefore, it must be injected subcutaneously. Injections are needed once or twice a day and should be administered at sites where the muscles are not used extensively in a sport or physical activity. Thus, for a student who is participating in track, the injection site should be the stomach rather than the legs. Another important consideration is the interaction between exercise and insulin dosage. There is an inverse relationship between the intensity of activity and the amount of insulin needed. A heightened level of activity necessitates a reduction in insulin dosage. The physical educator must alert the diabetic student to any planned changes in kind or intensity of activity; in anticipation, the student will have to make appropriate adjustments in insulin dosage and/or diet.

Generally, the student with diabetes tries to keep three factors constant: exercise, insulin usage, and diet. A variation in one or more of these factors may precipitate a physiological imbalance. For example, a reduction in calories and an increase in exercise intensity may lead to insulin shock; an increase in calories and a reduction in insulin can lead to diabetic coma. Please see box 15.1 for a description of the symptoms of diabetic coma and insulin shock.

A student with diabetes must constantly monitor the sugar level in the body. Several times a day the student will use a urine sample or blood test to evaluate the sugar level; the teacher should provide the time and privacy to conduct this test. A key to diabetes control is keeping the blood glucose level in the normal range. Close monitoring permits diabetics to determine how well they are managing their condition and is essential for safe participation in physical education or athletics.

There are some other factors for both the physical educator and the diabetic to consider. A student involved in an intense exercise or training program should increase food intake rather than altering insulin dosage. Carbohydrate loading is not appropriate for diabetics; rather, they should take in carbohydrates during exercise, because they lack the capacity for storing glycogen. As diabetics become older, circulatory problems may become a real concern. This should not exclude the school-age student from participating in contact or collision sports. Good skin care, however, should be encouraged by the physical educator. Table 15.1 provides the components for good foot care that a student with diabetes should follow (Conti & Chaytor, 1995).

Physicians today recognize how beneficial participating in all kinds of physical activity — including contact sports—can be for students with diabetes. We live in a society that is increasingly aware of the benefits of physical fitness; these benefits extend to students with diabetes. See the Track and Field application example describing the arrangements made for a track-and-field unit for a student with diabetes mellitus.

> **Box 15.1 Symptoms of Diabetic Coma and Insulin Shock**

Symptoms of *ketoacidosis* or *diabetic coma* are extreme thirst; dry mucous membranes; labored breathing; weak and rapid pulse; sweet, fruity breath odor; vomiting; and high sugar content in the urine. This condition is a first-rank medical emergency and should be treated at a hospital with dosages of insulin and intravenous fluids for a dehydrated state.

Symptoms of *insulin shock* or *hypoglycemia* are fatigue, excessive perspiration, hunger, double vision, tremor, and absence of sugar in the urine. If this condition is not in an advanced stage, the ingestion of fruit juice, a candy bar, honey, or a carbonated (not diet) drink should rectify the sugar imbalance. When shock is severe enough to result in unconsciousness, an injection of glucagon or dextrose is used to elevate the glucose level in the blood.

Table 15.1 Good Foot Care

Action	Reason
Inspect feet on daily basis	Look for redness, blisters, cuts, which can lead to infections
Avoid bare feet even for brief periods	Slippers should be worn at night and sandals in locker rooms to prevent getting any type of infection
Wash feet on a daily basis	Care should be taken to wash between toes where fungus can thrive
Trim toenails straight across	Ingrown toenails may lead to skin breaks and infections
Change socks every day and after every workout or physical education class	Skin irritations must be prevented

SEIZURE DISORDERS

Seizure disorders, convulsive disorders, and epilepsy are terms used to describe a condition of the brain that is characterized by recurrent seizures. These episodes are related to erratic electrochemical brain discharges. It is estimated that seizure disorders occur in 650 per 100,000 people in the general population (Bennett, 1989); the incidence is higher in families that have a history of such disorders. Possible causes of epilepsy are severe birth trauma, drug abuse, congenital brain malformation, infection, brain tumor, poor cerebral circulation, and head trauma. Between 50 percent and 70 percent of all cases are idiopathic, with no known cause and no structural damage to the nervous system. The Epilepsy Foundation of America claims that each year at least 200,000 Americans sustain seizure-producing head trauma.

Types

Several systems have been developed to categorize seizure disorders. One approach dichotomizes this condition into generalized nonconvulsive and general convulsive seizures. Another system classifies seizures as grand mal, petit mal, jacksonian, focal epilepsy, psychomotor, jackknife, and Lennox-Gastaul syndrome. The classification system endorsed by the Epilepsy Foundation of America has four categories: partial seizures, generalized seizures, unilateral seizures, and unclassified seizures.

Partial seizures originate from a localized area of the brain and usually are symptom specific. The traditional terms psychomotor and jacksonian would indicate seizures in this category. A jacksonian seizure begins as a localized seizure and may subsequently involve surrounding areas of the brain. It is manifested in jerky or stiff movements of an extremity, with tingling sensations in the limb. The seizure my start in a finger, spread to the hand, and finally affect the rest of the arm. Sometimes a partial seizure of this type becomes a generalized tonic-clonic (grand mal) seizure.

The complex partial seizure may have different types of symptoms, but generally purposeless behavior is in evidence. The person may exhibit a glassy stare, produce indiscernible speech, engage in purposeless wandering, and emit strange noises. After this type of seizure, a person may seem confused and is sometimes mistakenly thought to be drunk or on drugs.

Myoclonic, atonic, generalized tonic-clonic, and absence seizures are forms of generalized seizures. Absence seizures (petit mal), often found in children, usually last from 1 to 10 seconds. There are no overt manifestations, such as postural changes or interruption of activities, but there is a brief change in the level of consciousness. A teacher may mistake this change in consciousness for daydreaming or inattentiveness. This type of seizure rarely occurs during exercise. Myoclonic seizures are characterized by brief, involuntary quivers of the body, sometimes occurring in rhythmic fashion. An atonic seizure is sometimes called "drop attack" because the person falls to the floor and is unconscious for a few seconds. This brief episode may result in an injury from the fall.

A generalized tonic-clonic (grand mal) seizure is the most common type of seizure. In many people it is preceded by a warning or an aura, such as a dreamy feeling, nausea, a visual disturbance, or olfactory sensations. Sensations of this nature are probably the beginning of abnormal brain discharges.

A generalized tonic-clonic seizure may begin with a guttural cry, loss of consciousness and falling to the floor, boardlike rigidity changing to quivering and jerking, wild thrashing movements, foaming at the mouth, labored breathing, and incontinence. This series of events may last from two to five minutes; the person

Application Example: Track and Field

Setting:	High school physical education class
Student:	15 year old with insulin-dependent diabetes mellitus
Unit:	Beginning a track-and-field unit after a golf unit
Task:	Inform student of the change in intensity level of this new unit so the student may make appropriate adjustments in medication and food intake
Application:	The physical education teacher might include the following procedures:

- ▶ Appoint a buddy to monitor any signs that the student is having any type of diabetic reaction.
- ▶ Develop a prearranged signal so carbohydrates may be provided to the student.
- ▶ Check that the student has proper footwear and does not start class with a low blood sugar level. (Did the student have breakfast or lunch?)
- ▶ Meticulously take care of any type of injury.

then regains consciousness, may be somewhat confused, and may feel sleepy and fatigued.

The following factors seem to promote the occurrence of seizures:

Alcohol consumption

Constipation

Flashing lights of a certain velocity

Hyperventilation

Increase in blood alkalinity

Menstrual period

Psychological stress

While the cause of many seizure disorders has not been determined, treatment is highly successful. Approximately 80 percent of people with seizure disorders can control the conditions by using anticonvulsant drugs. Medications frequently prescribed for epileptics include phenytoin, phenobarbital, carbomazeprine, and primidone. Phenytoin (Dilantin) is prescribed for generalized tonic-clonic seizures because it does not produce the side effect of drowsiness. In females it may cause increased hair growth on the face and extremities. Phenobarbital (Luminal) is also used for generalized seizures but may produce hyperactivity with elementary school-age students. For partial seizures ethosuximide (Zarontin) and trimethadione (Tridione) are often prescribed. Refer to box 15.2 for a description of first aid measures needed for tonic-clonic seizures.

If a seizure occurs, it is important to make a detailed written account of it. This information is not only important for the school system's incident reports but also provides input that lets the physician evaluate the severity and duration of the person's seizures and adjust drug dosages, if necessary.

Physical Education and Students With Seizure Disorders

Participation in physical education by students with seizure disorders has evolved through several stages. At one time physical education for this population was relatively passive, with activities like croquet, golf, and bowling being recommended. The next stage was a more vigorous program, with most activities being permitted except contact sports. Today, all activities including contact sports are deemed appropriate; the only stipulation is that the student's seizures be controlled

▶ Box 15.2 First-Aid Measures for a Generalized Tonic-Clonic Seizure

If an epileptic has some warning or aura of an impending seizure, help the person into a supine position. Remove glasses and false teeth, if appropriate. Loosen tight-fitting clothes, and place a pillow or rolled-up cloth material under the head. If there is no warning, help the person into a lying position and follow the procedures just described; in addition, clear the area of hard objects. Do not attempt to restrain movements or force anything into the person's mouth, because tongue blades or other objects may break teeth and lacerate the mouth. If the mouth is open, some authorities advocate placing a soft object such as a clean, folded handkerchief between the teeth. Turn the head to provide an open airway and allow saliva to drain from the mouth. After the seizure subsides, the person should be told what happened, given reassurance, and, if tired, be allowed to rest.

and the activities be supervised (Lang, 1996). The possibility of a blow to the head causing a seizure does not seem to be a valid reason for exclusion. Another misconception is that physical activity will precipitate seizures; in reality, physical activity may elevate the seizure threshold. The extent to which seizures are controlled through medication is the key factor in the selection of physical education activities.

Frequently, students with seizure disorders do not exhibit normal levels of physical fitness and motor skill development. While progress has been made toward controlling seizures, social factors such as protective parents and the stigma associated with a seizure disorder contribute to a passive lifestyle. Physical educators may help these students become more actively involved in sports and physical activity.

A major concern of the physical educator is the student's safety. While students with seizure disorders may engage in most activities, certain precautions should be observed. To engage in aquatic activities, which are often an object of concern, students should have their seizures under control. For swimming, the buddy system should be standard operating procedure. Also, the physical educator should be sure that the medication is not causing a side effect that may compromise safety procedures. If seizures are not controlled or side effects pose a risk, then activities such as weight training, gymnastics, and rope climbing must be modified or excluded from the student's program. However, because 80 percent of all students with epilepsy have their seizures under control, most students with seizure disorders may engage in a varied and beneficial physical education program.

ASTHMA

The student with asthma may be apprehensive about participating in physical education class. The regular or adapted physical educator should help to alleviate this anxiety; unfortunately, some physical educators have been inept in handling situations with asthmatic students. The purpose of this section is to provide guidance concerning the involvement of asthmatic students in physical education classes.

There are 35 million people affected by asthma in the United States; approximately 50 percent of all cases begin in children under the age of 10. Twice as many males as females are affected by this condition.

While there are various theories concerning the types and causes of asthma, this condition is now considered an inflammatory process (Gaskin, 1994). The mechanisms for the obstructive symptoms are spasm of the muscular layer in the bronchial walls, swollen mucous membrane, and mucus secretions in the airways. All three conditions reduce the diameter of the bronchial passages and impede air flow (Rupp, 1996).

Systems for classifying asthmatic disorders are oriented toward causation. One system reflects a dichotomy of causes. Extrinsic or allergic asthma occurs in children or adults who have a history of allergic reactions. These people are allergic to such offending substances as dust, pollen, animal danders, mold spores, and certain drugs. Intrinsic or nonallergic asthma may be induced by excessive exercise or may be psychosomatic in nature. Some authorities contend that a person may suffer from a combination of both extrinsic and intrinsic types.

The student with asthma may present symptoms ranging from brief episodes of wheezing; shortness of breath; and coughing to breathlessness, which will cause the person to talk in one- to two-word sentences, to tighten neck muscles with inhalation, and to exhibit a blue or gray coloration of lips and nail beds. When a student with asthma has difficulty breathing, the physical educator should help establish an upright position with shoulders relaxed, advise the student to drink a lot of fluids, provide reassurance, and encourage the student to take appropriate medication if one has been prescribed for such occasions. The school nurse, parent, or physician should be notified if a prescribed medication does not seem to be effective.

Physical educators, especially, should be aware of a condition known as either exercise-induced asthma (EIA) or exercise-induced bronchospasm (EIB). Mahler (1993) estimates that 90 percent of persons with asthma experience EIA. Exercise of high intensity or duration seems to precipitate muscular contraction of the bronchial tubes, which results in an asthmatic attack. A commonly accepted hypothesis is that intrathoracic airways are cooled during exercise by air that has not been completely warmed. This is related to abnormal increases in the rate and depth of breathing, because the coolness affects airway mast cells, which liberate bronchoconstrictive substances.

The following factors influence the occurrence of EIA:

Absence of warm-up

Condition of air

Exercise intensity

Interval versus sustained exercise

Type of exercise

Use of preexercise medication

The implications of these factors for the physical education setting will be reflected in guidelines presented in the section called "Physical Education for Students with Asthma." Rupp (1996) contends that physical fitness improvement can help reduce the occurrence of EIA and reduce medication needs.

Medication and Asthma

Recent advances in medication markedly have helped to prevent EIA and have permitted persons with asthma to be active in physical education and sport. For general control of asthma and to facilitate exercise participation, anti-inflammatory therapy has emerged as the first line of treatment. Inhaled corticosteroids are frequently prescribed by physicians for all ages who have asthma (Disabella & Sherman, 1998). Cromolyn sodium and beta adrenergic agonists such as terbutaline are often the agents of choice for persons who have persistent problems with corticosteroids. These drugs should be administered 30 minutes to an hour before physical education class or sport participation. A word of caution must accompany the use of these pharmacological agents. First, there is concern among physicians that aerosols may be used to excess, and some recommend oral medication for children and certain adolescents. Excessive aerosol use can have deleterious effects; a relationship has been established between excessive use and an increase in mortality rate of asthmatics in Great Britain. Second, the use of certain drugs is prohibited in high-level competition. For example, use of isoproterenol or ephedrine would disqualify an athlete from Olympic competition.

Physical Education for Students With Asthma

Physical educators who plan to integrate the student with asthma into the physical education class should follow these guidelines:

▶Consult the school nurse and/or the records to ascertain the status of the student's asthmatic condition. If no up-to-date information is available, a conference should be scheduled with the appropriate individualized education committee. (See chapter 4 for a discussion of these committees.)

▶Together with the student, discuss feelings about exercise and collectively develop goals for physical education.

▶Conduct warm-up activities at the beginning of each class before bouts of vigorous activity. The purpose is to increase the body temperature until a mild sweat is developed. Warm-up might consist of walking, jogging, light mobilizing activities, and even some strengthening activities.

▶Classes should last from 30 to 40 minutes; shorter classes and interpolated rests should initially be used for the unfit student.

▶If possible, students with asthma should engage in physical activity four or five times a week.

▶Gradually increase the level of exercise intensity. With an interval training regime, exercise intensity may be elevated to 70 percent of maximum predicted heart rate. Rest intervals should be only brief enough to re-

duce the heart rate to 50 percent of maximum predicated heart rate.

▶Administer preexercise medication as prescribed by the physician.

▶Establish emergency procedures for coping with EIA episodes.

▶End each class with a cool-down period. At no time should vigorous activity be stopped abruptly. The student should at least walk around the gymnasium until the heart rate returns to within 20 beats per minute of the resting level.

▶Short-burst (anaerobic) or short-duration activities should be the predominant type of curricular activity for asthmatic students. For example, softball is a short-burst activity; soccer would not fall into this category.

▶While general strengthening exercises are valuable for the asthmatic, special emphasis should be placed on developing abdominal, trunk, and shoulder muscles. Gymnastics may serve as a valuable adjunct to this strength development phase.

▶Class participation should emphasize cooperative endeavors and enjoyment of team play rather than winning.

▶Aquatic activities provide many benefits. Conducted in a warm, damp environment, they promote control of breathing and involve numerous muscle groups.

▶Other activities that emphasize breath control may be incorporated into the student's curriculum. Karate and various forms of dance are valuable in this regard.

▶A healthful environment should be maintained. The physical educator should regularly clean and air gym mats; and students should be responsible for keeping their gym clothes, lockers, and gym shoes clean. Mats, shoes, and lockers harbor molds and dust.

▶While specific breathing exercises are a topic of controversy, activities that stress exhalation may be of value in a physical education class. Blowing Ping-Pong balls while on a gym scooter or in a swimming pool may help expiration and facilitate air flow out of the lungs. To help prevent hyperventilation, make sure students exhale twice as often as they inhale. Other activities that emphasize exhalation are balloon relays, blowing tissues in the air, and laughing.

▶Whether a student is on the gym floor, in the swimming pool, or on the softball field, provision should be made for disposing of coughed-up mucus. Tissues, some type of spittoon, or other convenience should be available for the asthmatic student.

▶A well-conducted physical education program should not only help students learn to pace themselves but also help them experience the joy of physical activity.

Exercise sessions should be omitted or markedly reduced on days when wheezing is a problem.

CANCER

Approximately 135,000 Americans die of cancer each year; it is second only to cardiovascular disease as a cause of death (Lee, 1995). While cancer is often associated with advancing age, it is a major cause of death among children from birth to early teens. Only accidents claim more victims among this age group.

Types

Cancer is usually categorized according to the tissue that is affected. Malignant tumors of connective, muscular, or bone tissue are called sarcomas; while neoplasms of the epithelial tissue that covers surfaces, lines cavities, and constitutes glands are referred to as carcinomas. A clear understanding has not been established concerning the mechanisms of cell division; cancer cells differ in size and multiply more rapidly than normal cells. This uncontrolled, rapid growth may originate at a primary site and spread, or metastasize, to other locations of the body. Certain substances such as asbestos, nitrogen mustard, and cigarette smoke have been identified as carcinogenic in nature. Among school-age students, leukemia and tumors of the central nervous system are the most frequently occurring forms of cancer. Bone tumors are more common in children than adults, with peak ages between 15 and 19 years.

Treatment of neoplasms involves one or a combination of these therapeutic approaches: surgery, radiation, and chemotherapy. Surgery is used to remove a bulky tumor, relieve pain, correct obstructions, and/or reduce pressure on surrounding structures. Radiation is applied to impede cell multiplication and destroy cancerous cells; it may reduce tumor mass and help with pain control. Ionizing radiation (gamma rays) and particle radiation (beta rays) are targeted at the cellular deoxyribonucleic acid (DNA). Radiation therapy may be delivered through external beam radiation or isotope implants. Chemotherapy involves a variety of drugs to inhibit tumor growth or impede metastasis. Chemotherapeutic agents inhibit cell growth by interacting with DNA, competing with metabolites, preventing cell reproduction by altering protein synthesis, and changing chemical susceptibility. While certain drugs may be effective in certain situations, chemotherapy may cause side effects of pain, anemia, loss of hair, vomiting, and dermatitis.

A holistic approach for the terminally ill person is hospice care, which includes comprehensive physical, psychological, social, and spiritual care. Good hospice care emphasizes coordinating efforts of health care staff, maintaining quality of life in a homelike environment, and providing emotional support for both patient and staff. In several large cities hospice programs have been established for children with leukemia.

Physical Education and the Student With Cancer

The involvement of the student with cancer in physical education is predicated on the following factors: the student's health status, the importance of physical activity as perceived by the oncologist, the kind of support available from parents, and the willingness of physical educators to cope with this type of student. The student with cancer would be eligible for the benefits of IDEA under the category of other health-impaired conditions.

Exercise as a maintenance or restorative technique for cancer patients holds promise. Buettner and Gavron (1981) reported that men and women who had a history of cancer benefited from an eight-week aerobic training program. When a sedentary control group was compared to the exercise group, the latter group exhibited significant improvement on five physiological measures. Another study compared an exercise group receiving chemotherapy for breast cancer with a control group (Winningham & MacVicar, 1985). Subjects in the exercise group exhibited more improvement on a graded exercise test than did the control group. Participants reported that their feelings of nausea decreased as the exercise session progressed. These studies and a review of work in this area lead Lee (1995) to conclude that exercise may enhance the quality of life for persons with cancer and may have a positive influence on the immune system.

While participation in exercise and physical education programs may be beneficial, each student's needs must be dealt with individually. The type of cancer, its status, and the general condition of the student determine the extent and intensity of the program. The physician must have input concerning the nature and intensity of activities. A person with lung cancer will definitely have respiratory restrictions, and a student with osteogenic sarcoma, a type of bone cancer, may not be allowed to jog or even walk because of a risk of fractures. While the physical well-being of a student with cancer cannot be minimized, psychological well-being is important as well. Being included into physical education and succeeding in this environment may greatly improve the student's self-image. To help minimize self-consciousness, the physical educator may let the student wear a ball cap or other head covering in class if hair loss has resulted from the medical treatment. Other variations in gym attire should be allowed when dermatological side effects are evident.

There may be days when chemotherapy or its side effects preclude active involvement in many activities. The following symptoms or situations are contraindications for exercise: pain in the legs or chest, unusual weakness, nausea during exercise, vomiting or diarrhea a day before class, sudden onset of labored breathing, dizziness or disorientation, and intravenous chemotherapy administered a day before class.

On other days the student may be able to engage in many physical education activities. Seventy percent of cancer patients report that a one-two punch of the malignancy and aggressive treatment reduces strength and energy. Patients are encouraged to increase their exercise level after high-dose chemotherapy (Sternberg, 1997).

Adaptations discussed in other sections of this book may be appropriate for students with certain types of cancer. The student with leukemia may profit from program modifications recommended for an anemic student or one with a cardiovascular problem. Adaptations for a student with Perthes disease or other orthopedic problems may be appropriate for the student with a bone tumor. A student with a tumor of the central nervous system may need the same adaptations as one who has cerebral palsy or muscular dystrophy. The key to physical education for the student with cancer, as for all students with unique needs, is the IEP.

CARDIOVASCULAR DISORDERS

The cardiovascular system, which comprises the heart, arteries, veins, and lymphatic system, may be considered the life-giving transportation system of the body. Disorders of this system appear during all stages of development and age periods. The following are major categories of cardiovascular disorders: congenital acyanotic defects, congenital cyanotic defects, acquired inflammatory heart disease, valvular heart disease, degenerative cardiovascular disorders, cardiac complications, and vascular disorders. An estimated 25 million Americans have some type of cardiovascular disease. While heart disease is the leading cause of death in people over 25, it is the sixth leading cause of death among people in the 15- to 25-year age range.

A student with a cardiovascular disease or problem may be covered by IDEA under the designation of "other health impaired." While IEPs including physical education may be developed for certain students with cardiac conditions, other students with this type of condition may be integrated into regular physical education without the benefit of an IEP. In both situations, the physical educator must know how best to involve students who are or have been afflicted with some type of cardiovascular problem. This section will orient the reader to some cardiovascular problems commonly found among school-age or preschool-age children and will recommend procedures and techniques for the delivery of physical education services to these students.

Rheumatic Heart Disease

A condition included in the broader designation of acquired inflammatory heart disease is rheumatic heart disease. Rheumatic fever is a systemic inflammatory disease of children that may occur following an un-treated or inadequately treated streptococcal infection. Rheumatic heart disease is the name given to the cardiac manifestations of rheumatic fever; it is estimated that 500,000 young people between the ages of 5 and 19 are affected by this condition.

Rheumatic fever, which usually follows a strep throat, may be a hypersensitive reaction to the streptococcal infection, in which antibodies produced to combat the infection affect specific tissue locations and produce lesions at the heart and joints. Approximately 1 to 3 percent of strep infections develop into rheumatic fever. Some common symptoms of rheumatic fever are polyarthritis, motor awkwardness (chorea), pancarditis (myocarditis, pericarditis, and endocarditis), skin rash, and subcutaneous nodules near tendons or bony prominences of joints. While most of the symptoms are transitory in nature, the destructive effect of rheumatic fever lies in pancarditis, which develops in up to 50 percent of cases. Endocarditis causes a scarring of the heart valves, which may result in a stenosis or narrowing of valve openings; this scarring may also prevent the valve leaflets from completely closing, which causes a regurgitation between heart chambers. Both situations cause the heart to work more intensely and, if the damage is quite severe, congestive heart failure may occur. The mitral valve is more often affected in girls and the aortic valve in boys.

Treatment of rheumatic fever and rheumatic heart disease begins with eradicating the streptococcal infection, relieving symptoms, and preventing recurrence of the strep throat or rheumatic fever. During the acute stage of the disease, penicillin is often prescribed along with aspirin. With active carditis, bed rest may be indicated for as long as five weeks. Following the acute stage, penicillin may be prescribed to prevent recurrence. With severe valve damage, repair or replacement may be the treatment of choice. The integration into physical education classes of the student who has had rheumatic fever or rheumatic heart disease will be discussed in the next section. The physical educator may help with a very important aspect of combating rheumatic fever or rheumatic heart disease, namely, preventive measures. If a student in a physical education class presents symptoms of a cold with a severe sore throat, the student should be referred to the school nurse or advised to obtain a throat culture. A strep throat must be treated with antibiotics to ensure that it does not lead to rheumatic fever.

Physical Education and Students With Cardiovascular Conditions

Determining the appropriate intensity level for students with cardiovascular disorders is the key to their integration into a physical education class. An affected student who is attending school should be involved in some type of physical education program. Numerous systems have been developed to classify individuals with cardiovascular disorders into groups that provide a baseline for activity selection. The system most frequently used to determine levels of intensity is based on metabolic equivalents (METs). Physicians often refer to this system in recommending exercise programs. The level of intensity, at least initially, should be based on the recommendation of the physician.

One MET is the equivalent of the basal oxygen requirement of the body at rest, or 3.5 milliliters of oxygen per kilogram of body weight per minute. If a person's maximum MET capacity is determined to be 7.0, the exercise prescription for a person with a heart condition is approximately 70 percent of this value or 4.9 METs. Using this information the physical educator may consult a table that recommends appropriate activities for different MET levels (table 15.2).

A new classification system is being implemented by physicians to correlate intensity of activity with types of activities (Mitchell, Haskell, & Raven, 1994.) This system may be used in physical education and sport settings. Table 15.3 presents this classification system. A perusal of this table indicates that the intensity level increases as one progresses from left to right or from top to bottom. The term dynamic exercise involves changes in muscle length and joint movement with rhythmic contractions. With these activities there are increases in oxygen consumption, cardiac output, and heart rate. Static exercise involves development of a relatively large intramuscular force with little or no change in muscle length or joint motion. With these activities, there will be increases in systolic, diastolic, and arterial pressure.

While physical education programs for students with cardiovascular problems must be personalized, the following suggestions should help in the implementation of each student's program:

1. Secure approval of the general framework of the physical education program (see chapter 3).
2. Gradually increase the level of intensity of all exercises and activities.
3. Stop all activity following pain in the sternal area, palpitation of the heart, cyanic appearance of the lips and nail beds, swelling of the ankles, or labored breathing.
4. Provide appropriate rest periods.
5. Reduce the intensity level of an activity if elementary children are squatting between activities or high school students are standing and breathing through their mouths.
6. Monitor pulse rate by comparing preexercise, exercise, and postexercise rates.
7. Reduce the intensity of certain exercises by having students perform them in a lying position.
8. Carefully monitor aquatic activities, which cause students to use a considerable number of muscle groups, thus placing greater demands on the cardiovascular system.
9. Reduce the highly competitive aspects of certain sport activities, which may elevate stress levels.
10. Modify programs according to climatic conditions: Hot and/or humid weather, for example, necessitates a reduction in intensity level.

Table 15.2 Metabolic Equivalents for Selected Activities

METs*	Activities
3-4	Archery
	Bowling
	Billiards
	Croquet
	Table tennis
5-6	Dancing (jitterbug, etc.)
	Golf (pulling cart)
	Roller skating
	Swimming (2 mph)
	Tennis (doubles)
7-8	Basketball (5-person teams)
	Cycling (10 mph)
	Running (12 minutes/mile)
	Tennis (singles)
	Touch football
9-10	Cycling (13 mph)
	Handball
	Racquetball
	Running (10 minutes/mile)
	Squash

*Metabolic equivalents (METs) are the body's basal oxygen requirement at rest.

Table 15.3   Classification of Activities for Physical Education

	Low static	Moderate static	High static
Low dynamic	Billiards Bowling Golf Riflery	Archery Equestrian	Field events (throwing) Gymnastics Karate/judo Rock climbing Weightlifting
Moderate dynamic	Baseball Softball Table tennis Tennis (doubles) Volleyball	Fencing Field events (jumping) Football Rugby Running (sprint) Synchronized swimming	Bodybuilding Wrestling
High dynamic	Badminton Cross-country skiing (classic technique) Field hockey Race walking Racquetball Running (long distance) Soccer Squash Tennis (singles)	Basketball Ice hockey Cross-country skiing (skating technique) Lacrosse Running (middle distance) Swimming Team handball	Cycling Rowing Speed skating

Adapted, by permission, from the American College of Cardiology, *Journal of the American College of Cardiology*, 1994, Vol. 24, 845-899.

11. Include a warm-up period. Although part of every student's program, it is even more important for students with cardiovascular problems.

12. Caution against extremes in temperature variations of ingested fluids.

13. Provide assistance in weight management. Added poundage places additional stress on the cardiovascular system.

14. Understand the possible interactions between medication and activity.

15. Modify intensity levels by the following means:

   ▶ Playing doubles in tennis and badminton and using the boundaries of the singles court

   ▶ Using courtesy runners in softball and kickball

   ▶ Allowing the ball to bounce more than the usual number of times in racquetball, volleyball, and tennis

   ▶ Reducing the velocity of balls by using a beach ball in place of an official volleyball and using Wiffle balls rather than regulation balls

ANEMIA

Anemia is a condition marked by a reduction in erythrocytes (red blood cells) or in the quality of hemoglobin, which is associated with oxygen transport throughout the body. A reduction in the oxygen-carrying capabilities of the blood necessitates increased cardiac output to compensate for the reduced amount of oxygen provided at the cellular level.

Characteristics of Anemia

Anemia may be an inherited condition or may be caused by hemorrhaging and dietary deficiencies. Hemorrhagic anemia may result from infections, severe trauma, postoperative or postpartum bleeding, and coagulation defects. Iron deficiency anemia is prevalent in young children, preadolescent boys, male and female athletes, pubescent girls who are experiencing their first menstrual periods, and premenopausal women (Browne, 1996). Iron is a main component of hemoglobin and is needed in the production of red blood cells. Lack of adequate stores of iron in the body leads to depleted erythrocyte mass and decreased hemoglobin concentration.

Sickle-cell anemia occurs most commonly, but not exclusively, in African-Americans. It results from defective hemoglobin, which causes red blood cells to have a sickle shape. This condition is linked to a recessive trait that produces defective hemoglobin molecules. About 1 in 10 African-Americans carries this abnormal gene; this situation is referred to as sickle-cell trait. If both parents have the trait, the chances are one in four that their child will have the disease. It is estimated that 1 in every 400 to 600 black children has sickle cell anemia. Hypoxia seems to affect the abnormal or defective hemoglobin molecules in the red blood cells, which become insoluble. An end result is an elongation or sickle-like appearance of these erythrocytes. This sickling effect may result in cell destruction and may impair circulation in capillaries and small blood vessels. A vicious cycle may be set up, with circulation impairment causing anoxic changes, which may lead to additional sickling and obstruction. To date, treatment consists mainly in management of symptoms, with hospitalization for severe aplastic crises. In these crises, blood transfusions, oxygen administration, and fluid ingestion are used. Symptoms accompanying sickle-cell anemia are jaundice, labored breathing, aching bones, chest pains, swollen joints, fatigue, and leg ulcers.

Sickle-cell trait present in 8 percent of African-Americans has been considered a benign condition. Individuals with this condition exhibit no anemia and manifest few or no symptoms. There is recent evidence that intense exercise, hot weather, and high altitudes or a combination of factors can evoke sickling with severe symptoms (Eichner, 1993). These three factors should be monitored in physical education settings and particularly with African-American students.

Physical Education for Students With Anemia

Many of the guidelines provided in the section dealing with cardiovascular disorders are applicable for the student with any type of anemia. The physical educator may be the first person to suspect that a student has anemia, observing lethargy, lack or loss of strength and endurance, irritability, and early onset of fatigue.

Excusing the anemic student from physical education is not an appropriate option, because lack of activity will aggravate the condition. It should be apparent to all concerned—parents, physicians, physical educators, and student—that a modified program of physical activity is in the student's best interest. Intensity levels in all activities must correlate with the student's physiological status. Initially, activities such as bowling and archery may be suitable. Conditioning may mean walking around the track and lifting light weights such as dumbbells. As the student's fitness improves through medical intervention and modified activity, the intensity level of activities may be elevated.

For the student with sickle-cell anemia, a coordinated effort between the physician and the physical educator is needed to design a safe and effective program. The physician must provide guidance on the level of intensity of physical fitness efforts, and the physical educator must select appropriate physical and motor activities. Even with motor activities that stress balance, hand-eye coordination, and agility, the physical educator must note any sign of cardiorespiratory distress such as rapid pulse; labored breathing; pale lips, tongue, or nail beds; and any complaint of pain. Underwater swimming is usually avoided, and jumping activities should not be excessive because of possible join inflammation. Both the student and the physical educator should monitor the condition of the student's skin, because skin ulcers may be a problem with this disorder.

HEMOPHILIA

Hemophilia is a hereditary bleeding disorder resulting from the inability of the blood to coagulate properly because of a deficiency of certain clotting factors. Hemophilia is the most common X-linked genetic disease and affects 1.25 in 10,000 live male births. Because of the genetic mode, females act as carriers; they have a 50 percent chance of transmitting the defective gene to each male child. There is also a 50 percent chance that each female child will likewise be a carrier of this defect.

Types

Hemophilia A, which affects more than 80 percent of all hemophiliacs, is caused by a deficiency of coagulation Factor VIII. Another type of hemophilia is sometimes called Christmas disease, or hemophilia B, resulting from a deficiency in Factor IX. The absence or deficiency of certain clotting factors is the reason for the vernacular designation, bleeder's disease.

Hemophilia may be classified as mild, moderate, or severe. The mild form may not appear until adolescence or adulthood. Its symptoms are a tendency to bruise, frequent nosebleeds, bleeding gums, and hematomas. Moderate hemophilia causes symptoms similar to those of the mild type, but bleeding episodes are more frequent, and there is occasional bleeding into the joints. Severe hemophilia is marked by spontaneous bleeding or severe bleeding after minor injuries. Bleeding into joints and muscle is more extensive than with the moderate type and causes pain, swelling, and extreme tenderness. Peripheral neuropathies may result from bleeding near peripheral nerves, with subsequent pain and muscle atrophy.

Physical Education for Students With Hemophilia

The student with hemophilia may derive both physical and social benefits from participating in physical education if appropriate precautions and modifications are in place. The student's physician should provide guidance on suitable physical education activities. Besides providing input on curriculum, the physician may wish to be informed (or instruct that the school nurse be informed) if certain injuries to the head, neck, or chest occur. Some of these injuries may require special blood factor replacement. The physician may alert the physical educator to watch for signs of severe internal hemorrhage, which may include severe pain or swelling in joints and muscles, joint stiffness, and abdominal pain. Students with hemophilia should wear a medical identification tag or bracelet at all times. Because blood transfusions carry the risk of infection with hepatitis, the physical educator may observe early symptoms of that disease: headache, vomiting, fever, nausea, pain over the liver, and abdominal tenderness. At no time should a student with hemophilia take aspirin, because this drug exacerbates the tendency to bleed.

If a student with hemophilia sustains some type of trauma, apply ice bags and pressure to the injured area and elevate the area if possible. The student should be restricted from activity for 48 hours after the bleeding is under control.

Activities that enhance physical fitness should be part of this student's program. Swimming is an excellent activity for enhancing cardiorespiratory endurance, muscular strength, and flexibility without subjecting the student to possible trauma. Jogging or fast walking may be used to develop aerobic capacities. If a student has had a problem with bleeding in a joint such as the knee, isometric rather than isotonic exercises may be used. Contact and collision sports such as football, basketball, and soccer are contraindicated. However, developing certain skill components of these sports, such as passing a football and shooting free throws, is desirable. Dual and individual sports such as tennis, archery, and golf are fine; racquetball is not an appropriate sport because of the risk of being hit by the ball.

ACQUIRED IMMUNODEFICIENCY SYNDROME

Acquired immunodeficiency syndrome (AIDS) is a very serious health disorder that has reached epidemic proportions. It is caused by an infection with human T-cell lymphotropic virus type III (HTLV-III), which is known as human immunodeficiency virus (HIV). As a result of HIV, people with AIDS have defective immune systems and cannot combat certain types of infections or rare malignancies. For each person with AIDS, there are five to seven individuals who develop symptoms called AIDS-related complex (ARC). These persons may be severely incapacitated or relatively well. Another category of individuals related to AIDS is the HIV carrier. While these people have the potential to infect other persons, they are capable of participating effectively in activities of daily living, including physical activity, without any evidence of the syndrome. It is estimated that within 5 to 7 years of infection 30 percent of the HIV carriers will develop a serious or life-threatening complication.

Characteristics of HIV Infection

The HIV virus has been found in such body fluids as blood, seminal fluid, vaginal secretions, saliva, and tears. It has not been documented that the last two fluids transmit this virus or that the disease can be airborne or spread by casual contact. Casual contact is defined as contact other than sexual contact or contact between mother and child during gestation or breast-feeding.

Two causal factors are associated with AIDS-infected children: birth to a mother who has this disorder and receiving blood or clotting factors containing HIV. Within the last decade there has been a dramatic rise in the incidence of HIV infection in infants and children (WHO, 1990). In 1999, the Centers for Disease Control and Prevention reported 297,137 persons living with AIDS. Researchers estimate that 65 percent of HIV-positive mothers in the United States transmit the virus to their offspring (Modlin & Saah, 1991). Authorities believe AIDS will soon be among the top five causes of death for children in the age range of one to four years. It is not clear if the number of elementary-age children infected with AIDS will appreciably increase. While blood screening techniques have improved, a certain number of hemophilic children may still acquire AIDS. The Centers for Disease Control has reported that 6 to 8 percent of certain adolescents may have HIV.

Physical Education and Students With AIDS

The physician has a critical role to play in determining the student's placement in the school setting and in the physical education class. Most state departments of health will allow a student to attend school if the student behaves acceptably (e.g., does not bite) and has no skin eruptions or uncovered sores. If any of these conditions are seen in physical education classes, the student should be removed from class immediately and sent to the appropriate school official. The same procedure should be followed if the student has a fever, a cough, or diarrhea.

The nature of the physical education program for students with AIDS depends on their physical capacities and motoric abilities. Children with AIDS manifest some of the following symptoms: opportunistic infections, chronic lung disease, failure to thrive, and encephalopathy. As this last symptom progresses, there is evidence of delays or loss of motor milestones and pyramidal tract problems. Cognitive impairments are more evident in children with AIDS than with the youngsters with ARC. Because some students with AIDS are hemophiliacs, the information provided in the sections on hemophilia and other cardiovascular problems is applicable to them (figure 15.1). If there is a bleeding episode in a physical education class because of an accident, an injury, or a situation related to a hemophilic condition, appropriate hygienic procedures should be followed. Blood or any body fluid should be treated with caution. Rubber gloves should be worn for cleaning up spills of any body fluid, and blood-soaked articles should be placed in leakproof bags for disposal or washing.

While appropriate hygienic practices are required, AIDS necessitates fewer adaptations and modifications for students than some other types of disabling conditions. Students who are HIV carriers can successfully engage in physical activity and are often involved in competitive sports. According to LaPerriere et al. (1994) there is increasing evidence that persons infected with HIV should remain physically active, because exercise is associated with improvements in mental health, immune system functioning, and neuroendocrine mechanisms. For example, progressive resistance exercises during the nonacute stage of AIDS did not exacerbate the disease but provided improved muscle function (Spence, Galantino, Mossbert, & Zimmerman, 1990). Persons who are HIV positive benefit from exercise training (Stringer, Berezovskaya, O'Brien, Beck, & Casaburi, 1998). In addition to the physical benefits, there are psychological benefits such as mood improvement and reduction in overall stress (Wagner, Rabkin, & Rabkin, 1998). The key to a physical education program for this type of student is carefully monitored activities.

STUDENTS AFFECTED BY SUBSTANCE ABUSE CONDITIONS

In the United States today, there are approximately 1.2 million females of childbearing age who are taking one or more of the following drugs: alcohol, cocaine, nicotine, or marijuana. The ingestion of these drugs in single or combination dosages increases the probability of producing an offspring with developmental difficulties (Sampson et al., 1997). A more alarming statistic is found in a survey by Adams (1988) who reported that 8 million of the 56 million women in their childbearing years are taking one or more of these drugs. It is estimated that between 2.8 and 4.6 of every 1,000 live births are present with fetal alcohol syndrome (Sampson et al., 1997). The effects of these drugs, particularly alcohol and cocaine, on offspring will be the focus of this section.

Photo by Mary Ann Carter. Used by permission

Figure 15.1 Ryan White was a student with hemophilia who received contaminated blood and developed AIDS. Before his illness, Ryan participated in most physical education activities.

Fetal alcohol syndrome (FAS) and fetal alcohol effects (FAE) were not truly diagnosed and given a label until 1973 (Niebyl, 1988). A combination of physical and mental defects that result from a mother ingesting excessive amounts of alcohol during pregnancy is referred to as FAS. Fetal alcohol effects describes a child who has been exposed to alcohol in utero but does not manifest all the symptoms of an individual with FAS. To be classified as a person with FAS, at least one characteristic must be found in each of these categories: retardation in growth before or after birth; facial anomalies such as epicanthic folds, flattened nasal bridge, or thin upper lip; and central nervous system dysfunction such as mental retardation, attention deficit disorder, or hyperactivity.

Characteristics of Fetal Alcohol Syndrome

A brief synopsis of a student with FAS will be presented to help gain better insight into this condition. From a physical perspective, these persons may exhibit altered facial features, limited range of motion in their joints, scoliosis, extra fingers and toes, hip dislocations, fusion of the radius and ulna, hollowed or depressed chest, ventricular septal defects, and poor performance of both fine and gross motor function. While the range of mental disabilities varies with different individuals, many are in the mild range of mental disabilities. In the social domain, they are also very much at risk. The mother-child relationship is often strained, which initially may preclude appropriate bonding; in later life these children are more prone to be physically and sexually abused and suffer neglect. There is also a higher risk of the child developing alcoholism than non-FAS persons. There is mounting evidence that FAS is not just a childhood disorder but is a long-term disorder affecting the adolescent and adult. While facial features are less distinctive after childhood, these individuals tend to be short of stature and microcephalic in nature. Academic functioning is often equivalent to a fourth grader, and maladaptive behavior such as distractibility, poor judgment, and problems perceiving social cues are evident (Steissguth, Aase, Clarren, Randels, LaDue, & Smith, 1991).

An estimated 10 to 20 percent of pregnant women are using drugs during the child's gestation. According to Chasnoff, Landeress, and Barrett (1990), 375,000 newborns are drug-exposed infants.

After marijuana, cocaine is the second most frequently used drug, with approximately 2 million Americans addicted to this substance (Howze & Howze, 1989). "Crack" cocaine babies are found to exhibit tremors, hypertonicity, hyperreflexia, seizures, mood swings, vomiting, and hypersensitivity (Kronstadt, 1991). Cocaine-exposed toddlers manifest behavior patterns that are less mature or appropriate in comparison to their nonaffected peers (Roding, Beckwith, & Howard, 1990). The frequency, type, and time of drug usage during gestation, as well as sociocultural factors, do affect the early development of crack babies (Cratty, 1990). Alcohol-related deficits are more severe in offspring of women over 30 years of age (Jacobson, Jacobson, & Sokol, 1996). Roding (1990) has observed with drug-exposed toddlers that their object manipulation skills are comparable to children with neurological problems. Church, Eldis, Blakley, and Banule (1997) have noted that these individuals have sensorineural and conduction hearing losses. The information in chapter 11 is applicable for students with FAS who manifest hearing losses. Attention deficit hyperactivity disorder has also been frequently associated with children with FAS (Coles et al., 1997). Chapter 11 provide insight into working with this type of student. There still is a need for research to investigate how these students fare in academic and physical education settings as they progress through elementary and high school.

Physical Education for Students Affected by Substance Abuse

The key to providing these students with successful physical education experiences is to develop individualized physical education programs. Because of the heterogeneity of their conditions, a teacher must consider each student individually. One student with FAS may be quite hypertonic in nature and should receive programming for this condition (which is similar to the individual with spastic cerebral palsy). Another student with FAS may exhibit hypotonicity; in this case, programming strategies similar to the hypotonic student with Down syndrome would be appropriate. Unfortunately, a certain percentage of these students will exhibit mental disabilities. Chapter 8 deals with this topic and provides guidance for this type of student affected by substance abuse. Extensive research has shown the following neuromuscular problems: delayed motor development, delayed postural reflex development, balance problems, coordination difficulties, and walking abnormalities (Osborn, Harris, & Weinberg, 1993). These are all areas that may be improved or enhanced in a physical education program. For young children with FAS early motor intervention programs have been found to counter developmental delays (Gianta & Streissguth, 1988). A real challenge for the physical educator will be to elicit a support system and a certain amount of continuity from the home environment. Many of these students come from single-parent homes, have been put into foster homes, and have a parent who is dealing with a substance abuse problem; these circumstances are not conducive to helping these students with their limitations or disabilities.

SUMMARY

The focus of this chapter has been certain conditions that, under IDEA, are designated as "other health impaired." Cancer, rheumatic heart disorders, diabetes, asthma, hemophilia, acquired immunodeficiency syndrome, and fetal alcohol syndrome are included in this category. Some students with these conditions (for example, students with asthma) may not have formal IEPs but may need certain modifications or adaptations in physical education. The real challenge in physical education is to integrate these students into the normal setting but provide programs for individualized needs.

Nondisabled Students and Adapted Physical Education

Paul R. Surburg

▷ ▷ ▷ ▷ **T**WO HIGH SCHOOL STUDENTS ARE participating in the Special Olympics Unified Sports program. During the first half of the basketball game, the student without a disability slightly sprains his ankle. Immediate care for this injury is administered. Rest, cold, compression, and elevation are immediately applied. During the second half of the game, one of the participants with mental retardation sustains a similar type of injury and the same type of care is provided. The following week both students were in their physical education classes. While these injuries were mild sprains, the physical education teacher followed the appropriate protocols, which are described in this chapter, to enhance full participation and recovery.

As noted in chapter 4, students without a designated disability but with unique physical education needs should have an individualized physical education program (IPEP). For example, a middle school student who has Osgood-Schlatter's condition may not be able to engage in all types of physical education activities. A person with a broken leg may be temporarily limited in the physical education setting (figure 16.1). People with long-term disorders will benefit from physical education programs that are modified to meet their present unique needs. Ironically, the information offered in this chapter to help nondisabled students may also be applicable to students with disabilities. A student with a deaf condition might break a leg, and information presented in the section on fractures would be useful for working with this student. Topics discussed under the heading of "Long-Term Disorders" are conditions that persons with disabilities may also exhibit. Thus, the content of this chapter is applicable to both types of students.

ACTIVITY INJURIES

Students may sustain activity or sports injuries in different settings. While some injuries may originate in a physical education class, more occur during free-time or recreational pursuits.

If an injury occurs in physical education class, immediate care should be provided in the RICE sequence: **Rest** should be given immediately to the injured part or joint. **Ice** or a cold application should be administered immediately and removed after 20 minutes. **Cold** may be reapplied in one to one and a half hours, depending on the extent of injury. Compression and **elevation** reduce internal bleeding and swelling. Unfortunately, in the recreational or free-time setting, these procedures are not always followed, and the injury may be exacerbated. Without the ice and compression, intra-articular pressure from the swelling may stretch structures such as ankle ligaments just as if a person had purposely twisted the joint. Some type of immobilization or rest is needed for musculoskeletal injuries; rest facilitates healing and reduces the risk of a prolonged recovery time.

Unless injured students are on an athletic team, there probably will be no type of rehabilitation service available to them. While physical educators cannot act in the capacity of athletic trainers, they may provide valuable assistance. If an injury is not being managed in an appropriate manner, the physical educator may recommend that the student see a physician or visit a sports medicine clinic. Many hospitals provide sports medicine services; these clinics or departments are staffed by sport physical therapists or athletic trainers.

If the injured student is progressing normally toward recovery, the physical education class may provide an opportunity for exercises or activities that will ameliorate the condition. The Brockport Physical Fitness Test is health related and criterion referenced. This test may be used to assess aerobic functioning, body composition, musculoskeletal functioning, and flexibility development, following recovery from activity injuries.

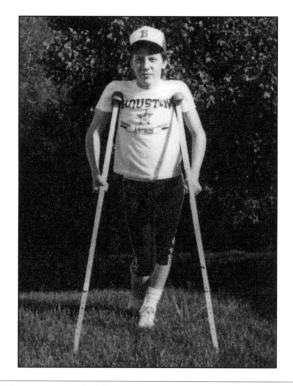

Figure 16.1   Student with a temporary limitation.

Ankle

Ankle sprains are a risk for anyone who actively engages in sports or physical activity. Jumping and other movements may cause a person to roll over on the ankle and stretch the medial or lateral side of this joint. Eighty-five percent of all ankle sprains are of the inversion type, in which the ligaments on the lateral side of the ankle are stretched.

Table 16.1 provides an exercise protocol for inversion sprains after primary care and treatment have been administered. While eversion sprains are less frequent, this injury tends to be more serious and entails a longer recovery period. Most of the exercises listed in table 16.1 could be used for an eversion sprain; the eversion exercises, being the mechanism of injury for this type of sprain, should be eliminated.

While the primary focus of this chapter is on activity and exercise, the reader should be aware that other rehabilitative measures may have been or are currently being used. Hydrotherapy and/or cryotherapy are used to treat sprained ankles; taping the ankle is another mode that may be used for participation in physical education class. Sports medicine practitioners differ on the value, duration, and methods of taping or strapping. If ankle taping has been prescribed by sports medicine personnel, compliance by the student should be encouraged. The physical educator, however, should not make a practice of taping ankles, for this could set a precedent for a time-consuming practice.

Knee

A variety of conditions and situations involving the knee may affect physical education performance. Table 16.2 provides basic rehabilitation programs for common knee problems. Please keep in mind both persons with and without disabilities may experience these knee problems. Several terms in this table will be briefly described. Chondromalacia patella refers to a degeneration or softening of the posterior surface of the patella and/or the femoral cartilage. Quad sets are isometric contractions of quadriceps femoris muscle groups. Six-second contraction of this muscle group constitutes a repetition. Straight leg lifts involve having a student assume a supine position and flex one leg at the hip while keeping the knee in complete extension. Short arc exercises are isotonic knee extension exercises done from 30 or 40 degrees of knee flexion to full or 0 degrees of knee extension. For chondromalacia patella, 30 to 0 degrees is appropriate, while for anterior cruciate ligament problems 90 to 45 degrees of motion is recommended. There is truly an individualized exercise program for each of these problems. A protocol that may be used in the latter stages of a rehabilitation program is the daily adjustable progressive resistance exercise (DAPRE), developed by Knight (1979). This strength development (table 16.3) program provides precise modification between sets of an exercise bout and between workouts.

Shoulder

The shoulder is composed of several major joints: sternoclavicular, acromioclavicular, scapulocostal, and glenohumeral. In many instances, activity-related strains or overuse syndromes involve the glenohumeral joint. Rotator cuff impingement syndrome, tendinitis, bursitis, and other glenohumeral problems may benefit from a general mobilizing and conditioning program for this joint (table 16.4). This program, however, is contraindicated for students suffering from anterior glenohumeral dislocation. In its chronic form, this condition is sometimes referred to as a "trick shoulder."

Mobilization for this condition consists primarily of adduction and/or internal rotation exercises. During the immobilization stage, isometric exercises are the exercises of choice; following immobilization the exercise regimen may progress from isometric exercises, such as pulling against rubber tubing, to pulley or free-weight exercises that emphasize adduction and internal rotation. It should be noted that external rotation and abduction are movements associated with the mechanism of injury.

An important phase of many rehabilitation programs is the development of proprioception and kinesthesia. There is evidence (Smith & Bronolli, 1989) that after a glenohumeral dislocation kinesthetic problems are evident. Because of this position-sense deficit, a comprehensive rehabilitation and conditioning program should include proprioceptive neuromuscular facilitation (PNF) exercises (Voss, Knott, & Iona, 1986) and other exercises such as moving the arm into functional positions without the benefit of sight.

Types of Exercises

Terms commonly used to describe lower-extremity exercises are open and closed kinetic chain exercises. Both types of exercises are used in rehabilitation and physical fitness programs. They are implemented to develop strength, flexibility, and proprioception. Open kinetic chain exercises are nonweight bearing with the distal end (foot) free to move, and they generally involve motion in one joint. Seated or prone on a bench to increase the strength of the quadriceps muscles or the hamstrings constitutes examples of open kinetic chain exercises.

Closed kinetic chain exercises are weight bearing in nature with the distal segment in contact with a supporting surface and several joints involved in the execution of movement. A half deep knee bend would be an example of a closed kinetic chain exercise. This type of exercise is often suggested for persons with certain types of knee injuries, such as anterior cruciate ligament problems and patellofemoral problems. It has been suggested that closed kinetic exercises put less strain on the ligament and these surfaces (Fitzgerald, 1997).

The question that naturally arises is: What is the best type of exercise? The answer is that both open and closed

Table 16.1 Rehabilitation Protocol for a Moderate Inversion Sprain of the Ankle

Stage	Activity	Purpose
I. Control and decrease foot swelling and pain	A1. Flexion, extension, and spreading of toes 2. Exercises for noninvolved leg and upper extremities 3. Crutch walking with touch weight bearing	A1. Work on intrinsic muscles and certain muscles that go across the ankle 2. Keep rest of body in good condition 3. Involve ankle in a minimal amount of motion, yet replicate a normal gait pattern as closely as possible
II. Begin restoration of strength and movements	B1. Ankle circumduction movements 2. Toe raises 3. Eversion exercises—isometric then isotonic 4. Achilles tendon stretch in sitting position with toes in, out, and straight ahead 5. Shift body weight from injured side to uninjured side	B1. Involve ankle in the four basic movements of this joint 2. Begin to develop plantar flexors 3. Reinforce side of ankle that has been stretched 4. Improve dorsiflexion and slight inversion and eversion 5. Begin to retain proprioception
III. Restore full function to ankle	C1. Ankle circumduction 2. Achilles tendon stretching in standing position 3. Eversion exercises 4. Plantar, dorsiflexion, and inversion exercises	C1. Continue to improve range of motion 2. Work on dorsiflexion 3. Improve strength for protection 4. Develop main muscle groups of ankle
IV. Restore full function of ankle	D1-4. As in previous stage 5. Tilt-board exercises 6. Walk-jog 7. Jog faster, stop 8. Run and sprint 9. Jog figure eights 10. Run figure eights 11. Cutting—half speed 12. Cutting—full speed 13. Run Z-shaped patterns 14. Backwards running	D1-4. As in previous stage 5. Work on proprioception 6. Develop functional and sport-specific activities

Table 16.2 Rehabilitation Exercise Programs for Common Knee Problems

Diagnosis	Rehabilitation areas to stress	Method
Chondromalacia patella, subluxating patella s/p[1] surgery for chronic dislocations or chronic irritative processes	Strengthen the quads, particularly VMO[2], without putting additional stress on patellofemoral surface Achieve or maintain full ROM[3] Strengthening of hamstrings, abductors, adductors, and lower-leg musculature	Quad sets, straight lifts, terminal extensions, lifts, short arcs with progressive resistance ROM and strengthening of hamstrings, abductors, and adductors can be done conventionally
Anterior cruciate deficient knees, chronic/acute; s/p casting; s/p reconstruction, intra-articular or extra-articular	Achieve full ROM Emphasis is placed on strengthening secondary stabilizers to the anterior cruciate	Active ROM exercises Hamstring strength aims to equal quad strength of the nonaffected leg, done with programs of isometrics; isotonic with concentric and eccentric contractions; isokinetics Quad strength to be elevated by use of isometrics at 90°, 60°, 30°, and full extension; isotonic resistance done only in 90°-45° flexion
Medial collateral ligament sprains, chronic/acute, s/p reconstruction; s/p immobilization	Special emphasis should be placed on strengthening quads and adductors General ROM Strengthen hamstrings, adductors, and lower-leg musculature	Achieve or maintain full ROM Quad sets Straight leg lifts with progressive resistance in hip flexion and adduction Isotonics for quads and hamstrings Isokinetics for quads and hamstrings

[1]s/p = Status/post
[2]VMO = Vastus medialis obliquus muscle
[3]ROM = Range of motion

kinetic chain exercises should be part of a rehabilitation and physical fitness program. If one thinks of a very basic movement such as walking there are both open (swing phase) and closed (stance phase) components to this motor skill.

Open and closed kinetic chain exercises may apply to the area of proprioception and kinesthesis. One could think of closed kinetic chain exercises as a means to enhance proprioception. The tilt-board exercise described in the next section is a closed kinetic chain exercise. Exercises such as PNF may help position sense and are often done as open kinetic chair exercises. For

more information on PNF, consult a chapter by this author (Surburg, 1999) written as part of *The Brockport Physical Fitness Training Guide*. Kinesthetic and proprioceptive defects are not only present after an injury but are problems with students who have various types of physical and learning disabilities.

Selected Exercises

This section describes selected exercises for rehabilitation programs (see tables 16.1 through 16.4) that may

Table 16.3   Daily Adjustable Progressive Resistance Exercise (DAPRE)

Set	Weight	Repetitions
1	50% of working weight	10
2	75% of working weight	6
3	100% of working weight	Maximum
4	Determined by repetitions done in third set[*]	Maximum number determines working weight for next session

Working weight adjustments	
Repetitions during third set[*]	**Working weight for fourth set**
0–2	Decrease 5 to 10 lb
3–4	Decrease 0 to 5 lb
5–7	Keep the same
8–10	Increase 2.5 to 5 lb
More than 10	Increase 5 to 10 lb
Repetitions during fourth set	**Working weight for next session[*]**
0–2	Decrease 5 to 10 lb
3–4	Keep the same
5–7	Increase 2.5 to 7.5 lb
8–10	Increase 5 to 10 lb
More than 10	Increase 10 to 15 lb

[*]From Knight, K. (1979). Rehabilitating chondromalacia patellae. *The Physician and Sportsmedicine*, 1, 147- 148.

be unfamiliar. Table 16.1 refers to tilt-board exercises. The apparatus used with this exercise, which is available commercially or could easily be constructed, is basically a circle of three-fourths inch plywood that is one foot in diameter (figure 16.2a). Attached to the center of the board is half of a pool ball or similar wooden ball. Standing with both feet on the board, the student attempts to balance on the board. This apparatus could be used to develop proprioceptive capabilities following an injury to the lower extremity. This exercise would be appropriate for balance and proprioceptive training of students with and without disabilities.

Terminal extension exercises are listed as strength exercises for quadriceps development in the exercise programs for common knee problems (see table 16.2). Ideally, these exercises are done with a knee extension machine rather than a weight boot or weights wrapped

around the ankle. Weights around the ankle cause a traction or pulling effect on the knee, which could stretch ligaments. Terminal extension exercises are initiated with the knee completely extended and resistance applied to the extremity (figure 16.2b). In subsequent exercise sessions, isotonic contractions are started with 5 degrees of extension. Over a number of sessions, the degree of extension is increased until the person can extend against resistance through 25 degrees of motion.

In table 16.4 Codman's exercise is listed as an exercise to mobilize the shoulder. In this exercise the participant bends over at the waist to achieve 90 degrees of trunk flexion and holds on to a chair or the end of a table. In this position the arm of the affected side should hang in a relaxed state. The person initiates motion first in a flexion-extension direction, then adduction-abduction, and finally circumduction. All of these motions are pendular in nature without benefit of muscular contractions at the shoulder joint. Progressions in this exercise include wider circumduction motions and holding 2.5- and 5-pound weights as motions are performed (figure 16.2c). As with the tilt board for proprioceptive development, Codman's exercise may be used to enhance range of motion and flexibility of students with certain types of disabilities.

LONG-TERM DISORDERS

This section will deal with several conditions that are classified as long-term disorders. The designation implies a condition or a problem that will last longer than 30 days. This time span is, to an extent, an arbitrary designation, because a third-degree ankle sprain

Table 16.4 Shoulder Exercise Protocol for the Glenohumeral Joint

Stage	Activity	Purpose
I. During period of immobilization	A1. Isometric contractions of major muscle groups of shoulder	A1. Reduce muscle atrophy
	2. Isotonic wrist exercises of involved extremity	2. These muscles may be kept in condition without involving shoulder muscle
	3. General conditioning exercises for the other three extremities	3. These muscle groups should be kept in good condition
II. Mobilization of the shoulder	B1. Work on moving the shoulder through abduction and external rotation	B1. Begin to gain appropriate range of motion
	2. Codman's exercise[1]	2. Begin to enhance four motions of the shoulder
	3. Wall-climbing exercise	3. Help to develop abduction and external rotation
	4. Continue conditioning of other extremities	4. Improve body fitness
III. Development of the shoulder	C1. Isotonic exercises that involve shoulder flexion, extension, abduction, adduction, medial rotation, and lateral rotation	C1. Develop muscles that cause specific motions
	2. Specific exercise involvement: bench press, pullovers, push-ups, parallel bar dips	2. Develop muscles for aggregate muscle action
	3. PNF[2] exercises, replicate arm positions without the use of sight	3. Improve kinesthesis
	4. Resistance movements that replicate sport activities	4. Conform to the principle of specificity

Note: Not to be used with glenohumeral dislocations.
[1]Explained in text under the heading "Selected Exercises."
[2]PNF = Proprioceptive Neuromuscular Facilitation

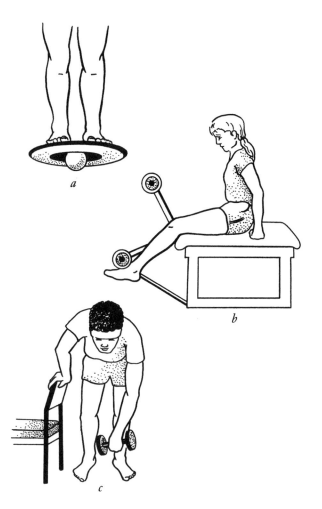

For physical educators, the student with a fracture presents two challenges: developing a program for a student immobilized in a cast and providing assistance after removal of the cast to integrate the student into the regular physical education program.

The first challenge must be dealt with from the perspective that the unaffected three-fourths of the extremities have normal movement. A student with a broken arm has no problem concerning ambulation and can easily maintain a high level of cardiovascular fitness. Strength development of three extremities can be pursued with only minimal modification or adaptation. With a broken arm, certain lifts such as the bench press would have to be eliminated, but development of the triceps of the nonaffected arm could be accomplished through other exercises, such as elbow-extension exercises. Involvement of the affected arm should be predicated on recommendations from the physician and on good judgment. For example, any type of isometric exercise involving muscles immobilized in a cast should be approved by the physician. Exercises, isotonic or isometric, involving joints above or below the cast area should also have physician approval; this, however, does not mean that exercises are contraindicated for these joints.

Participation in physical education activities is based on the nature of the activities in the curricular unit. While a track unit may mean little restriction for the student with a broken arm, a unit on gymnastic activities may require considerable restriction. For the person with a broken radius of the nondominant arm, a badminton unit will need very little modification. An archery unit, while less demanding as a vigorous activity, does require use of both upper extremities; however, the student could use a crossbow, which needs dexterity of one arm and could be mounted on a camera tripod. When unit activities preclude participation because of a fracture, physical fitness of the remaining three extremities may be the focus of the student's involvement in physical education class.

Once the cast is removed, the curricular focus should be on integrating the student into normal class activities. Part of this integration process may be to help develop range of motion, flexibility, strength, and muscular endurance in the affected limb. This assistance may be very important for students from low socioeconomic levels who may not have the benefit of appropriate medical services. Most students, regardless of socioeconomic background, can benefit from a systemic reconditioning program. Activities and exercises described in the earlier section entitled "Activity Injuries" may be incorporated into this program. The student's total integration into the regular physical education program is dependent on a group of factors: the nature of the fracture, the extent of immobilization, the duration of the reconditioning period, and the nature of unit activities.

Figure 16.2 (a) Rehabilitation exercise programs for common knee problems, (b) daily adjustable progressive resistance exercise (DAPRE), and (c) shoulder exercise protocol for the glenohumeral joint.

(discussed earlier in this chapter under "Activity Injuries") may not be totally rehabilitated within 30 days. Three conditions—fractures, Osgood-Schlatter's condition, and weight control—will be the primary focus here, but certain principles and procedures relevant to these conditions may also be applied to the integration into regular physical education classes of students with other long-term conditions.

Fractures

While most bones may be fractured in various types of accidents, bones in the upper and lower extremities are often fractured in activity-related accidents, for example, in landing on an outstretched arm. While the focus of this chapter is activity, readers should be aware that fractures may result from other situations, such as child abuse and pathologic bone-weakening conditions like cancer.

Application Example: Weight Training

Setting:	Middle school physical education class
Student:	13-year-old student with Osgood-Schlatter's condition
Task:	Develop a suitable strength program for a weight-training unit
Application:	The physical educator might include the following modifications:

- ▶ Use the daily adjustment progressive resistance exercise for all waist-up exercises and the uninvolved leg.
- ▶ For the leg with Osgood-Schlatter's condition, work on strength exercises for the ankle and hip.
- ▶ Do flexion and extension range of motion exercises for the knee with Osgood-Schlatter's condition.

Osgood-Schlatter's Condition

A long-term condition that often presents a dilemma for the physical educator is Osgood-Schlatter's condition. Not a disease, as some books describe it, the condition involves incomplete separation of the epiphysis of the tibial tubercle from the tibia. Whether Osgood-Schlatter's condition is due to singular or plural causes, it affects primarily males ages 10 to 15 and females ages 8 to 13 (Krauss, Williams, & Catterall, 1990). The treatment varies from immobilization in a cast to restriction of explosive extension movements at the knee, such as jumping and kicking (Wall, 1998). Variation in treatment depends on the severity of the condition and the philosophy of the attending physician concerning its management.

The physical educator should help the student with this condition during both its stages: the acute stage of involvement and the recovery stage. Because 60 to 75 percent of all cases are unilateral, affected students may be considered to have normal status for three-fourths of their extremities. In essence, the approach discussed in the earlier section on fractures may be applied here. The comparison is applicable not only from a programmatic standpoint but also from a causality perspective, because with Osgood-Schlatter's condition there is a type of avulsion or fracture of bone from the tibial tuberosity. Involvement of the affected knee in activity must be based on a physician's recommendation. For example, certain physicians may approve isometric contractions of the quadriceps and stretching of the hamstrings in the affected extremity. Ankle exercises may also be considered appropriate for the affected limb. When all symptoms have disappeared and the physician has approved full participation, the student should begin a general mobilizing program in physical education class. Development of strength and flexibility of the affected limb should be part of the student's physical education program. Exercises such as straight-leg raises, short-arc exercises and wall-slide arc are used to improve quadriceps strength. Stretching of the quadriceps and hamstrings should be part of the student's fitness program (Meisterling, Wall, & Meisterling, 1998). The physical educator should evaluate (and improve where needed) the student's gait pattern following the occurrence of Osgood-Schlatter's condition. Consider the Weight Training application example centered around a student with Osgood-Schlatter's condition.

Weight Control Problems

Many people associate weight control problems with the obese person going for the fourth serving at an all-you-can-eat smorgasbord. In reality there are two types of weight control problems: overweight and underweight. Both problems pose serious threats to a student's health. This problem may be the only condition a student must cope with, or weight control may be an accompanying condition or syndrome of a disability as defined by the public laws.

Underweight

Only recently have the terms anorexia nervosa and bulimia become familiar to the general public. School staff and faculty, including physical educators, should realize that a coordinated effort among parents, students, physicians, and school personnel is needed to deal with these problems.

Anorexia nervosa is a preoccupation with being thin that is manifested in willful self-starvation and may be accompanied by excessive physical activity. These individuals are 15 percent below what is considered normal for their weight and height, have a distorted body image, and females often develop primary or secondary amenorrhea. There is also what physicians call the female athlete triad. Eating disorders, amenorrhea, and osteoporosis constitute this triad. Increases in stress fractures are related to the third element in this triad (Rencken, Chestnut, & Drinkwater, 1996). Ninety percent of all cases of anorexia nervosa involve females, usually between the ages of 12 and 19 with a mean age onset of 16 years (Wichmann & Martin, 1993). During the early stages, both parents and student may be unaware that the condition is developing. As it progresses, the student becomes emaciated and hungry but denies the existence of the problem. Anorexia nervosa is not to be regarded lightly, for mortality rates range from 5 to 15 percent among diagnosed anorexics. Death is the result of circulatory collapse or cardiac arrhythmias due to electrolyte imbalance.

Bulimia is a condition associated with anorexia nervosa; it involves obsessive eating with ritualistic purging of ingested food by means of self-induced vomiting or laxatives. This condition is not a problem exclusive to girls and female athletes. Male wrestlers engage in these practices to make their weight classes (Oppliger et al., 1993). This practice also leads to electrolyte imbalance, impaired liver and kidney functioning, stomach rupture, tooth decay, and esophagitis.

The goals of treatment to promote weight gain in these conditions are simplistic in nature but challenging in execution. Hospitalization in a medical or psychiatric unit may be needed to initiate treatment, with a brief stay of two weeks or a longer period of several months to a year. Treatment consists of behavior modification, vitamin and mineral supplements, appropriate diet, and psychotherapy for the student and family. Low self-esteem, guilt, and anxiety are often part of the student's underlying problems.

The physical educator may be one of the first to recognize these problems. Anorexia or bulimia students present a profile of being compliant high achievers. A preoccupation with thinness, apparent weight loss, and an increasing involvement in aerobic activities may be signs that referral to an appropriate health professional is needed. Other signs and symptoms are amenorrhea; dry hair and skin; cold, discolored hands and feet; lightheadedness; and difficulty concentrating (Joy, et al., 1997). Whatever the stage of the student's disorder, physical education activities must be monitored to ensure an appropriate level of exertion. A caloric deficit—more calories being used than are taken in—should not be allowed to develop through or in conjunction with physical education activities. Fitness enhancement should be a gradual process, with strength gains achieved before cardiovascular endurance is attempted. Precautions or exercise contraindications for students with cardiovascular conditions, to be discussed in a subsequent section, are applicable in this situation. Dual and individual sports, some with certain modifications, are of suitable intensity and good for facilitating social interaction. A priority with the anorexic or bulimic student should be enhancement of self-esteem.

Overweight

A major health concern today is the number of overweight and obese students. Overweight is generally regarded as 10 percent over the appropriate weight, based on height and somatotype. Obesity is generally defined as at least 20 percent over appropriate weight. However, there are gender differences: Females may be considered obese when they exceed the appropriate weight by 30 to 35 percent; for males the range is 20 to 25 percent. One in seven children is overweight; for teenagers the ratio is one in nine (Centers for Disease Control and Prevention, 1994). A recent study reported that 33 percent of adults are in the obese category. This is a 25-percent increase in about a 10-year period (Backburn, et al., 1994). Both physical educators and adapted physical educators must cope with and provide suitable physical education experiences for overweight and obese students.

The cause of this condition is multifaceted in nature. Hypothalamic, pituitary, and thyroid dysfunction are causes of endocrine obesity. While endocrine disorders may cause excessive weight gain, fewer than 10 percent of all cases of obesity can be attributed to this cause. In the case of the hypothalamic dysfunction, damping of the hunger sensation does not take place, and a person feels hungry after consuming sufficient calories. Certain medications such as cortisone and adrenocortical steroids may produce the side effect of appetite elevation, with concomitant weight gain. There seems to be a set of emotional factors that may result in overweight and obesity. Some people eat as a means of compensation or to reduce feelings of anxiety; others, however, eat excessively when content and happy.

Demographic variables related to obesity are gender, age, and socioeconomic level. Females are more likely than males to be obese at all age levels. The critical age range for obesity is between 20 and 50 years, and obesity is more prevalent among people from lower socioeconomic levels.

An increase in the number (hypertrophic obesity) or size (hyperplastic obesity) of the fat cells that make up adipose tissue may lead to obesity. There are two periods during which there are rapid increases in fat cell production: the third trimester of pregnancy through the first year of life and the adolescent growth spurt. Between these two periods, the number of fat cells increases gradually. There is some evidence that vigorous exercise during childhood may reduce the size and number of fat cells (Saltin & Rowell, 1980).

Strategies for Weight Control

The physical educator is one member of a team who may help a student with a weight control problem. Just as exercise alone cannot remediate a weight problem, the physical educator alone cannot solve this problem. A team effort is needed, with the physician overseeing medical and dietary matters, the parents providing appropriate diet and psychological support, and the physical educator selecting the exercises and activities best for the obese or overweight student.

There are basically three ways to lose weight: diet, exercise, and a combination of the two. Diet alone is the most common method used by adults and is often the most abused method. The criterion frequently used to judge success with this approach is how quickly the maximum number of pounds can be lost. A crash diet may even trigger a starvation reaction, which causes the basal metabolic rate to diminish and may increase metabolic efficiency during physical activity. In this way, the body counteracts certain effects of the crash diet and does not lose the desired weight. A more gradual approach is a reduction of 500 calories a day, which in a week equals 3,500 calories—the number of calories needed to lose a pound of fat. Under the direction of a physician, an obese student may be on a diet that reduces intake by more than 500 calories per day.

Several tips or keywords may be provided to students to help the diet phase of the weight control program. The keywords are pyramid, plastic, fast foods, and drinks. Students should be reminded that the bottom part of the food pyramid contains the foods of choice such as fruits, vegetables, and starches. Students should be advised to avoid eating anything wrapped in plastic such as candy and potato chips. Lunch should be a bag lunch that includes fruits and vegetables and not food from fast-food restaurants. Finally, the students should be cautioned about what they drink. Water should be selected over diet soda, juices should be consumed in moderation, and milk should be fat free (Sallis, 1998).

Exercise helps with weight control, but it is a slow, difficult process if used as the sole method for weight reduction. The dual action of a sensible diet and physical activity is a healthy and productive approach to weight control. A reduction of 500 calories per day and an addition of 400 calories of exercise yields a safe 900-calorie deficit each week (Clark, 1995).

Physical Education and the Student With an Overweight Problem

A well-designed physical education program for overweight and obese students may contribute to increased caloric expenditure. There are, however, certain limitations or problems the physical educator must recognize and cope with in developing such a program.

There are barriers to exercise that the physical educator must overcome to implement a successful physical activity program. The following barriers have been proposed by Parr (1996):

Lack of motivation

Lack of time

Lack of access to facilities or equipment

Previous negative experience

Weight

Poor balance

Anxiety

Discomfort and pain

The physical education teacher may provide experiences and advice to deal with these barriers. The first barrier may be removed by providing evidence of successful control by assessing the percent of weight lost rather than the pounds that were lost. The teacher can celebrate the progress that is being made. Time in each physical education period may be devoted to weight-control activities. Being in a physical education class provides access to both facilities and equipment to enhance a weight control program. An atmosphere of commitment to students' goals and positive feedback from teacher and peers can counter previous negative experiences. As noted in the next paragraph, there are many activities that a student may engage in that are suited to students with excessive weight. The physical education teacher may work on balance and proprioceptive activities and provide activities such as water aerobics that minimize weight on joints and reduce reliance on balance skills. The physical educator should ease a person into a physical activity program and provide activities that are enjoyable. This is the best antidote for anxiety. Appropriately selected activities that reduce intensity, duration, and weight bearing situations will minimize or eliminate discomfort and pain associated with exercise.

The physical education program should be developed to provide successful experiences for the obese student. Activities that require lifting or excessively moving the body weight will not result in positive experiences. Gymnastic activities, distance running, rope climbing, and field events such as long jump may need extensive modification for the obese student. Gradual enhancement of endurance capabilities should be part of the student's program. Fast walking, bicycle riding, and certain swimming pool activities may help to develop aerobic endurance. Items from *The Brockport Physical Fitness Test* may be used to evaluate health-related components of fitness. Preferred general standards of body fat for children and adolescents range from 10 to 20 percent for males and from 17 to 25 percent for females (Winnick & Short, 1999). Aquatic activities are usually deemed appropriate for many types of special populations; with the obese student, activities such as water calisthenics may be quite appropriate, because buoyancy may reduce stress on joints such as the knees, ankles, and feet. Excessive buoyancy may be counterproductive, as Sherrill (1998) points out, because this force may keep parts of the body out of the water, thus impeding the execution of certain swimming strokes.

Many of the typical units covered in a physical education class will need considerable modification for students who are obese. A basketball or football unit may center on developing certain fundamental skills such as passing, catching, kicking, and shooting. Softball may include the development of fundamental skills and may involve modification of some rules, for example, allowing for courtesy, or pinch, runners. Dual and individual sports with modifications such as boundary or rule changes (i.e., the ball may bounce twice in handball) are appropriate activities. Golf, archery, and bowling need no modification, while tennis and racquetball may be feasible only in doubles play.

All curricular experiences should be oriented toward helping obese students develop a positive attitude toward themselves and toward physical activity. Physical education experiences—whether doing a caloric analysis of energy expenditures, learning to drive a golf ball, or being permitted to wear a different type of sport clothing than the typical gym uniform—should help obese students cope with their condition and should contribute directly or indirectly to the solution of the weight control problem. Finally, any strategy to improve self-image will help these students deal with their situation. Likewise, any strategy that changes the other students' attitudes toward those with weight problems will facilitate integration of obese and overweight students into the social environment.

In the final analysis the focus of a physical education class must be on suitable physical activity. Short, McCubbin, and Frey (1999) provide helpful physical activity guidelines for weight loss in *The Brockport Physical Fitness Training Guide*.

SUMMARY

This chapter has addressed conditions that both students with disabilities (as defined by IDEA) and students without these types of conditions may experience. While activity injuries and long-term disorders may be found among students with and without disabilities, these conditions should not preclude participation in physical education class or physical activity outside of the school setting. Suggestions for appropriate physical activity and physical fitness experiences have been provided for students with activity injuries, long-term disorders, and weight control problems.

III

Developmental Considerations

This part of the book focuses on early childhood development and services for youngsters with disabilities at the youngest ages. Chapters 17 through 20 discuss motor development (chapter 17); perceptual-motor development (chapter 18); adapted physical education for infants and toddlers (chapter 19); and preschoolers and children in early elementary programs (chapter 20).

The information presented in chapters 17 and 18 is intended to review and build on foundational knowledge related to motor development classes taken in a student's professional preparation program. Chapter 19, which deals with the infant and toddler population, opens with a discussion of the role of teachers in programs for infants and toddlers. Following this, topics include assessment, goals, and objectives for programs, along with recommendations for interacting with infants, toddlers, and their families. Chapter 20 provides information related to assessment, program objectives, and developmentally appropriate teaching approaches and activities.

17

Motor Development

David L. Gallahue

▷

▷

▷

DO YOU REMEMBER WHEN YOU were young, about four or five, and wanted to complete what now is a simple task, to just tie your shoelaces? Back then, however, it was a monumental accomplishment. First, there were the fine motor requirements of the task itself (T). Second, there were the individual differences in the rate of learning among you and your preschool playmates (I). Finally, there were environmental factors, such as the fact that your mom or dad may have dressed you in shoes with Velcro fasteners, and you had little need to learn how to tie laces (E). Motor development and the learning of new movement skills involves for all of us, able-bodied and disabled, a transaction among the requirements of the specific task, with a variety of personalized factors within the individual and the environment. Keep the letters T, I, E in mind as you read this chapter, focusing on how the task, the individual, and the environment combine to determine the sequence, rate, and extent of learning any movement skill.

For years the topic of motor development has been of considerable interest to physical educators in general and adapted physical educators in particular. Knowledge of the developmental process lies at the very core of education that is adapted to meet the needs of the individual, whether it be in the classroom, gymnasium, swimming pool, or on the playing field. Without sound knowledge of the learner's individual level of development, teachers can only guess at appropriate educational techniques and intervention strategies to be used to maximize one's learning potential. Educators who are developmentally based in their instruction incorporate learning experiences geared to the specific needs of their students. They reject the all too frequent textbook ideal of students all being at the same level of development at given chronological ages. In fact, one of the most valuable outcomes of studying human development has been less reliance on the concept of age appropriateness and more attention to the concept of individual appropriateness.

In a very real sense, adapted physical education is developmental education. Program content and instructional strategies, by their very nature, are designed to meet individual developmental needs. Unfortunately, human development is frequently studied from a compartmentalized viewpoint. That is, the cognitive, affective, and motor domains are viewed, by many, as unrelated entities. Although valid perhaps from the standpoint of basic research, such a perspective is of little value when it comes to understanding the learning process and devising appropriate intervention strategies. It is essential for teachers of students with developmental disabilities to be knowledgeable about the normal process of development in order to have a baseline for comparing the individuals with whom they are dealing. The totality and integrated nature of the individual must be recognized, respected, and accommodated in the educational process. This chapter focuses on defining motor development and describing the categories of human movement. Developmental

theory is briefly examined from the viewpoints of Dynamic Systems Theory and the Phases of Motor Development, two popular theoretical frameworks. The chapter concludes with common principles that emerge from the neuromaturational viewpoint of motor development.

MOTOR DEVELOPMENT DEFINED

Development is the continuous process of change over time, beginning at conception and ceasing only at death. **Motor development**, therefore, is progressive change in movement behavior throughout the life cycle. Motor development involves continuous adaptation to changes in one's movement capabilities in the never-ending effort to achieve and maintain motor control and movement competence. Such a perspective does not view development as being compartmentalized into specific cognitive, affective, and motor domains, nor does it view development as being stage-like or age dependent. Instead, a life-span perspective suggests that some aspects of one's development can be conceptualized into domains, as being stage-like and age related, while others cannot. Furthermore, the concept of achieving and maintaining competence encompasses all developmental change, both positive and negative.

Motor development may be studied both as a process and as a product. As a process, it may be viewed from the standpoint of underlying factors that influence both the motor performance and movement capabilities of individuals from infancy through old age. As a product, motor development may be studied from a descriptive or normative standpoint and is typically viewed in broad time frames, phases, and stages.

Currently, Dynamic Systems Theory (Kamm, Thelen, & Jensen, 1990; Thelen & Smith, 1993) is popular among developmentalists as a means of better understanding the processes of development. On the other hand, the Phases of Motor Development (Gallahue & Ozmun, 1998) serve as a descriptive means for better understanding and conceptualizing the products of development. Both will be briefly discussed, but first let's look at the categories of human movement.

CATEGORIES OF MOVEMENT

Both the process and products of motor development are revealed through changes in one's movement behavior across the life span. All of us, infants, children, adolescents, and adults, are involved in learning how to move with control and competence in response to the daily movement challenges we face. We are able to observe developmental differences in motor behavior by observing changes in body mechanics and motor performance scores. In other words, a "window" is provided through which the individual's actual movement behavior can be observed.

Observable movement takes many forms and can be grouped into categories. One technique classification scheme involves three movement categories: stability, locomotion, and manipulation and combinations of the three (Gallahue, 1996; Gallahue & Ozmun, 1998). Stability is the most basic form of movement, and it is present to a greater or lesser extent in all movement. A stability movement is any movement that places a premium on gaining and maintaining one's equilibrium in relation to the force of gravity. Gaining control of the muscles of the head, neck, and trunk is the first stability task of the newborn. Sitting with support, sitting unaided, and pulling oneself to a stand are important stability task of the normally developing infant. Standing without support; balancing momentarily on one foot; and being able to bend and stretch, twist and turn, and reach and lift are all important stability tasks of childhood through old age.

The locomotor movement category refers to movements that involve a change in location of the body relative to a fixed point on the surface. To walk, run, jump, hop, skip, or leap is to perform a locomotor task. In our use of the term, activities such as a forward or backward roll may be considered to be both locomotor and stability movements—locomotor, because the body is moving from point to point, and stability, because of the premium placed on maintaining equilibrium in an unusual balancing situation.

The manipulative movement category refers to both gross and fine motor manipulation. The tasks of throwing, catching, kicking, and striking an object are all considered to be gross motor manipulative movements. Activities such as sewing, cutting with scissors, and typing are fine motor manipulative movements.

A large number of our movements involve a combination of stability, locomotor, and/or manipulative movements. In essence, all voluntary movement involves an element of stability. It is for this reason that stability is viewed as the most basic category of movement, and it is absolutely essential for progressive development in the other two categories. The individual with cerebral palsy, for example, is frequently encumbered in his walking gait because of difficulty in negotiating the stability problems of static and dynamic balance inherent in independent walking.

MOTOR DEVELOPMENT AS A DYNAMIC SYSTEM

Theory should undergird all research and science. The study of motor development is no exception. Motor development theory is based on the observation of individuals without disabilities as they acquire and refine movement skills. If applied wisely, this information provides insight into the influence of various developmental disabilities on the learning of new movement skills, thereby permitting adoption of appropriate instructional strategies.

To be of practical benefit, developmental theory must be both descriptive and explanatory. It is important to know about the products of development in terms of what people are typically like during particular age periods (description). It is equally important, however, to know what causes these changes (explanation). Many motor developmentalists are now looking at explanatory models in an attempt to understand more about the underlying processes that actually govern development. Dynamic Systems Theory is popular among many (Kamm, Thelen, & Jensen, 1990; Caldwell & Clark, 1990; Thelen, 1989; Thelen & Smith, 1994).

In brief, the term dynamic conveys the concept that developmental change is nonlinear and discontinuous rather than linear and continuous. Because development is viewed as nonlinear, it is seen as a discontinuous process. That is, individual change over time is not necessarily smooth and hierarchical, and it does not necessarily involve moving toward ever-higher levels of complexity and competence in the motor system. Individuals, particularly those with disabling conditions, are encumbered by impairments that tend to impede their motor development. For example, children with spastic cerebral palsy are frequently delayed in learning to walk independently. When independent walking is achieved, the gait patterns will be individualized and achieved at a point in time appropriate for each. Although, by definition, development is a continuous process, it is also a discontinuous process. In other words, from a dynamical perspective, development is viewed as a "continuous-discontinuous" process. The dynamics of change occur over time but in a highly individual manner using multiple pathways influenced by a variety of critical factors within the system. Children develop a movement skill to meet the specific goal of a task through exploration, discovery, and selection in the process of learning (Thelen, 1995; Gallahue, 1996). When learning new movement skills, children utilize a variety of strategies and adopt one that personally works best for them in achieving the goal of the task.

The term **systems** conveys the concept that the human organism is self-organizing and composed of several subsystems. It is self-organizing in that, by their very nature, humans are inclined to strive for motor control and movement competence. It is the various subsystems operating within the requirements of the movement task (i.e., performance demands movement pattern formation and degrees of freedom), coupled with the biology of the individual (i.e., anatomical, physiological, mechanical, and perceptual factors) and the conditions of the learning environment (i.e., environmental context, opportunity for practice, encouragement, and instruction) that actually determine the rate, sequence, and extent of development. In other words, it is not some preprogrammed universal plan that unfolds on an inflexible schedule.

Dynamic Systems Theory is unlike Neuromaturation Theory, which views maturation of the central nervous system as the principal component responsible for changes in motor behavior. Dynamic Systems Theory views all subsystems as potential contributors to changes in motor behavior. Using our example of children with spastic cerebral palsy, they will, as a self-organizing system, develop their individually unique gait patterns in response to their capabilities in terms of meeting the achievement demands of the walking task.

Dynamic Systems Theory attempts to answer the *why*, or process, questions that result in the observable product of motor development. That is: What are those enabling factors (termed affordances) that encourage or promote developmental change, and what are those inhibiting factors (constraints) that serve to restrict or impede development? For children with cerebral palsy, constraints are neurological and biomechanical in nature. Affordances may include assisted support, handholds, encouragement, and guided instruction.

For years, developmentalists have recognized the interactive role of two primary systems on the developmental process: heredity and environment. Many developmentalists now, however, have taken this view one step further in recognizing that the specific demands of the movement task itself actually transact with the individual (i.e., hereditary or biological factors) and the environment (i.e., experience or learning factors) in the development of stability, locomotor, and manipulative movement abilities. Such a transactional model implies that factors within various subsystems of the task, the individual, and the environment not only interact with one another but also have the potential for modifying and being modified by the other, as one strives to gain motor control and movement competence (figure 17.1).

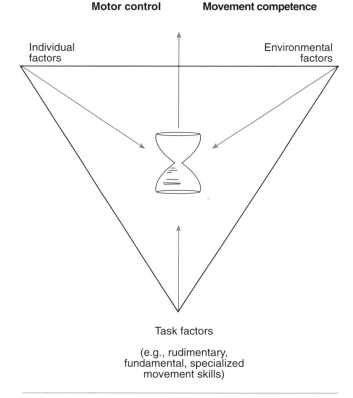

Figure 17.1 A transactional model of causality in motor development and movement skill acquisition.

Both the processes and the products of motor development should constantly remind us of the individuality of the learner. Individuals have their own timetables for the development and extent of skill acquisition. Although our "biological clocks" are rather specific when it comes to the sequence of movement-skill acquisition, the rate and extent of development are individually determined and dramatically influenced by the specific performance demands of the task itself. Typical age periods of development are just that—typical, and nothing more. Age periods merely represent approximate time ranges during which certain behaviors may be observed for the mythical "average" individual. Overreliance on these time periods would negate the concepts of continuity, specificity, and individuality in the developmental process; and these time periods are of little practical value when working with individuals with developmental disabilities.

THE PHASES OF MOTOR DEVELOPMENT

If movement serves as a "window" for viewing motor development, then one way of studying development is by examining the typical sequential progression in the acquisition of movement abilities. The Phases of Motor Development (figure 17.2) and the developmental stages within each phase (table 17.1) serve as a useful descriptive model for this study (Gallahue & Ozmun, 1998).

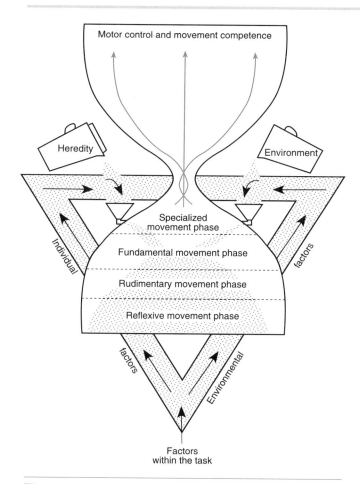

Figure 17.2   The hourglass: Gallahue's life-span model of motor development.

Reprinted, by permission, from D.L. Gallahue and J.C. Ozmun, 1997, *Understanding motor development*, 4th ed. (Boston: McGraw-Hill).

Reflexive Movement Phase

The very first movements of an infant are reflexive. Reflexes are involuntary, subcortically controlled movements. Through reflex activity the infant gains information about the immediate environment. The infant's reactions to touch, light, sounds, and changes in pressure trigger involuntary movements. These movements, coupled with the increasing cortical sophistication in the early months of life, play an important role in helping the child learn more about her body and the outside world and are typically referred to as primitive reflexes and postural reflexes.

Primitive reflexes are information gathering, nourishment seeking, and protective responses. Postural reflexes resemble later voluntary movements and are used to support the body against gravity or to permit movement. Tables 17.2 and 17.3 provide a summary of common primitive and postural reflexes, respectively. The reflexive movement phase may be divided into two overlapping stages, as described in the following paragraphs.

Information Encoding Stage

The information encoding (gathering) stage of the reflexive movement phase is characterized by observable involuntary movement during the fetal period until about the fourth month of infancy. During this stage lower brain centers are more highly developed than the motor cortex and are essentially in command of fetal and neonatal movement. These brain centers are capable of causing involuntary reactions to a variety of stimuli of varying intensity and duration. During this stage, reflexes serve as the primary means by which the infant is able to gather information, seek nourishment, and seek protection through movement.

Information Decoding Stage

The information decoding (processing) stage begins around the fourth month of postnatal life. There is a gradual inhibition of many reflexes as higher brain centers continue to develop. Lower brain centers gradually relinquish control over skeletal movements and are replaced by voluntary movement mediated by the motor area of the cerebral cortex. The decoding stage replaces sensorimotor activity with perceptual-motor behavior. That is, the infant's development of voluntary motor control involves processing sensory stimuli with stored information, not merely reacting to stimuli.

One means of diagnosing possible central nervous system disorders in the infant is through observation and reflex testing. The complete absence of a reflex is usually less significant than a reflex that remains too long. Other evidence of possible damage may be found in a reflex that is too strong or too weak. Also, a reflex that elicits a stronger response on one side of the body than on the other may indicate central nervous system dysfunction. For example, an asymmetrical tonic neck reflex that shows full arm extension on one side of the body and only weak extensor tone when the other side is stimulated may provide evidence of damage.

Examination of reflexive behaviors in the neonate provides the physician with a primary means of diagnosing central nervous system integrity in full-term, premature, and at-risk infants. Furthermore, these examinations serve as a basis for intervention by the physical and occupational therapists and by the adapted physical education specialist working with individuals displaying pathological reflexive behavior.

Rudimentary Movement Phase

The first forms of voluntary movement are rudimentary. Rudimentary movements are maturationally determined behaviors seen in the normally developing infant from birth to about age two. They are heavily influenced by heredity and characterized by a highly predictable sequence that is resistant to change under normal conditions. The rate at which these abilities appear will, however, vary from child to child and relies on biological, environmental, and task-specific factors. The rudimentary movement abilities of

Table 17.1 The Phases and Stages of Motor Development

Phase of motor development	Approximate age periods of development	The stages of motor development
Reflexive movement phase	In utero to 4 months 4 months to 1 year	Information encoding stage Information decoding stage
Rudimentary movement phase	Birth to 1 year 1 to 2 years	Reflex inhibition stage Precontrol stage
Fundamental movement phase	2 to 3 years 4 to 5 years 6 to 7 years	Initial stage Elementary stage Mature stage
Specialized movement phase	7 to 10 years 11 to 13 years 14 years and up	Transition stage Application stage Lifelong utilization stage

Reprinted, by permission, from D.L. Gallahue and J.C. Ozmun, 1997, *Understanding motor development*, 4th ed. (Boston: McGraw-Hill).

Table 17.2 Sequence of Emergence of Selected Primitive Reflexes of the Newborn

Reflex	Onset	Inhibition	Stimulus	Behavior
Moro	Birth	Third month	Supine position—sudden loud noise causes rapid or sudden movement of infant's head	Stimulation will result in extension of the infant's extremities, followed by a return to a flexed position against the body.
Tonic neck (asymmetrical)	Birth*	Sixth month	Supine position—neck is turned so head is facing left or right	Extremities on side of body facing head position extend, those on side opposite flex.
Tonic neck (symmetrical)	Birth*	Sixth month	Supported sitting—flexion or extension of infant's head	Extension or flexion of neck will result in extension of arms and flexion of legs.
Grasping	Birth	Fourth to sixth month	Supine position—stimulation of the palm of the hand or the ball of the foot	Stimulation will result in a grasping action of the fingers or toes.
Babinski	Birth	Sixth month	Supine position—stimulation by stroking the sole of the foot	Stimulation will result in extension of the toes.
Sucking	Birth	Third month	Supine or supported sitting—stimulus applied directly above or below the lips	Touching area of mouth will result in a sucking action of the lips.

*Not seen in all children.

Table 17.3 **Sequence of Emergence of Selected Postural Reflexes of the Newborn**

Reflex	Onset	Inhibition	Stimulus	Behavior
Labyrinthine righting	Second month	Twelfth month	Supported upright position—tilting of trunk forward, rearward, or to side	Infant will attempt to keep head in an upright position by moving head in opposite direction from tilt
Supportive reactions	Fourth month (arms) Ninth month (legs)	Twelfth month	Prone or upright supported—movement of the child's extremities toward a surface	Extension of the extremities to a position of support
Pull-up	Third month	Fourth month	Upright sitting supported by hands—tilting of child from side to side and front and back	Infant will flex arms in an attempt to maintain equilibrium
Stepping	Second week	Fifth month	Supported upright position—infant is held in upright position and soles of feet are allowed to touch a surface	Definite stepping action of only the lower extremities
Crawling	Birth	Fourth month	Prone unsupported position—stimulus is applied to sole of one foot	Crawling action exhibited by both the upper and lower extremities
Swimming	Birth	Fifth month	Prone held over water—infant is held over or in water	Swimming movements elicited in both the upper and lower extremities

the infant represent the basic forms of voluntary movement required for survival. See table 17.4 for a descriptive profile of selected rudimentary stability, locomotor, and manipulative abilities.

The rudimentary movement phase may be subdivided into two stages that represent progressively higher orders of motor control and movement competence. These stages are discussed in the following paragraphs.

Reflex Inhibition Stage

The reflex inhibition stage of the rudimentary movement phase begins at birth. Although reflexes dominate the newborn's movement repertoire, the infant's movements are increasingly influenced by the developing cortex. Development of the cortex and the lessening of certain environmental constraints cause several reflexes to gradually disappear. Primitive and

postural reflexes are replaced by voluntary movement behaviors. At the reflex inhibition level, voluntary movement is poorly differentiated and integrated. That is, the neuromotor apparatus of the infant is still at a rudimentary stage of development. Movements, though purposeful, appear uncontrolled and unrefined. If, for example, the infant desires to make contact with an object, there will be global activity of the entire hand, wrist, arm, shoulder, and even trunk. In other words, the process of moving the hand into contact with the object, although voluntary, lacks control.

Precontrol Stage

At around one year of age, normally developing children begin to bring greater precision and control to their movements. The process of differentiating between sensory and motor systems and integrating perceptual and motor information into a more meaningful and congruent whole takes place during the precontrol stage. The rapid development of higher cognitive processes as well as motor processes makes for rapid gains in rudimentary movement abilities during this stage. Children learn to gain and maintain their equilibrium, to manipulate objects, and to locomote throughout their environments.

The maturational process may partially explain the rapidity and extent of development of movement con-trol during this stage, but the growth of motor proficiency is no less amazing.

Fundamental Movement Phase

The fundamental movement abilities of early childhood are an outgrowth of the rudimentary movement phase of infancy. Fundamental movements are generally viewed as basic movement skills such as walking, running, throw-

Table 17.4 Developmental Sequence of Selected Rudimentary Movement Abilities

Movement pattern	Selected abilities	Approximate age of onset
Control of head and neck	Turns to one side Turns to both sides Held with support Chin off contact surface Good prone control Good supine control	Birth 1 week First month Second month Third month Fifth month
Control of trunk	Lifts head and chest Attempts supine-to-prone position Success in supine-to-prone roll Prone to supine roll	Second month Third month Sixth month Eighth month
Sitting	Sits with support Sits with self-support Sits alone Stands with support	Third month Sixth month Eighth month Sixth month
Standing	Supports with handholds Pulls to supported stand Stands alone	Tenth month Eleventh month Twelfth month
Horizontal movements	Scooting Crawling Creeping Walking on all fours	Third month Sixth month Ninth month Eleventh month
Upright gait	Walks with support Walks with handholds Walks with lead Walks alone (hands high) Walks alone (hands low)	Sixth month Tenth month Eleventh month Twelfth month Thirteenth month
Reaching	Globular ineffective Definite corralling Controlled	First to third month Fourth month Sixth month
Grasping	Reflexive Voluntary Two-hand palmar grasp One-hand palmar grasp Pincer grasp Controlled grasping Eats without assistance	Birth Third month Third month Fifth month Ninth month Fourteenth month Eighteenth month
Releasing	Basic Controlled	Twelfth to fourteenth month Eighteenth month

Reprinted, by permission, from D.L. Gallahue and J.C. Ozmun, 1997, *Understanding motor development*, 4th ed. (Boston: McGraw-Hill).

ing, and catching, which are building blocks for more highly developed and refined movement skills. This phase of motor development represents a time in which young children are actively involved in exploring and experimenting with the movement capabilities of their bodies. It is a time for discovering how to perform a variety of basic stabilizing, locomotor, and manipulative movements, first in isolation and then in combination with one another. Children who are developing fundamental patterns of movement are learning how to respond with motor control and movement competence to a variety of stimuli. Tables 17.5, 17.6, and 17.7 provide a descriptive overview of the typical development sequence of several fundamental stability, manipulative, and locomotor movements.

Several researchers and assessment instrument developers have attempted to subdivide fundamental movements into a series of identifiable sequential stages (McClenaghan & Gallahue, 1978; Haubenstricker & Seefeldt, 1986; Robertson & Halverson 1984; Gallahue & Ozmun, 1998). For the purposes of our model, we

Table 17.5 Sequence of Emergence of Selected Fundamental Stability Abilities

Movement pattern	Selected abilities	Approximate age of onset
Dynamic balance Dynamic balance involves maintaining one's equilibrium as the center of gravity shifts.	Walks 1-inch straight line	3 years
	Walks 1-inch circular line	4 years
	Stands on low balance beam	2 years
	Walks on 4-inch wide beam for a short distance	3 years
	Walks on same beam, alternating feet	3–4 years
	Walks on 2- or 3-inch beam	4 years
	Performs basic forward roll	3–4 years
	Performs mature forward roll*	6–7 years
Static balance Static balance involves maintaining one's equilibrium while the center of gravity remains stationary.	Pulls to a standing position	10 months
	Stands without handholds	11 months
	Stands alone	12 months
	Balances on one foot 3–5 seconds	5 years
	Supports body in basic three-point inverted positions	6 years
Axial movements Axial movements are static postures that involve bending, stretching, twisting, turning, and the like.	Axial movement abilities begin to develop early in infancy and are progressively refined to a point where they are included in the emerging manipulative patterns of throwing, catching, kicking, striking, trapping, and other activities	2 months to 6 years

*The child has the developmental "potential" to be at the mature stage. Actual attainment will depend on task, individual and environmental factors.

Reprinted, by permission, from D.L. Gallahue and J.C. Ozmun, 1997, *Understanding motor development*, 4th ed. (Boston: McGraw-Hill).

Table 17.6 Sequence of Emergence of Selected Fundamental Manipulative Abilities

Movement pattern	Selected abilities	Approximate age of onset
Reach, grasp, release Reaching, grasping, and releasing involve making successful contact with an object, retaining it in one's grasp, and releasing it at will.	Primitive reaching behaviors	2–4 months
	Corralling of objects	2–4 months
	Palmar grasp	3–5 months
	Pincer grasp	8–10 months
	Contolled grasp	12–14 months
	Controlled releasing	14–18 months
Throwing Throwing involves imparting force to an object in the general direction of intent.	Body faces target, feet remain stationary, ball thrown with forearm extension only	2–3 years
	Same as above but with body rotation added	3.6–5 years
	Steps forward with leg on same side as the throwing arm	4–5 years
	Boys exhibit more mature pattern than girls	5 years and over
	Mature throwing pattern*	6 years
Catching Catching involves receiving force from an object with the hands, moving from large to progressively smaller balls.	Chases ball; does not respond to aerial ball	2 years
	Responds to aerial ball with delayed arm movements	2–3 years
	Needs to be told how to position arms	2–3 years
	Fear reaction (turns head away)	3–4 years
	Basket catch using the body	3 years
	Catches using the hands only with a small ball	5 years
	Mature catching pattern*	6 years
Kicking Kicking involves imparting force to an object with the foot.	Pushes against ball; does not actually kick it	18 months
	Kicks with leg straight and little body movement (kicks *at* the ball)	2–3 years
	Flexes lower leg on backward lift	3–4 years
	Greater backward and forward swing with definite arm opposition	4–5 years
	Mature pattern (kicks *through* the ball)*	5–6 years
Striking Striking involves sudden contact to objects in an overarm, sidearm, or underhand pattern.	Faces object and swings in a vertical plane	2–3 years
	Swings in a horizontal plane and stands to the side of the object	4–5 years
	Rotates the trunk and hips and shifts body weight forward	5 years
	Mature horizontal pattern with stationary ball	6–7 years

*The child has the developmental "potential" to be at the mature stage. Actual attainment will depend on environmental factors.

Reprinted, by permission, from D.L. Gallahue and J.C. Ozmun, 1997, *Understanding motor development*, 4th ed. (Boston: McGraw-Hill).

Table 17.7 Sequence of Emergence of Selected Fundamental Locomotor Abilities

Movement pattern	Selected abilities	Approximate age of onset
Walking Walking involves placing one foot in front of the other while maintaining contact with the supporting surface.	Rudimentary upright unaided gait	13 months
	Walks sideways	16 months
	Walks backward	17 months
	Walks upstairs with help	20 months
	Walks upstairs alone—follow step	24 months
	Walks downstairs alone—follow step	25 months
Running Running involves a brief period of no contact with the supporting surface.	Hurried walk (maintains contact)	18 months
	First true run (nonsupport phase)	2–3 years
	Efficient and refined run	4–5 years
	Speed of run increases, mature run*	5 years
Jumping Jumping takes three forms: (1) jumping for distance; (2) jumping for height; and (3) jumping from a height. It involves a one- or two-foot takeoff with a landing on both feet.	Steps down from low objects	18 months
	Jumps down from object with one-foot lead	2 years
	Jumps off floor with both feet	28 months
	Jumps for distance (about 3 feet)	5 years
	Jumps for height (about 1 foot)	5 years
	Mature jumping pattern*	6 years
Hopping Hopping involves a one-foot takeoff with a landing on the same foot.	Hops up to 3 times on preferred foot	3 years
	Hops from 4 to 6 times on same foot	4 years
	Hops from 8 to 10 times on same foot	5 years
	Hops distance of 50 feet in about 11 seconds	5 years
	Hops skillfully with rhythmical alteration, mature pattern*	6 years
Galloping The gallop combines a walk and a leap with the same foot leading throughout.	Basic but inefficient gallop	4 years
	Gallops skillfully, mature pattern*	6 years
Skipping Skipping combines a step and a hop in rhythmic alteration.	One-footed skip	4 years
	Skillful skipping (about 20%)	5 years
	Skillful skipping for most*	6 years

*The child has the developmental "potential" to be at the mature stage. Actual attainment will depend on environmental factors.

Reprinted, by permission, from D.L. Gallahue and J.C. Ozmun, 1997, *Understanding motor development*, 4th ed. (Boston: McGraw-Hill).

Application Example: Improvement of Throwing and Catching Skills

Setting: An adapted physical education class

Students: A group of students with mental retardation

Task: The adapted physical education teacher is interested in improving these students' throwing and catching skills. The objective is for them to be able to use these fundamental manipulative skills more effectively in a variety of play, game, and sport activities.

Application: Assess their present skill levels in overhand throwing and ball catching, and use this information when engaging them in individually appropriate skill development activities. Use the four-step approach that follows:

- ▶ **Preplan** a partner throwing and catching activity to create a situation for observing the students' present developmental skill levels (initial, elementary, mature).

- ▶ **Assess** each student's present skill level from a location where unobtrusive observation is possible. While they are playing catch, do a total-body configuration analysis by watching each student's entire throwing or catching pattern. After identifying students at less than the mature level, do a segmental analysis by breaking the movement pattern down and observing individual body segments.

- ▶ **Plan and implement** a series of individually appropriate movement lessons designed to bring lagging body parts in line with more advanced ones.

- ▶ Take time to **evaluate students** and revise lessons. Use ongoing observation and evaluation to determine if they have made progress in throwing and catching. Make modifications in subsequent lessons as appropriate.

will view the entire fundamental movement phase as having three separate but often overlapping stages, namely, the initial, elementary, and mature stages.

Initial Stage

The initial stage of a fundamental movement phase represents the child's first goal-oriented attempts at performing a fundamental skill. Movement itself is characterized by missing or improperly sequenced parts, markedly restricted or exaggerated use of the body, and poor rhythmical flow and coordination. In other words, the spatial and temporal integration of movement is poor during this stage.

Elementary Stage

The elementary stage involves greater control and better rhythmical coordination of fundamental movements. The temporal and spatial elements of movement are better coordinated, but patterns of movement are still generally restricted or exaggerated. Children of normal intelligence and physical functioning tend to advance to the elementary stage primarily through the process of maturation. Many individuals, adults as well as children, fail to get beyond the elementary stage in many fundamental patterns of movement.

Mature Stage

The mature stage within the fundamental movement phase is characterized by mechanically efficient, coordinated, and controlled performances. The majority of available data on the acquisition of fundamental movement skills suggest that normally developing children can and should be at the mature stage by age five or six in most fundamental skills. Manipulative skills, however, which require trading and intercepting moving objects (catching, striking, volleying), develop somewhat later because of the sophisticated visual-motor requirements of these tasks. Consider the application example on Improvement of Throwing and Catching Skills.

Specialized Movement Phase

The specialized phase of motor development is an outgrowth of the fundamental movement phase. Instead of continuing to be closely identified with learning to move for the sake of movement itself, movement now becomes a tool that is applied to a variety of specialized movement activities for daily living, recreation, and sport pursuits. This is a period when fundamental stability, locomotor, and manipulative skills are progressively refined, combined, and elaborated on so that they may be used in increasingly demanding situations. The fundamental movements of hopping and jumping, for example, may now be applied to jumping rope, performing folk dances, and performing the triple jump (hop-step-jump) in track and field.

The onset and extent of skill development within the specialized movement phase depend on a variety of task, individual, and environmental factors. Task complexity; individual physical, mental, and emotional limitations; and environmental factors such as opportunity for practice, encouragement, and instruction are but a few. There are three identifiable stages within the specialized movement phase.

Transitional Stage

Somewhere around the seventh or eighth year of life, children commonly enter a transitional movement skill stage. They begin to combine and apply fundamental movement skills to the performance of specialized skills in sport and recreational settings. Walking on a rope bridge, jumping rope, and playing kickball are examples of common transitional skills. These skills contain the same elements as fundamental movements, but greater form, accuracy, and control of movement are now required. The fundamental movement skills that were developed and refined for their own sake during the previous phase now begin to be applied to play, game, and daily living situations. Transitional skills are simply an application of fundamental movement patterns in somewhat more complex and specific forms.

Application Stage

From about age 10 to age 13, interesting changes take place in skill development. During the previous stage, children's limited cognitive abilities, affective abilities, and experiences, coupled with a natural eagerness to be active, caused the normal focus (without adult interference) on movement to be broad and generalized to "all" activity. During the application stage, increased cognitive sophistication and a broadened experience base enable them to make numerous learning and participation decisions based on a variety of factors. Children begin to make conscious decisions for or against participation in certain activities. These decisions are based, in large measure, on how they perceive the extent to which factors within the task, within themselves, and within the environment either enhance or inhibit chances for personal enjoyment and success.

Lifelong Utilization Stage

The lifelong utilization stage typically begins around age 13 and continues through adulthood. The lifelong utilization stage represents the pinnacle of the process of motor development and is characterized by the use of one's acquired movement repertoire throughout life. The interests, competencies, and choices made during the previous stage are carried over to this stage; further refined; and applied to a lifetime of daily living, recreational, and sports-related activities. Factors such as equipment and facility availability and physical and mental limitations affect this stage. Among other things, one's level of activity participation will depend on talents, opportunities, physical condition, and personal motivation. One's lifetime performance level may range anywhere from the Paralympics and Special Olympics, to intercollegiate and interscholastic competition, to participation in organized or unorganized play, and important daily living skills such as toileting, grooming, and self-help.

In essence, the lifelong utilization stage represents a culmination of all preceding phases and stages and should be viewed as a lifelong process. One of the primary goals of adapted physical education programs is to help individuals to a point that they become happy, healthy, contributing members of society. The hierarchical development of the continuum of movement abilities embodied in the Phases of Motor Development serves as a primary means for achieving this goal. Only when we recognize that the progressive development of movement abilities in a developmentally appropriate manner is imperative to the balanced motor development of infants, children, adolescents, and adults will we begin to make significant contributions to the total development of the individual.

PRINCIPLES OF MOTOR DEVELOPMENT

The development and refinement of movement patterns and skills are influenced in complex ways. Both the process and product of one's movement are rooted in their unique genetic and experiential background, coupled with the specific demands of the movement task itself. Any study of motor development would be incomplete without a discussion of several of these influencing factors. The unique genetic inheritance that accounts for our individualities can also account for our similarities in many ways. One of these similarities is the trend for human development to proceed in an orderly, predictable fashion. A number of principles of motor development seem to emerge from this predictable pattern of development.

Developmental Direction

The principle of developmental direction was first formulated by Arnold Gesell (1954) as a means of explaining increased coordination and motor control as being a function of the maturing nervous system. Through observation, Gesell noted an orderly, predictable sequence of physical development that proceeds from the head to the feet (cephalocaudal) and from the center of the body to its periphery (proximodistal). The principle of developmental direction has, however, come into some criticism during the last few years and should not be viewed as operational at all levels of development nor in all individuals. The observation of tendencies toward distinct developmental directions may not be an exclusive function of the maturing nervous system as originally hypothesized, but it may be due, in part, to the demands of the specific task itself. For example, the task demands of independent walking are considerably greater than those for crawling or creeping. There is less margin for error in independent walking than there is in creeping and, in turn, crawling. In other words, it is mechanically easier to crawl than it is to creep, and to creep than it is to walk. Therefore, the apparent cephalocaudal progression in development may not be simply due to maturation of the nervous system but also to the specific performance demands of the task itself.

Rate of Growth and Development

One's growth rate follows a characteristic pattern that is universal for all and resistant to external influence. Even the interruption of the normal pace of growth is compensated for by a still unexplained process of self-regulatory fluctuation (Gesell, 1954) that comes into operation to help the growing child catch up to his agemates. For example, a severe prolonged illness may markedly limit a child's gain in height and weight, but upon recovery from the illness there will be a definite tendency to catch up if the condition does not persist and treatment is administered. The same phenomenon is seen with low-birth-weight infants. Despite this low weight at birth, there is still a tendency to catch up to the characteristic growth rate of one's agemates in a few years. Restricted opportunity for movement and deprivation of experience have been repeatedly shown to interfere with children's abilities to perform movement tasks that are characteristic for their particular age level. The effects of this deprivation of sensory and motor experience can sometimes be overcome when nearly optimal conditions are established for children (McGraw, 1939). The extent to which children will be able to catch up to their peers, however, depends on the duration and severity of deprivation and the age of the children, coupled with their individual genetic growth potential.

Differentiation and Integration

The coordinated and progressive intricate interweaving of neural mechanisms of opposing muscle systems into an increasingly mature relationship is characteristic of the developing child's motor behavior. Termed reciprocal interweaving by Gesell (1954), there are two different but related processes associated with this increase in functional complexity: differentiation and integration.

The process of differentiation is associated with the gradual progression from the gross globular (overall) movement patterns of infants to the more refined and functional movements of children and adolescents as they mature. For example, the manipulative behaviors of the newborn in terms of reaching, grasping, and releasing objects are quite poor. There is little control of movement, but as the child develops, control improves. The child is able to differentiate between various muscle groups and begins to establish control. Control continues to improve with practice until we see the precise movements of cursive writing, cutting with scissors, building with blocks, and playing the violin.

The process of integration refers to bringing various opposing muscle and sensory systems into coordinated interaction with one another. For example, the young child gradually progresses from ill-defined corralling movements when attempting to grasp an object to more mature and visually guided reaching and grasping behaviors. This differentiation of movements of the arms, hands, and fingers, followed by integration of the use of the eyes with the movements of the hand to perform eye-hand coordination tasks, is crucial to normal development.

Developmental Variability and Readiness

The tendency to exhibit individual differences is crucial. Each person is unique with his or her own timetable for development. This timetable is a combination of a particular individual's heredity and environmental influences. Although the sequence of appearance of developmental characteristics is predictable, the rate of appearance may be quite variable. Therefore, strict adherence to a chronological classification of development by age is without support or justification.

The average ages for the acquisition of all sorts of developmental tasks, ranging from learning how to walk to gaining bowel and bladder control, have been discussed in the professional literature and the daily conversation of parents and teachers for years. These average ages are just that and nothing more. They are merely approximations and are meant to serve as convenient indicators of developmentally appropriate behaviors in the normally developing individual. It is common to see deviations from the mean of as much as six months to one year or more in the appearance of numerous movement skills. The tendency to exhibit individual differences is closely linked to the principle of readiness, and it helps to explain why some individuals are ready to learn new skills when others are not.

Readiness refers to conditions within the task, the individual, and the environment that make a particular task appropriate for an individual to master. Physical and mental maturation, motivation, prerequisite learning, and an enriching environment all influence readiness. At this juncture we simply do not know how to identify precisely when one is ready to learn a new movement skill. However, research suggests that early experience in a movement activity before the individual is ready is likely to have minimal benefits (Magill, 1998).

It is critical to recognize and respect the concepts of developmental variability and readiness, both among individuals and within individuals. We simply cannot deal with individuals with developmental disabilities on the basis of chronological age or grade level and expect to be successful in movement skill acquisition and fitness enhancement.

Critical and Sensitive Learning Periods

The principle of critical or sensitive learning periods is closely aligned to readiness, and it revolves around observation that there are certain time frames when an individual is more sensitive to certain kinds of stimulation. Normal development in later periods may be hindered if the child fails to receive the proper stimulation during a critical period. For example, inadequate nutrition, prolonged stress, inconsistent parenting, or a lack of appropriate learning experiences may have a more negative impact on development if introduced early in life rather than at a later age. The principle of critical periods also has a positive side. It suggests that appropriate intervention during a specific period of time tends to facilitate more positive forms of development than if the same intervention occurs later.

Current views of the critical period hypothesis reject the notion that there are highly specific time frames in which one must develop motor skills (Seefeldt, 1975). There are, however, periods during which development of certain skills is more easily accomplished. These are referred to as sensitive periods. A sensitive period is a broader time frame for development and is susceptible to modification. Learning is a phenomena that can continue throughout life.

Phylogeny and Ontogeny

Many of the rudimentary abilities of the infant and the fundamental movement abilities of young children are considered to be phylogenetic. That is, they tend to appear automatically and in a predictable sequence within the maturing child. Phylogenetic skills are resistant to external environmental influences. Movement skills such as the rudimentary manipulative tasks of reaching, grasping, and releasing objects; the stability tasks of gaining control of the gross musculature of the body; and the fundamental locomotor abilities of walking, jumping, and running are examples of phylogenetic skills.

Ontogenetic skills, on the other hand, are those that depend primarily on learning and environmental opportunities. Skills such as swimming, bicycling, and ice skating are considered to be ontogenetic, because they do not appear automatically within individuals but require a period of practice and experience and are influenced by one's culture. Phylogeny and ontogeny need to be reevaluated in light of the fact that many skills, heretofore considered phylogenetic, can be influenced by environmental interaction.

SUMMARY

The acquisition of motor control and movement competency is an extensive process beginning with the early reflexive movements of the newborn and continuing throughout life. The process by which an individual progresses from the reflexive movement phase, through the rudimentary and fundamental movement phases, and finally to the specialized movement skill phase is influenced by factors within the task, the individual, and the environment.

Reflexes and rudimentary movement abilities are largely maturationally based. They appear and disappear in a fairly rigid sequence deviating only in the rate of their appearances. They do, however, form an important base that fundamental movement abilities are developed on.

Fundamental movement abilities are basic movement patterns that begin developing around the same time that the child is able to walk independently and move freely through the environment. These basic locomotor, manipulative, and stability abilities go through a definite, observable process from immaturity to maturity. A variety of stages within this phase have been identified for a number of fundamental movements; these are the initial, elementary, and mature stages. Attainment of the mature stage is influenced greatly by opportunities for practice, encouragement, and instruction in an environment that fosters learning. These same fundamental skills will be elaborated on and refined to form the specialized movement abilities so highly valued for recreational, competitive, and daily living tasks.

The specialized movement skill phase of development is in essence an elaboration of the fundamental phase. Specialized skills are more precise than fundamental skills. They often involve a combination of fundamental movement abilities and require a greater degree of exactness in performance. Specialized skills have three related stages. From the transition stage onward, children are involved in the application of fundamental movement skills and their purposeful utilization in play, games, sport, and daily living tasks. If the fundamental abilities used in a particular activity are not at the mature stage, the individual will have to resort to the use of less mature patterns of movement.

Principles of development emerge as we study the process of growth and motor development. These principles serve as an avenue for theory formulation. Dynamic Systems Theory and the Phases of Motor Development serve as helpful means for conceptualizing both the process and the product of motor development. Individuals with developmental disabilities are more like their agemates than they are unlike them. Therefore, it is essential that we understand and apply principles of normal development in the education of all individuals.

Chapter 18

Perceptual-Motor Development

Joseph P. Winnick

▶
▶
▶
▶

MARGARET IS A TOTALLY BLIND four-year-old who would like to move independently in her preschool gymnasium. Her teacher has encouraged Margaret to orient herself to play areas according to sound cues associated with each area. In essence, she is encouraging Margaret to develop her auditory perceptual abilities to help compensate for her loss of sight. Do you feel that Margaret's teacher should help her develop her auditory perceptual abilities? If yes, which components of auditory perception should be developed, and what are some physical education activities that could help?

The ability to learn and function effectively is affected by perceptual-motor development. Perceptual-motor ability permits the human to receive, transmit, organize, integrate, and attach meaning to sensory information and to formulate appropriate responses. These responses are necessary for the individual to move and to learn while moving in a variety of environments; thus, they have direct or indirect impact in physical education and sport.

Ordinarily, perceptual-motor development occurs without the need for formal intervention. In other instances, perceptual-motor abilities need attention, because they have not developed satisfactorily. For example, deficits related to perceptual-motor ability are often named as characteristics of persons with learning disabilities. These deficits may include poor spatial orientation, poor body awareness, immature body image, clumsiness or awkwardness, coordination deficits, and poor balance. The higher incidence of perceptual-motor deficits among people with cerebral palsy or mental retardation is well known. Perceptual-motor experiences are particularly important in cases where sensory systems are generally affected but residual abilities may be enhanced; and in cases where perceptual-motor abilities must be developed to a greater degree to compensate for loss of sensory abilities. People with visual and/or auditory disabilities exemplify these situations.

In this introductory section it is important to comment on the influences of perceptual-motor programs. In the 1960s and early 1970s, such programs were strongly advocated and supported because of the belief that they led to a significant improvement in academic and intellectual abilities. Research recently conducted has not supported this notion. On the other hand, research indicates clearly that perceptual-motor abilities, as measured by various tests, may be attained through carefully sequenced programs (Winnick, 1979). For example, it is clear that balance, a perceptual-motor ability that is basic to movement skill, may be enhanced through systematic training. Since these perceptual-motor abilities are fundamental to many motoric, academic, and daily living skills, the nurturing and/or remediation of these skills are vital and are relevant to adapted physical education.

After reading this chapter it should be clear that all the movement activities experienced in physical education are perceptual-motor experiences. When perceptual-motor abilities require nurturing or when they have developed inadequately, there may be a need to plan programs to enhance their attainment. This chapter is designed to serve as a resource for planning and program implementation.

OVERVIEW OF THE PERCEPTUAL-MOTOR PROCESS

In order to implement perceptual-motor programs most effectively, it is helpful to have an understanding of the perceptual-motor process. A simplified four-step schematic of perceptual-motor functioning is presented in figure 18.1.

Sensory Input

The first step in the perceptual-motor process, sensory input, involves receiving energy forms from the environment and from within the body itself, as sensory stimuli, and processing this information for integration by the central nervous system. Visual (sight), auditory (hearing), kinesthetic (movement), vestibular (balance), and tactile (touch) sensory systems provide raw data for the central nervous system. These sensory systems serve to gain information that is transmitted to the central nervous system through sensory (afferent) neurons.

Sensory Integration

The second step in the perceptual-motor process involves sensory integration. Present and past sensory information is integrated, compared, and stored in short- and/or long-term memory. An important phase occurs as the human organism selects and organizes an appropriate motor output based on the integration. The resultant decision becomes part of long-term memory, which is transmitted through the motor (efferent) mechanisms.

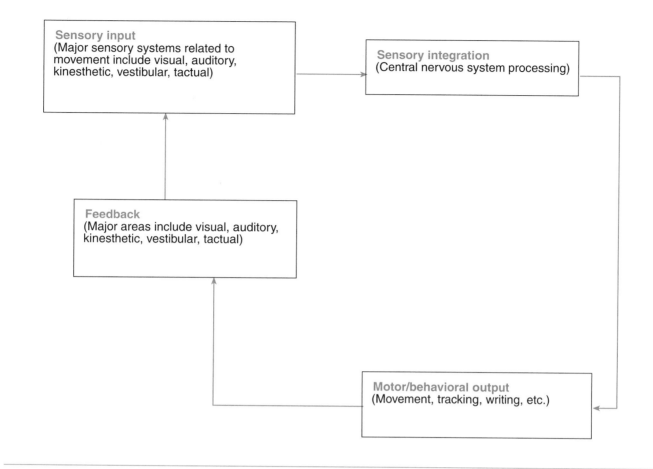

Figure 18.1   A simplified model of the perceptual-motor process.

Motor/Behavioral Output and Feedback

The third major step in the perceptual-motor process is **motor/behavioral output**. Overt movements and/or behaviors occur as a result of decisions from the central nervous system. As output occurs, information is also continually fed back as sensory input about the nature of the ongoing response by the human organism. This *feedback* constitutes step four and serves as sensory information to continue the process. As with sensory input, feedback in movement settings is usually kinesthetic, tactual, visual, or auditory. During feedback, the adequacy or nature of the response is evaluated or judged. If it is judged inadequate, adjustments are made; if it is successful, adjustments are not required.

Terms Associated With the Process

Terms associated with the perceptual-motor process are used in a variety of ways. The following are definitions of some terms used in this chapter.

Perception is the monitoring and interpretation of sensory data resulting from the interaction between sensory and central nervous system processes. Perception occurs in the brain and enables the individual to derive meaning from sensory data. **Perceptual-motor development** is the process of enhancing the ability to integrate sensory stimuli arising from or relating to observable movement experiences. It involves the ability to combine kinesthetic and tactual perceptions for the development of other perceptions, the use of movement as a vehicle to explore the environment and to develop perceptual-motor abilities, and the ability to perceive tactually and kinesthetically. Perception occurs as sensory information is interpreted or given meaning. Because perceptual-motor includes both an individual's interpretation and response to sensory stimulation, it requires cognitive ability. On the other hand, **sensorimotor activity** occurs at a subcortical level and does not involve meaning, interpretation, or cortical-level functioning. Sensorimotor activity is characterized by motor responses to sensory input. The sensory integration process results in perception and other types of sensory data synthesis. Thus, perception is one aspect of sensory integration.

During the 1980s and continuing into the 1990s, there has developed in physical education literature a view on perception that appears to have implications for perceptual-motor development. This view, known as the direct or ecological approach to perception, is based on the writings of J.J. Gibson (1977, 1979). This view emphasizes that perception is specific to each individual and that the environment is perceived directly in terms of its utility or usefulness for the perceiver. Persons perceive the environment in terms of the actions they can exert on it, that is, the affordances provided by the environment. For example, children may perceive a chair predominantly in terms of their ability to crawl under it, while adults perceive it as an object to sit on (although they may also recognize its other possibilities). Advocates of this orientation feel that perceptual deficits may be defined in terms of inadequate perception of affordance. Thus, perception may become a prime candidate in the search for potential rate limiters (factors that impede performance) in children with movement problems (Burton, 1990).

An Example of a Perceptual-Motor Skill

The perceptual-motor skill of batting a pitched softball can be used to illustrate these basic introductory concepts. As the pitched ball comes toward the plate, the batter focuses on the ball and tracks it. Information about speed, direction, spin, and other flight characteristics is picked up (stimulus reception) by the visual system (sensory input) for further processing. The information is transmitted via sensory neurons to the central nervous system, where it undergoes sensory integration. Because of various characteristics in the environment, the nature of the sensory information, and past experience, the incoming object is perceived as a softball (perception) to be struck (perceptual motor). The batter's past experience influences the ability to fixate and track the softball and processes information about how to hit it. How to hit it involves comparative evaluations in which, for example, the arc of the ball is compared with that from previous instances when the act was performed.

The batter decides on the appropriate swing (decision about motor behavior) on the basis of earlier steps in the process. This decision becomes part of the long-term memory, to be used for future reference. During the pitch, the brain is constantly kept informed about the position of the bat and the body; the brain uses this information to enhance the overt behavior of swinging the bat. Once the nature of the motor behavior or response has been determined, messages are sent to appropriate parts of the body to initiate the response (motor output response). As the batter moves and completes the task, the information is provided on which to judge whether the response is successful or inadequate (feedback). If the pitch is missed, the batter may decide that adjustments are necessary in similar future instances; this information is stored in long-term memory and serves as a basis for learning. Perceptual-motor functioning is concerned with the entire process depicted in figure 18.1 and described in this example; batting is thus one type of perceptual-motor skill.

If the ecological perspective on perception is applied to this example, the ball will be perceived by each observer in terms of ability to hit the ball or to hit the ball to a particular place on the field, given one's body size and skill, rather than a focus on the ball's velocity, spin, and so on. Thus, individuals will perceive the ball in terms of their ability to hit it, and based on this perception, they will decide what to do and respond accordingly.

Perceptual-Motor Deficits

Because figure 18.1 depicts perceptual-motor processing, it is a useful reference for breakdowns in the process: breakdowns at the input, integration, output, and feedback sites. A breakdown at the *input* site may occur for a variety of reasons. For example, individuals with sensory impairments like blindness or deafness may not be able to adequately pick up visual or auditory information from the environment, with the result that this information does not appropriately reach the central nervous system. Students who are mentally retarded or autistic may not attend to, and thus may not receive, relevant information. People with neuromuscular impairments may be inhibited by the lack of appropriate kinesthetic, vestibular, or tactual information basic to quality input.

Sensory *integration* may be affected by factors such as mental retardation or neurological conditions that impair the functioning of the central nervous system. Also, sensory integration may be influenced by the quality of information received and the ability to process sensory input. Motor *output* can be affected by inappropriate functioning at the input and integration steps as well as by disabilities that influence movement. Conditions associated with cerebral palsy, muscular dystrophy, and other neurological or orthopedic impairments are examples. Breakdowns at the *feedback* site can result from factors that affect input, integration, and output and any additional factors that bear on the ability to modify or correct behavior. The clumsy child, for example, may lack body awareness because of faulty kinesthetic perception; this would impair the adequacy of feedback.

Related to the ecological approach, Burton (1987, 1990) has suggested that the influence of perception be examined more closely when movement problems occur. Consistent with this view, it is suggested that perceptual judgments be assessed using actual performance as a criterion to determine if the motor outcomes are due to faulty perception. The game of golf provides a good example of this concept. In golf, the club used for a particular shot is selected on the basis

Table 18.1 An Analysis of Prominent Perceptual-Motor Need and Deficit Areas

Disability	Prominent need and deficit areas
Visual disability (visual impairments)	Need to focus on the development of residual visual perceptual abilities and to help the child compensate for visual perceptual-motor deficits by enhancing auditory, vestibular, tactual, and kinesthetic perception. Give particular attention to input and feedback steps in the perceptual-motor process.
Auditory (deaf and hard of hearing)	Need to focus on the development of residual hearing and vestibular abilities (if affected) and help the child compensate by enhancing development associated with sensory systems that are intact. Give particular attention to input, integration, and feedback steps in the perceptual-motor process.
Haptic disabilities (primarily, children who are clumsy; children with orthopedic, neuromuscular, or neurological impairments)	Need to focus on the development of vestibular, kinesthetic, and tactual perception and to integrate motor experiences with visual and auditory perception. There may be a particular need to focus on input, motor response, and feedback steps.
Mental or affective disabilities (children with mental retardation, emotional disturbance, etc.)	Needs and focus on needs determined by assessment of perceptual-motor abilities. Involvement throughout the perceptual-motor process may exist.

of distance, height desired, and so on. If, after selecting a club, the golfer makes a shot short of the target, it is possible that this inadequate performance resulted from selecting the wrong club (faulty perception of shot requirements) rather than from poor skill. For example, the golfer may have underestimated the distance, wind resistance, and so on. Although the shortness of the shot may have been due to other reasons, the possibility for faulty perception exists and should be assessed.

Many of the factors causing perceptual-motor deficits are associated with student disabilities seen in the school environment. Due to the influence on perceptual-motor development, physical educators may need to develop programs to nurture development or remediate performance. The nature of an individualized program will depend on the cause of the perceptual-motor breakdown, the student's perception and abilities, and the purpose of the program. In the case of a student who is totally blind, for example, it may be necessary to focus on heightening auditory perceptual-motor components to improve orientation to school grounds and to focus on kinesthetic percep-

tion to enhance efficient movement in the environment. Table 18.1 presents an analysis of prominent perceptual-motor need and deficit areas as a function of specific impairments.

Sensorimotor Stimulation

As stated earlier, perceptual-motor deficits may be due to sensory input and processing problems. Physical educators who have reviewed and analyzed research and literature on the value of sensory stimulation (Cratty, 1986a, b; Sherrill, 1998) agree with this author that sensorimotor stimulation enhances development and learning. There is also agreement in the literature that acceptable levels of sensorimotor development generally occur without the need for special intervention by professionals. However, programs including sensorimotor activity are sometimes recommended for individuals with disabilities, especially those exhibiting severe disabilities. Sensorimotor stimulation is designed to overcome sensorimotor developmental delays or deficits due to factors such as inadequate central nervous system development or functioning, diseases, disabilities, and disabilities, and opportunities for development. Programs designed to nurture sensory systems have been conducted by various professionals, including educators as a part of early childhood and/or physical education, movement therapists as a part of movement education programs, and occupational therapists as a part of sensory integration programs. Goals of programs established by professionals, influence purposes and approaches in programs. This author recommends sensorimotor stimulation to enhance adequate perceptual-motor and motor development.

Programs designed for sensorimotor nurturing emphasizing movement have several identifiable commonalties. First, they are closely associated with the major sensory modalities associated with movement (tactile, kinesthetic, vestibular, visual, and auditory). Second, they involve and focus on subcortical activity (i.e. activity not dependent on cerebral cortex and cortical tracts). A key distinction between sensorimotor activity and perceptual-motor activity is that the former is subcortical and the latter includes cognition. Third, in program implementation, there is considerable overlap in sensorimotor and perceptual-motor activities. In fact, perceptual-motor functioning is dependent on sensorimotor functioning, because sensory information is processed, integrated, and organized in the cerebral cortex and this activity is key to perceptual-motor functioning.

Sherrill (1998) indicates that the factors that affect sensorimotor function include muscle and postural tone, reflexes and postural reactions, and sensory input systems.

In this regard, this chapter is delimited to an introduction of activities associated with the stimulation of sensory systems. Hopefully, the information presented in table 18.2 will serve as a good beginning resource for physical educators called on to provide or coordinate with programs designed to nurture sensory input systems. Components as well as sample activities typically associated with such programs are presented in the table. The components include tactile, kinesthetic, vestibular, visual, and auditory integration. Although activities are grouped with a single component, it must be realized that most activities involve and stimulate two or more components as they are conducted. These components correspond closely to the perceptual-motor components to be discussed subsequently in the chapter.

Facilitating Development

The teacher has an important role in nurturing or remediating perceptual-motor abilities. The exact role and teaching styles to be employed should vary with the characteristics of the learner. Since perceptual-motor abilities appear to develop optimally between the ages of two and seven, indirect teaching styles like movement exploration and guided discovery are generally appropriate. More direct teaching styles may be effective as children reach ages six and seven.

Table 18.2 **Typical Components and Activities Associated With Sensorimotor Development**

Components	Typical activities
Tactile integration	Water activities; massage; stroking; partner activities; bare-footed activities; movement education activities; activities performed on various surfaces; handling objects of various textures; crawling through tunnels; tactually discriminating between familiar and unfamiliar objects
Kinesthetic integration	Naming body parts; moving body parts for a purpose; active movements that develop a knowledge of body parts, their position in space, and the ability to innervate them (time, space, force, flow of movement)
Vestibular integration	Rocking and cradling in arms, cribs, or chairs; simple bouncing activities on spring-type or trampoline equipment; scooter and vestibular board activities; simple swinging activities using hammocks or other simple swing-type supplies; therapy ball activities; sitting and standing postures; nonlocomotor movements
Visual integration	Activities involving object manipulation including ball handling, in which speed, distance, size, color, and mass are modified; recognition and tracking of objects; looking; sorting; fundamental visual-motor activities; finding objects
Auditory integration	Activities involving sound recognition, auditory discrimination, sound localization, auditory figure ground; fundamental motor activities; listening and responding to auditory stimuli; making sounds and talking; activities involving auditory memory

Burton (1987) suggested two implications for teaching to enhance perceptual skill that may have important potential for enhancing perceptual-motor development. First, he emphasized that teachers should provide *purposeful movement*, since it is most motivating and encourages attention to the information in the environment. Movements are most purposeful when they are performed in natural settings as a means to an end rather than as the end itself. Key to the provision of purposeful movement is to select an activity involving an objective beyond the actual movement itself. For example, assume that the teacher wishes to improve the accuracy of kicking a playground ball. A purposeful movement would be to kick the ball into a goal in a soccer lead-up activity or game. This is more motivating to the student than kicking a ball for the sake of kicking a ball.

A second teaching implication is for students to become more accurately attuned to *affordances in the environment* (Burton, 1987). One way to do this is by encouraging students to make perceptual judgments and to assess the accuracy of their judgments. Applied to a golf example, the question posed may be, "Can I hit the ball to the green using the nine iron?" In another example, a question posed may be, "Can I pass the ball to a teammate without it being intercepted by a defensive player?" In these situations the congruency between perception and action or movement is evaluated. The accuracy of one's perception of the environment or the affordances available is evaluated. In situations like these, the influence of perception on motor performance may be evaluated to determine whether poor performance results from faulty perception.

Although technically there are many sensory systems associated with perceptual-motor development, the remainder of this chapter discusses visual, auditory, kinesthetic, and tactual perception.

VISUAL PERCEPTUAL-MOTOR DEVELOPMENT

Visual perceptual-motor abilities are important in academic, physical education, and sport settings. In the academic setting, visual perceptual abilities are used in writing, drawing, reading, spelling, and arithmetic. In physical education and sport, they are important for catching, throwing, and kicking objects, playing tag, balancing, running, and performing fundamental movements. Age-appropriate visual perceptual-motor abilities are built on visual acuity, which affects the ability to see, fixate, track, and so forth, and thus is required for the input step. On the basis of input, the individual develops the abilities or components of visual perceptual-motor development associated with central nervous system processing and output. Components closely associated with movement include visual figure-ground perception, spatial relations, visual constancy, and visual-motor coordination.

Figure-Ground Perception

Figure-ground perception involves the ability to distinguish a figure from its background and give meaning to the forms or the combination of forms or elements that constitute the figure. It requires the ability to differentiate and integrate parts of objects to form meaningful wholes and to appropriately shift attention and ignore irrelevant stimuli. Visual figure-ground perception is called on when students are asked to pick out a specific letter of the alphabet from a field of extraneous items. In sport, figure-ground perception is clearly demonstrated in baseball, because a batter must distinguish a white ball from a background in attempting to hit it. Students with inadequate perception may exhibit difficulties in differentiating letters, numbers, and other geometric forms; combining parts of words to form an entire word; or sorting objects. In physical education, figure-ground perception is required in games that depend on tracking moving objects, observing lines and boundaries, and in activities requiring concentration on relevant stimuli. These include activities in which children move under, over, through, and around perception boxes, tires, hoops, or playground equipment, as well as activities in which they follow or avoid the lines and shapes associated with obstacle courses, geometric figures, maps, mazes, hopscotch diagrams, or footprints.

Spatial Relationships

The perception of spatial relationships means locating objects in space relative to oneself (egocentric localization) and locating objects relative to one another (objective localization). Egocentric localization, often referred to as perception of position in space, is demonstrated as youngsters attempt to move through hoops without touching them. Objective localization is seen as a player attempts to complete a pass to a guarded teammate.

Spatial relationship, which affects virtually all aspects of academic learning, involves knowing direction, distance, and depth. Position in space is basic to the solution of reversal or directional problems (such as the ability to distinguish *d*, *p*, and *q*; 36 and 63; *saw* and *was*; *no* and *on*). Perception of spatial relationship also encompasses temporal ordering and sequencing. People who have difficulty putting objects in order will have difficulty in various academic areas, including arithmetic sequencing problems (performing operations in correct order). Some authors have contended that spatial awareness is preceded by body awareness and that the awareness of relationships in space grows out of an awareness of relationships among the parts of one's own body.

Visual Perceptual Constancy

Perceptual constancy is the ability to recognize objects despite variations in their presentation. It entails recognizing the sameness of an object, although the object may in actuality vary in appearance, size, color, texture, brightness, shape, and so on. For example, a football is recognized as having the same size even when seen from a distance. It has the same color in daylight as in twilight and maintains its shape even when only its tip is visible. Development of perceptual constancy involves seeing, feeling, manipulating, smelling, tasting, hearing, naming, classifying, and analyzing objects. Inadequate perceptual constancy affects the recognition of letters, numbers, shapes, and other symbols in different contexts. Physical education and sport provide a unique opportunity for the nurturing of constancy, because objects are used and manipulated in a variety of ways and are viewed from many different perspectives.

Visual-Motor Coordination

Visual-motor coordination is the ability to coordinate vision with body movements. It is the aspect of visual perceptual-motor ability that combines visual with tactual and kinesthetic perception; thus, it is not an exclusively visual ability. Although coordination of vision and movement may involve many different parts of the body, eye-hand and eye-foot coordination are usually the most important in physical education and sport activities. Effective eye-limb coordination is also important in academic pursuits, such as cutting, pasting, finger painting, drawing, tracing, coloring, scribbling, using the chalkboard, and manipulating clay and toys. It is particularly important in writing. Effective eye-limb coordination is also necessary for such basic skills as putting on and tying shoes, putting on and buttoning clothes, eating or drinking without tipping glasses and plates, and using simple tools.

Using and Developing Visual Perceptual-Motor Abilities

A wide variety of experiences in physical education and sport call on and may be used to stimulate visual perceptual-motor abilities. Although motor activities are not generally limited to the development of one specific ability, some activities are especially well suited for figure-ground development. These include rolling, throwing, catching, kicking, striking, dodging, and chasing a variety of objects in a variety of ways; moving under, over, through, and around perception boxes, tires, hoops, geometric shapes, ropes, playground equipment, pieces of apparatus, and other "junk"; following or avoiding lines associated with obstacle courses, geometric shapes, maps, mazes, hopscotch games, or grids; stepping on or avoiding footprints,

Figure 18.2   Playing in a tub of balls is an activity that helps students learn spatial concepts, colors, shapes, and texture.

stones, animals, or shapes painted on outdoor hardtops or floors; imitating movements as in Leapfrog, Follow the Leader, or Simon Says; and doing simple rope activities, including moving under and over ropes and jumping rope. An example of a program set up for a child for developing visual figure-ground skill can be found in the Perceptual-Motor Skill Development application example.

Spatial relationships are involved in trampolining, tumbling, swimming, rope jumping, rhythms and dance, obstacle courses, and the like. Activities particularly useful in helping the individual to develop spatial abilities include moving through tunnels, tires, hoops, mazes, and perception boxes. Elementary games like dodgeball, tag, and Steal the Bacon, in which one must locate objects in space relative to oneself (egocentric localization) or relative to one another (objective localization), foster perception of spatial relationships as well (figure 18.2).

Visual-motor coordination is clearly important in physical education and sport. Games that include throwing, catching, kicking, and striking balls and other objects are among those activities requiring such coordination. Age-appropriate games are highly recommended because they enhance the purposefulness of movement.

AUDITORY PERCEPTUAL-MOTOR DEVELOPMENT

Age-appropriate auditory perceptual-motor abilities are built on auditory acuity. The ability to receive and transmit auditory stimuli as sensory input is the foundation of the development of auditory figure-ground perception, sound localization, discrimination, temporal

Application Example: Perceptual-Motor Skill Development

Setting: A classroom teacher asks the physical education teacher to suggest motoric activities to help develop visual figure-ground activities for a six-year-old student named Jimmy.

Student: This six-year-old boy has a learning disability involving serious difficulty with visual figure-ground differentiation and integration.

Application: The physical educator suggests the following activities:

▶ While using a scooter, have Jimmy follow roads in his community drawn on the gym floor.

▶ Involve Jimmy in partner activities in which soft, large balls and balloons are rolled, tossed, and stopped.

▶ Have Jimmy walk, hop, or jump on and off shapes, various spots, hula hoops, or lines drawn on a mat or on the floor.

auditory perception, and auditory-motor coordination. Educators should give much attention to auditory perception when working with students who have sensory impairments.

Auditory Figure-Ground Perception

Auditory figure-ground perception is the ability to distinguish and attend to relevant auditory stimuli against a background of general auditory stimuli. It includes the ability to ignore irrelevant stimuli (such as those in a noisy room or a room in which different activities are conducted simultaneously) and to attend to relevant stimuli. In situations where irrelevant stimuli are present, people with inadequate figure-ground perception may have difficulty concentrating on the task at hand, responding to directions, and comprehending information received during the many listening activities of daily life. They may not attend to a honking horn, a shout, or a signaling whistle. Their problems in physical education and sport may occur on occasions when beginning, changing, or ending activities are signaled through sound.

Auditory Discrimination

Auditory discrimination is the capacity to distinguish different frequencies, qualities, and amplitudes of sound. It involves the ability to recognize and discriminate among variations of auditory stimuli presented in a temporal series and also involves auditory perceptual constancy. The latter is the ability to recognize an auditory stimulus as the same under varying presentations. Auditory discrimination thus involves the ability to distinguish pitch, loudness, and constancy of auditory stimuli. People with inadequate auditory discrimination may exhibit problems in games, dances, and other rhythmic activities that depend on this ability.

Sound Localization

Sound localization is the ability to determine the source or direction of sounds in the environment. Sound localization is used on the basketball court to find the open player who is calling for the ball, and it is basic to goal ball, in which blindfolded players attempt to stop a ball emitting a sound. Sound localization is vital to orientation and mobility for people with visual impairments. They often need to further develop their abilities to locate sound and may do so by reacting to various stimuli (e.g., bells, voices, horns).

Temporal Auditory Perception

Temporal auditory perception involves the ability to recognize and discriminate among variations of auditory stimuli presented in time. It entails distinguishing rate, emphasis, tempo, and order of auditory stimuli. Individuals with inadequate temporal auditory perception may exhibit difficulties in rhythmic movement and dance, singing games, and other physical education activities.

Auditory-Motor Coordination

Auditory-motor coordination is the ability to coordinate auditory stimuli with body movements. This coordination is readily apparent when a person playing goal ball reaches for a rolling ball (ear-hand coordination) according to where the player believes the ball is located. Linking auditory and motor activities is also demonstrated when a person responds to a beat in music (ear-foot coordination) or to a particular cadence when football signals are called out. Auditory-motor coordination is evident as a skater or gymnast performs a routine to musical accompaniment.

Development of Auditory Perceptual-Motor Abilities

Physical education and sport offer many opportunities to develop auditory perception. Participants may follow verbal directions or perform activities in response to tapes or records; the activities may be suggested by the music itself. For example, children may walk, run, skip, or gallop to a musical beat; they may imitate trains, airplanes, cars, or animals, as suggested by the music. Dances and rhythmic activities with variations in the rate and beat are useful, as are games and activities in which movements are begun, changed, or stopped in response to various sounds. Blind or blindfolded children may move toward or be guided by audible goal locators, or they may play with balls that have bells attached to them. Triangles, drums, bells, sticks, or whistles may direct children in movement or serve as play equipment. A teacher conducting such activities should minimize distracting stimuli and vary the tempo and loudness of sound. It may be necessary to speak softly at certain times so that the participants must concentrate on listening.

PROPRIOCEPTION

Proprioception encompasses those perceptual-motor abilities that respond to stimuli arising within the organism. These include sensory stimuli arising from muscles, tendons, joints, and vestibular sense receptors. Such abilities emphasize movement and are discussed within the categories of kinesthetic perception and balance.

Kinesthetic Perception

It is apparent even to the casual observer that we use information gained through auditory and visual receptors to move within and learn from the environment. Just as we know a sight or sound, we also have the ability to know a movement or body position. We can know an action before executing it or feel the correctness of a movement. The awareness and memory of movement and position are kinesthetic perceptions. They develop from impulses that originate from the body's proprioceptors. Because kinesthetic perception is basic to all movement, it is associated with visual-motor and auditory-motor abilities.

Like all perceptions, kinesthetic perceptions depend on sensory input (including kinesthetic acuity) provided to the central nervous system. The central nervous system, in turn, processes this information in accord with the perceptual-motor process. Certain diseases and conditions may cause kinesthetic perception to be impaired. For example, in the case of a person who has had an amputation, all sensory information that normally would be processed by a particular extremity could be missing. Cerebral palsy, muscular dystrophy, and other diseases or conditions affecting the motor system may result in a pattern of input or output that is different from that of an individual without disabilities. A youngster with a learning disability may have difficulty selecting appropriate information from the many sources in the organism. Inadequate kinesthetic perception may manifest itself in clumsiness due to lack of opportunity for participation in motoric experiences. Abilities closely associated with kinesthetic perception are body awareness, laterality, and verticality.

Body Awareness

Body awareness is an elusive term that has been used in various ways by writers representing different but related disciplines. Used here, body awareness is a comprehensive term that includes body schema, body image, and body concept or knowledge. Body schema is the most basic component and is sometimes known as the sensorimotor component, because it is dependent on information supplied through activity of the body itself. It involves awareness of the body's capabilities and limitations, including the ability to create appropriate muscular tensions in movement activities, and awareness of the position in space of the body and its parts. At basic levels, body schema helps the individual know where the body ends and external space begins. Thus, an infant uses feedback from body action to become aware of the dimensions and limitations of the physical being and begins to establish separateness of the body from external surroundings. As body schema evolves, higher levels of motor development and control appear and follow a continuous process of change throughout life.

Body image refers to the feelings one has about one's body. It is affected by biological, intellectual, psychological, and social experiences. It includes the internal awareness of body parts and how they function. For example, people learn that they have two arms and two legs or two sides of the body, and sometimes they work in combination and other times they function independently. Intellectual, social, and psychological factors enter into the perception of oneself as fast, slow, ugly, beautiful, weak, strong, masculine, feminine, and so forth.

Body concept, or body knowledge, is the verbalized knowledge one has about one's body. It includes the intellectual operation of naming body parts and the understanding of how the body and its parts move in space. Body concept is developed by experiences involving body schema and body image.

The importance of movement experiences for the stimulation and nurturing of body awareness and the importance of body awareness for movement are obvious. Movement experiences that may serve in the developmental years to enhance body awareness include those in which parts of the body are identified, named,

Figure 18.3   Balance activities on a large ball stimulate postural adjustments and body awareness.

pointed to, and innervated: imitation of movements, balance activities, rhythmic or dance activities, trampoline and scooter board activities, mimetic activities, movement and exploration, swimming games, activities conducted in front of a mirror, stunts and tumbling, and a variety of exercises. Virtually all gross motor activities involve body awareness at some level.

Laterality and Verticality

Laterality and verticality refer to internal awareness of right and left and up and down, respectively. Laterality is the internal awareness of the sides of the body and their differences. Verticality refers to an internal awareness of up and down. Development of verticality is also believed to be enhanced through experimentation in upper and lower parts of the body. Laterality and verticality should be conceived as very much related to body awareness. Laterality and verticality are significantly included in many physical education activities, and nurturing these abilities enhances their successful development and performance. Examples include most balance, locomotor, and object control activities.

Balance

As mentioned previously, proprioception includes sensation pertaining to vestibular sense reception. The vestibular apparatus provides the individual with information about the body's relationship to gravitational pull and thus serves as the basis for balance or equilibrium. Vestibular sense perception combines with visual, auditory, kinesthetic, and tactual information to enhance attainment of static and dynamic balance.

It is well known that balance is a key element in the performance of movement activities. Many activities may be used to nurture or remediate balance during the perceptual-motor developmental years. These include activities conducted on tilt boards, balance boards, and balance beams. Mimetic activities, stunts and tumbling, and a variety of games may also be used to develop balance (figure 18.3).

TACTUAL PERCEPTION

Tactual perception is the ability to interpret sensations from the cutaneous surfaces of the body. Whereas kinesthetic perception is internally related, tactual perception is externally related and responds to touch, feel, and manipulation.

Through these aspects of the tactile system, the human experiences various sensations that contribute to a better understanding of the environment. For example, tactual perception enables one to distinguish wet from dry, hot from cold, soft from hard, and rough from smooth. The importance of tactual perception is evident as a student who is blind feels a lacrosse stick to understand what it is, tries to stay on a cinder track while running, or learns to move through the environment using cane travel. For all youngsters, learning is enhanced when they touch, feel, hold, and manipulate objects. The term *soft* becomes more meaningful and tangible as youngsters feel something soft and distinguish it from something hard.

Gross motor activities in physical education and sport offer many opportunities to use tactual perception. Relevant activities include those involving contact of the hands or the total body with a variety of surfaces. Tactual perception combines with kinesthetic sensations as youngsters crawl through a tunnel, walk along a balance beam, jump on a trampoline, climb a ladder, wrestle, or tumble. Individuals may walk barefoot on floors, lawns, beaches, balance beams, or mats or in swimming pools; they may climb ropes, cargo nets, ladders, and playground equipment. Swimming activities are particularly important because of the unique sensations that water provides.

SUMMARY

Perceptual-motor development is a process of enhancing the ability to integrate sensory stimuli arising from or relating to observable movement experiences. It is associated with sensory systems. This chapter discusses sensorimotor stimulation and perceptual-motor development. It focuses on how movement activities are involved in and can be used for sensorimotor stimulation and perceptual-motor development and thereby contribute to development.

Infants and Toddlers

Cathy Houston-Wilson

▶
▶
▶

MARIA, AN EARLY CHILDHOOD ADAPTED physical education teacher at the United Cerebral Palsy Center for Young Children, sets up her motor room with incline mats, soft balls, scarves, and tunnels to accommodate infants and toddlers with special needs and their caregivers. She has specific objectives for each child: weight bearing for Adam, balancing and ambulating for Molly, and visual tracking for Juan. The children and their caregivers enter the motor room and immediately begin acting on the equipment that has been set out for them. Maria greets the children individually as they enter and talks with their caregivers about the progress the children are making and what objectives have been developed for the day with the equipment that has been set up. Maria and the caregivers follow the young children's lead as they begin to play with the equipment.

The purpose of this chapter is to present an overview of adapted physical education as it relates to infants and toddlers with unique needs. Specifically, the following topics will be covered: the role of teachers in early childhood adapted physical education; goals and objectives of early childhood motor programs for young children with and without unique needs; and the importance of families in the development of the young child.

LEGISLATION

Infants and toddlers, caregivers, and adapted physical education? Do these groups belong together, you may ask. The answer is yes. As a result of federal legislation, particularly PL 99-457 the Education of the Handicapped Amendments of 1986, now reauthorized as PL 105-17 the Individuals with Disabilities Education Act (IDEA) Amendments of 1997, infants and toddlers with developmental delays or at risk for developmental delays are provided early intervention services to enhance their development and to minimize their potential for delays. According to federal legislation (PL 105-17), infants and toddlers with disabilities are defined as individuals under three years of age who are experiencing developmental delays in one or more of the following functional areas: cognitive development; physical development; communication development; social or emotional development; adaptive development; or who have a diagnosed physical or mental condition that has a high probability of resulting in developmental delay. Infants and toddlers with disabilities may also include, at a state's discretion, at-risk infants and toddlers.

In order to determine if infants and toddlers experience developmental delays and/or are "at risk" for developmental delays, states develop criteria to be met. Based on these criteria, decisions are made as to who is eligible to receive early intervention services. Definitions of developmental delays vary depending on criteria set by individual states. For example, in New York state, a developmental delay would be established if the child demonstrates any of the following:

▶ A 12-month delay in one functional area

▶ A 33-percent delay in one functional area and a 25-percent delay in each of two areas

▶ A score of at least 2 standard deviations below the mean in one functional area or a score of at least 1.5 standard deviations below the mean in each of two functional areas; or informed clinical opinion of the multidisciplinary team for those children in which a standardized score is either inappropriate or cannot be determined because of the child's age, condition, or type of instruments available (Regulations of the Commissioner of Health, 1993)

Early childhood experts agree that determining eligibility for early intervention services for children who are so young is a complex process. A variety of methods for determining eligibility are therefore recommended, including direct observation, play-based assessment, standardized tests or developmental inventories, clinical opinion, and, most importantly, reports by parents and/or caregivers (Early Childhood Intervention Council of Monroe County, 1994). In addition to assessing the child, a family may choose to be involved in what is known as a *family assessment*. With this process, the family identifies their "concerns, priorities and resources in relation to maximizing the potential for their child's development" (Early Childhood Intervention Council of Monroe County, 1994, p. 7).

To determine if the child is experiencing a developmental delay in one or more of the functional areas listed above, a comprehensive multidisciplinary assessment must be conducted on the child. Tests used to determine unique needs must be valid and administered by trained professionals from the multidisciplinary team. Members of the multidisciplinary team include the parents; a service coordinator; advocates; professionals from at least two different disciplines, for example, a speech and language specialist; an occupational therapist; or a teacher of adapted physical education; and any other individual with an interest in the child. The team members engage in various forms of data collection to determine the strengths and needs of the child. Once a developmental delay has been established, an Individualized Family Service Plan (IFSP) is developed by the team members. The IFSP is a written document that contains the following information:

▶ The child's present level of performance in the five functional areas of development

▶ A statement of the family's resources, priorities, and concerns relating to enhancing the development of the child

▶ A statement of major outcomes to be achieved and the criteria, procedures, and timelines used to determine the degree to which progress is being made and whether modifications or revisions are necessary

▶ A statement of specific early intervention services necessary to meet the unique needs of the child and his family, including frequency, intensity, and method of delivery

▶ A statement of the natural environments in which early intervention services shall be provided, including a justification of the extent, if any, to which the services will not be provided in a natural environment (the child's home or day care)

▶ The projected dates for initiation of services and the anticipated duration of the services

▶ The identification of the service coordinator from the profession most immediately relevant to the child's or family's needs who will be responsible for the implementation of the plan and coordination with other agencies and persons

▶ The steps to be taken to support the transition of the toddler with a disability to preschool or other appropriate services (IDEA, 1997)

THE ROLE OF TEACHERS OF EARLY CHILDHOOD ADAPTED PHYSICAL EDUCATION

Where do teachers of early childhood adapted physical education fit into this process? Infants and toddlers with developmental delays, whether they be cognitive, physical, or emotional, may also demonstrate psychomotor delays. Psychomotor delay creates a major disadvantage to the child, since movement serves as an important basis from which children initially learn. For example, if children lack the ability to maintain an upright position or ambulate across a room, they are unable to interact with their environments in a meaningful way. Teachers of early childhood adapted physical education are in a unique position to develop and implement appropriate motor programs that can stimu-

late and enhance the movement abilities of infants and toddlers with special needs. Teachers of early childhood adapted physical education may serve as valuable members of the multidisciplinary team by conducting motor assessments and/or observing motor behavior, thereby helping to select goals and objectives to enhance the child's development. They may provide direct teaching services or serve in a consultant role to early intervention providers and parents. As resource consultants they can provide appropriate activities that can stimulate development in areas beyond the motor domain, including cognitive, social or emotional, communication, and adaptive development.

In summary, teachers of early childhood adapted physical education conduct assessments and teach and/or serve as consultants to caregivers of infants and toddlers with special needs. Since the ability to conduct valid assessments is essential to determining eligibility for services, the following section provides information that can help teachers of early childhood adapted physical education to accurately test and assess the motor needs of infants and toddlers with developmental delays.

ASSESSMENT

As noted, assessment serves as the primary means of determining eligibility for and implementing early intervention services. There are various assessment techniques used to determine the status of infants and toddlers. These techniques and examples of each are provided in the following sections.

Screening

Prior to a comprehensive assessment, infants and toddlers may first be screened to determine if there is a probable delay. Upon completion of the screening procedure, recommendations are made regarding the need for further evaluation. One example of a motor screening test is the Milani-Comparetti Motor Development Screening Test for Infants and Young Children (1987). The Milani-Comparetti was designed to test the development of children from approximately birth to two years of age. The majority of the test items are scored within the first 16 months of a child's life; however, the test provides the most detailed information regarding children between the ages of 3 and 12 months. The test assesses primitive reflexes, righting reactions, protective reactions, equilibrium reactions, postural control, and active movement. Another popular screening test is the Denver II Developmental Screening Test (Frankenburg, et al., 1992). The Denver II assesses children from birth to six years of age in the areas of personal/social development, language, fine and gross motor ability, and adaptive behavior. The Denver II relies heavily on the input of caregivers as well as on direct observation of developmental tasks to determine performance level. Teachers of early childhood adapted physical education should have a solid background in motor development so that these screening tests and subsequent assessments can be administered with relative ease.

Standardized Assessment

One way to determine the unique needs of infants and toddlers with special needs is through the use of standardized assessments. Standardized assessments are often used because they lend themselves to the determination of developmental status, are technically sound, and are relatively easy to administer by educators. Standardized assessment instruments may be norm and/or criterion referenced. Norm-referenced assessments allow testers to compare the child's performance against others of similar age and characteristics, while criterion-referenced tests compare the child's performance against preestablished criteria. Tests that are both norm and criterion referenced help not only with identification of unique needs but also with program planning and implementation.

One example of a motor assessment instrument that is both norm and criterion referenced is the Peabody Developmental Motor Scales (Folio & Fewell, 1983). This test is designed for young children from birth to seven years and utilizes two scales, the gross motor scale and the fine motor scale. The gross motor scale contains items related to reflexes, balance, nonlocomotor, locomotor, and receipt and propulsion of objects. The fine motor scale contains items related to grasping, hand use, eye-hand coordination, and manual dexterity. In addition to producing motor assessment data, the Peabody test provides activity cards to aid in the development and implementation of appropriate goals and objectives to meet the unique needs of the child.

Another example of a standardized test that is both norm and criterion referenced is the Brigance: Inventory of Early Development (Brigance, 1999). The Brigance is unique in that it contains both a screening test and formal assessment. Once the child has been screened and a delay is suspected, the child is further evaluated with the formal assessment. The Brigance is designed for children from birth to seven years. Areas that are assessed include motor skills, self-help skills, speech and language, general knowledge/comprehension, and early academic skills. The Brigance tends to be one of the most widely used comprehensive assessments, because it targets all areas of functional development, generating a valid picture of the child's current level of performance.

Curriculum-Based Assessment

Curriculum-based assessment has also become a popular means in which to generate data relative to a child's present level of performance. Curriculum-based assessment takes place in natural environments and is based on a predetermined set of curriculum objectives (Bailey & Wolery, 1989). The Carolina Curriculum for Infants and Toddlers with Special Needs (Johnson-Martin, Jens, Attermeier, & Hacker, 1991) and the Hawaii Early Learning Profile (HELP) Checklist Ages Birth to Three Years (Furuno, O'Reilly, Hosaka, Inatsuka, Zeisloft-Falbey, & Allman, 1988), as well as the Hawaii Early Learning Profile (HELP) Strands (Parks, 1992), are all examples of curriculum-based assessment. The Carolina Curriculum was developed specifically for infants and toddlers from birth to 24 months, while the HELP series was developed for infants and toddlers from birth to 36 months. Both tests assess the child in five functional areas of development, including cognitive skills, communication/language, social-emotional adaptation, fine motor skills, and gross motor skills. Each area assessed is embedded in a naturally occurring activity. Based on the assessment data, a profile of the child is developed. Because these assessments are linked to a curriculum, there is a smooth transition from the assessment phase to the intervention phase. Both the Carolina Curriculum and the HELP series provide activities to develop skills assessed.

Transdisciplinary Play-Based Assessment

The final form of assessment that will be presented in this chapter is transdisciplinary play-based assessment (Linder, 1993). Transdisciplinary play-based assessment differs from traditional forms of assessment in several ways. Traditional forms of assessment typically involve experts in various domains determining the strengths and needs of the child for a particular domain. For example, a teacher of adapted physical education may provide data related to motor abilities, while a speech and language teacher may provide information regarding communication. These experts then come together and report their findings and generate a comprehensive picture of the youngster's current level of performance. Transdisciplinary play-based assessment, also known as arena assessment, calls for a play facilitator to interact with the child, her parents, and a peer. The interactions are based on a set of criteria that are observed unobtrusively in both structured and unstructured play environments by representatives from various disciplines who are knowledgeable about all areas of development (Linder, 1993). During the interactions, observations of cognitive, social-emotional, communication and language, and sensorimotor development are assessed. Based on the observations, developmental level, learning style, interaction patterns, and other relevant behaviors are analyzed and recorded onto observation sheets, which are later transferred to summary sheets during post-observation sessions

(Linder, 1993). The role of the parents in transdisciplinary play-based assessment cannot be overemphasized. Parents are involved in the process from start to finish by completing a developmental checklist, directly interacting with the child during the assessment, and developing the IFSP. This form of assessment has many benefits; most notably it helps to provide an accurate picture of the youngster's present level of performance, because data are collected in natural environments with caregivers not only present but directly involved in the assessment process. Youngsters are typically more at ease and perform more as they normally would with this type of assessment. In addition, by utilizing a play facilitator, the child is exposed to only one professional individually assessing the child rather than the traditional three to five members of an assessment team. Teachers of early childhood adapted physical education can easily serve as play facilitators, since their primary means of developing motor abilities is through the use of play.

Since well-prepared teachers of adapted physical education should be able and willing to provide assessment services, they may serve as valuable members of the multidisciplinary team responsible for direct screening and assessment of motor behaviors for infants and toddlers with disabilities. In addition to their abilities to provide screening and assessment data, teachers of early childhood adapted physical education can help infants and toddlers achieve their motor objectives by providing developmentally appropriate activities and environments. The next section identifies goals and objectives of early childhood motor programs for infants and toddlers.

GOALS AND OBJECTIVES IN MOTOR PROGRAMS FOR INFANTS AND TODDLERS

Infants and toddlers learn by experiencing their environments in several ways. Infants and toddlers experience their environments through their senses (seeing, hearing, tasting, smelling, and feeling), through reciprocal adult-child interactions, and through movement actions and reactions (Bredekamp & Copple, 1997). For example, when an infant cries, a natural reaction on the part of the parent or caregiver is to interact with the child to determine his needs. Similarly, if a child swats at a mobile (i.e., movement action), the natural reaction is for the mobile to move. These constant interactions between the child, the environment, and those within the environment serve as the basis for cognitive, affective, and psychomotor development of infants and toddlers.

Within these early years, adult-child interactions need to be warm and positive so that infants and toddlers can develop a sense of trust in the world and feel-

Figure 19.1   Ample toys that are similar in nature should be readily available.

ings of self-competence (Bredekamp & Copple, 1997). As trust is established, infants and toddlers become receptive to new experiences. Teachers of early childhood adapted physical education are in a unique position to provide safe and secure motor environments in which infants, toddlers, and their caregivers can discover and explore the world around them. Thus, a primary goal of motor programs for infants and toddlers is to develop a sense of trust in both their caregivers and their environments.

A second goal of motor programs for infants and toddlers is to aid in the development of independence. While infants may not be ready to function independently of their caregivers, they should be provided opportunities to engage in isolated activities independently from time to time. Toddlers, however, will need guidance and support as they begin to release their total dependence on their parents or caregivers and learn to function more independently. This need for independence helps shape positive psychosocial behaviors later in life. Play is seen as crucial to the development of the child's independence, because it allows children to function within their own boundaries not the boundaries that have been established by others (Gallahue & Ozmun, 1995). As youngsters attempt new skills they need to feel successful so that they will continue to attempt either the same skill, leading to more refined movement, or new skills. Self-initiated repetition is one way to foster this independence (Bredekamp & Copple, 1997). Teachers of early childhood adapted physical education will be responsible for developing environments that allow for ample opportunities to practice already learned skills as well as new skills. For example, if a toddler finds rolling a ball stimulating, a variety of balls with various shapes and sizes should be available to the child. It is important to keep in mind that, in terms of play, toddlers are egocentric and should not be expected to share. Enough equipment that is the same or similar in nature should be available to the children (figure 19.1).

A third goal of motor programs for infants and toddlers relates to providing opportunities for active movement. Typically developing infants and toddlers will naturally progress from reflexive movement to rudimentary movement. Rudimentary movement includes the following three categories: stability, locomotion, and manipulation (Gallahue & Ozmun, 1995). Stability, the ability to maintain control of the head, neck, and trunk, serves as the basis for locomotion and manipulation. The enhanced motor ability that comes with rudimentary movement such as crawling, creeping, and walking allows infants and toddlers to move freely within their environments. Once there, the ability to reach, grasp, and release allows infants and toddlers to then make meaningful contact with objects within the environment. Together, these movement interactions help to shape the overall development of infants and toddlers and lead to more refined movement forms, known as fundamental movement. Providing activities and objects within the environment that elicit movement is the role of teachers of early childhood adapted physical education.

In summary, motor programs for typically developing infants and toddlers should seek to provide a safe nurturing environment in which trust and independence can flourish. It should also provide opportunities for motor activities that allow infants and toddlers to utilize their newly found motor skills, such as crawling, walking, grasping, and releasing, so that they may have meaningful contacts with persons and objects within their environments. These combined goals realized through a series of objectives will help to shape not only motor development in infants and toddlers but also their cognitive and social development.

Infants and toddlers with unique needs, however, may have specific motor delays that may inhibit their abilities to move freely and interact with their environments. While the goals and objectives previously discussed are applicable to all infants and toddlers, the following section identifies unique motor needs that teachers of early childhood adapted physical education may encounter, as well as strategies to enhance the development of children with unique needs.

GOALS AND OBJECTIVES IN MOTOR PROGRAMS FOR INFANTS AND TODDLERS WITH UNIQUE NEEDS

While typically developing infants and toddlers move from the stage of reflexive movement to rudimentary movement in a smooth integrated fashion, infants and toddlers with special needs may demonstrate unique motor problems that may benefit by intervention. Garwood and Sheehan (1989) delineate infant and toddlers with unique needs into two risk categories,

established risk and biologically and/or environmentally at risk. While the following list is not comprehensive, established-risk conditions include children with cerebral palsy, Down syndrome, hydrocephalus, pediatric AIDS, muscular dystrophy, seizure disorders, or spina bifida. Biological risk factors include low birth weight, respiratory difficulties, prematurity, central nervous system disorders, and prenatal alcohol and/or drug abuse by the mother. Environmental risk factors include poverty, low maternal education or age, family instability, low social support systems, and weak parent-infant bonding (Eason, 1995).

Cowden, Sayers, and Torrey (1998) identify several areas of emphasis as motor programs for infants and toddlers with unique needs are developed. The first deals primarily with increasing muscle tone and strength. Infants and toddlers who are lacking in muscle tone are said to be hypotonic. Disabilities that are associated with hypotonicity include Down syndrome, muscular dystrophy, or metabolic disorders (Cowden, Sayers, & Torrey, 1998). Infants with low muscle tone will often demonstrate delays in primitive reflex integration, especially with the tonic neck group of reflexes (Cowden, Sayers, & Torrey, 1998). Teachers of early childhood adapted physical education should provide physical assistance as needed to move the child through various strength-enhancing activities. Strength-control activities should occur in four positions: prone, supine, side lying, and upright (Eason, 1995). In the prone position, the child is developing greater control of the head and trunk. These prerequisite skills will aid in the development of locomotor movements such as crawling and creeping. A typical prone position stimulus that elicits head control involves squeaking or rattling a favorite toy above the child's head. The child will attempt to reach for the toy and as a result the head will be lifted from the floor. Control of the trunk can be realized by encouraging and assisting, if necessary, the child to roll from stomach to back. Again, a favorite toy just out of reach will motivate the child to roll over. After the child has accomplished the task, it is important to allow the child to interact with the toy. Once head and trunk control have been established, crawling and creeping positions should be maintained (figure 19.2).

Activities in the supine and side-lying positions also stimulate muscle strength. However, it is important to note that for children with extremely high or low muscle tone, supine positions should be avoided and activities should be done in the side-lying position (Eason, 1995). A typical supine or side-lying activity involves offering the child a toy just slightly out of reach. The child will then reach for the toy (extend arms), obtain it, and bring the toy into midline (flex arms). These types of activities can be repeated in several different ways by using a variety of equipment to stimulate the child's interest. Finally, activities that enhance

Figure 19.2   A young child bearing weight on arms and knees.

strength can be attained in upright positions. For example, in a sitting position, children can interact with caregivers and their environments. Any activity that maintains the child's interest, such as Peek-a-Boo, done over and over again often stimulates a child and fosters an upright position. As the child gains additional strength, standing and finally walking will be achieved.

The second area of motor emphasis for infants and toddlers with special needs may be to decrease muscle tone and enhance reflex integration. Increased muscle tone, also known as hypertonicity, results from delayed reflex integration. Infants and toddlers with spastic cerebral palsy will exhibit hypertonicity. The lack of reflex integration interferes with typical movement skills, thus creating substantial motor delays. Activities that involve relaxation techniques, massage, therapy exercise balls, and appropriate positioning help to minimize and alleviate hypertonicity. Figure 19.3 illustrates a young child being placed in a prone position on a therapy ball with legs slightly separated and toes pointed outward. The child is gently rocked forward and backward. This position on the therapy ball helps to normalize muscle tone by relaxing the muscles.

Stimulation and development of the sensory motor system are also areas of emphasis for infants and toddlers with unique needs. Cowden, Sayers, and Torrey (1998), and Winnick in chapter 18 of this book identify five components of the sensory system and provide activities to enhance each. These components include vestibular integration, visual-motor control, auditory discrimination, tactile stimulation, and kinesthetic and spatial awareness. Vestibular integration activities are needed so that youngsters can "right" themselves, that is, maintain an upright position. In-

fants and toddlers who lack the ability to right themselves can easily be injured and demonstrate delayed locomotion and balance abilities. Therapy balls are useful in helping to enhance postural reactions. With the child lying over the ball, the ball is rolled forward so that the child extends arms in a protective fashion. Activities that utilize incline mats, balance boards, and scooters, as well as a variety of other movement activities, are all appropriate for enhancing postural reactions, since they stimulate opportunities for movement in a variety of positions and situations. The ability to track objects, discriminate between shapes and sizes, and distinguish objects from a background are all under the guise of visual-motor control (figure 19.4). Activities that can enhance children's visual-motor control should be incorporated into the repertoire of early childhood motor programs. Simple tasks such as tracking a rolling ball, matching shapes, and walking on overlapping geometric shapes are all appropriate for enhancing visual-motor control.

Auditory discrimination (i.e., the ability to distinguish sounds) can be enhanced by having children search for sounds in a room, move the body to the rhythm of music, and follow simple directions. Some children may be hypersensitive to touch or lack reactions to touch. In these instances, carefully sequenced tactile stimulation activities can be incorporated into the program. Allowing the child to interact with a variety of textured equipment such as scooters, mats, balance boards, and ropes will prove helpful in the child's ability to distinguish hard and soft objects. Physically stroking a child with various textures in a manner that is tolerated by the child is also helpful. Finally, kinesthetic and spatial awareness can be enhanced by providing activities that allow the child to move through space, such as obstacle courses with tunnels, mats, and scooters. Table 18.2 on page 286 provides additional activities to enhance sensorimotor development.

Figure 19.3   Child on a therapy ball.

Figure 19.4 Tracking a slow-moving scarf is one way to enhance visual-motor control.

The ability to manipulate objects may also be a need area of emphasis for infants and toddlers with unique needs. The ability to reach, grasp, hold, and release is crucial for acquiring self-help skills such as eating and for making meaningful contacts within the environment such as swinging on a swing.

In summary, areas of emphasis associated with infants and toddlers with unique needs may include increasing muscle tone and strength, decreasing muscle tone and enhancing reflex integration, stimulating the sensory motor system, and enhancing manipulative abilities. The following section describes developmentally appropriate interactions parents, caregivers, and teachers of early childhood adapted physical education should embrace in implementing programs.

DEVELOPMENTALLY APPROPRIATE INTERACTIONS WITH INFANTS AND TODDLERS

Traditional formats for teaching physical education obviously do not apply to teaching infants and toddlers. Teachers of early childhood adapted physical education are primarily responsible for setting up an environment that is intriguing to youngsters and their parents or caregivers. Direct teaching is typically not recommended with such young children; rather, indirect approaches to attain goals such as discovery and exploration are embedded within the learning environment. Infants and toddlers should be free to explore and engage in self-initiated repetition as much as desired. Teachers and caregivers should utilize indirect teaching strategies such as following the child's lead in activities that are initiated by the child rather than the teacher or caregiver. The adult watches and interacts at the child's level by describing what the child is doing

rather than telling the child what he should be doing (McCall & Craft, 2000). For example, if a young child is interacting with blocks, the caregiver can label the blocks by their colors or shapes with statements such as "Shannon is stacking the red block." Adults also play a role in providing choices for young children. A child should be able to choose to play in any of several suitable environments; for example, a motor room may contain push toys, stacking toys, balls, and tunnels. Any of these activities would be suitable, as each has specific embedded values. Tunnels and balls allow children to enhance their gross motor abilities, while push toys and stacking toys enhance fine motor abilities. Activity participation will have naturally occurring consequences, that is, if a child climbs to the top of the slide, the natural consequence would be to slide down. These activities contribute to an understanding of the environment and sensorimotor development. Refer to the Developing an Appropriate Motor Environment application example, which describes a developmentally appropriate motor environment.

As noted, typically parents or caregivers will accompany the young child in the motor environment; thus, teachers of early childhood adapted physical education need to respect and embrace the role of parents in their programming. The following section provides tips for interacting with parents or caregivers to effectively enhance the development of the child.

INTERACTING WITH FAMILIES

Teachers of early childhood adapted physical education for infants and toddlers should understand that their programs should be based on a family-centered philosophy and that goals and objectives are developed not in isolation of the family, but rather in conjunction with the family. One of the basic underlying assumptions of the development of the IFSP is that infants and toddlers are dependent on their families for survival (McGonigel, 1991); thus, the role of families in early intervention cannot be minimized. Bennett, Lingerfelt, and Nelson (1990) developed family-centered principles that teachers of early childhood adapted physical education may adopt as effective methods for interacting with families. The first principle deals with basing intervention efforts on the needs and aspirations identified by the family. Fiorini, Stanton, and Reid (1996) identify several questions one may ask regarding motor development such as "What types of activities would you like your child to be doing now that she is not doing at the present time?" "What worries you most about your child's motor ability?" "What are you most excited about in terms of your child's motor ability?" This process enables parents to set the goals and objectives they feel are important to their child's development. The role of the early childhood adapted physical educator is to help prioritize the goals.

Application Example: Developing an Appropriate Motor Environment

Setting: Motor room at a day-care facility

Students: Toddlers with and without unique needs

Task: Setting up a developmentally appropriate motor environment that facilitates the movement of the young children

Application: The environment contains the following areas:

- ▶ Push toys for fine motor development
- ▶ Stacking toys for fine motor development
- ▶ Balls for gross motor development
- ▶ Tunnels for gross motor development

The teacher allows the children to interact within the environment and follows their lead, as they move from one piece of equipment to another. Children with special needs may need to be brought to various areas if they are unable to ambulate independently. Choice of area should be encouraged.

The second family-centered principle is based on expanding and developing the family's repertoire of skills and competencies. Teachers of early childhood adapted physical education can consult with and demonstrate appropriate motor interactions that parents or caregivers can model. As parents or caregivers engage their child in developmentally appropriate activities and see gains in motor development, confidence in their abilities to do so is strengthened.

Finally, the most important family-centered principle is communication. Teachers of early childhood adapted physical education need to maintain constant open communication with parents or caregivers. The development of the young child is based on the combined efforts of parents and supportive professionals whose major role is to guide the family in the development of their child not to direct the development of the child. With a strong collaborative partnership, infants and toddlers with special needs and their families can benefit from early intervention services, and the child's development can greatly be enhanced.

SUMMARY

This chapter has provided an overview of the role of teachers of early childhood adapted physical education as it relates to infants and toddlers with unique needs. As noted, teachers of early childhood adapted physical education may be involved in early intervention services for infants and toddlers with unique needs by serving as a member of the multidisciplinary team convened to determine unique needs, conduct valid assessments of the child's motor abilities, develop appropriate motor goals and objectives for the IFSP, and implement motor programs. In some instances, teachers of early

childhood adapted physical education may serve as consultants to caregivers or parents on providing appropriate motor programs.

While all infants and toddlers can benefit from developmentally appropriate motor programs, infants and toddlers with unique needs will have specific goals and objectives embedded in their activities. Goals may include increasing muscle tone and strength, decreasing muscle tone and enhancing reflex integration, stimulating the sensory motor system, and enhancing manipulative abilities. A variety of suggested activities were presented to assist teachers of early childhood adapted physical education realize these goals; however, the reader is encouraged to seek out recommended resources outlined in the resources for this chapter at the end of the book to further develop motor programs for infants and toddlers with unique needs.

Finally, this chapter has highlighted the importance of family in the lives of infants and toddlers. It is suggested that teachers of early childhood adapted physical education embrace three family-centered principles when working with infants and toddlers with unique needs. These family-centered principles include basing intervention efforts on the needs and aspirations identified by the family, expanding and developing the family's repertoire of skills and competencies, and maintaining open communication.

Teachers of early childhood adapted physical education are in a unique position to help minimize further delays of infants and toddlers with unique needs by creating movement environments that allow children to discover, explore, and interact with their environments. These interactions not only enhance motor development but cognitive and social development as well.

Chapter 20

Early Childhood Adapted Physical Education

Lauriece L. Zittel ◆ Cathy Houston-Wilson

▶

▶

▶

▶

▶

▶

MR. TODD AND MS. TYLER are elementary physical education specialists at Park Ridge Elementary School. Recently, their district instituted an inclusive early childhood program for preschoolers. Young children ages three to five with disabilities and/or developmental delays from the community who have been identified as needing special education services are enrolled in this program. The district combined this special education program with an already existing preschool program housed in the elementary school. Federal legislation (Individuals with Disabilities Education Act [IDEA]) requires that all preschool-age children with diagnosed disabilities or developmental delays receive special education programming. Additionally, those children experiencing motor delays are entitled to receive adapted physical education services at the discretion of individual states. The teachers decided that rather than just providing instruction to the children with developmental delays, they would design a preschool physical education program for all of the young children in this new program. Having taken graduate courses in early childhood adapted physical education at the local university, both teachers are familiar with developmentally appropriate practices for teaching young children and are eager to "test out" their newly learned skills. The challenge, as they recognize it, will be to design a preschool program that will build a skill foundation for the children as they prepare to enter the elementary physical education program.

The early childhood physical education specialists in this scenario are not alone. It is not uncommon to find preschool classrooms for children with developmental delays in an elementary school building. Many physical educators have had the experience of teaching children with disabilities in kindergarten through the third grade, but the addition of the preschool-age population may be new to them. The purpose of this chapter is to present readers with information about accurately assessing the abilities of young children, the developmental differences between preschool-age children and those entering elementary school, planning for instruction, and developmentally appropriate teaching approaches.

IDENTIFYING YOUNG CHILDREN WITH DEVELOPMENTAL DELAYS

More than a decade ago federal legislation, now known as PL 105-17 the Individuals with Disabilities Education Act Amendments, was passed to enhance opportunities for young children with special needs. Children ages three through nine who are experiencing developmental delays are to be provided early educational opportunities and may qualify for these services if they are "(1) . . . experiencing developmental delays, as defined by the State and as measured by appropriate diagnostic instruments and procedures, in one or more of the following areas: physical development, social or emotional development, or adaptive development; and (2) Who, by reason thereof, needs special education and related services" (Federal Register, 1999, p. 1241). Each state has established its own eligibility requirements in order

to determine if a child qualifies for services. For example, New York identifies a child with a disability as one who exhibits "a significant delay or disorder in one or more functional areas related to cognitive, language and communication, adaptive, social-emotional, or motor development which adversely affects the student's ability to learn" (*Regulations of the Commissioner of Education*, 1998). Identification is based on a comprehensive evaluation administered by a multidisciplinary team. The results of the evaluation are compared to eligibility criteria that have been established by individual states or school districts to signify if a delay exists. In New York, eligibility criteria is

(a) a 12 month delay in one or more functional area(s); or (b) a 33 percent delay in one functional area, or a 25 percent delay in each of two functional areas; or (c) if appropriate standardized instruments are individually administered in the evaluation process, a score of 2.0 standard deviations below the mean in one functional area, or a score of 1.5 standard deviations below the mean in each of two functional areas. (Regulations of the Commissioner of Education, 1998, p. 4)

In the preschool and early childhood years (ages three to nine), a noncategorical approach is used to classify students as eligible to receive special education services. This noncategorical approach in the early childhood years allows for programming that is based on the functional area(s) of development of each child rather than disability-specific programming. Federal legislation also requires that instruction be provided to these youngsters in the least restrictive and most natural learning environment. Thus, inclusive programs designed to accommodate children both with

Box 20.1 **Examples of Motor Development Delays**

- Absence or delay of motor development milestones
- Hypertonia
- Hypotonia
- Abnormal reflex persistence
- Decreased range of motion
- Spinal misalignment
- Extremity asymmetry
- Inability to bear weight
- Decreased balance or equilibrium
- Poor head control
- Inability to transition from one position to another
- Decreased quality of movement
- Poor ocular motor control
- Poor fine motor control
- Poor perceptual processing abilities
- Lack of play skills

and without unique needs should be the norm rather than the exception. Teachers of physical education are therefore faced with the challenge of determining motor delays and structuring motor programs that will meet the needs of young children with varying levels of abilities. Box 20.1 provides examples of motor development delays or disorders that infants and toddlers may experience, which may continue as they enter early childhood programs. These delays could be a result of an ineffective or atypical approach to motor learning that interferes with the acquisition and/or generalization of motor skills, atypical posture or movement patterns that are not explained by an established diagnosis, or sensory dysfunction (Early Childhood Intervention Council of Monroe County, 1994b).

Accurately assessing these youngsters' abilities will enhance the planning and implementation of early childhood adapted physical education programs. Children with diagnosed developmental delays, as well as specific disabilities, present individual challenges for instructors. The next section presents appropriate assessment procedures that can assist teachers of early childhood adapted physical education in determining the functional abilities of the young children in their classrooms. Identifying a gross motor developmental delay is an essential component in determining a child's eligibility to receive an individualized adapted physical education program. Planning and implementing devel-

opmentally appropriate physical education instruction for children with gross motor delays is dependent on the accuracy of the assessment results.

ASSESSMENT OF PERFORMANCE

Making accurate decisions about children's individualized program needs requires an understanding of their current abilities. Professional guidelines and federal legislation recommend that assessment information come from a variety of measures and sources. Additionally, each child should be observed in a variety of settings. A complete picture of a young child's present level of performance (in all areas of learning) can be drawn from a group of individuals most familiar with the child's routines and behaviors. A multidisciplinary team should collect information while observing the child in structured settings as well as unstructured play environments. Members of the multidisciplinary team should include classroom teachers, adapted and/or regular physical educators, therapists, parents/guardians, and others who see the child on a regular basis. These individuals will develop the child's Individualized Education Program (IEP) and participate as part of the IEP planning committee as outlined in federal guidelines (IDEA, 1997).

The assessment process may be completed using a formal, standardized procedure or an informal, play-based procedure (Linder, 1993), the difference lying primarily in the level of intrusiveness (Hills, 1992). Prior to selecting an instrument to assess young children, the purpose of testing should be clear (Zittel, 1994). Will the results be used to document a developmental delay, or is the information needed for planning and teaching? Norm-referenced, standardized tests are administered to determine a child's gross motor developmental level and provide comparison information of same-age children without delays. However, standardized, norm-referenced tests are often not program specific. Knowing that a young child is performing 1.5 standard deviations below the norm is not helpful for determining specific teaching objectives. Criterion-referenced instruments typically provide more information about delays in specific skill areas (locomotor, object control, etc.) to assist with program planning.

Standardized Formal Testing

Selecting standardized test instruments that are both norm- and criterion-referenced will assist adapted physical education specialists in making program eligibility decisions as well as in providing instructional information. Instruments that incorporate flexible testing procedures (i.e., equipment selection and testing environments) will be more sensitive to the testing characteristics of young children with disabilities and therefore have a more positive impact on the accuracy of test results. Good examples of standardized, norm- and criterion-referenced tools used to assess the gross-motor skills of young children include the Brigance: Diagnostic Inventory of Early Development—Revised (Brigance, 1991), the Peabody Developmental Motor Scales, second edition (PDMS-2) (Folio & Fewell, 2000) and the Test of Gross Motor Development, second edition (TGMD-2) (Ulrich, 2000). Each of these assessment instruments is designed to assist early childhood physical education specialists in determining a child's gross-motor developmental level as well as to provide information for instructional programming. Additionally, each of these tools will assist specialists in developing goals and objectives/benchmarks for the child's individualized education program. The testing procedures outlined for the Brigance, the PDMS-2, and the TGMD-2 provide test administrators with flexibility in the way the test environment is structured, what equipment is selected, and how instructions are provided. For example, the PDMS-2 test manual suggests that a station-testing format be used for evaluating a number of young children at one time. And, the materials used to administer the PDMS-2 are those commonly found in preschool or primary programs and are familiar to most children.

Federal guidelines encourage early childhood teachers to use data from multiple observations and measures for gathering information to make decisions about performance ability. Information collected in standardized settings should be combined with observations made in natural play environments.

Informal Testing

The authenticity of data collected in natural settings provides instructors with useful information about the child's preferences and abilities. The ability to write accurate instructional objectives and structure effective teaching environments is maximized when teachers observe and record behaviors of children during typical games and activities and watch them interact with age-appropriate equipment. Criterion-referenced skill checklists are used in curriculum-based assessment and provide information about the child's abilities in the context of a predetermined set of curriculum objectives (Bailey & Wolery, 1989).

In the area of early childhood adapted physical education, two curriculum-based tools are recommended here to assist teachers of physical education in securing performance assessment data. SMART START: Preschool Movement Curriculum (Wessel & Zittel, 1995) and I CAN Primary Skills K-3—Revised (Wessel & Zittel, 1998) are both programs that include skill checklists that teachers can use to collect information for individualizing movement programs. In both programs the authors have organized the skill checklists into a **LOOP model**. The SMART START preschool checklists include **L**ocomotor (jumping, hopping, etc.), **O**rientation (body awareness, imitative expressive movements, etc.), **O**bject Control (throwing, kicking, etc.), and **P**lay Skills (parachute play, tricycle riding, etc.). The I CAN K-3 skill checklists also include the fundamental skill areas (locomotor, orientation, and object control), but instead of play skills, **P**ersonal-Social participation (problem solving, self-control/respect, etc.) skills are assessed. Participation checklists were added to the I CAN K-3 curriculum because of a need expressed by teachers. The LOOP organization, in both of these early childhood programs, allows instructors to assess children according to specific skill areas that are relevant for both the preschool and elementary years. Additionally, the model gives instructors the freedom to select skill objectives from different areas for different units of instruction. For example, a preschool adapted physical educator may choose to assess three locomotor skills (vertical jumping, hop, gallop) and two play skills (parachute play, tricycle riding) to organize one six-week unit of instruction.

The Carolina Curriculum for Preschoolers with Special Needs (Johnson-Martin, Attermeier, & Hacker, 1990) and the Assessment, Evaluation, and Programming System for Infants and Children (AEPS) (Bricker & Pretti-Frontczak, 1996) are multidomain, curriculum-based programs designed specifically for preschool-age children with disabilities or those that are at risk for developmental delays. The Carolina assesses children in five learning domains and provides a curriculum sequence in each area. The gross-motor domain assesses skills in locomotion, stair climbing, jumping, balance, balls, and outdoor equipment. The AEPS includes a functional skill assessment across six learning domains and a curriculum to provide programming and teaching suggestions for instructors. The gross-motor domain includes two skill strands. Strand A is Balance and Mobility in Standing and Walking and focuses on stair climbing. Strand B is called Play Skills and includes five goals in the following areas: jumps forward; runs avoiding obstacles; bounces, catches, kicks, and throws; skips; and rides and steers a two-wheel bicycle.

Teacher-made checklists and rubrics can also be useful methods for assessing motor and play skill performance. Standards of performance can be written to address the unique motor abilities of children, and progress on those standards can be used to evaluate improvement. These methods give teachers the freedom to collect performance information on skills or behaviors that may not be found in published checklists. Additionally, task analyzing skills for children with severe disabilities will give early childhood teachers the opportunity to assess functional abilities by individualizing the critical elements in a checklist. For example, a teacher may be assessing the galloping abilities of preschool-age children while they pretend that yardsticks are their horses. One child in the class with cerebral palsy uses a gait trainer and, with the horse attached, is working on stepping. In this example, the teacher may use the SMART START galloping checklist for some of the children in the class and create his own stepping checklist to evaluate the ability of the child with cerebral palsy.

Performance data, gathered formally and/or informally, should provide a direct link to designing an individualized intervention plan. The activities a teacher uses should be designed to address each child's gross-motor strengths and challenges. Teachers of early childhood adapted physical education should become familiar with appropriate curriculum designs and instructional strategies that are both age and individually appropriate.

OBJECTIVES OF EARLY CHILDHOOD PROGRAMS

The development of early childhood physical education programs should be aligned with quality practices in early childhood and early childhood special education. Young children who are experiencing delays in their motor development should receive opportunities and instruction that parallel what their same-age peers receive but that are modified to address individual challenges. Preschool movement programs should provide children with the opportunity to explore and act on objects in their physical environments. The preschool years give instructors the opportunity to guide children through games and activities in order to build a skill foundation. It follows then that the early elementary years (kindergarten to third grade) allow the teacher to integrate the knowledge and skills that children have acquired and begin to focus on refining fundamental skills that will be required for more advanced games and activities. The importance of seeing the developmental connection in the early childhood years is critical for physical education curriculum development. Moving toward mastery of the basic fundamental movements and skills and beginning to integrate those skills into games and activities are processes.

Activity environments designed to provide instruction for young children with developmental delays and those with specific disabilities should be individualized according to assessment information. Arbitrarily selecting games and activities because they "seem fun" and the children "appear to enjoy them" is not necessarily in line with good practice. Specifically, learning environments should reflect the strengths and challenges

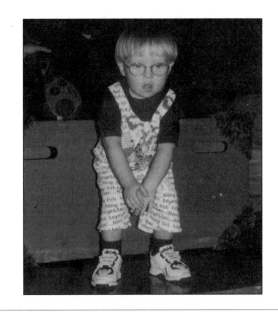

Figure 20.1   A young child attempting the objective of jumping. The platform from which he is jumping is wide and sturdy to allow him to confidently engage in the targeted skill.

that have been identified during the assessment process and written on the IEP as instructional objectives. Instruction is based on a good understanding of each child's present level of performance. An activity setting should be carefully planned to build on what children already know as well as to facilitate the acquisition of new skills (figure 20.1).

Developmental theorists have supported instruction that encourages children to explore and manipulate their environments in order to construct meaning (Piaget, 1952; Vygotsky, 1978). Individualizing instruction for each child in the class is the challenge faced by teachers providing early childhood adapted physical education. An understanding of the child's developmental abilities (physically, socially, and cognitively) as well as the effect that a certain disability may play on this development will have to be considered.

Developmental Differences Between Preschoolers and Primary-Age Children

The cognitive and social developmental status of a four-year-old differs from that of a six-year-old. As children develop cognitively and socially, they incorporate their movement strategies in different ways. It is critical for teachers providing adapted physical education services to understand age-related developmental differences in order to construct appropriate learning environments for children who exhibit delays in one or more areas of learning.

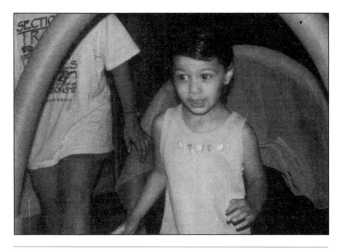

Figure 20.2 Interesting equipment, such as tunnels, encourage students to explore their environments while stimulating their learning.

Developmentally appropriate movement environments designed for preschool-age children will differ from those planned for kindergarten and elementary school children. A "watered down" kindergarten curriculum presented to children in preschool is not appropriate. Games, activities, and equipment that are meaningful to a four-year-old may be of little interest to a seven-year-old and vice versa. For example, preschool-age children love to climb and move through cardboard boxes and tunnels. Figure 20.2 shows a young girl intrigued by the "magical tunnel" made of foam domes and scarves. With a little creativity and imagination, teachers of early childhood physical education can create stimulating and motivating learning environments. A refrigerator box that has holes cut for climbing and hiding may lure a preschooler to explore and move for a long time. Preschool-age children are intrigued with new spaces and the opportunity to explore these seemingly simple environments. On the other hand, a seven-year-old would find these activities simplistic and boring. They would be much more interested and challenged by moving under and through a parachute lifted by classmates. A child in first or second grade (six to seven years old) will be challenged by activities that encourage a higher level of problem solving. They have more of an ability to reason and logically integrate thoughts than their younger counterparts. For a three- or four-year-old, a parachute activity that includes anything more than moving the parachute up and down is often frightening and unpredictable. The National Association for the Education of Young Children (NAEYC) (Bredekamp & Copple, 1997) provides guidelines for developmentally appropriate practices in early childhood and discusses the differences between preschool- and primary-age children in their physical, social, cognitive, and language development. Teachers providing adapted physical education should keep in mind that the cognitive and social development of young children cannot be ignored while developing goals and objectives in the psychomotor domain. The interplay between each of these functional areas of learning and an individual child's development within each area has to be considered when planning movement environments and instruction.

Developmental Considerations for Young Children With Disabilities

The effect of a specific disability on the communication, social, cognitive, and/or physical development of a child must be recognized prior to planning instruction. How a child's disability will impact motor learning and performance is essential for the development of an appropriate physical education program. Young children with orthopedic impairments, for example, may begin independently exploring their physical environments by using a walker, wheelchair, or crutches but may also require specific accommodations in order to benefit from age-appropriate activities. Instructors should be aware of physical barriers that exist in the activity setting and design the environment in a way that encourages interactions with peers and equipment. Assistive devices that allow children with orthopedic impairments to initiate tasks that are both physically and intellectually challenging should be made available to these youngsters in order to facilitate independence. Young children with specific delays in social interaction, for example, children with autism, will require modifications in the manner in which games and activities are introduced and delivered. Small- or large-group activities may be difficult for children with autism and practicing motor skills may need to be completed in social environments that offer solitary or parallel play options. For young children with autism, interaction with others may not be the best instructional approach or least restrictive environment for learning new skills. Children with cognitive delays, such as those with mental retardation, would benefit from an environment that is consistent, predictable, and repetitive. As shown in figure 20.3, a predictable environment is set up in such a way that the young child knows where to throw the ball. The cardboard box cut into the shape of a triangle also demonstrates ways in which cognitive concepts such as shapes can be embedded in the learning activity. Physical educators need to be aware of the various characteristics of young children with disabilities and plan activities and environments accordingly.

PLANNING FOR INSTRUCTION

Curriculum, including assessment and instruction, should be designed according to what is known about how children learn (Bredekamp & Rosegrant, 1992).

Figure 20.3 Consistent and predictable environments facilitate learning among children with disabilities.

What is age appropriate must be balanced with what is individually appropriate when designing physical education instruction for young children with disabilities and/or developmental delays. The three Cs of curriculum design—Content, Construction, and Contact (Wessel & Zittel, 1995, 1998)—can serve as a guide for teachers of adapted physical education who are working toward attaining this balance for preschool- and primary-age children. First, the skills selected as instructional *content* must be considered. The instructional focus of a preschool physical education program may include community-based, neighborhood activities like tricycle riding and pulling a wagon, whereas it may be appropriate to teach primary-age children how to use a jump rope. The selection of appropriate content is dependent on how well the teacher has examined assessment information and understands the developmental differences between children of the same or similar chronological age. Second, *construction* of a teaching environment must be carefully planned. How the teacher of adapted physical education constructs or arranges the physical environment and how activities are introduced will differ for preschool- and primary-age children. Given the developmental premise that a three-year-old differs from a seven-year-old, the manner in which children of different ages interact in a physical environment will vary. Physical environments should be designed to facilitate interactions for the purpose of skill development. The organization of the physical environment, as well as how and when activities are introduced, must be considered. Finally, a critical consideration in planning for instruction is the thought given to strategies that maximize the *contact* a young child has with equipment and peers versus their contact time with adults (teachers). Young children, regardless of developmental level, must be given the opportunity to explore their physical environments in order to develop impressions about their world. Young children with disabilities and/or developmental delays may require additional prompting during exploration so they are not denied opportunities to interact with equipment and peers. Instructors working with young children have the primary responsibility of facilitating interactions within the movement environment versus providing direct instruction. Table 20.1 provides strategies physical educators can use to facilitate effective teacher and environmental interactions.

Organization of the curriculum may vary in different preschool and primary grade programs. Teachers of early childhood adapted physical education may choose to set up an instructional unit to focus on a certain fundamental skill area at a certain time of the year or introduce skills in a particular sequence. For example, a specialist may design a unit that targets locomotor skills at the beginning of the school year. All games and activities during that unit will include the ongoing assessment and teaching of skills like hopping, jumping, and galloping. Other instructors may choose to organize instructional material around themes (Clements, 1995). Often, the theme approach is used in collaboration with classroom teachers and the instructional concepts they are teaching. For example, a physical education Day at the Zoo or Day at the Circus provides children with the opportunity to practice a variety of fundamental and motor fitness skills while interacting within a familiar theme. The curricular vehicle used to organize instruction (units or themes) will often depend on the instructor's style of teaching and her comfort level with this younger age group. The number of skill objectives taught and the time needed to learn each objective will depend on the severity of the child's delay. However, what should never be compromised in any early childhood program are the principles underlying appropriate instruction for this young age group.

DEVELOPMENTALLY APPROPRIATE TEACHING APPROACHES

The following sections will be used to describe developmentally appropriate teaching approaches that should be used by teachers for planning and delivering preschool- and primary-age adapted physical education programs.

Preschool-Age Children

Preschool classrooms typically include 12 to 18 children, with one teacher and an assistant. Classrooms for young children with severe developmental delays and specific disabilities often have fewer children. One

Table 20.1 Key Indicators in Assisting Teachers to Be Effective Facilitators

Teacher interaction	Effective implementation
Teacher as observer	▶ Pursue student involvement ▶ Collect data on the needs and interests of the children ▶ Monitor activity environment to ensure success
Teacher as facilitator	▶ Allow for choice making (assist when necessary) ▶ Maximize opportunities to practice skills with child-directed repetition ▶ Maintain physical proximity and provide support as needed ▶ Model activity behavior, challenge present performance level, guide or redirect to alternative activity, if necessary ▶ Encourage development of positive social skills (helping, sharing, negotiating)

Environmental interaction	Effective implementation
Equipment	▶ Use familiar and meaningful objects mixed with novel, challenging objects ▶ Arrange objects according to spatial and safety constraints to allow for active activity involvement ▶ Provide enough materials for multiple trials without waiting
Peers	▶ Create a motivating environment in which children can model appropriate skill performance; demonstrate alternative plan activities and encourage task persistence

Modified, with permission, from J.A. Wessel and L.L. Zittel, 1995, *SMART START: Preschool movement curriculum designed for children of all abilities* (Austin, TX: Pro-Ed, Inc.).

challenge in providing instruction for this age group is the fact that many preschool programs have multi-age classrooms, meaning that 3, 4, and 5 year olds are taught in the same class. As has been stated earlier in this chapter, planning instruction will require teachers to consider the developmental status of each child in the class regardless of chronological age and/or how a specific disability may impact that development. Although ages and abilities may differ for children in the same class, the approach to intervention should remain similar for this age group. Developmentally appropriate practice in preschool physical education emphasizes the role of teacher as one of guide or facilitator. Teachers structure the environment with specific objectives in mind, for example, throwing and kicking, and guide students toward these movement objectives. Environments are adapted to allow for a variety of choices to meet the objectives and to maximize active participation (Avery, Boos, Chepko, Gabbard, & Sanders, 1994). This intervention strategy, known as child initiated, differs from the teacher-directed style of instruction, in which the teacher focuses on one task at a time and students move as a group from one activity to the next on the teacher's signal. It would be developmentally inappropriate to expect preschool children to respond to such a teacher-directed format. Refer to the Child-Directed,

Teacher-Facilitated lesson application example for a sample child-directed lesson plan.

One example of a child-directed approach to early childhood education is known as Activity-Based Intervention (ABI) (Bricker & Woods Cripe, 1992). ABI is designed to embed each child's IEP goals and objectives within activities that are naturally motivating to young children. Different strategies (equipment, space, etc.) can be used by teachers to design activity areas that are motivating for children and encourage interactions with their environments to ensure that skill practice takes place. Young children with specific disabilities and/or developmental delays will benefit from structured movement environments that incorporate the following principles:

Child-directed versus teacher-directed learning

Experience with both novel and familiar equipment

Exposure to peer models

Opportunity for choice

Self-initiated exploration

The following scenario is used to highlight how each of these principles of best practice can be seen within one preschool physical education activity setting.

SARA AND HER CLASSMATES ARE on their way to physical education class, where they will soon discover that today is a wheel toy adventure. As they enter the activity area, they see tricycles, ride toys, wagons, and scooters scattered throughout. They also notice that jump ropes have been laid down to create pathways, cardboard trees have been spaced throughout the activity area, and arches held up with orange cones have been scattered around as obstacles to move under and through. As Mr. Todd welcomes the children into the activity environment, he announces, "Let's take a ride through the park today."

As the children scurry into the activity area to select their modes of transportation, Mr. Todd takes notice of Sara. He knows that Sara has autism, and, upon reviewing her IEP, he is aware of the fact that she has a gross-motor IEP objective focused on pulling an object around obstacles and a social skill objective focused on initiating social interactions. He notices that Sara gets on a ride toy and begins to maneuver through the "park." As Sara crosses the room, she slows down to take notice of a teddy bear sitting in a wagon. Sara knows that she has been given a ride in a wagon before and decides that she will give the teddy bear a ride. Mr. Todd sees that Sara has directed her attention toward the wagon. Knowing that pulling an object is one of her objectives, he begins to interact with Sara telling her how nice it is that she is giving the teddy bear a ride. Sara continues to pull her wagon through the pathways and around the cardboard trees. Mr. Todd is able to facilitate social interactions (another one of Sara's IEP objectives) by suggesting that Sara ask Maria if she would like to go for a ride. Once Maria is in the wagon, Mr. Todd suggests that Sara ask Maria where she would like to go. Maria's request(s) will encourage Sara to move through her physical environment encountering a variety of inanimate obstacles in the park as well as other children moving with wheel toys.

A structured movement environment should allow children to direct the process by which they manipulate the physical environment. Child-directed learning isn't synonymous with free play. Teachers of preschool adapted physical education are responsible for designing environments that will motivate children to initiate practice on specific skills outlined in their individualized plans. Interaction with equipment and practice on certain IEP objectives will be far more probable and children will persist with the task much longer if the interaction is initiated by the young child. The teacher then becomes a facilitator within the movement environment. The teacher follows the child's lead and facilitates challenges based on the choices the child has made. The intensity with which children practice a skill or explore a new task will depend on how interested they are in the task itself. The fact that Sara recognized the teddy bear sitting in the wagon and related that to her own previous experience was enough to interest her in the task and encourage further exploration. Mr. Todd was able to accommodate the interests and abilities of different children in his class by using multiple pieces of equipment and structuring his activity environment around one theme. The opportunity to view peer models and initiate social interactions was built into the structure of the activity. This activity design now gives Mr. Todd the opportunity to facilitate the childrens' viewing of peer models. For example, he may now prompt Erik to follow Sara while riding his tricycle.

Primary-Age Children

Primary-grade classes, including kindergarten through third grade, typically vary in size depending on school system policies. A newer approach in primary education is to incorporate multi-age class groupings. In this structure, kindergarten and first grade students may be educated in the same classroom. Designing developmentally appropriate instruction for this age group can only follow if a good assessment of individual abilities has been completed. As children grow and mature physically, discrepancies in movement abilities between children in the same age range may become more evident. However, young children with disabilities and/or developmental delays between six and eight years of age will have an interest in participating in the same physical activities as their same-age, typically developing counterparts. Planning instructional content for this age group should focus around fundamental movement skills (locomotion, object control, perceptual motor) and build on the rudimentary skills learned in preschool. For some children this will mean refining specific skills and for others it will be a time to begin to integrate fundamental skills into organized games and activities. Integrating fitness concepts within the curriculum will also become important. Regardless of the curricular focus, it is critical that children learn to move and enjoy movement at this age to increase the possibility that they will continue to be physically active as they mature.

Application Example: Child-Directed, Teacher-Facilitated Lesson

Setting: The early childhood teacher has selected the following content objectives: body space awareness (under, over, through) and jump down.

Issue: What environmental construction would be appropriate to ensure child-directed learning?

Application: The goal in this setting is to minimize teacher interaction and maximize each child's interaction with peers and equipment. The teaching/learning environment should contain multiple opportunities to practice the lesson objectives. Children will choose equipment and teachers will facilitate learning by providing verbal prompts, demonstrations, physical assistance, or redirection as necessary. The following equipment will prompt child-directed activity:

▶ Tunnels, cardboard boxes, and archways with scarves attached will prompt movement through and under

▶ Bolsters covered with mats will prompt climbing over

▶ Various platforms or steps with pictures of letters, numbers, smiley faces, or bugs will prompt stepping up and jumping down

Designing instructional settings for primary-age students of differing abilities should combine movement exploration and guided-discovery techniques with specific skill practice. With an exploration style of teaching, the teacher selects the instructional materials to be used and designates the area to be explored (Pangrazzi, 1998). Rather then having the environment already set up with embedded goals and objectives, students choose a piece of equipment and figure out various ways to interact with the equipment. Teachers may offer directives such as "Get a hoop and see how many ways you can make it spin." This style of teaching takes advantage of children's desires to move and explore. It emphasizes self-discovery, which is a necessary and important part of learning, and allows children to note variations in movement forms and equipment usage. However, when using this teaching style, the instructor should avoid praising students for their creative movements too early, because it may lead to imitative or noncreative behavior (Pangrazzi, 1998).

Guided discovery is used when there is a predetermined choice or result that the teacher wants students to discover (Pangrazzi, 1998). With this approach to teaching, students are presented with a variety of methods to perform a task and then are asked to choose the method that seems to be most efficient or that works best. Students with disabilities may often find that their movement choices may differ from what typically developing peers choose as the appropriate movement form. This discovery reinforces the fact that just because a movement form is different does not mean that it is incorrect.

Children should also have the opportunity to use fundamental skills in low-organization-type games. Equipment and instruction should be modified, if nec-

essary, to accommodate the physical, sensory, behavioral, and/or cognitive abilities of each child in the class. The activity environment should be organized to accommodate the varying learning styles (visual, kinesthetic) of children in the class (Coker, 1996). General principles guiding the design of developmentally appropriate adapted physical education lessons for primary grade students are as follows:

▶ A variety of teaching styles should be used to individualize instruction for the variety of learning styles (visual, auditory) present in one classroom.

▶ Varied equipment (e.g., different size rackets) should be available to allow for student choice.

▶ There should be rule flexibility in tasks to be completed to encourage creativity and problem solving.

▶ There should be varied classroom designs (activity stations, small group, large group) and the opportunity for peer observation and interaction.

The scenario on page 313 provides an example of how the principles can be used to accommodate children in a first grade classroom.

ACTIVITIES

Developmentally appropriate activities for preschool- and primary-age students are those that are selected, designed, sequenced, and modified to maximize learning and active participation. Fundamental motor skills and patterns form the basis for higher-level movement sequences and skills and should therefore be emphasized in early childhood motor programs. Activities that

▷ LOGAN IS SELECTED AS THE line leader by his classmates as they prepare to leave for physical education. Ms. Tyler greets the children outside of their classroom and walks with them to the gymnasium. As they walk together, Ms. Tyler explains to the children that they will continue to work on striking activities today. Once inside the gymnasium door, the children see a familiar sight. Ms. Tyler has set up five activity stations and each has a different task to complete. Ms. Tyler asks the children, "What types of implements do we use to strike?"

The children's responses to Ms. Tyler's question are varied. As they look over the gymnasium, they are reminded of some answers. "We strike with bats," says Breanna. "We can use Styrofoam rackets," yells Mia. "I like the hockey sticks," exclaims Patrick. Ms. Tyler explains that everyone will have an opportunity to use all of the implements today as they spend time at each activity station. She informs the children that they will select an implement out of the station bin and begin practice at one station until they hear the sound of the drum. At that time, they will put their implements back in the bin and follow the floor arrows to the next station, select their new implements, and begin the task. Ms. Tyler has asked Logan to begin at the racket/balloon station. As Ms. Tyler designates other stations for children in the class, Logan turns to Sammy and whispers, "That's my favorite!" Ms. Tyler is familiar with Logan's IEP and has written his gross-motor objectives. Striking is a focus for Logan in addition to working on increasing his static and dynamic balances. He will be working on both today. Logan has spina bifida and uses braces on his legs. He ambulates slowly but is proud of the fact that, over the summer months, before first grade began, he stopped using his walker for most activities. The activities that Logan will participate in today are functional for him. He will work on striking just like the rest of his classmates, but this activity will also address his balance goals as he swings different implements and moves between stations.

Ms. Tyler has constructed enough activity stations to accommodate this class of 22 children and given them the opportunity to explore different movements with different striking implements. The equipment used to accomplish the skill objective in a few of the stations has been changed, but for the most part the students are familiar with the activity structure and they have been practicing their striking for several classes now. This activity environment allows Ms. Tyler to individualize instruction, provide specific skill feedback to children at different times, and modify equipment and tasks to accommodate the learning abilities of all of the children in this class. Ms. Tyler knows that in her next period first grade class she will see Matt. He has a severe mental disability and Ms. Tyler will use a **location station** approach for him. This means that she will utilize the same environmental structure to practice striking, but when the drum sounds Matt will remain in the same balloon-striking station and his cross-age peer model will assist him in selecting another implement to strike the balloon. The rest of the class will rotate stations. All of this has been planned by Ms. Tyler, because she is familiar with the children she teaches and their ability levels. So, for now she admires Logan's stability as she watches him swing the racket to hit the suspended balloon to his classmate Melissa. As Melissa tracks the balloon and gets ready to hit it back, she exclaims, "Here it comes, Logan!"

develop and enhance body awareness and perceptual motor skills should also be incorporated into the program. Utilizing the strategies presented in this chapter, physical educators can select games and activities to meet IEP goals and objectives that are both age and developmentally appropriate and presented in a way that fosters child-directed interactions. The SMART START (1995) and the I CAN Primary Skills K-3 (1998), both created by Wessel and Zittel, provide a variety of games and activities that can be used in early childhood physical education programs and that can be easily adapted to youngsters with unique needs. Box

20.2 provides a sample of games and activities commonly used in developmentally appropriate early childhood physical education classes.

SUMMARY

The increase in early childhood motor programs has brought new challenges to physical educators. Children as young as three years old may be, at the discretion of states, included in adapted physical education programs designed to meet their unique needs. While some physical educators may have experience in working with older

> Box 20.2 **Sample Games and Activities in Early Childhood Physical Education**

Animal Tracks—Scatter footprints on the floor. Have children walk, run, gallop, or skip around the tracks. Challenge students to jump on the "tracks" that animals have left in the snow.

Cross the River—Use ropes to form rivers, varying the distance of the rivers. Have children walk, run, gallop, or skip around the rivers. Challenge children to jump over rivers.

Follow the Leader—The leader performs an action and children mimic the leader. Leaders are switched periodically. Encourage students to try various movement forms.

Balancing Fun—Children are challenged to balance on various body parts and to balance various forms of equipment (i.e., beanbag) on body parts.

Falling Rocks—Children roll (i.e., logroll, forward roll, backward roll) down incline mats, pretending they are falling rocks.

Beach Ball Pass—Divide children into two groups separated by a line in the middle. Have children continually push balls to the other side encouraging them to push to different people each time.

Kick to the Target—Divide children into two groups separated by a suspended rope with various targets attached (i.e., milk jugs, detergent bottles). Have children continually kick at the target.

Strike a Ball—Suspend balls from a rope and provide children with various striking implements (i.e., paddle, bats) and encourage students to strike a target.

Tunnels and Bridges—Make tunnels and bridges with ropes, cones, and hoops. Have children ride scooters or tricycles through the tunnels and over the bridges.

Base Travel—Scatter bases in various formats. Have children ride scooters or tricycles around the bases.

students with developmental delays or disabilities, the opportunity to work with young children presents new and exciting challenges. This chapter has provided information to assist teachers of early childhood adapted physical education; it has provided strategies and tools necessary for the development of appropriate goals, objectives, and activities for young children with unique physical education needs. Specifically, information related to assessment, program planning, and program implementation has been highlighted. With an understanding of what constitutes developmentally appropriate intervention, teachers of adapted physical education should be able to positively influence the movement abilities of young children. Intervention that is planned and delivered in settings that challenge and reinforce motor learning in a positive manner provides young children with opportunities to experience success and enjoy movement.

Activities for Individuals With Unique Needs

Part IV includes separate chapters on physical fitness (chapter 21); body mechanics and posture (chapter 22); rhythms and dance (chapter 23); aquatics (chapter 24); team sports (chapter 25); individual, dual, and adventure sports (chapter 26); winter sports (chapter 27); and the enhancement of wheelchair sport performance (chapter 28). Although the content of each chapter is influenced by the nature of the activities discussed, there are several common threads. To the extent relevant and appropriate, the chapters identify skills, lead-up activities, modifications, and variations associated with the activity in question. In many instances, these include modifications and variations used in established organized sport programs. In fact, some chapters provide information on organized sport programs. Finally, chapters provide information helpful for the enhancement of education in a variety of settings—particularly inclusive settings.

In these chapters, the reader's attention is directed to the many techniques that can be used in adapting activities. These include, but are not limited to, modifying or reducing skill or effort demands of an activity; providing physical assistance or assistive devices; modifying facilities, equipment, or rules; and modifying activities to emphasize abilities rather than disabilities. These and other techniques have many interesting applications. Examples include striking stationary rather than moving objects, using hand rather than foot cranks or pedals to move a tricycle, and running or responding to a sound cue instead of a visual cue.

The importance of the use of wheelchairs in physical education and sport is recognized by including an entire chapter (chapter 28) on ways of enhancing performance when using wheelchairs.

Although this part is last in the book, it is of great importance, because it applies much information presented earlier to the content associated with physical education and sport. It focuses on how the content of a program is adapted or modified to meet unique needs. It is this part of the book that will serve as a great resource to service providers on the job, since it provides answers on how they may modify an activity for a specific youngster with unique needs to be active, "physically educated," and to work toward self-actualization.

Physical Fitness

Francis X. Short

▷ **T**WO TEACHERS, MR. BARNETT, A social studies teacher, and Mrs. Novak, a physical education teacher, were in the teachers' lounge.

Mr. Barnett asked, "What are you working on?"

▷ "I'm planning some fitness activities for my third period class."

"Isn't that the class that has a few students in wheelchairs? I thought physical fitness was just for athletes or students interested in bodybuilding."

▷ "Well," said Mrs. Novak, "some of my students with disabilities *are* athletes, but the objectives of my fitness program are more health related than sport related."

"Health related? I haven't heard of too many junior high school students dying of heart attacks lately," said Mr. Barnett, half jokingly.

▷ "Well, that's right," said Mrs. Novak with a smile, "but good fitness and physical activity habits should be established early, and, besides, health-related fitness, especially for youngsters with disabilities, is not just about reducing the risk of disease. Having good health-related fitness also means having the independence to perform many important day-to-day skills like pushing a wheelchair uphill or even getting dressed. In fact, fitness may actually be more important to students with disabilities than it is for the general student body.

"By the way," continued Mrs. Novak, "*your* 'general student body' looks like it could stand some health-related fitness."

▷ "Hey ....," said Mr. Barnett while sucking in his stomach.

Whether it is necessary to sit independently or to successfully complete a marathon, physical fitness is critical to the development of persons with disabilities. Activities for daily living (including self-help skills), job requirements, recreational opportunities, and reducing the risk of developing health problems all require certain levels of physical fitness. This chapter examines definitions, the concepts of personalizing physical fitness, patterns of physical activity that can enhance fitness, and recommendations for improving health-related physical fitness. The chapter is written primarily from a "noncategorical" perspective; that is, most of the information is about fitness in general (with implications for adapted physical education) and is not specific to any one category of disability. Disability-specific fitness information can be found in other chapters of this book.

DEFINITION AND COMPONENTS OF PHYSICAL FITNESS

Caspersen, Powell, and Christenson (1985) defined physical fitness as a "set of attributes that people have or achieve that relates to the ability to perform physical activity" (p. 129). Physical activity, in turn, has been defined as any bodily movement produced by skeletal muscle resulting in a substantial increase over resting energy expenditure (Bouchard & Shephard, 1994). Caspersen et al. further suggest that the components of physical fitness can be categorized into two groups: one related to health and the other to skills that are necessary for athletic ability. These authors identified the health-related components as cardiorespiratory endurance, muscular endurance, muscular strength, body composition, and flexibility. The skill-related components of physical fitness include agility, balance, coordination, speed, power, and reaction time. Throughout this chapter primary attention is given to development of health-related aspects of physical fitness, but teachers and coaches are reminded that developing the skill-related components is also necessary to foster certain types of physical activity (e.g., sport performance).

To completely understand what is meant by "health-related physical fitness," we must first understand the nature of health. Health has been defined as "a human condition with physical, social, and psychological dimensions, each characterized on a continuum with positive and negative poles. Positive health is associated with a capacity to enjoy life and to withstand challenges; it is not merely the absence of disease. Negative health is associated with morbidity and, in the extreme, with premature mortality" (Bouchard & Shephard, 1994, p. 84). Winnick and Short (1999) suggest that aspects of health having a potential relationship to physical fitness could be broadly categorized as either physiological or functional. Physiological health relates to the organic well-being of the individual. Indices of physiological health include traits or capacities that are associated with well-being, absence of a disease or a condition, or low risk of developing a disease or condition. Functional health relates to the physical capability of the individual. Indices of functional health include the ability to perform important tasks independently and the ability to independently sustain the performance of those tasks.

Given that information, a definition of health-related physical fitness originally proposed by Pate (1988) (with slight modification) seems most appropriate:

Health-related fitness refers to those components of fitness that are affected by habitual physical activity and relate to health status. It is defined as a state characterized by (a) an ability to perform and sustain daily activities and (b) demonstration of traits or capacities that are associated with a low risk of premature development of diseases and conditions related to movement.

As this definition suggests, relationships exist among health-related physical fitness, physical activity, and health. These relationships are depicted in figure 21.1. In essence each of these concepts can influence, and be influenced by, each of the others. Increases in physical activity and in physical fitness can both contribute to positive health status. Increases in physical activity usually will result in improved physical fitness, while improved physical fitness likely will affect the potential types (exercise, sport) and patterns (frequency, intensity, duration) of physical activity available to the individual. Finally, reductions in health status (i.e., negative health) typically will reduce participation in physical activity and also restrict progress in physical fitness.

Although health-related fitness has traditionally been associated with five components (see Caspersen et al., 1985), Winnick and Short (1999) more recently categorized health-related physical fitness into three components: aerobic functioning, musculoskeletal functioning, and body composition. Each of these components, along with relevant subcomponents and possible test items, is discussed below. Most of the test items offered as examples come from the Brockport Physical Fitness Test (BPFT). The BPFT is a health-related, criterion-referenced fitness test that is appropriate for youngsters with disabilities. It is described in chapter 5.

Aerobic Functioning

Aerobic functioning refers to that component of physical fitness that permits one to sustain large-muscle, dynamic, moderate-to-high intensity activity for prolonged periods of time. Aerobic functioning can be estimated by measuring aerobic capacity (an index of physiological health) and/or aerobic behavior (an index of functional health). Aerobic capacity refers to the highest rate of oxygen that can be consumed by exercising and ordinarily is the preferred measure of aerobic functioning. Field tests of aerobic capacity include a 1-mile run and the 20-meter multistage shuttle run (i.e., the PACER). Estimating aerobic capacity in a field setting, however, is not always possible for students with disabilities, because the estimates are based on equations developed on people without disabilities. Aerobic behavior provides an alternative measure of

Figure 21.2   Musculoskeletal functioning can contribute to functional health.

aerobic functioning and refers to the ability to sustain physical activity of a specific intensity for a particular duration. Field tests of aerobic behavior include measuring the length of time one can exercise in a target heart rate zone (e.g., the Target Aerobic Movement Test). Aerobic behavior can be measured in any youngster who can sufficiently elevate the heart rate through physical activity.

Musculoskeletal Functioning

Musculoskeletal functioning is a component of physical fitness that combines elements of muscular strength, muscular endurance, and flexibility/range of motion (figure 21.2). Combining these elements speaks to their relationship, especially when designing physical fitness programs. Improving range of motion of a joint in a youngster with a disability, for instance, may require improving the extensibility of the agonistic muscle while improving the strength of the antagonistic muscle. Muscular strength is the ability of the muscles to produce a maximal level of force over a short period of time. We can measure strength by using dynamometers (e.g., grip strength) or by recording the maximum amount of weight that can be lifted in a single repetition. Muscular endurance refers to the ability to sustain submaximal levels of force over an extended period of time. Curl-ups and flexed arm hang are examples of field-based tests of muscular endurance. Flexibility is conceptualized as the extent of movement possible in multiple joints while performing a functional movement. The back-saver sit-and-reach and the shoulder stretch are measures of flexibility. Range of motion is defined as the extent of movement possible in a single joint. An objective measure of range of motion can be obtained through goniometry; the Target Stretch Test provides a subjective alternative (see appendix C).

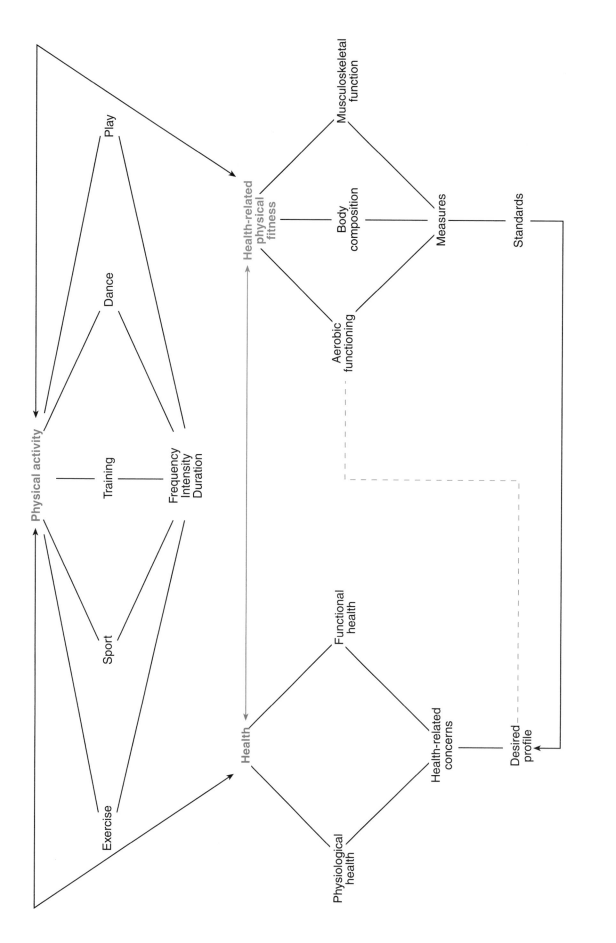

Figure 21.1 Relationships among physical activity, health, and health-related fitness.
Reprinted, by permission, from J.P. Winnick and F.X. Short, 1999, *The Brockport physical fitness test manual*, (Champaign, IL: Human Kinetics), 10.

Body Composition

Body composition shows the percentage of one's body weight that is fat versus the percentage that is muscle, bone, connective tissue, and fluids, or indicates the appropriateness of one's body weight to height. Skinfold measures can be used in field settings to estimate percent of body fat, while body mass index (a weight to height ratio) is often used to assess the appropriateness of one's weight.

PERSONALIZING PHYSICAL FITNESS

When programming for students with disabilities, the physical educator must first ask, "Physical fitness for what purpose?" Objectives may vary widely from student to student. Fitness objectives for a student with a disability can be influenced by a number of factors including, but not necessarily limited to, present level of physical fitness, functional motor abilities, physical maturity, age, nature of the disability (and how the disability affects activity selection), student interests and/or activity preferences, and availability of equipment and facilities. In an adapted physical education program, objectives can range from fitness for the execution of rudimentary movements (e.g., sitting, reaching, creeping) to fitness for the execution of specialized movements (sports skills, recreational activities, vocational tasks) to fitness for the pursuit of a healthier lifestyle (better daily functioning and reduced risk of acquiring diseases or conditions).

With the help of the student, as appropriate, teachers should design personalized fitness programs. In personalizing physical fitness, certain elements of the fitness program are emphasized for each student. For instance, for each student the teacher might ask the following questions:

▶ What are the highest priority fitness needs for this student (including any health-related concerns) and what would be the objectives of the program (or personalized "profile")?

▶ Which component(s) of health-related or skill-related physical fitness should be targeted for development given the identified needs and objectives?

▶ Which areas of the body will be trained? (For students with physical disabilities, the teacher must decide whether or not to train affected body parts.)

▶ Which tests will be used to measure physical fitness for this student, and what standards will be adopted for evaluative purposes?

The Brockport Physical Fitness Test, for instance, provides opportunities for teachers to personalize the test according to the needs of the student. Test users have the option of adopting a set of tests and standards identified for students with a particular disability or modifying the recommended test battery and/or standards based on the unique fitness needs of the student.

PATTERNS OF PHYSICAL ACTIVITY

Once the objectives of the personalized program have been established and baseline testing has been conducted, the teacher must arrange "patterns" of physical activity for the student in such a way that fitness will be improved and objectives attained. As depicted in figure 21.1, patterns of physical activity can be arranged by manipulating the variables of frequency (how often?), intensity (how hard?), and duration (how long?) of various types of activity (games, sports, exercise, and dance).

Monitoring appropriate levels of frequency and duration is fairly straightforward. Frequency generally is expressed in terms of the number of times a day or days in a week that activity is performed (e.g., three to five days per week; twice a day for five to seven days per week). Duration is simply monitored by timing the length of the activity period (6 to 10 seconds, 20 to 40 minutes) or counting the number of repetitions of an exercise.

Tracking intensity, particularly in a field setting, however, is somewhat less objective and sometimes a bit more difficult. Different indices of intensity are used with different components of fitness. For instance, for muscular strength and endurance, intensity could be estimated from the degree of exertion or the amount of resistance (e.g., a certain percentage of the maximum weight that can be lifted one time). In the case of flexibility, intensity could be a function of perceived discomfort or length of a stretch (e.g., touching toes rather than ankles). The intensity of activity selected to improve either aerobic functioning or body composition might be measured by heart rate, ratings of perceived exertion (RPE), or estimated metabolic equivalents (METs) (table 21.1). Heart rate monitors are becoming increasingly popular in public schools and provide an easily determined, objective measure of intensity. Pulse rate taken manually by either teachers or students, of course, provides an inexpensive alternative to the heart rate monitor. Maximal heart rate is usually estimated from the simple formula 220 − age (e.g., the estimated maximal heart rate for a 10–year-old is 210).

RPE provides a subjective, but nevertheless valid, measure of intensity whereby participants gauge effort by sensations such as perceived changes in heart rate, breathlessness, sweating, muscle fatigue, and lactate accumulation (Arnold, Ng, & Pechar, 1992). Participants then translate these sensations to a numerical scale such as one suggested by Borg (1998) and shown in table 21.2. Borg's RPE scale has good utility for adapted

Table 21.1 **Estimating Activity Intensity by Three Methods**

Intensity	Method		
	% Maximal heart rate	RPE	METs
Very light	<35	<10	<2
Light	35-54	10-11	2-4
Moderate	55-69	12-13	5-7
Vigorous	70-89	14-16	8-10
Very vigorous	>90	17-19	>10
Maximal	100	20	12

Adapted from U.S. Department of Health and Human Services. *Physical Activity and Health: A Report of the Surgeon General.* Atlanta, GA: U.S. Department of Health and Human Services, Centers for Disease Control and Prevention, National Center for Chronic Disease Prevention and Health Promotion.

physical education programs and has been successfully in research projects employing subjects with mental retardation, asthma, spinal cord injuries, and cerebral palsy.

MET values represent multiples of resting energy expenditure. For instance, when an individual participates in an activity that is four METs, the body is using four times more oxygen than it does at rest. MET values reflect an "average" person's energy expenditure for a particular activity (e.g., running a 10-minute mile = 10.2 METs, playing basketball = 8.3 METs). In adapted physical education programs, the use of METs to estimate intensity has the most relevance for able-bodied youngsters with disabilities (e.g., mental retardation, learning disabilities, visual impairments) and the least relevance for those students who have physical disabilities (cerebral palsy, spinal cord injuries, and amputations), since the theoretical "average" person does not have a physical disability.

In the following section, suggestions are made for arranging patterns of physical activity to improve specific components or subcomponents of health-related physical fitness. Modifications for youngsters with disabilities also are discussed as appropriate.

PHYSICAL ACTIVITY RECOMMENDATIONS FOR HEALTH-RELATED PHYSICAL FITNESS

As suggested by figure 21.1, physical activity can influence the development of health-related physical fitness as well as impact health status. This section explores the relationship between physical activity and health-related physical fitness (also review, however, box 21.1, which

discusses physical activity and health). Recommendations for developing health-related fitness follow under the headings aerobic functioning, body composition, muscular strength and endurance, and flexibility/range of motion. Adjustments to those recommendations for youngsters with disabilities, as well as some suggested activities, also are included.

Table 21.2   **Borg's Ratings of Perceived Exertion Scale**

6	No exertion at all
7	Extremely light
8	
9	Very light
10	
11	Light
12	
13	Somewhat hard
14	
15	Hard (heavy)
16	
17	Very hard
18	
19	Extremely hard
20	Maximal exertion

Reprinted, by permission, from G.A. Borg, 1998, *Borg's perceived exertion and pain scales* (Champaign, IL: Human Kinetics), 31.

Box 21.1 Physical Activity and Health

The idea that regular participation in physical activity can enhance an individual's health status is now firmly established. Regular physical activity can reduce the risk of premature mortality and the risk of acquiring coronary heart disease, hypertension, colon cancer, and diabetes mellitus; it also appears to reduce depression and anxiety, improve mood, and enhance ability to perform daily tasks (U.S. Department of Health and Human Services, 1996).

As a result of this relationship, the Centers for Disease Control and Prevention (CDC) and the American College of Sports Medicine (ACSM) published joint recommendations that all Americans accumulate at least 30 minutes of moderate-level physical activity on most, preferably all, days of the week (Pate et. al, 1995). Perhaps the most significant aspect of this recommendation is the notion that moderate-level activity is of sufficient intensity to act as a buffer against certain diseases and conditions. In contrast, physical activity recommendations for improving physical fitness generally require more vigorous levels of intensity. Furthermore, the CDC/ACSM recommendation suggests that *any* moderate-level activity can enhance health status. Traditional activity recommendations for the development of physical fitness, on the other hand, have focused on specificity of training and have been dominated by one form of activity, namely exercise.

The Council for Physical Education for Children (COPEC, n.d.) has suggested that the CDC/ACSM recommendation be modified for children. In defense of a modified recommendation, COPEC cites some significant differences when comparing children to adults, including a relatively short attention span, a tendency toward concrete rather than abstract thought, normal activity patterns that are more intermittent than continuous, a weaker relationship between physical activity and physical fitness, and the possibility that participation in more intense activities will be perceived as too difficult, thus leading to withdrawal from physical activity. The COPEC guidelines include the following:

▶ Elementary school children should accumulate *at least 30 to 60 minutes* of age-appropriate and developmentally appropriate physical activity on all, or most, days of the week.

▶ An accumulation of *more than 60 minutes, and up to several hours per day*, of age-appropriate and developmentally appropriate activities is *encouraged* for elementary school children.

▶ Some of the childrens' activity each day should be in periods lasting 10 to 15 minutes or more that includes moderate to vigorous activity.

Children and adolescents with disabilities should be encouraged to meet either the CDC/ACSM or COPEC recommendations for physical activity. Those who participate in structured physical fitness development programs will have a "head start," because many of the recommendations for fitness development are more rigorous than either the CDC/ACSM or COPEC recommendations especially with regard to intensity. Nevertheless, those implementing adapted physical education programs may find the CDC/ACSM and COPEC guidelines more appropriate for certain students than traditional "exercise prescriptions" for fitness for a few reasons. First, it may be difficult or impossible to measure fitness in some students with disabilities. Secondly, since the CDC/ACSM and COPEC recommendations tend to be less intense, they may be more attainable by some youngsters with disabilities. Finally, a wider range of possible activities is available to meet the CDC/ACSM or COPEC guidelines than is usually recommended for fitness development.

Aerobic Functioning

Frequency, intensity, and duration guidelines for improving the aerobic functioning of school-age youngsters are summarized in table 21.3. In essence, children are encouraged to be active on all, or most, days of the week for at least 30 to 60 minutes each day. The intensity of activity ranges from "deemphasized" for younger children to moderate (and sometimes vigorous) for older children. Adolescents may reduce their frequencies and durations somewhat in exchange for higher levels of intensity (moderate to vigorous). Higher levels of intensity usually will be necessary to improve aerobic capacity.

Many youngsters with disabilities can meet these guidelines without modification. Some adjustments, however, may be appropriate in certain situations. Ordinarily, frequency will not have to be adjusted; as with nondisabled students, students with disabilities should be encouraged to participate regularly in aerobic activity. Greater periods of recovery from physical activity, however, may be necessary for youngsters with

Table 21.3 Guidelines for Developing Aerobic Functioning

Group	Frequency	Intensity	Duration
Adolescents (13-17)	3-5 days/week	55-90% HR max (~115-180 beats/min) 12-16 RPE 5-10 METs	20-60 min/day (accumulated: >10 min/bout)
Older children (10-12)	4-7 days/week	55-70% HR max* (~115-145 beats/min) 12-13 RPE* 5-7 METs*	30-60+ mins/day (accumulated, intermittent)
Younger children (6-9)	4-7 days/week	Deemphasized; participation is encouraged	30-60+ mins/day (accumulated, intermittent)
Adjustments for youngsters with disabilities	No change unless disability can be exacerbated by regular activity	Reduce as a function of fitness level; adjust THRZ for persons using arms-only activity and for persons with SCI quadriplegia	Accumulate more intermittent activity or reduce total time if necessary

*These values represent moderate physical activity; ideally this level will be exceeded to vigorous levels at times.
Reprinted, by permission, from F.X. Short, J. McCubbin, and G. Frey, 1999, Cardiorespiratory endurance and body composition. In *The Brockport physical fitness training guide*, edited by J.P. Winnick and F.X. Short (Champaign, IL: Human Kinetics), 13-37.

neuromuscular diseases, arthritis, or other disabilities requiring greater periods of rest. Similarly, youngsters with disabilities should strive to attain the same duration guidelines as their nondisabled peers. When students with disabilities cannot maintain continuous activity for the recommended length of time, the first alternative would be to accumulate more intermittent activity (i.e., shorter bouts of activity more often to achieve the recommendations) before electing to reduce the duration guidelines.

Adjustments to intensity likely will be necessary when youngsters have a history of inactivity. For these students, activity may be more appropriately conducted at lighter intensities, particularly if the frequency and duration guidelines can be achieved. Intensity should be gradually increased over time. If intensity is being measured via heart rate, it will be necessary to adjust the target heart rate zone (THRZ) for students who use their arms to propel wheelchairs, for those who engage in other "arms-only" forms of activity, or for those with spinal cord injuries (SCI). Subtracting 10 beats per minute from the THRZ values given in table 21.3 constitutes a reasonable adjustment in intensity for arms-only activity.

A wide range of activities can be used to meet the aerobic functioning guidelines in table 21.3. Teachers should select age-appropriate developmental activities

for children (ages 6 to 12). Included in this group might be jump rope activities, relay races, obstacle courses, climbing activities (e.g., jungle gyms or monkey bars), active lead-up games, active games of low organization, and rhythmic activities including creative dance and "moving to music."

Adolescents (ages 13 to 18) generally are ready for more sport-related activities and should be exposed to activities that have a lifetime emphasis. Appropriate activities might include fast walking, jogging or running, swimming, skiing, racket sports, basketball, soccer, skating, cycling, rowing, hiking, parcourse (fitness trail) activities, and aerobic dance or aerobic aquatics.

Many youngsters with disabilities can participate in most of the activities listed in the previous paragraphs, although some modifications may be necessary. As with all good teaching in physical education, teachers should "start with the student" and select, modify, or design appropriate activities rather than start with an activity and hope that it is somehow appropriate for the student. Youngsters in wheelchairs or those who are blind often require alternative activities (or activity modification) to meet the aerobic functioning guidelines. Some ideas for youngsters using wheelchairs might include slalom courses, free wheeling (e.g., "jogging" in a wheelchair), speed bag work (i.e., rhythmically striking an overhead punching bag), arm ergometry, seated aerobics, and active wheelchair sports (e.g., sledge hockey, basketball, track, rugby, and team handball). Students with visual impairments might participate in activities such as calisthenics, rowing, stationary or tandem cycling, wrestling or judo, step aerobics, aerobic dance, swimming, or track.

Body Composition

This section of the chapter will focus on weight-loss strategies associated with physical fitness. When people fall outside the appropriate range of body composition, most do so because they are too fat (or are too heavy for their heights). Readers should remember, however, that some youngsters may be outside the healthy range because they are too lean (or are too light for their heights). Excessive leanness is also a health-related concern. When physical educators believe that a student's excessive leanness may be due to inadequate diet, an eating disorder, or the possible existence of a medical condition, the school's medical staff should be consulted.

Strategies for improving body composition (i.e., for weight loss) are similar to those associated with aerobic functioning. In most cases, weight-loss recommendations can be achieved by following the guidelines for improved aerobic functioning (see table 21.3). In fact, there would be little reason to ever adjust those recommendations for achieving weight-loss in children. In the case of adolescents (and adults), however, there may be circumstances where weight-loss might be a more important goal than improved aerobic functioning. In such cases, the aerobic functioning guidelines could be adjusted slightly to meet weight-loss recommendations for adolescents (and adults). These weight-loss recommendations would be to participate in moderate-level activity, 4 to 7 days per week, for 30 to 60 or more minutes per day. When compared to the aerobic functioning guidelines for adolescents, the recommendations for weight loss generally are less intense but more frequent and, at least potentially, for longer durations.

Another weight-loss recommendation is to attain a minimal energy expenditure of at least 1,000 kilocalories (kcals) per week. If MET estimates are available, kilocalories can be calculated from the following equation:

$$\text{kcals/min} = \text{METs} \times 3.5 \times \text{body weight (kg)} / 200$$

As mentioned earlier in the chapter, published MET estimates may not be particularly appropriate for youngsters with physical disabilities. This is due to differences in mechanical efficiency that may exist between nondisabled participants and those with physical disabilities. Estimating kilocalories expenditure for students with physical disabilities, however, is still possible. Let us say that Eddie is a 14-year-old, 150-pound (68 kilogram) boy with spastic paraplegia. He uses forearm crutches to walk and run. He enjoys playing floor hockey, and his dad devised a way to attach the blade of a floor hockey stick to the end of one of his crutches so that he could play in his regular physical education class. Since there are no MET estimates for youngsters with cerebral palsy playing floor hockey with forearm crutches (and a modified stick), energy cost will have to be estimated in a different way. Eddie's teacher, for instance, could determine Eddie's average heart rate during the game. (This most easily could be done with a monitor but could also be done by taking manual pulse rates periodically.) If Eddie averaged 155 beats per minute over the course of the game, he would have been working at about 75 percent of his maximal heart rate (220 – 14 = 206; 155 / 206 = .75). By consulting table 21.1, the teacher can see that Eddie is at the low end of "vigorous" activity. The low end of vigorous intensity roughly corresponds to 8 METs (the range for vigorous intensity is 8 to 10 METs). If 8 METs is used to estimate the energy cost of Eddie's activity, his energy expenditure in kilocalories per min would be:

$$\text{kcals/min} = 8 \times 3.5 \times 68 / 200 = 9.5$$

To expend 1,000 kilocalories for the week, Eddie will need to participate in floor hockey (or some other 8-MET activity) for 105 minutes (1,000 / 9.5 = 105). If he plays twice a week, he could meet the 1,000 kilocalories goal if he played for 53 minutes each time (105 / 2 = 52.5), or he could play three times per week for 35 minutes each time to reach 1,000 kilocalories.

Less intense activities will require greater frequencies and/or durations to meet the goal. For instance, if Eddie played table tennis (perhaps a 5-MET activity for him), his energy expenditure would be about 6.0 kilocalories per min. At that rate, he would need to accumulate about 167 minutes of activity to expend 1,000-kilocalories in a week (1,000 / 5 = 167). If he played for 30 minutes each time, he would have to play table tennis (or some other 5-MET activity) between 5 and 6 days per week to achieve the kilocalories goal (167 / 30 = 5.6).

Muscular Strength and Endurance

Exercise is traditionally the type of physical activity selected to improve muscular strength and muscular endurance. Recommendations generally call for an activity pattern characterized by performing 8 to 10 different dynamic (i.e., isotonic or isokinetic) exercises at least twice a week with at least 1 day of rest between each exercise. It is also recommended that 1 set of 8 to 12 repetitions maximum be performed when strength is the primary goal, and 12 to 15 repetitions maximum when endurance is being targeted. (The intensity of the activity is a function of the amount of resistance that can be overcome within the range of maximum repetitions.)

Exercise tends to be the preferred mode of activity for strength and endurance, because muscle groups can be more easily isolated and intensity more easily monitored. For gains in strength and endurance to be made,

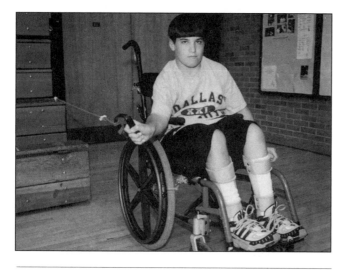

Figure 21.3   Elastic exercise bands help to overload this student's muscles, which increases his muscular strength and endurance.

muscles must be overloaded; that is, they must work at a greater level than normal. To achieve overload, resistance of some kind (e.g., gravity, one's body weight, free weights, exercise machines, elastic bands, medicine balls, weighted cuffs) usually is necessary (figure 21.3). When resistance is used, a very light load (or possibly no load at all) is recommended initially, followed by a progressive increase in resistance as exercise skill is mastered and as muscles develop. The level of resistance generally should never require a maximum exertion to attain a single repetition (i.e., one repetition maximum) in prepubescent children.

The exercise recommendations discussed in this section are appropriate for a wide range of individuals including many youngsters with disabilities. Some youngsters with disabilities may need to work at lighter intensities (lower resistance) but might still be able to meet the frequency and repetition guidelines given earlier. Others might benefit from a reduction in the number of separate exercises recommended per day (down from 8 to 10) or from a pattern that increases the frequency of exercise but reduces the intensity and duration. Programs for youngsters with medical conditions should be developed with medical consultation.

Although exercise may be the preferred type of activity for enhancing strength and endurance, other types of activities certainly can contribute. In fact, children likely will benefit from a wide range of developmentally appropriate activities. Activities that require children to move their own weights or the weight of an object against gravity might be most appropriate. Examples include climbing on monkey bars, jungle gyms, ropes, or cargo nets; crawling through obstacle courses; pushing a wheelchair up a ramp; propelling scooter boards with legs or arms or by pulling on a rope; and emphasizing fundamental movements that require power such as throwing, kicking, jumping, leaping, and hopping.

Flexibility and Range of Motion

As with muscular strength and endurance, exercise is usually the preferred activity mode for improving flexibility and range of motion. Again, properly selected exercises can be used to isolate muscles or muscle groups, and intensity can be monitored or controlled. When exercises are used by teachers to improve flexibility/range of motion (ROM), general recommendations call for performing exercise sessions at least three days per week. Each exercise should be conducted for a period of 10 to 30 seconds and repeated 3 to 5 times per session. The intensity of a flexibility exercise is judged by a feeling of mild discomfort; that is, muscles should be stretched to a point where a slight pulling or "burning" sensation is felt and held there for 10 to 30 seconds.

Although some stretching activities are generally recommended at the start of a physical activity program for warm-up, specific flexibility/ROM training will be more successful when conducted toward the end of the activity session. Muscular and collagenous tissues are more likely to be warm toward the end of the session and be more receptive to stretching (Surburg, 1999). There are a number of stretching techniques that can be used to improve flexibility/ROM including passive stretching, active-assisted stretching, active stretching, and proprioceptive neuromuscular facilitation (PNF).

In passive stretching the person is not actively involved in the exercise. The muscle is stretched by some outside force (e.g., a weight, sandbag, gravity, machine, therapist). Passive stretching is usually used in rehabilitative settings when individuals are weak, when muscles are paralyzed, or when exercise protocols require sustained stretches. Nevertheless, passive stretching can be used in educational settings as well. Physical educators who wish to employ passive stretching techniques should consult qualified professionals.

Active-assisted stretching combines the efforts of the individual and a partner assistant (e.g., teacher, therapist, aide). The individual stretches the muscle as far as she can, and then the partner assists the movement through the full (or functional) range of motion.

Active stretching is characterized by the participant moving the joint through its full (or functional) range of motion without outside assistance. Active stretching exercises have been categorized as static (slow stretches to a point of mild discomfort and holding for a time generally ranging from 6 seconds to 45 seconds) and ballistic (a "momentum exercise" where "bouncing" or twisting movements are used to elongate a muscle for a brief time) (Surburg, 1999). Although ballistic stretching may replicate certain sport-related movements, it generally is not a recommended technique for improving flexibility/ROM. The preferred method is static stretching.

Proprioceptive neuromuscular facilitation is an exercise system designed to improve strength, coordination, and kinesthesia, as well as flexibility (Surburg, 1999). Surburg (1999) has described five PNF techniques that have been used specifically to increase flexibility/ROM:

▶ Rhythmic stabilization: The participant performs rhythmical, alternating, isometric contractions of agonists and antagonists.

▶ Contract-relax: A partner or teacher passively stretches the muscle to elongation; the participant contracts the muscle against resistance provided by the partner/teacher until the body part has returned to its original/resting position; the muscle is relaxed for five seconds prior to the next repetition (five to six repetitions).

▶ Hold-relax: A partner or teacher passively stretches the muscle to elongation; the participant then performs a six-second isometric contraction in that position followed by a five-second rest.

▶ Hold-relax-contract: This stretch is the same as hold-relax except that following the isometric contraction of the target muscle (i.e., the muscle to be stretched) the participant contracts the opposite muscle group isotonically and without resistance.

▶ Contract-relax-contract: This stretch is similar to contract-relax but adds an isotonic contraction of the antagonist following the isotonic contraction of the target muscle (the agonist).

Most youngsters with disabilities will benefit from one or more of the stretching techniques just described. The general recommendations for frequency, intensity, and duration described at the beginning of this section also may be appropriate when youngsters have unique flexibility/ROM needs; however, some modification in these recommendations may be necessary. Although intensity (stretches to a point of mild discomfort) ordinarily would remain the same, flexibility/ROM training for youngsters with unique needs likely should consist of greater frequency (exercise sessions conducted two to three times daily) and duration (individual stretches lasting 10 minutes or more). For longer stretches (more then five minutes) it would be appropriate to limit the number of exercise repetitions to one (recommendations generally call for three to five repetitions per exercise for shorter stretches).

General Considerations

When developing programs of physical fitness, the physical educator should be aware of students' initial levels of fitness and select activities accordingly. In all cases the procedure should be to start slowly and progress gradu-

ally. Students should be taught to warm up prior to a workout and cool down afterward. The physical educator should motivate students to pursue higher levels of fitness by keeping records, charting progress, and presenting awards. Selecting enjoyable activities will also help to maintain interest in physical fitness; for instance, charting a class's cumulative running distances on a map for a "cross country run" will be more interesting and motivating than simply telling them to "run three laps." The physical educator should also be a good role model for students; this includes staying fit and participating in class activities whenever possible. Finally, the physical educator should view physical fitness as an ongoing part of the physical education program and not just one unit of instruction. Even though different units will be taught throughout the year, activities within a unit (exercises, games, and drills) should be arranged to enhance, or at least maintain, physical fitness.

Developing structured physical fitness programs for students with more severe disabilities may be difficult or impossible to achieve. In these situations teachers probably should focus more on increasing the students' physical activity rather than their physical fitness. In this approach teachers should identify movement experiences that are possible and enjoyable for the youngsters and motivate the youngsters to increase their levels of participation. Increases in physical activity likely will lead to increases in physical fitness, even if physical fitness cannot be validly assessed (see chapter 5 for information on measuring physical activity). The Physical Fitness application example presents a fitness plan for a girl with a thoracic spinal cord injury.

SUMMARY

Physical fitness is critical to the person with a disability. In addition to improved performance, health, and appearance, high levels of fitness can foster independence, particularly among individuals with physical disabilities. The goals of a fitness program for individuals with unique needs depend on the type and severity of the disability and current levels of physical fitness. Increasing fitness for developing physiological or functional health and for improving skill and/or performance are all reasonable goals in adapted physical education. Fitness programs should be "personalized" to meet the goals of each student. Teachers must understand that, with few exceptions, students with disabilities will exhibit a favorable physiological response to training. In many cases the recommendations for frequency and duration of exercise do not differ significantly from those made for the nondisabled student. The mode, or type of activity, and intensity, however, frequently must be modified to provide the student with an appropriate workout.

Application Example: Physical Fitness

Setting: A 15-year-old girl with a thoracic spinal cord injury is in an inclusive physical education class. One of the class goals is to improve the students' physical fitness. Present level of performance (i.e., baseline data) was established using the Brockport Physical Fitness Test. Improvements in aerobic functioning and body composition (i.e., weight loss) were identified as the primary fitness objectives.

Issue: What strategies should be used to help the student achieve these goals?

Application: The teacher might recommend or pursue the following strategies:

- ► Use ACTIVITYGRAM (or some other self-report instrument to monitor activity levels).
- ► Encourage the student to achieve the Surgeon General's recommendation to engage in at least moderate-level activity—with a minimum heart rate of 103 beats per minute ($220 - 15 = 205 \times .55 = 112.75 - 10$ for arm activity $= 102.75$)—for at least 30 minutes per day on most, preferably all, days of the week.
- ► Counsel student to work within a target heart rate zone of 134 to 164 (i.e., 70 to 85 percent of predicted maximum heart rate with a 10-beat adjustment for arm-only activity) for at least 20 minutes a day for 3 days of the week, in pursuit of the Surgeon General's recommendation.
- ► If the student has difficulty achieving these guidelines, reduce criteria for intensity (i.e., heart rate) but maintain frequency and duration (i.e., days and minutes).
- ► Recommend enjoyable aerobic-based activities for use in and outside of class, such as wheelchair slaloms, free wheeling, speed bag work, swimming, arm ergometry, seated aerobics and dance, scooter board activities, and wheelchair sports.
- ► Consult with parents on the nature of the program; encourage their support at home and possibly include a dietary component with their help.

Body Mechanics and Posture

Luke Kelly

▶ **D**ENISE, A MIDDLE SCHOOL PHYSICAL education teacher, finds the following note from the guidance department in her mailbox at the end of the day on Wednesday:

"Starting Monday, you will have a new student Sally Britton in your fifth period class. Ms. Britton has some type of medical condition called scoliosis, and her parents Mr. and Mrs. Britton would like to meet with you prior to Monday. The parents apparently have some concern about Sally's brace and the middle school physical education curriculum. Call us if you have any questions."

How would you respond to this situation?

Denise vaguely remembers something about scoliosis being a type of posture problem, but she does not remember any of the specific implications for physical education. She goes to her office to find her adapted physical education textbook, but as usual someone has borrowed it and not returned it. Realizing that she needs to find some information quickly so that she can contact the parents, Denise goes to the school media center (it was called the library last year). She finds an open computer and double clicks on the Netscape icon and then on the Infoseek search button. She types in "scoliosis" and presses the Enter key. The first document returned branches her to the Southern California Orthopedic Institute home page on scoliosis where Denise gets a quick review of the nature, cause, and common treatments for scoliosis (an abnormal lateral curvature of the spine). Denise clicks on a few more suggested Web sites to learn about the latest information on braces for scoliosis and restrictions for activity. Having updated her memory, Denise proceeds to call Sally's parents and set up a meeting.

In the above scenario, Denise encountered a common postural deviation. The term good posture implies appropriate alignment of body segments and body mechanics. Optimal body mechanics are desirable because they allow individuals to maximize their efficiency and minimize the strain placed on the body—both of which are important for safe and successful performance in physical education and sport. Posture refers to appropriate body mechanics, whether the body is stationary or moving; it should not be thought of only as correct body alignment in standing or sitting. "Good posture" is a relative term. There are acceptable variations among individuals due to differences in body build and composition, and there are marked deviations that clearly require professional attention and remediation. The purpose of this chapter is to provide the physical educator with the knowledge and skills to identify postural deviations and, where appropriate, to remediate them in physical education.

POSTURAL DEVIATIONS

Mild postural deviations are quite common and can often be remediated through proper instruction and practice in physical education. It is estimated that 70 percent of all children have mild postural deviations and that 5 percent have serious ones. The prevalence of serious postural deviations is, unfortunately, much greater among students with disabilities.

Poor posture can result from any one or a combination of factors such as ignorance, environmental conditions, genetics, physical and/or growth abnormalities, or psychological conditions. In many cases children are unaware that they have poor posture, because they do not know what correct posture is and how their posture differs. In other cases, postural deviations can be traced to simple environmental factors such as poorly fitting shoes. In students with disabilities, poor posture can be caused by factors affecting balance (e.g., visual impairments), neuromuscular conditions (e.g., spina bifida and cerebral palsy), or congenital defects (e.g., bone deformities and amputations). Finally, poor posture can occur as a result of attitude or self-concept. Students who have a poor body image or who lack confidence in their abilities to move tend to display defensive postures that are characterized by poor body alignment.

Physical educators should play an important role in the identification and remediation of postural deviations in all students. The physical educator is often the one educator in the school with the opportunity and background to identify and address postural problems. Unfortunately, postural screening and subsequent remediation are overlooked in many physical education programs. This is ironic because the development of kinesthetic awareness and proper body mechanics is fundamental to teaching physical education and is clearly within the domain of physical education as defined in IDEA.

Several excellent general posture screening tests are available that can be used easily by physical educators and involve minimal preparation and equipment to administer. Two examples, referenced at the end of this book, are the Posture Grid (Adams & McCubbin, 1991), and the New York Posture Rating Test (New

York State Education Department, 1966). Posture screening should be an annual procedure in all physical education programs. Particular attention should be paid to children with disabilities because of their generally higher incidence of postural deviation. The appropriate school personnel, as well as the parents or guardians of all children identified as having present and potential postural problems, should be informed of and requested to pursue further evaluation of all these students. The instructor can remediate most mild postural deviations within the regular physical education program by educating the children about proper body mechanics and prescribing exercises that can be performed both in and out of class. Sample exercises are described at the end of this chapter.

The purpose of the remainder of the chapter is to review some of the specific postural deviations that occur in the spinal column and lower extremities. Implications for adapted physical education, as well as procedures for assessment and remediation, will be discussed for each deviation.

SPINAL COLUMN DEVIATIONS

Viewed from the back, the spinal column should be straight with no lateral (sideways) curves. Any lateral curvature in the back is abnormal and is referred to as scoliosis. Viewed from the side, the spinal column has two mild curves. The first natural curve occurs in the thoracic region, where the vertebrae are concave forward (curving slightly in a posterior or outward direction). An extreme curvature in this region is abnormal and is known as kyphosis. The second natural curve occurs in the lumbar section, where there is mild forward convexity (inward curvature) of the spine. An extreme curvature in this region is also abnormal and is called lordosis. An exaggerated lumbar curve in lower elementary-age children is natural but should disappear by the age of eight.

Scoliosis

Lateral deviations in the spinal column are generally classified according to whether the deviation is structural or nonstructural. Structural deviations are generally related to orthopedic impairments and are permanent or fixed changes in the alignment of the vertebrae that cannot be altered through simple physical manipulation, positioning, or exercise. Nonstructural, or functional, deviations are those in which the vertebrae can be realigned through positioning and/or removal of the primary cause—such as ignorance or muscle weaknesses—and remedied with practice and exercise.

Structural scoliosis is also frequently classified according to the cause of the condition. Although scoliosis has many possible causes, the two most common are labeled idiopathic and neuromuscular. Idiopathic means that the cause is unknown. Neuromuscular means that the scoliosis is the result of nerve and/or muscle problems.

Structural idiopathic scoliosis occurs in approximately 2 percent of all school-age children. The onset of scoliosis usually occurs during the early adolescent years, when children are undergoing a rapid growth spurt. This form of scoliosis is characterized by an S-shaped curve, usually composed of a major curve and one or two minor curves. The major curve is the one causing the deformity. The minor curves, sometimes referred to as secondary or compensatory curves, usually occur above and/or below the major curve and are the result of the body's attempt to adjust for the major curve. Although both genders appear to be equally affected by this condition, a greater percentage of females have the progressive form that becomes more severe if not altered. The cause of this progressive form of scoliosis is unknown, but there is some evidence that suggests a possible genetic link in females.

A second type of structural scoliosis, more commonly found in children with severe disabilities, is caused by neuromuscular problems. This form of scoliosis is usually characterized by a C-shaped curve. In severe cases, this form of scoliosis can lead to balance difficulties, pressure on internal organs, and seating problems (pressure sores) for students in wheelchairs.

Nonstructural scoliosis can be the result of a number of causes and can be characterized by either an S- or a C-curve. The primary causes can be broadly classified as either skeletal or muscular. An example of scoliosis with a skeletal cause would be a curve that has resulted from one leg being shorter than the other. An example of scoliosis with a muscular cause would be a curve that has resulted because the muscles on one side of the back have become stronger than the muscles on the other side and have pulled the spinal column out of line. Fortunately, nonstructural scoliosis can usually be effectively treated by identification and correction of the cause—that is, inserting a lift in the child's shoe to equalize the length of the legs, or strengthening and stretching the appropriate muscle groups in the back (see the Correcting Postural Deviations application example).

Assessment of Scoliosis

Early identification is extremely important for both structural and nonstructural scoliosis so that help can begin and the severity of the curve can be reduced. Scoliosis screening should be conducted annually for all children, particularly from the 3rd through the 10th grade, when the condition is most likely to occur. If a child is suspected of having scoliosis, the parents or guardians, as well as other appropriate school personnel, should be notified and further evaluation conducted.

Application Example: Correcting Postural Deviations

Situation: During the annual posture screening evaluation of her 4th grade students, the physical educator notices that one of her students has a mild lateral curvature to the right when viewed from behind. When the teacher asks the student to hang from a pull-up bar, the curve disappears.

Student: This 10-year-old boy is very coordinated and active in many sports. His favorite sport is baseball, and he was the star pitcher last year. The physical educator deduces that this student has a nonstructural postural deviation probably caused by his baseball pitching, which overdeveloped and tightened the back muscles on the right side.

Application: On the basis of the information presented here, the physical educator takes the following actions:

▶ Develops an exercise routine designed to stretch the muscles on the right side of the upper back and strengthen the muscles on the left side of the upper back.

▶ Contacts the parents and informs them that their son has a mild, nonstructural postural deviation of his back. She tells them that at this time the deviation can be corrected by a routine of exercises. She asks for their assistance in encouraging their son to regularly perform the exercises and in monitoring the remediation of the deviation. She also asks them to identify a reward their son can earn for doing his exercise routine.

▶ Meets with the student and explains the problem and how it can be corrected by a regular exercise routine. The teacher demonstrates the exercises and has the student perform them a few times to ensure they were done correctly.

▶ Develops a progress chart she and the student can use to monitor the degree of the curvature and the student's exercise compliance. She gives the student a copy of this chart each month to share with his parents.

Students suspected of having scoliosis should be monitored frequently, approximately every three months, to ascertain if the condition is progressing. A scoliosis assessment, which can be performed in less than a minute, involves observing the student shirtless. The assessments should be done individually and by an assessor of the same gender as the children being assessed, because children of these ages are usually self-conscious about the changes occurring in their bodies and about being seen undressed.

To perform a scoliosis assessment, check the symmetry of the child's back while the child is standing and then when the child is bent forward. First, with the child standing erect, observe from a posterior view for any differences between the two sides of the back, including the following points (figure 22.1, rows A-D):

1. Does the spinal column appear straight or curved?

2. Are the shoulders at the same height, or does one appear higher than the other?

3. Are the hips at the same horizontal distance from the floor, or does one appear higher than the other?

4. Is the space between the arms and trunk equal on both sides of the body?

5. Do the shoulder blades protrude evenly, or does one appear to protrude more than the other?

Then ask the child to perform the Adam's position, bent forward at the waist to approximately 90 degrees (figure 22.2). Examine the back from both a posterior and an anterior view for any noticeable differences in symmetry, such as curvature of the spine or one side of the back being higher or lower than the other, particularly in the thoracic and lumbar regions.

Care and Remediation of Scoliosis

The treatment of scoliosis depends on the type and the degree of curvature. As mentioned earlier, nonstructural scoliosis can frequently be corrected when the cause is identified and the condition remediated through a program of specific exercises and body awareness. With structural scoliosis, the treatment varies according to the degree of curvature. Children with mild curvatures (less than 20 degrees) are usually given exercise programs to keep the spine flexible and are monitored on a regular basis to make sure the curves are not becoming more severe.

Children with more severe curves (20 degrees to 40 degrees) are usually treated by means of braces or orthotics, which force the spine into better alignment

POSTURE RATING CHART

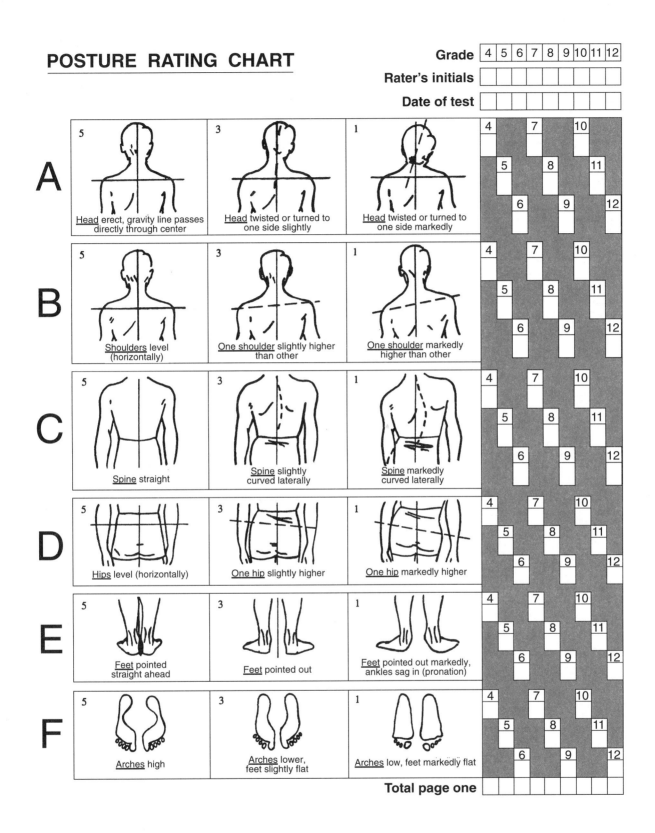

Figure 22.1   The New York State Posture Rating Chart.

From *New York State Physical Fitness Test for Boys and Girls Grades 4-12.* by New York State Education Department, 1966, Albany, N.Y.
Copyright by New York State Education Department. Reprinted by permission.

(continued)

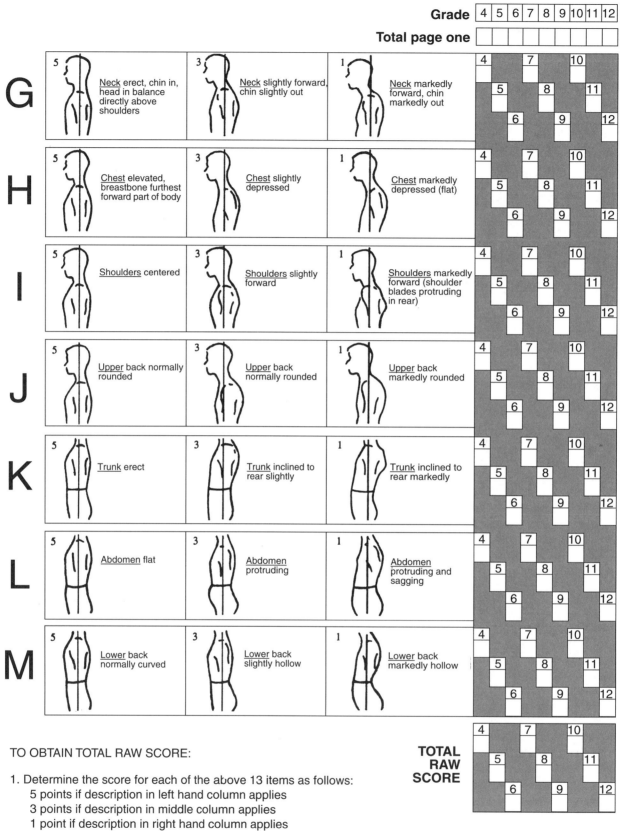

G Neck erect, chin in, head in balance directly above shoulders (5) | Neck slightly forward, chin slightly out (3) | Neck markedly forward, chin markedly out (1)

H Chest elevated, breastbone furthest forward part of body (5) | Chest slightly depressed (3) | Chest markedly depressed (flat) (1)

I Shoulders centered (5) | Shoulders slightly forward (3) | Shoulders markedly forward (shoulder blades protruding in rear) (1)

J Upper back normally rounded (5) | Upper back normally rounded (3) | Upper back markedly rounded (1)

K Trunk erect (5) | Trunk inclined to rear slightly (3) | Trunk inclined to rear markedly (1)

L Abdomen flat (5) | Abdomen protruding (3) | Abdomen protruding and sagging (1)

M Lower back normally curved (5) | Lower back slightly hollow (3) | Lower back markedly hollow (1)

TOTAL RAW SCORE

TO OBTAIN TOTAL RAW SCORE:

1. Determine the score for each of the above 13 items as follows:
   - 5 points if description in left hand column applies
   - 3 points if description in middle column applies
   - 1 point if description in right hand column applies
2. Enter score for each item under proper grade in the scoring column
3. Add all 13 scores and place total in appropriate space

Figure 22.1   *(continued)*.

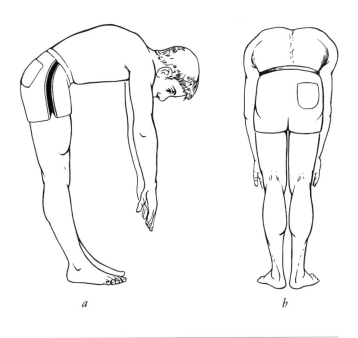

a   *b*

Figure 22.2   Illustration of the Adam's position: *(a)* side view, *(b)* back view.

and/or prevent it from deviating further. The Milwaukee and Charleston Twisting braces shown in figure 22.3 are the two most effective and commonly used braces for the treatment of scoliosis. To see the latest changes to the Milwaukee brace, consult the electronic resources referenced for this chapter at the end of the book. A number of more cosmetic braces, developed in recent years, are made of molded orthoplast and are custom fitted to the individual. These braces have been found to be effective in treating mild and moderate curves. One of the major advantages of the orthoplast braces is that they are less conspicuous and tend to be worn more consistently. These braces must be worn continuously until the child reaches skeletal maturity—in many cases, for four to five years. The brace can be removed for short periods for activities such as swimming and bathing. The treatment of scoliosis in persons who are wheelchair-bound may also involve modifying the chair to improve alignment and to equalize seating pressures. For the latest information on braces used to treat scoliosis, check the electronic resources listed for this chapter at the end of the book.

In extremely severe cases of scoliosis, where the curve is greater than 40 degrees or does not respond to bracing, surgery is employed. The surgical treatment usually involves fusing together the vertebrae in the affected region of the spine by means of bone grafts and the implantation of a metal rod. Following surgery, a brace is typically worn for about a year until the fusion has solidified.

Kyphosis and Lordosis

Abnormal concavity forward (backward curve) in the thoracic region (kyphosis) and abnormal convexity forward (forward curve) in the lumbar region (lordosis) are usually nonstructural and the result of poor posture (see figure 22.1, rows G-M). These deformities are routinely remediated by exercise programs designed to tighten specific muscle groups and stretch opposing muscle groups and by education designed to make students aware of their present posture, proper body mechanics, and the desired posture.

The physical educator can assess kyphosis and lordosis by observing children under the conditions described previously for scoliosis screening but from the side view. Look for exaggerated curves in the thoracic and lumbar regions of the spinal column. Kyphosis is usually characterized by a rounded appearance of the upper back. Lordosis is characterized by a hollow back appearance and a protruding abdomen.

Structural kyphosis, sometimes referred to as Scheuermann's disease or juvenile kyphosis, is similar in appearance during the early stages to the nonstructural form described earlier, but it is the result of a deformity in the shape of the vertebrae in the thoracic region. While the cause of this vertebral deformity is unknown, it can be diagnosed by x-rays. This form of kyphosis is frequently accompanied by a compensatory lordotic curve. The prevalence of this deformity is not known, but it appears to affect both genders equally during adolescence.

Early detection and treatment of structural kyphosis through bracing can result in effective remediation of the condition. The treatment typically involves wearing a brace continuously for one to two years until the vertebrae reshape themselves. A variety of braces and orthotic jackets have been developed for the treatment of this condition. The Milwaukee brace used for the treatment of scoliosis is considered one of the more effective braces for treating this form of kyphosis.

Implications for Adapted Physical Education

As discussed earlier, physical educators can play a major role in screening for postural deviations in the spinal column. Children identified as having mild, nonstructural postural problems should receive special instruction and exercises to remediate their problems. Following are several general guidelines that should be considered whenever a physical educator is designing, implementing, or monitoring an exercise program to correct postural deviations.

▶ Establish and follow policies and procedures for working with students suspected of having structural or serious postural deviations of the spinal column.

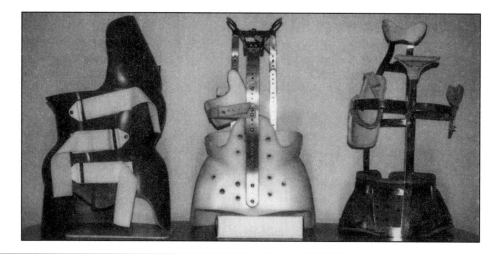

Figure 22.3 The Milwaukee and Charleston Twisting braces are commonly used in the treatment of scoliosis.

▶In an exercise program to remediate a postural deviation, the general objective is to strengthen the muscles used to pull the spinal column back into correct alignment and to stretch or lengthen the muscles that are pulling the spinal column out of alignment. The stretching program should be performed at least twice a day, and muscle-strengthening exercises should be performed at least every other day. A number of different exercise and stretching routines should be developed to add variety and keep the exercise routine interesting. Setting the exercise routines to music and establishing reward systems are also recommended, especially for students who are not highly motivated to exercise.

▶ All exercise and activity programs should begin and end with stretching exercises, with the greatest emphasis on static stretching. Stretches should each be performed 5 times and held for a count of 15 seconds.

▶ The exercise program should be initiated with mild, low-intensity exercises that can be easily performed by the children. The intensity of the exercises should be gradually increased as the children's strength and endurance increase.

▶ In most cases, individual exercise programs should be initiated and taught in adapted physical education. After the student has learned the exercise routine, it can be performed in the regular physical education class and monitored by the regular physical education teacher. First, explain to the student the nature of the postural deviation being addressed, the reasons that good posture is desirable, and the ways in which the exercises will help. The exercises should then be taught and monitored until it is clear that the child understands how to perform them correctly. The importance of making the student aware of the difference between the present posture and the desired posture cannot be overemphasized. Many children with mild postural deviations are simply unaware of the problem and, therefore, do not even try to correct their postures. Mirrors and videotapes are useful media for giving students feedback on posture. When working with students with visual impairments, the teacher will need to provide specific tactile and kinesthetic feedback to teach them the feeling of the correct postures. Cratty (1971) has described a tactile posture board, composed of a series of movable wooden pegs projecting through a vertical board, which can be placed along a student's spine to provide tactual feedback related to both postural deviations and desired postures.

▶ Children should follow their exercise programs at home on the days they do not have physical education. Some form of monitoring system, such as a log or progress chart, should be used. Periodically the children should be evaluated and given feedback and reinforcement to motivate them to continue working on their exercise programs.

▶ Exercises that make the body symmetrical are recommended. The use of asymmetric exercises, especially for children being treated for scoliosis, should be used only following medical consultation.

▶ When selecting exercises to remediate one curve (i.e., the major curve), take care to ensure that the exercise does not foster the development of another curve (i.e., the minor curve).

▶ The exercise program should be made as varied and interesting as possible to maintain the student's motivation and involvement. Alternating between routine exercises and activities like swimming and rowing will usually result in greater compliance with the program. Motivation is an even greater concern with students with mental disabilities, who may not comprehend why they need to exercise or why better posture is desirable; they will require more frequent feedback and reinforcement. Showing students random Polaroid snapshots is a good technique for keeping their attention on their postures and rewarding them when they are displaying the desired postures.

▶ Students wearing braces such as the Milwaukee brace can exercise and participate in physical education, although activities that cause trauma to the spine (e.g., jumping and gymnastics) may be contraindicated. As a general rule, the brace will be self-limiting.

Recommended Exercises

Using the guidelines just presented and drawing on an understanding of the muscles involved in a spinal column deviation, a physical educator should be able to select appropriate exercises and activities. The following lists include sample exercises for the upper and lower back that can be used in the remediation of the three major spinal cord deviations discussed in this chapter. These exercise lists provide examples and are far from complete. Several resources that provide additional exercises and more detailed descriptions of their performance are listed for this chapter at the end of the book.

Sample Upper-Back Exercises

The following sample exercises can be used to remediate nonstructural deviations in the upper spine.

▶ Symmetrical swimming strokes such as the backstroke and the breaststroke. If a pool is not available, the arm patterns of these strokes can be performed on a bench covered with a mat. Hand weights or pulley weights can be used to control the resistance.

▶ Rowing using either a rowboat or a rowing machine. The rowing action can also be performed with hand weights or pulley weights.

▶ Various arm and shoulder lifts from a prone position on a mat. Small hand weights can be used to increase resistance.

▶ Hanging from a bar. This is a good stretching exercise.

▶ Lateral (sideways) trunk bending from either a standing or a kneeling position. Forward bending should be avoided.

Sample Lower-Back Exercises

The following sample exercises can be used to remediate nonstructural deviations in the lower back.

▶ Any form of correctly performed sit-ups commensurate with the student's ability. Emphasis should be placed on keeping the lower back flat and performing the sit-ups in a slow, continuous action, as opposed to performing a large number of repetitions. Raising of the hips and sudden jerky movements should not be allowed.

▶ Pelvic tilt. This can be done from a supine position on a mat or standing against a wall.

▶ Alternating or combined knee exchanges (bringing the knee to the chest) from a supine position on a mat.

▶ Doing a bicycling action with the legs from a supine position on a mat.

▶ Leg lifts. Any variation of leg lifts is appropriate as long as the lower back is kept flat and pressed against the floor.

New techniques and treatments for spinal column postural deviations are constantly being developed. To find the latest information, it is recommended that teachers periodically consult the electronic resources for this chapter at the end of the book.

LOWER-EXTREMITY AND FOOT DISORDERS

Deviations in the alignment of the hips, legs, and feet are common and can result in pain, loss of efficiency, and other postural problems. The correct alignment of the hips, legs, and feet is shown in figure 22.4a. The assessment of postural alignment of the lower extremities is included in the general postural screening instruments presented in the first part of this chapter.

The purpose of this section is to review some of the common deviations found in the hips, legs, and feet. Many of these problems are related and frequently occur together. As with the spinal column deviations discussed earlier, deviations in the alignment of the lower extremities can be either structural or nonstructural. Structural deviations are usually caused by skeletal or neuromuscular problems, which may be congenital or acquired. Nonstructural deviations are typically the result of muscle imbalances that, if not treated, may eventually become structural deviations.

Hip Deformities

Four common deformities occur at the hip joint. The first two (coxa valga and coxa vara) are the result of the angle between the head of the femur and the shaft of the femur and can occur in anyone. The third and fourth deformities involve hip dislocations, which are caused by congenital abnormalities in the formation of the hip joint or are acquired because of either trauma or gradual deformation of the hip joint.

Coxa Valga

The alignment of the upper leg is the result of the way the head of the femur articulates with the hip socket and the angle between the head of the femur and the shaft of the femur. In coxa (hip) valga (out), the angle between the head of the femur and the shaft is greater than normal, which results in the upper leg appearing abducted or bowed. The increased angle also causes the affected leg to be longer. This condition is structural and usually congenital and cannot be affected by increasing the individual's awareness of the deviation, exercise, or activity in physical education.

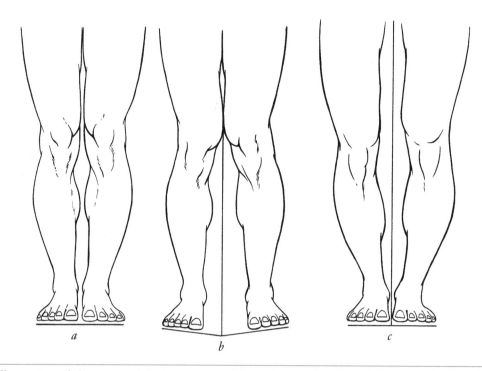

Figure 22.4   Illustration of alignment of the lower extremities: *(a)* normal, *(b)* knock-knees, and *(c)* bowlegs.

Coxa Vara

This deformity is the opposite of coxa valga. In coxa vara, the angle between the head of the femur and the shaft is smaller than normal, which results in the upper leg appearing bent inward or in a knock-kneed appearance. This condition, more common than coxa valga, can be either congenital or acquired. The acquired form, which occurs most frequently in males during the adolescent growth spurt, is usually caused by trauma or dislocation of the hip. If identified early, the condition is usually treated by several weeks of abstention from weight bearing. If the condition is allowed to progress, bracing and surgery may be required to correct it.

Hip Dislocations

Hip dislocation is the situation in which the head of the femur is separated from the hip socket. Hip dislocations can be classified as either congenital or acquired. Congenital dislocations, sometimes referred to as developmental hip dislocations, result from deficits in prenatal development and abnormal birth conditions. The cause of this condition is unknown, but there is some relatively strong evidence of a hereditary link. The condition is more prevalent in females and occurs most frequently in the left hip.

Depending on the severity of a congenital hip dislocation, it may be detected immediately after birth or, in mild cases, not until the child begins to walk. Other observable symptoms are decreased range of adduction of the hip on the affected side and asymmetrical fat folds on the upper legs. If the condition is not detected early, older children exhibit additional symptoms of exaggerated lordotic and scoliotic curves, pain, and fatigue. Treatment of congenital hip dislocations varies with the degree of the dislocation and the age at which it is detected. In mild cases involving very young children, dislocation can be successfully treated with splints. In more severe cases, a combination of casts, traction, and even surgery may be required.

In most cases, congenital hip dislocations will have been treated and remediated before the child enters school and will require no special attention in physical education. Occasionally, a case will be encountered where the condition was detected late, and the child needs attention in the area of postural training and conditioning of the muscles in the hip region. The physical educator should consult with the appropriate medical personnel to coordinate the child's physical education objectives with the objectives being worked on in rehabilitation.

Acquired hip dislocations can be either acute or gradual. Acute dislocations are caused by trauma or injury to the hip. Gradual dislocations are caused by a progressive deformation of the hip joint and are common in many children with neuromuscular conditions, such as cerebral palsy and spina bifida, who spend a majority of their time sitting in wheelchairs. These children are born with normal hip joints that are gradually deformed because of unbalanced muscle forces placed on the joint. Hip dislocations in these children frequently occur with very little associated pain. When the hip is dislocated, the affected leg will appear shorter and will be rotated inward in a scissoring position. Physical educators should be aware of these signs and of the potential for this problem in the children with neuromuscular disabilities they serve. The physical educator should consult regularly with the medical personnel on the status of students with potential hip dislocations and on physical education activities that might aggravate this condition and that, therefore, should be avoided.

Hip dislocation conditions can be treated prior to the actual dislocation by surgery that cuts the muscles, tendons, and/or nerves of the muscles exerting the inappropriate force on the joint. This usually involves the adductor muscles and tendons and the obturator nerve. After the hip has actually been dislocated, more extensive surgery involving restructuring of the hip joint is required.

Knee Deformities

A number of alignment problems referred to as knee deformities are the result of muscle imbalances, compensatory actions resulting from hip deviations, or abnormalities in the lower leg (tibia and fibula).

Knock-Knees

In this deviation the lower legs bow outward, with the result that the ankles are forced apart when the knees are touching (see figure 22.4b). This condition is usually nonstructural and common in very young children, and it frequently corrects itself. In young children, the deviation can be treated through bracing. If not treated, the deformity will eventually become structural. Knock-knees is a common postural deviation in obese children. Poor alignment of the knee results in a disproportionate amount of weight being borne by the medial aspect of the knee and predisposes the joint to potential strain and injury. Physical educators should carefully consider this point when selecting physical education and/or athletic events for students with this condition. Activities that increase the possibility of trauma to the knees—such as jumping from heights, running on hard or uneven surfaces, and/or games and sports where the knees could be hit laterally—should be carefully evaluated.

Bowlegs

Bowlegs, or genu (knee) varum (inward), is the condition in which there is an inward bowing of the legs, resulting in the knees being separated when the ankles are touching (see figure 22.4c). The bowing can occur in either the femur or the tibia but is most common in the tibia. One or both legs can be affected. This condition is a common nonstructural problem in many young children, usually correcting itself by the time a child reaches the age of three. If the condition has not corrected itself by that age, it must be treated with braces or it will most likely become a permanent structural deformity. This structural deformity is frequently accompanied by compensatory deformities in the feet. By school age, bowlegs cannot be remediated by activity or exercises in physical education. The physical educator's major responsibility, when the condition is suspected in the lower elementary grades, is to refer the child to a physician for possible treatment through bracing. Most children with this deformity, although sometimes appearing awkward in their movements, can successfully participate in all regular physical education activities.

Tibial Torsion

Tibial torsion is the result of an inward rotation of the lower leg (tibia). This condition can result in toeing in. The condition, caused by a muscle imbalance that twists the tibia inward, frequently occurs in young children who are in a non-weight bearing position for a prolonged period as a result of injury or illness. When it is not contraindicated by the injury or illness, children should be encouraged to bear weight on their legs each day. Attention should be focused on maintaining the legs in proper alignment and stretching the muscles that pull the tibia inward. If identified early in children, this deformity can be treated through bracing. In more severe cases a combination of surgery and bracing may be required.

Knee Flexion Deformity

Knee flexion deformity is common in children with neuromuscular disabilities, such as cerebral palsy, who are confined to wheelchairs. It is characterized by the legs being permanently bent or contracted in a sitting position. This condition can be painful, makes the legs harder to manage, and prevents the children from using standing tables or attempting ambulation even with the aid of braces and crutches. The condition is initially treated with splints and typically requires surgical lengthening of the hamstring tendons in which their insertions are repositioned. Physical educators can assist in preventing this condition by encouraging wheelchair-bound students to move their knees through the full range of motion. This might mean having the students leave their wheelchairs and perform a stretching routine on a mat while the other class members are doing other stretching exercises.

Foot Deformities

Although foot deformities can be caused by skeletal and neuromuscular abnormalities, the majority are caused by compensatory postures required to offset other postural misalignments in the legs, hips, and spine. As a result, the feet are typically the most abused structure in the body. Most foot deformities in children are identified and treated before they enter school. Occasionally, mild deformities can go unnoticed by parents and will be identified by the physical educator when the student complains of pain or avoids certain activities in physical education.

Physical educators should review the medical files and be aware of any students who have foot deformities. In most cases, the students will require no special consideration and will be able to participate normally in the regular physical education setting. When appropriate, the physical educator may need to monitor specific students to make sure they are wearing their braces and/or orthotics and performing any exercises prescribed by their physicians. Several of the common foot deformities found in children are described in the following paragraphs to provide physical educators with a basic understanding of the conditions and how they are treated.

Clubfoot

Clubfoot, or talipes, refers to a number of deformities in the foot in which the foot is usually severely twisted out of shape. This condition is usually congenital or acquired as the result of a neuromuscular condition. The term talipes is usually followed by one or more descriptors indicating the nature of the deformity: equinus (toe walking caused by tight heel cords), calcaneus (opposite of equinus and caused by loose heel cords, resulting in the foot being flexed), varus (toes and sole of the foot are turned inward, causing the individual to walk on the outside edge of the feet), and valgus (toes and sole of the foot are turned outward, causing the individual to walk on the inside edge of the feet). Mild forms of the conditions are treated with braces and orthopedic shoes. More severe forms require a combination of corrective surgery and braces.

Pronation

Pronation is a foot deformity in which the individual walks on the medial (inside) edge of the feet. The condition is frequently accompanied by toeing out. Pronation is usually acquired and can be remediated, if identified early, through corrective shoes and exercises.

Flatfoot

Flatfoot, or pes planus, may be acquired or congenital and may be the result of a fallen or flat longitudinal arch. Flatfoot is considered a postural deviation only when it is acquired as a result of poor body mechanics. In such cases, the structure of the foot is altered, which reduces its mechanical efficiency in absorbing and dis-

tributing force. This postural deviation, in turn, can cause pain in the longitudinal arch and result in other postural deviations as well. This condition is common in children with visual impairments and children who are obese. Treatment can involve prescriptive exercises and orthotics (inserts in the shoes).

Hollowfoot

Hollowfoot, or pes cavus, the opposite of flatfoot, is characterized by an extremely high longitudinal arch. This condition is usually congenital and is frequently associated with clubfoot. As in pes planus, the change in alignment of the foot due to the extremely high arch reduces the foot's ability to absorb and distribute force. The condition can be treated with orthopedic shoes and, in severe cases, with surgery.

Morton's Toe

Morton's toe refers to a deformity caused by a fallen metatarsal arch. The fallen arch puts pressure on surrounding nerves, which makes this condition very painful. The condition is caused by a disproportional amount of weight and stress being placed on the ball of the foot over a prolonged period of time. The treatment usually involves identifying and removing the cause, introducing appropriate exercises, and inserting an arch support in the shoe.

Hallux Valgus

Hallux valgus is a condition in which the big toe is bent inward toward the other toes. The condition is caused by pressure forcing the big toe inward, usually as a result of poorly fitting shoes. If the condition persists a bunion forms, and eventually a calcium deposit builds up on the head of the first metatarsal. Treatment involves removing the bunion and calcium deposit and fitting the individual with appropriate shoes.

SUMMARY

All children can benefit from good posture and body mechanics to safely and efficiently participate in physical education and athletics. Poor posture and body mechanics predispose children to potential injury. Physical educators, given the nature of their position and preparation, can play an important role in the identification and remediation of postural deviations. Postural screening should be an annual event in all physical education programs. Awareness and understanding of proper body mechanics should be taught to all children so that they may monitor and evaluate their own postures.

23

Rhythm and Dance

Ellen M. Kowalski

> **A**NDREA AND KEVIN HAD THE greatest difficulty in physical education of all the third graders in class. One day, during a lesson on ball-handling skills, the physical educator decided to try a different approach by incorporating the fun of rhythm and dance. Rather than just telling the children to keep practicing, she put on popular music that had a good steady beat and had the class clap, tap, and jump to the rhythm. Then she challenged the class to bounce-catch or dribble to match the rhythm of the music. After a little while Andrea and Kevin both came running up very excited saying, "This is fun! Watch me! I can do it!"

The purpose of this chapter is to provide the reader with an understanding of the value of teaching rhythm and dance to persons with disabilities. This chapter introduces basic elements of rhythm and movement and provides progressions for teaching rhythm and dance in the physical education curriculum. This chapter also provides ideas for developing a curriculum that is age appropriate and designed for the correct stage. The chapter also includes suggestions for modifying rhythm and dance instruction for students with various disabling conditions.

BENEFITS OF TEACHING RHYTHM AND DANCE

Many children like Andrea and Kevin, able-bodied and disabled, have difficulty coordinating their movements rhythmically. Almost 80 percent of the school-age population are uncomfortable moving to music, following a visual movement demonstration, and expressing themselves rhythmically (Weikart, 1989). Rhythm is a basic element inherent in all aspects of life, from the beating of the human heart to the rhythm inherent in movement skills. Few teachers realize the contribution that rhythm and dance activities can make to the psychological and motor development of children and to the long-term well-being of adults (Schmitz, 1989). Rhythm is the connecting thread that enhances the development of skills and abilities, develops comfort with movement, enhances self-esteem, and promotes social interaction. Dance is a moving, feeling art that allows for self-discovery, creative teaching, and cooperation among students (Walden, 1999). Additionally, all children learn through various modalities (visual, auditory, tactile, kinesthetic). A developmentally appropriate program of rhythmic activities can help nurture and develop auditory, visual, and tactile/kinesthetic decoding abilities, basic timing, creative movement, and language (Weikart, 1989).

All students, regardless of exceptionality have special needs. Dance does not discriminate between age or ability level, high functioning or low, ambulatory or not. The medium of rhythm is particularly valuable to children with disabilities because it provides children of all ability levels the opportunity to participate in activities that are educational, recreational, and remedial. "Dance is all inclusive—respecting individuals for what they bring to the moment; their abilities, and their strengths . . ." (Schwartz, 1989, p. 49). By integrating students in a common environment and experience, attitudes, behavior, and self-control can be positively affected (Wolf & Launi, 1996).

Many children with disabilities display developmental delays in self-esteem, body and spatial awareness, and coordination and exhibit deficits in spatial awareness, motor sequencing, and timing. Linked to theoretical concepts of sensory integration (Ayres, 1972), dance provides an excellent medium for developing perceptual-motor skills and spatial awareness, offering opportunities to explore the abilities and limitations of the physical self. Through dance, children facilitate the development of accurate body image, kinesthetic awareness, position and movement in space, and balance (Boswell, 1993) and posture (Woodard, Lewis, Koceja, & Surburg, 1996).

Dance provides a connection between the mental and physical self. Children's perceptions and feelings about themselves, their identities, and self-esteem are largely influenced by how well their bodies move (Marsh & Shavelson, 1985). The qualities of dance—its elements of discipline, perseverance, and teamwork—can often bring out unexpected responses in individuals like no other form of activity. Particularly beneficial for individuals with learning disabilities, behavior disorders, and emotional disturbance, rhythm and dance focuses on children becoming comfortable with their bodies and, therefore, plays a crucial role in the developing self-concept.

Children with disabilities often have difficulty with self-expression, creative movement, and interpretation. Dance, by the nature of its definition, encourages children to express themselves through movement. Dance is a living, moving language that offers a medium for self-discovery, self-expression, and creativity despite limitations imposed by a disability (Pesetsky & Burack, 1984). It is a wonderful medium to encourage expressive movement, because, unlike many other activities in physical education, there is no right or wrong (Joyce,

1980). Dance needs no words thus serving as an excellent way to help develop communication and expression for children who are nonverbal or have limited verbal capacity. Creative movement is important to help all students, especially those with disabilities, investigate new movement patterns and explore their bodies' capacities for movement (Riordan, 1989). Through creative dance activities, those who have little physical movement or language can connect with their inner thoughts and feelings; they can learn to express these feelings and find alternative ways to interact with others and the world around them in a nonthreatening environment.

In addition to being a lifetime activity, at the most basic level, rhythm and music can be used to facilitate movement. Sometimes it is difficult to motivate children with disabilities to move, especially those at severe and profound levels. Rhythm and music can be powerful motivators to encourage all hesitant children into the gymnasium and to maximize movement during instructional time.

National Influences

Currently, state and national influences in education have stimulated change in physical education curricula at all levels. Standards set on the national (National Association for Sport and Physical Education, 1993) and state (New York State Education Department, 1995) levels emphasize that students need to develop skills and knowledge necessary to establish and participate in fitness and maintain personal health. These national and state standards clearly identify the importance of all students developing an understanding and respect for differences among individuals thus supporting the need for inclusive activities. In addition to these standards, the Adapted Physical Education National Standards (National Consortium for Physical Education and Recreation for Individuals with Disabilities, 1995) and the definition of physical education in the Individuals with Disabilities Education Act (IDEA) of 1997 encourage that Individualized Education Plan (IEP) development and instruction focus on lifetime, *functional* skills. Rhythm and dance instruction is perfectly suited to facilitate these current state and national standards. Dance is a functional, inclusive, lifetime activity that enables individuals to enjoy movement and to laugh with others in a leisure activity. Children need to develop basic rhythmic skills that allow them to enjoy interactive, social activities with family and friends in a leisure setting, especially as they reach adulthood.

Adapted Dance and Dance Therapy

Both educators and dance therapists use rhythm and dance in their programs for individuals with disabilities. However, a distinction needs to be made between adapted dance and dance therapy. Although adapted dance can be an art form, an educational modality, and can be therapeutic in nature, it cannot be considered therapy. Until the middle 1960s, rhythmic movement and dance for individuals with disabilities was referred to as dance therapy, with no differentiation made between terms. The American Dance Therapy Association (ADTA), formed in 1966, defines dance/movement therapy as "the psychotherapeutic use of movement as a process which furthers the emotional and physical integration of the individual. Dance therapy is distinguished from other utilizations of dance . . . by its focus on the nonverbal aspects of behavior and its use of movement as the process for intervention" (Sherrill, 1998). Dance therapy is a specific treatment modality used as a nonverbal psychotherapy with individuals exhibiting psychological, emotional, and behavioral problems. Though similar to adapted dance, dance therapy is more closely aligned with physical or occupational therapy and can only be conducted by a certified dance therapist (Sherrill & Delaney, 1986). Adapted dance, which parallels the definition of adapted physical education, refers to rhythmic movement instruction designed or modified to meet the unique needs of individuals with disabilities. "The purpose of adapted dance . . . is to facilitate self-actualization, particularly as it relates to understanding and appreciation of the body and its capacity for movement" (Sherrill, 1998, p. 411). Two types of dance pedagogies are often used by professionals involved in dance/movement pedagogy: creative and structured dance. Whereas creative dance techniques encourage movement exploration and free response, structured dance provides easily defined goals because of its objective and sequential construction. Based on the participants and goals and objectives of the class, both styles have an appropriate place in the adapted physical education curriculum (Roswal, Sherrill, & Roswal, 1988).

USING RHYTHM AND DANCE IN THE PHYSICAL EDUCATION PROGRAM

Rhythmical movement experiences are important to the motoric development of children with disabilities. Dance is an integral part of physical education and should be given a major emphasis in the curriculum. Unfortunately, this is not always the case. Many physical educators have a limited background in dance and are uncomfortable teaching rhythmic activities. Consequently, these teachers tend to include few rhythm and dance experiences in their programs or eliminate them all together. As a result, students lack adequate experiences for rhythm and dance to become a natural part of their movement repertoire. Compounding the problem, teachers often unknowingly introduce dance activities at too high a level, beginning with movements that are too complex for their students' ability levels. Dance "steps" that involve the coordination of several body parts require a level of integration, kinesthetic awareness, and motor sequencing that is often difficult for children to perform.

To help diminish these problems, dance and rhythmic activities need to be threaded throughout the entire physical education curriculum and not used only as a two-week unit. There needs to be a change in perception, that is, the use of rhythm and dance as a teaching tool and not just to motivate. With well-designed and developmentally sequenced rhythm and dance activities, participation in rhythmic and dance experiences can be therapeutic, successful, and enjoyable.

Participation in rhythm and dance activities can often be frustrating for all individuals, regardless of age. Participation in rhythm and dance activities can be especially frustrating for children with disabilities in an integrated setting. Because the growth and maturation of motor patterns are often different in children with disabilities, basic motor skills may not develop fully or may develop as splinter skills—which are "particular perceptual or motor acts that are performed in isolation and to not generalize to other areas of performance (Auxter, Pyfer, & Heuttig, 1997, p. 582)—resulting in difficulty when performing rhythmic movements and dance steps. Unfortunately, minimal exposure such as a two-week dance unit does not develop the necessary skills.

Rhythm Awareness: The Beginning

Helping individuals feel comfortable with their bodies is an important goal of rhythmic activities. If children have basic rhythm awareness, they are much more likely to participate in and enjoy rhythmic activities in physical education classes and in social settings. Inherent in almost any piece of music or rhythmical sequence, rhythm involves three components: beat (the underlying pattern found in music or rhythmic sequence), tempo (speed), and accent (emphasis). At the most basic level, rhythmic movement requires the ability to effectively use time and space. Many children with disabilities lack rhythm awareness and thus are unable to respond naturally to pulse beats or to the time intervals between beats (Sherrill & Delaney, 1986). Rhythmic competency, or basic timing, is composed of two abilities: beat awareness and beat competency (Weikart, 1989). Beat awareness is defined as the ability to feel and express the steady beat of a rhyme, song, or recorded musical selection using nonlocomotor movements. Beat competency involves the ability to walk to a beat in self-space or general space. Weikart (1989) suggests a basic teaching progression to help

students attain a basic level of rhythmic competency. The first task, being able to "feel" the beat, involves hearing and moving to a steady beat such as rocking, patting, or clapping. One aspect of feeling the beat is being able to distinguish between the big beat and the little beat. The big beat involves tapping, clapping, or moving *every other* beat whereas the little beat involves tapping to every beat. Having students move to both the big and little beats allows teachers to modify challenges in coordination, balance, timing, and control to meet the needs of the class. Refer to the Rhythms application example for ideas about teaching the difference between the big beat and the little beat.

The second task involves organizing and repeating two nonlocomotor movements to the beat. Children may be able to tap their knees to a beat but not be able to bend their knees and clap to a beat. The third task involves walking to a beat. The addition of mobility requires greater integration and coordination of body parts. Beat awareness activities involve simple movements to a beat creating a bonding to rhythm by providing a link between sound and movement (Bornell, 1989; Weikart, 1989).

Following simple guidelines can assist teaching rhythmic competency to all age groups and ability levels, including individuals with disabilities.

Begin with the individual. It is challenging enough to follow a beat by oneself without attempting to coordinate one's movements with someone else's. Many children with mobility impairments or temporal perception problems experience difficulty moving to an external rhythm (Krebs, 1990). Allow children to move to their own rhythms, initially. Children then imitate a steady beat created by the teacher using simple movements such as clapping or rhythm instruments. When using music, be sure to reinforce the heavy or even beat, which is much easier to hear and move to.

Activities should begin with nonlocomotor movements—sitting by yourself, with a partner, or with objects such as balls or beanbags. When standing, use movements such as bending, twisting, rocking, or walking in place. Eventually incorporate marching in place, walking, and other locomotor movements into rhythm activities.

Initially, use whole-body movements (rocking, bending at waist), individual body parts (one arm), same body part on both sides of the body (bend both knees/elbows), and alternating body parts (lift right arm, then left). Eventually incorporate two or more body parts, either sequentially or simultaneously. For example, bending knees to a beat is much simpler rhythmically than bending knees and clapping.

Once basic beat awareness is obtained individually, students can work on performing movements with a partner in which they have to coordinate their own movements with the movements of another. Activities include children clapping hands (double patty cake) and

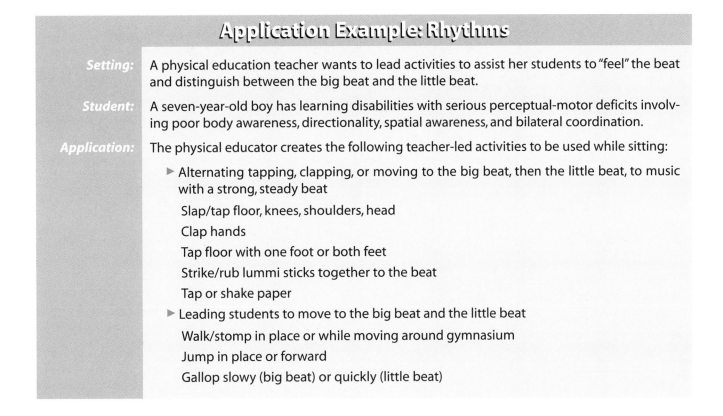

Application Example: Rhythms

Setting: A physical education teacher wants to lead activities to assist her students to "feel" the beat and distinguish between the big beat and the little beat.

Student: A seven-year-old boy has learning disabilities with serious perceptual-motor deficits involving poor body awareness, directionality, spatial awareness, and bilateral coordination.

Application: The physical educator creates the following teacher-led activities to be used while sitting:

▶ Alternating tapping, clapping, or moving to the big beat, then the little beat, to music with a strong, steady beat

Slap/tap floor, knees, shoulders, head

Clap hands

Tap floor with one foot or both feet

Strike/rub lummi sticks together to the beat

Tap or shake paper

▶ Leading students to move to the big beat and the little beat

Walk/stomp in place or while moving around gymnasium

Jump in place or forward

Gallop slowy (big beat) or quickly (little beat)

mirroring each other's movements to a beat or sitting in a choo-choo train formation tapping the beat on the back of the person in front of them. To increase the level of difficulty, teachers add equipment such as beanbags, balls, or ribbon sticks to the rhythmical movement. Children must coordinate the movement of an external object with the movement of their bodies. When adding equipment, teachers should begin with activities involving nonlocomotor movements first and then add locomotor movements.

Incorporating Elements of Movement in Instruction

Although developing beat competency and rhythm awareness should be the first objective, rhythmic and dance activities can also focus on the elements of movement. All dance is composed of elements of movement (Laban, 1963): space (shape, level, size, pathway), time (beat, accent, pattern), force (light or strong), and flow (free or bound). Laban (1963) grouped movement into 16 basic themes. Each theme represents a movement idea that corresponds to the progressive unfolding of movement in the growing child. Both creative and modern dance expand on the elements of movement to allow all students to test the limits of their bodies' capabilities, especially those who are motorically challenged. By focusing on the elements of movement through creative dance and movement education (exploration and guided discovery), teachers can enhance

body image and spatial awareness, language development through associations between words and movement, creativity, and expression of feelings and emotions. Preston (1963) simplified and reorganized Laban's themes into seven movement themes for organizing educational dance content. Building on beat awareness, these themes provide teachers guidelines for developing rhythm and dance progressions and activities. A valuable tool in an integrated setting, these themes can assist the teacher to individualize activities, allowing each student to work at his own level. These seven themes are referred to throughout the remaining portions of the chapter:

Theme 1—Awareness of the body (total-body actions and actions of individual body parts)

Theme 2—Awareness of weight and time (contrasting qualities)

Theme 3—Awareness of space (areas, directions, levels, pathways, and extensions)

Theme 4—Awareness of flow of movement (use of space and time)

Theme 5—Awareness of adaptation to partners and small groups (simple forms of relationships)

Theme 6—Awareness of body (emphasis on elevation, body shapes, and gestures)

Theme 7—Awareness of basic effort actions (rhythmic nature of time, weight, and space)

Table 23.1 Suggested Dance Progression by Theme and Age

Key:

___   Theme is developmentally appropriate in physical education program.

****   Theme may be appropriate but used less frequently.

.........More advanced themes to increase complexity in a physical education experience.

Theme	Age in years									
	3	4	5	6	7	8	9	10	11	12
1. Awareness of body	___	___	___	___	___	****	****	****	****	****
2. Awareness of weight and time	___	___	___	___	___	___	****	****	****	****
3. Awareness of space				...	___	___	___	___	****	****
4. Awareness of flow					...	___	___	___	___	****
5. Awareness of adaptation to partners and small groups			...	___	___	___	___	___	___	
6. Awareness of the body			...	___	___	___	___	___	___	
7. Awareness of the basic effort actions			...	___	___	___	___	___	___	

Adapted, by permission, from B.J. Logsdon, et al., 1984, *Physical education for children: A focus on the teaching process*, 2nd ed. (Philadelphia: Lea & Febiger), 161.

Once children learn the basic content of a theme, experiences may be designed to combine previous themes with the those newly learned. Modified to incorporate the early childhood age group, table 23.1 outlines a dance program that continues through age 12 (Logsdon et al., 1984).

Introducing Simple Dance Steps

As children develop basic rhythm awareness, teachers can begin to introduce dances involving simple steps and movement sequences. No matter what the age group or the disability, teachers should follow a basic progression when teaching a dance, especially when introducing a new dance. Classes should be given the opportunity to listen to the entire piece of music first to give them a feeling of the rhythm and tempo; then they can be asked to tap or clap to the beat. Without music first, teachers should teach dance steps by having children verbally rehearse/label each step. Verbal rehearsal/labeling involves active learning where verbal labels are spoken in the same timing in which the movement is performed. Documented as an effective teaching strategy (Kowalski & Sherrill, 1992; Weiss & Klint, 1987), verbal rehearsal/labeling assists encoding and integration of informatio, improving the ability to plan, sequence, and perform a motor pattern. Children with disabilities are often motorically awkward, displaying difficulties in movement integration, motor memory, and motor sequencing. For children who have difficulty encoding and decoding visual and auditory information and then translating it into smooth coordinated movement, verbal rehearsal creates a concrete link between movement and thought and enhances motor sequencing and motor memory.

One such progression for enhancing motor sequencing and memory is the Say and Do method (Weikart, 1989), designed to teach dance by connecting language and movement. Verbal labels identifying the body part, locomotor movement, or motion are spoken in the same timing in which the movements are performed. The four-step language process begins with no movement and no music, just chanting the movement labels out loud to a rhythm set by the teacher; no music, chanting out loud while simultaneously performing the movements; adding music, chanting while simultaneously performing the movements; and thinking the words while simultaneously performing the movements to music.

DEVELOPMENTALLY AND AGE-APPROPRIATE RHYTHM AND DANCE ACTIVITIES

Rhythm and dance activities provide children with disabilities wonderful opportunities to become involved with their able-bodied peers. To ensure successful and enjoyable participation, activities and dances should be selected that are appropriate for the age level not the disability level. Selecting appropriate activities requires careful consideration of the group's developmental level as well as accurate assessment of each individual's skill level. With young children, teachers must be careful that movement activities designed for the age group are not too difficult, resulting in repeated failure (Weikart, 1989). Although older, adolescents and adults may still need to experience and develop the basic levels of rhythmic competency. Although the focus and objectives remain the same, it is extremely important that selected rhythmic activities are appropriate for the age group. Because of developmental level, selecting age-appropriate activities for persons with severe and profound disabilities is particularly difficult. Even so, nursery songs such as "Ring Around the Rosie" or "The Wheels on the Bus" are unacceptable to use with older students and adults.

Early Childhood

Rhythmic activities for children ages three to five should emphasize body awareness (theme 1) and fundamental spatial concepts (themes 2 and 3). Especially for children with perceptual-motor deficits, all rhythmic activities should combine chanting with movement. For example, chanting and tapping various body parts (e.g., head, head, head, head; knees, knees, knees, knees) or performing simple movements (push, push, push, push) helps create a strong word-movement association. Action songs such as "Head, Shoulders, Knees, and Toes" are popular for teaching body part names and locations or simple movement concepts about space, effort, and time (bend/straight, big/small). Teachers can also create action songs to familiar tunes (e.g., "Skip to My Lou") by substituting words (shake, push, tap, jump) relevant to the action or body part being taught. On a simpler level, single bilateral symmetrical movement, such as moving both feet or hands in a single movement (Weikart, 1989), is recommended because, at this level, bilateral integration is yet unrefined; even simple integrated movements are often too complex. Young children need and enjoy activities that are simple and extremely repetitive in nature. Use activities that are primarily nonlocomotor, and allow each child to perform in her own space. Locomotor skills should be limited to simple walks or marches. Most importantly, whether using music or a simple rhythm instrument, action song, or chant, teachers should not be overly concerned with students moving "correctly" to an external beat but should allow them to move to their own rhythms (Weikart, 1989). There is nothing sadder than when a child is concentrating so hard on the tempo that the joy of movement is lost (Sherrill, 1998).

Primary

Rhythmic activities for children ages five to seven expand on activities taught in early childhood. Although

activities are still primarily focused on body and space awareness (themes 1, 2, 3) and are repetitive in nature, teachers can begin to utilize combined movements (themes 1, 4, 6) in their rhythmic activities. Important for children with disabilities, rhythmic activities help develop awareness of variation of movement force and flow (theme 4) and discrimination between even and uneven rhythm and various tempos (theme 2). Action songs and rhythmic chants remain as a foundation for rhythmic activities involving isolated body parts (themes 2, 3) and total-body movement (themes 1, 4, 6, 7). Increased coordination and ability to combine rhythm with locomotor and axial movement allow activities to contain a greater variety of movement and utilize words that involve total-body coordination rather than body parts. Percussive instruments (tambourines, drums, and shakers) or other equipment (ribbon sticks, scarves, beanbags, lummi sticks) are an excellent way to increase coordination requirements and bridge rhythmic movement from nonlocomotor (sitting) to locomotor skills (marching or galloping) (figure 23.1). Simple dances such as Hands Up or the Macarena, which teach basic dance steps and movements, should be introduced as early as possible. Individual, or nonpartner, dances (e.g., Hokey Pokey, Alley Cat, and the Bunny Hop) are the easiest to start with because they do not require coordination of one's own movements with the movements of another. Nonlocomotor dances such as the Macarena or Hands Up can be easily performed. Rhythmic activities can and should be designed to stimulate imagination and creative movement (Jay, 1991). Creativity can be enhanced by stimulating imaginative play (themes 1, 2, 3) such as singing action songs; imitating people, animals, and things; or acting out nursery rhymes, poems, stories, and songs (Krebs, 1990).

Figure 23.1 Led by teacher Sally Ayres, children explore rhythm with scarves.

Late Childhood and Early Adolescence

Children 8 to 11 years old demonstrate increased levels of attention and cognitive functioning, balance, coordination, perceptual-motor abilities, and social interaction. Children are ready to be challenged by adding the use of objects such as balls, hoops, ribbon sticks, and ropes to rhythmic activities. Socially, children should begin to work cooperatively with partners and in small groups. Line dances, simple folk and square dances, and activities using improvisation (theme 6) help provide older children with the fundamentals to learn more complex dance movements (Krebs, 1990) that are important to successful participation in integrated social settings later in life.

Adolescence and Young Adulthood

Adolescence brings a shift in the demands of the school environment, where social behavior becomes a central focus of a student's life. These social demands are no different for adolescents with disabilities. Interest is on the "cool" way to dress or learning the latest dance everyone is doing to the most recent popular music. Rhythmic programs during the adolescent years should be predominantly geared toward teaching social dance skills and lifetime fitness activities.

More than any other age, adolescents with disabilities need to be provided opportunities to participate in age-appropriate, integrated activities. Teachers need to keep up with current trends in music and social dances (e.g., hip hop, The Electric Slide, Achy Breaky Heart), often seen at school dances and family and social gatherings. Adolescents with disabilities may not have the opportunity to learn these current dances because of motoric difficulty or lack of exposure. More fundamental than learning current dance steps, students need to feel good about expressing themselves rhythmically. Frequently seen at school dances and social gatherings is freestyle dance. Freestyle dance, which is no dance in particular, is simply moving rhythmically to music in a way that fits the music and reflects the individual. Students can create their own dance styles by taking a few simple movements, performing them to the beat, and then changing the direction, emphasis, or tempo (themes 3, 4, 6, 7).

With a basis of rhythmic competency and basic dance steps learned, teachers can help students with disabilities learn to express themselves through movement and develop their own unique styles. Rhythmic programs should also include aerobic dance, now popular in many physical education programs. Aerobics, which combines simple locomotor and nonlocomotor activity, is fun, social, and easily modified for an integrated group. Currently, there are several adapted aerobics tapes available.

Though more formal than freestyle or aerobic dancing, adolescents can also enjoy integrated rhythmic activities through folk, square, and ballroom dance. Folk and square dances are social dances in which both children and adults can participate. Folk and square dance are not only popular worldwide but provide valuable insight into many cultures and customs around the world. Currently, ballroom dance, especially swing, is experiencing a "rebirth" in popularity among young adults. Whether it be freestyle, aerobic, folk, square dance, or swing, adolescents with disabilities can be provided with a variety of opportunities to participate in age-appropriate, integrated activities and feel good about themselves.

MODIFYING RHYTHM AND DANCE ACTIVITIES FOR PERSONS WITH DISABILITIES

With only slight modification, all individuals of all ages, can enjoy and benefit from rhythm and dance activities. One of the most significant benefits of teaching rhythmic activities and dance to individuals with disabilities is that it can be used to work with all levels of movement control. Whether using locomotor or nonlocomotor movements, teachers must be able to recognize movement complexity and then modify the challenge. Weikart (1989) identifies important elements of rhythmic movement that provide the basis for increasing or decreasing rhythmic complexity of selected activities and dances. Through adjusting these elements, teachers can modify games and activities so that all students are challenged and able to actively participate. These elements are as follows:

- ▶ Movement in which the hands touch the body is easier than movement in which they do not.
- ▶ Most nonlocomotor movements are easier than locomotor movements.
- ▶ Movement without an object is easier than movement with an object.
- ▶ Movement without a partner or group is easier than movement timed with a partner or group.
- ▶ Bilateral movements are easier than movements using one or alternating sides of the body.
- ▶ Symmetrical movements (paired body parts moving in the same way) are easier than asymmetrical movements (body parts moving in different ways).
- ▶ Single, repetitive movements (e.g., eight slaps each on the floor, then the knees) are easier than sequenced movements (single touch floor, knees, shoulder, head).
- ▶ Movement to a slower beat is easier than to a faster beat.

For students who have difficulty processing multipart directions and sequencing movements, teachers should break dance steps into small movement phrases and teach them separately first, and then in combination. Teachers may initially need to slow the tempo down or manually assist students' movements to help them feel the beat or tempo. Although specific modifications and techniques are associated with various disabling conditions, in general, teachers should follow a few basic guidelines: keep movement sequences short and within the comprehension level and physical tolerance of the group; place less emphasis on verbal explanation and greater emphasis on manual guidance; and select music containing a strong, steady underlying beat.

When teaching students with mental retardation, it is important for instruction to be multisensory and concrete, allowing multiple opportunities for students to understand directions. Because of the abstract nature of language, understanding verbal directions is often difficult. Emphasis should be on visual and manual guidance to help students learn movement sequences. Prompt with light touch or physical guidance through movement to assist remembering or feeling the movement. Footprints and arrows on the floor, numbers on cones, enlarged pictures and diagrams, as well as visual demonstration, focus attention and clarify directions. Props such as pieces of clothing, hoops, balls, elastic bands, yarn balls, balloons, ribbons, sandpaper, and scarves in activities are useful to help students focus their energy and develop concepts through concrete experience (Sherrill & Delaney, 1986; Silk, 1989). For example, teachers can use hula hoops as a prop for stretching overhead or scarves for experiencing lightness of movement (Silk, 1989). When teaching action songs, simple melodies that are short with frequently repeated phrases should be selected (Krebs, 1990). Similar to introducing a dance, words to the melody should be taught first, followed by the movements or steps without partners or groups.

Children with learning disabilities often have difficulty planning, organizing, sequencing, and remembering movement because of problems in information processing and motor planning (Lazarus, 1990). For children exhibiting motor awkwardness, rhythmic activities should focus on developing temporal perception and body localization (Schmitz, 1989). It is important to incorporate basic rhythm awareness using simple bilateral movements that contain a low level of coordination. If processing delays are apparent, providing for maximum repetition and focusing children on relevant cues through verbal labeling are crucial. Children with learning disabilities should be taught to verbalize their movements in all activities thus helping them to encode movement sequences by connecting language and movement at basic levels of processing (Kowalski & Sherrill, 1992). For children displaying social imperception, using activities that focus on expressive movement helps children learn to interpret and express feelings accurately. Pantomime and mimetic activities, as well as creative dance, can also be instrumental in the development of appropriate social behavior.

Children with behavior disorders and mental illness characteristically have poor body image, low self-esteem, poor self-control, lack of trust in others, difficulty identifying and expressing feelings, and poor interpersonal skills. According to Bannon (1994) these children tend to be disconnected from their bodies, as well as from the world around them. Children who have been brought up in an unsafe environment often have experienced fear and anxiety and thus build defenses to protect themselves. Although they may understand and appreciate their bodies and derive great pleasure from exploration of space, they still may be unable to use movement to convey their needs, wishes, feelings, and moods to others (Sherrill & Delaney, 1986).

Often used with emotional disturbance and mental disabilities, dance therapy is an effective therapeutic intervention because it focuses on nonverbal communication through movement and allows children to reflect their thoughts and feelings in a nonthreatening way. Although it is not dance therapy, teachers can apply principles of dance therapy in rhythmic activities to help students with emotional problems develop a better perspective about themselves and their feelings. Through creative expression and modern dance, activities should primarily focus on the expression of feelings rather than on specific steps or movements. Whenever possible, the school guidance counselor and/or psychologist can assist in facilitating students to express their feelings.

Wolf and Launi (1996) established an innovative, integrated program using modern dance instruction; they suggest that a questionnaire asking students how they express their feelings is an effective technique to focus students prior to beginning a dance unit or project. Questions might include, for example, "How do you usually express your feelings?" "Which emotion do you have the most trouble expressing?" "What kind of dance/music do you enjoy?" Dances that encourage children to use their bodies in different ways, experiment with opposites in movement, and explore movement through space should be created. Initially, teachers should mimic the students' movements rather than initiating movement. Eventually students can begin to reflect the movement patterns of others.

Because students with hearing impairments may be unable to hear music, it does not mean they cannot dance. Although hearing does make it easier, profoundly deaf and hearing-impaired students can learn to dance as well as their hearing peers. Selecting music that is high frequency should be avoided because many students will be unable to hear it. Music should be amplified and contain a heavy bass. In addition, placing speakers face down on a wood floor may help to increase the intensity of the vibrations and bass tones. When integrating hearing and deaf students in dance activities, it is important that all information is presented visually (Hottendorf, 1989). All forms of communication should be utilized: written directions on posterboard and blackboards; basic signs, gestures, and pantomime; and the spoken word. Teachers should always face the students while giving instructions and allow them to place themselves where they can see best. Walking around and talking makes it extremely difficult for students to lip-read. The rhythm of music should be consistently demonstrated visually by beating a drum, clapping hands, tapping legs, or counting with the fingers. Teachers should demonstrate dance steps with their backs to the students, making it easier to mimic right and left movements (Krebs, 1990).

When teaching students with visual impairments, rhythmic activities should be modified by replacing visual cues with tactile and/or kinesthetic cues. Totally blind students, paired with a partially sighted or sighted partner, can acquire a kinesthetic sense of force, flow, timing, and space concepts by placing their hands on the partner's shoulders, arms, or hips. Teachers and students can mark dance steps and patterns by moving their hands on the floor. Dances can be modified so that either contact is maintained continuously (circle, line, square) or dancers are totally separate from each other in their own defined spaces (Krebs, 1990). Complex dance movements should be replaced with simpler forward, backward, and side steps. Teachers can enhance residual vision by using brightly colored tape to mark position and indicate direction or space and establish boundaries by adding surface textures to the floor (thin mats, taping down newspaper or butcher paper). Especially when teaching students with visual impairments, it is important to make sure that the sound system and acoustics are clear and any distracting or competing noises are eliminated. For example, when leading/cueing aerobics, music that is too loud acts more as a distracter to the instructor's voice than a motivator.

Characteristically, children with mobility impairments (e.g., cerebral palsy, muscular dystrophy, spina bifida) have difficulty performing movements smoothly and with control. When teaching rhythmic activities and dances, independent movement and control should be emphasized. To modify activities the tempo may need to be slowed, movements simplified or substituted, and timing of coordinated steps adjusted (Harris, 1989; Krebs, 1990) to allow students to focus on controlling a movement or holding a position. For students unable to push themselves, being pushed by a volunteer will not give the dancer anything more than a nice ride. Actively engage the student by substituting upper-body movements (moving head, bending body, clapping/shaking hands, raising/lowering arms, touching thighs,

raising feet) to the same number of counts as steps (Caler & Cronk, 1997). For ambulatory students, walking and nonlocomotor movements (swaying, swinging arms, balancing) can be substituted for dance steps that are too fast or complex. Students who use wheelchairs may substitute upper-body, arm, and head movements for leg movements and rolling for locomotor skills (Kindel, 1986). Teachers can also actively engage the students by discussing their ideas for modifying the activity for them to be successful (Caler & Cronk, 1997). Although limited motorically, physically challenged students need to be given opportunities for creative expression as much their able-bodied peers. Rather than being limited to traditional activities such as wheelchair square dancing, students should be encouraged to explore and create new movements or patterns made possible by the chair (Riordan, 1989).

THEATRICAL DANCE AND PERFORMANCE PROGRAMS

Beginning during childhood and adolescence, many students become interested in more formalized and expressive forms of dance (tap, ballet, jazz, and modern). Students with disabilities need to experience formal dance, both as a participant as well as an observer (Boswell, 1989; Schmitz, 1989). It is important for children to be exposed to role models with disabilities who demonstrate high levels of skill, not only in sport, but in dance as well. For example, within Gallaudet University for the Deaf, a professional-level dance company performs ballet, modern dance, and jazz.

Currently, there is an emerging culture in dance created by a growing participation of dancers and performers with disabilities in professional dance. An increasing number of dance companies across the United States are featuring artists with disabilities both as soloists and in integrated dance pieces. Through their performances, improvisational dance companies such as *Light in Motion* from Seattle, Washington, and performing artists like Bill Shannon (figure 23.2) are redefining the parameters of dance and broadening public and professional perceptions of what dance is (Walden, 1999).

Although dancers with disabilities are important in serving as role models, it is important for children of all abilities to observe and experience disabled and integrated dance programs. For example, Wolf and Launi (1996) created an integrated dance program where children with different exceptionalities performed for the entire school. Another example is Wheelchair Dance, as defined by the International Sports Organization for the Disabled, which is practiced in more than 40 countries on various levels: rehabilitative, recreational, and competitive; folk, ballroom, formation, and creative. Integrated wheelchair dance can be performed in sev-

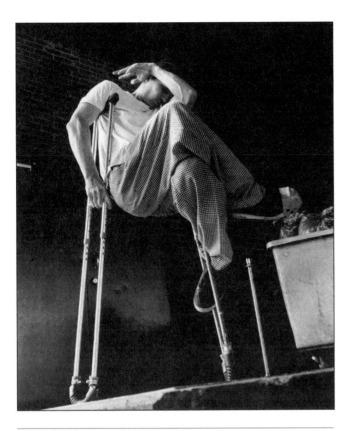

Figure 23.2   Through his improvisational work, performance artist Bill Shannon redefines perceptions of dance.

eral forms. Couple dancing can be practiced in two forms: duo dancing where two wheelchair users dance together or combi dancing where a wheelchair user is dancing with a nondisabled partner. Finally, there is group dancing where wheelchair users dance alone or together with nondisabled partners (Krombholz, 1997). Regardless of form or level, social-integrative wheelchair dancing clearly demonstrates how the movements of the wheelchair and the steps of the "pedestrian" are coordinated so that the couple dances in harmony.

Recognizing this new emerging art form, VSAarts, in a partnership with the American Association for Active Lifestyles and Fitness (AAALF), and the National Dance Association (NDA) have gathered authors experienced in the fields of dance education, disabilities, and curriculum development to develop a resource guide entitled *DanceConnects*. Created by a design team including dancers with disabilities, *DanceConnects* provides the philosophical foundations of dance and disability as well as dance content and methodologies necessary to facilitate the evolution of inclusive dance. This text, geared for practitioners of all levels, is organized in a user-friendly format designed to assist educators to include all types of disabilities in the dance experience. Still in its development, more information on this program can be obtained by contacting VSAarts at 800-933-8721.

SUMMARY

Rhythmic movement and dance are valuable forms of communication and creative expression that enrich one's life in many ways and contribute to the motor and psychological development of a child. Children with disabilities can benefit greatly from a modified dance curriculum where activities are designed to meet the unique needs of the individual. Benefits of rhythmic and dance activities include development of mind/body connections, body image, and spatial awareness; improvement of mobility, strength, coordination, and flexibility; and enhancement of self-image and the development of social skills (Schmitz, 1989). Rhythm awareness and basic rhythm skills, such as locomotor and nonlocomotor movements, are prerequisite to successful participation in many games and activities taught in integrated settings. Now more than ever, national and state standards are emphasizing the need to teach all students lifetime, functional skills. Today, with the trend toward full inclusion continuing in education, the value and significance of rhythm and dance have a greater role in physical education. If persons with disabilities are to truly be integrated with their able-bodied peers, physical educators must ensure opportunities for reaching their full potentials by threading dance and rhythmic activities throughout the entire curriculum.

Aquatics

Monica Lepore

▷ **J**ACK IS AN EIGHT-YEAR-OLD boy needing full physical assistance to participate in his regular physical education class. Jack's parents have asked the school district for a physical education assessment to determine if Jack is benefiting from his current support and placement. His parents have noticed Jack exhibits more independence when placed in a flotation device in his backyard pool than anywhere else. Therefore, they have asked that a swimming component be part of the assessment.

During the land portion of the physical education assessment, it was evident Jack is unable to participate in physical activities without adult intervention. He was not able to consistently perform voluntary movements against gravity and could not raise his heart rate unless physically assisted. In contrast, during the pool assessment, Jack was able to raise his heart rate by 40 beats using a head/neck flotation device without teacher intervention. In addition, he continually moved his arms and legs for nine minutes without prompting.

Therefore, it was decided by Jack's individualized education planning committee that he should receive adapted aquatic instruction at the expense of the school district in addition to regular physical education instruction.

This scenario demonstrates that adapted aquatic instruction can complement a land-based adapted physical education program. The purpose of this chapter is to assist the reader in identifying benefits of aquatic instructional programs, illustrate the best practices in adapted aquatics, and provide information for meeting the needs of students with disabilities in aquatic programs.

BENEFITS OF ADAPTED AQUATICS

Aquatics provides physical fitness and motor skill development within a physical education program for children with disabilities. In the vignette above, Jack's parents are within their legal rights to request swimming as part of their son's Individualized Education Program (IEP), since aquatics is listed as a component of physical education under PL 105-17 (Individuals with Disabilities Education Act). School districts have to realize that aquatics is not a luxury nor is it a therapeutic (related) service. **Adapted aquatics** means modifying the teaching environment, skills, facilities, equipment, and instructional strategies for persons with disabilities. It includes aquatic activities of all types such as instructional and competitive swimming, small craft, water aerobics, and skin and scuba diving (AAHPERD-AAALF, 1996).

Physical educators, school administrators, parents, related service personnel, and special education teachers must be educated to the benefits of aquatics and its role in a child's education. The physical and psychosocial benefits of aquatics for children with disabilities are more pronounced and significant than for individuals without disabilities. Many people whose disabilities impair mobility on land can function independently in an aquatic environment without the assistance of braces, crutches, walkers, or wheelchairs due to the buoyancy afforded by water. Although adapted aquatics does not focus on therapeutic water exercise, warm water facilitates muscle relaxation, joint range of motion, and improved muscle strength and endurance (Skinner & Thompson, 1983). Swimming strengthens muscles that enhance postural stability necessary for locomotor and object control skills. Water supports the body, enabling a person to possibly walk for the first time thus increasing strength for ambulation on land. Adapted aquatics activities also enhance breath control and cardiorespiratory fitness. Blowing bubbles, holding the breath, and inhalation/exhalation during swimming strokes improve respiratory function and oral motor control, aiding in speech development (Martin, 1983) (figure 24.1).

Figure 24.1 Breath control exercises help improve oral motor control.

Benefits are not limited to the physical realm. Water activities that are carefully planned and implemented meet individual needs by providing an environment that contributes to psychosocial and cognitive development. As an individual with a physical disability learns to move through the water without assistance, self-esteem and self-awareness improve. Moreover, the freedom of movement made possible by water not only boosts morale but also provides the incentive to maximize potential in other aspects of rehabilitation (Skinner & Thompson, 1983).

The motivational and therapeutic properties of water provide a stimulating learning environment. Some instructors have even reinforced academic learning in adapted aquatics, successfully reinforcing cognitive concepts (American Red Cross, 1977). For example, instructors have centered games and activities around math, spelling, reading, and other concepts. Participants may count laps, dive for submerged plastic letters, or read their workouts. These activities also help participants improve judgment and orientation to the surrounding environment.

GENERAL TEACHING SUGGESTIONS

Each person is unique, and individualization is the key to safe, effective, and relevant programming. Therefore, it should never be assumed that all characteristics associated with disabilities are endemic to each person with that diagnosis. Generalizations serve merely to present a wide scope of information that may pertain to swimmers with any particular disability. Each swimmer should be taught sufficient safety and swimming skills to become as safe and comfortable as possible during aquatic activities. Choice and presentation of skills should be tailored to meet the specific needs of each individual (Lepore, Gayle, & Stevens, 1998).

Prior to instruction the teacher must gather information from written, oral, and observational sources. In addition to reading previous records and interviewing the swimmer and significant others, it is imperative that an aquatic assessment be conducted to determine present level of functioning. General instructional suggestions include writing long-term goals and short-term performance objectives, task analyzing aquatic skills, determining proper lift and transfer methods, establishing communication signals, and developing holding and positioning techniques to facilitate instruction.

Teaching basic safety skills first, such as mouth closure, rolling over from front to back, changing directions, recovering from falling into the pool, vertical recovery from front and back positions, and holding onto the pool wall (figure 24.2), will help to alleviate fear of more difficult skills. A balanced body position in the water is an important prerequisite for skills. The

Figure 24.2   Prerequisite skills such as wall climbing help alleviate the fears associated with more difficult tasks.

aquatic instructor must experiment with horizontal and vertical rotation as well as appropriate placement of arms, legs, and head to teach the development of proper buoyancy, balance, and water comfort in relation to the unique physical characteristics of the student.

Some students with developmental delays and/or cerebral palsy may exhibit primitive reflexes that mimic fear responses. Aquatic instructors should be knowledgeable in the areas of human development and reflex patterns to discern between the two. This knowledge can also aid in eliciting voluntary movement patterns through proper positioning (Priest, 1995). See the section on individuals with cerebral palsy in this chapter for more practical tips on positioning and working around reflex retention.

Finally, presenting swimming cues in a concise manner, connected to something that the person already is familiar with, strengthens learning. Since swimming takes place in such a unique setting, swimmers with disabilities need cues that refer to situations or things they already know or know how to do. An example of this is the use of the phrase "move your hands like you are opening and closing curtains" to depict the movement of the hands during treading water or sculling.

AQUATIC ASSESSMENT

Individualized instructional planning begins with defining which skills a participant needs to learn and assessing the present level of performance in those skills. Prior to performing the assessment, an instructor should determine the skills that will be assessed. To help prioritize, questions such as these should be asked of the participant or caregiver: What is the participant interested in learning? What are important safety skills for the participant to acquire? Where will the participant use the skills outside of class? What are same-age peers performing in aquatics? What equipment does the family have available to them?

What are the medical, therapeutic, educational, and recreational needs of the participant? Afterward, it is important to determine what activities and concepts are common to a few of the areas; these are the priorities. Assessment items that will determine the present level of performance in these skills should be developed (Block, 1994).

Swimming instructors usually use checklists or a rubric to determine the extent of aquatic skills that an individual possesses (figure 24.3). One norm-based assessment exists (Conatser, 1995), but most instructors use either a curriculum-based or ecologically based assessment. Curriculum-based assessment items include skills that the swimmer needs to function effectively within an integrated class. The American Red Cross Progressive Swim Levels I-VII (American Red Cross, 1992), YMCA swim levels (YMCA of the USA, 1999), and the SwimAmerica (SwimAmerica, n.d.) skills programs are examples of curricula from which a swim instructor would draw skills for a curriculum-based assessment checklist. The Data Based Gymnasium Program (Dunn, Morehouse, & Fredericks, 1986) is another example of a curriculum model utilizing task-analyzed aquatic skills that can be employed for assessment purposes.

An ecologically based aquatic assessment may also be considered. Ecologically based assessments include skills needed for an individual's current and future environment. Aquatic skill assessments of this nature may include components of the curriculum-based assessment but also include individual skills not addressed in the regular curriculum, such as entering and exiting the pool area, dressing, using appropriate language in a swim group, performing stretching exercises before swimming, knowing how to swim in a circle, using a flotation device, or clearing the mouth of water. These are skills that need to be learned but would not usually appear within a regular swim curriculum. Ecologically based assessments should be used to assess all important areas of the aquatic experience and should be developed on the basis of individual needs.

ADAPTING SWIMMING SKILLS

Before adapting skills to meet an individual's needs, an instructor must first look at why the skill is needed and how and where it will be used. Some swimmers want to pass the competencies for the American Red Cross Swim Levels, some may want to improve cardiorespiratory functioning, and yet others may want to enter a swim meet. These differing purposes for performing the front crawl might cause an instructor to take a different approach to adapting strokes and other aquatic skills. The most important considerations in adapting strokes should be these:

▶ What are the physical constraints of the disability?
▶ What is the most efficient way to propel through the water, given the constraints?
▶ What movements will cause or diminish pain and/or injury?
▶ What adaptations can be made that will make the stroke or skill as much like the nonadapted version as possible?
▶ What equipment is available to facilitate the skill?

The instructor may need to

▶ adjust the swimmer's body position by adding flotation or light weights,
▶ change the propulsive action of the arms and/or legs, or
▶ adapt the breathing pattern.

Adjusting the swimmer's body position is typical for people who have disabilities such as cerebral palsy, stroke, traumatic brain injury, spina bifida, obesity, limb loss, muscular dystrophy, polio, or traumatic spinal cord injury. Due to variations from the norm, with regard to muscle mass and body fat in many individuals with physical disabilities, the center of gravity and center of buoyancy may be atypical. Finding an efficient body position and experimenting with a variety of flotation devices and weights, including scuba diving and ankle weights, inflatable arm floaties, foam swim noodles, rescue tubes, lifejackets, and ski belts, are imperative.

The swimmer's arm and leg actions may need adjustment also. Typical efficient propulsive action may not be feasible due to contractures, muscle atrophy, or missing limbs. Adaptations, such as changing the ideal 'S' curve of the front crawl arms to a modified 'C' or 'J', should be experimented with. Lower-body propulsive adaptations might include bending the knees more during flutter kicking, using the scissors kick while breathing in the front crawl, or using fins while doing the butterfly dolphin kick during any of the strokes.

Breathing patterns can be changed from one-side breathing to alternate-side breathing, front breathing, rolling over to back to breathe, or using a snorkel. Swimmers can be taught explosive breathing, breathing using mouth only, or breathing using a closed throat technique.

Nonphysical adaptations include developmentally appropriate progressions, frequent practice of skills, detailed task analysis, verbal and visual cues, repeating directions, and altering the skill objective.

ORIENTATION TO WATER

Acquisition of aquatic skills is based on the learner's readiness to receive the skill, readiness to understand the goal, opportunities to practice at a challenging but manageable level, and ability to receive feedback. Orientation to water focuses on the readiness of the learner

Skill Evaluation Chart

Student: ______________________________

Scoring Codes:
O = Cannot perform
X = Independent
V = Verbal Cues
PC = Picture Cues
G = Gesture Cues
P = Physical Cues
*More than one code may describe a step (e.g. V, PC)

Target steps or skills	Dates							
I. Entries								
Ladder entry								
Side roll in								
Jump: shallow								
Jump: deep								
Dive: kneel								
Dive: compact								
Dive: stride								
Dive: front								
II. Exits								
Ladder								
Pull-up—side								
III. Water orientation								
Washes face								
Puts chin in water								
Puts mouth in water								
Puts mouth and nose in water								
Puts face in water								
Puts whole body in water								
Blows bubbles								
Blows bubbles with face in water								
Blows bubbles lying on front with face in water								
Blows bubbles with full body underwater								
Bobs 5 times in shallow water								
Bobs 10 times in shallow water								
Bobs 5 times in deep water								
Bobs 10 times in deep water								
IV. Front propulsion								
Pushes off side with face out of water								
Pushes off side with face in water								
Pushes off side with face in water and kicks								
Arm stroke while walking								
Arm stroke, with underwater recovery—5 feet								
Arm stroke, with underwater recovery— face in 5 feet								
Arm stroke, with underwater recovery, face in — kicking 10 feet								
Arm stroke with overwater recovery—10 feet								

Figure 24.3 Aquatic activity achievement checklist. *(continued)*

From *Adapted Physical Education and Sport*, third edition by Joseph P. Winnick, 2000, Champaign, IL: Human Kinetics.

Arm stroke with kick—20 feet									
Front crawl with rhythmic breathing to front—20 feet									
Front crawl with breathing to side—20 feet or more									
V. Breast stroke									
Push off in streamlined position for beginning breaststroke									
Breaststroke with arms—on deck									
Breaststroke arms while standing in water									
Breaststroke arms over a noodle—30 feet									
Breaststroke kick correctly on deck—5 times									
Breaststroke kick over a noodle—30 feet									
Breaststroke combined arms and kick—30 feet									
VI. Back propulsion									
Back float—5 seconds									
Back glide off wall with a noodle									
Back glide—10 feet									
Back glide with kick—20 feet									
Back glide with finning or sculling—10 feet									
Back crawl arms—on deck									
Back crawl arms over a noodle									
Back crawl arms with kick—20 feet									
VII. Side propulsion									
Side stroke glide									
Side stroke legs—on deck									
Side stroke legs over a noodle—20 feet									
Side stroke arms—on deck									
Side stroke arms over noodle—20 feet									
Side stroke—30 feet									

Figure 24.3   *(continued)*.

From *Adapted Physical Education and Sport*, third edition by Joseph P. Winnick, 2000, Champaign, IL: Human Kinetics.

and other psychological and physiological factors. Physiological factors are those in which anatomical and physiological variations in an individual's body affect how and what a person learns. This includes how disability and medication affect each body system. A swimmer may not be neurologically ready to perform a skill due to brain damage, lack of central nervous system maturity, or a developmental delay. When the instructor understands the affect of a disability on learning and provides developmentally appropriate skill progressions, learning is augmented (Langendorfer & Bruya, 1995).

Psychologically, each person is unique and learns at an individual rate, depending on a number of psychological factors. Individuals with disabilities may have psychological characteristics that hinder the acquisition of aquatic skills. Some psychological factors such as anxiety and cognitive readiness are factors to examine before developing instructional strategies.

Anxiety stems from fear and inhibits mental adjustment to the aquatic environment. Although mental adjustment takes time for new or frightened swimmers, it may be even more difficult for persons with disabilities. Poor breath control due to oral muscle dysfunction, asthma, and high- or low-muscle tone limits the ability to develop rhythmic breathing and breath holding. These and other issues, such as not being able to grasp and hold the pool gutter, leave individuals with some disabilities at a high risk of having fear and anxiety control their openness to learning. Some factors that may cause a swimmer to be anxious include fear of drowning, past frightening water experiences, submerging unexpectedly and choking on water, fear reinforced by warnings (e.g., "Don't go near that water or you will drown"), capsizing in a boat, being knocked down by a wave, or feelings of insecurity caused by poor physical ability or unfamiliar surroundings (Lepore, Gayle, & Stevens, 1998).

Fear stimulates physiological responses, such as heightened muscle tone, increased involuntary muscle movements, and inability to float. Fear is a powerful emotion that may lead to poor self-respect and insecurity, preventing participation in aquatics (Moran, 1961). Helping participants get past fear and anxiety to practice aquatic skills that will make them water safe is an initial step in teaching swimming. When participants are free of fear, they are free to learn. The following tips will facilitate implementation of a fear-reduction program:

▶ Do not ridicule or exhibit impatience with fearful reactions.

▶ Use patience without pampering.

▶ Gently guide; don't force.

▶ Explain everything in a calm, sympathetic, matter-of-fact voice.

▶ Progress from step to step gradually.

▶ Use noncompetitive activities.

▶ Encourage practice of breath control at home (Moran, 1961; Hicks, 1988).

▶ Assess an individual's readiness with the Aquatic Readiness Assessment tool (Langendorfer & Bruya, 1995). Areas to assess with this tool are water entry, breath control, buoyancy and body position, arm actions, leg actions, and combined movements.

Fear is diminished when the aquatic instructor and swimmer easily communicate. In addition to communication skills, a thorough understanding of proper participant positioning, guiding, and supporting is essential. Proper methods of transferring, touching, and supporting participants in the locker room, on the pool deck, and in the pool will also develop relationships based on trust. Knowing how to use and work all the adapted equipment, wheelchairs, and flotation devices provides an atmosphere of efficiency and safety that makes everyone feel comfortable. Likewise, holding someone with a firm and balanced grip (figure 24.4), as close as safety and comfort allow, communicates care and establishes trust and rapport (Lepore, Gayle, & Stevens, 1998). In addition, an environment in which the instructor exhibits a consistent personality, provides discipline methods that are flexible but consistent, uses caring verbal assurances, and provides balanced and controlled physical handling promotes trust, security, and mental adjustment.

The aquatic instructor should use fun activities instead of drills to promote a more comfortable atmosphere. Games, music, and props will help a fearful student become more ready to accept the aquatic setting. Activities such as a flower hunt with plastic flowers inserted into the gutters at various intervals helps acclimate the fearful student in a nonthreatening way. Other activities include square and social dance and physical education activities such as cooperative musical chairs using hoops, land games, and activities such as basketball and sponge tossing to inflatable tubes. These activities build on what a person is familiar with, and the aquatic instructor can progress from there.

FACILITY AND EQUIPMENT CONSIDERATIONS

Facilities and equipment must be accessible and safe, as well as lend themselves to successful and satisfying experiences, for participants and instructors alike. Familiarity with the Americans with Disability Act (ADA) guidelines for accessibility, state and local health codes for aquatic facilities, and resources for equipment and supplies that facilitate aquatic participation help provide quality swimming experiences.

Facilities

Facility discussions should include information about the locker room, pool deck, and pool itself. The locker room can be a place of frustration for individuals with disabilities. Factors such as inadequate lighting for individuals with visual impairment and combination-only lockers that impede independence for those with arthritis do not motivate individuals to use a facility. Other factors inhibiting independence include benches cemented into the floor in front of lockers; shower area ledges or lips that limit access for participants in wheelchairs; and lack of Braille signs on lockers, entrances, and exits. Since accessibility guidelines in relation to physical education, recreation, and aquatics facilities have been published, it is easier now to know exactly what is appropriate (Grosse & Thompson, 1993).

Facility design must enable participants to make transitions between the locker area, pool deck, and water. Due to differing abilities among participants, facilities may need more than one mode for safe and dignified entrances and exits. A variety of facility designs lend themselves to safe access to the pool. Wet ramps, dry ramps, and gradual steps with handrails are examples of built-in methods of transferring into the pool. Dry ramps are constructed into the pool deck outside of

Figure 24.4 Firm, balanced holding positions alleviate fear.

the pool with a transfer wall (figure 24.5). A wet ramp connects the deck directly to the water. Gradually sloping steps are a helpful adaptation for many participants using the pool. In addition to the structure of a pool, the pool temperature must be compatible to the groups it serves. In general, the majority of children with disabilities perform better with a pool temperature between 86 degrees and 90 degrees. Air temperature should be approximately four degrees higher.

Equipment

Proper equipment and supplies are even more important for classes serving individuals with disabilities than for the general population. Six basic reasons to use adapted equipment and supplies include entrance and exit requirements, safety, support, propulsion, fitness, and motivation (Crawford, 1988; Heckathorn, 1980).

Safe entrances and exits are crucial to accessible swim instruction. In addition to the method of entry afforded by the facility design, lifts, portable ramps, stairs, and ladders are important items for transferring into and out of the water when equipment is not built into the facility. Lifts often provide primary access to pools for individuals with severe orthopedic disabilities. Such equipment includes pneumatic systems, water-powered systems, mechanical lifts, and fully automated electrical lifts. The National Center on Accessibility (NCA, 1996) recommends that pool lifts that facilitate independent usage be used. Independent usage is most facilitated when hand controls are located at the front edge of the seat, are operational with one hand, do not require tight grasping, and require five pounds or less of force to operate.

If the pool has a wet ramp, a movable floor, or zero depth entry as the means of entry and exit, and an aquatic chair with push rims must be provided. The aquatic chair should measure 17 inches above the deck, be 19 inches wide at the seat, and have footrests and armrests that can be moved out of the way (NCA, 1996).

For individuals who have good upper-body function, but cannot negotiate stairs or ladders due to lower-body involvement, the Transfer Tier can facilitate more independent pool access. Participants transfer from a wheelchair onto the upper step and then lower themselves into the pool step by step; they reverse the process to exit.

Safety equipment is a mandatory part of an adapted aquatic program. Safety equipment includes typical rescue equipment as well as such items as a floor covering to decrease slipping, closed-cell foam mats for use during seizures, and transfer mats to cover pool gutters.

Equipment used for support is the most prominent equipment typically seen in an adapted aquatic program. Support equipment that is useful in an adapted swim program may include personal flotation devices (PDFs), foam noodles, Wet Vests, sectional rafts, and flotation collars. A large variety of flotation devices,

Figure 24.5   A dry ramp provides access to a transfer ledge.

including PFDs, water wings, pull buoys, dumbbell floats, and sectional rafts, give an extra "hand" when working with individuals who are dependent on others to stay above the water. Flotation devices ensure safety, eliminate fear, provide support, and help participants maintain a level position in the water (Heckathorn, 1980). Because flotation devices help to support, stabilize, and facilitate movement, they open a new world to individuals with mobility impairments, allowing freedom of movement not possible on land.

Although flotation devices are useful, they may pose concerns. For example, they may impair independence if swimmers rely on them too long after they should have progressed to independent, unaided swimming; and they may not provide the balance and support that is claimed because they are typically tested on nondisabled individuals (Dunn, 1981). Therefore, if students use flotation devices for support, proper supervision must be provided even if the PFDs are Coast Guard approved. In general, when using flotation devices, most of the flotation must center over the lungs and upper chest not around the stomach or solely across the back (Shurte, 1981).

Propulsion equipment affords a person with a disability the ability to move in ways he may not be able to on land. Forward movement in the water is affected by a swimmer's physical ability, body shape, and efficiency of swim stroke (Anderson, 1992). The first step to efficient propulsion is to devise flotation or other support to put the body in the most streamlined and balanced position possible. If the participant is still having difficulty with propulsion, try other devices. Hand paddles and fins increase surface area and press against the water for propulsive efficiency. Those who have part or all of their arms missing may be able to use a swimming hand prosthesis or Plexiglas paddles attached to the residual stump. For specific ideas about using these devices, see Paciorek and Jones (1994) and Summerford (1993).

An increased interest in water fitness has resulted in a greater diversity of fitness supplies. Underwater treadmills, aquacycles, water workout stations, and aqua-exercise steps provide cardiovascular conditioning, muscle toning, and strength training. Water fitness participants also use supportive and resistive equipment and supplies in the water that are handheld, pushed, or pulled including finger and hand paddles, balance bar floats, upright flotation vests and wraps, aquashoes, webbed gloves, waterproof ankle and wrist weights, workout fins, buoyancy cuffs, water-ski belts, aquacollars, and water jogging belts.

Motivational equipment provides age-appropriate experiences for children. The developmental levels, interests, and attention spans of children require a different approach to aquatic instruction and recreation. Attractive, brightly colored equipment; nontoxic and sturdy supplies; toys; flotation devices; and balls will help enhance instructional strategies that focus on fun. Other devices include swim belts, bubbles, and squares, many of which come with modules to increase or decrease flotation. Water logs, also known as water noodles or woggles, are hefty, flexible buoyant logs that encourage water exploration and kicking in a fun way.

MEETING UNIQUE NEEDS OF PARTICIPANTS

To meet the needs of a variety of individuals in providing safe, effective, and relevant aquatic opportunities, it is necessary to know certain unique attributes or characteristics of learners. However, it is important not to assume that these attributes or characteristics apply to every individual in an identified category. Suggestions not specific to the aquatic setting and additional basic teaching tips for each disability are included in other chapters of this book.

Individuals With Cerebral Palsy

Individuals with cerebral palsy (CP) exhibit a wide variety of skills due to type and severity of CP and the different body parts affected. Due to this diversity, the following tips in box 24.1 are suggested for teaching swimming to individuals with CP.

▶ **Box 24.1** **Suggestions for Teaching Swimming to Individuals With Cerebral Palsy**

Make stroke adaptations based on a limited range of motion. Try having the individual use an underwater versus out-of-water recovery of the arms, especially for the front crawl.

Maintain water temperature between 86 to 90 degrees Fahrenheit and air temperature 4 degrees higher than the water temperature.

While the participant is in the supine position, guard against temporary submersion of the face because a weak cough will not clear water from the throat.

Consider hand paddles for participants with wrist flexion contractures.

Develop strokes executed in the back lying position thus eliminating the need for head control with rhythmic breathing.

Have the participant wear a ski belt or rescue tube across the chest and under the armpits (with closing clip on back) to elevate chest and face area.

For individuals with primitive reflex retention, consider allowing participants to wear a flotation collar to hold their heads above water.

For individuals with primitive reflex retention, the following suggestions apply:

▶ Keep in mind that sudden noises, movements, or splashing may cause sudden reflex activity, possibly causing the participant to lose a safe position. Maintain a position at or near the participant's head to prevent sudden submersion.

▶ Neck hyperextension or turning of the head to the side may affect arm and leg control in persons with reflex retention. Encourage a full body roll for breathing or the use of a snorkel.

▶ Avoid quick movements and sudden hands-on and hands-off movements. Slow movements and a steady touch are best with persons who have high muscle tone.

▶ Be aware of sudden spastic movements during transfers in and out of the pool. Have adequate personnel during transfers and use a mat under the transfer area.

▶ Encourage participants to flex their heads slightly while on their backs. When their heads are in extension and they are laying on their backs, the mouths tend to open and their arms tend to extend.

▶ Keep the participant stable, as unstable positions in the water or a feeling of falling causes the body to stiffen, the arms and legs to involuntarily extend and flex, and the mouth to open.

▶Use positions that inhibit reflexes, such as a neutral or slightly tucked chin position and the head in midline of the shoulders. Hips and knees should be slightly flexed.

▶ Use symmetrical activities as much as possible (both sides of the body doing the same thing at the same time) such as breaststroke, elementary or inverted backstroke, finning, or sculling.

▶ Use caution with the scissors kick and the flutter kick as these tend to promote the crossed extension reflex, causing scissoring of the legs. If scissoring occurs, place a comfortable piece of cushioning between the knees during swimming.

Individuals With Intellectual Disabilities

Problems with memory and comprehension are usually the primary disability for many people with mental retardation, traumatic brain injury, severe learning disabilities, stroke, autism, prenatal exposure to drugs, and pervasive developmental disability. The suggestions in box 24.2 focus on these two issues.

> ### Box 24.2 Suggestions for Teaching Swimming to Individuals With Intellectual Disabilities

> ▶ Tactile cues are often the best (i.e., tapping a person on the shoulder to cue breathing at the right time).

> ▶ Use a Plexiglas clipboard and a grease pencil to list tasks that the participant must accomplish. Place the clipboard list by the pool edge, and, as each task is completed, encourage the participant to check it off.

> ▶ Use numbered lap counters to keep track of laps.

> ▶ Use a kitchen timer to know when to leave or how much time to spend on a task.

> ▶ Use basic orientation questions at each session such as "Where is the best place to enter?"

> ▶ Have the participant practice the skills in as many situations as possible.

> ▶ Inform participants daily of class expectations.

> ▶ Emphasize and repeat safety directions.

> ▶ Never minimize any safety issues.

Individuals Who Are Deaf and Hard of Hearing

As members of a swim class, deaf participants need aquatic instructors who either sign or make provisions for an interpreter. Although interpreters are important for communication, the instructor should learn basic signing to develop a personal relationship with a student. In general, deaf participants are integrated into regular programs, but, if deafness is secondary to a physical or mental disability, other placements may be warranted. The tips and suggestions in box 24.3 are useful in the aquatic setting for participants who are deaf or hard of hearing.

> ### Box 24.3 Suggestions for Teaching Swimming to Individuals Who Are Deaf

> ▶ When using an interpreter, look at the participant while speaking.

> ▶ In outdoor pools, face participants away from the sun during directions.

> ▶ Be aware of the glare of the light on the water, which reduces visibility.

> ▶ Avoid demonstrating at the same time as talking since it is hard to watch the demonstration and the interpreter at the same time.

> ▶ Provide a dry place to store hearing aids.

> ▶ When showing water safety videos, use a closed caption decoder.

> ▶ Use e-mail to distribute notes, rules, and safety information in advance.

> ▶ Since auditory emergency signals are not useful, establish visual signs to get attention.

Individuals With Orthopedic Disabilities

Although individuals with orthopedic disabilities reflect a wide variety of characteristics, there are similarities that can be considered within the aquatic environment. Balance, buoyancy, body position, and range of motion may be affected. Individuals with arthrogryposis, amputations, spina bifida, spinal cord injuries, osteogenesis imperfecta, traumatic brain injury, stroke, spinal cord injury, orthopedic disabilities, multiple sclerosis, muscular dystrophy, and myasthenia gravis might benefit from the teaching tips provided in box 24.4.

> **Box 24.4 Suggestions for Teaching Swimming to Individuals With Orthopedic Disabilities**

▶ Use water tables or tot docks for rest areas.

▶ Look for ways to streamline the body, such as changing head position or attaching flotation devices or weights to lower or raise body position. Achieve a balanced body position by experimenting within proper safety limits.

▶ Check skin for abrasions before and after swimming because of decreased sensation.

▶ Encourage use of aquashoes to decrease lesions caused by transferring and scraping feet when swimming.

▶ Be aware that muscle spasms and strange sensations may sometimes interrupt the aquatic session.

▶ Become knowledgeable about proper assistance in taking off and putting on braces and other orthotic devices.

▶ Alter stroke mechanics as necessary due to uneven muscle strength and abnormal centers of gravity and buoyancy. Change strokes as little as possible from normal efficiency. If necessary, use smaller range of motion or sculling arm movements with participants.

▶ If upper-body impairment causes difficulty in lifting the head to breathe, a participant should use a mask and snorkel or roll over onto the back to breathe. Initially, teach the back crawl or elementary backstroke.

▶ Ensure that all excretion collection bags are emptied before swimming.

▶ Allow the individual to wear a neoprene vest or wet suit to keep warm in cooler pools.

▶ Provide assistance for balance problems while on deck.

Individuals Who Are Seizure Prone

Individuals with seizure disorders need aquatic instructors who have a plan of action in case of a seizure incident. Current practice suggests that steps be taken to ensure that the person having the seizure has an open airway and is protected from physical injury caused by contact with other people or objects. When in doubt, always activate the emergency medical system (EMS). The following section describes how to manage a seizure effectively.

The first-aid objectives for assisting an individual having a seizure in the pool are to keep the individual's face above the water, to maintain an open airway, and to prevent injury by providing support with a minimal amount of restraint. One position that meets these objectives is to stand low in the water behind the individual's head and place the individual in a supine position. Then support the individual under the armpits, shoulders, and head. Remember to provide only the support needed to keep the participant's face out of the water, as unnecessary restraint may cause injury to both participant and rescuer. Remove the person from the water when it is safe to do so. The natural qualities of the water provide buoyancy and support during a seizure if the individual is kept away from the pool edge, equipment, and other persons. However, it is important not to allow the participant to remain in the pool if the seizure lasts for more than several minutes, continues in rapid succession, if injury or hypothermia are imminent, or if the person needs CPR.

If the participant requires removal from the pool during a seizure, several rescuers or aides can lift the participant from the water (figure 24.6). One type of lift requires several rescuers standing on one side of the individual, rolling the individual toward their chests, and laying the individual on a mat or towels on the side of the pool. Render first aid for any injuries and contact EMS, if necessary. The participant's medical or participation form should indicate the exact protocols for care in the event of a seizure. Refer to the tips in box 24.5 for working with a person who is seizure prone.

Individuals Who Have Visual Impairment and Blindness

Individuals with visual impairment and blindness do not have specific precautions relative to the aquatic environment that are significantly different from any other physical activity instructional setting. The suggestions in box 24.6 will enhance aquatic instruction of individuals with visual impairments or blindness.

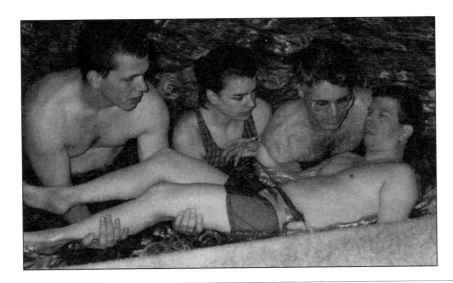

Figure 24.6   Three individuals are needed to transfer a person having a seizure out of a pool.

> **Box 24.5 Suggestions for Teaching Swimming to Individuals Who Are Seizure Prone**

- Obtain medical clearance and a list of any contraindicated activities.
- Fill out an appropriate incident report following a seizure.
- Maintain supervision during aquatic activities.
- Factors provoking seizure onset include playing games of holding breath for "as long as you can" as well as hyperventilation before underwater swimming; excessive drinking of pool water, which can lead to hyperhydration or hyponutremia; hyperthermia; and excessive looking into the sun.
- Discuss scuba diving with participants and their physicians before attempting deep dives.
- Be aware that some seizure medications increase photosensitivity. When outdoors it may be important to swim in the early evening. Use sunscreen or wear T-shirts.

> **Box 24.6 Suggestions for Teaching Swimming to Individuals Who Are Blind or Visually Impaired**

- Instructors wear a brightly colored Lycra shirt or tights to draw attention to their leg or arm movements while demonstrating to those who have residual vision.
- Use of lane lines for swimming laps and practicing strokes is helpful.
- Use auditory signals such as a radio (if it can be heard over the noise of a typical pool) to identify the deep end of the pool.
- Place a water sprinkler attached to a hose near the end of the pool to signal that the swimmer is nearing the end of the pool.
- Use a tennis ball impaled on a long folding cane to tap the heads or shoulders of swimmers to convey they are coming to the end of the pool.
- Provide opportunities to experience the environment in and around the pool with a sighted friend to help orientation and mobility.

SWIMMING AS A COMPETITIVE SPORT

Many opportunities await an individual with a disability to participate in sport, but several conditions need to be realized. First, individuals with disabilities need to know about the opportunities. Second, they need the skill prerequisites to participate. Third, they may need adapted equipment.

Competitive swim opportunities can be found by contacting various swim teams that compete under the guidelines of USA Swimming and by contacting disabled sports organizations (DSOs). DSOs that have swimming as competitive events include the USA Deaf Sports Federation; Disabled Sports, USA; Dwarf Athletic Association of America; Special Olympics International; United States Association for Blind Athletes; United States Cerebral Palsy Athletic Association; and Wheelchair Sports, USA. Competitive swim events also exist in local integrated swimming clubs where adapted rules are provided (Anderson, 1992). Those interested can train individually and compete in integrated events or segregated DSO meets. Competitive diving is offered by the Special Olympics and by the USA Deaf Sports Federation.

Aquatic instructors need to teach prerequisite skills to swimmers who want to participate in competitive swim events. These skills require more precision and independence than needed in noncompetitive experiences. Starts, turns, and stroke mechanics must be taught, and athletes must be prepared for sprint or long-distance events. Many individuals with disabilities train in a one-on-one situation with a coach or swim instructor. Some train on integrated U.S.A. sanctioned swimming teams, high school teams, or college teams, while others prefer to participate in segregated disability sports teams. There are pros and cons to each situation: Competitive swim training in which individuals with disabilities are integrated with individuals without disabilities often lack proper adapted equipment and qualified staff, while segregated disability sports opportunities are not often found outside of major cities. Developmental programs may be offered at special summer camps or sporadically at rehabilitation centers or special schools.

OTHER AQUATIC ACTIVITIES

Individuals with disabilities of all ages enjoy water sports as much as their nondisabled counterparts. Water sports provide outlets for persons with disabilities to participate in aquatic recreational opportunities with their peers, families, and community members. With legal mandates for accessibility, more chances exist for participation in instructional, recreational, and competitive water sports. Activities such as water skiing, scuba diving, snorkeling, and boating can serve as avenues for increasing independence and normalization.

Water Skiing

Prerequisites to water skiing include consultation with a swimmer's physician, acquisition of basic swim skills, and knowledge of using a PFD. All skiers should practice using a PFD for support and buoyancy in a controlled environment before using a PFD in open water. The driver of the boat, the observer, and the skier should agree beforehand on communication techniques, whether they be hand or head movement signals, to make the activity safe for all.

To make skiing easier for the beginner and for those with disabilities, equipment modifications must be made, especially for those with lower-extremity involvement. A ski bra is one piece of equipment that keeps the skis together for those with leg weakness or paraly-

sis. A knee board, ski biscuit (inflatable inner tube with a floor cover), or specially designed sit-ski can accommodate the skier who cannot stand up. Two popular sit-skis are the KAN SKI and Wake Jammer. These feature high seatbacks, an aluminum seat tube or "cage," and quick-release tow rope attachments and foot bindings on a wide- or regular-width ski. Outriggers are also available for the novice or skier with severe balance impairment. Information about adapted water skiing equipment is provided by Paciorek and Jones (1994).

Scuba Diving and Snorkeling

Traditionally, scuba diving was not a sport open to individuals with disabilities, but during the last decade, scuba and snorkeling have become part of a nucleus of adventure-based activities offered to individuals with numerous disabilities. Before beginning training, the instructor and diver need to discuss specific water access and entry techniques from the pool, beach, or boat (Petrofsky, 1995; Robinson & Fox, 1987). Once in the water, however, no architectural barriers prevent interaction with nature, and mobility is enhanced by a minimal amount of gravity. A wide variety of traditional scuba training programs are available that provide diver certification, such as National Association for Underwater Instructors (NAUI), Professional Association of Dive Instructors (PADI), and the Young Men's Christian Association (YMCA). Also available to meet the more specific needs of individuals with disabilities who may find the general scuba courses not adapted enough is the Handicapped Scuba Association (HSA), founded in 1981. Unlike the more traditional scuba certification programs, HSA uses a multilevel credential that classifies divers according to physical performance standards regardless of type of disability. Level A consists of diving students who can care for themselves and others, level B are students who need partial support, and level C are students who need full support.

While all agree that certified divers should possess requisite knowledge and skills for a safe and successful experience, controversy surrounds the diving community in regard to medical clearance and certification. Scuba diving has been generally accepted for most individuals with orthopedic, sight, and hearing disabilities. However, secondary disabilities such as limited breathing capacity; osteoporosis; poor circulation; temperature regulation disorders; psychological conditions; and medical conditions such as seizure disorders, insulin-dependent diabetes, and asthma present a real concern for physicians and dive instructors (Lin, 1987; Paciorek & Jones, 1994; Petrofsky, 1995). Presently, the only sound advice is for the prospective diver with a disability to consult with a physician experienced in hyperbaric medicine, to use caution when diving, and to be conservative.

Some modifications to equipment might include pressure gauges that have Braille numbers or that emit auditory signals, divers tethered together, hand paddles or swim mitts, diving boots, low-volume masks, octopus regulators, jacket-type buoyancy compensators, flexible vented fins, Velcro on wet suits, and diver propulsion vehicles for those who cannot propel themselves (Paciorek & Jones, 1994; Jankowski, 1995).

Boating

Boating activities are enjoyed by all but are especially good for the individual with lower-body involvement because paddling, rowing, and sailing emphasize upper-body strength, allowing participation with nondisabled peers and family members. The national governing body for canoeing and kayaking is the American Canoe Association (ACA), which sponsors the Disabled Paddler's Committee and dedicates itself to promoting canoeing and kayaking as lifetime recreational activities for individuals with physical disabilities.

In addition to canoeing and kayaking, sailing opportunities have expanded rapidly during the last decade through new programs and adapted boats for individuals with disabilities. The United States Sailing Association promotes sailing at all levels. One of the first adapted sailing programs in the United States, the Lake Merritt Adapted Boating Program of the Office of Parks and Recreation in Oakland, California, began in 1981. Glo Webel, boating programs coordinator, pioneered the development of sailing facilities for individuals with disabilities and has coauthored a text entitled, *Open Boating: A Handbook* (Webel & Goldberg, 1982).

Another pioneer and innovator in sailing is Harry Horgan, founder of Shake-A-Leg of Newport, Rhode Island. His boat design, with its adapted seating, proved to be successful, and participants consider it to be the benchmark of modified sailing vessels. In addition, the National Ocean Access Project, now associated with Disabled Sports, USA, has continued to improve accessibility of sailing vessels by modifying traditional designs, such as with the Kaufman "Drop-in Seat" and by otherwise customizing boats for sailors with disabilities. Competitive sailing opportunities continue to grow nationally and were showcased at the 1995 International Special Olympics Games and at the 1996 Atlanta Paralympic Games. Elite sailors with disabilities strive to participate in the U.S. Disabled Sailing Team (Lepore, Gayle, & Stevens, 1998).

The United States Rowing Association is the national governing body for rowing. Its Adaptive Rowing Committee has developed educational materials and is dedicated to promoting rowing among individuals with disabilities and providing the same spectrum of opportunities as are enjoyed by nondisabled rowers (Tobin, 1990).

Safety and risk management are concerns for everyone in boating, but some individuals with disabilities need to take extra precautions. Aquatic instructors can become certified through the American Canoe Association or the level I Coaching Program, available through U.S. Rowing. Webre and Zeller (1990) suggest that safety planning of any boating class should include swim skill assessment of participants, considerations for accessibility to the boating site, review of medical information, and considerations involved with any medical condition. In addition, assessing what the participant can do on land and determining what medical information needs to be shared with others in the group in relation to an emergency action plan are crucial.

The amount of responsibility a paddler or rower should have depends on functional ability. It is important to test balance, stability, and buoyancy of the boat with paddlers or rowers and to test equipment before undertaking a river or lake trip. Other elements of safety include planning for embarkation and disembarkation, instructor-to-student ratio, and—as with all water sports—an emergency action plan. To determine which boat, method, and paddle are most appropriate, consider the participant's balance, grip strength and endurance, coordination, and upper-extremity range of motion. Consider, too, how much sight and hearing the person possesses, the ability to make decisions, and knowledge of cause and effect.

Water orientation should include instruction in safety, personal rescue, and using a PFD. After the water orientation, boat orientation may begin on land, move into a pool and then calm outdoor water, and finally progress to moving to open water. Boat orientation should take into account terminology that is understandable to the participant, exploration of the boat by blind participants, entry and exit procedures, and propulsion and steering techniques. It is at this time that participants and instructors must work together to modify equipment through trial and error, based on knowledge of available commercial equipment.

Entry and exit procedures can be modified in several ways. For example, a modification may be as simple as the instructor standing in the water stabilizing a boat or two assistants helping to lift a boater onto a transfer mat from the dock to a boat. If the river or lake bed is firm enough, it may be possible to push a water wheelchair into shallow water for water entries, having assistants help lift and transfer.

Further modifications of equipment to enhance propulsion techniques could be printing the words "right" and "left" on the opposite paddle blades on a double-blade paddle or on the inside of the boat to help a paddler with an intellectual impairment, painting the inside of the boat with nonslip paint, using suction cup bath mats on the bottom or seats of the boat, having various paddle lengths available, and having the participant use rubber or leather palm gloves for a better grip.

Commercial equipment for seating and gripping is available, such as a custom-made seat, ensolite on the seat to protect those with skin problems, mitts for grasping, and sling-back seats to help with sitting balance. There are several models of canoes and kayaks to choose from, and the instructor should analyze the participant's ability and intended use prior to suggesting a boat. A sea kayak is the most stable of kayaks, and a rowboat is the most stable small craft; therefore, they are both good for beginners. Specially designed boats such as the Row Cat II, which is an "untippable" catamaran that is easy to board and propel, and the Modified Alden Ocean Shell provide extra stability for the paddler or rower with a disability. The Modified Alden Shell is made by Special Sports Corporation of Knoxille, Tennessee, and Row Cat II is made by Martin Marine Company of Kittery Point, Maine (Lepore, Gayle, & Stevens, 1998).

INCLUSION IN AQUATIC ACTIVITIES

Including an individual with a disability into an aquatic activity with nondisabled peers requires the teacher to review the results of the individual skill assessment and to look at the goals of the program, class, or activity in which the student will be placed. There is no doubt that the aquatic setting is unlike the gymnasium and sport field settings. Even if the student is included in a regular physical education environment on land, several questions must be answered by the individualized education planning team before the student begins an aquatic program within an inclusive setting. Typical questions may be: How many of the participant's targeted goals and objectives match what will take place in the general aquatic program? Can the participant follow rules and guidelines within the regular aquatic program so as not to compromise the safety of all? Is there an age-appropriate class available? Does the placement provide an emotionally and physically safe environment? Is the ultimate goal of the placement to be able to participate in aquatic activities in an integrated setting? Does the placement meet other goals in addition to instructional goals, for example, recreational or therapeutic goals? Refer to the Aquatics application example for a practical situation describing the inclusion of a student with disabilities.

Participants should have a minimal level of basic skill competencies and possess several prerequisite skills for successful experiences in an integrated class. Participant prerequisites might include such factors as fundamental social, cognitive, and aquatic readiness skills vital to inclusive group integrity and learning. Medical and health conditions are also issues. Lepore, Gayle, and Stevens (1998) contend that some medical and health conditions, such as the following, may warrant a more segregated setting.

▶ Open sores such as decubitus ulcers
▶ Uncontrolled seizures leading to emergency removal from the pool and causing clearing of the pool for each seizure incident

Application Example: Aquatics

Setting: A child with multiple physical disabilities has been included in the regular sixth grade aquatics class during her physical education period. The class is learning how to dive, but the child does not have the prerequisite skills to participate.

Student: This 12-year-old girl has no cognitive disabilities but has spastic cerebral palsy and uses an electric wheelchair. She has head control and can close her mouth in response to splashing water. She can also hold onto the pool gutter and use her arms to do a modified elementary backstroke.

Application: The adapted aquatic instructor suggests the following modifications:

▶ An aide and/or additional aquatic instructor who are trained in adapted aquatics must be made available.

▶ While the rest of the class is practicing kneeling or standing dives, the student will practice sitting on the pool edge with maximal support while in a lifejacket. With an aide in the water, the assistant on the pool deck will help the student fall into the pool and recover on her back.

▶ During diving practice, she can work on surface dives in the deep end with an aide to assist her to plunge under the water and then to help her recover onto her back.

▶ Have the student work on her IEP aquatic goals in the shallow end.

▶ Encourage her to work on various diving tasks but not the dive itself, such as streamlined body position and pike or tuck position.

▶ Tracheotomy tubes or ventilator dependency that may require shallow water, qualified health care professionals, heavily grounded electrical cords, and calm water with no splashing.

▶ Neuromuscular conditions that require a specific water temperature, which may not be available in the general aquatic facility

▶ Neurological conditions that require gradual change from water to air temperature due to inadequate thermoregulation systems

▶ High susceptibility to infection, needing more sterile environments

▶ Allergies to chlorine, requiring pools with alternative chemicals

▶ Behavior disorders, such as uncontrolled aggression, compromising the safety of others

▶ Hemophilia, possibly requiring calm water, limited bumping into other participants and equipment, and modified pool temperatures due to arthritic conditions

▶ Detached retinas requiring the need to avoid projectiles such as balls and any other bumping of the head and face

Aquatic skill prerequisites are necessary for success in aquatic classes. Even if the tasks are as simple as holding the pool gutter, closing the mouth when someone splashes, or not drinking pool water, these skills might be necessary for success and safety in the regular class. Support services often need to be provided to assist with skill prerequisites.

Unlike land-based physical activities, sometimes an individual with a disability cannot participate in aquatic groups with same-age nondisabled peers due to lack of ability. For example, if the entire instructional unit is taking place in the diving well and the individual is overly fearful, the caregiver and individual must communicate with the instructor about what is needed, desired, and feasible.

Ways that inclusion can be enhanced include the following:

▶ Providing an alternative activity to one that might be inappropriate

▶ Using an instructor aide to repeat directions or provide physical support

▶ Providing a temporary segregated program

▶ Working with an adapted aquatic instructor in another area of the pool

▶ Using peers who are trained as water safety aides or adapted aquatic aides who can be helpful in providing specific assistance

SUMMARY

Aquatic instruction is an important part of a physical education program for individuals with disabilities. The benefits go beyond the physical and reach into the psychological and social realms. Physical educators should be familiar with the many possibilities afforded by water to advocate for aquatic instruction within the physical education program for individuals with disabilities.

Although the most important aspect of an adapted aquatic program is the instructor's knowledge, skill, and attitude, a critical role is played by facilities and equipment for the comfort, safety, support, and, ultimately, the achievements of participants. Fortunately, physical barriers to participation in aquatic programs of individuals with disabilities are disappearing with advances in technology to better assist and support individual needs. As these barriers disappear and adapted facilities, equipment, and supplies become more widely available, individuals with disabilities are afforded more opportunities to participate with their able-bodied peers.

Chapter 25

Team Sports

David L. Porretta

▶ JOE, A 15-YEAR-OLD, has just moved into a new school district and is to receive regular physical education with his chronological age peers. Joe is functioning academically at grade level, is socially well adjusted, and has a mobility impairment that requires him to use a wheelchair. Joe's physical education teacher Mr. Bailey was consulted as to whether or not Joe could safely and

▶ effectively participate in regular physical education. Mr. Bailey believes that students with disabilities should be included with nondisabled peers whenever possible. Knowing the way in which the secondary curriculum was now structured at the high school, Mr. Bailey knew that some accommodations would need to be made, especially since many of the units were team sports. While

▶ unsure as to how these accommodations would be made, Mr. Bailey would do his best, since he was committed to teaching *all* students.

In this chapter, variations and modifications of selected team sports are presented. These variations and modifications are designed to help physical education teachers and coaches provide the best possible programs for students with disabilities.

GETTING INVOLVED

Team sports are a popular way for individuals with disabilities to become involved in physical activity. In elite or inclusive settings, individuals with disabilities have excelled and continue to excel in amateur as well as professional team sports, and many interscholastic and recreational sport programs encourage participation of persons with disabilities on an integrated basis.

Fully integrated sport is especially encouraged for people with auditory impairments. In fact, as early as the late 19th century people with auditory impairments were excelling in sport with nonimpaired persons. For example, William "Dummy" Hoy was a major league baseball player from 1888 to 1902 and was noted as the first person with deafness to become a superstar in the game, as well as the person to invent hand signals used by umpires for calling balls and strikes. Kenny Walker (professional football) and Curtis Pride (professional baseball), both deaf, are examples of persons who have excelled in sport. Both get signals from managers/coaches and team players while on the field. In other sports like floor hockey, basketball, and volleyball, which are played in a relatively small area, very few modifications may need to be made. In volleyball, the beginning or ending of play may be signaled by an official pulling the net. On the other hand, sports played out-of-doors on a large field may require somewhat more modification. For instance, in football and soccer, flags and hand gestures can supplement whistles as signals. For players with deafness, a bass drum on the sideline may signal the snap of the ball instead of the quarterback's verbal cadence. There are teams composed entirely of players with deafness who compete against nondisabled individuals.

A number of organizations now provide sport programs for athletes with disabilities. The American Athletic Association for the Deaf (AAAD), one such organization, offers competition solely for those with hearing impairments. Team events include volleyball, soccer, and basketball, in which athletes are classified according to gender and degree of hearing loss. These sports are regularly featured at the World Games for the Deaf and follow international sports federation rules with some minor adjustments. Because few modifications are needed for people with hearing impairments to effectively participate in team sports, the focus in the remainder of this chapter will be on people with other types of disabilities.

Other sport organizations, such as the National Beep Baseball Association (NBBA), Special Olympics, and the United States Cerebral Palsy Athletic Association (USCPAA), have been formed to meet the needs of individuals with disabilities for segregated sport competition. However, Special Olympics does promote team sports competition by persons with mental retardation in totally inclusive settings. This program is known as Unified Sports. Beep baseball, goal ball, quad rugby, and wheelchair softball are relatively new team sports that have been designed for players with disabilities.

Only the significant modifications of each sport are presented in this chapter. Information regarding a more detailed description of the rules and regulations for each sport can be obtained from each sponsoring sport organization.

BASKETBALL

Basketball is a popular activity in both physical education and sport programs. It incorporates the skills of running, jumping, shooting, passing, and dribbling. Varying or modifying skills, rules, and/or equipment can allow individuals with disabilities to effectively participate in the game. Generally, most ambulatory persons can participate in basketball with few or no modifications. However, those with severe mental disabilities or mobility problems may need greater modifications, for example, wheelchair basketball for people in wheelchairs.

Game Skills

Important basketball game skills include shooting, passing, and dribbling. Selected modifications are provided for each skill.

Shooting and Passing

Bounce passing is advised for partially sighted players because the sound of the bounce lets them know from what direction the ball is coming. Bounce passing also provides more time for players with unilateral upper-limb involvement to catch the ball. One-hand shots and passes should be encouraged for players who have upper-limb impairments. Players in wheelchairs find the one-hand pass useful for long passes; when shooting at the basket, however, they often prefer the two-hand set shot (especially for longer shots) because both arms can put more force behind the ball. For people with ambulation difficulties, a net placed directly beneath the basket during shooting practice facilitates return of the ball. Players with upper-limb involvement may find it helpful to trap or cradle the ball against the upper body when trying to catch a pass.

Dribbling

For players with poor eye-hand coordination or poor vision, dribbling can be performed with a larger ball. For those having poor body coordination, it may be necessary to permit periodic bouncing while running or walking, although they can dribble the ball continually when standing still. Players in wheelchairs will need to dribble to the left or the right of the chair and carry the ball in their laps when wheeling.

Lead-Up Games and Activities

Lead-up games and activities are important prerequisites in learning to play basketball. Selected lead-up games and activities are presented here.

Horse

Two or more players may play this shooting game, competing against each other from varying distances from the basket. To begin the game, a player takes a shot from anywhere on the court. If the shot is made, the next player must duplicate the shot (type of shot, distance, etc.). Failure to make the shot earns that player the letter "H." If, however, the second player makes the shot, an additional shot may be attempted from anywhere on the court for the opponent to match. Players attempt shots that they feel the opponent may have difficulty making. The first person to acquire all of the letters H-O-R-S-E loses.

Circle Shot

This activity involves shooting a playground ball in any manner to a large basket approximately 1.1 meters (45 inches) high from six different spots on the floor, ranging from approximately .64 meters (2 feet) to 1.5 meters (5 feet) away surrounding the basket. Two shots are attempted from each spot, for a total of 12 shots. The player's score is the number of shots made.

Other Activities

Other basketball lead-up activities may include bouncing a beach ball over a specified distance and shooting or dropping a playground ball into a large barrel or container.

Sport Variations and Modifications

Basketball is an official sport of the Dwarf Athletic Association of America (DAAA). The only modification to the game is that players use a slightly smaller ball (the size used in international play by women) for better dribbling and shooting control.

In Special Olympics competition, the game follows rules developed by the Federation Internationale de Basketball Amatuer (FIBA) for all multinational and international competition (Special Olympics, 1996). The significant modifications are as follows:

▶ Fouls are called only in rough contact.

▶ A smaller basketball, 72.4 centimeters (28.5 inches) in circumference and .51 kilograms (18 ounces) to .57 kilograms (20 ounces) in weight, may be used for women's and junior division play.

▶ A shorter basket, 2.4 meters (8 feet), may be used for junior division play.

▶ Players may take two steps beyond what is allowable after dribbling.

The National Wheelchair Basketball Association (NWBA) (1998) has also modified the game for wheelchair users (figure 25.1). Following are some of the major rule modifications:

▶ The wheelchair is considered part of the player.

▶ Players must stay firmly seated in the chair at all times.

▶ An offensive player shall not remain for four seconds in the key.

▶ Dribbling consists of simultaneously wheeling the chair and dribbling the ball (however, the player may not take more than two consecutive pushes without bouncing the ball).

▶ Three or more pushes results in a traveling violation.

▶ No player of the team with a throw-in the front court shall enter the free-throw lane until the throw-in starts.

▶ Personal fouls are charged to players who block, push, charge, or impede the progress with either their bodies or wheelchairs.

Photo courtesy of Rick Swauger. Printed by permission.

Figure 25.1   Wheelchair basketball.

Skill Event Variations and Modifications

Special Olympics offers individual skills competition in shooting, dribbling, and passing for individuals with lower ability levels, not for athletes who can already play the game. Scores for all three events are added together to obtain a final score. The shooting competition is called spot shot and measures the athlete's skill in shooting a basketball. Six spots are marked on the basketball floor, three spots to the left of the basket and three spots to the right of the basket. The athlete attempts two shots from each of the six spots. The first six shots are taken from the right of the basket, and the second six shots are taken from the left of the basket. Points are awarded for every field goal made. The farther the spot is from the basket, the higher the point value. For any shot that hits the backboard and/or rim and does not go into the basket, one point is scored. The athlete's score is the sum of all 12 shots.

The 10-meter event requires the athlete to dribble with one hand as fast as possible for a distance of 10 meters (32 feet 9 inches). If control of the ball is lost, the athlete can recover the ball. If, however, the ball goes outside of the designated 1.5 meter (4 foot 9 inch) lane, the ball may be retrieved or a back-up ball placed outside of the lane 5 meters (16 feet 4 inches) from the start of the event may be picked up. Points are awarded depending on how long it takes to dribble the entire 10 meters (32 feet 9 inches). A one-second penalty is

added for each illegal (e.g., two-handed) dribble. Two trials for this event are allowed. The athlete's score is the best of the two trials.

In the target pass event, the athlete must stand within a 3-meter (9-foot 8-inch) square and pass the ball in the air to a 1-meter (3-foot 3-inch) square target 1-meter (3-foot 3-inch) from the floor from a distance of 2.4 meters (7 foot 8 inch). Five attempts are allowed. The athlete receives three points for hitting the inside of the target; two points for hitting the lines of the target; one point for hitting the wall but no part of the target; and one point for catching the ball on the return from the wall. The final score is the sum of all five passes.

Other Variations and Modifications

Game rules may be simplified by reducing the types of fouls players are allowed to commit. A playground ball may be used, and the basket can be lowered and/or enlarged. The game area may be restricted to half court for players with mobility impairments, such as those using lower-limb prostheses. Shorter play periods and frequent substitutions may be incorporated into the game for players with cardiac or asthmatic conditions. Lighter-weight balls can be used by those with insufficient arm strength.

FLOOR HOCKEY

The game of floor hockey is gaining popularity in physical education and sport programs across the country. Game skills include stickhandling, shooting, passing, checking, and goalkeeping. With appropriate variations and modifications, the game can be played by most individuals with disabilities.

Game Skills

Important floor hockey game skills include stickhandling, shooting, passing, checking, and goalkeeping. Selected modifications are provided for each skill.

Stickhandling

The key to successful stickhandling is being able to keep the head up. Keeping the head up allows the player to attend to the field of play and opponents rather than the puck. To facilitate better stickhandling, the stick blade can be enlarged for players with motor-control problems or visual impairments. The size and length of the stick are important factors for some individuals. Lighter sticks should be available for smaller players and those with muscle weaknesses, and shorter sticks should be made available for those in wheelchairs. Players with crutches may use the crutch as a stick to strike the puck as long as sufficient balance can be maintained. Those with crutches or leg braces can hit the puck more successfully from the stationary position. Players with unilateral upper-limb deficiencies or amputations can control the stick with the nonimpaired limb, because sticks are light in weight. However, for those with poor grip or poor arm or shoulder strength, the stick can be secured to the limb with a Velcro strap. Players in wheelchairs may need to stress passing or shooting rather than dribbling past opponents. When moving the chair, they usually place the stick in their laps.

Shooting and Passing

Shooting and passing in floor hockey requires quickness, accuracy, and the ability to shoot the puck when moving. Making a pass is a difficult skill and requires good timing and eye-hand coordination. In the initial stages of learning this skill, the player should make passes from a stationary position to a player who is also in a stationary position. More advanced stages of passing may include passing the puck from a stationary position to a moving player. When passing or shooting on goal, players with visual impairments can push the puck with the stick rather than using a backswing before striking the puck. They will find playing with a larger, brightly colored puck very helpful. It is also helpful if a coach or a sighted teammate calls to them when the puck is passed in their direction.

Checking

Checking requires the player to gain control of the opponent's puck. Here, the player positions the stick under the opponent's stick and attempts to lift the opponent's stick away from the puck. Once the opponent's stick is raised, the puck can be controlled. Players should first practice checking in a stationary position in a slow manner and then gradually increase in the speed of movement. The highest level of checking is when both players are moving.

Goalkeeping

Goalies try to keep the puck from going into the goal and, therefore, need to make quick movements. They may stop the puck a number of different ways—using either their sticks or their feet, blocking it with their bodies, or catching it. Once the puck is controlled it may be put back into play by using the stick or by throwing it into the playing area. Goalies need to wear a face mask, pads, and glove and use a larger stick. Players with asthma or poor cardiorespiratory endurance levels can be successful at the goalie position because little running is needed.

Lead-Up Games and Activities

Lead-up games and activities are important prerequisites to learning floor hockey. Selected lead-up games and activities are provided.

Stop the Puck

Three players face the net at a distance of 5 meters (16 feet 4 inches). One player is positioned in front of the net and the other two players position themselves 3 meters (9 feet 8 inches) to the left and right of the middle player. Each of the three players has two pucks. On command, players, in rapid succession, shoot their pucks at the goal attempting to score. The goalie makes as many saves as possible.

Puck Dribble Race

The player starts the race from behind a starting line. On command, the player dribbles the puck as quickly as possible in a forward direction for a distance of 10 meters (32 feet 8 inches).

Other Activities

Additional activities may include pushing a puck to the goal as fast as possible using a shuffleboard stick, kicking the puck with the feet as fast as possible to the goal, dribbling the puck as quickly as possible around a circle of cones, and shooting a sock stuffed in the shape of a ball (sock ball) with a poly hockey stick into a large box turned on its side.

Sport Variations and Modifications

As played under Special Olympics rules (Special Olympics, 1999), floor hockey is very similar to ice hockey. The game can be played on any safe, level, properly marked surface with minimum dimensions of 12 meters by 24 meters (40 feet by 80 feet) and maximum dimensions of 15 meters by 28 meters (49 feet by 92 feet). Additional official floor dimensions (e.g., center circle, goal line, face-off circles) can be found in the *Winter Sports Rules* (Special Olympics, 1999). Six players compose a team (one goalkeeper, two defenders, and three forwards). All players wear helmets and distinctive team markings. Players use sticks that resemble broom handles, except for the goalie, who uses a regulation ice hockey stick. The goalie must wear a mask and helmet; pads and gloves can be used. All other players must wear helmets with protective cages and shin guards; pads, gloves, and mouth guards can also be worn. The puck is a circular felt disc (approximately 20 centimeters [8 inches] in diameter) with a 10-centimeter (4-inch) hole in the center. The end of the stick is placed in the hole to control the puck. Face-offs, offsides, and minor and major violations are part of the game. Games consist of three nine-minute periods of running time, with a one-minute rest between periods. For penalties, "frozen" puck situations, time-outs, and line changes the clock is stopped.

Roller hockey, very similar to floor hockey, is played in Special Olympics summer competition. Roller hockey is played on any flat, properly marked surface. The official dimensions for the court can be found in the *Summer Sports Rules* (Special Olympics, 1996). The game uses a soft, flexible plastic ball, 24.75 centimeters (9.75 inches) in circumference. All skaters must wear helmets, and the goalie must wear a face mask. Shin guards, knee pads, gloves, and mouthpieces can be worn. The stick is constructed of durable plastic with a rounded blade, and players may use only their sticks to maneuver the ball toward the goal. The object of the game is to shoot the ball into the opponent's goal cage and secure a goal. Teams are composed of five players each (including the goalie). The game is composed of two eight-minute periods with a three-minute rest between periods.

Skill Event Variations and Modifications

Special Olympics offers individual skills competition for floor hockey in shooting (two events), passing, stickhandling, and defense. A final score is determined by adding the scores of all five events. Shooting competition has two events, "Shoot Around the Goal" and "Shooting for Accuracy." In the shooting around the goal event, the athlete takes one shot from five different locations around the goal with each location being 6 meters (19 feet 7 inches) from the goal. The athlete has a 10-second time limit to shoot all the pucks. One puck will be at each location before the athlete starts shooting. Each puck that goes into the goal is worth five points. The score is the total of the five shots. In the shooting for accuracy event, the athlete takes five shots from directly in front of the goal from a distance of 5 meters (16 feet 4 inches). The goal is divided into the following point sections: five points for a shot entering the goal in either of the upper two corners, three points for a shot entering the goal in either of the lower two corners, two points for a shot entering the goal in the upper-middle sections, one point for a shot entering the goal in the lower-middle section, and zero points for a shot not entering the goal.

The passing event requires that the athlete make 5 passes from behind a passing line located 8 meters (26 feet) from 2 cones placed 1 meter (39 inches) apart. The athlete is awarded five points each time the puck passes between the two cones, while three points are awarded whenever the puck hits a cone and also crosses the line. The total score is the sum of the scores for the five passes made.

In the stickhandling event, the athlete stickhandles the puck in a figure-eight fashion past 6 cones 3 meters (9 feet 8 inches) apart for a distance of 21 meters (68 feet 8 inches) and then shoots the puck at the goal. The time elapsed from the beginning of the event to the shot on goal is subtracted from 25. One point is also subtracted for each cone missed, and five points are added to the score if the goal is made.

In the defensive event, the athlete gets 2 attempts to steal the puck from 2 opponents who try to keep it away from the athlete within a 12-meter (39-foot 4-inch) square area. Fifteen seconds on each attempt is given. Each steal is worth 10 points. If the puck is not stolen, the athlete may score up to 20 points for pressing opponents, trying to stick check the opponent with the puck, trying to stay between opponents, and touching the puck but not gaining control.

Other Variations and Modifications

In order to accommodate players with varying abilities, the game may be played on a smaller playing surface and with larger or smaller goals. To make the game less strenuous, a Wiffle ball, a sock ball, a yarn or Nerf ball, or a large reinforced beanbag may be used. If players have impaired mobility, increasing the number of players on each team may be helpful. Body contact can be eliminated for players with bone or soft-tissue conditions like juvenile rheumatoid arthritis or osteogenesis imperfecta. Penalties may be imposed on players who deliberately bump into opponents in wheelchairs.

FOOTBALL

Football uses the skills of passing, catching, kicking, blocking, and tackling. Most individuals can participate effectively in football or some variation of it. The regulation game of tackle football can be played by people with mild impairments who are in good physical condition. A small number of players possessing partial sight can and do participate on high school and collegiate teams. In fact, persons who are legally blind can be successful placekickers. However, people with total blindness will be unable to play the regulation game. Those with unilateral amputations either below the knee or above or below the elbow will be able to participate in regulation football as long as their prostheses do not pose a safety risk. Flag or touch football is more commonly offered in physical education and recreation sport programs. Most individuals with mental impairments, except those with severe or profound impairments, can be safely integrated with other players. Individuals with significant physical impairments may find modified or less integrated participation more suitable.

Game Skills

Important football game skills include passing, catching, and kicking. Selected modifications are provided for each skill.

Passing

Players with partial sight are able to pass the ball as long as distances are short and receivers wear bright-colored clothing. For players with poor grip strength, pronounced contracture of the hand or wrist, or on crutches, a softer and smaller ball may be used to facilitate holding and gripping. People in wheelchairs will be able to effectively pass the ball if they have sufficient arm and shoulder strength. Wheelchair users will find it almost impossible to perform an underhand lateral pass while facing the line of scrimmage; therefore, this type of pass must be performed while facing the receiver.

Catching

Players with partial or no available sight as well as those in wheelchairs should be facing the passer when attempting to catch the ball. Instead of trying to catch with their hands, players should be encouraged to cradle the ball with both hands or to trap the ball in the midsection. The ball should be passed from short distances without great speed; a foam ball should be used for safety purposes. Players who have unilateral arm deformities should catch the ball by stopping it with the palm of their nonimpaired hands and trapping it against their bodies. Wheelchair users will be able to catch effectively if the ball is thrown accurately; using a chair limits one's catching range.

Kicking

Players with partial or no available sight may be encouraged to practice punting without shoes so they can feel the ball contacting their feet. In learning this punt, players should be instructed to point their toes (plantar flex the foot) while kicking. A player with unilateral arm amputation can punt the ball by having it rest in the palm of the nonimpaired hand. A punting play may begin with the player already holding the ball instead of with a snap from the center.

Lead-Up Games and Activities

Lead-up games and activities are important prerequisites to learning the game of football. Selected lead-up games and activities are provided.

Kickoff Football

The game is played on a playground, 27.4 meters (30 yards) by 54.8 meters (60 yards), by 2 teams with approximately 6 to 8 players each. The object of the game is to return the kickoff as far as possible before being touched. The football is kicked off from the center of the field. The player with the ball returns it as far up the field as possible without being touched by opposing team members. The team returning the ball may use a series of lateral passes to advance it; forward passes are not permitted. Play stops when the ball carrier is touched. The other team then kicks off from the middle of the field. The winner is the team advancing farthest up the field.

Football Throw Activity

A player attempts to pass a football from a distance of 9.1 meters (30 feet) through the hole of a large rubber tire suspended 1.2 meters (4 feet) from the ground on a rope. Ten attempts are given; the player's score is the number of successful passes out of 10.

Other Activities

Other lead-up activities may include placekicking or punting the ball for distance, centering the ball to a target for accuracy, performing relays in which players hand the ball off to each other, and guessing the number of throws or kicks it will take to cover a predetermined distance.

Sport Variations and Modifications

Wheelchair football has steadily gained in popularity since it began in 1948. For more than 15 years the Santa Barbara (California) Parks and Recreation Department/Adapted Programs has been sponsoring wheelchair football tournaments at a national level. Because of its growing popularity, the Universal Wheelchair Football Association (UWFA) was formed in 1997 to promote the game. Wheelchair football is played on any hard, flat surface. The boundaries for a standard basketball court work well. The degree of physical contact is determined by how hard players block/tackle other players. Some players, through mutual consent, like to play more aggressively. With few exceptions, the game is very similar to touch football. Team size varies, but six players per side are preferred because confusion can occur if a larger number of players on each team participate. A standard foam (Nerf, Poof) football is used, and players can participate in manual chairs, motorized chairs, or scooters. Major rule modifications ("Play Ball," 1997) are as follows:

▶ A first down is from the foul line on the basketball court to half court or any equivalent distance identified (if a parking lot is used, the distance for a first down is two parking spaces).

▶ A "kick" is simulated by throwing the ball down the field.

▶ When "punting," the offensive team must alert the defensive team before play begins.

▶ The defensive team must loudly and clearly count off three seconds before rushing the quarterback, except for the one blitz that is allowed per series of downs.

▶ Contact behind the opponent's rear axle is considered clipping.

Skill Event Variations and Modifications

Various skill events can be offered. A catching event may require a player to run a specified pattern (e.g., down and out) and catch the ball. Five attempts are given, with the total number of catches constituting the player's score.

In a field goal kicking event, players attempt to placekick a football over a rope suspended 2.4 meters (8 feet) from the ground between two poles 15.2 meters (50 feet) apart. Kicks may be attempted from 4.6 meters (5 yards), 9.1 meters (10 yards), 13.7 meters (15 yards), 18.3 meters (20 yards), or 22.9 meters (25 yards) from the rope. Ten kicks are given, and players may kick from any or all of the five distances. Points are awarded according to the distance kicked. The points are as follows: 1 point for a 4.6-meter kick, 2 points for a 9.1-meter kick, 3 points for a 13.7-meter kick, 4 points for an 18.3-meter kick, and 5 points for a 22.9-meter kick. The total number of points from 10 successful kicks is the player's score.

Other Variations and Modifications

Simplified game situations that include only the performance of specific game skills can be used. For example, the game may be played allowing only passing plays. The field can be both shortened and narrowed, and the number of players on each team can be reduced. In addition, first-down yardage can be reduced to less than 9.1 meters (10 yards). The game may also be played with a kickball or volleyball by people of low skill. Individuals with arm or leg deformities as well as those with visual impairments can play most line positions.

SOCCER

Soccer is included in many physical education and sport programs. Skills in soccer include running, dribbling, kicking, trapping, heading, and catching (goalie only) the ball. Because of its large playing area and continuous play, soccer requires stamina. However, the game can be varied or modified so that persons with disabilities can participate in any setting. For those with very minimal impairments (e.g., mild learning disability), no modifications in the game are necessary. When regular sport competition is not possible, USCPAA and Special Olympics provide competition between players with disabilities.

Game Skills

Important soccer game skills consist of kicking, trapping, heading, and goalkeeping. Selected modifications are provided for each skill.

Kicking and Trapping

Whether for dribbling, passing, or shooting, kicking is of paramount importance in the game of soccer. People with upper-limb amputations can learn to kick the ball effectively, but they may have difficulty with longer kicks because the arms are normally abducted and extended to maintain balance. Individuals with unilateral lower-limb amputations may use a prosthesis for support when passing or shooting the ball. Players may be unable to kick the ball effectively with a prosthetic device and therefore may wish to play in a wheelchair. Wheelchair users will be able to dribble the ball by using the footrests of their chairs to contact the ball and push it forward. They can also be allowed to throw the ball, because they do not have use of their lower limbs.

Most players can effectively learn to trap the ball. Foam balls are good to use with players who are hesitant to have the ball hit their bodies. Persons in wheelchairs may have some difficulty trapping, because the sitting position impedes the reception of the ball on their chests and abdomens. However, some players learn to trap the ball in their laps.

Heading

Most players can learn to head the ball successfully, although heading should not be encouraged for players with conditions such as brain injury or atlantoaxial instability. Players with mental or visual impairments may find using a balloon or beach ball helpful for learning to head, because these balls are soft and give players time to make body adjustments prior to ball contact. Players with upper-limb amputations can be very effective in heading the ball. Players in wheelchairs will be able to head the ball effectively as long as it comes directly to them. However, the distance the ball can be headed will be limited because the chair's backrest and the sitting position limit the player's ability to exert force on the ball.

Goalkeeping

Goalkeeping requires that the goalie be able to catch (or at least trap the ball with the hands) and kick the ball. It also requires that the goalie be able to react quickly to the ball. Because playing this position requires little cardiorespiratory endurance, many players with limited endurance can successfully play goalie. Those in wheelchairs can catch effectively as long as upper-limb involvement is minimal. Most players with one arm will find it difficult to catch while in the goalie position. In this case, they should be encouraged to slap, trap, or strike the ball. For some players with limited mobility, reducing the size of the goal can be helpful.

Lead-Up Games and Activities

Lead-up games and activities are important prerequisites to learning the game of soccer. Selected lead-up games and activities are provided.

Line Soccer

The game may be played on a playground with 2 teams, preferably of 8 to 10 players each. Players on each team stand beside each other (.64 meters or about 2 feet from the person next to them). Both teams face each other from a distance of about 8 meters (26 feet 2 inches). Players on each team try to kick a soccer ball below shoulder level past their opponents. After each score, players on both teams rotate one position to the right. A team scores one point each time the ball passes the opponents' line, and one point is scored against a team that uses their hands to stop the ball.

Accuracy Kick

This activity involves kicking a playground ball into a goal area 1.5 meters (5 feet) wide (with flagsticks at each end) from a distance of 3 meters (9 feet 8 inches). A player is allowed three kicks from either a standing or a sitting position. The player receives three points each time a ball is kicked into the goal, two points each time the ball hits a flagstick but does not pass through the goal, and one point each time the ball is kicked in the direction of the goal but does not reach the goal. Following three kicks, players' scores are compared.

Other Activities

Additional activities may include heading a beach ball into a large goal area from a short distance, dribbling a soccer ball in a circular manner around stationary players as fast as possible, throwing the ball in-bounds for distance, keeping a balloon in the air by kicking it, punting a soccer ball for distance, and scooter soccer.

Sport Variations and Modifications

Modifications in the sport of soccer have been introduced by several sport organizations. Disabled Sports, USA (DS/USA) sponsors soccer competition for players with amputations. Players who are missing part or all of one leg participate on forearm crutches, and the crutch is treated as an extension of the arm.

Sport modifications have also been introduced by the USCPAA. That organization has developed rules for competition in both indoor wheelchair soccer and seven-a-side soccer (USCPAA, 1997). In indoor wheelchair soccer, players are assigned to one of four divisions: Division I includes athletes with a brain injury as governed by USCPAA; Division II includes athletes with a les autres disability; Division III includes athletes with a spinal cord disability as governed by Wheel-

chair Sports, USA (WS/USA); Division IV includes athletes with an amputation as governed by DS/USA. Within each of these divisions, athletes are further classified according to whether they use a power or manual chair for mobility. Teams are composed of six on-court players (one of which is a goalie). The following are some modifications applied to indoor wheelchair soccer:

- ►The game should be played on a gymnasium floor with boundaries not less than 15.2 meters (50 feet) wide and 28.6 meters (94 feet) long or not more than 15.2 meters (50 feet) wide and 30.5 meters (100 feet) long.
- ►A 25.4-centimeter (10-inch) circumference rubber playground ball is used.
- ►The goal measures 1.7 meters (5 feet 6 inches) high, 1.5 meters (5 feet) wide, and .9 meters to 1.5 meters (3 to 5 feet) deep.
- ►Penalty shots and power plays are used.
- ►Goalies may leave the goal area with the ball or to gain possession of the ball.
- ►Players may not touch the playing surface while in possession of the ball.
- ►The wheelchair, a limb, and/or any part of the body can be used to move the ball.
- ►Dribbling the ball with one or both hands simultaneously is permitted.
- ►A maximum of three seconds is permitted for a player to hold or maintain possession of the ball before attempting a pass, dribble, or shot.
- ►Unnecessary roughness and holding or ramming into another wheelchair result in penalties.

Players who ambulate (classes 5 through 8) are eligible to play seven-a-side soccer. Rules generally follow Federation Internationale de Football Association (FIFA) standards. Along with the seven-player limit, some modifications are as follows:

- ►One class V or class VI player must be a member of the team, and one of these players must be on the field at all times.
- ►No offside rule is applied.
- ►An underhand throw-in is permitted.
- ►Teams may consist of male and female players.
- ►The field dimensions are 75 meters (82 yards) by 55 meters (60 yards).

In Special Olympics, the sport is played as either 11-a-side or 5-a-side soccer, and rules follow FIFA standards (Special Olympics, 1996). There are no major modifications for 11-a-side soccer. The following modifications, among others, are applied for five-a-side soccer:

▶ The field dimensions must be a maximum of 50 meters (164 feet) by 35 meters (114 feet 9 inches) and a minimum of 40 meters (131 feet) by 30 meters (98 feet 4 inches). The smaller field is recommended for lower-ability teams.

▶ The goal must be 4 meters (13 feet) by 2 meters (6 feet 6 inches).

▶ A ball over the sideline results in a kick-in by a player from the opposing team to the player who last touched it.

The game of soccer has also very recently been modified for persons in power wheelchairs ("Play Ball," 1997). The game, originated in Canada, is played on an indoor court (regulation basketball court) with a 45.7- to 60-centimeter (18- to 24-inch) physioball. The strategy is similar to able-bodied rugby. Bumpers on footrests and anti-tip bars on chairs are mandatory because of safety and ball-control issues. To score, the ball must travel through a goal at the end of each court.

Skill Event Variations and Modifications

Special Olympics offers individual skills competition in dribbling, shooting, and running and kicking for athletes with lower-ability levels. Athletes perform each of the three events twice, and all scores are then added for a total score. In the dribbling event, the player dribbles the ball 15 meters (49 feet 2 inches) while staying in a 5-meter (16-foot 4-inch) wide lane into a 5-meter (16-foot 4-inch) finish zone. The finish zone is marked with cones. The clock is stopped when both the player and the ball are stopped inside the finish zone. If players dribble over the finish zone, they must dribble it back in the zone in order to finish. If the ball runs over the sideline, the referee places another ball in the center of the lane opposite the point at which the ball went out. The time it takes to do this is converted into points by a point system. The maximum number of points that can be obtained is 60 and the minimum is 10, less a deduction of 5 points each time the ball runs over the sideline or a player touches the ball with his hands.

In the shooting event, the player walks, jogs, or runs forward a distance of 2 meters (6 foot 6 inch) and then kicks a stationary ball into a 4-meter (13-foot 1-inch) wide by 2-meters (6-foot 6-inch) deep goal from a distance of 6 meters (19 feet 7 inches). Once the kick is made, the player returns to the starting line. A total of 5 kicks are allowed, and each successful kick is worth 10 points.

In the run and kick event, the player stands 4 meters (13 feet 1 inch) from 4 balls (1 to the left, 1 to the right, 1 in front, and 1 in back). The player begins by running to any ball and kicks it 2 meters (6 feet 6 inches) through a target gate 2 meters (6 feet 6 inches) wide made of cones. This continues until all four balls have been kicked. The total time in seconds from when the player starts to when the last ball is kicked is recorded. The time is then converted into points. The maximum number of points is 50, and the minimum number is 5. In addition, a bonus of five points is added for each ball kicked successfully through the target.

Other Variations and Modifications

For a simplified game, fewer than 11 players can participate and field dimensions can be reduced. For players with low stamina, a partially deflated ball (which does not travel as fast as a fully inflated ball) can be used, and frequent rest periods, substitutions, or time-outs can be incorporated into the game. A soft foam soccer ball can be used; in some cases, a cage ball can replace a soccer ball. Players with upper-limb deficiencies may be allowed to kick the ball in-bounds on a throw-in. Additional players may be situated along the sidelines to take throw-ins for their teams. Penalty kicks can be employed for penalties occurring outside of goal areas. To avoid mass conversion on the ball, players can be required to play in specific zone areas.

SOFTBALL

Softball uses the skills of throwing, catching, fielding, hitting, and running. With certain modifications, the game can be played in an integrated setting by most persons with disabilities, although players with visual or mobility impairments may find competing in a segregated setting more appropriate. The National Wheelchair Softball Association (NWSA) sponsors competition for athletes in wheelchairs, NBBA sponsors competition for players with blindness, Special Olympics sponsors competition for players with mental retardation, and the DAAA sponsors competition for players with dwarfism. The DAAA follows American Softball Association rules with no modifications.

Game Skills

Important softball game skills consist of throwing, catching, fielding, and batting. Selected modifications are offered for each skill.

Throwing

People with visual impairments will throw with better accuracy if the person receiving the ball communicates verbally with the thrower. For players with small hands or hand deformities, the use of a smaller or foam ball is recommended. Individuals with cerebral palsy, because of control problems, may find using a slightly heavier ball more advantageous. Individuals with lower-limb disability will be able to throw the ball quite well. However, they will have difficulty throwing for distance because body rotation may be limited. A player with one upper limb will be able to throw the ball without much difficulty.

Catching and Fielding

Players with visual impairments will more easily learn to catch if a large, brightly colored ball is initially rolled or bounced. A beep baseball, described in this section under the heading "Sport Variations and Modifications," will be most helpful. Persons using wheelchairs and those with crutches or with braces may wish to use a large glove. Players with one upper limb will be able to catch with one hand as long as eye-hand coordination is well developed. A glove is helpful to most players with one arm who have a remaining segment of the affected limb, provided they have mastered the technique of catching the ball and then freeing it from the glove for a throw. (After the catch, the player removes the glove with the ball in it by placing it under the armpit of the limb segment; the hand is then quickly drawn from the glove to grasp the ball for the throw.) This technique has been used by Jim Abbott, a major league pitcher.

An oversize glove can facilitate fielding for players with poor eye-hand coordination. Fielders should face the direction from which the ball is being hit, and players with visual impairments should be encouraged to listen for a ground ball moving along the ground. Players with assistive devices or in wheelchairs can be paired with sighted, able-bodied players for assistance in fielding. While a fielder in a wheelchair should be able to intercept the ball independently, the assisting player can retrieve it from the ground after interception and hand it to the player with a disability for the throw.

Batting

Players with one upper limb will be able to bat as long as the nonimpaired limb possesses enough strength to hold and swing the bat. To facilitate hitting, the player may use a lighter bat, grasping it not too close to the handle. Players in wheelchairs or on crutches must rely more on arm and shoulder strength for batting, because they will be unable to shift their body weight from the back leg to the front leg to provide power for the swing. Plastic bats with large barrels will be helpful for people with poor arm and grip strength or poor eye-hand coordination. In this case, a large Wiffle ball should also be used.

Lead-Up Games and Activities

Lead-up games and activities are important prerequisites to learning the game of softball. Selected lead-up games and activities are provided.

Home Run Softball

The game is played on a softball field with a pitcher, a catcher, a batter, and one fielder. The object of the game is for the batter to hit a pitched softball into fair territory and then run to first base and return home before the fielder or pitcher can get the ball to the catcher. The batter is out when three strikes are made, a fly ball is caught, or the ball reaches the catcher before the batter returns home.

Lead-Up Team Softball

The game is played in any open area, with six players constituting the team. Players position themselves in any manner approximately 3.6 meters (12 feet) from each other. To begin play, the first player throws the ball to the second player. Each player attempts to catch the softball and throw it (in any manner) accurately to the next player. The sixth player, on catching the ball, attempts to throw it to a 1-meter-square (3-foot by 3-foot) target from a distance of 4 meters (13 feet). Following that throw, players rotate positions until each player has had an opportunity to throw the ball to the target.

Other Activities

Additional activities may include throwing beanbags in an underarm manner through a hoop suspended from the floor, hitting balls for distance from a batting tee, punching a volleyball pitched in an underarm manner, keeping a balloon in the air by hitting it with a plastic stick, and batting a ball suspended from the ceiling or a tetherball pole.

Sport Variations and Modifications

Beep baseball is designed for athletes with visual impairments. It is sanctioned by the NBBA. The object of the game is for the batter to hit a regulation 16-inch audio softball equipped with a special sound-emitting device and to reach base before an opposing player fields the ball. (The Telephone Pioneers of America distribute the balls nationwide. To obtain one, contact the local telephone company and request the address of that organization.) Teams are composed of six players who play. Each team may have two additional sighted players on the roster; they may play blindfolded only when no other player with an impairment is available to play. The sighted players function as the pitcher and catcher while on offense. The pitcher throws the ball in an underarm motion to the batter in an attempt to "give up" hits. The pitcher must give two verbal cues to the batter before the pitch. These are "ready" and "pitch" (or "ball"). The catcher assists the batters by positioning them in the batter's box and also retrieves pitched balls. On defense, one or both sighted players (spotters) stand in the field and assist their six teammates in fielding the ball by calling out the number or name of the defensive player closest to the ball. The spotters cannot field balls themselves. If a hit ball presents a chance of injury to a player, the spotter may yell a warning. Also, a spotter may knock down an unusually hard-hit ball headed directly toward a player; however, a run will be awarded to the offensive team. A batter gets four strikes before being called out. A batter may allow one ball to go by without penalty. Any additional pitched balls that are not swung at are called strikes. Each side has three outs per inning; there are six innings to an official game unless more innings are needed to break a tie.

Upon hitting the ball beyond the foul line, the batter runs to one of two cone-shaped bases (the Telephone Pioneers of America also distribute bases nationwide) that are 48 inches high. Bases contain battery-powered, remotely controlled buzzers; before the ball is hit, the umpire determines which base buzzer is to be activated by giving a hand signal to the base operator. To score a run, the batter must touch the appropriate base before an opposing player cleanly fields the ball. However, if the opposing player fields the ball before the batter reaches base, the batter is out (figure 25.2).

The game of softball has been modified for players in wheelchairs by the NWSA (figure 25.3). The game is played under official rules for 40.6-centimeter (16 inch) slow-pitch softball with the following major modifications (NWSA, 1996):

▶ Manual wheelchairs with foot platforms must be used.

▶ The field is a smooth, level surface of blacktop or similar material.

▶ Second base is composed of a 1.2-meter (4-foot) diameter circles; first and third bases are composed of 1.2-meter (4-foot) semicircles.

▶ Teams are balanced by a point system.

▶ Neither a hitter nor fielder (when playing the ball) can have any lower extremity in contact with the ground.

▶ Fielders cannot play the ball with any lower extremity in contact with the ground.

▶ Teams must have a player with quadriplegia on the team and in active play.

▶ Each team is composed of 10 players.

The game of softball is also modified for Special Olympics play (Special Olympics, 1996). Official events include slow-pitch team competition and tee-ball competition. Both events follow Federation Internationale de Softball (FIS) and National Governing Body (NGB) rules for slow-pitch softball. The following are some modifications that are used for the slow-pitch game:

▶ The distance from home plate to the pitching rubber can be modified to a distance of 12.2 meters (40 feet).

▶ Ten players play defense at any one time.

▶ An extra player may be used. If one is used she must play the entire game. The extra player may be substituted for at any time.

▶ If an extra player is used, all 11 players must bat, and any 10 are allowed to play defense.

▶ If the batter has two strikes and fouls off the third pitch, the batter is out.

▶ The catcher must wear a face mask and batter's helmet.

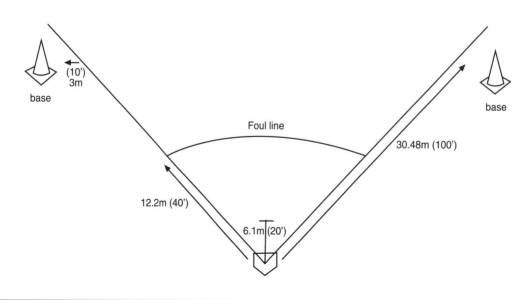

Figure 25.2　Beep baseball field.

Figure 25.3   Player participating in a wheelchair softball game.

Skill Event Variations and Modifications

Special Olympics offers individual skill competition designed for athletes with lower abilities (not for athletes that can already play the game) in four events: base race, throwing, fielding, and hitting. The scores for each of the four events are added together to obtain the athlete's final score. In the base-running event, the player must start at home plate; run around the bases, which are positioned 19.8 meters (65 feet) apart; and return home as fast as possible. The time needed to run the bases is subtracted from 60 to determine the point score. A five-second penalty is given for each base missed or touched in an improper order. The best of two trials is counted.

In the throwing event, the object is to throw a softball as far and accurately as possible. Two attempts are given, and the player's score is the distance of the longest throw (measured from the restraining line to the point where the ball first touches the ground). The score reflects the throwing distance in meters minus the error distance (the number of meters the ball landed to the left or right) of a perpendicular throwing line marked from the restraining line. A distance-throwing event (without an accuracy component) is offered in competition sponsored by Wheelchair Sports, USA (WS/USA). Similar events sponsored by USCPAA include a soft shot and an Indian club throw for distance. USCPAA junior events include a softball throw.

The fielding event requires the player to stand between 2 cones 3 meters (10 feet) apart and catch a total of 10 ground balls (5 attempts per trial for a total of 2 trials) thrown by an official from 19.8 meters (65 feet) away. The throw from the official must hit the ground

before traveling 6.1 meters (20 feet). The athlete can move aggressively to the thrown ball. Five points are received for catching the ball in the glove or trapping it against the body but off the ground, two points are received for a ball that is blocked, and no points are received for a missed ball.

The hitting event requires the player to bat for distance by hitting a softball off a batting tee. Three attempts are allowed, and the longest hit is the player's score. The distance is measured in meters from the tee to the point where the ball first touches the ground. If the score falls between meters, the score is rounded down to the lower meter.

Other Variations and Modifications

To accommodate players' varying ability levels, the number of strikes a batter is allowed can be increased. In some cases, fewer than four bases can be used, and distances between bases can be shortened. Half innings may end when 3 outs have been made, 6 runs have been scored, or 10 batters have come to bat. In addition, lighter-weight and large-barreled bats may be used. For players with poor eye-hand coordination, like those with cerebral palsy or traumatic brain injury, the ball may be hit from a batting tee. A larger ball or restricted-flight softball, which travels a limited distance when hit, may be used. For players with more severe impairments, the ball may be rolled down a groove or tube-like channel when they are at bat. A walled or fenced area is recommended for players with mobility impairments so that distances to be covered can be shortened. In addition, a greater number of players on defense may be allowed, especially if players have mobility problems. The game can also be modified so that it is played in a gymnasium with a Wiffle ball and bat.

VOLLEYBALL

Volleyball is a popular game included in most programs. Game skills include serving, passing, striking, and spiking the ball. Most persons with disabilities can be integrated into the game. However, for players with severe mental disabilities or those with significant visual or mobility impairments, the game may require modifications.

Game Skills

Important volleyball game skills consist of serving and striking. Selected modifications are provided for each skill.

Serving

Players with disabilities can learn to serve quite effectively. It is helpful to begin with an underhand serve; the nondominant hand is beneath the ball, while the dominant hand (fisted) strikes the ball in an underhand motion. Very young players or those with insufficient arm and shoulder strength can move closer to the net. As players develop coordination, they can switch to the overhand serve. Players with one functional arm can serve overhand effectively by tossing the ball into the air with the nonimpaired arm and then hitting it with the same arm. Wheelchair users are able to perform both the underhand and overhand serves, although for the underhand serve, it is important to be in a chair without armrests.

Striking

Players with visual impairments can competently hit the ball with two hands if the ball is first allowed to bounce. This gives the player more time to visually track the ball. Because of limited mobility, players in wheelchairs—like those on crutches—will need to learn to return the ball with one hand. However, players in wheelchairs can use both hands to return the ball within their immediate area. As these players become more adept in predicting the flight of the ball, they will be able to make a greater percentage of returns.

Lead-Up Games and Activities

Lead-up games and activities are important prerequisites to learning the game of volleyball. Selected lead-up games and activities are provided.

Keep It Up

The game is played by teams forming circles approximately 4.6 meters (15 feet) to 6.1 meters (20 feet) in diameter. Any number of teams, of about six to eight members each, may play. To begin the game, a team member tosses the volleyball into the air within the circle. Teammates, using both hands, attempt to keep hitting the ball into the air (it must not hit the ground). A player may not strike the ball twice in succession. The team that keeps the ball in the air the longest scores one point, and the team with the most points wins the game.

Serving Activity

A player hits a total of 10 volleyballs, either underhand or overhand, over a net and into the opposite court. Point values are assigned to various areas within the opposite court, with areas farther away from the net having higher values. The player's score is the point total for all 10 serves.

Other Activities

Additional activities may include setting a beach ball or large balloon to oneself as many time as possible in succession, serving in the direction of a wall and catching the ball as it returns, and spiking the ball over a net that is about 30.8 centimeters (1 foot) higher than the player.

Sport Variations and Modifications

Disabled Sports, USA (DS/USA), DAAA, and Special Olympics offer team sport competition in volleyball. Volleyball competition is governed internationally by International Sports for the Disabled (ISOD) and the International Paralympic Committee (IPC). Players, especially for DS/USA competition, are classified according to functional ability. For example, players with amputations are classified according to the site and degree of amputation. Two modifications of the sport under DAAA auspices consist of a slightly lowered net and court dimensions that are used by ISOD for seated competition. It is interesting to note that sitting volleyball was first introduced in the Netherlands as early as the mid-1950s. Since then, sitting volleyball has grown into a very popular sport. The World Organization Volleyball for Disabled (WOVD, 1997), affiliated with both the ISOD and IPC, has established both standing and sitting rules for the game for both amputations and les autres (other locomotor disabilities). Table 25.1 illustrates the basic differences between sitting and standing volleyball.

Special Olympics volleyball competition is based on Federation Internationale Volleyball (FIVB) and the rules of each individual country's NGB. Some modifications of the game (Special Olympics, 1996) are as follows:

▶ The ball can be hit with any part of the body on or above the waist.

▶ The serving area can be moved closer to the net but no closer than 4.5 meters (14 feet 9 inches).

▶ A lighter-weight ball no more than 226.4 grams (8 ounces) can be used.

Table 25.1 **Rule Differences Between Sitting and Standing Volleyball**

Game characteristics	Sitting	Standing
Court dimensions	10 m (32 ft 8 in.) × 6 m (19 ft 7 in.)	18 m (59 ft) × 9 m (29 ft 5 in.)
Attack lines	2 m (6 ft 5 in.) from the middle of the center line	3 m (9 ft 8 in.) from the middle of the center line
Net size	6.5 m (21 ft 7 in.) × .8 m (2 ft 6 in.)	9.5 m (31 ft 2 in.) × 1 m (3 ft 3 in.)
Net height	1.15 m (3 ft 8 in.) (men) 1.05 m (3 ft 4 in.) (women)	2.43 m (7 ft 9 in.) (men) 2.24 m (7 ft 3 in.) (women)
Equipment	Players are not allowed to sit on thick materials.	NA
	Players are not allowed to wear orthopedic appliances.	Players are allowed to wear orthopedic appliances.
	Position of players is determined by the position of their buttocks.	Position of players is determined by the position of their feet contacting the ground.
	Players are not allowed to lift their buttocks from the court when carrying out any type of attack hit.	NA
Play	When serving, the server must be in the service zone and the buttocks must not touch the court.	When serving, the server's feet must be in the service zone and must not touch the court.
	Net contact is permitted with leg(s) if action of playing the ball is outside of attack zone.	Net contact with any part of the body is a fault only when the action of playing the ball is in the front zone.
	Touching the opponent's court with a hand/foot/feet/leg(s) is permitted provided player does not interfere with opponent.	Touching the opponent's court with a foot or feet is permitted; contact with any other body part is forbidden.
	Front-row players are allowed to block the opponent's service.	It is a fault to block the opponent's serve.

Skill Event Variations and Modifications

Special Olympics competition includes three skill events: overhead passing (volleying), serving, and passing (forearm). These events are designed for athletes with lower-ability levels, not for those who can already play the game. Scores obtained in each of the three events are added together to obtain a final score. For the overhead passing event, the player stands 2 meters (6 feet 7 inches) from the net and 4.5 meters (14 feet 9 inches) from the sideline on a regulation size court. A tosser provides the player with 10 2-handed underhand tossed balls from the backcourt 4 meters (13 feet) from the baseline and 4.5 meters (14 feet 9 inches) from the sideline in the left back

position. The player sets the tossed ball to a target (a person standing 2 meters [6 feet 7 inches] from the net and 2 meters [6 feet 7 inches] from the front left sideline position). If any toss is not high enough for the player to set, it is repeated. The peak of the arc of each set should be above net height. The height of each set is measured. One point is awarded for setting the ball 1 meter (3 feet 3 inches) above the athlete's head; 3 points are awarded for setting the ball above net height; and 0 points are given for illegal contact, a ball that goes lower than head height, or over the net outside the court. The final score is the sum of all the points awarded for each of 10 attempts.

In the forearm passing event, the athlete stands on a regulation court at the right back position 3 meters (9 feet 10 inches) from the right sideline and 1 meter (3 feet 3 3/4 inch) from the baseline. A two-hand overhead toss is made by a tosser standing on the same side of the net in front center court 2 meters (6 feet 7 inches) from the net. The athlete returns the toss with a forearm pass to a target—a person standing on the same side of and 2 meters (6 feet 7 inches) from the net and 4 meters (13 feet 1 1/2 inches) from the side away from the tosser. Varying point values are marked on the front court. This is repeated with the athlete at the left back position. To receive the maximum number of points, the peak of the arc of the pass must be at least net height. A ball landing on a line is assigned the higher point value. One point is received if the ball passes below net height. The final score is determined by adding together the five attempts from both the left and right sides.

Serving competition requires the athlete to serve a ball into the opponent's side of the court. That court is divided into three areas of equal size, and a point value is assigned to each area. One point is awarded for a serve landing in the area of the opponent's court closest to the net, three points are awarded for a serve landing in the middle third area, and five points for a serve landing in the area closest to the opponent's end line. For serves that land on a line, the athlete receives the higher point value. The final score is the total number of points made in 10 serves.

Other Variations and Modifications

Volleyball is easily modified for most players with disabilities. Most often, court dimensions are reduced, the net is lowered, and the serving line is brought closer to the net to accommodate the varying abilities of players, especially for players in wheelchairs. Balls may be permitted one bounce before players attempt to return them over the net, or an unlimited number of hits by the same team may be allowed before the ball is returned. Players with arm or hand deficiencies can be allowed to carry a hit or return. To serve, players may throw the ball over the net rather than hitting it, and they may catch the ball before returning it. Players may have greater success by using a large, colored beach ball or a foam ball. Players with mobility problems, such as those using crutches or walkers, may play the game from a seated position, or the number of players on each team may be increased.

GOAL BALL

Goal ball, a sport invented more than 50 years ago in Europe, was created primarily for sport and rehabilitation for post-World War II blind veterans. The game was first introduced to world competition at the 1976 Paralympics in Toronto, Canada. Goal ball requires players to use auditory tracking, agility, coordination, and team-mindedness skills. The game is played in a silent arena where blindfolded players attempt to score goals by rolling a ball across an opponent's goal line. Game skills are throwing, shot blocking, and ball control. Goal ball follows rules established by the International Blind Sports Association (ISBA).

Males and females compete separately. To remove any advantages for players possessing partial sight, all players are blindfolded, even those who are totally blind. Instead of actually blindfolding players, eyeshades in the form of blacked-out swim goggles are used. A hard rubber ball 76 centimeters (30 inches) in circumference and weighing approximately 1.25 kilograms (2.75 pounds) is used. Each ball contains bells that allow players to track it during play. Information on obtaining goal balls may be obtained from the United States Association for Blind Athletes, 33 N. Institute, Colorado Springs, CO 80903.

Because of the nature of the game, many players wear protective padding. Those who wear pads generally cover the knees, elbows, and hips. Pads are usually of the volleyball and ice hockey types.

Coaches are not permitted to communicate with their players outside of halftime or during official time-outs. Spectators must also remain silent so players may hear the ball. However, the rules permit communication between players in the form of talking, finger snapping, and/or tapping on the floor.

The game consists of two seven-minute periods, and halftime is three minutes long. Running time is not used; rather, the clock is stopped at various points in the game (e.g., a scored goal). Three 45-second team time-outs are allowed during regulation play. Whistles are used to communicate clock times to players. Each team is allowed a total of six players. Three players are on the court at any one time, and each team is allowed three substitutions per game.

Players must remain within their respective play zones. Boundaries are marked with white tape approximately 5.1 centimeters (2 inches) wide. Various textured tape is used so that players can distinguish boundaries and zones (figure 25.4).

Play begins with a throw by a designated team. During the game, the ball must touch the floor at least once in the neutral area as well as in the team or landing areas. If not, the throw will be disallowed and the ball will be turned over to the defensive team. The ball may be passed twice before each throw on goal.

All three players may play defense. Defensive players may assume a kneeling, crouching, or lying position to contact the ball, but they cannot assume a lying position on the playing surface until the ball has been thrown by an opponent. Defenders may move laterally within their team area. However, they cannot rush forward into the throwing area to intercept the ball except to follow a deflection. A player must throw the ball within eight seconds after defensive control has been gained.

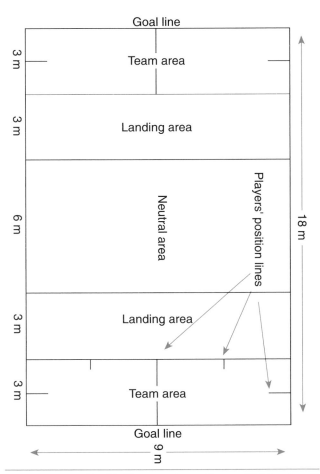

Goal line

3 m

Team area

3 m

Landing area

6 m

Neutral area

Players' position lines

18 m

3 m

Landing area

3 m

Team area

Goal line

9 m

Figure 25.4   Goal ball arena. According to IBSA rules all measurements are to be within +/- .05 meters.

Should a personal or team penalty be assessed to a team, the offending team must defend a penalty throw. In such a case the player committing the penalty (personal penalty) is the only player to defend against the penalty throw. In the case of a team penalty, the player who made the last throw before the penalty was awarded will defend the throw. If a team penalty is awarded before a throw has been taken, the player to defend the throw will be at the coach's discretion.

The USABA encourages and promotes goal ball development camps across the country and provides technical assistance and professional support to members in regions throughout the country who are interested in offering these camps. The camps are designed to assist new players in learning the game.

Game Skills

Important goal ball skills consist of throwing, blocking, and ball control. Selected modifications are offered for each skill.

Throwing

Throwing can be accomplished easily by most players. The ball is thrown in an underhand manner so that it rolls along the ground. Individuals with poor upper-body strength or poor motor control may need to use a lighter ball. Players with one arm can effectively throw the ball, while players with orthopedic impairments who use scooters may need to push the ball with both hands along the ground rather than throwing it in an underhand motion. In other cases, players on scooters can "throw" the ball by striking it with a sidearm motion when it is located at their sides.

Blocking and Ball Control

Blocking and ball control are essential skills because defensive players must stop the ball from entering the goal area. Once the ball is blocked, it is brought under control with the hands so that a throw can be made. It may be helpful to have players with mental retardation see the ball when learning the skill of blocking so that they can more effectively coordinate body movement with the sound of the ball. Players with amputations can wear prosthetic devices to assist in blocking, as long as the ball does not damage the device or vice versa.

Lead-Up Games and Activities

Lead-up games and activities are important prerequisites to learning the game of goal ball. Selected lead-up games and activities are provided.

Heads-Up

This game involves one-on-one competition with players positioning themselves in their half of the arena. A player is given possession of the ball, with the object of scoring a goal. The ball must be continuously rolled as the offensive player moves about; the player cannot carry the ball. The defensive player can take control of the ball from the offensive player by deflecting or trapping a shot on goal. The first player to score three goals wins the game.

Speedy

This game involves two teams of three players each who play within an enclosed area. Players on each team assume a crouching or kneeling position and form a triangle with a distance of 3.7 meters (12 feet) between players. Players position themselves on small area rugs and face the middle of the triangle. On command, a designated player rolls a goal ball as quickly as possible to the player on the immediate right, who controls the ball and rolls it to the next player on the right, and so on. In the event the ball is not controlled by a player to whom it was rolled, that player must retrieve the ball and return to the area rug before rolling to the next player. Each time the ball returns to the player who started the game, a point is scored. The team scoring the most points in one minute wins the game. Teams compete one at a time so that players can hear the ball.

386 Adapted Physical Education and Sport

Other Activities

Other activities may include throwing for accuracy to the goal and passing the ball as quickly as possible between two players for a specified period of time.

Sport Variations and Modifications

Goal ball has quickly gained popularity in the United States and is an official competitive sport of the USABA. People with other types of disabilities as well as people without impairments can participate in the game as long as blindfolds are worn.

Other Variations and Modifications

The game may be modified by increasing or decreasing the size of the play arena and the number of players on a side. People with mobility problems may play on scooter boards, and for those with poor upper-arm and shoulder strength, a lighter-weight ball may be used.

QUAD RUGBY

Quad rugby (internationally known as *wheelchair rugby*), originally called murderball, was developed in the latter part of the 1970s by wheelchair athletes from Manitoba, Canada. In 1981 it was introduced in the United States and quickly gained popularity. As such, the United States Quad Rugby Association (USQRA) was formed in 1988 to promote and regulate the game at both the national and international levels. USQRA is a member of Wheelchair Sports, USA (WS/USA), an umbrella organization for wheelchair sports. To further advance the sport, the USQRA and the Paralyzed

Veterans Association (PVA) sponsor many instructional clinics each year around the country. Today, quad rugby is the fastest-growing wheelchair sport in the world. Stoke Mandeville in England was the site of the first international competition in 1990. And, in 1996, it was played at the Paralympics as an exhibition sport. The International Wheelchair Rugby Federation (IWRF) holds a World Championship every four years, and in 2000 the sport has full medal status at the Paralympics. Wheelchair athletes participating in a quad rugby competition can be found in figure 25.5.

The game is designed for players with quadriplegia who might be prevented from participating in sports such as wheelchair basketball. As such, persons who exhibit upper-limb limitations such as spinal cord injuries, cerebral palsy, spina bifida, and other les autres conditions are eligible to compete. Generally, quad rugby is a combination of American football, ice hockey, and wheelchair basketball. Teams are composed of four on-court players. In order to equalize competition, players are classified according to their functional abilities by a point system. Players are provided a classification number from one of seven classifications, ranging from .5 (most impaired) to 3.5 (least impaired). The player classified as .5 has function comparable to C5 quadriplegia, whereas a player classified as 3.5 has function comparable to a C7-C8 incomplete quadriplegic. The USQRA follows the rules (with very minor changes) set forth by the IWRF. A complete copy of the IWRF rules and regulations with USQRA modifications can be obtained from WS/USA, 3595 E. Fountain Blvd., Suite L-1, Colorado Springs, CO 80910.

The game is played on a regulation basketball court. Goal lines (at each end of the court) are identified by rubber cones placed at the end of each line. In front of each goal line is a key area 8 meters (26 feet 2 inches) by 1.75 meters (5 feet 7 inches). The object of the game

Photo courtesy of Quad Rugby Today. Printed by permission.

Figure 25.5   International quad rugby competition.

is to carry the ball (regulation volleyball) over the opponent's goal line. Players try to gain possession of the ball by forcing a bad pass or a violation. The team with the greatest number of goals at the end of the game wins. At no time can all four defensive players be in the restricted area. Full chair contact is allowed, and, because of the nature of the game, any form of hand protection may be used; however, whatever is used cannot be harmful to other players (e.g., hard or rough material).

An unlimited number of pushes can be made by the ball handler; however, the ball must be either passed or bounced within 10 seconds; if it is not, a turnover is assessed. A player has 15 seconds to advance the ball into the opponent's half of the court. While only 3 defensive players can be in the key area at any one time, all 4 offensive players are allowed in the key area, but only for 10 seconds. Players committing personal fouls must serve time in a penalty box. As a result, the opposing team has a power play. Yilla and Sherrill (1994) provide a detailed description of offensive and defensive strategies.

Game Skills

Important quad rugby game skills consist of wheelchair mobility, throwing, and catching. A brief description of each skill is provided.

Wheelchair Mobility

Chair mobility is essential in the game of quad rugby. Mobility skills consist of maneuvering the chair with the ball, picking, and sprinting. In order to maneuver effectively, the player must place the ball securely in his lap. Maneuverability consists of being able to move forward and backward as fast as possible and being able to turn the chair quickly to the left or right. Players also must be able to bounce and pass the ball while the chair is in motion. Players set picks in order to gain offensive advantage. This is where an offensive player maneuvers her chair into position to the side or behind a defensive player guarding an offensive teammate. This creates an offensive advantage in that the offensive player (not the one creating the pick) moves or drives around the teammate creating the pick. Players having the ability to quickly start from a stopped position will have an advantage over other players, especially when having to sprint down the court. Sprinting up and down the court places the offensive player in a better position to score and places the defensive player in an advantageous position to defend against a score.

Throwing and Catching

Throwing (accuracy and distance) and catching are essential game skills in the game of quad rugby. The nature of the game requires that the ball be advanced,

and throwing (passing) the ball is the quickest way to do it. Depending on ability level, a one-hand or two-hand throw can be done. Those with more severe upper-limb impairments may need to throw two handed. The skill of catching needs to be mastered while the chair is both moving and stationary. Of course, the ability to catch while the chair is in motion is an advantage. For players who do not have full control of the hands, catching (trapping) the ball can be done with closed fists, wrists, and/or forearms.

Lead-Up Games and Activities

Lead-up games and activities are important prerequisites to learning the game of quad rugby. Selected lead-up games and activities are provided.

Mobility Relay

Relay teams are composed of three or more players. Three cones are placed at 4-meter (13-foot 1-inch) intervals from a starting line. The distance from the starting line to the third cone is 12 meters (39 feet 4 inches). The object of the relay is to wheel forward as fast as possible in a figure-eight fashion around each of the three cones. Once the third cone is rounded, the player returns to the starting line by wheeling backward. The first team to successfully complete the relay wins.

Accuracy Pass Activity

A passing line is marked on a court floor 5 meters (16 feet 4 inches) from a stationary, empty wheelchair situated on the court. The start line is situated 3 meters (9 feet 8 inches) behind the passing line. The total distance between the start line and the unmanned wheelchair is 8 meters (26 feet 2 inches). On command the player begins moving forward and passes the ball to the wheelchair (facing the player) before crossing the passing line. The player gets a total of five throws. The final score is the number of times the player hits the wheelchair with the ball in the air (as opposed to rolling it on the ground). Should the player go over the passing line before the ball is thrown, a fault is called and the pass does not count.

Other Activities

Recently, Yilla and Sherrill (1998) validated a battery of quad rugby skills tests. These skills consist of sprinting, passing (distance and accuracy), picking, and maneuvering. Each of these skills can be used as part of practice for the competitive sport, or they can be used as lead-up activities in physical education classes.

INCLUSION

During physical education and, at times, in athletics, players with disabilities can be included in regular or adapted sports of basketball, floor hockey, volleyball,

Application Example: Football Inclusion

Setting:	Interscholastic freshman football
Students:	Players with mild mental retardation and/or specific learning disabilities
Application:	To accommodate the learning needs of these players, the coach could, among other things:

- ▶ require players to play only on "special teams" (e.g., kick-off, punt return) so that they will not need to know multiple play assignments,

- ▶ use task analysis for the learning of new, more complex skills/assignments,

- ▶ allow players to wear a wristband to remind them of their play assignments,

- ▶ have teammates remind them of their play assignments while in the huddle, and

- ▶ teach new plays and/or skills at the beginning of practice as opposed to the end of practice.

football, soccer, and softball. The games of goal ball and quad rugby were designed specifically for persons with certain defined disabilities. However, they may be played by persons with disabilities as well as nondisabled participants. As long as certain techniques are applied, most players with disabilities can be successfully and safely integrated into the sports previously identified. Numerous modifications and variations have already been presented in this chapter. In addition, there are other techniques that have been successfully implemented. One technique is to match abilities and positions; teachers and coaches should attempt to assign specific positions on the basis of ability levels of players. In football, players with mild mental retardation possessing good catching skills could play end positions, while others with good speed could play the backfield. Players with upper-arm impairments can serve as placekickers in football. In specific instances, a physical disability can be used to advantage. For example, Tom Dempsey, whose partial amputation of the kicking foot allowed for a broader surface with which to kick the ball, was a very successful placekicker in the National Football League.

Integration can also be enhanced by teaching to the players' abilities. When teaching or coaching players with mental retardation, one should emphasize concrete demonstrations more than verbal instructions. Verbal instructions, when used, should be short, simple, and direct as when coaching first and third base in softball. In football and basketball, a few simple plays that have been overlearned can encourage success. In sports like football or basketball, the player holding the ball can shout or call out to help a partially sighted player with ball location. Additional examples for promoting inclusion in the game of football can be found in the Football Inclusion application example.

Modification of equipment can also foster integration. The teacher or coach can place audible goal loca-

tors in goal areas in sports such as basketball, soccer, and floor hockey for players with visual impairments. Goals, when used, can be brightly painted or covered with tape. For example, the crossbar and goalposts in soccer could be brightly painted. In games played on an indoor court, such as floor hockey or basketball, mats may be placed along the sidelines to differentiate the playing surface from the out-of-bounds area.

Special Olympics now promotes competition with nondisabled persons, provided that Special Olympics athletes have demonstrated the ability to participate in team sports. This ability includes not only the attainment of certain skills but also teamwork and team strategy (Special Olympics, 1996). Special Olympics provides a detailed handbook that covers the philosophy of Unified Sports, research and evaluation results, and field implementation, as well as sports rules for basketball, bowling, soccer, softball, and volleyball.

SUMMARY

This chapter described a number of popular team sports included in physical education and sport programs. Team sports included in competition sponsored by sport organizations such as AAAD, DAAA, NBBA, DS/ USA, NWBA, NWSA, Special Olympics, USABA, and USCPAA as well as team sports pertaining to various international sport organizations (e.g., IWRF, UWFA, WOVD) were also discussed. Game skills, along with variations and modifications specific to each sport, were identified. Also presented were lead-up games and activities, as well as rules and strategies, corresponding to those found in programs for individuals without disabilities. Finally, suggestions for integrating persons with disabilities into team sports were offered in addition to the many modifications and variations presented earlier to facilitate inclusion.

Individual, Dual, and Adventure Sports and Activities

E. Michael Loovis

▷

▷   **K**EVIN APPROACHED THE PHYSICAL EDUCATION instructor at a local college and inquired about signing up for the beginning tennis course. The instructor indicated that it was appropriate, and Kevin appeared at the first class. After the normal introductory lecture, students were paired up for

▷   some basic drills. Initially, only one student would participate with Kevin. Apparently everyone thought Kevin was incapable of doing well in tennis. It soon became obvious that Kevin was quite capable and other students soon wished to be paired with him. Kevin picked up the forehand and backhand strokes quickly. He did have some difficulty with the serve, but even there he was ahead

▷   of most of the class. Kevin completed his competency testing before anyone else in the class and was patiently waiting for the in-class tournament to begin. When it did, Kevin quickly dispatched all opponents! He obviously had become the best player in the class, regardless of his need to

▷   perform while seated in a wheelchair.

This chapter examines a variety of individual, dual, and adventure activities and sports in which persons with unique needs can participate successfully. Participation will be analyzed from two perspectives: within the context of sanctioned events sponsored by sport organizations and as part of physical education programs in the schools. In terms of organized competition, discussion will be limited to the rules, procedural modifications, and adaptations that are in use and that are the only approved vehicle for participation. The remainder of the chapter will be a compendium of modifications or adjustments for a variety of activities and sports. The incorporation of each sport by major sports organizations will be discussed, if applicable.

TENNIS

Because of the nature of the game and the availability of variations and modifications, tennis or some form of it can be played by all persons except the most severely disabled. Because the game is played as a singles or doubles event, the skill requirements (both psychomotor and cognitive) can be modified in numerous ways to encourage participation. Regardless of the variations or modifications, the basic objective remains the same: to return the ball legally across the net and to prevent one's opponent from doing the same.

Sport Skills

Typically, tennis can be played quite adequately with the ability to perform only the forehand and backhand strokes and the serve. For the ground strokes, good footwork (for ambulatory players) or effective wheelchair mobility along with good racket preparation— moving the racket into the backswing well in advance of the ball's arrival—is fundamental to execution. Under normal circumstances, movement into position to return the ball and racket preparation are performed simultaneously.

Lead-Up Activities

Adams and McCubbin (1991) describe an elementary noncourt lead-up game, Target Tennis, that is appropriate for wheelchair-bound as well as ambulatory students and that can be played indoors with limited space. Basically, the players position themselves behind the end-zone line, which is 10 feet from a target screen. The screen has five 10-inch-diameter openings, which serve as the targets. The player tosses a tennis ball into the air and, with an overhand swing, attempts to bounce the ball midway between the end-zone line and the target, so that the ball goes through one of the five openings. A bonus serve is permitted for every point scored. No points are awarded if the ball bounces twice.

Special Olympics (1998) provides four developmental events that can serve as lead-up activities: *target stroke, target bounce, racket bounce,* and *return shot.* In target stroke, the athlete is given 10 attempts to drop hit the ball within the boundaries of the opponent's singles court; one point is awarded for each successful hit. Target bounce consists of having the athlete bounce a tennis ball on the playing surface using one hand; the score is the highest number of consecutive bounces in two trials. In racket bounce, the athlete bounces the ball off the racket face as many times consecutively as possible; the score is the most consecutive bounces in two trials. In return shot, the athlete attempts to return a ball that has bounced once over the net and into the opponent's singles court; one point is awarded for each successful hit and the greatest number of successful consecutive attempts in two rounds are counted.

Sport Variations and Modifications

In 1980 the National Foundation of Wheelchair Tennis (NFWT) was founded to develop and sponsor competition in that sport. In 1981 the Wheelchair Tennis Players Association (WTPA) was formed under the aegis of the NFWT, with the purpose of administering the rules and regulations of the sport. A Wheelchair

Tennis Committee within the United States Tennis Association (USTA) was approved in 1996. As recently as 1998, the WTPA merged with the USTA's Wheelchair Tennis Committee, making the USTA the governing body for wheelchair tennis. The rules for wheelchair tennis are the same as for regular tennis, except that the ball is allowed to bounce *twice* before being returned (Parks, 1997). The first bounce must land in-bounds, while the second bounce may land either in-bounds or out-of-bounds.

The USTA sanctions the following divisions: men's and women's open, A and B, men's and women's E (novice), and boy's and girl's junior. The E Quad division was established for individuals with limited power, mobility, and strength in at least three limbs as a result of accidents, spinal cord injuries, or other conditions. Also included in this division are walking quads, power wheelchair users, and triple amputees.

Special Olympics (1998) offers the following events: singles, doubles, Unified Sports doubles, and an individual skills contest. The latter consists of the racket bounce, "Ups," forehand volley, backhand volley, forehand groundstroke, backhand groundstroke, serve-deuce court, serve-advantage court, and alternating groundstrokes with movement. A player's final score is the cumulative scores of all nine events.

Other Variations and Modifications

If mobility is a problem, the court size can be reduced to accommodate persons with disabilities. This might be accomplished by having able-bodied players defending the entire regulation court while persons with disabilities defend one-half of their court. It could likewise be accomplished by permitting players to strike the ball on the second bounce. Variations in the scoring system can facilitate participation. An example is scoring by counting the number of consecutive hits, which, in effect, structures the game as cooperative rather than competitive. If the player with a disability has extremely limited mobility, then the court could be divided into designated scoring areas, with those closest to that player receiving higher point values. Racket control may be a concern for some students because the standard tennis racket may be too heavy. There are several solutions to this problem, including shortening the grip on the racket, using a junior-size racket, or substituting a racquetball racket. In the case of an amputee, the racket can be strapped to the stump, provided there is a functional stump remaining to allow for effective leverage and racket use.

If mobility and/or racket preparation are problematic, then reduction of court size, at least initially, will assist in learning proper racket positioning and stroking because footwork will be minimal. If still unsuccessful, the player can be placed in the appropriate stroking position with the shoulder of the nonswinging arm perpendicular to the net. At this point, all that is necessary is to move into and swing at the ball.

In serves, the ball is routinely tossed into the air by the hand opposite the one holding the racket. In preparation for striking the ball at the optimal height, the racket is moved in an arc from a position in front of the body down to the floor and up to a position behind the back. At this point, the racket arm is fully extended to strike the ball as it descends from the apex of the toss. For some individuals who lack either the coordination or strength to perform the serve as described, an appropriate variation is to bring the racket straight up in front of the face to a position in which the hand holding the racket is approximately even with the forehead or slightly higher. Although serving in this manner reduces speed and produces an arc that is considerably higher than normal, it does allow for serving on the part of some persons who might otherwise not learn to serve correctly. To accomplish the toss, a single-arm amputee may grip the ball in the racket hand by extending the thumb and first finger beyond the racket handle when gripped normally and hold the ball against the racket. The ball is then tossed in the air and stroked in the usual way. A double-arm amputee having the racket strapped to the stump uses a different approach. The ball lies on the racket's strings, and, with a quick upward movement, the ball is thrust into the air to be struck either in the air or after it bounces. In the case of Quad tennis, another individual may drop the ball for an E player who is unable to serve in the conventional manner (Parks, 1997).

TABLE TENNIS

Like tennis, all persons can play table tennis or some version of it except the most severely disabled. It, too, can be played in singles or doubles competition, and, although the requisite skills are less adjustable than in tennis, the mechanical modifications that are available make this sport quite suitable for persons with disabilities. Regardless of the variations and modifications, the basic objective of the game remains the same: to return the ball legally onto the opponent's side of the table in such a way as to prevent the opponent from making a legal return.

Sport Skills

As in tennis, the basic strokes are the forehand and backhand. Unlike tennis, the serve is not a separate stroke. A serving player puts the ball in play with either a forehand or backhand stroke. The ball is required to strike the table on the server's side initially before striking the table on the receiver's side. In addition, servers must strike the ball outside the boundary at their end of the court.

Lead-Up Activities

Appropriate for use in physical education programs is an adapted table tennis game that originated at the Children's Rehabilitation Center, University of Virginia Hospital. Two or four people can play this game, called surface table tennis. This game involves hitting a regulation table tennis ball so that it moves on the surface of the table and passes through a modified net. The net is constructed from two pieces of string attached one-half inch apart to the top of official standards and two or three pieces of string three-fourths of an inch apart at the bottom (Adams & McCubbin, 1991). At the start of play, the ball is placed on the table. A player strikes it so that it rolls through the opening in the net into the opponent's court. Points are awarded to the player who last made a legal hit through the net. Points are lost when a player hits the ball over the net either on a bounce or in the air, when the ball fails to pass through the net, or when a player hits the ball twice in succession.

Another lead-up game, Corner Ping-Pong, was developed at the University of Connecticut (Dunn & Fait, 1997). It is played in a corner, with an area 6 feet high and 6 feet wide on each side of the corner. One player stands on either side of the center line. The server drops the ball and strokes it against the floor to the forward wall. The ball must rebound to the adjacent wall and then bounce onto the floor of the opponent's area. If the server fails to deliver a good serve, one point goes to the opponent. The ball may bounce only once on the floor before the opponent returns it. The ball must be stroked against the forward wall within the opponent's section of the playing area so that it rebounds to the adjacent wall and onto the floor in the server's area. Failure to return the ball means a point for the server. Scoring is similar to that for table tennis. Each player gets five consecutive serves. A ball that is stroked out-of-bounds is scored as a point for the other player. Game is 21 points, and the winner must win by 2 points.

Special Olympics provides three developmental events that can serve as lead-up activities. They are *target serve*, *racket bounce*, and *return shot* (Special Olympics, 1998). In target serve, the athlete serves five balls from the right side and five balls from the left side of the table; a point is awarded for each ball that lands in the correct service area. In racket bounce, the athlete bounces the ball off the racket face as many times consecutively as possible in 30 seconds; the score is the most consecutive bounces in two trials. Return shot involves attempting to return a tossed ball to the feeder's side of the table; one point is awarded if the ball is successfully returned, while five points are earned if the ball lands in one of the service boxes. The athlete attempts to return 5 balls with a maximum of 25 points possible.

Sport Variations and Modifications

Table tennis is included as a sport in competitions offered by Wheelchair Sports, USA (WS/USA); the United States Cerebral Palsy Athletic Association (USCPAA); Disabled Sports, USA (DS/USA); the Dwarf Athletic Association of America (DAAA); the United States of America Deaf Sports Federation (USADSF) through its affiliate, the U.S. Deaf Table Tennis Association, Inc.; and the Special Olympics. For the most part, competition is based on the rules established by the International Table Tennis Federation and International Paralympic Committee (IPC). Some modifications are permitted. For example, DAAA (1998) permits the use of a riser or an elevated platform. In WS/USA and USCPAA competitions, the following rules apply (WS/USA, 1996):

▶Competitors' feet and footrests may not touch the floor.

▶Serve shall be called a let if the ball leaves the table by either of the receiver's sides; if on bouncing on the receiver's side, the ball returns in the direction of the net; or if the ball comes to rest on the receiver's side of the playing surface.

▶The playing surface shall not be used as a support with the free hand while playing the ball; a player may use the playing surface to restore balance after a shot has been played as long as the table does not move.

▶Strapping is permitted only below the knee.

▶The playing area may be reduced; however, it should not be less than 8 meters long and 7 meters wide.

▶There are no exceptions to the playing rules for players who stand.

Special Olympics (1998) sanctions the following events: singles, doubles, mixed doubles, wheelchair competition, individual skills contest, Unified Sports doubles, and Unified Sports mixed doubles. The individual skills contest is comprised of five events including hand bounce, racket bounce, forehand volley, backhand volley, and serve. Scores from all five events are added together for a final score.

Other Variations and Modifications

Several assistive devices are available for individuals with severe disabilities such as muscular dystrophy and other disorders that weaken or affect the shoulder. A ball-bearing feeder provides assistance for shoulder and elbow motion by using gravity to gain a mechanical advantage and makes up for a loss of power resulting from weakened muscles. The bihandle paddle, which consists of a single paddle with handles on each side, was designed to encourage greater range of motion for participants in adapted table tennis. Its greatest asset is increased joint movement resulting from the bilateral nature of hand and finger positioning. A strap-on paddle has been designed for players with little or no functional finger flexion or grasp. The paddle is attached to the back of the hand with Velcro straps. The major disadvantage is that it precludes use of a forehand stroking action. The table tennis cuff is useful with individuals who have limited finger movement and grip strength. It consists of a clip that attaches to a metal clamp on the paddle's handle; a Velcro strap then holds the cuff securely to the hand (Adams & McCubbin, 1991). Another device, the space ball net, can replace the paddle for players who are blind (Dunn & Fait, 1997). This device, held in two hands, consists of a lightweight metal frame that supports a nylon lattice or webbing. The net is large enough and provides an adequate rebounding surface to make participation feasible by persons who are blind. Additionally, a special table has been designed for players who are visually impaired or blind. It is all white, and, when combined with a special orange ball containing metal beads, accommodates players who have visual impairments.

ANGLING

The American Casting Association (ACA) is the governing body for tournament fly and bait casting in the United States. It sets the rules by which eligible casters may earn awards in registered tournaments. None of the contemporary sports organizations for disabled people sponsors competition in angling, and it is likewise not found very often in physical education programs. It is more likely to be used as a recreational sport.

Lead-Up Activities

Angling requires the mastery of casting and other fishing skills. In terms of lead-up activities, there are at least two casting games that deserve mention. Skish involves accuracy in target casting at various distances. Each participant casts 20 times at each target. One point is awarded for each direct hit (plug landing inside target or similar goal). Three targets can be used simultaneously to speed up the game, with players changing position after each has cast 20 times at a target. The player with the greatest number of hits at the end of 60 casts is the winner. The second game, speed casting, is a variation of skish. Each player casts for five minutes at each target. The player with the highest score at the end of 15 minutes is the winner (Adams, Daniel, McCubbin, & Rullman, 1982).

Variations and Modifications

Major assistive devices for use in angling have become very popular in recent years. For example, the freehand recreation belt (figure 26.1), a specially designed

Figure 26.1 Freehand recreation belt.

harness into which one inserts rod and reel, permits reeling and fighting fish with one hand. Other devices include the Ampo Fisher I, Van's EZ cast, Batick Bracket, and Handi-Gear. There are also several lines of electronic fishing reels (Adams & McCubbin, 1991).

ARCHERY

Under normal circumstances, shooting the longbow is a six-step procedure. The steps, in order of occurrence, are assuming the correct stance, nocking the arrow, drawing the bowstring, aiming at the target, releasing the bowstring, and following through until the arrow makes contact with the target. One or more of these steps may be problematic and may require some modification in the archer's technique.

Sport Variations and Modifications

Target archery is an athletic event sponsored by USCPAA, WS/USA, and DS/USA. These groups observe the rules established by the Federation of International Target Archery, with certain modifications. The following are adjustments that may be employed to encourage participation by individuals in wheelchairs including athletes from USCPAA and les autres athletes (WS/USA, 1996):

▶ An adjustable arrowrest and arrowplate and a draw check indicator on the bow may be used, provided they are not electric or electronic and do not offer any additional aid in aiming.

▶Archers may use an adaptive equipment archery body support.

▶Only archers with quadriplegia may have the bow bandaged or strapped into the hand; archers with bow-arm disability may use an elbow or wrist splint, and they may also have a person load their arrows into the bow.

▶Quadriplegic archers may use compound bows; if they use a mechanical release aid with recurve and compound bows, they compete in the adaptive equipment division.

Other Variations and Modifications

Several assistive devices are available to aid the archer who has a disability. These include the bow sling, commercially available from most sports shops, which helps stabilize the wrist and hand for good bow control; the below-elbow amputee adapter device (figure 26.2), which is held by the terminal end of the prosthesis and requires a slight rotation of the prosthesis to release the string and the arrow; the wheelchair bowstringer, which consists of an in-ground post with two appropriately spaced bolts, around which the archer who has a disability places the bow to produce enough leverage to string it independently; and the elbow brace, which is used to maintain extension in the bow arm when the archer has normal strength in the shoulder but minimal strength in the arm, possibly because of contractures (Adams & McCubbin, 1991). Additionally, the vertical bow set can accommodate individuals with bilateral upper-extremity involvement (Wiseman, 1982).

Other program adjustments include the use of a crossbow with the aid of the tripod assistive device for bilateral upper-extremity amputees. Various telescopic sights are also commercially available for the partially sighted archer.

Although the United States Association for Blind Athletes (USABA) does not sponsor competition in archery, several modifications can facilitate participation by individuals with visual impairments (Hattenback, 1979). These modifications are as follows:

▶Using foot blocks to ensure proper orientation with the target

▶Placing an audible goal locator behind the target to aid in directional cueing

▶Using a brightly colored target for partially sighted individuals

▶Placing balloons on the target as a means of auditory feedback

BADMINTON

Special Olympics (1998) sponsors competition in badminton but only as a demonstration sport. Badminton was added as a new medal sport by DAAA in 1998. Competition is based on the rules of the International Badminton Federation (IBF). The game is, however, ideally suited for individuals with disabilities and is played routinely in physical education class.

Sport Skills

Badminton can be played quite adequately using only the forehand and backhand strokes and the underhand serve. Beyond these strokes, development of the clear, smash, drop shot, and drive will depend on the participant's ability.

Lead-Up Activities

There are at least two lead-up games that deserve mention. Loop badminton (Dunn & Fait, 1997) is played with a standard shuttlecock, table-tennis paddles, and

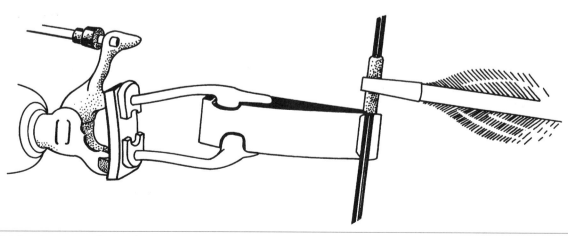

Figure 26.2   Amputee adapter device for archery.

Application Example: Individual, Dual, and Adventure Sports and Activities

Setting:	Middle school physical education class
Student:	7th grade student with muscular dystrophy who uses a wheelchair and is able to push himself
Unit:	Badminton
Task:	Serving the shuttlecock
Application:	The physical educator might make reasonable accommodations in the following ways:

> ▶ Modify regular equipment to make it lighter or more manageable
>
> ▶ Permit serving into a larger court area
>
> ▶ Lower the net
>
> ▶ Shorten distance of serve

a 24-inch loop that is placed on top of a standard 46 inches in height. The object of the game is to hit the shuttlecock through the loop, which is positioned in the center of a rectangular court 10 feet long and 5 feet wide. Scoring is the same as in the standard game of badminton. Loop badminton is well adapted for individuals with restricted movement who wish to participate in an active game that requires extreme accuracy. A second modified game is called balloon badminton (Adams & McCubbin, 1991). In this game a balloon is substituted for the shuttlecock, and table-tennis paddles are used instead of badminton rackets. The game can also be played by people with visual impairments if a bell is placed inside the balloon to aid in directional cueing.

Special Olympics provides three developmental events that serve as lead-up activities. They are *target serve*, *target stroke*, and *return serve* (Special Olympics, 1998). In target serve, the participant has 10 chances to hit the shuttlecock within the boundaries of the opponent's singles court; one point is awarded for each successful hit. In target stroke, the athlete has 10 chances to hit the shuttlecock to the opponent's empty court; one point is awarded for each successful hit. Return serve involves attempting to return a serve to anywhere in the opponent's court; one point is awarded for each successful return up to a maximum of 10 points.

Sport Variations and Modifications

Special Olympics sanctions the following events: singles, doubles, Unified Sports doubles, mixed doubles, Unified Sports mixed doubles, and individual skills contests. The latter is unique to Special Olympics and includes six events as follows: hand feeding, racket feeding, the "Ups" contest, forehand stroke, backhand stroke, and serve. A final score is determined by adding the scores of all six events. In Special Olympics competition, the following rules apply to wheelchair athletes (Special Olympics, 1998):

▶ Athletes have the option of serving an overhand serve from either right or left serving area.

▶ The serving area is shortened to half the distance.

DAAA (1998) permits the following rules modifications:

▶ Sidearm serving is allowed.

▶ No overhead serves are permitted.

Other Variations and Modifications

Some standard modifications are routinely used (see the Individual, Dual, and Adventure Sports and Activities application example). These modifications include, but are not limited to, reducing court size, strapping the racket to the stump of the double-arm amputee, and using Velcro on the butt end of the racket and on the top edge of the cork or rubber base of the shuttlecock to aid in retrieval (Weber, 1991). Several assistive devices are used to facilitate participation in badminton. The extension-handle racket involves splicing a length of wood to the shaft of a standard badminton racket. This is helpful for a wheelchair player or one with limited movement. Another device is the amputee serving tray (figure 26.3). Attached to the terminal end of the prosthesis, it permits easier serves and promotes active use of the prosthesis.

BOWLING

Bowling is an activity in which both ambulatory and wheelchair-bound people can participate with a high degree of success. Usually the ambulatory bowler demonstrates a procedure that incorporates the following actions: approach (which may be modified if lower-extremity involvement exists); delivery, including the swinging of the ball; and release. A wheelchair-bound bowler will eliminate the approach and will either

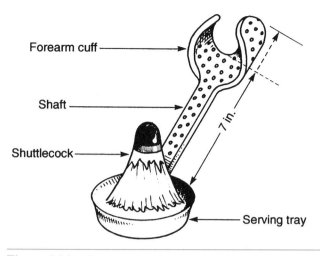

Figure 26.3 Amputee serving tray.

Adapted, by permission, from R.C. Adams and J.A. McCubbin, 1991, *Games, sports, and exercise for the physically handicapped*, 4th ed. (Philadelphia: Lea & Febiger), 218.

perform the swing and release independently or will use a piece of adapted equipment to assist this part of the procedure. Two organizations that have made a significant impact on the lives of bowlers with disabilities are the American Blind Bowlers Association (ABBA) and the American Wheelchair Bowling Association (AWBA).

Lead-Up Activities

Special Olympics (1998) sponsors two developmental events that qualify as lead-up activities. These two events are target roll and frame bowl.

Target Roll

This event consists of rolling two 2-pound bowling balls in the direction of 2 regulation bowling pins positioned on a standard bowling lane that is modified to equal half its normal length. Participants bowl five frames using the standard scoring systems of the American Bowling Congress (ABC).

Frame Bowl

In this event the bowler rolls 2 frames using 30-centimeter-diameter plastic playground balls. The object is to knock down the greatest number of plastic bowling pins from a traditional 10-pin triangular formation. The lead pin is set five meters from a restraining line. Bowlers may either sit or stand and may use either one or both hands to roll the ball; the ball must be released behind the restraining line. Pins that are knocked down are cleared between the first and second rolls; pins are reset for each new frame. A bowler's score equals the number of pins knocked down in two frames. Five bonus points are awarded when all pins are knocked down on the first roll of a frame, and two bonus points are awarded when all remaining pins are knocked down on the second roll of the frame.

Sport Variations and Modifications

Bowling is a sanctioned event in the competitions of USCPAA and Special Olympics. Both follow the rules of the ABC. The USCPAA (1997) has developed the following procedures and/or rules for its competition:

▶There are four divisions: two chute and two nonchute.

▶Nonchute divisions include class 3 through 8 bowlers who do not use specialized equipment; the retractable ball handle is permitted in these divisions.

▶Chute divisions are for class 1 through 6 bowlers using specialized equipment. There are two divisions: closed chute, for class 1 and 2 bowlers who need assistance with equipment (wheelchair, chute, or ball); and open chute, for classes 3 through 6 bowlers who do not require assistance with their equipment.

▶All bowlers must be able to bowl each ball independently within a 60-second period.

Special Olympics (1998) sponsors competition in bowling and follows the rules established by the ABC and the Women's International Bowling Congress. Modified rules for use in Special Olympics are as follows:

▶Ramps and other assistive devices are permitted for singles competition only.

▶Bowlers using ramps shall compete in separate divisions. There are two classifications of ramp bowling: unassisted and assisted.

▶Bowlers are permitted to bowl up to three consecutive frames.

Other Variations and Modifications

Ambulatory bowlers use several types of assistive devices. The handle grip bowling ball, which snaps back instantly upon release, is ideal for bowlers with upper-extremity disabilities and for individuals with spastic cerebral palsy, especially those who have digital control difficulties. The ABC has approved the handle grip ball for competitive play. Upper-extremity amputees who use a hook can utilize an attachable sleeve made of neoprene to hold and deliver a bowling ball. The sleeve can be made to compress by use of a spring and to expand for release with the identical action used to open the conventional hook. Stick bowling, which is similar to use of a shuffleboard cue, was designed for individuals with upper-extremity involvement, primarily grip problems. The AWBA permits stick bowling in its national competitions, provided the bowlers apply their own power and direction to the ball.

Wheelchair-bound bowlers have several assistive devices that facilitate participation. Although not approved by the AWBA for national competition, ramp or chute bowling has become extremely popular with bowlers who are severely disabled. The counterpart of stick bowling for the wheelchair-bound bowler is the adapter-pusher device, which was originally designed for wheelchair bowlers lacking sufficient upper arm strength to lift the ball. The handlebar-extension accessory, used in conjunction with the adapter-pusher device, assists ambulatory bowlers who lack sufficient strength to lift the ball. Also available is the bowling ball holder ring (third arm), a device that attaches to the wheelchair arm and holds a ball while the bowler wheels down the approach lane.

Although USABA does not sponsor competition in bowling, the International Blind Sports Federation (IBSF) sponsors Nine Pin Bowling or Skittles. There are several modifications that can enhance participation in the sport by people with visual impairments. These modifications are as follows:

▶ The use of a bowling rail for guidance, which is the standard method employed by most bowlers in the ABBA National Blind Bowling Championship Tournament

▶ The use of an auditory goal locator placed above or behind the pins

▶ A scoring board system that tactually indicates the pins that remain standing after the ball is rolled

FENCING

Sir Ludwig Guttmann introduced fencing as a competitive event for people with disabilities in 1953 as part of the Stoke Mandeville Games in Stoke Mandeville, England. It was included in the Paralympic Games in Rome in 1960. The objective in fencing is to score by touching the opponent's target while avoiding being touched. If fencing is included in physical education, a number of variations and modifications make achievement of the sport's primary objective feasible for individuals with disabilities.

Sport Variations and Modifications

Competition in fencing for wheelchair-bound individuals is normally conducted according to the rules of the IPC. In competition the following classes are recognized: 1A, 1B, class 2, class 3, and class 4. This continuum represents competitors with no sitting balance and a severely affected fencing arm to competitors with good sitting balance and support legs and a functional fencing arm, respectively. Where appropriate, fencers in the above classes with significant loss of grip or control of the sword hand may bind the sword to the hand with a bandage or similar device (IPC, 1992-1993). Among the modified rules that have been written to ensure equal opportunity for all participants in wheelchair fencing, the following are important to note:

▶ A fencing frame must be used; the fencer with the shortest arms determines the length of the playing area.

▶ Fencers cannot purposely lose their balance, leave their chairs, rise from their seats, or use their legs to score a hit or to avoid being hit; the first offense is a warning, with subsequent offenses penalized by awarding one hit for each occurrence; accidental loss of balance is not penalized.

▶ The legs and trunk below the waist are not valid target areas in épée; the target area for foil and saber is exactly the same as in able-bodied competition— torso excluding arms, legs, and head and waist up including arms and head, respectively.

Other Variations and Modifications

Fencing is ordinarily conducted on a court measuring 6 by 40 feet; however, to accommodate wheelchair participants, it is suggested that either the dimensions of the standard court be changed to 8 by 20 feet or that a circular fencing court with a diameter of 15 to 20 feet be used (Adams et al., 1982). Fencers who are blind will require a smaller, narrower court, which may conceivably be equipped with a guide rail (Dunn & Fait, 1997).

Basically, the only piece of adaptive equipment is the lightweight sword, which permits independent participation, especially for individuals with upper-extremity disabilities. In cases where no modification is necessary, the epee is recommended for ease of handling rather than the foil or saber (Orr & Sheffield, 1981).

HORSEBACK RIDING

The North American Riding for the Handicapped Association (NARHA), founded in 1969, is the primary advisory group to riders with disabilities in the United States and Canada. NARHA does not sponsor competition; however, it advises therapeutic, recreational, and competitive riding programs. It, likewise, certifies riding instructors, accredits therapeutic riding facilities, and provides guidelines for operating safe programs.

Sport Skills

The Cheff Center in Augusta, Michigan, and the EQUEST program in Dallas, Texas, are two of the largest instructor-training programs for therapeutic riding in the United States. Horseback riding involves, among other skills, mounting, maintaining correct positioning on the mount, and dismounting. At the Cheff Center, students with disabilities receive a six-phase lesson. The phases include *mounting, warm-up, riding instruction, exercises, games,* and *dismounting* (McCowan, 1972). Special consideration should be given to the selection and training of horses used for therapeutic riding programs (Spink, 1993). Horses should be suitably sized, that is, small, because children are less likely to be fearful of smaller animals. In addition, smaller horses permit helpers to be in a better position for assisting unbalanced riders; the shoulder of the helper should be level with the middle of the rider's back.

Lead-Up Activities

Several possibilities exist for using games in the context of horseback-riding instruction. The origin of some of these games is pole bending, which is common in Western riding. It is used to teach horses how to bend and to teach riders how to compensate during the bending maneuver. One game that encourages stretching of the arms involves placing quoits over the poles; this activity is conducted in relay fashion, with two- or three-member teams competing. Another game consists of throwing balls into buckets placed on a wall or pole. The traditional game of Green Light, Red Light can be played as a way of reinforcing certain maneuvers, such as halts, which are taught to riders.

Sport Variations and Modifications

Special Olympics and USCPAA each have their own equestrian competition or show. Les autres athletes participating in equine competition follow USCPAA rules.

Special Olympics (1998) offers the following events: dressage; English equitation; stock seat equitation; Western riding; working trails; Gymkhana events including pole bending, barrel racing, figure-eight stake race, and team relays; drill teams of twos or fours; prix caprilli; showmanship at halter/bridle classes; and Unified Sports team and drill relays. There are eight divisions to which riders are assigned, based on a Rider Profile that is completed by the coach for each rider prior to any competition. The divisions are C-S, C-I, B-SP, B-S, B-IP, B-I, AP, and A. Distinctions among divisions range from C-S, requiring a leader (horse handler) and one or two sidewalkers to act as spotters; to A, where a rider is expected to compete with no modifications to national governing body (NGB) rules. Special Olympics (1998) has designated the following specific rules:

▶ Riders who must wear other footwear as the result of a physical disability must submit a physician's statement with their entries; English tack-style riders must use either Peacock safety stirrups, S-shaped stirrups, or Devonshire boots; Western tack-style riders must use tapaderos or other approved safety stirrups.

▶ All riders must wear protective SEI-ASTM or BHS-approved helmets with full chin harness.

▶ Riders may use adaptive equipment without penalty but must in no way be attached to the horse or saddle.

The USCPAA offers competition in the following events: dressage including walk, beginning walk/trot, and intermediate walk/trot, training level 1, first level 1, and classes in pairs to music and musical freestyle; obstacle including walk, intermediate walk/trot, and walk/trot/canter. The USCPAA (1997) has the following specific rules:

▶ No equipment is permitted that would in any way affix a rider to a horse or saddle, with the exception of Velcro and rubber bands.

▶ Leaders and/or sidewalkers are permitted; leaders may walk only beside the horse; no aide is permitted to speak to a rider during an event; coaches or family members are not allowed to act as aides for their riders.

▶ For dressage tests, riders with visual impairments may either have callers or use beepers for purposes of location.

Other Variations and Modifications

Mounting is the single most important phase of a riding program for individuals with disabilities. Because for some people with disabilities the typical method of mounting is impossible (i.e., placing the left foot in the stirrup, holding onto the cantle, and springing into the saddle), there are alternatives based on the rider's abilities. Several basic types of mounting procedures are used by riders with disabilities; these range from totally assisted mounts, either from the top of a ramp or at ground level, to normal mounting from the ground (McCowan, 1972).

Once mounted, individuals with disabilities have available to them numerous pieces of special equipment that can make riding an enjoyable and profitable learning experience. One commonly used item is an adapted rein bar, which permits riders with a disability in one arm to apply sufficient leverage on the reins with the unaffected arm to successfully guide the horse; it is faded as soon as the rider learns to apply pressure with the knees. Another adaptation is the Humes rein, consisting of large oval handholds fitted on the rein; these handholds allow individuals with disabilities involving the hands to direct the horse with wrist and arm movements. Body harnesses are used extensively in programs for riders with the disabilities. They consist of web belts approximately four inches wide with a leather handhold in the back, which a leader can hold onto to help maintain a rider's balance. Most riders who have disabilities also use the Peacock stirrup (figure 26.4a), which is shaped like a regular stirrup except that only one side is iron while the other side has a rubber belt attached to top and bottom. This flexible portion of the stirrup releases quickly in case of a fall, reducing the chance that a foot could get caught. The Devonshire boot (figure 26.4b) is used frequently if a

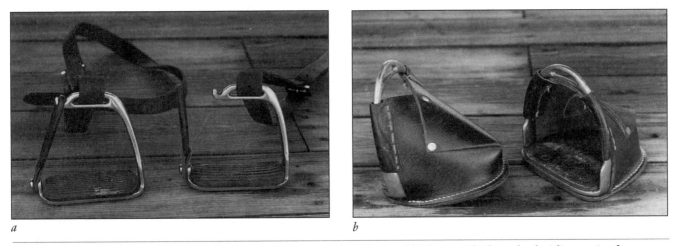

a   *b*

Figure 26.4   Special stirrups like the *(a)* Peacock stirrup and *(b)* Devonshire boot make horseback riding easier for people with disabilities.

rider has tight heel cords or weak ankles. Designed much like the front portion of a boot, it prevents the foot from running through the stirrup, and consequently it encourages keeping the toes up and heels down, which can be invaluable if heel-cord stretching is desirable.

GYMNASTICS

Gymnastics has enjoyed considerable popularity in recent decades because of its visibility in the Olympic Games. As a result, individuals with disabling conditions have likewise begun to participate in gymnastic programs where opportunities have been available. For example, people with orthopedic involvements can participate and have participated in gymnastics programs for able-bodied persons (Winnick & Short, 1985).

Sport Skills

Beyond possessing the physical attributes necessary to participate in gymnastics (e.g., strength, agility, endurance, flexibility, coordination, and balance), participants must learn to compete either in one or more single events or in all events, referred to as the all-around. Under normal conditions men compete in the following events: pommel horse, rings, horizontal bar, parallel bars, and floor exercise. Women compete in balance beam, uneven parallel bars, vaulting, and floor exercise. In both men's and women's competitions, participation in all events qualifies athletes for a chance to win the all-around title.

Lead-Up Activities

The closest thing to lead-up activities related to gymnastics can be found in the Special Olympics Sports Skills Program and in select developmental events that are offered as part of the Special Olympics Games. In its sports skills program manual, *Gymnastics* (Special Olympics, n.d.), general conditioning exercises with emphasis on flexibility and strength are recommended. Doubles tumbling and balance stunts are also suggested. As an introductory experience, educational gymnastics, which uses a creative, problem-solving approach, can be used to teach basic movement concepts. This could eventually enable participants to compete in more advanced forms of gymnastic competition.

Sport Variations and Modifications

Gymnastics is offered in the Special Olympics Games. Events for men include vaulting, parallel bars, pommel horse, horizontal bar, rings, and floor exercise. Women's events include vaulting, uneven parallel bars, balance beam, and floor exercise. Both men and women can compete in the all-around competition. Mixed-gender events include vaulting, wide beam, floor exercise, and tumbling with an all-around competition. Only women compete in rhythmic gymnastics in the following events: ribbon, ball, rope, hoop, and an all-around competition consisting of the four events. Exceptions are levels A and B, which are coeducational and which are performed while sitting. There are also group routines for females that are performed with anywhere from four to eight athletes.

Each of the individual competitions (i.e., mixed gender, men's, and women's) has a Unified Sports event for each of the events that comprise the specific category. No significant rule modifications are required, and each participant's performance is judged according to the rules established for that event by the International Gymnastics Federation (FIG) and the NGB (Special Olympics, 1998).

Special Olympics (1998) rules to ensure equitable competition include the following:

▶ Gymnasts with visual impairments have the option of performing the vault with no run, one step, two steps, a multiple bounce on the board (with hands starting on the horse) or a two- or three-bounce take-off; audible cues may be used in all routines.

▶ In floor exercise, the coach may signal gymnasts with a hearing impairment to begin their routines.

▶ Gymnasts using canes or walkers may have a coach walk onto the floor and remove (and replace) walkers and other aids as needed without any deduction of points.

▶ Gymnasts with visual impairments may use audible cues during floor exercise; music may be played at any close point off the mat, or the coach may carry the music source around the perimeter of the mat.

▶ Gymnasts in levels A and B perform their rhythmic gymnastics routines while seated; gymnasts who are blind can have audible cues during competition; athletes who are deaf can have a visual cue to start with the music without penalty.

Other Variations and Modifications

Few modifications are used in gymnastics competition. If gymnastics is used in physical education programs, all of the modifications observed by the sports organizations in the conduct of their competitions would be valid.

WRESTLING

The sport of wrestling requires considerable strength, balance, flexibility, and coordination. If individuals with a disability possess these characteristics and if they can combine a knowledge of specific techniques with an ability to demonstrate them in competitive situations, then there is no reason that wrestling cannot be a sport in which persons with disabilities experience success.

Sport Skills

Wrestling consists of several fundamentals and techniques that are essential for success. These include *takedowns, escapes and reversals, breakdowns and controls*, and *pin holds*. When learned and performed well, these maneuvers assist in accomplishing the basic objective in wrestling, which is to dominate opponents by controlling them and holding both shoulders to the mat simultaneously for one second.

Lead-Up Activities

The development of specific lead-up activities for wrestling has apparently not been an area of creative activity. If lead-up activities are desirable, then one can conceivably use certain traditional elementary physical education self-testing activities that have some relationship to wrestling. Two such activities are listed in the following paragraphs.

Hand Wrestling

While standing, two people face each other and grasp right hands; each person raises one foot off the ground. On signal, each attempts to cause the other to touch either the free foot or hand to the ground.

Indian Leg Wrestle

Two people lie side by side facing in opposite directions. Hips are adjacent to the partner's waist. Inside arms and legs are hooked. Each person raises the inside leg to the count of three; on the third count they bend knees, hook them, and attempt to force the partner into a backward roll.

Sport Variations and Modifications

Both USABA and USADSF sponsor wrestling competitions. The USABA competition takes place in one division known as the Open Division and is contested according to international freestyle rules as interpreted by the International Blind Sports Federation (IBSF). International competition is held at the following weight categories for men: up to 119.5, 127.86, 138.88, 152.11, 167.54, 187.39, 213.8, and 275.57 pounds. Weight categories for women are up to 101, 112, 123, 136.5, 149.5, and 163 pounds. The following modification has been instituted to render conditions more suitable for athletes with visual impairments:

> *Opponents begin the match in the neutral standing positions with gentle overlapping of each wrestler's hand over the hand of the opponent that is directly opposite; right to left and left to right hands. When contact is broken the match is interrupted and restarted in the neutral position at mat center. (IBSF, 1997)*

Both Greco-Roman wrestling (which prohibits holds below the waist and use of the legs in attempting to take opponents to the mat) and freestyle wrestling are sanctioned events in competitions governed by USADSF. These events are conducted according to the rules established by the International Federation of Wrestling.

Competition in judo is also sanctioned by USABA. It is contested according to the rules of the International Judo Federation as interpreted by the IBSF. In the Paralympic Games, World Championships, and Regional Championships, competition for all weight classifications shall be combined for classes B1, B2, and B3. There are individual men's and women's competitions and men's and women's team competitions.

Other Variations and Modifications

Wrestling is not for everyone. For people with disabilities who want to attempt this sport, there are several modifications that can be used. For those with lower-extremity difficulties that prevent ambulation, all maneuvers should be taught from the mat with emphasis on arm technique. Bilateral upper-extremity involvement will probably restrict participation in all but leg wrestling maneuvers. After removal of prostheses, single-arm amputees can participate with emphasis placed on arm maneuvers.

TRACK AND FIELD

Because all major sports organizations for individuals with disabilities offer competitive opportunities in track and field, this section will focus on the rules modifications that have been enacted in order to make participation maximally available. For each organization the track portion will be discussed first, followed by the field events. No attempt will be made to examine sport skills, variations and modifications, and lead-up activities, except as related to Special Olympics.

Wheelchair Sports, USA

Track-and-field competitions are governed by the rules of The Athletic Congress (TAC). The WS/USA sponsors a classed division in track-and-field events. Classes include T1, T2A, T2B, T3, T4, F1, F2, F3, F4, F5, F6, F7, and F8 (standing and sitting). Classes T1 and T2 compete in 100-, 200-, 400-, 800-, 1,500-, and 5,000-meter races, as well as a 4 x 100-meter circular relay. Classes T3 and T4 also compete in a 10,000-meter race and a 4 x 100- and a 40 x 400-meter circular relay. T3 and T4 also contest 10-, 15-, and 20-kilometer races as well as half and full marathons. In field events all classes compete in the discus, shot put, and javelin, except class F1, which does not put the shot and which substitutes the club throw for the javelin event. Class F2 can choose between the club throw and javelin. Additionally, all classes compete in the pentathlon, which consists of five individual events. F1 competes in a 100-meter race, club throw, a 400-meter race, discus, and an 800-meter race. Classes F2 and F3 compete in shot put, javelin, discus, and 100- and 800-meter races. Classes F4 through F8 substitute 200- and 1,500-meter distances. There is also a junior division that contests mostly similar events but also includes races at shorter distances such as 40 and 60 meters as well as a slalom event.

The WS/USA (1996) has designated the following specific rules:

▶ Wheelchairs shall have at least two large wheels and one small wheel with only one round handrim for each large wheel; the use of chain-driven or geared equipment is not permitted in any sanctioned WS/USA competition; only hand-operated mechanical steering devices are permitted.

▶ Batons are not exchanged in relay races; the takeover shall be a touch on any part of the body of the outgoing competitor within the takeover zone.

▶ Athletes must ensure that no part of their lower limbs can fall to the ground or track during an event; if used, strapping must be of a nonelastic material.

▶ Approved hold-down devices can be used to stabilize competitors' chairs in field events; classes F1 to F6 must have at least one part of the upper leg or buttock in contact with the cushion or seat until the implement is released; F7 and F8 competitors are permitted lifting as long as one foot is in contact with the ground inside the throwing circle.

▶ Competitors in the club throw conform to the rules for discus as outlined in the rules of the International Amateur Athletic Federation (IAAF).

United States Association for Blind Athletes

IAAF rules are employed in track-and-field competitions. Within the United States Association of Blind Athletes (USABA) structure, there are three visual classifications: B1, B2, and B3. The following events are provided for males and females across all three classes: 100-, 200-, 400-, 800-, 1,500-, 5,000- and 10,000-meter races. Men and women in all classes also run a marathon. There are two relays for men and women: a 4 x 100 meter and a 4 x 400 meter with combined visual classes. Both men and women in all three classes compete in long jump, high jump, triple jump, discus, javelin, and shot put. There are pentathlons that are contested by all classes and by men and women. Both men and women compete in the long-jump, discus, and 100-meter events. Men throw the javelin, while women put the shot. Men run a 1,500-meter event, and women run an 800-meter race. Two divisions of youth events, junior and intermediate, are also contested. The IBSF (1997) has determined that the following rule modifications are necessary to provide more suitable competition for athletes with visual impairments:

▶ Class B1 sprinters may run the 100-meter race with the help of not more than two callers, one of whom must remain behind the finish line; the second caller, if one is used, has no restriction on position taken but may not cross the finish line ahead of the athlete.

▶ Guides are allowed for B1 and B2 in 200-meter races through marathon events. When guides are used in events between 200 and 800 meters, there is an allowance of two lanes per competitor.

▶ Competitors may also decide what form guidance will take. They may choose an elbow lead, a tether, or run free; at no time will the guide push or pull the competitor, nor will the guide ever precede the athlete; the runner may receive verbal instructions from the guide.

▶ Acoustic signals (a caller) are permitted for B1 and B2 athletes in field events.

▶ Class B1 high jumpers may touch the bar as an orientation prior to jumping; B2 jumpers are permitted to place a visual aid on the bar.

▶ Class B1 and B2 shot put, discus, and javelin throwers may enter the throwing circle or runway (run-up track) only with the assistance of a helper, who must leave the area prior to the first attempt.

United States of America Deaf Sports Federation

Track competition for men includes races at standard distances from 100-meter through 25-kilometer road racing. It also includes 110- and 400-meter hurdles, 3,000-meter steeplechase, and 20-kilometer walk. Along with the standard field events, the United States of America Deaf Sports Federation (USADSF) provides competition in pole vaulting and hammer throw. Women's competition in track and field parallels that described for women in USABA with one exception—the 100-meter hurdle.

Special Olympics

Special Olympics offers a greater number and diversity of track-and-field events than any other sport organization for people with disabilities. Included in the list of possible events that can be offered at a sanctioned competition are 100-, 200-, 400-, 800-, 1,500-, 3,000-, 5,000-, and 10,000-meter races. There are walking races of 100, 400, and 800 meters; women compete in 100-meter hurdles, while men compete in 110-meter hurdles. Additionally, there are 4 x 100- and 4 x 400-meter relays. In field competition the following events are contested: long jump, high jump, shot put, and pentathlon. Each of these events is mirrored with Unified Sports events with the exception of the hurdle events. There are also track-and-field events for athletes in wheelchairs. Long-distance events combining racing and walking have expanded to include 1,500-, 3,000-, 5,000-, and 10,000-meter competitions. There is also a 15,000-meter walking event, a half marathon, and a full marathon. Unified Sports competitions include 1,500-, 3,000-, 5,000-, 10,000-, and 15,000-meter walking and running events. A 15,000-meter walking event is contested along with a half and full marathon. IAAF rules are employed in competitions sanctioned by Special Olympics (1998). Modifications to those rules include the following:

▶ In running events a rope or bell can be utilized to assist athletes who are visually impaired; a tap start can be used only with an athlete who is deaf and blind.

▶ In race-walking events, athletes are not required to maintain a straight support leg while competing.

▶ In the softball throw, athletes can use any type of throw.

Special Olympics (1998) comes closer than the other organizations to describing lead-up activities. It does this through the provision of 17 developmental events including 25- and 50-meter dashes or walks; the 10-, 25-, and 50-meter assisted walk; softball throw; 10- and 25-meter wheelchair races; a 30-meter wheelchair slalom; a 4 x 25-meter wheelchair shuttle relay; 30- and 50-meter motorized wheelchair slaloms; a 25-meter motorized wheelchair obstacle race; a tennis ball throw for distance; and the standing long jump.

United States Cerebral Palsy Athletic Association

Competition sanctioned by the United States Cerebral Palsy Athletic Association (USCPAA) is governed by rules established by TAC. Events are contested via an 8-class system and consist of races as short as 60-meter (weave) for class 1 athletes in electric wheelchairs up to 3,000 meters and cross country running for classes 5 through 8. There are 4 x 100-meter and 4 x 400-meter open relays for classes 2 through 8, including wheelchair and ambulant events. The following events constitute the field portion: shot put, discus, javelin, club throw, and long jump. Additional events include the *precision throw, soft shot, distance kick, high toss,* and *medicine ball thrust.* There is also a pentathlon for classes 3 through 8. Modifications of rules (USCPAA, 1997) used to ensure equitable competition are as follows:

▶ Class 5 athletes who use canes or crutches must use their assistive devices in a manner such that they make contact with the surface of the track a minimum of 1 time approximately every 10 meters.

▶ Athletes in wheelchair relays must make personal contact with their team members to complete a successful changeover. This contact can be on any part of the outgoing teammate; either the incoming or outgoing competitor may initiate the tag within the change zone.

The USCPAA (1997) incorporates the following additional modifications in its field events:

▶ An attendant or approved holding device may secure the chair in place; however, neither an attendant nor the apparatus may be inside the throwing area.

▶ The soft shot (5-inch-diameter cloth weighing a maximum of 6 ounces), precision throw, high toss, and soft discus (Spongedisc) are used for class 1 only.

▶Distance kick and medicine ball thrust are offered for class 2 athletes who cannot engage in routine throwing events. In the distance kick a 13-inch playground ball is placed on a foul line; the competitors initiate a backswing and then kick the ball forward while remaining seated in their chairs. Distance of the kick is the criterion. In the medicine ball thrust, a 6-pound medicine ball is used. Competitors may not kick the ball; rather, the foot must remain in contact with the ball throughout the entire movement until release.

Dwarf Athletic Association of America

The Dwarf Athletic Association of America (DAAA) sanctions the following track events: 20-meter run for children under 7 years of age (futures); 20-meter run for juniors, 7 to 9 years; 40-meter run for juniors, 7 to 9 and 10 to 12 years; and a 60-meter run for juniors, 10 to 12 and 13 to 15 years; and athletes over 40 years of age (master). There is also a 100-meter open race and a 4 x 100-meter relay. The rules of TAC and wheelchair competition are typically adhered to. Field events that are contested in DAAA competition include shot put, discus, and javelin in open and shot put and discus in master's classes. Juniors (13 to 15) may participate in shot put and discus. Other juniors and futures events include softball throw, flippy flyer (soft discus), and tennis ball throw for futures (DAAA, 1998).

GOLF

Golf has been an event in the Special Olympics since 1995. It is also an activity that can be effectively included in physical education programs for individuals with disabilities. The sport of professional golf achieved infamous recognition in 1998. Casey Martin, a golfer with a chronic and debilitating circulatory disorder in his leg, sued the Professional Golf Association (PGA) under the provisions of the Americans with Disabilities Act to traverse the course using a cart. He won the right to use a cart in PGA events.

Sport Skills

Golf, as it is normally played, requires a person to grasp the club and address the ball using an appropriate stance. Being able to swing the golf club backwards, then forward through a large arc including follow-through are also requisite tasks.

Lead-Up Activities

An appropriate lead-up activity is miniature golf. This popular version of golf is quite suited to persons with disabilities. For many, this may represent the extent to which the golf experience is explored. Holes should range from 8 to 14 feet from tee mat to hole with a width of three feet, which accommodates reaching a ball lying in the center of the course from a wheelchair.

Sport Variations and Modifications

Special Olympics (1998) has created rules based on the *Rules of Golf* as written by the Royal and Ancient Golf Club of St. Andrews. Official events include an individual skills contest (level 1), an alternate shot team play competition (level 2), an individual stroke play competition of 9 holes (level 3), an individual stroke play competition of 18 holes (level 4), and a Unified Sports team play (level 5). Individual skills contests are designed to train athletes to compete in basic golf skills. Competition is held in *short putting, long putting, chipping, pitch shot, iron shot,* and *wood shot.* The alternate shot team play competition involves pairing one Special Olympics athlete with one golfer without mental retardation who serves as a coach and mentor. The format is a nine-hole tournament that is played as a modified four-person scramble. Level 3 enables athletes to play in regulation 9-hole golf competition, while level 4 is designed to play in 18-hole competitions. Level 5 or Unified Sports team play is designed to provide the Special Olympics athlete an opportunity to play in a team format with a partner without mental retardation but with similar ability.

Other Variations and Modifications

Because of various limitations experienced by persons with disabilities, the essential sport skills are often problematic. Dunn and Fait (1997) have detailed many practical considerations necessary for successful participation by golfers who possess disabilities. These include using powered carts for those who lack stamina to walk around the golf course, but who can physically play the game; having a player whose right arm is missing or incapacitated play left-handed, or vice versa; providing a chair for players who cannot balance on one crutch or who are unable to stand (Longo, 1989) (those using a chair or sitting in a wheelchair should have the chair turned so they are facing the ball); and eliminating the preliminary movement of the club (waggle) for blind golfers, because this could produce an initial malalignment of the club with the ball. Additionally, information about distance to the hole can be provided by tapping on the cup or by telling golfers how far they are positioned from the cup, and, for some wheelchair players, using extra long clubs is necessary to clear the foot plates. The Putter Finger is an assistive device that consists of a molded rubber suction cup designed to fit on the grip end of any putter. It is used to retrieve the ball from the hole (Adams & McCubbin, 1991). J. H. Huber (personal communication, January 1971) developed another adaptation that enables golfers who are blind to practice independently. Three pieces of material, all of which produce a different sound when struck, are hung 15 to 20 feet in front of golfers while they practice indoors. Golfers are instructed about the positions of the different pieces of material and the sound made by each. Because feedback about the direction of the ball's line of flight is available, they can determine whether the ball went straight, hooked, or sliced. The **golf chirper** (Cowart, 1989) is used to develop independent putting skills; it serves as a cup locator and audio feedback device. The **amputee golf grip** developed by Synergetic Muscle-Powered Prosthetic Systems fits any standard prosthetic wrist. It permits full rotation during backswing, squared club face at impact, and complete follow-through.

POWERLIFTING

Powerlifting has developed over the years as an extremely popular sport for people with disabilities. For example, in 1992, 153 participants from 34 countries took part in the Barcelona Paralympic competition. In this chapter powerlifting as a sport is distinguished from routine weight training.

Sport Skills

As administered by the International Powerlifting Federation, the sport includes two lifts: the *bench press* and the *squat*. In IPC competition, the sport is restricted to the bench press. Participants are classified by weight; however, braces and other devices are not counted in the total weight. Wheelchair Sports, USA makes adjustments to recorded weight according to the site of an amputation.

Lead-Up Activities

Special Olympics (1998) offers three events that provide meaningful competition for athletes with lower-ability levels: *push-up*, *sit-up*, and *exercycle*. The **modified push-up** is executed in the kneeling position. As many push-ups as possible are performed in 60 seconds; legal push-ups consist of lowering the head and upper back to the floor, touching the chin to the floor, and returning to the starting position. The athlete may not receive assistance, and push-ups do not have to be performed continuously. Sit-ups are performed in the supine position with knees bent and feet held flat on the floor. The athlete folds arms across the chest with hands grasping opposite shoulders; one of the elbows must touch the participant's knee or thigh for the sit-up to be legal. One's score consists of the number of complete sit-ups in 60 seconds. In the exercycle, the athlete sits on the bike with feet on the pedals. Assistance may

be provided to stay on the bike but not for pedaling. The athlete begins pedaling at the sound of the starting whistle and pedals a distance of one kilometer; the score is the amount of time it takes to pedal the distance.

Variations and Modifications

DAAA, WS/USA, USCPAA, USABA, and Special Olympics offer competitive powerlifting programs as a demonstration event. The IBSF (1997) sanctions three events: bench press, squat, and deadlift. Special Olympics (1998) offers bench press, deadlift, and squat. It also offers two combination events: bench press and deadlift; and bench press, deadlift, and squat. Special Olympics also offers Unified Sports competition that parallels the previously mentioned events. WS/USA provides competition in powerlift press and bench press. Each organization has specific rules that accommodate its athletes. Some of the more significant modifications are as follows:

- A safety device engineered to protect lifters against the **clasp knife reflex** (exaggerated stretch reflex) is mandatory in all sanctioned events (USCPAA, 1997).
- Strapping the legs above the knees to the bench is permissible as long as it is done with the strap provided by the organizing committee (WS/USA, 1996).
- A lifter who has a physical disability may be strapped to the bench either between the naval and nipples and/or between the knees and ankles (Special Olympics, 1998).
- Athletes may bench press with legs straight or with knees bent and feet flat on the bench (DAAA, 1998).

CYCLING

Cycling, whether bicycle or tricycle, is a useful skill from the standpoint of a lifelong leisure pursuit. It can likewise be a strenuous sport that is pursued for its competitiveness. To compete, participants must develop a high level of fitness and learn effective race strategy.

Sport Skills

Under most circumstances cycling requires the ability to maintain one's balance on the cycle and to execute a reciprocal movement of the legs to turn the pedals. Technological advances have enabled persons to cycle who would never have thought previously about cycling as a leisure pursuit or as a competitive event.

Sport Variations and Modifications

The USADSF sponsors 3 events, which include the 1,000-meter sprint, a road race, and a time-trial race on the road. The USCPAA sponsors events in four divisions including divisions 1 and 2 for tricycles and division 3 and 4 for bicycles. The tricycle events include a 3- to 10-kilometer time trial and a 10- to 40-kilometer road race for classes 2, 5, and 6. The bicycle events include a 10- to 20-kilometer time trial and a 35- to 75-kilometer road race for classes 5 through 8. Classes 5 through 8 also compete in a velodrome or cycling track event, a flying 200, a pursuit 3-k/4-k, and a Kilo. Hand-propelled tricycles are not permitted in USCPAA competitions (USCPAA, 1997). The IBSF (1997) offers four event categories. These include road races, track races, individual pursuit, and sprints. Within each category there are races for men, women, and mixed tandems. Road races are 110 to 35 kilometers for men, 60 to 70 kilometers for women, and 65 to 80 kilometers for mixed teams. Track races are 1,000 meters for men and mixed teams and 500 meters for women. Individual pursuit events are 4 kilometers for men and 3 kilometers for women and mixed teams. Sprint competitions are contested at 4 laps of 250 meters for men, women, and mixed teams. With few exceptions the rules for IBSF cycling are the same as those for the United States Cycling Federation. The primary exception is that the pilot (front rider) in tandem riding events must be sighted with a group 1 permit; the stoker (back rider) can be from any vision class and must have a group 2 permit.

Special Olympics offers the following events: 500-meter time trial; 1-, 5-, and 10-kilometer time trials; and 5-, 10-, 15-, 25-, and 40-kilometer road races. Unified Sports cycling events parallel the competitions metioned above. All events are governed by the rules established by the International Federation of Amateur Cycling (Special Olympics, 1998).

Other Variations and Modifications

Riding a bicycle can be a difficult task. Individuals with impaired balance or coordination may require some adaptation. Three- and four-wheeled bicycles with or without hand cranks can facilitate cycling for disabled persons. If riding a two-wheeled bicycle is the desirable approach, then training wheels suitable for full-size adult bikes can be constructed. Additionally, tandem cycling can be used in cases where total control of the bicycle is beyond the ability of the person with a disability, for example, the visually impaired. The Quickie Kidz Bike (figure 26.5) is a made-to-order, hand-cranked bicycle/wheelchair available in single or three to five-speed models. It is available from Sunrise Medical, Inc.

BOCCIE

Boccie, the Italian version of bowling, is generally played on a sand or soil alley 75 feet long and 8 feet wide. The

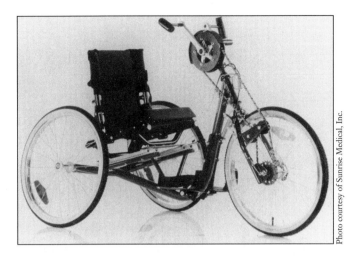

Photo courtesy of Sunrise Medical, Inc.

Figure 26.5   Quickie Kidz Bike.

playing area is normally enclosed at the ends and sides by boards that are 18 inches and 12 inches high, respectively.

Boccie made its initial appearance in the Paralympics at the 1992 Barcelona Games. Boccie is the only Paralympic sport in which men and women compete together in all events.

Sport Skills

The game requires that players roll or throw wooden balls in the direction of a smaller wooden ball or "jack." The object is to have the ball come to rest closer to the "jack" than any of the opponent's balls. To do this, players try to roll balls to protect their own well-placed shots, while knocking aside their opponent's balls.

Lead-Up Activities

The Empire State Games for the Physically Challenged have adopted a new game, crazy bocce, as a demonstration activity. This game consists of throwing two sets of four wooden balls alternately into various-size rings for specified point totals. Three smaller rings sit inside one large ring, which is 13 feet in circumference. Points are awarded only if the ball remains inside one large ring. If the ball lands inside the large ring (but not in any of the smaller rings), one point is awarded. If the ball lands inside the small blue or red ring, two points are earned. Landing inside the small yellow ring nets three points. The game is usually played with the large ring in a small wading pool (figure 26.6). The large ring can also be attached to swimming pool sides, using a suction cup attachment that is provided. The game can also be played in the snow, on the beach, on the lawn, and on carpet. Crazy bocce is enjoyed by young and old alike.

Figure 26.6   Playing crazy bocce.

Sport Variations and Modifications

Both individual and team boccie are sanctioned events in the national competition of USCPAA, DAAA, and the Special Olympics. A minor adjustment to established rules permits the use of ramps or chutes by DAAA athletes. In the Paralympic Games, athletes eligible for competition are those classified as C1 and C2 according to the Cerebral Palsy International Sports and Recreation Association (CP-ISRA). Athletes are combined in two groups designated as sitting and standing. Les autres athletes follow the rules established by USCPAA. Major modifications to the rules in either USCPAA (1997) or DAAA (1998) sanctioned events are as follows:

► The court is laid out on a tile or wood gymnasium floor or asphalt surface and measures 12.5 by 6 meters; stools, chairs, or other sitting devices are permitted in the thrower's box during matches.

► An assistant is allowed to adjust ramps/chutes and player's chair position within the throwing box; however, all direction for adjustments must be initiated by the player; assistive devices should not contain any mechanical device, such as a spring-loaded device, that would aid in propulsion.

► C1 players who have difficulty holding or placing the balls can receive assistance from one aide; however, they must throw, kick, strike, push, or roll the ball independently; players may use more than one assistive device during a match only after the referee has indicated it is their turn to throw.

► All balls must be thrown, rolled, pushed, struck, or kicked into the court; use of a head pointer, chin lever, or pull lever is acceptable.

There are six divisions of play: four individual, one pairs, and one team. Each division is played by competitors of either sex.

Other Variations and Modifications

There are several ways to modify boccie for participation by people with disabilities. A major concern is a lack of sufficient strength to propel the ball toward the jack. In such cases, substitution of a lighter object, such as a Nerf ball or balloon, or reduction of the legal court size would facilitate participation. Another area of concern is upper-extremity involvement, which could prohibit rolling or throwing the ball. This concern can be overcome if the individual is permitted to kick the ball into the target area or perhaps, as in regular bowling, to use a bowling cue/stick.

ADVENTURE ACTIVITIES

Another programmatic area that has gained considerable momentum over the past 25 to 30 years is the adventure curriculum. Perhaps the best-known program is Project Adventure, started by R. Lentz in 1971. Project Adventure uses a sequence of activities that encourage the development of individual and group trust, cooperation, confidence, courage, independence, and competence. These themes or goals are achieved through *trust activities, cooperative games, initiative problems, rope course elements,* and *high ropes courses.* Over the years it became clear that these experiences would benefit individuals with disabilities just as they benefit those without disabilities. In 1992, with the encouragement of Project Adventure, Havens published

Photography courtesy of NICAN (Australia). Taken from a film by Outward Bound Australia.

Figure 26.7 Mountaineering descent by wheelchair.

Bridges to Accessibility. The major theme of this text is the provision of integrated adventure experiences that are accessible, rather than adapted—for persons with all abilities.

Persons with disabilities engage in adventure activities such as canoeing (Wachtel, 1987), white-water rafting (Roswal & Daugherty, 1991), kayaking (Kegel & Peterson, 1989), and backpacking (Huber, 1991). Still others participate in activities that are considered by many to be high-risk, nontraditional experiences, at least for persons with disabilities. These include rock climbing (Roos, 1991) (figure 26.7), mountain biking, mountain trekking, and mountaineering.

Hal O'Leary, who is the Recreation Program Director at the National Sports Center for the Disabled (NSCD) in Winter Park, Colorado, indicated that white-water rafting for the individual with disabilities

is not much different than it is for persons without disabilities (personal communication, October 1998). In the case of wheelchair users, the chair is placed on the floor of the raft, which is contrary to where one is normally positioned. Also, air mattresses or inflatable seats can be fitted in the boat to provide support and comfort. For persons with severe disabilities, the NSCD uses "beanbag" chairs to position individuals to participate comfortably in the rafting experience.

Huber (1991) reported on the successful through-hike of the Appalachian Trail by Bill Irwin. Irwin lost his sight to an eye disease at the age of 28. He made the hike from Springer Mountain, Georgia, to Mount Katahdin, Maine, in a little more than eight months, accompanied only by his guide dog.

Rock climbing (Roos, 1991) is becoming a favorite sport of individuals with disabilities. Mark Wellman was the first paraplegic to climb the sheer 3,000-foot granite face of El Capitan in Yosemite National Park. Eric Weihenmayer was the first person with blindness to climb El Capitan; he likewise was the first to climb the 20,320-foot summit of Denali in Alaska. More recently, mountaineering has become the talk of the world of sport for individuals with disabilities with the successful climb of Mount Everest in 1998 by Tom Whittaker, who has a below-the-knee amputation. The equally successful 1998 All Abilities Trek accompanied Whittaker's summit. A group of five people with disabilities and seven able-bodied trekkers attempted to reach the Mount Everest base camp located at 17,000 feet. These "Kripples in the Kumboo," as they call themselves, became the first group of people with disabilities to reach base camp; it took 41 days. Many of these activities are conducted in a one-day format. Also available is the adventure or wilderness trip that may last for days or weeks. Several organizations provide separate wilderness experiences for persons with disabilities, while others offer integrated experiences for people of all abilities. The more prominent organizations providing these experiences include the following:

▶ SPLORE—Special Populations Learning Outdoor Recreation and Education, 27 West 3300 South, Salt Lake City, UT 84115; 801-484-4128.

▶ C.W. HOG—Cooperative Wilderness Handicapped Outdoor Group, Idaho State University, Student Union, Box 8128, Pocatello, ID 83209; 208-236-3912.

▶ Wilderness Inquiry II—1313 Fifth Street Southeast, Suite 327A, Minneapolis, MN 55414; 612-379-3858.

▶ P.O.I.N.T.—Paraplegics On Independent Nature Trips, 4144 N. Central Expy., Ste. 515, Dallas, TX 75204; 214-827-7404.

►B.O.E.C.—Breckenridge Outdoor Education Center, 917 Airport Rd., P.O. Box 697, Breckenridge, CO 80424; 970-453-4676.

►Bradford Woods—5040 State Road 67 North, Martinsville, IN 46151; 765-342-2915.

INCLUSION

The variations and modifications that have been highlighted in this chapter reflect what is considered good practice in physical education as well as in sanctioned sports programs (e.g., limiting the play areas is an adaptation technique used in several sports such as tennis, badminton, and boccie). Individual, dual, and adventure sports and activities provide a unique opportunity for encouraging inclusion of people with disabilities with their able-bodied peers. From elementary school through high school, variations and modifications can be utilized to alter a sport in subtle ways (e.g., maintaining physical contact while wrestling). As a result of this approach, students with disabilities can participate and/or compete in inclusive physical education and sport programs, and, not only derive the benefits of instruction in activities that are themselves normalizing, but also receive that instruction in the least restrictive environment. Because of the reduced temporal and spatial demands of most of the individual, dual, and cooperative sports and activities, there is every reason to believe that success in activities such as those highlighted in this chapter will be readily attainable within accessible programs in inclusive settings.

As it relates to inclusion in adventure activities, the Cooperative Wilderness Handicapped Outdoor Group (C.W. HOG) at Idaho State University uses the Common Adventure principle: All participants engage in planning, decision making, expense, and execution of the experience, regardless of their ability levels.

SUMMARY

For the most part, this chapter presented those individual and dual sports that are currently available as parts of the competitive offerings of the major sport organizations serving athletes who have disabilities. In each case, the particular skills needed in the able-bodied version of the game or sport were detailed. Additionally, lead-up activities were suggested, as well as variations and modifications for use in competitive sport or for use in physical education programs. It is worth noting that the rules governing sport for individuals with disabilities have undergone subtle changes since 1995. National governing bodies and international federations have changed rules in ways that permit fewer and fewer modifications. Space limitations prevented discussion of other games and sports such as riflery and air pistol, shuffleboard, darts, and billiards; information on these activities can be found in Adams and McCubbin (1991). Also highlighted were accessible adventure activities, including brief mention of the "Common Adventure" principle.

Chapter 27

Winter Sport Activities

Luke Kelly

▶

I T'S 7:00 A.M., GOING TO be partly sunny, high around 15 with winds out of the west today," blares the DJ on the radio as Andrew and Emily drive into the sunrise toward the mountains.

"It is going to be a great day for skiing," Emily says.

▶

"Yeah," says Andrew, "I hope they have the black diamond run open. I want to get some big air off that run of moguls near the top."

"You can't get big air," says Emily. "You're lucky if you don't get taken out by a tree the way you ski. Do you want to make a little wager who can get down black diamond the fastest?" asks Emily.

▶

"No way," responds Andrew. "You are not going to get another free lunch off me." As they approach the lodge, Andrew says, "You want me to drop you off here and then I'll go park?"

"That would be great," says Emily. "Get my chair set up, and then I will take our skis and get in line for the lift tickets." Andrew gets Emily's wheelchair out of the trunk and sets it up. While Emily

▶

transfers from the car seat to her wheelchair, Andrew gets the skis off the roof.

"Are you sure you are going to be able to carry your skis and mine?" asks Andrew.

"No problemo," says Emily. "Just don't get lost parking the car and make me pay for the lift tickets."

▶

The reason Andrew would not bet Emily on a race down the black diamond run was because she is a world-class sit-skier. Emily "broke her back" in a skiing accident when she was 13 and lost the use of both of her legs. Emily took up sit-skiing after her injury because she loved to ski and wanted an activity she could do outdoors with her friends. She lived in Colorado, so skiing was the natural choice for the activity.

▶

T he purpose of this chapter is to introduce the reader to a number of winter sport activities that can be included in physical education and sport programs for persons with disabilities. It is not within the scope of this chapter to cover in detail how each winter sport skill should be taught. Instead, general guidelines are provided, along with a brief description of each winter sport activity and some specific adaptations for various participants with different disabilities.

VALUE OF WINTER SPORTS

A major goal of physical education for both students with and without disabilities is to provide the knowledge, skills, and experiences they need to live healthy and productive lives. At the completion of their school physical education programs, students should have the basic physical fitness and motor skills required to achieve this goal. It would be logical to assume that the emphasis on various sport skills in the school curriculum would reflect the students' needs in terms of carryover value and the likelihood of continuing participation after the school years. However, one area, that of winter sport skills, is frequently underrepresented in the physical education and sport curriculum. This is a serious omission for all students and especially students with disabilities. In many parts of the country, the winter season is the longest season during the school year. Winter sport activities provide opportunities for individuals with disabilities to

▶ maintain or improve their physical fitness levels,

▶ participate in many social/community recreation activities, and

▶ pursue athletic competition.

Failure to provide youngsters with disabilities with winter sport skills limits their recreational options during the winter months, which, in turn, may both affect their fitness and isolate them from many social activities and settings.

Individuals with disabilities, given proper instruction and practice, can pursue and successfully participate in a variety of winter sports such as alpine skiing (downhill), cross-country skiing (nordic), ice skating, ice picking, sledding, curling, and hockey. Instructional programs for individuals with disabilities should be guided by equal attention to safety, motivation (fun), and skill development. Safety concerns should encompass the areas of physical and motoric readiness, appropriate clothing and equipment, and instructor qualifications (Leonard & Pitzer, 1988).

ALPINE SKIING

Downhill skiing is a winter sport in which most individuals with disabilities can participate with little or no modification. Skiing frees many individuals with disabilities from many of the limitations that hinder their mobility (figure 27.1) on land and allows them to move with great agility and at great speeds. For many indi-

Figure 27.1   A paraplegic uses a mono-ski and outriggers to practice chairlift unloading under the guidance of Mark Andrews and Dara Kuller of Massanutten Adaptive Ski School, Harrisonburg, Virginia.

viduals with physical, mental, and sensory impairments, skiing offers a unique opportunity to challenge their environment.

The key to learning to ski is controlling one's weight distribution and directing where the weight is applied on the surface (edges) of the skis. The goal of any introductory ski program is to provide students with the basic skills needed to enjoy and safely participate in the sport. The basic skills of downhill skiing can be grouped into six categories:

▶Independence in putting on and taking off one's equipment
▶Independence in using rope and chairlifts
▶Falling and standing
▶Walking (sidestepping, herringbone)
▶Stopping (wedge, parallel)
▶Turning (wedge, parallel)

Instruction

It is strongly recommended that actual ski instruction be preceded by a conditioning program and the development of basic skills such as falling and standing. When the actual ski instruction begins, the skill sequence must be matched to the needs and abilities of the learners to both ensure safety and maximize enjoyment. While independent recovery (standing back up on one's skis) is ultimately a required skill for independent skiing, it may not be appropriate to concentrate on this skill during early learning. For many individuals with disabilities, learning to stand up on skis after falling is very strenuous and often a frustrating experi-

ence. Students who are made to master this skill first are likely not to experience much success or fun and will soon become disenchanted with the idea of learning to ski. Initial instruction should focus on learning actual skiing skills such as a wedge stop, and the instructor should provide assistance to compensate for the lack of other skills, such as the ability to independently recover from falls. This form of instruction will provide students with confidence and some of the thrills of moving on skis (figure 27.2). As skill and enjoyment increase, the students will become more motivated to work on mastering the other essential skills, such as independent recovery.

Assistive Devices

A number of assistive devices have been developed to offset some of the limitations imposed by various disabilities or to compensate for the general low fitness and/or poor motor coordination found in many individuals with disabilities. The most commonly used device is the ski-bra (figure 27.3), which is mounted to the tip of the skis and serves two primary functions. First, it stabilizes the skis while still allowing them to move independently. Second, it assists the skier in positioning the skis in a wedge position, which improves balance and facilitates stopping and turning. The ski-bra can be used as a temporary learning device for any skier (e.g., mentally retarded, visually impaired, orthopedically impaired) during the early stages of learning to ski to assist with balance and control of the skis. The ski-bra may also be used as a permanent assistive device for individuals with lower-extremity orthopedic impairments who lack sufficient strength and/or control of their lower limbs.

Canting wedges are another common modification used to assist skiers with disabilities. Small, thin wedges are placed between the sole of the ski boot and the ski.

Figure 27.2   A quadriplegic learns to use a bi-ski tethered by Mark Andrews of Massanutten Adaptive Ski School, Harrisonburg, Virginia.

Photo courtesy of David Burton, Kludge Children's Rehabilitation Center, University of Virginia. Printed with permission.

Figure 27.3   Photograph of a ski-bra.

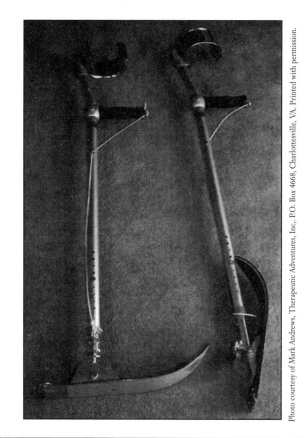

Photo courtesy of Mark Andrews, Therapeutic Adventures, Inc., P.O. Box 4668, Charlottesville, VA. Printed with permission.

Figure 27.4   Outriggers. The left outrigger is in the down position for skiing, and the right outrigger is in the up position and can be used as a crutch.

The wedges adjust the lateral tilt of the boot and subsequently affect the distribution of the weight over the edges of the skis. Canting wedges are commonly used to assist skiers who have trouble turning to one side or the other.

Outriggers (figure 27.4) are common assistive devices used by skiers with amputations and other orthopedic impairments who require additional support primarily in the area of balance. The outriggers are made from a Lofstrand crutch with a short ski attached to the bottom. The ski on the end of the crutch can be placed in a vertical (up) position and used as a crutch or positioned in a horizontal position for use as an outrigger. **Three-track** and **four-track** skiing are common terms used to describe the use of outriggers. Three-track skiing refers to individuals who use only one ski and two outriggers—for example, single-leg amputees. Four-track skiing refers to those who use two skis and two outriggers.

Sit-skiing may be more accurately described as a form of sledding, but it is included here because it is performed on ski slopes and is the method used by skiers with paraplegia and quadriplegia to ski. Sit-skiing involves the use of a special sled (figure 27.5). The person with paraplegia is strapped into the sled, which contains appropriate padding and support to hold the

skier in an upright sitting position. The top of the sled is covered by a water-repellent nylon skirt to keep the skier dry. The bottom of the sled is smooth, with a metal runner or edge running along each side. Skiers control the sled by shifting their weight (over the edge) in the direction they want to go. A single kayak-type pole or two short poles can be used by the skier to assist in balancing and controlling the sled. Special mittens are available to allow individuals with limited grip strength to hold on to the poles. For the protection of both the sit-skier and other skiers on the slope, the beginner sit-skier should always be tethered to an experienced ski instructor. Modified sit-skis are also available and can be used by persons with paraplegia for cross-country skiing. Consult the electronic resources at the end of this book to find the latest advances in sit-skiing equipment.

Ski instructors use various forms of physical assistance to help and guide skiers with different disabilities. Instructors must be able to provide sufficient physical assistance during early learning to ensure both safety and success. Providing physical assistance to a moving beginner skier requires specific skills that must be learned and perfected. As the skills of a skier with a disability increase, the instructor must also know how to gradually fade out the physical assistance to verbal cues and eventually to independence.

Photo courtesy of Mark Andrews, Therapeutic Adventures, Inc., P.O. Box 4668, Charlottesville, VA. Printed with permission.

Figure 27.5   Sit-skiing sleds. View of an athlete with a spinal cord injury in a stationary sit-ski.

In addition to the more universal assistive devices described previously, there are numerous other devices that have been created to address specific needs of skiers with disabilities. Special prosthetic limbs, for example, have been developed to allow single- and double-leg (mono-ski) amputees to ski. Many of these devices are homemade by ski instructors trying to help specific individuals. Watching a national ski competition for individuals with disabilities will clearly demonstrate that there is no limit to the devices that can be created to assist skiers who have various disabilities.

CROSS-COUNTRY SKIING

Cross-country skiing in recent years has become a very popular winter sport. It is an excellent physical fitness and recreational activity. Because both the arms and the legs are used in cross-country skiing, this activity develops total-body fitness. Two other advantages of the sport are that it costs nothing after the initial equipment is purchased and that it can be done almost anywhere (e.g., golf courses, parks, open fields). The major disadvantage of cross-country skiing for many individuals with disabilities, when compared to downhill skiing, is that the skier must create the momentum to move. This difference eliminates many individuals with more severe orthopedic impairments who lack either the strength or the control to generate the momentum needed to cross-country ski. However, because the activity is performed on snow and does not require that the feet actually be lifted off the ground, many individuals with cerebral palsy who have difficulty walking (shuffle gait) can successfully cross-country ski.

A complete cross-country skiing outfit (skis, poles, shoes, and gaiters) is relatively inexpensive. Waxless (fish-scale or step pattern) skis are recommended over wax skis for beginners. Waxless skis require no maintenance or preparation prior to use, and they provide more than sufficient resistance and glide for learning and enjoying cross-country skiing.

Initial instruction should take place in a relatively flat area with prepared tracks. Most beginning cross-country skiers tend to simply walk wearing their skis, using their poles for balance. This, unfortunately, is incorrect and very fatiguing. The key in learning to cross-country ski is getting the feel of pushing back on one ski while the weight is transferred to the front foot and the front ski is slid forward. Instructors should focus on demonstrating this pattern and contrasting it with walking. A very effective technique is to physically assist the beginning skier through this pattern so that the learner can feel what it is like. This works particularly well with skiers with mental and visual impairments. Cross-country skiers with visual impairments must be accompanied by sighted partners who usually ski parallel to them and inform them of upcoming conditions (e.g., turns, changes in grades). The forward push-and-glide technique is the preferred pattern for most skiers with disabilities, as opposed to the more strenuous and skill-demanding skating technique used by world-class nordic skiers.

COMPETITIVE SKIING FOR INDIVIDUALS WITH DISABILITIES

Skiing for individuals with disabilities is sponsored by a number of sport associations that conduct local, state, and national skiing competitions. Three of the largest and most prominent sponsors of ski competitions are Special Olympics, which sponsors competitions for individuals with mental retardation; Disabled Sports, USA (DS/USA), which sponsors national skiing competitions for individuals with disabilities; and the United States Association for Blind Athletes (USABA), which sponsors an annual national competition for skiers with visual impairments.

Special Olympics (Special Olympics, 1999) sponsors local, state, and national ski competitions. Competition is offered in both alpine and nordic events. The alpine events include downhill, giant slalom, and slalom races. The nordic events include the 500-meter, 1-kilometer, 3-kilometer, 5-kilometer, 7.5-kilometer, and 10-kilometer races, as well as a 3 x l-kilometer or a 4 x 1-kilometer relay race. Athletes are classified for competition into one of three classes—novice, intermediate, or advanced—on the basis of preliminary time trials in each event. There are no age or gender divisions. Special Olympics also offers developmental (noncompetitive/participation) alpine and nordic events. The developmental alpine events include a 10-meter glide and a 10-meter ski walk. The nordic developmental events include a 10-meter pole walk (no skis), a 10-meter ski walk (no poles), a glide event, and a 30-meter snowshoe race.

DS/USA sponsors the National Ski Championships each year. The nationals are preceded by a series of regional meets where athletes must qualify for the nationals. The national meets involve competition in three categories: alpine (downhill, slalom, giant slalom), nordic (5-kilometer, 10-kilometer, 15-kilometer, 20-kilometer, 30-kilometer, biathlon, and relays), and sit-skiing (as listed for alpine and nordic). Athletes are classified according to the site and severity of their disabilities and the type of adapted equipment used in skiing (DS/USA, 1997). Skiers with orthopedic impairments are divided into 12 classes, described briefly in the next section.

Classification of Skiers With Orthopedic Disabilities

The following is a list of classifications for individuals with orthopedic disabilities:

Class L1—Disability of both legs, skiing with outriggers and using two skis or on one ski using a prosthesis

Class L2—Disability of one leg, skiing with outriggers or poles and on one ski

Class L3—Disability in both legs, skiing on two skis with poles

Class L4—Disability of one leg, skiing on two skis with poles

Class L5—Disability of both arms or hands, skiing on two skis with no poles

Class L6—Disability of one arm or hand, skiing on two skis with one pole

Class L9—Disability of a combination of arm and leg, using equipment of their choice

Class L11—All athletes who meet the Group 1 Class in sledge sports or International Stoke Mandeville Games Federation (ISMGF) 2/3 (above T10 inclusive) and all athletes who meet the Group 1, 2, 3 Classes with major disability of one arm

Class L12—All athletes who meet the Group 2 and 3 Classes in sledge sports or ISMGF 3/4/5 (below T10 and through L1, double above-knee amputations, and other neurological conditions)

Skiers with visual impairments are divided into three classes on the basis of visual acuity with maximum correction. These classifications are used by both USABA and DS/USA for their ski competitions.

Classification of Skiers With Vision Disabilities

The following are the three most common classification for individuals with visual disabilities:

Class B1—Totally blind, can distinguish light and dark, but not shapes

Class B2—Partially sighted. Best correctable vision up to 20/600 and/or visual field of five degrees

Class B3—Partially sighted. Best correctable vision from 20/600 to 20/200 and/or field of vision from 5 to -20 degrees

The USABA, for its national ski competition, offers giant slalom and downhill alpine events as well as 5-kilometer, 10-kilometer, and 25-kilometer nordic events (USABA, 1999). Separate competitions are offered for each gender within each classification; there are no age divisions. Sighted guides are used in all of the events to verbally assist the skiers who are visually impaired. USABA sponsors approximately five nordic skiing training camps each year to promote skiing for individuals with visual impairments.

In addition to the DS/USA classifications just described, there are 3 age divisions: ages 0 through 16 (juniors), ages 17 through 39, and age 40 and over (seniors). Separate competitions are offered for each gender in each age division except when there are not enough participants of one gender to compose a heat. The same classifications apply for both alpine and cross-country skiing. The only difference between the men's and women's events is that females are limited to the 5-kilometer and 10-kilometer cross-country events.

Sit-skiing competition is conducted only in the United States, and, therefore, athletes competing in this category are not classified according to the international classification system. Sit-skiers are classified into one of two groups. Group 1 is composed of athletes with disabilities in the lower limbs, with injury between T5 and T10 inclusive. (Athletes with higher injuries, above T5, typically are not able to sit-ski.) Group 2 is for athletes with all other disabilities resulting from injury below T10 and conditions such as spina bifida, amputation, cerebral palsy, polio, and muscular dystrophy.

ICE SKATING

Ice skating is another inexpensive winter sport that is readily accessible in many regions of the country. Most individuals with disabilities who can stand and walk independently can learn to ice skate successfully. For those who cannot, a modified form of ice skating, ice picking, is available. While skating is common in many areas on frozen lakes and ponds or water-covered tennis courts, the preferred environment for teaching ice skating is an indoor ice rink. An indoor rink offers a more moderate temperature and a better quality ice surface, free from the cracks and bumps commonly found in natural ice. Ice rinks frequently can be used by physical education programs during off times such as daytime hours on weekdays.

Properly fitting skates are essential for learning and ultimately enjoying ice skating. Ice skates should be fitted by a professional experienced in working with and fitting individuals with disabilities. Either figure or hockey skates can be used. The important consideration is that the skates provide good ankle and arch support so that the skater's weight is centered over the ankles and the blades of the skates are perpendicular to the ice when the skater is standing.

As discussed earlier in this chapter, instruction should be guided by safety and success. The greatest obstacle in learning to ice skate is the fear of falling. Although falling while first learning to skate is inevitable, steps can and should be taken to minimize the frequency and severity of the falls and, consequently, the apprehension. At the same time, the early stages of learning must be associated with success, which gives learners confidence that they will be able to learn to skate. It is recommended that padding be used around the major joints most likely to hit the ice during a fall. Knee and elbow pads reduce the physical trauma of taking a fall and also provide a form of psychological security that alleviates the fear of falling. When teaching individuals with disabilities to ice skate, football pants with knee, hip, and sacral pads along with elbow pads have been found to be very beneficial during the early stages of learning (see the Overcoming Fear application example).

The locomotor skill of ice skating is very similar to walking. The weight, the center of gravity, is basically transferred in front of the base of support and from side to side as the legs are lifted and swung forward to catch the weight. The back skate is usually rotated outward about 30 degrees to provide some resistance to

Application Example: Overcoming Fear

Setting: A physical education class is starting an ice hockey skills unit. Most of the students can skate forward well and are working on speed, changing direction, and skating backwards.

Student: John is an 11-year-old with mild mental retardation who recently moved to Michigan from Florida. He has slightly below-average coordination for his age but generally is willing to try new skills. John has recently become a big MSU hockey fan.

Issue: John had a negative experience with ice skating the first time he tried it and now is extremely fearful and unwilling to even put skates on.

Application: Based on the above information and a meeting with John's parents, the physical educator decided on the following strategies:

► Meet individually with John to discuss his fears of ice skating and to explain how these fears will be addressed.

► Contact the local university and borrow some official Michigan State hockey pads and a helmet of appropriate size

► John will start using a padded Hein-A-Ken skate aid to give him confidence. He will then transition to using a hockey stick as a balance aid.

► An aide was prepared by the physical educator and assigned to work with John during the initial lessons to ensure he was successful and to prevent any new negative experiences.

sliding backward as the weight is transferred to the forward skate. Because success during the early lessons is essential, one-on-one instruction from an experienced instructor is highly recommended.

The primary aid used in teaching ice skating is physical assistance. Some individuals with orthopedic and neuromuscular impairments may benefit from the use of polyproplylene orthoses to stabilize their ankles. Ankle-foot orthoses are custom made and can be worn inside the skates. The most universal skating aid is the Hein-A-Ken skate aid, which is simply a walker that has been modified to be used on ice (Adams & McCubbin, 1991). This device does not interfere with the skating action of the legs, and it provides the beginning skater with a stable means of support independent from the instructor. The skate aid can be used for temporary assistance during the early stages of learning for students who need a little additional support or confidence; it can also be a more permanent assistive device for skaters with more severe orthopedic impairments. It is beneficial to add some foam padding to the top support bar in the front of the skate aid to further reduce the chance of injury from falls. If skate aids are not available, chairs can be used in a similar fashion.

Ice picking is a modified form of ice skating in which the participant sits on a sledge, a small sled with blades on the bottom, and uses small poles (picks) to propel the sledge on the ice (figure 27.6). Ice picking can be performed by almost anyone and is particularly appropriate for individuals who only have upper-limb control (e.g., paraplegics or those with spina bifida). Ice picking is an excellent activity for developing upper-body strength and endurance. All skating activities and events (speed skating and skate dancing) can be modi-

fied and performed in sledges. Because both able-bodied individuals and individuals with disabilities can use the equipment, ice picking offers a unique way to equalize participation and competition in integrated settings.

Special Olympics sponsors (Special Olympics, 1999) ice skating competitions in two categories: figure skating and speed skating. The figure-skating events include singles, pairs, and ice dancing; the speed-skating events include the 100-, 300-, 500-, 800-, 1,000- and 1,500-meter races. For each event, athletes are divided into three classifications—novice, intermediate, and advanced—on the basis of preliminary performance and time trials. Developmental (noncompetitive/participation) ice-skating events are also offered. These include the slide for distance, the 10-meter assisted skate, the 10-meter unassisted skate, and the 30-meter slalom.

SLEDDING AND TOBOGGANING

In snowy regions of the country, sledding and tobogganing are two common recreational activities that are universally enjoyed by both children and adults. Many individuals with disabilities, however, avoid these activities because they lack the simple skills and confidence needed to successfully take part in them. The needed skills and confidence can easily be addressed in a physical education program. Given proper attention to safety and clothing, almost all children with disabilities can participate in sledding and tobogganing. Sleds and toboggans can be purchased and/or rented at minimal cost. Straps and padding can be added to commercial sleds and toboggans to accommodate the specific needs occasioned by individuals with different disabilities. Even individuals with the most severe disabilities can experience the thrill of sledding or tobogganing when paired with an aide who can control and steer the sled.

HOCKEY

Ice hockey is a popular winter sport in the northern areas of the United States and is the national sport of Canada. The game is played by two teams who attempt to hit a puck into the opposing team's goal using their hockey sticks. Hockey is a continuous, highly active, and exciting sport. Because of these features, numerous modifications and adaptations have been made to hockey to accommodate players with disabilities. Some common modifications are as follows:

▶ Using soft plastic balls, plastic pucks, or doughnut-shaped pucks instead of the traditional ice hockey pucks

▶ Using shorter and lighter sticks made of plastic, which are more durable, easier to handle, and less harmful to other players

Photo courtesy of Mark Andrews. Printed with permission.

Figure 27.6   Sledges and picks used in sledge hockey.

▶ Changing the surface to a less slippery surface like a gymnasium floor or tennis court

▶ Changing the boundaries, the number of players per team, and the length of the playing periods to accommodate the ability of the players

▶ Changing the size of the goals

Modifications can easily be made to enable sticks to be held by players with physical impairments or used from wheelchairs by using tape and Velcro. A wide range of abilities can be accommodated in a game if the teams are balanced and the players' abilities are matched to the various positions.

Special Olympics sponsors local, state, and national competition in two versions of hockey: floor hockey and poly hockey. These two games are basically the same except for the sticks and pucks used in each. In floor hockey, a stick similar to a broomstick with a vinyl coating on the end is used in conjunction with a doughnut-shaped puck. In poly hockey, plastic sticks similar to regular hockey sticks are used along with a plastic puck shaped like a regulation puck. The goalkeeper in both versions uses a regular hockey goalie stick. Although there are some minor variations in the skills related to each version due to the differences in equipment, the basic skills are the same for both games.

Sledge hockey is a modified form of ice hockey played on sledges. The only difference from the regulation game of ice hockey is that the game is played from a sledge and the puck is struck with a modified stick called a pick (see figure 27.6). The pick is approximately 30 inches in length. On one end it has metal points that grip the ice and allow the athlete to propel the sledge. The other end, called the butt, is rubber coated. The butt end of the pick is held while the sledge is being propelled. When the athlete wants to hit the puck, the hand is slid down the shaft of the pick to cover the spiked end, and then the butt end of the pick is used to strike the puck.

Sledge hockey is an excellent recreational and fitness activity. Using sledges is also an ideal way of equating able-bodied students and students with orthopedic impairments in the same activity. Canada is the leader in developing sledge hockey. For more information on sledge hockey, consult the electronic resources provided for this chapter at the end of this book.

Logical modifications should be made to the regulation game of hockey to accommodate beginners, such as reducing the playing area, increasing the number of players on each team, increasing the size of the goal, playing without goalkeepers, or changing the size or type of puck (e.g., substituting a playground ball). The goal of all modifications should be to maximize participation and success in the basic sledge and hockey skills while gradually progressing toward the regulation game.

CURLING

Curling is a popular recreational activity and sport in Europe and Canada. The playing area is an ice court 46 yards long and 14 feet wide, with a 6-foot circular target, called a house, marked on the ice at each end. The game is played by two teams of four players, with pieces of equipment called stones (a kettle-shaped weight 36 inches in circumference and weighing approximately 40 pounds, with a gooseneck handle on top). A game is composed of 10 or 12 rounds called heads; a round consists of each player delivering (sliding) two stones. Players on each team alternate delivering stones until all have been delivered. After each stone is delivered, teammates can use brooms to sweep frost and moisture from the ice in front of the coming stone to keep it straight and allow it to slide farther. At the end of a round, a team scores a point for each stone they have closer to the center of the target than the other team. The team with the most points at the end of 10 or 12 rounds is the winner. If the score is tied, an additional round is played to break the tie.

Curling can easily be modified to accommodate individuals with just about any disability. The distance between the houses and the weight of stones can be reduced to facilitate reaching the targets. The size of the targets can also easily be increased to maximize success. Audible goal locators can be placed on the houses to assist players with visual impairments. The sweeping component of the game may be difficult to modify to include players who are nonambulatory or have visual impairments. In these cases, mixed teams could be formed of players with different disabilities and/or combining nondisabled players and players with disabilities so that each team has a few members who could do the sweeping. Finally, assistive devices like those used in bowling (ramps and guide rails) could be used to help players with more severe disabilities deliver the stones. For the latest information on curling, check the electronic resources provided for this chapter at the end of this book.

SUMMARY

Winter sports, in general, are excellent all-around activities. They develop motor skill, strength, and physical fitness, while at the same time provide participants with functional recreational skills they can use for the rest of their lives. For many individuals with disabilities, winter sports performed on snow and ice allow them to move with agility and speed not possible under their own power on land. Winter sports, therefore, should be an essential component in the physical education and sport programs of all students and, especially, for students with disabilities. For this reason, activities have been discussed in this chapter with particular focus on ways to modify them for people with unique needs.

Enhancing Wheelchair Sport Performance

▶ ▷ ▷

THE BALL WAS LOW. The guard saw that, and although he was 6 inches shorter than the forward, if he could just get "up" enough, he could tip it away. His system rose the needed inches and he clutched the ball as he returned to the ground. Out of the corner of his eye he saw his forward, guarded, but open, at the other end of the court. He sent the ball the length of the court but again, the ball was low. The forward pivoted to the left and just "rose" enough to tip the ball from the marauding center. As the opposing center challenged the ball, the forward performed his patented fall-away shot. As his body fell to the floor, and the clang of wheelchairs rang, and the official's whistle blew, he saw the ball pass through the basket … good for two points.

Yes, "as the clang of wheelchairs rang." This is the National Wheelchair Basketball Association as it enters the 21st century. This example of the combination of body and wheelchair into a performance "system" has opened up elite disability sport to a new dimension, far from that envisioned by early pioneers. There is now a functional perspective in wheelchair sports that places disability second to the demands of the specific sport. This transition has been spearheaded by the athletes themselves, and they have been instrumental in developing many of the innovations in equipment and technique that have occurred.

Coupled with innovations in equipment and technique has been increased specialization by the athletes. It was once possible for an athlete to compete at World Championship level in a number of sports, but this is no longer the case. Sport for athletes with disabilities now parallels the able-bodied world, and to reach the highest levels, athletes must find the sport (and often the event or position) for which they are best suited. Suitability for specific sports is most commonly determined by interest, disability type, body size and shape, and the psychological makeup of the athlete. In order to maximize wheelchair sport performance, event selection should be combined with a systems approach to combining the athlete and the wheelchair for competition. Taking all of these changes and advances into consideration, this chapter covers the process of selecting both a sport and an appropriate chair, as well as training and performance considerations.

THE ATHLETE AND THE WHEELCHAIR: A SYSTEMS APPROACH

Conceptually, the athlete and the wheelchair combined can be viewed as a performance system. This reflects the functional model of disability sport now prevalent in both the design of wheelchairs and the functional classification systems employed in elite disability sport competition. Much of the early research available in wheelchair sports examined *either* the athlete *or* the wheelchair. Many of these studies also examined wheel-

chair users outside of their preferred environment, that is, their competitive wheelchairs and employed instruments such as ergometers that had scientific but not ecological validity. In appreciating the functional development in wheelchair sports, it is necessary to understand the importance of a systems approach. A systems approach incorporates the combination of athlete and wheelchair into a performance *system* that is defined by the needs of the specific sport. This chapter will examine performance enhancements for the wheelchair athlete, provide a brief description of the equipment, and then examine the combination of the athlete and the wheelchair into a performance system.

The Athlete

Success in wheelchair sports requires that an athlete be suited to meet the performance considerations of that sport. For example, success in basketball requires height for the forward and center positions and speed and agility in the guard positions. In the main, athletes with predominantly fast-twitch muscle fibers should focus on sprint events, and those with slow twitch should focus on the endurance events. Those with the psychological profile that includes subsuming the self into a team mold should, of course, focus on team events; those with an individual orientation should focus on the appropriate individual sport. While this is somewhat of a truism in able-bodied sports, in the past, wheelchair athletes have not focused on fulfilling their personal goals by selecting the appropriate challenge. After identification of the appropriate event, the athlete needs to focus on meeting the demands of that particular sport by identifying a training regime.

Training

Wheelchair users respond to physical training in a similar, but not identical, manner to each other and to the able-bodied population (Shephard, 1990; Wells & Hooker, 1990). In addition to developing a foundation of health-related fitness (see chapter 21) when identifying the training requirements in a given sport, the athlete should develop levels of fitness associated with performance for the particular sport. The athlete should also consider factors such as individual orientation to specific sports and body anthropometry. Performance-related fitness includes components such as movement, coordination, agility, power, speed, and balance (Gallahue & Ozmun, 1998). These components are applicable to wheelchair performance and should be developed in a sport-specific manner. Ultimately, the trained athlete will demonstrate improved performance in the sport under examination (Curtis, 1981a). Note that because agility and balance require the combination of the athlete with the wheelchair, they are addressed in a subsequent section.

Training regimens for athletes with disabilities parallel those for athletes who are able-bodied and, in many cases, appear identical. Where modifications are necessary, it is usually due to the smaller active muscle mass and to the lack of alternate modes of training available. Runners can train their legs through running, jumping, stair climbing, cycling, or weight training. Athletes who use wheelchairs are usually limited to weight training, pushing their chairs, or arm cranking (if they have access to this specialized equipment). This makes repetitive overuse injuries more prevalent and makes the incorporation of rest into the training plan critical.

Underpinning most athletic performances is the development of a sound cardiovascular "base." However, training the cardiovascular system presents unique problems to the wheelchair athlete. Cardiovascular endurance is produced by stressing the heart and respiratory system through the use of major-muscle groups, which expend large amounts of energy over prolonged periods of time. For able-bodied athletes, running, cycling, and swimming use the large muscles of the trunk and lower limbs and are excellent modes of exercise. The wheelchair athlete, however, is limited to using the relatively small muscles of the arm and, in some cases, the muscles of the trunk. This smaller working muscle mass places lower demands on the heart and lungs and makes cardiovascular endurance training more difficult. However, training rollers, arm-crank ergometers, and upper-body exercise (UBE) systems that are modalities used in cardiovascular endurance programs of elite wheelchair athletes can be employed in this respect.

While a cardiovascular base is essential, power more directly relates to the performance demands made in the anaerobic sports exemplified on the court (tennis, basketball, and quad rugby). Because of the small muscle mass involved in wheelchair propulsion and the asymmetry of the propulsion movements, systematic strength and flexibility training is critical. Stretching, both before and after exercise, may be more important for athletes with disabilities than for the able-bodied (Curtis, 1981b). Weight training will develop strength and, indirectly, power and is a major component in the regime of the elite athlete. It should be noted that the wheelchair athlete propels the wheelchair using a relatively small range of motion at the shoulder and elbow, and this action is asymmetrical in that the forces of extension (during propulsion) are far greater than the muscular forces used in recovery. This asymmetry can lead to muscle imbalance around the shoulder joint, which in turn may cause serious overuse injury and postural problems (Curtis, 1981b). An important component of the weight-training program for wheelchair users should, therefore, be development of the posterior (back) muscles. A simple rule of thumb is to pair the muscles in the training program (i.e., biceps and triceps) and to include free-weight exercises that require the athlete to be face down (prone) on the work bench. Exercises for the latissimus dorsi and the trapezius muscles of the back are examples of the muscles that should be targeted in a weight program if such exercises are not contraindicated. Attention to appropriate stretching practices will also improve weight-training regimes (Curtis, 1981b); and, where there is a strength imbalance in the muscles due to the disability, stretching can reduce problems such as contracture. The next section examines other disability-specific sports medical concerns.

Medical Concerns

Wheelchair athletes face several disability-specific sports medicine problems, of which the most important are those associated with thermal regulation. Damage to the spinal cord presents problems at both high and low ambient temperatures because impairment of sensory nerves means that athletes are often unable to feel heat, cold, or pain. Thus, in cold weather, they receive no sensory warning that body extremities (usually the feet) are becoming frozen. Coupled with reduced blood flow to the inactive feet, frostbite becomes an ever-present danger against which the athlete must guard.

At high ambient temperatures, the problem is damage to the nerves that initiate and control sweat production. The problem is particularly severe in athletes with quadriplegia, many of whom have little or no body-sweat production and thus no way of reducing their core temperatures. The provision of shade, adequate drinking fluids, and wet towels for surface-temperature reduction can help alleviate this problem.

A particularly pernicious medical problem that has surfaced in wheelchair sport is the life-threatening but deliberate precipitation of autonomic dysreflexia by athletes with quadriplegia, a process referred to as "boosting." Autonomic dysreflexia is a medical condition characterized by hypertension, piloerection, headaches, and bradycardia and associated with very high levels of catecholamine. Autonomic dysreflexia is unique to individuals with spinal cord injury above the major splanchnic outflow at the sixth thoracic vertebrae.

Athletes with quadriplegia believe that "boosting" increases their athletic performances, and experimental evidence (Burnham, et al., 1993) supports this view. National and international sports groups are aware of both the use of "boosting" and its dangers, and the practice is now banned in Paralympic competition.

The Wheelchair

Developments in wheelchair design that match the chair to the demands of the sport have led to a multiplicity of choices for the athlete wishing to enhance performance (LaMere & Labanowich, 1984 a, b, c). It is no longer feasible at the elite level to expect a general wheelchair to perform to the competitive demands of a sport. Just as the able-bodied athlete wears different shoes for different sports, the elite wheelchair athlete uses different wheelchairs for different sports. Nevertheless, there are some commonalties in wheelchair design that generalize across the spectrum of competitive wheelchairs.

The Wheelchair Frame

The wheelchair frame performs one major function. It holds the other components—the seat, the mainwheels, and the front wheel(s)—in their proper relative positions. Perhaps the most important design consideration in building a wheelchair frame is to make it as light and as stiff as possible. It needs to be light weight so that the athlete has to propel as little weight as possible during athletic performance. It needs to be rigid so that the energy that the athlete applies to the wheelchair is used to drive the chair rather than to bend and deform the frame. Frame flexing absorbs energy directly, but because the wheel alignment of the wheelchair changes as the frame flexes, additional energy is lost when the mainwheels do not point straight ahead in the direction of travel. In addition, the frame must be matched to the body size and shape of the athlete and will vary dependent on the performance considerations of the sport.

Wheels

The rear wheels of racing wheelchairs are larger in diameter and narrower in cross-section than those used in other sport chairs because of the differing demands of the activity. Wheels for court chairs have to be able to withstand rotational torque and, in many cases, contact with other wheelchairs. Therefore, some athletes in contact sports such as basketball and quad rugby forgo lightness for additional robustness in the mainwheel and have stronger, cross-spoked wheels that can withstand these extra forces.

For racing, unless the athlete is very small, the mainwheels are usually high-quality racing bicycle

Figure 28.1   Basketball performance system.

wheels in the European 700C or North American 27-inch size. For road racing, the narrowest possible tires (19 millimeters or less) are used, while for track racing there appears to be some benefit to using wider 23- to 28-millimeter tires. Since additional weight in the wheels of a wheelchair slows the athlete down twice as much as additional weight in the frame of the chair, racing wheels should be as light, strong, and rigid as possible. Although athletes have experimented with both disk and three-spoke airfoil wheels, there have been no demonstrated performance gains over traditionally spoked wheels, probably because of their greater weight and poorer performance in cross-wind conditions. The trend for front wheel(s) has been to use as large a wheel as the rules allow, and 18- and 20-inch (.46- and .51-meter) diameter front wheels are common.

For court chairs like those used in the basketball game pictured in figure 28.1, the size of mainwheels ranges from 24 inches to the maximum permissible 26 inches, dependent on the position played (Yilla, La Bar, & Dangelmaier, 1998). The preferred front wheel now tends to be the skateboard-type available at regular hobby shops.

Number of Wheels

A wheelchair remains stable as long as the center of gravity of the athlete plus the chair remains inside the wheelchair's base of support. The base of support is the area of ground marked by the points at which the wheels contact the surface. In four-wheeled designs, the base of support is rectangular with the base a little narrower at the front than at the rear, while in three-wheeled designs, the base of support is triangular. This means that, as the weight of the athlete moves forward (as he leans forward to cut down air resistance), the center of gravity gets nearer to the edge of the base of support of the three-wheeled chair, and the chair becomes less stable. This lack of stability can be a problem for less-experienced athletes or for some court sports, but for those who can handle them, three-wheeled chairs are faster. Three-wheel designs are faster because there is less resistance to passage over the ground for three rather than four wheels (Higgs, 1992a), and three-wheeled designs also have considerably less wind resistance than four-wheelers, under most wind conditions (Higgs, 1992b).

Note that a performance consideration in many court sports is the necessity to extend laterally forward to the right or the left, and this may require the extra stability inherent in a four-wheel design. Caution should be exercised, therefore, in selecting a three-wheel chair for basketball or quad rugby where stability is at a premium. For tennis and racing, elite athletes almost exclusively now use a three-wheel configuration. An additional wheel for front-to-rear stability can be attached to the back of the chair and is now popular in tennis, basketball, and quad rugby.

Mainwheel Alignment

To allow the wheelchair to roll with the least resistance, it is critical that the mainwheels point straight ahead. If the mainwheels point slightly outward (toe-out) or slightly inward (toe-in) they significantly slow down the wheelchair. O'Reagan, Thacker, KauzIarich, Mochel, Carmine, and Bryant (1981) showed that, for some tires, toe-in or toe-out of as little as three degrees increased rolling resistance tenfold. Since the front wheels (castors) are free to move to allow turning to take place, they are not subject to the same toe-in/toe-out problems as mainwheels. They do, however, increase rolling resistance greatly when their bearings become worn.

Camber Angle

Sports chairs are cambered to allow for superior turning and/or ease of pushing. The mainwheels of a racing wheelchair are cambered to allow for maximum application of force to the pushrim. With the wheels cambered, the hands fall naturally to the handrim. With court chairs, the camber greatly enhances maneuverability. The majority of athletes use camber angles between 6 and 12 degrees, with 8 to 10 degrees being most popular.

Seat Height

All other factors being equal, the most effective seat height is a function of the athlete's trunk and arm length and of the handrim size that is selected. Higgs (1983) reported that in athletes at the 1980 Paralympic Games, superior performances in racing events were recorded by those with lower seats. Experimental work by Traut (1989) showed greater propulsion efficiency when a "relatively low" seat position was used. Experimental work by Meijs, Van Oers, Van de Woude, and Veeger (1989) and by Van de Woude, Veeger, and Rozendal (1990) showed that there was a relationship between the elbow angle (when the athlete was sitting upright in a general sport wheelchair with hands placed on the top center of the handrim) and propulsion efficiency. Their results showed that efficiency was greatest when the elbow angle was 80 degrees and that the energy cost of sitting too high in the chair was greater than the penalty paid for sitting too low. However, performance considerations (height in basketball or the post position in quad rugby) offset some propulsion considerations.

Additional sport-specific considerations for wheelchairs follow. Note that there is a fundamental difference in the configuration of the racing wheelchair relative to court chairs.

Specific Considerations for Racing Wheelchair

As can be seen by the racer in figure 28.2, the structure of the racing wheelchair has changed fundamentally since Bob Hall first pushed in the Boston Marathon. In fact, Goosey and Campbell (1998) concluded that wheelchair design (combined with disability) may be more important factors in pushing efficiency than propulsion techniques. Specific considerations for racing wheelchairs involve the weight distribution of the athlete in the wheelchair, which can be modified by the anterior/posterior seat position, the size of the handrims, and the use of accessories. Detailed explanations of these considerations follow.

Anterior-posterior seat position

Little is known about the optimum anterior-posterior position of the wheelchair seat, although this position affects both the chair's stability and the effectiveness of application of force to the handrim. If the athlete is too far toward the rear of the wheelchair, there is a tendency for the chair to become unstable (particularly when going uphill) and for the athlete to "flip" out the back. A rear seat position makes it difficult for the athlete to apply force to the front of the handrim where the most effective application of driving force can be made. There is, to date, little research evidence on optimum anterior/posterior seating, and positioning for elite athletes is usually achieved by trial and error.

Courtesy of SPORTS 'N SPOKES/Paralyzed Veterans of America.

Figure 28.2   Racing performance wheelchair.

Handrims

The handrims are the point at which the energy produced by the wheelchair athlete is transmitted to the wheelchair, and as such they are critical to producing optimal performance. The three most important aspects of the handrim are its diameter, its width, and the material with which it is covered.

Handrim diameter

The handrim acts like the gearing for the wheelchair. If a small-diameter handrim is used, the athlete has selected a "high" gear that produces poor acceleration but a high top speed. Conversely, if a large-diameter handrim is used, the benefit is greater acceleration at the cost of a lower top speed. In general, stronger athletes are able to effectively push smaller-diameter handrims. Thus the optimum handrim diameter is a function of the size of the athlete, the relative importance of acceleration and top speed in the race being run, and the strength of the athlete. Most handrims are between 14 and 16 inches (.35 and .41 meter) in diameter, and athletes usually experiment to determine what works best for them.

Handrim width

If the handrim is made of relatively wide tubing, it is easier to grasp, which makes starts and uphill climbing easier. On the other hand, it is possible that narrower tubing would encourage higher wheeling speeds because the athlete is more likely to "strike" the handrim rather than hold it and push. In the absence of research studies, athletes determine their optimum handrim diameter by trial and error.

Handrim material

The handrim covering is of great importance because it is this material that the hand strikes during propulsion. If it is too smooth or shiny, the hand will slip when power is applied. For this reason, a number of materials have been used for handrim covers. Many racers also apply adhesives to increase handrim traction. Although the frictional grip of the handrim is important, it is only half of the hand-wheelchair interface, and the hand covering used by the athlete is of equal or greater importance. Most athletes wear gloves that have been sculptured to their exact requirements by the application of hundreds of layers of adhesive tape. This glove-and-tape combination provides protective cushioning and high, instant grip between the hand and the handrim. Again, the athlete is encouraged to experiment with different materials to find the combination that meets his or her needs.

Accessories

Racing wheelchair accessories are almost as numerous as the wheelchair athletes who use them, but almost all racing wheelchairs incorporate at least a steering device, a compensator, and a computer. With downhill racing speeds reaching more than 40 mph (64 kph), the need for a steering mechanism to help the athlete negotiate corners is obvious, and the usual steering device is a small handle attached directly to the front-wheel mounting. This lever can be moved left or right to steer the wheelchair, although steering only occurs when the lever is held in place. Once released, the front wheel returns (under spring action) to a neutral, straight-ahead position. This process is called active steering, since turning only occurs when steering input is applied by the racer. In addition to this active steering mechanism, the wheelchair also incorporates a compensator.

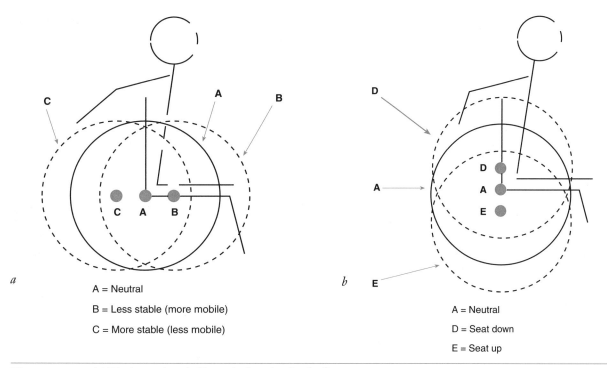

A = Neutral

B = Less stable (more mobile)

C = More stable (less mobile)

A = Neutral

D = Seat down

E = Seat up

Figure 28.3   *(a)* Horizontal and *(b)* vertical mainwheel adjustment.

The purpose of the compensator is to permit small, long-term adjustments to the direction in which the chair moves, and it is most important in road racing. Most road races are held on public roads that are designed with a high crown along the midline, with the road falling away for drainage toward the curb. A chair propelled along the crown would go straight, but a chair wheeled near the curb for safety would be moving forward on a sideways sloping surface. The front of the chair would be constantly "falling away" from the crest of the crown, and the chair would tend to steer into the curb. A compensator applies a small offset to the front wheel to allow the chair to move straight ahead without the athlete having to make constant small corrective steering adjustments. The compensator also stabilizes the wheelchair in the event of surface irregularities.

Bicycle computers are now relatively inexpensive and are almost universally found on racing wheelchairs. These computers provide essential information on distance traveled, cadence, and top and average speeds. This feedback is essential for developing and maintaining accurate training and racing logs that facilitate performance enhancement.

Specific Considerations for Court Chairs

Because of similarities in configuration, it is easy to assume that a wheelchair that is appropriate for everyday use is also sufficient for sport. While a general wheelchair may be sufficient for recreational physical activity, when examining enhancing wheelchair sport performance, a first step is to realize the importance of utilizing sport-specific equipment. The following paragraphs contain some general guidelines that should help

when selecting a wheelchair for court sports. Yilla (1997) provided a more detailed explanation on selecting a wheelchair specifically for basketball, and much of this information transfers to the selection of other court chairs.

Selecting a Wheelchair for Court Sports

If the athlete is inexperienced, advice should be sought from someone who has experience in court sports and understands the athlete's function level at the position played. This will help give the athlete insight into the functional demands of the sport. If this is a first court chair, it should be purchased from a reputable manufacturer.

There are now a bewildering number of options in performance wheelchairs and in their designs; many are experimental. It is advisable to avoid experimental designs until the athlete is comfortable with the performance demands of the sport. The athlete's first, sport-specific performance chair should be adjustable both for height and for point of balance so that it can be modified (figure 28.3, a-b). However, because of weight and performance considerations, the athlete should avoid wheelchairs that have too many adjustments.

Wheelchairs Without Wheels

Functionality is the focus of performance wheelchairs, and these chairs can diverge from the traditional concept of the wheelchair to the point that these "wheelchairs" no longer have wheels! In particular, throwing chairs for field events, mono-skis for snow skiing, and mono-skis for water skiing are all "wheelchairs" without wheels (figures 28.4 and 28.5).

Figure 28.4   A throwing chair for field events.

Wheelchairs without wheels are the latest sport wheelchair development. They are designed to provide a rigid, stable base from which the athlete can perform optimally. Throwing chairs are firmly anchored to the ground during the throw; there is no need for wheels, and, as can be seen in figure 28.4, they are no longer part of the throwing chair design. The heavy metal frame is designed to provide stability for the athlete and to provide anchor points so that the chair can be tied down to prevent it from moving during throws. The seat is built as high above the ground as the rules allow, and since cushioning absorbs some of the power of the throw, the seat is usually hard. Sit-skis and mono-skis provide a similar sport-specific approach, and, in meeting the functional demands of the activity, the wheels have become superfluous.

THE SYSTEMS APPROACH— COMBINING THE ATHLETE AND THE WHEELCHAIR

The process of combining the athlete and the wheelchair into a sport "system" will vary dependent on the specific sport. However there are some general principles that can be applied when fitting the wheelchair itself. Additionally, other performance considerations are broken down into the division between racing wheelchairs and court chairs.

Fitting the Wheelchair to the Athlete

Proper fitting of the wheelchair to the athlete is critical for high levels of athletic performance. Most manufacturers provide retail experts who are experienced in measuring athletes for performance wheelchairs.

In fitting the frame, the two most critical dimensions are the width and the relative positions of the seat and wheels. If the frame is too narrow for the athlete, there will be insufficient clearance between the wheels and the athlete; this results in the wheel rubbing the athlete's body. This both slows the chair and produces frictional injury to the athlete. If the frame is too wide, the handrim will be difficult to reach and even more difficult to push effectively.

The relative positioning of the seat and the mainwheels is dependent on the material in the wheelchair, the sport to be played, the athlete's weight, and the athlete's fitness level. Therefore, an adjustment on the wheelchair for this relative position is essential no matter how experienced the athlete. Because different manufacturers present this adjustment option in different ways, this should be an important factor when selecting the chair. Refer to the Enhancing Wheelchair Sports Performance application example for a list of considerations to keep in mind while helping athletes find the chair that suits them.

System Considerations for Racing Wheelchairs

There are a number of system considerations that apply to racing wheelchairs. The following section identifies *propulsion techniques* and how to *overcome negative forces* as important considerations in developing an athlete's wheelchair racing system.

Propulsion Techniques in Track and Road Racing

Coupled with the evolution of the racing wheelchair has been the development of ever-more efficient propulsion techniques (Higgs, 1985). A six-phase technique (figure 28.6) is most frequently used, although not all athletes use each phase with the same degree of effectiveness. A subsequent analysis by O'Connor, Robertson, and Cooper (1998) led the authors to conclude that there is a need for coaches to become more knowledgeable concerning appropriate wheelchair propulsion techniques.

Figure 28.5   A mono-ski for water skiing.

The basic stroke

The **propulsion cycle** starts with the hands drawn up as far above and behind the handrim as is possible given the seating position and flexibility of the athlete. The hands are then accelerated as rapidly and forcefully as possible (acceleration phase) (see points A on figure 28.6) until they strike the handrim. The moment of contact is the impact energy transfer (point B on figure 28.6) phase, during which the kinetic energy stored in the fast-moving hand is transferred to the slower-moving handrim. With the hand in contact with the handrim, there is a force application, or push, phase (points C on figure 28.6), and this continues until the hands reach almost to the bottom of the handrim. During the force application phase, most of the propulsion comes from the muscles acting around the elbow and shoulder. As the hands reach the bottom of the handrim, the powerful muscles of the forearm are used to pronate the hand, which allows the thumb to be used to give a last, powerful "flick" to the handrim. This last flicking action is reversed by a few athletes who use supination in the rotational energy transfer phase (points D on figure 28.6) to flick the handrim with the fingers rather than the thumb. Immediately following the rotational energy transfer, the hands leave the handrim during the castoff phase (see point E on figure 28.6). Here it is important that the hand be moving faster than the handrim as it pulls away since a slower hand will act as a brake to the wheelchair. Often athletes will use the pronation or supination of the rotational energy transfer phase to accelerate the hands and arms and thus allow them to be carried up and back under ballistic motion. This upward and backward motion is called the backswing phase (points F on figure 28.6) and is used to get the hands far enough away from the handrim to allow them to accelerate forward to strike the handrim at high speed at the start of the next stroke.

This basic propulsion stroke is modified by the terrain over which the athlete is wheeling, by the tactics of the race, and by the athlete's level of disability. On uphill parts of a course, the athlete shortens the backswing and acceleration phases so as to minimize the time that force is not applied to the handrim, when the chair could roll backwards. Tactically, the athlete is either wheeling at constant speed or is making an attack and needs to accelerate. The basic stroke described here is used at steady speed, while during bursts of acceleration the major change in stroke takes place during the backswing. At steady speeds the backswing is a relatively relaxed ballistic movement in which the velocity at castoff is used to raise the hand to its highest and most rearward position. This relaxed backswing is efficient and allows a brief moment of rest during each stroke. During acceleration, however, the major change in stroke dynamics is to increase the number of strokes from approximately 80 per minute to more than 120 per minute. This is achieved by a rapid reduction in the time taken for a more restricted backswing.

The start

In short races, the start is critical, and a good start can provide the margin of victory. Critical aspects of the start are the upper-body action and modifications to the first few strokes. Newton's third law, the law of action and reaction, confirms that if the upper body is thrown forward at the starting gun there will be a reaction of the lower part of the body (plus the front of

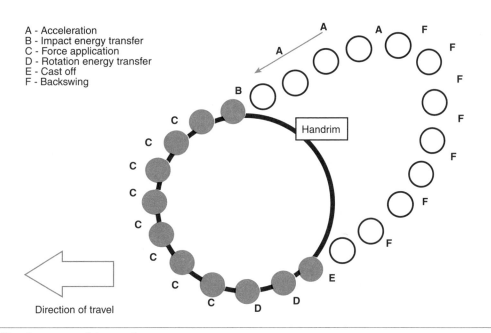

Figure 28.6   Six-phase propulsion cycle.

A - Acceleration
B - Impact energy transfer
C - Force application
D - Rotation energy transfer
E - Cast off
F - Backswing

Handrim

Direction of travel

Application Example: Enhancing Wheelchair Sports Performance

Setting: A community-based junior wheelchair sports program

Student: A 16-year-old junior wheelchair basketball player needs recommendations to refine his *individualized transition program* to incorporate adult wheelchair sports. The player is "tall" and has played the center/forward positions and wishes to purchase his own wheelchair.

Issue: What considerations should be taken into account in making recommendations to this athlete?

Application: Considerations for this athlete center around equipment, physical fitness, and individual skills.

▶ Equipment considerations

Athlete's height, the desire to play a certain position, athlete's classification level.

Adjustability for height and point of balance. (Because the player has a desire to be a center/forward, the adjustability should include being able to maximize the seat height to about 21 inches.)

System considerations such as strapping and mobility in the wheelchair.

A reputable manufacturer.

▶ Individual physical fitness

A strength-training program that targets the upper-body muscles in paired groups (i.e., biceps and triceps).

A cardiovascular conditioning program that utilizes an arm crank ergometer or, preferably, a training roller.

▶ Individual skills targeted

Wheelchair mobility skills both with and without the basketball.

Shooting skills both stationary and moving.

Passing skills both stationary and moving.

Studying the sophisticated strategies involved in the adult game.

the wheelchair) rising up to meet it. Then, as the upper body pushes down on the handrim and rocks back up, the legs and the wheelchair will return to the ground, and only then will the chair start to move forward. This is clearly inefficient, and for this reason it is critical that athletes start with their bodies as far forward as possible and, at the gun, the arms drive the wheelchair forward, while the athlete tries to prevent the chair from rising. In this way maximum energy is transferred to the forward motion of the wheelchair.

The stroke is also modified during the start. Since the wheelchair is stationary, the hands grip the handrim (rather than striking it), and for the first few strokes the arc of pushing is very restricted with as rapid as possible a recovery. The key is to get three or four short, hard strokes before making the transition to a striking rather than pushing stroke.

Retarding Forces and Overcoming Them

While the athlete provides the energy to drive the wheelchair forward, the twin retarding forces of rolling resistance and aerodynamic drag act to slow it down.

When propulsive forces are greater than resistance, the wheelchair accelerates, and when the retarding forces are greater, the chair is slowed. Obviously, reductions in rolling resistance and aerodynamic drag translate directly into higher wheeling speeds and improved athletic performance.

Rolling resistance

On a hard, smooth surface, the majority of rolling resistance of a wheelchair wheel occurs at the point where the tire is in contact with the ground. As the tire rotates, each part is compressed as it passes under the hub and is in contact with the road surface; it then rebounds as it begins to rise again, and contact with the surface is broken. Not all the energy used to compress the tire is recovered on the rebound, and the energy loss (called hysteresis) is the major determinant of rolling resistance. Obviously, the thicker and less flexible the wall of the tire, the greater the hysteresis energy loss. If the tire wall can be prevented from flexing, then energy losses will be reduced, and this is exactly what happens when the air pressure in the tire is increased.

Experiments have confirmed (Higgs, 1993) the theoretical expectation that, on a hard, smooth surface, rolling resistance will decrease as tire pressure increases. Thus, the road-racing athlete should inflate tires to the highest safe pressure for best performance.

Empirical work also suggests that rolling resistance decreases as the diameter of the wheel increases, and for this reason athletes use the largest wheels permitted by the rules. Experience also suggests that wider tires have lower rolling resistance than similarly constructed narrow tires. But since they also have higher aerodynamic drag, it is not clear if they offer higher overall performance in road racing.

On the track, hysteresis energy losses still occur in the tires when they are compressed and allowed to rebound, but the tires also cut into the surface of the track and produce surface deformation energy losses. If tire pressures are very high, the hysteresis energy losses in the tire will be low, but the hard inflation pressure will cut deeply into the track, and surface deformation losses will be high. Conversely, at low tire pressures the "flatter" tire will spread the weight load of athlete and chair over a greater area and reduce surface deformation losses. At the same time, however, the tire hysteresis energy losses will increase. The best athletic performance will occur at the tire pressure that minimizes the combined energy losses from both hysteresis and surface deformation. Experiments have been conducted (Higgs, 1993) that indicate that moderate tire pressures of approximately 90 pounds per square inch give the best results on synthetic athletic tracks, and that the penalty for overinflation of the tire is greater than the penalty for underinflation. The same series of experiments also indicated that wider, 28-millimeter (1.10-inch) tires had lower rolling resistance and better track performance than similarly constructed 20-millimeter (.78-inch) tires.

Rolling resistance of racing wheelchairs is also affected by the mainwheel's camber angle, although the relationship is complex. While the evidence concerning rolling resistance and camber angle is complex, findings with regard to longitudinal wheel alignment could not be clearer. Wheels that are not parallel and pointing straight ahead dramatically increase the rolling resistance of a wheelchair. Athletes should do everything in their power to check and adjust alignment prior to every important race, particularly in light of the damage to racing wheelchair frames caused by poor handling by airline baggage handlers and others.

Aerodynamic drag

The problems of aerodynamic drag of racing wheelchairs and athletes are unique in the field of sport because of the relatively low speeds at which events take place. Coutts and Schutz (1988) calculated that races on the track take place at average speeds between 3 and 7 meters/second (6.65 and 15.75 mph). While race times of wheelchairs have dramatically improved since 1988, wheelchair times are still considerably slower than the speeds found in cycling. This creates special low-speed aerodynamic conditions.

Aerodynamic drag is caused by two separate but interrelated forces called **surface** drag and **form** drag. Surface drag is caused by the adhesion of air molecules to the surface of an object passing through it, and it is very powerful at low speeds. Form drag, on the other hand, is caused by the difference in air pressure between the front and the back of an object, which in turn is created by the swirls and eddy currents formed as the wheelchair and athlete pass through the air.

For wheelchair racers the problem is that smooth surfaces increase surface drag while decreasing form drag, and, at the speeds at which wheelchair track events take place, there is uncertainty as to which type of surface gives the best overall result.

Some aspects of aerodynamic drag reduction are beyond doubt, and those are to reduce both surface and form drag by minimizing the drag, producing areas of the wheelchair and athlete's clothing. To reduce surface drag there is a need to reduce the exposed surfaces of the chair and athlete. When an athlete's trunk is resting on her knees, the air is unable to flow over the top surface of the thigh and the front surface of the chest, and this reduces the surface area on which molecules of air can exert their retarding influence. Form drag is mostly reduced by ensuring that all of the exposed components of the wheelchair are streamlined and that the chair and athlete have the lowest possible frontal area. Frontal area is the area of the body and chair perpendicular to the flow of air; this is what is seen in silhouette when looking head-on at an approaching athlete. Tipping the upper body forward (like the position taken by a racing cyclist) and eliminating loose clothing are the two major ways in which frontal area is reduced. Hedrick, Wang, Moeinzadeh, and Adrian (1990) have shown that frontal area can be reduced by up to 44 percent by adjustment of body position.

Drafting

Because aerodynamic drag represents approximately 40 percent of the force acting to slow down a wheelchair racer, methods of cheating the wind pay considerable dividends. The single most effective way in which drag can be reduced is by the process of **drafting**. Drafting occurs when a wheelchair follows closely behind another wheelchair, which acts as a wind deflector. The second wheelchair may experience aerodynamic drag forces less than half of those that would occur under nondrafting conditions. In still air or in a headwind, the wheelchair that follows gains greatest benefit when it follows directly behind the lead chair and as closely as possible. If a choice must be made between dropping back farther but remaining directly behind the lead chair or staying closer and moving out to one side, the evidence (Pugh, 1971) is overwhelming that it is better to stay directly behind. At the end of long races, the energy saved by drafting can be a critical determinant of race outcome, and modern race tactics confirm this. Frequently teams will work together, taking turns at both leading and drafting, so that their overall performance will be increased.

System Considerations for Court Wheelchairs

There are two fundamental properties that affect the system considerations for court wheelchairs (Yilla, La Bar, & Dangelmaier, 1998). These are the horizontal (anterior/posterior) positioning and the vertical positioning (see figure 28.3, a-b) of the chair. Appropriate adjustments of these will facilitate the development of the balance (vertical adjustment) and agility (horizontal adjustment) components of performance-related fitness.

Horizontal Positioning

Horizontal positioning of the mainwheels affects the mobility of the chair. The further forward the mainwheel from a hypothesized neutral position (see figure 28.3a, position A), the more maneuverable the chair (see figure 28.3a, position B). A backward tilt to the seat rail can also be incorporated; this shifts the center of gravity of the system to the rear and further promotes mobility. Unfortunately, the further forward the mainwheel relative to the center of gravity, the more likely it is that the chair will raise and tilt up. Therefore, the forward placement of the mainwheel is restricted by the athlete's ability to force the front end of the chair down. This ability is a function of abdominal and lower-body strength and therefore a function of the athlete's disability level. Therefore, disability level has been a limiting factor in how far forward the mainwheels could be placed, with more restricted athletes positioning their chairs in the more stable position, C, in figure 28.3a. This has been somewhat mitigated by a recent change in rules in basketball, quad rugby, and tennis that allows for a "fifth" wheel. This fifth wheel can be placed on the rear of the chair, provided it stays within the area contained by the mainwheel. This wheel offsets the tendency of the chair to tilt up if the mainwheel is set forward of the neutral position (position B, figure 28.3a), thus affording more restricted athletes the capacity to move their rear wheels forward, thereby improving their mobility.

Vertical Positioning

Vertical positioning of the mainwheel affects the height at which the athlete sits and the "system's" center of gravity. This fundamentally affects the chair's handling properties. Again using a hypothetical neutral position (figure 28.3b, position A), the lower down the athlete sits relative to this neutral position (figure 28.3b, position D), the more maneuverable is the wheelchair. Therefore, all other things being equal, the wheelchair athlete should sit as low as possible. However, performance considerations place a premium on height for certain positions, such as center in basketball and the post position in quad rugby. When playing these positions, it is advantageous to position the mainwheel to maximize sitting height (figure 28.3b, position E). However, as the center of gravity is higher, this will make the chair less maneuverable. This can be offset somewhat by using larger mainwheels but this will, in turn, reduce start speeds that are vital for court sports.

Therefore, when enhancing wheelchair sport performance on the court, the athlete should identify the functional aspects of the game and their roles or positions. This will in part be dependent on the disability level of the athlete. After identifying those roles, athletes should select the wheelchair setup that will improve functionality within those roles. It is stressed that the positioning of the mainwheel will fundamentally affect the performance characteristics of the chair. Finally, with the athlete combined to the wheelchair in a performance system, the athlete should address the skill development essential for success in the specific sport.

Skill Development

The skill aspects of the sport are critical to the elite athlete's program. Common to skill tests in court sports is speed (which is dependent on power, which is dependent on strength), and maneuverability with the target object, whether it be a basketball, volleyball (as used in quad rugby), or tennis racket. Other sport-specific skills can be obtained from the research literature. Skill tests have been developed for wheelchair basketball, quad rugby, and tennis (Brasile, 1986; Brasile & Hedrick, 1996; Yilla & Sherrill, 1998; Moore & Snow, 1994).

Instructional materials that focus on the skills and strategies involved in many wheelchair sports are also available (Hedrick, Byrnes, & Shaver, 1989; Moore & Snow, 1994). Again, the systems approach should be incorporated, with athletes practicing their skills in their competitive systems that include a sport-specific wheelchair, strapping, bracing, and an applicable prosthesis.

As pioneer Sir Ludwig Guttmann stated, "it is no exaggeration to say that the paraplegic [sic] and his [sic] chair have become one, in the same way as a first-class horseman and his [sic] mount (Weisman & Godfrey, 1976). This is the essence of the system approach to enhancing wheelchair sport performance.

THE FUTURE

Technological advances and improved training techniques have changed the look of wheelchair sport dramatically. The opportunities are now seemingly endless. New sports are constantly being adapted to meet the needs of those with disabilities, and, with each new sport, comes the need for specialized equipment designed to meet that sport's unique demands.

The demands of each sport and the needs of athletes involved in the sport must be systematically assessed, and suitable equipment must then be designed and built. The combining of the athlete and the wheelchair into a sport-specific, functional, performance system is the trend in elite wheelchair athletics, with even greater equipment and technique specialization anticipated as the wave of the future.

SUMMARY

This chapter has dealt with information for enhancing performance in wheelchair sports. It dealt with the appropriate selection of a sport by the athlete, the athlete's individual training regime, and some sport medicine issues faced by athletes with disabilities. Basic types of sport wheelchairs were identified (racing and court), as were specific performance considerations that influence the selection of the appropriate chair. The chapter concluded with an examination of the combination of the athlete and the chair to create a performance system. Again, racing performance considerations (stroke analysis, starting techniques, overcoming rolling resistance, and drafting) were separated from court performance considerations (adjusting the chair and individual skill development).

Appendix A:
Definitions Associated With the Individuals With Disabilities Education Act

There are several definitions associated with infants, toddlers, and children with disabilities. To a great extent, the definitions used in this book are based on those from IDEA. Those definitions are summarized here.

INFANTS AND TODDLERS WITH DISABILITIES

The term "infant and toddler with a disability" means an individual under three years of age who needs early intervention services because the individual

1. is experiencing developmental delays, as measured by appropriate diagnostic instruments and procedures in one or more of the areas of cognitive development, physical development, communication development, social or emotional development, and adaptive development; or
2. has a diagnosed physical or mental condition that has a high probability of resulting in developmental delay; and may also include, at a state's discretion, at-risk infants and toddlers (Individuals with Disabilities Education Act Amendments of 1997 (PL 105-17) U.S.C. 1400 (1997).

The term "at-risk infant or toddler" means an individual under three years of age who would be at risk of experiencing a substantial developmental delay if early intervention services were not provided to the children with disabilities.

CHILDREN WITH DISABILITIES

The term "children with disabilities" means those children having mental retardation; hearing impairments including deafness, speech, or language impairments; visual impairments including blindness; serious emotional disturbance; orthopedic impairments; autism; traumatic brain injury; other health impairments; specific learning disabilities; deaf-blindness; or multiple disabilities, and who because of those impairments need special education and related services.

The term "children with disabilities," for children ages three through nine, may, at a state's discretion, include children

► who are experiencing developmental delays, as defined by the state and as measured by appropriate diagnostic instruments and procedures, in one or more of the following areas: physical development, cognitive development, communication development, social or emotional development, or adaptive development; and
► who, for that reason, need special education and related services (Office of Special Education and Rehabilitative Services (OSE/RS). 34 CFR 300 [1998]).

The terms used in this definition are defined as follows:

1. "Autism" means a developmental disability significantly affecting verbal and nonverbal communication and social interaction, generally evident before age three, that adversely affects a child's educational performance. Other characteristics often associated with autism are engagement in repetitive activities and stereotyped movements, resistance to environmental change or change in daily routines, and unusual responses to sensory experiences. The term does not apply if a child's educational performance is adversely affected primarily because the child has a serious emotional disturbance.
2. "Deaf-blindness" means concomitant hearing and visual impairments, the combination of which causes such severe communication and other developmental and educational problems that they cannot be accommodated in special education programs solely for children with deafness or children with blindness.
3. "Deafness" means a hearing impairment that is so severe that the child is impaired in processing linguistic information through hearing, with or without amplification, that adversely affects a child's educational performance.

4. "Hearing impairment" means an impairment in hearing, whether permanent or fluctuating, that adversely affects a child's educational performance but that is not included under the definition of deafness.

5. "Mental retardation" means significantly subaverage general intellectual functioning, existing concurrently with deficits in adaptive behavior and manifested during the developmental period, that adversely affects a child's educational performance.

6. "Multiple disabilities" means concomitant impairments (such as mental retardation-blindness, mental retardation-orthopedic impairment, etc.), the combination of which causes such severe educational problems that they cannot be accommodated in special education programs solely for one of the impairments. The term does not include deaf-blindness.

7. "Orthopedic impairment" means a severe orthopedic impairment that adversely affects a child's educational performance. The term includes impairments caused by congenital anomaly (e.g., clubfoot, absence of some member, etc.), impairments caused by disease (e.g., poliomyelitis, bone tuberculosis, etc.), and impairments from other causes (e.g., cerebral palsy, amputations, and fractures or burns that cause contractures).

8. "Other health impairment" means having limited strength, vitality, or alertness, due to chronic or acute health problems such as a heart condition, tuberculosis, rheumatic fever, nephritis, asthma, sickle cell anemia, hemophilia, epilepsy, lead poisoning, leukemia, or diabetes, that adversely affects a child's educational performance.

9. "Serious emotional disturbance" is defined as follows:

(a) The term means a condition, exhibiting one or more of the following characteristics over a long period of time and to a marked degree, that adversely affects a child's educational performance.

1. An inability to learn that cannot be explained by intellectual, sensory, or health factors.

2. An inability to build or maintain satisfactory interpersonal relationships with peers and teachers.

3. Inappropriate types of behavior or feelings under normal circumstances.

4. A general pervasive mood of unhappiness or depression.

5. A tendency to develop physical symptoms or fears associated with personal or school problems.

(b) The term includes schizophrenia. The term does not necessarily apply to children who are socially maladjusted, unless it is determined that they have a serious emotional disturbance.

10. "Specific learning disability" means a disorder in one or more of the basic psychological processes involved in understanding or in using language, spoken or written, that may manifest itself in an imperfect ability to listen, think, speak, read, write, spell, or to do mathematical calculations. The term includes such conditions as perceptual disabilities, brain injury, minimal brain dysfunction, dyslexia, and developmental aphasia. The term does not apply to children who have learning problems that are primarily the result of visual, hearing, or motor disabilities; mental retardation; emotional disturbance; or environmental, cultural, or economic disadvantage.

11. "Speech or language impairment" means a communication disorder, such as stuttering, impaired articulation, a language impairment, or a voice impairment, that adversely affects a child's educational performance.

12. "Traumatic brain injury" means an acquired injury to the brain caused by an external physical force, resulting in total or partial functional disability or psychosocial impairment, or both, that adversely affects a child's educational performance. The term applies to open or closed head injuries resulting in impairments in one or more areas, such as cognition; language; memory; attention; reasoning; abstract thinking; judgment; problem solving; sensory, perceptual, and motor abilities; psychosocial behavior; physical functions; information processing; and speech. The term does not apply to brain injuries that are congenital or degenerative, or brain injuries induced by birth trauma.

13. "Visual impairment including blindness" means an impairment in vision that, even with correction, adversely affects a child's educational performance. The term includes both partial sight and blindness.

Appendix B: Adapted Sport Organizations

With the advances made in utilizing the Internet to access and disseminate information, many of the national governing bodies and multisport and unisport disability sports organizations have developed Web sites. Many of these provide links to other sites related to disability issues and adapted sport. Web sites and other information are listed below. Readers should realize that these have a history of frequent change.

Multisport Organizations

Disabled Sports, USA
451 Hungerford Dr., Suite 100
Rockville, MD 20805
301-217-9838 (phone)
301-217-0968 (fax)
E-mail: information@dsusa.org
Web site: **http://www.dsusa.org/**

Dwarf Athletic Association of America
418 Willow Way
Lewisville, TX 75067
972-317-8299 (phone)
972-317-8299 (fax)
E-mail: jfbda3@aol.com

Special Olympics
1325 G Street, NW, Suite 500
Washington, DC 20005-3104
202-628-3630 (phone)
202-824-0200 (fax)
E-mail: SOImail@aol.com
Web site: **http://www.specialolympics.org**

United States Association for Blind Athletes
33 North Institute Street
Colorado Springs, CO 80903
719-630-0422 (phone)
719-630-0616 (fax)
E-mail: usaba@usa.net
Web site: **http://www.usaba.org/**

United States Cerebral Palsy Athletic Association
25 West Independence Way
Kingston, RI 02881
401-848-2460 (phone)
401-848-5280 (fax)
E-mail: uscpaa@mail.bbsnet.com
Web site: **http://www.uscpaa.org**

USA Deaf Sports Federation
3607 Washington Blvd., Suite 4
Ogden, UT 84403-1737
801-393-7916 (phone) TTY: use your state relay service
801-393-2263 (fax)
E-mail: usadsf@aol.com
Web site: **http://www.usadsf.org**

The United States Olympic Committee
Disabled Sports Services Dept.
One Olympic Plaza
Colorado Springs, CO 80909-5760
719-578-4958 or 4818 (phone)
719-578-4976 (fax)
719-447-8773 (TTY)
Web site: **http://www.usoc.org/**

Wheelchair Sports, USA
3395 E. Fountain Blvd., Ste L-1
Colorado Springs, CO 80910
719-574-1150 (phone)
719-574-9840 (fax)
E-mail: wsusa@aol.com
Web site: **http://www.wsusa.org/**

Unisport Organizations

Aquatics

Aqua Sports Association for the Physically Challenged
9052-A Birch St.
Spring Valley, CA 91977
619-589-0537 (phone)
619-589-7013 (fax)

USA Swimming Adapted Swimming
Web site: **http://www.usa-swimming.org/adapted/index.html**

U.S. Wheelchair Swimming, Inc.
c/o Wheelchair Sports, USA
3595 E. Fountain Blvd., Ste. L-1
Colorado Springs, CO 80910
719-574-1150 (phone)
719-574-9840 (fax)

Archery

Wheelchair Archery, USA*
c/o Wheelchair Sports, USA
3595 E. Fountain Blvd, Ste. L-1
Colorado Springs, CO 80910
719-574-1150 (phone)
719-0574-9840 (fax)

Athletics

Wheelchair Athletics of the USA*
2351 Parkwood Rd.
Snellville, GA 30039
770-972-0763 (phone)
770-985-4885 (fax)

Basketball

National Wheelchair Basketball Association*
Charlotte Institute of Rehabilitation
c/o Adaptive Sports/Adventures
1100 Blythe Blvd.
Charlotte, NC 28203
704-355-1064 (phone)
704-466-4999 (fax)
Web site: **http://www.nwba.org**

Bowling

American Blind Bowling Association
411 Sheriff St.
Mercer, PA 16137
414-421-6400 (phone)
414-421-1194 (fax)

American Wheelchair Bowling Association
6264 North Andrews Ave.
Ft. Lauderdale, FL 33309
954-491-2886

National Deaf Bowling Association
2208 Gateway Oakes Dr. #192
Sacramento, CA 95833
916-564-8328 (phone)
800-735-2922 (relay)

Cycling

United States Handcycling Federation
207-443-3063
E-mail: info@ushf.org
Web site: **http://www.ushf.org**

Equestrian

American Competition Opportunities for Riders
With Disabilities (ACORD), Inc.
5303 Felter Road
San Jose, CA 95132
408-261-2015 (phone)
408-261-9438 (fax)

North American Riding for the Handicapped
Association
P.O. Box 33150
Denver, CO 80233
800-369-RIDE
E-mail: narha@narha.org
Web site: **http://www.narha.org**

Flying

Freedom's Wings
1832 Lake Ave.
Scotch Plains, NJ 07076
908-232-6354

International Wheelchair Aviators
PO Box 2799
Big Bear City, CA 92314
909-585-9663 (phone)
909-585-7156 (fax)

Football

Universal Wheelchair Football Association
UC Raymond Walters College
Disability Services Office
9555 Plainfield Rd.
Cincinnati, OH 45236-1096
513-792-8625 (phone)
513-792-8624 (fax)

Golf

Association of Disabled American Golfers
PO Box 280649
Lakewood, CO 80228-0649
303-922-5228 (phone)
303-922-5257 (fax)

Amputee Golfer Magazine
National Amputee Golf Association
P.O. Box 23285
Milwaukee, WI 53223-0285
800-633-6242 (phone)
414-376-1268 (fax)
Web site: **http://www.amputee-golf.org**

Hockey

American Sledge Hockey Association
10933 Johnson Ave. So.
Bloomington, MN 55437
612-881-2129

Quad Rugby

United States Quad Rugby
101 Park Place Circle
Alabaster, AL 35007
205-868-2281 (phone)
205-868-2283 (fax)
Web site: **http://www.quadrugby.com**

Road Racing

Achilles Track Club
42 West 38th Street
New York, NY 10018
212-354-0300
E-mail: achillestc@aol.com
Web site: **http://www.geocities.com/~achillestc**

Crank Chair Racing Association
3294 Lake Redding Dr.
Redding, CA 96003-3311
530-244-3577

Sailing

Access to Sailing
6475 East Pacific Coast Hwy
Long Beach, CA 90803
562-499-6925 (phone)
562-437-7655 (fax)

Scuba Diving

Handicapped Scuba Association International
1104 El Prado
San Clemente, CA 92672
714-498-6128
E-mail: hsahdq@comp.com
Web site: **http://ourworld.compuserve.com/
homepages/hsahdq/index.htm**

Shooting

National Rifle Association Disabled Shooting
Services
11250 Waples Mill Road
Fairfax, VA 22030
703-267-1495

National Wheelchair Shooting Federation*
102 Park Ave.
Rockledge, PA 19046
215-379-2359

Skiing

Ski for Light, Inc.
1400 Carole Ln.
Green Bay, WI 54313
920-494-5572
E-mail: pagsj@aol.com

U.S. Disabled Ski Team
P.O. Box 100
Park City, UT 84060
801-649-9090

Soccer

American Amputee Soccer Association
Web site: **http://www.ampsoccer.org/**

Bay Area Outreach and Recreation Program
830 Bancroft Way
Berkeley, CA 94710
510-849-4663 (phone)
510-849-4616 (fax)
Web site: **http://www.borp.org**

International Amputee Football Federation
Web site: **http://www.ampsoccer/iaff/index.htm**

International Paralympic Committee (7-a-side soccer
rules)
Web site: **http://info.lut.ac.uk/research/paad/ipc/
soccer/rules.html**

United States Amputee Soccer Association
c/o Disabled Sports, USA–Northwest
117 E. Louisa St., #202
Seattle, WA 98102
260-467-5157
E-mail: USAmpSoccer@dsusa.org
Web site: **http://www.dsusa.org/~dsusa/
usamputeesoccer/**

Softball

National Beep Baseball Association
c/o Jeanette Bigger
2231 West First Street
Topeka, KS 66606
E-mail: info@nbba.org
Web site: **http://www.nbba.org**

National Wheelchair Softball Association
1616 Todd Ct.
Hastings, MN 55033
612-437-1792

Table Tennis

American Wheelchair Table Tennis Association*
23 Parker St.
Port Chester, NY 10573
914-937-3932

Team Handball

United States Deaf Team Handball Association
Gene Duve
510-862-2907
E-mail: geneduve@aol.com

Tennis

International Wheelchair Tennis Federation
Bank Lane, Roehampton
London SW15 5XZ, England
011-44-181-878-6464 (phone)
011-44-181-392-4741 (fax)
Web site: **http://www.itftennis.com**

National Foundation of Wheelchair Tennis
940 Calle Amanecer, Suite B
San Clemente, CA 92673
714-361-3663 (phone)
714-361-6603 (fax)
E-mail: NFWT@aol.com
Web site: **http://www.nfwt.org**

Water Skiing

Water Skiers With Disabilities Association
American Water Ski Association
799 Overlook Dr.
Winter Haven, FL 33884
800-533-2972

Weightlifting

United States Wheelchair Weightlifting Federation*
39 Michael Pl.
Levittown, PA 19057
215-945-1964

*National governing body (NGB) of Wheelchair Sports, USA

Other Resources Related to Adapted Sport

Australian Sports Commission
Web site: **http://www.ausport.gov.au**
Provides extensive information on coaching athletes with disabilities.

Director for Disabled Sports (BlazeNet)
Center for Sports Medicine, Science and Technology
Georgia State University
125 Decatur Street
Atlanta, GA 30303
404-651-1928 (phone)
404-651-1929 (fax)
E-mail: blazenet@gsu.net
Web site: **http://education.gsu.edu/blazenet**
The Georgia State University Center for Sports Medicine, Science and Technology is dedicated to interdisciplinary research, teaching, and service in science and technology fundamental to sports medicine and physical activity, with a major program emphasis on persons with disability. The center operates BlazeNet (funded by a grant from the U.S. Disabled Athletes Fund), an information and referral service for sports, recreation, and physical activity programs and services available in Georgia for persons with physical disabilities.

D-Sport Online Magazine
Web site:
http://www.portal.cal/~igregson/dsport.html
Quarterly publication focusing on disability sport issues.

International Paralympic Committee
Web site: **http://www.paralympic.org**
Provides information related to the happenings of the International Paralympic Committee.

Minnesota Association for Adapted Athletics
John Bartz
Minnesota State High School League
2100 Freeman Blvd.
Brooklyn Center, MN 55430
651-560-2262
or
George Hanson
Department of Children, Family and Learning
1500 Highway 36 West
Roseville, MN
651-582-8365
This program offers the first and most comprehensive program of high school athletic opportunities for students with disabilities.

Palaestra
Challenge Publications, Ltd.
PO Box 508
Macomb, IL 61455
309-833-1902 (phone/fax)
E-mail: challpub@macomb.com
Web site: **http://www.palaestra.com**
Publication related to adapted sport, physical education, and recreation for people with disabilities.

Sports 'N Spokes
2111 E. Highland Avenue, Ste. 180
Phoenix, AZ 85016-9611
602-224-0500
Web site:
http://www.sns-magazine.com/sns/default.htm
Sports magazine for wheelchair users.

One Stop
Web site: **http://monticello.avenue.gen.va.us/disabledsports/home.html**
Provides information and links to many adapted sport organizations.

Wheelchair Racing Resource Page
Web site: **http://www.execpc.com/~birzer/**
Provides up-to-date information and links related to wheelchair racing, organizations, equipment, scholarships/grants, and sports camps.

Worldsport.com
Web site: **http://www.worldsport.com/**
Provides information on hundreds of sports worldwide.

World T.E.A.M. Sports
2108 South Boulevard, Ste 101
Charlotte, NC 28203
704-370-6070 (phone)
704-370-7750 (fax)
Web site: **http://www.worldteamsports.org**

Selected International Disability Sport Organizations

Canadian Wheelchair Basketball Association
1600 James Naismith Dr.
Gloucester, Ontario
K1B 5N4 Canada
613-841-1824 (phone)
613-841-5151 (fax)
E-mail: cwba@cwba.ca
Web site: **http://www.cwba.ca/**

International Wheelchair Basketball Federation
Robert J. Szyman, Secretary General
5142 Ville Maria Lane
Hazelwood, MO 63042–1646
314-209-9006 (phone)
314-739-6688 (fax)
E-mail: iwbfsecgen@aol.com
Web site: **http://www.iwbf.org/**

International Wheelchair Aviators
PO Box 2799
Big Bear City, CA 92314
909-585-9663 (phone)
909-585-7156 (fax)

Blind Sport New Zealand
Web site: **http://www.blindsport.org.nz/**

Canadian Deaf Sports Association
1600 James Naismith Drive, STE 303
Gloucester, ON
K1B 5N4 Canada
613-748-5789 (TTY/voice)
613-748-5706 (fax)

Canadian Wheelchair Sports Association
1600 James Naismith Dr.
Gloucester, Ontario
K1B 5N4 Canada
613-748-5685 (phone)
613-748-5722 (fax)

International Committee on Silent Sports
Web site: **http://www.ciss.org/**

International Blind Sports Association
Web site: **http://www.ibsa.es**

International Paralympic Committee
Adenauerallee 212
D-53113 Bonn
Germany
49-228-2097 200 (phone)
49-228-2097 209 (fax)
E-mail: info@paralympic.org
Web site: **http://www.paralympic.org**

International Paralympic Committee links to international organizations
Web site: **http://info.lboro.ac.uk/research/paad/ipc/general/int-orgs.html**
This site provides information on the major international disability sport organizations.

International Wheelchair Tennis Federation
Bank Lane, Roehampton
London SW15 5XZ, England
011-44-181-878-6464 (phone)
011-44-181-392-4741 (fax)
Web site: **http://www.itftennis.com**

British Disabled Water Ski Association
The Tony Edge National Centre
Heron Lake
Wraysbury
Middlesex TW19 6HW
United Kingdom
Web site: **http://www.bdwsa.org.uk/**

Wheelchair Sports
Web site: **www.wsusa.org**
This is a great source of information on global wheelchair sports events and much more. In this site, you will find information on all aspects of wheelchair sport and related topics, from future sports events to authorized merchandise and on-line donations.

Olympic Village, Guttman Road
Aylesbury, Bucks HP21 9PP, UK.
+44 (0) 01268 436179 (phone)
+44 (0) 01296 436484 (fax)
Web site: **http://193.129.121.27/wsw/home.htm**

Appendix C: Brockport Physical Fitness Test

This appendix contains a brief description of the test items included in the Brockport Physical Fitness Test. Although all 27 test items are presented here, a test battery for a particular individual generally includes 4 to 6 items. Test selection guidelines are included in the test manual. For a full description of the test, please see Winnick and Short, *The Brockport Physical Fitness Test*, Human Kinetics, 1999.

AEROBIC FUNCTIONING

▸PACER Test (20 meters and modified 16 meters)—At the sound of a tape-recorded beep, youngsters run from one line to another, either 20 meters or 16 meters away. They must arrive at the second line prior to the next beep (initially a nine-second interval). The time between beeps gradually decreases over the length of the test so students will find it increasingly difficult to keep up with the pace the longer the test goes on. The test score is the number of laps completed on pace.

▸Target Aerobic Movement Test—Youngsters engage in any type of activity that will elevate their heart rates into a target heart rate zone. They then attempt to maintain those elevated heart rates for 15 continuous minutes.

▸One-Mile Run/Walk—Youngsters attempt to complete a one-mile distance as quickly as possible.

BODY COMPOSITION

▸Skinfold Measures—Skinfold calipers are used to determine the youngster's skinfold thickness in order to estimate body fat percentage. Measures are taken at one of the following site options: triceps (only), triceps plus calf, or triceps plus subscapular.

▸Body Mass Index—Height and weight measures are used in a ratio to determine if an individual is either overweight or underweight for his height.

MUSCULOSKELETAL FUNCTIONING

Muscular Strength/Endurance

▸Trunk Lift—From a prone position with hands under thighs, participants attempt to lift their chins up to 12 inches from the mat by arching the back.

▸Dominant Grip Strength—Youngsters squeeze a grip dynamometer as hard as possible with their "favorite" hands.

▸Bench Press—From a supine position on a bench, youngsters attempt to repeatedly lift a 35-pound barbell from the chest to a straight-arm position above the chest. Boys are limited to 50 repetitions and girls to 30 repetitions.

▸Push-Up—Initially participants lie prone on a mat with hands placed under the shoulders (palms flat on the mat), elbows at 90 degrees, legs straight, and toes tucked. The participant then pushes up so that the arms (and back) are straight and the body weight is supported completely by the hands and toes. Youngsters attempt to complete as many push-ups as possible by performing one push-up every three seconds.

▸Isometric Push-Up—Participants attempt to hold the up position for the push-up for a given period of time.

▸Seated Push-Up—Participants who are wheelchair users (paraplegic) attempt to lift their buttocks and posterior thighs off the seats of their wheelchairs by pushing up from the armrests or tires of the chairs with their hands and arms. An alternative is to lift the buttocks off a mat using seated push-up blocks.

▸Dumbbell Press—From a seated position, youngsters attempt to repeatedly lift a 15-pound dumbbell from shoulder height to a straight-arm position directly above the shoulder. Both boys and girls are limited to 50 repetitions.

▸Reverse Curl—Participants (with a spinal cord injury and quadriplegia) attempt to lift a one-pound weight from lap level to shoulder level with a tenodesis grasp and elbow flexion.

▸40-Meter Push/Walk—Youngsters with certain mobility problems attempt to cover at least 40 meters in 60 seconds while maintaining a low heart rate (i.e., below moderate-level intensity).

▸Wheelchair Ramp Test—Youngsters in wheelchairs try to negotiate a standard ANSI ramp.

▸Curl-up—Youngsters lie in a supine position with knees bent and feet flat on the mat; arms are straight at the side with palms down and fingers at the edge of a four-and-a-half-inch-wide cardboard strip. Youngsters lift their upper backs off the mat until the fingers slide to the far edge of the strip and then return to the starting position. Youngsters perform as many curl-ups as possible (up to 75) by doing 1 curl-up every 3 seconds.

▸Curl-Up (modified)—Identical to the curl-up except there is no cardboard strip. Instead, youngsters place their hands on the top of their thighs and slide them to the kneecaps during the curl-up.

▸Flexed Arm Hang—Participants grasp (palms forward) an overhead bar (feet off the floor) with elbows bent and chin above the bar and attempt to hold that position for as long as possible.

▸Extended Arm Hang—Participants grasp (palms forward) an overhead bar (feet off the floor) with elbows straight and attempt to hold that position for up to 40 seconds.

▸Pull-Up—Participants grasp (palms forward) an overhead bar (feet off the floor) with elbows straight. They then attempt to repeatedly lift the body with the arms until the chin is above the bar.

▸Pull-Up (modified)—Using a special apparatus, students lie in a supine position and grasp a bar an arm's length above their chests. While keeping heels on the ground and back straight, the participants pull their bodies toward the bar until the chin passes an elastic band placed seven to eight inches below the bar. Participants attempt to perform as many modified pull-ups as possible.

Flexibility/Range of Motion

▸Back-Saver Sit-and-Reach—The youngster places one foot against a sit-and-reach box with a straight leg while the other, leg is bent at the knee with the foot flat on the floor. One hand is placed on top of the other, and the youngster attempts to reach as far across the top of the box as possible while maintaining the straight leg. The test is repeated with the opposite leg.

▸Shoulder Stretch—Participants attempt to touch the fingertips of their two hands behind their backs. The right hand reaches over the right shoulder between the scapulae while the left hand is brought up the back from the waist by bending the elbow. The test is repeated with the opposite arms.

▸Apley Test (modified)—Youngsters attempt to touch with one hand one of three landmarks given here in descending order of difficulty: superior angle of the opposite scapula, top of the head, and the mouth. The test is repeated with the opposite hand.

▸Thomas Test (modified)—Youngsters lie supine on a table and pull one knee to their chests while the tester evaluates the length of the opposite hip flexors by observing the extent of "lift" present in the opposite leg. The test is repeated with the other leg.

▸Target Stretch Test—Participants demonstrate their maximum movement extent for a variety of single-joint actions (e.g., wrist extension, shoulder abduction, elbow extension, forearm supination), and testers estimate the extent of movement from pictorial criteria.

THE BROCKPORT PHYSICAL FITNESS TEST KIT

The *Brockport Physical Fitness Test Kit* provides a complete package for fitness testing for youths with physical and mental disabilities. It includes the following:

- ▸ The comprehensive test manual that explains development of the test and testing procedures
- ▸ A training guide to assist you in improving your students' fitness
- ▸ *Fitness Challenge*, the companion software that makes test use much easier
- ▸ A video that demonstrates clearly how to use the test with this population
- ▸ Curl-up strips
- ▸ Skin caliper
- ▸ PACER audio CD/cassette

References and Resources

CHAPTER 1

References

Amateur Sports Act of 1978 (PL 95-606), 36 U.S.C. 371 (1978).

Brown v. Board of Education of Topeka, Kansas 347 U.S. 483 (1954).

Buell, C.E. (1983). *Physical education for blind children*. Springfield, IL: Charles C Thomas.

Committee on Adapted Physical Education. (1952). Guiding principles for adapted physical education. *Journal of Health, Education and Recreation, 23*, 15.

Gannon, J.R. (1981). *Deaf heritage: A narrative history of deaf America*. Silver Spring, MD: National Association for the Deaf.

Individuals with Disabilities Education Act (IDEA) Amendments of 1997 (PL 105-17), 20 U.S.C. 1400 (1997).

Mills v. Board of Education of the District of Columbia, 348 F. Supp. 966 (1972).

Office of Special Education and Rehabilitative Services (OSE/RS), 34 CFR 300 (1998).

Olympic and Amateur Sports Act in the Omnibus Appropriations Bill (PL 105-277), *Federal Register*. October 21, 1998.

Pennsylvania Association for Retarded Children v. Commonwealth of Pennsylvania, U.S. District Court, 343 F. Supp. 279 (1972).

Rehabilitation Act Amendments of 1992 (PL 102-569) Section 102 (p) (32) (1992).

Sherrill, C. (1998). *Adapted physical activity, recreation and sport: Crossdisciplinary and lifespan*. Madison, WI: WCB McGraw-Hill.

U.S. Department of Education. (1998). To assure the free appropriate public education of all children with disabilities: Twentieth annual report to Congress on the implementation of the Individuals with Disabilities Education Act. Washington, DC: Author

Resources

Written

Block, M.E. (1995). Americans with disabilities act: Its impact on youth sports. *Journal of Education, Recreation and Game, 66*(1), 28-32. Summarizes major parts of the act and answers questions on how it affects youth sports.

DePauw, K.P., & Gavron, S.J. (1995). *Disability and sport*. Champaign, IL: Human Kinetics. Comprehensive book on sport and individuals with disabilities. Topics include historical perspectives, sport organizations and sport structures, sport opportunities, and issues related to sport and disability.

French, R., Henderson, H., Kinnison, L. & Sherrill, C. (1998). Revisiting Section 504, physical education and sport. *Journal of Physical Education, Recreation and Dance, 69*(7), 57-61. Discusses Section 504 and its implications for physical education, including least restrictive environment, free and appropriate education, procedural safeguards, evaluation, and disciplinary procedures.

Minnesota Association for Adapted Athletics. Information can be obtained from John Bartz, Minnesota State High School League, 2100 Freeway Boulevard, Brooklyn Center, MN 55430, phone 651-763-2262 and from George Hanson, Department of Children, Family, and Learning, 1500 Highway 36 West, Roseville, MN 55113, 651-582-8365.

Paciorek, M.J., & Jones, J.A. (1994). *Sports and recreation for the disabled*. Carmel, IN: Cooper Publishing Group. A resource guide on sports and recreational activities for individuals with disabilities. Helpful in identifying sport organizations and the sport opportunities they provide.

Winnick, J.P., Auxter, D., Jansma, P., Sculli, J., Stein, J., & Weiss, R.A. (1980). Implications of Section 504 of the Rehabilitation Act as related to physical education instructional, personnel preparation, intramural, and interscholastic/intercollegiate sport programs. In J.P. Winnick & F.X. Short (Eds.), *Special athletic opportunities for individuals with handicapping conditions*. Brockport, NY: SUNY College at Brockport (ERIC Ed 210 897) or *Practical Pointers*, 3(11)1-20. Provides a full position paper related to Section 504 of the Rehabilitation Act of 1973.

Audiovisual

Physical Activity for All: Professional Enhancement Program (PAFA). (1999). This is a digital video disc, read-only memory (DVD-ROM) with modules on 12 adapted physical education topics, including movement science foundations; legal and professional aspects; physical education: inclusive settings; physical education: special settings; family, community, and school; uniqueness (unique attributes of learners); collaboration; assessment; curriculum and instruction; principles and practices of physical fitness; play; leisure and sports. Running time is approximately six hours.

Electronic

Challenge Publications. Web site: **www.palaestra.com**. This is the home of *Palaestra*, published by Challenge Publications, PO Box 508, Macomb, IL 61455.

Human Kinetics. Web site: **http://www.humankinetics.com**. This is the home of the *Adapted Physical Activity Quarterly*, published by Human Kinetics Publishers, Inc., 1607 N. Market Street, Box 5076, Champaign, IL 61825-5076.

Paralyzed Veterans of America. Web site: **www.pva.org/** This is the Web site of *Sports 'N Spokes*, published by the Paralyzed Veterans of America, 2111 East Highland Avenue, Suite 180, Phoenix, AZ 85016-4702.

CHAPTER 2

References

Block, M.E. (1994). *A teacher's guide to including students with disabilities in regular physical education*. Baltimore, MD: Paul H. Brookes.

Craft, D.H. (1996). A focus on inclusion in physical education. In B. Hennessy (Ed.), *Physical education sourcebook*. Champaign, IL: Human Kinetics.

Resources

Written

Active Living Alliance for Canadians with a Disability. (1994). *Moving to inclusion.* Gloucester, Ontario, Canada: Author. A curriculum consisting of nine books in English and French. Each book provides ideas for individualization for individuals with a specific disability. Available from Canadian Association for Health, Physical Education, and Recreation, 1600 James Naismith Drive, Gloucester, Ontario K1B 5N4, Canada; phone 613-748-5639.

Auxter, D., Pyfer, J., & Huettig, C. (1997). *Principles and methods of adapted physical education and recreation* (8th ed.). St. Louis: Mosby. This book describes and effectively applies community-based programming to adapted physical education and sport. It also includes an excellent chapter on facilities and equipment.

Block, M.E. (1995). Americans with Disabilities Act: Its impact on youth sports. *Journal of Physical Education, Recreation and Dance,* 66(1), 28-32. This article presents an overview of the Americans with Disabilities Act (ADA) as it pertains to youth sports. It also presents questions and answers on how the ADA directly affects youth sports.

Block, M.E. (1996). Implications of federal laws and court cases for physical education placement of students with disabilities. *Adapted Physical Activity Quarterly,* 13(2), 127-152. Examines federal laws regarding inclusion and least restrictive environment, court cases regarding inclusion, and the application of these federal laws and court decisions to physical education placement.

Craft, D.H. (Ed.). (1994). Inclusion: Physical education for all. *Journal of Physical Education, Recreation and Dance,* 65(1), 22-56. Periodical providing a special issue on inclusion, including information on making curricular modifications, promoting equal status relationships among peers, teaching collaboratively with others, research on inclusion, ideas on infusion, and experiences implementing inclusion in two schools.

Craft, D.H. (1996). A focus on inclusion in physical education. In B. Hennessy (Ed.), *Physical education sourcebook.* Champaign, IL: Human Kinetics. This manuscript was written to define and clarify the meaning of inclusion, review legislation affecting inclusive physical education, and provide strategies for teaching in an inclusive manner.

Sherrill, C. (1998). *Adapted physical activity, recreation, and sport: Cross disciplinary and lifespan.* Madison, WI: WCB McGraw-Hill. Includes a checklist for evaluating school district needs related to adapted physical education.

Winnick, J.P. (1994). Rating scale for adapted physical education. This rating scale can be used as one self-assessment instrument on which to base evaluation of a school's adapted physical education program. The scale presents criteria statements reflecting guidelines implicitly suggested in this chapter. The rating scale can be found in the electronic instructor's guide for this book. For more information contact Joseph Winnick, Department of Physical Education and Sport, SUNY, College at Brockport, Brockport NY, 14420.

Winnick, J.P., Auxter, D., Jansma, P., Sculli, J., Stein, J., & Weiss, R.A. (1980). Implications of Section 504 of the Rehabilita-tion Act as related to physical education instructional, personnel preparation, intramural, and interscholastic/intercollegiate sport programs. *Practical Pointers,* 3(11), 1-20.

Audiovisual

Texas Woman's University Presents—The Survival Series: Inclusion Techniques in Physical Education (Texas Woman's University), videotape. Department of Kinesiology, Texas Woman's University, Box 23717 TWU Station, Denton, Texas 76204. This resource includes three videotapes demonstrating inclusion strategies for physical education. The running time varies with each videotape.

CHAPTER 3
References

Davis, R., & Ferrara, M. (1996). Athlete classification: An exploration of the process. *Palaestra,* 12(2), 38-44.

Department of Health, Education, and Welfare. (1977). Nondiscrimination on basis of handicap. *Federal Register,* 42(86), 22676-22702.

Federal Register, May 4, 1977, PL 93-112, the Rehabilitation Act of 1973, Section 504.

Federal Register, June 4, 1997, PL 105-17, Individuals with Disabilities Education Act Amendments of 1997.

Federal Register, October 21, 1998, PL 105-277, Olympic and Amateur Sports Act in Omnibus Appropriations Bill.

Olympic and Amateur Sports Act (1998), Omnibus Appropriations Bill.

Paciorek, M.J., & Jones, J.A. (1994). *Sports and recreation for the disabled.* Carmel, IN: Cooper Publishing Group.

Rehabilitation Act of 1973. *Revisions of 1998, 29 U.S.C. Chapter 16.*

Richter, K.J., Adams-Mushett, C., Ferrara, M.S., & McCann, B.C. (1992). Integrated swimming classification: A faulted system. *Adapted Physical Activity Quarterly,* 9, 5-13.

Sherrill, C. (1998). *Adapted physical activity, recreation and sport* (5th ed.). Dubuque, IA: WCB/McGraw-Hill.

USOC Constitution. (1998). Article VII, Section 1 (H). Web site: **http://www.usoc.org/c&b/constitution.html**.

Winnick, J.P. (1987). An integration continuum for sport participation. *Adapted Physical Activity Quarterly,* 4, 157-161.

Winnick, J.P. Auxter, D., Jansma, P., Sculli, J., Stein., & Weiss, R.A. (1980). Implications of Section 504 of the Rehabilitation Act as related to physical education instructional, personnel preparation, intramural, and interscholastic/intercollegiate sport programs. *Practical Pointers,* 3(11), 1-20.

Resources

A list of sport organizations for people with disabilities can be found in appendix B.

Written

Block, M.E. (1995). Americans with Disabilities Act: Its impact on youth sports. *Journal of Physical Education, Recreation and Dance,* 66(1), 28-32. This article provides information on the legal rights for participation in youth sports for children with disabilities.

Coaching athletes with disabilities. Australian Sports Commission. A variety of educational packages are available on sport and physical activity for people with disabilities through the Willing and Able program and Coaching Athletes with a Disability (CAD) scheme. The Willing and Able program is designed to help teachers and community leaders include young people with disabilities in regular physical education and sport programs. The CAD scheme focuses on assisting coaches to develop individualized training programs for athletes with disabilities to improve their performances in their chosen sports. CAD manuals outline the terminology, structure of sport, inclusion methods, and coaching principles needed to coach athletes with a variety of disabilities. This series of extensive manuals and videos can be purchased by contacting the Australian Sports Commission Disabilities Program, P.O. Box 176, Belconnen ACT 2616, Australia (02) 6214 1792; e-mail: DEP@ausport.gov.au; Web site: **http://www.ausport.gov.au/**

Curtis, K., McClanahan, S., & Hall, K. (1986). Health, vocational and functional states in spinal cord injured athletes and non-athletes. *Archives of Physical Medicine and Rehabilitation, 67*, 862-865.

DePauw, K.P., & Gavron, S.J. (1995). *Disability and sport.* Champaign, IL: Human Kinetics. Provides readers with an understanding of the historical context for sport today and trends for the future, an awareness of sport modifications, and multitude of sport opportunities available for individuals with disabilities.

French, R., Henderson, H., Kinnison, L., & Sherrill, C. (1998). Revisiting section 504, physical education, and sport. *Journal of Physical Education, Recreation and Dance, 69*(7), 57-63. Describes the legal ramifications of Section 504 of the Rehabilitation Act of 1973, regarding accessibility of sports for people with disabilities.

Paciorek, M.J., & Jones, J.A. (1994). *Sports and recreation for the disabled.* Carmel, IN: Cooper Publishing Group. A resource guide providing information on 53 sports and recreation activities for people with disabilities. Information provided on adapted equipment and manufacturers, disabled sport organizations, and national governing bodies.

Special Olympics Sports Skills Guides. Special Olympics, Inc., Washington, DC. These guides describe a developmental approach of teaching the many sports offered through Special Olympics. The guides are excellent for use in physical education by students with and without mental impairments.

Audiovisual

America's Best, videotape. Message Makers, 1217 Turner, Lansing, MI 48906; 517-482-3333. The story of the Dwarf Athletic Association of America.

Conviction of the Heart: Paralympic School Program (K-6/7-12), curriculum and videotape. (1996). U.S. Disabled Athlete Fund, 2015 S. Park Place, Suite 180, Atlanta, GA 30339. A two-part curriculum for grades K-6 and 7-12, providing an introduction to the Paralympic Games and disability awareness of Paralympic athletes.

Disabled Sports Team-American Champions, videotape. (1996). Message Makers, 1217 Turner, Lansing, MI 48906; 517-482-3333. Tells the story of the United States Disabled Sports Team at the 1996 Paralympic Games.

USOC Disabled Sports Promo, videotape. USOC Media Services, One Olympic Plaza, Colorado Springs, CO 80909. Describes the disabled sports programming coordinated through the United States Olympic Committee.

World Triumph—1996 U.S. Paralympic Team, videotape. (1996). Available from Message Makers, 1217 Turner, Lansing, MI 48906; 517-482-3333. Tells the story of the triumph of the 1996 U.S. Paralympic Team at the Atlanta Paralympic Games.

CHAPTER 4
References

Department of Education. (1998). *Federal Register,* 34 CFR Ch. III, July 1, 1998.

French, R., Henderson, H., Kinnison, L., & Sherrill, C. (1998). Revisiting Section 504, physical education and sport. *Journal of Physical Education, Recreation and Dance, 69*(7), 57-63.

Individuals with Disabilities Education Act Amendments of 1997 (PL 105-17), 20 U.S.C. 1400 (1997).

New York State Education Department. (1998). *Part 200-Students with disabilities.* Vocational and Educational Services for Individuals with Disabilities, Albany, NY: Author.

New York State Education Department. (1995). *Test access and modifications for individuals with disabilities.* Vocational and Educational Services for Individuals with Disabilities, Albany, NY: Author.

Resources
Written

Houston-Wilson, C., & Lieberman, L. (1999). The individualized education program in physical education: A guide for regular physical educators. *Journal of Physical Education, Recreation and Dance, 70*(3), 60-64. Summarizes IEPs for physical education teachers. Attention is given to assessment for the IEP and writing the IEP.

Zirkel, P. (1993). *Section 504 and the schools.* Horsham, PA: LRP Publications. This resource manual comes in a loose-leaf notebook format and addresses issues related to 504 and the education of students with disabilities. It comes with bibliographic references and sample forms.

Electronic

IDEA '97. Web site: **www.ed.gov/offices/OSERS/IDEA/the_law. html**. Sponsored by the U.S. Department of Education, this site includes opportunities to obtain copies of the law, information on regulations related to the law, recent happenings pertaining to IDEA, frequently asked questions, and other information.

Idea practices. Web site: **www.ed.gov/offices/OSERS/IDEA/the_law.html**. Sponsored by the Council for Exceptional Children (CEC) and other organizations to promote the intent of IDEA through effective institutional practice and collaboration. Includes information on IDEA, including opportunities to read or print the law and regulations, Department of Education news, a summary of important court cases, and IDEA resources.

PE Central. Web site: **http://pe.central.vt.edu/adapted**. Summarizes IEP development for physical educators as well as other topics in adapted physical education and links.

CHAPTER 5

References

American Alliance for Health, Physical Education and Recreation. (1976a). *Special fitness test for mildly mentally retarded persons*. Washington, DC: Author.

American Alliance for Health, Physical Education and Recreation. (1976b). *Youth fitness test*. Washington, DC: Author.

American Alliance for Health, Physical Education, Recreation and Dance. (1980). *The health-related physical fitness test*. Reston, VA: Author.

Block, M., Lieberman, L., & Connor-Kuntz, F. (1998). Authentic assessment in adapted physical education. *Journal of Physical Education, Recreation and Dance*, 69(3), 48-55.

Burton, A., & Miller, D. (1998). *Movement skill assessment*. Champaign, IL: Human Kinetics.

Cooper Institute for Aerobics Research. (1999a). *Prudential FITNESSGRAM*. Dallas, TX: Author.

Cooper Institute for Aerobics Research. (1999b). *Teacher FITNESSGRAM software*. Dallas, TX: Author.

Davis, W., & Burton, A. (1991). Ecological task analysis: Translating movement theory into practice. *Adapted Physical Activity Quarterly*, 8, 154-177.

Folio, M., & Fewell, R. (2000). *Peabody developmental motor scales* (2nd ed.). Austin, TX: Pro-Ed.

Hensley, L. (1997). Alternative assessment for physical education. *Journal of Physical Education, Recreation and Dance*, 68(7), 19-24.

Houston-Wilson, K. (1995). Alternate assessment procedures. In J. Seaman (Ed.), *Physical best and individuals with disabilities: A handbook for inclusion in fitness programs*. Reston, VA: AAHPERD.

Johnson, L., & Londeree, B. (1976). *Motor fitness testing manual for the moderately mentally retarded*. Washington, DC: AAHPER.

Melagrano, V. (1996). *Designing the physical education curriculum*. Champaign, IL: Human Kinetics.

Meyer Rehabilitation Institute. (1992). *Milani-Comparetti motor development screening test for infants and young children: A manual*. Omaha: Author.

Safrit, M.J. (1990). *Introduction to measurement in physical education and exercise science*. St. Louis: Times Mirror/Mosby.

Short, F., & Winnick, J. (1999). *The Brockport physical fitness test technical manual. Fitness challenge software*. Champaign, IL: Human Kinetics.

Special Olympics, Inc. (1995-99). *Special Olympics sports skills program guides*. Washington, DC: Author.

Ulrich, D. (2000). *Test of gross motor development* (2nd ed.). Austin, TX: Pro-Ed.

Welk, G., & Wood, K. (2000). Physical activity assessment: A practical review of instruments and their use in the curriculum. *Journal of Physical Education, Recreation and Dance*, 71(1), 30-40.

Weston, A., Petosa, R., & Pate, R. (1997). Validation of an instrument for measurement of physical activity in youth. *Medicine and Science in Sports and Exercise*, 29(1), 138-143.

Winnick, J., & Short, F. (1999a). *The Brockport physical fitness test manual*. Champaign, IL: Human Kinetics.

Winnick, J., & Short, F. (Eds.). (1999b). *The Brockport physical fitness training guide*. Champaign, IL: Human Kinetics.

Resources

Written

Brigance Diagnostic Inventory. (1999). Curriculum Associates, Inc., 153 Rangeway Rd., PO Box 2001, North Billerica, MA 01862. Composite scores for gross motor skills provided for youngsters up to age seven. Web site: **www.curriculumassociates.com**

Bruininks-Oseretsky Test of Motor Proficiency. (1978). American Guidance Services, 4201 Woodland Rd., Circle Pines, MN 55014. Norm-referenced test of motor ability for youngsters ages 4 1/2 to 14 1/2. Web site: **www.agsnet.com**

Denver Developmental Screening Test (Denver II). (1990). Denver Developmental Materials, Inc., PO Box 6919, Denver, CO 80206. Developmental milestones for normal gross motor development; age norms up to six years.

Ohio State SIGMA. (1979). Mohican Publishing Co., PO Box 295, Loudonville, OH 44842. Criterion-referenced test of fundamental movements; designed primarily for use with students with mental retardation.

President's Challenge. (1997). President's Council on Physical Fitness and Sports, Washington, DC 20001. Standards for the 85th percentile are presented for five measures of fitness; students ages 6 to 17 who meet the criteria for all five items qualify for the Presidential Physical Fitness Award. Other award programs are available including one for health-related physical fitness. Web site: **www.surgeongeneral.gov/ophs/pcpfs.htm**

U.S. Department of Health and Human Services. (1996). *Physical activity and health: A report of the Surgeon General*. Atlanta, GA: U.S. Department of Human Services, Center for Disease Control and Prevention, National Center for Chronic Disease Prevention and Health Prevention.

Audiovisual

Brockport Physical Fitness Test Video, videotape. American Fitness Alliance, Youth Fitness Resource Center, PO Box 5076, Champaign, IL 61825-5076. Discusses and demonstrates the proper techniques for administering the 27 items in the Brockport Physical Fitness Test. Some background information also is provided. Running time is approximately 30 minutes.

Electronic

See chapter 21 for information on software to support FITNESSGRAM and the Brockport Physical Fitness Test.

CHAPTER 6

References

Active Living Alliance for Canadians with a Disability. (1994). *Moving to inclusion*. Gloucester, Ontario, Canada: Author.

American Alliance for Health, Physical Education, Recreation and Dance. (1999a). *Physical Best Activity Guide—Elementary*. Champaign, IL: Human Kinetics.

American Alliance for Health, Physical Education, Recreation and Dance. (1999b). *Physical Best Activity Guide—Secondary*. Champaign, IL: Human Kinetics.

American Alliance for Health, Physical Education, Recreation and Dance and Cooper Institute for Aerobics Research. (1995). *You stay active*. Dallas, TX: Cooper Institute for Aerobics Research.

Auxter, D., Pyfer, J., & Huettig, C. (1997). *Principles and methods of adapted physical education and recreation* (7th ed.). St. Louis: Mosby.

Balan, C.M., & Davis, W.E. (1993). Ecological task analysis: An approach to teaching physical education. *JOPERD*, 64 (9), 54-81.

Bidabe, D.L. (1995). *M.O.V.E.* Bakersfield, CA: Kern County Superintendent of Schools.

Block, M., Zeman, R., & Henning, G. (1997). Pass the ball to Jimmy! A success story in inclusive physical education. *Palestra*, 13(3), 37-41.

Bruininks, R.H. (1978). *Bruininks-Oseretsky test of motor proficiency*. Examiners Manual. Circle Pines, MN: American Guidance Service.

Burton, A.W., & Miller, D.E. 1998. *Movement skill assessment*. Champaign, IL: Human Kinetics.

Davis, W.E., & Burton, A.W. (1991). Ecological task analysis: Translating movement behavior into practice. *Adapted Physical Activity Quarterly*, 8, 154-177.

Dunn, J. (1997). *Special physical education: Adapted, individualized, developmental*. Madison, WI: Brown & Benchmark.

Dunn, J., Morehouse, J., & Fredericks, H. (1986). *Physical education for the severely handicapped: A systematic approach to a data based gymnasium* (2nd ed.). Austin, TX: Pro-Ed.

Gallahue, D.L. (1993). *Developmental physical education for today's children*, (2nd ed.). Madison, WI: Brown & Benchmark.

Hellison, D.R. (1985). *Goals and strategies for teaching physical education*. Champaign, IL: Human Kinetics.

Houston-Wilson, C., Dunn., J.M., van der Mars, H., & McCubbin, J. (1997). The effect of peer tutors on motor performance in integrated physical education classes. *Adapted Physical Activity Quarterly*, 14, 298-312.

Karp, J., & Adler, A. (1992). *ACTIVE*. Bothell, WA: Author.

Kelly, L.E. (1989). *Project ICAN-ABC*. Charlottesville, VA: University of Virginia.

Mosston, M., & Ashworth, S. (1990). *The spectrum of teaching styles: From command to discovery*. White Plains, NY: Longman.

Nichols, B. (1996). *Moving and learning: The elementary school physical education experience* (3rd ed.). St. Louis: Times Mirror/Mosby.

Sherrill, C. (1998). *Adapted physical activity, recreation and sport: Cross disciplinary and lifespan* (5th ed.). Boston: WCB/McGraw-Hill.

Special Olympics sport skill guides (1985). Washington, DC: Special Olympics.

Vodola, T. (1976). *Project ACTIVE maxi-model: Nine training manuals*. Oakhurst, NJ: Project ACTIVE.

Webster, G.E. (1993). Effective teaching in adapted physical education: A review. *Palaestra*, 9(3), 25-31.

Werder, J.K., & Bruininks, R.H. (1988). *Body skills: A motor development curriculum for children*. Circle Pines, MN: American Guidance Service.

Wessel, J.A. (1979). *I CAN—Sport, leisure and recreation skills*. Northbrook, IL: Hubbard.

Wessel, J.A., & Kelly, L.E. (1985). *Achievement-based curriculum development in physical education*. Philadelphia: Lea & Febiger.

Wessel, J.A., & Zittel, L.L. (1998). *I CAN primary skills: K-3* (2nd ed.). Austin, TX: Pro-Ed.

Wessel, J.A., & Zittel, L.L. (1995). *SMART START preschool movement curriculum*. Austin, TX: Pro-Ed.

Winnick, J.P., & Short, F.X. (1999a). *The Brockport physical fitness test manual*. Champaign, IL: Human Kinetics.

Winnick, J.P., & Short, F.X. (Eds.). (1999b). *The Brockport physical fitness training guide*. Champaign, IL: Human Kinetics.

Resources

Written

Active Living Alliance for Canadians with a Disability. (1994). *Moving to inclusion*. Gloucester, Ontario, Canada: Author. A curriculum consisting of nine books in English and French. Each book provides ideas for individualization for individuals with a specific disability. Available from Canadian Association for Health, Physical Education, and Recreation, 1600 James Naismith Drive, Gloucester, Ontario K1B 5N4, Canada; phone 613-748-5639.

American Alliance for Health, Physical Education, Recreation and Dance (AAHPERD). (1999). *Physical Best Educational Program*. Champaign, IL: Human Kinetics. A comprehensive health-related physical fitness curriculum. Materials available include: Elementary and Secondary Activity Guides, Teachers Guide, Instructor Video, and access to workshops and certification through AAHPERD. All materials available from Youth Fitness Resource Center, P.O. Box 5076, Champaign, IL 61825-5076; phone 800-747-4457; Web site: **http://www.americanfitness.net**

Bibabe, D.L. (1995). *M.O.V.E.* Bakersfield, CA: Kern Superintendent of Schools. A top-down activity-based curriculum to teach individuals with profound multiple disabilities functional activities. For additional information and training locations contact M.O.V.E. International, 1300 17th St., City Center, Bakersfield, CA 93301; phone 800-397-MOVE; e-mail: moveint@fc.kern.org; Web site: **http://www.move-international.org**

Gallahue, D.L. (1993). *Developmental physical education for today's children*. Madison, WI: Brown & Benchmark. Discusses the development process from the prenatal period through age 12. Treats psychomotor, cognitive, and affective factors influencing the motor development of children, and also describes teaching behaviors and styles that promote effective teaching of students with special needs. Helpful in working with developmentally delayed children.

Karp, J., & Adler, A. (1992). *ACTIVE*. Bothell, WA: Author. Curriculum includes instructional programs for individuals with mental retardation, learning disabilities, orthopedic impairments, sensory impairments, eating disorders, and breathing problems. Included are normative as well as criterion-referenced tests in the areas of motor ability, nutrition, physical fitness, and posture. Materials are available from Joe Karp, 20214 103rd Place NE, Bothell, WA 98011-2455.

Magill, R.A. (1993). *Motor learning: Concepts and applications* (4th ed.). Dubuque, IA: Brown & Benchmark. Provides an understanding of the scientific foundations of motor learning while stressing practical applications. Offers an introduction to motor learning followed by information on both the learner and the learning environment.

Mosston, M., & Ashworth, S. (1986). *Teaching physical education* (3rd ed.). Columbus, OH: Merrill. Provides a comprehensive discussion of teaching styles as well as excellent examples of their application.

Sherrill, C. (1998). *Adapted physical activity, recreation and sport: Cross disciplinary and lifespan* (5th ed.). Dubuque, IA: WCB/McGraw-Hill. Comprehensive, excellent source of information on the philosophy of humanism applied to adapted physical education and the developmental approach to teaching. Provides examples of teaching to meet individual needs through task analysis, activity modification, and activity analysis.

Werder, J.K., & Bruininks, R. H. (1988). *Body skills: A motor development curriculum for children.* Circle Pines, MN: American Guidance Service. A boxed motor development curriculum for children ages 2 to 12 years. Provides a systematic procedure for assessing, planning, and teaching gross motor skills. Scores on assessment tools correlate to the Bruininks-Oseretsky Test of Motor Proficiency. Both the Body Skills curriculum and the Bruininks-Oseretsky Test can be obtained from American Guidance Service (AGS), P.O. Box 99, Circle Pines, MN 55014-1796; phone 800-323-2560.

Wessel, J.A., & Zittel, L.L. (1998). *I CAN primary skills: K-3* (2nd ed.). Austin, TX: Pro-Ed. A spiral-bound user-friendly teacher resource for use with students in grades K-3 physical education classes. The curriculum provides program goals, objectives, assessments, and instructional activities that are matched to desired student outcomes. Includes locomotor and rhythmic skills, object control skills, personal-social participation skills, and health-related physical fitness components. All materials are available from Pro-Ed, 8700 Shoal Creek Blvd., Austin, TX, 78757-6897; phone 512-451-3246.

Winnick, J.P., & Short, F.X. (Eds.). (1999). *The Brockport physical fitness training guide.* Champaign, IL: Human Kinetics. Provides guidelines for the development of health-related physical fitness of children with disabilities.

Electronic

Adapt-talk provides an e-mail opportunity to discuss adapted physical education pedagogy challenges with peers and professionals worldwide. It is a subdivision of PE-Talk and is sponsored by Sport Time International. To subscribe to this service, send an e-mail (type subscribe in message) to adapt-talk-digest request@lists2.sportime.com

PE Central. Web site: **http://pe.central.vt.edu**. PE Central is an online resource for adapted physical educators. It includes information on lesson ideas, frequently asked questions, and information about a diversity of professional topics.

CHAPTER 7

References

Arnold, L.E., Gadow, K., Pearson, D.A., & Varley, C.K. (1998). Stimulants. In S. Reiss & M.G. Aman (Eds.), *Psychotropic medication and developmental disabilities: The international consensus handbook.* Washington, DC: American Association on Mental Retardation.

Baumeister, A.A., Sevin, J.A., & King, B.H. (1998). Neuroleptics. In S. Reiss & M.G. Aman (Eds.), *Psychotropic medication and developmental disabilities: The international consensus handbook.* Washington, DC: American Association on Mental Retardation.

Coleman, M.C. (1992). *Behavior disorders: Theory and practice* (2nd ed.). Needham Heights, MA: Allyn & Bacon.

Cullinan, D., Epstein, M.H., & Lloyd, J.W. (1991). Evaluation of conceptual models of behavior disorders. *Behavioral Disorders, 16*, 148-157.

Dunn, J.M., & Fredericks, H.D.B. (1985). The utilization of behavior management in mainstreaming in physical education. *Adapted Physical Activity Quarterly, 2*, 338-346.

Dunn, J.M., Morehouse, J.W., & Fredericks, H.D.B. (1986). *Physical education for the severely handicapped: A systematic approach to a data based gymnasium.* Monmouth, OR: Teaching Research.

ERIC/OSEP Special Project. (1997). *Research connections in special education.* Reston, VA: The ERIC Clearinghouse on Disabilities and Gifted Education/The Council for Exceptional Children.

Hallahan, D.P., & Kauffman, J.M. (1997). *Exceptional children: Introduction to special education* (6th ed.). Englewood Cliffs, NJ: Prentice Hall.

Hellison, D.R. (1995). *Teaching responsibility through physical activity.* Champaign, IL: Human Kinetics.

Hellison, D.R., & Templin, T.J. (1991). *A reflective approach to teaching physical education.* Champaign, IL: Human Kinetics.

Jones, M.M. (1998). *Within our reach: Behavior prevention and intervention strategies for learners with mental retardation and autism.* Reston, VA: Division of Mental Retardation and Developmental Disabilities of the Council for Exceptional Children.

Kalachnik, J.E., Leventhal, B.L., James, D.H., Sovner, R., Kastner, T.A., Walsh, K., Weisblatt, S.A., & Klitzke, M.G. (1998). Guidelines for the use of psychotropic medication. In S. Reiss & M.G. Aman (Eds.), *Psychotropic medication and developmental disabilities: The international consensus handbook.* Washington, DC: American Association on Mental Retardation.

Kazdin, A.E. (1989). *Behavior modification in applied settings* (Rev. ed.). Homewood, IL: The Dorsey Press.

Lavay, B.W., French, R., & Henderson, H.L. (1997). *Positive behavior management strategies for physical educators.* Champaign, IL: Human Kinetics.

Maslow, A.H. (1970). *Motivation and personality* (2nd ed.). New York: Harper & Row.

Mendler, A.N. (1992). *How to achieve discipline with dignity in the classroom.* Bloomington, IN: National Educational Service.

Parrish, J.M. (1997). Behavior modification. In M.L. Batshaw (Ed.), *Children with disabilities* (4th ed.). Baltimore, MD: Paul H. Brookes.

Premack, D. (1965). Reinforcement theory. In D. Levine (Ed.), *Nebraska symposium on motivation.* Lincoln: University of Nebraska Press.

Redl, F. (1952). *Controls from within.* New York: Free Press.

Rinck, C. (1998). Epidemiology and psychoactive medication. In S. Reiss & M.G. Aman (Eds.), *Psychotropic medication and developmental disabilities: The international consensus handbook.* Washington, DC: American Association on Mental Retardation.

Sherrill, C. (1998). *Adapted physical education, recreation, and sport: Crossdisciplinary and lifespan* (5th ed.). Boston: WCB/McGraw-Hill.

Sovner, R., Pary, R.J., Dosen, A., Geyde, A., Barrera, F.J., Cantwell, D.P., & Huessy, H.R. (1998). Antidepressants. In S. Reiss & M.G. Aman (Eds.), *Psychotropic medication and developmental disabilities: The international consensus handbook.* Washington, DC: American Association on Mental Retardation.

Walker, J.E., & Shea, T.M. (1988). *Behavior management: A practical approach for education* (4th ed.). Columbus, OH: Merrill.

Werry, J.S. (1998). Anxiolytics and sedatives. In S. Reiss & M.G. Aman (Eds.), *Psychotropic medication and developmental disabilities: The international consensus handbook.* Washington, DC: American Association on Mental Retardation.

Resources

Written

French, R., Henderson, H.L., & Horvat, M. (1992). *Creative approaches to managing student behavior.* Park City, UT: Family Development Resources. Helps teachers manage behavior effectively to assure that learning can occur. The first part of the book is dedicated to a clear exposition of the major principles of operant conditioning, enabling the reader to comprehend the second section, which analyzes the most common behavioral profiles identified in physical education, recreation, and coaching and suggests prevention and intervention strategies.

French, R., & Lavay, B. (Eds.). (1990). *A manual of behavior management techniques for physical educators and recreators.* Kearney, NE: Educational Systems Associates. Provides an overview of the research and discussion of effective behavior management techniques related to persons with disabilities. Readings address topics such as justification and importance of behavior management, critical analysis, research design, general techniques and strategies, and techniques applied to specific populations.

Audiovisual

Almost Everything You Ever Wanted to Know About Motivating People: Maslow's Hierarchy of Needs, videotape. (1987). Salenger Educational Media, 1635 12th Street, Santa Monica, CA 90404. Basic needs such as food, shelter, security, recognition, and achievement are reviewed in light of Maslow's hierarchy of needs; their meaning for motivation in organizational settings is analyzed and illustrated using dramatized incidents. Running time is approximately 16 minutes.

Catch'em Being Good: Approaches to Motivation and Discipline, videotape. (1990). Research Press, Box 31775, Champaign, IL 60820. This film presents methods for helping children with emotional, behavioral, and academic problems and shares the more rewarding application of positive discipline based on warm teacher-child interaction. Running time is approximately 30 minutes.

Painting a Positive Picture: Proactive Behavior Management, videotape. (1994). Indiana Bureau of Child Development, 402 E. Washington Street, Indianapolis, IN 46204. This video examines and illustrates how to manage children's behavior, while maintaining a perspective between discipline and punishment. Scenarios portraying young children in child/day care situations are presented. Running time is approximately 28 minutes.

Teaching People With Developmental Disabilities, videotape. (1988). Research Press, Box 31775, Champaign, IL 60820. This film is a hands-on training program illustrating how to use task analysis, prompting, reinforcement, and error correction to teach functional skills to individuals with mild to moderate developmental disabilities. Running time is approximately 88 minutes.

Three Approaches to Psychotherapy. III. Pt. 2., videotape. (1986). Psychological & Educational Films, PMB #252, E. Coast Highway, Corona Del Mar, CA 92625. This video highlights Dr. Donald Meichenbaum illustrating his methods of cognitive behavior modification therapy. Running time is approximately 47 minutes.

CHAPTER 8

References

Accardo, P.J. & Capute, A.J. (Eds.). (1998). Mental retardations. *Mental Retardation and Developmental Disabilities Research Reviews, 4* (1), 2-5.

American Association on Mental Retardation. (1992). *Mental Retardation: Definition, classification, and systems of supports* (9th ed.). Washington DC: Author.

Burack, J.A., Hodapp, R.M., & Zigler, E. (Eds.). (1998). *Handbook of mental retardation and development.* Cambridge, United Kingdom: Cambridge University Press.

Cunningham, C. (1987). *Down syndrome: An introduction for parents* (Rev. ed.). Cambridge, MA: Brookline.

Dunn, J.M. (1997). *Special physical education: Adapted, individualized, developmental* (7th ed.). Madison, WI: Brown & Benchmark.

Dunn, J.M., Morehouse, J.W., & Fredericks, H.D.B. (1986). *Physical education for the severely handicapped: A systematic approach to a data based gymnasium.* Austin, TX: Pro-Ed.

Eichstaedt, C.B., Wang, P.Y., Polacek, J.J., & Dohrmann, P.F. (1991). *Physical fitness and motor skill levels of individuals with mental retardation: Mild, moderate, and individuals with Down syndrome: ages 6 to 21.* Normal: Illinois State University Printing Services.

Hagberg B., & Kyllerman, M. (1983). Epidemiology of mental retardation—A Swedish survey. *Brain Development, 5,* 441-449.

Hensley, L.D. (1997). Alternative assessment for physical education. *Journal of Physical Education, Recreation and Dance, 68*(7), 19-24.

Karp, J., & Adler, A. (1992). *ACTIVE.* (Available from J. Karp, 20214 103rd Place NE, Bothell, WA 98011-2455).

Kern County Superintendent of Schools Office. (1995). *M.O.V.E.: Mobility opportunities via education.* Bakersfield, CA: Author.

Londeree, B.R., & Johnson, L.E. (1974). Motor fitness of TMR vs. EMR and normal children. *Medicine Science and Sport, 6,* 247-252.

Loovis, E.M., & Ersing, W.F. (1979). *Assessing and programming gross motor development for children.* Cleveland Heights, OH: Ohio Motor Assessment Associates.

Lund, J. (1997). Authentic assessment: Its development and applications. *Journal of Physical Education, Recreation and Dance, 68*(7), 25-28, 40.

McLaren, J., & Bryson, S.E. (1987). Review of recent epidemiological studies of mental retardation: Prevalence, associated disorders, and etiology. *American Journal of Mental Retardation, 92,* 243-254.

Murphy, C.C., Boyle, C., Schendel, D., Decoufle, P., et al. (1998). Epidemiology of mental retardation in children. *Mental Retardation and Developmental Disabilities Research Reviews,* 4(1), 6-13.

Official Special Olympics summer sports rules: 1996-1999 (Rev. ed.). (1997). Washington DC: Special Olympics.

Piaget, J. (1952). *The origins of intelligence in children.* New York: International Universities Press.

Rarick, G.L., Dobbins, D.A., & Broadhead, G.D. (1976). *The motor domain and its correlates in educationally handicapped children.* Englewood Cliffs, NJ: Prentice Hall.

Rarick, G.L., & McQuillan, J.P. (1977). *The factor structure of motor abilities of trainable mentally retarded children: Implications for curriculum development.* (DHEW Project No H23-2544). Berkeley, CA: Department of Physical Education, University of California.

Scott, K.G. (1988). Theoretical epidemiology: Environment and life-style. In J.F. Kavanagh (Ed.), *Understanding mental retardation.* Baltimore, MD: Paul H. Brookes.

Special Olympics sports skills program guides (1995-1999). Washington, DC: Special Olympics.

U.S. Department of Education. (1998). *Twentieth annual report to Congress on the implementation of the Individuals with Disabilities Education Act* (DOE Publication No. 1998-616-188/90444). Washington, DC: U.S. Government Printing Office.

Vodola, T.M. (1978). *Developmental and adapted physical education: A.C.T.I.V.E. motor ability and physical fitness norms: For normal, mentally retarded, learning disabled, and emotionally disturbed individuals.* Oakhurst, NJ: Township of Ocean School District.

Winnick, J.P., & Short, F.X. (1999a). *The Brockport physical fitness test manual.* Champaign, IL: Human Kinetics.

Winnick, J.P., & Short, F.X. (Eds.). (1999b). *The Brockport physical fitness training guide.* Champaign, IL: Human Kinetics.

Yeargin-Allsopp, M., Murphy, C.C., Cordero, J.F., et al. (1997). Reported biomedical causes and associated medical conditions for mental retardation among 10-year-old children, metropolitan Atlanta, 1985 to 1987. *Developmental and Medical Child Neurology, 39,* 142-149.

Zigman, W., Silverman, W., & Wisniewski, H.M. (1996). Aging and Alzheimer's disease in Down syndrome: Clinical and pathological changes. *Mental Retardation and Developmental Disabilities Research Reviews,* 2(2), 73-76.

Resources

Written

Eichstaedt, C.B., & Lavay, W. (1992). *Physical activity for individuals with mental retardation: Infancy through adulthood.* Champaign, IL: Human Kinetics. P.O. Box 5076, Champaign, IL 61825-5076; phone 217-351-5076. A comprehensive book that covers movement competency from infancy throughout adulthood for persons with mental retardation.

Special Olympics. (1995-1999). *Sports skills program guides.* Washington, DC: Special Olympics, Inc., 1325 G St. NW, #500, Washington, DC 20005; phone 202-628-3630. A series of sport-specific instructional manuals including long-term goals, short-term objectives, skill assessments, task analyses, teaching suggestions, progression charts, and related information.

Special Olympics, Inc. *Spirit: The Magazine of Special Olympics.* Washington, DC: Special Olympics, Inc., 1325 G. St. NW, #500, Washington, DC 20005; phone 202-628-3630. Published quarterly by Special Olympics to promote its aims and programs and to provide information about Special Olympics to its participants, volunteers, and others interested in Special Olympics.

Audiovisual

And Then Came John, videotape. (1988). New York, NY: Filmakers Library, 124 East 40th Street, New York, NY 10016; phone 212-808-4980. Movingly documents the importance of family and community support in the development of persons with disabilities.

Fred's Story, videotape. (1996). Gilman, CT: Pennycorner Press in Association with Storyline Motion Pictures. Program Development Associates, 5620 Business Ave., Suite B, Cicero, NY 13039; phone 800-543-2119. An older gentleman tells of decades in a large institution, then a move to his own apartment. He talks about living in the training school and being on his own. His reflections on life inside are intercut with contemporary footage expressing the ideals of the institution. Running time is 27 minutes 20 seconds.

Special Friends, videotape. (1994). New York, NY: Filmakers Library, 124 East 40th Street, New York, NY 10016; phone 212-808-4980. Depicts the 19-year friendship between two young women with Down syndrome. They enjoy themselves immensely, and, because of their friendship, each attains an impressive level of independence. Family support is the key ingredient in the success of this friendship. Shows that people with mental retardation have the same needs as other adults: a job, a place to live, and relationships. Running time is 38 minutes.

The Spirit of Special Olympics, videotape. (1997). Washington, DC: Special Olympics, Inc., 1325 G. St. NW, #500, Washington, DC 20005; phone 202-628-3630. Orientation to the Special Olympics movement. Provides background and insight into Special Olympics as well as the myriad of volunteer opportunities. Running time is 20 minutes

Electronic

American Association on Mental Retardation. Web site: **http://www.aamr.org**. Information is provided about AAMR's definition of mental retardation, about the mental retardation and disabilities field, and current events and resources.

The ARC of the United States. Web site: **http://www.thearc.org**. Home of The ARC of the United States, the country's largest voluntary organization committed to the welfare of individuals with mental retardation and their families. It provides government, research, and program services reports; fact sheets; and publications catalog. It also offers The ARC's positions on issues that affect people with mental retardation and their families.

National Down Syndrome Congress. Web site: **http://www.ndscenter.org/index2.htm**. Site of the national advocacy organization for Down syndrome. Provides information on health concerns, position statements on issues, bibliographies, and resources for parents of children with Down syndrome.

National Fragile X Foundation. Web site: **http://www.nfxf.org**. Provides the latest information on Fragile X syndrome, resources and materials for educators and for families of individuals with Fragile X syndrome.

Presidents Committee on Mental Retardation. Web site: **http://www.acf.dhhs.gov/programs/pcmr**. Site of the committee that advises the president and secretary of Health and Human Services on matters relating to programs and services for persons with mental retardation. Offers the organization's seven national goals, publications, and state and national resources.

Special Olympics, Inc. Web site: **http://www.specialolympics.org**. Provides information on Special Olympics worldwide programs, upcoming games and competitions, individual profiles of Special Olympics athletes, and how to get involved.

CHAPTER 9
References

Barkley, R. (1997). Attention-deficit/hyperactivity disorder, self-regulation, and time: Toward a more comprehensive theory. *Developmental and Behavioral Pediatrics*, 18, 271-279.

Boucher, C.R. (1999). *Students in discord: Adolescents with emotional and behavioral disorders*. Westport, CT: Greenwood.

Bruininks, V.L., & Bruininks, R.L. (1977). Motor proficiency and learning disabled and nondisabled students. *Perceptual and Motor Skills*, 44, 1131-1137.

Brunt, D., Magill, R.A., & Eason, R. (1983). Distinctions in variability of motor output between learning disabled and normal children. *Perceptual and Motor Skills*, 57, 731-734.

Cantwell, D., & Baker, L. (1991). Association between attention deficit-hyperactivity disorder and learning disorders. *Journal of Learning Disabilities*, 24(2), 88-94.

Cinelli, B., & DePaepe, J.L. (1984). Dynamic balance of learning disabled and nondisabled children. *Perceptual and Motor Skills*, 58, 243-245.

Craft, D. (1996). A focus on inclusion in physical education. In B. Hennessy (Ed.), *Physical education teacher resource handbook*. Champaign, IL: Human Kinetics.

Delong, R. (1995). Medical and pharmacologic treatment of learning disabilities. *Journal of Child Neurology*, 10, (Suppl 1), S92-S95.

Fiore, T.A., Becker, E.A., & Nero, R.C. (1993). Educational interventions for students with attention deficit disorder. *Exceptional Children*, 60(2), 163-173.

Harvey, W., & Reid, G. (1997). Motor performance of children with attention-deficit hyperactivity disorder: A preliminary investigation. *Adapted Physical Activity Quarterly*, 14, 189-202.

Haubenstricker, J.L. (1983). Motor development in children with learning disabilities. *Journal of Physical Education, Recreation and Dance*, 53, 41-43.

Individuals with Disabilities Education Act. (1997). *Individuals with Disabilities Education Act Amendments of 1997*. Washington, DC: U.S. Government Printing Office.

Kahn, L.E. (1982). *Self-concept and physical fitness of retarded students as correlates of social interaction between retarded and non-retarded students*. Unpublished doctoral dissertation, New York University.

Levine, M. (1990). *Keeping a head in school*. Cambridge, MA: Educators Publishing Service.

Moats, L.C., & Lyon, G.R. (1993). Learning disabilities in the United States: Advocacy, science, and the future of the field. *Journal of Learning Disabilities*, 26(5), 282-294.

Rapp, D. (1992). *Is this your child? Discovering and treating unrecognized allergies*. New York: Morrow.

Raskind, M., Goldberg, R., Higgins, E., & Herman, K. (1999). Patterns of change and predictors of success in individuals with learning disabilities: Results from a twenty-year longitudinal study. *Learning Disabilities Research and Practice*, 14(1), 35-49.

Reason, R. (1999). ADHD: A psychological response to an evolving concept (Report of a working party of the British Psychological Society). *Journal of Learning Disabilities*, 32(1), 85-91.

Riccio, C.A., Hynd, G.W., Cohen, M.J., & Gonzalez, J.J. (1993). Neurological basis of attention deficit hyperactivity disorder. *Exceptional Children*, 60(2), 118-124.

Rief, S. (1993). *How to reach and teach ADD/ADHD children: Practical techniques, strategies, and interventions for helping children with attention problems and hyperactivity*. West Nyack, NY: Center for Applied Research in Education.

Sherrill, C., & Pyfer, J.L. (1985). Learning disabled students in physical education. *Adapted Physical Activity Quarterly*, 2, 283-291.

Sutaria, S.D. (1985). *Specific learning disabilities: Nature and needs*. Springfield, IL: Charles C Thomas.

Tomasi, S., & Weinberg, S. (1999). Classifying children as LD: An analysis of current practice in an urban setting. *Learning Disabilities Quarterly*, 22(1), 31-42.

Torgeson, J.K. (1980). Conceptual and educational implications of the use of efficient task strategies by learning disabled children. *Journal of Learning Disabilities* 13, 364-371.

U.S. Department of Education. (1997). *To assure the free appropriate public education of all children with disabilities. 19th annual report to Congress on the implementation of the Individuals with Disabilities Education Act*. Washington, DC: U.S. Government Printing Office.

Zentall, S.S. (1993). Research on the educational implications of attention deficit hyperactivity disorder. *Exceptional Children*, 60(2), 143-153.

Resources

Written

Bishop, P., & Beyer, R. (1995). Attention deficit hyperactivity disorder (ADHD): Implications for physical educators. *Palaestra*, 11(4), 39-46. Shares useful teaching suggestions for the physical education setting.

Boucher, C.R. (1999). *Students in discord: Adolescents with emotional and behavioral disorders*. Westport, CT: Greenwood. A resource helpful to teachers wishing to better understand the behaviors of adolescents with emotional and behavioral disorders. Two chapters specifically discuss the social and emotional issues that can accompany LD and ADHD. The many scenarios based on real teenagers are helpful in understanding the behaviors described and teaching suggestions offered.

Decker, J., & Voege, D. (1992). Integrating children with attention deficit disorder with hyperactivity into youth sport. *Palaestra*, 8(4), 16-20. Contains the "Youth Sport Participation Profile" of potential use to coaches, parents, and physical educators who seek to include children with ADHD in youth sports.

Kurcinka, M.S. (1992). *Raising your spirited child*. New York: Harper Perennial. A guide for parents whose child is more intense, sensitive, perceptive, persistent, energetic. The positive, reassuring tone coupled with practical suggestions makes this a particularly helpful resource for parents of children with ADHD.

Rief, S. (1993). *How to reach and teach ADD/ADHD children: Practical techniques, strategies, and interventions for helping children with attention problems and hyperactivity*. West Nyack, NY: Center for Applied Research in Education. A practical resource for classroom teachers, but many of the ideas can be applied to teaching physical education.

Audiovisual

ADHD: Inclusive Instruction and Collaborative Practices, videocassette. (1995). Paul H. Brookes, P.O. Box 10624, Baltimore, MD 21285-0624; phone 800-638-3775; fax 410-337-8539; Web site: **www.pbrookes.com**. This video outlines seven key elements to success with students with ADHD. Practical teaching strategies are illustrated in real classroom teaching settings.

How to Help Your Child Succeed in School: Strategies and Guidance for Parents of Children With ADHD and/or Learning Disabilities, videocassette. (1997). Paul H. Brookes, P.O. Box 10624, Baltimore, MD 21285-0624; phone 800-638-3775; fax 410-337-8539; Web site: **www.pbrookes.com**. The focus is on the key for success—a strong partnership in education between home and school. Topics include developing reading, writing, math skills; building organizational and study skills; and coping with learning disabilities.

Electronic

Children with Attention Deficit Disorders (CH.A.D.D.). Web site: **http://www.chadd.org**. 8181 Professional Place, Suite 201, Landover, MD 20785; phone 800-233-4050; fax 301-306-7090. Provides information on conferences, legislation, news releases, local chapters, and research studies.

Division on Learning Disabilities (DLD) within the Council for Exceptional Children (CEC). Web site: **http://www.cec.sped.org**. 1920 Association Drive, Reston, VA 22091-1589; phone 703-620-3660; fax 703-264-9494. This site includes the DLD mission statement, events calendar, resources, and publications.

Learning Disabilities Association of America (LDA). Web site: **http://www.ldanatl.org**. 4156 Library Road, Pittsburgh, PA 15234-1349; phone 412-341-1515; fax 412-344-0224. Includes LDA bulletins, upcoming events, position statements, resources, and publications.

CHAPTER 10

References

Achenbach, T.M., Howell, C.T., Quay, H.C., & Conners, C.K. (1991). National survey of problems and competencies among four- to sixteen-year-olds: Parents' reports for normative and clinical samples. *Monographs of the Society for Research in Child Development*, 56(3), 1-133, serial no. 225.

American Psychiatric Association. (1994). *Diagnostic and statistical manual of mental disorders* (4th ed.). Washington, DC: Author.

Batshaw, M.L., & Conlon, C.J. (1997). Substance abuse: A preventable threat to development. In M.L. Batshaw (Ed.), *Children with disabilities*. Baltimore, MD: Paul H. Brookes.

Bauer, A.M. (1991). Drug and alcohol exposed children: Implications for special education for students identified as behaviorally disordered. *Behavioral Disorders*, 17, 72-79.

Breen, M.J., & Altepeter, T.S. (1990). *Disruptive behavior disorders in children: Treatment-focused assessment*. New York: Guilford Press.

Carpenter, S.L., & McKee-Higgins, E. (1996). Behavior management in inclusive classrooms. *Remedial and Special Education*, 17, 195-203.

Christopher, J., Boucher, J., & Smith, P. (1993). Symbolic play in autism: A review. *Journal of Autism and Developmental Disabilities*, 23, 281-307.

Collier, D., & Reid, G. (1987). A comparison of two models designed to teach autistic children a motor task. *Adapted Physical Activity Quarterly*, 4, 226-236.

Connor, F. (1990). Combating stimulus overselectivity: Physical education for children with autism. *Teaching Exceptional Children*, 23, 30-33.

Dunn, J.M., Morehouse, J.W., & Fredericks, H.D.B. (1986). *Physical education for the severely handicapped: A systematic approach to a data based gymnasium*. Monmouth, OR: Teaching Research.

Elliot, Jr., R.O., Dobbin, A.R., Rose, G.D., & Soper, H.V. (1994). Vigorous, aerobic exercise versus general motor training activities: Effects on maladaptive and stereotypic behaviors of adults with both autism and mental retardation. *Journal of Autism and Developmental Disorders*, 24, 565-576.

Federal Register. (1997, October 22). *Assistance to states for the education of children with disabilities, preschool grants for children with disabilities, and early intervention program for infants and toddlers with disabilities; Proposed rules*. Washington, DC: Department of Education.

Hallahan, D.P., & Kauffman, J.M. (1997). *Exceptional learners: Introduction to special education*. Boston: Allyn & Bacon.

Hawley, T.L., & Disney, E.R. (1992). *Crack's children: The consequence of maternal cocaine abuse (Social Policy Report, Volume 6, Number 4)*. Ann Arbor, MI: Society for Research in Child Development.

Hellison, D.R. (1995). *Teaching responsibility through physical activity*. Champaign, IL: Human Kinetics.

Hellison, D.R., & Georgiadis, N. (1992). Teaching values through basketball. *Strategies*, 5, 5-8.

Hellison, D.R., & Templin, T.J. (1991). *A reflective approach to teaching physical education*. Champaign, IL: Human Kinetics.

Henley, M. (1997). Six surefire strategies to improve classroom discipline. *Learning*, 26, 43-45.

Kauffman, J.M. (1997). *Characteristics of emotional and behavioral disorders of children and youth*. Upper Saddle, NJ: Prentice Hall.

Levinson, L.J., & Reid, G. (1993). The effects of exercise intensity on the stereotypic behaviors of individuals with autism. *Adapted Physical Activity Quarterly*, 10, 255-268.

Mangus, B.C., & Henderson, H. (1988). Providing physical education to autistic children. *Palaestra*, 4, 38-43.

Manjiviona, J., & Prior, M. (1995). Comparison of Asperger syndrome and high-functioning autistic children on a test of motor impairment. *Journal of Autism and Developmental Disorders*, 25, 23-29.

Mauk, J.E., Reber, M., & Batshaw, M.L. (1997). Autism and other pervasive developmental disorders. In M.L. Batshaw (Ed.), *Children with disabilities*. Baltimore, MD: Paul H. Brookes.

McEachin, J.J., Smith, T., & Lovaas, O.I. (1993). Long-term outcome for children with autism who received early intensive behavioral treatment. *American Journal of Mental Retardation*, 97, 359-372.

Mesibov, G.B., & Shea, V. (1997). *From theoretical understanding to educational practice*. New York: Plenum.

Miller, A., & Eller-Miller, E. (1989). *From ritual to repertoire: A cognitive-developmental systems approach with behavior-disordered children*. New York: Wiley.

Miyahara, M., Tujii, M., Hori, M., Nakanishi, K., Kageyama, H., & Sugiyama, T. (1997). Brief report: Motor incoordination in children with Asperger syndrome and learning disabilities. *Journal of Autism and Developmental Disorders*, 27, 595-603.

Morin, B., & Reid, G. (1985). A quantitative and qualitative assessment of autistic individuals on selected motor tasks. *Adapted Physical Activity Quarterly*, 2, 43-55.

National Center for Education Statistics. (1998). *The condition of education, 1998*. Washington, DC: Department of Education.

Nelson, J.R., Smith, D.J., Young, R.K., & Dodd, J.M. (1991). A review of self-management outcome research conducted with students who exhibit behavioral disorders. *Behavioral Disorders*, 16, 169-179.

Quay, H.C. (1986). Classification. In H.C. Quay & J.S. Werry (Eds.), *Psychopathological disorders of childhood*. New York: Wiley.

Quay, H.C., & Peterson, D.R. (1987). *Manual for the revised behavior problem checklist*. Coral Gables, FL: Authors.

Quill, K., Gurry, S., & Larkin, A. (1989). Daily life therapy: A Japanese model for educating children with autism. *Journal of Autism and Developmental Disorders*, 19, 625-635.

Reid, G., Collier, D., & Cauchon, M. (1991). Skill acquisition by children with autism: Influence of prompts. *Adapted Physical Activity Quarterly*, 8, 357-366.

Rhodes, W.C., & Doone, E.M. (1992). One boy's transformation. *Journal of Emotional and Behavioral Problems*, 1, 10-15.

Rosenberg, M.S., Wilson, R., Maheady, L., & Sindelar, P.T. (1997). *Educating students with behavior disorders*. Boston: Allyn & Bacon.

Rosenthal-Malek, A., & Mitchell, S. (1997). Brief report: The effects of exercise on the self-stimulatory behavior and positive responding of adolescents with autism. *Journal of Autism and Developmental Disorders*, 27, 193-202.

Rossi, R.J. (Ed.). (1994). *Schools and students at risk*. New York: Teachers College Press.

Sherrill, C. (1998). *Adapted physical activity, recreation and sport: Crossdisciplinary and lifespan*. Boston: WCB/McGraw-Hill.

Sinclair, E. (1998). Head start children at risk: Relationship of prenatal drug exposure to identification of special needs and subsequent special education kindergarten placement. *Behavioral Disorders*, 23, 125-133.

Singer, L., Arendt, R., Farkas, K., Minnes, S., Huang, J., & Yamashita, T. (1997). Relationship of prenatal cocaine exposure and maternal postpartum psychological distress to child development outcome. *Development and Psychopathology*, 9, 473-489.

Southam-Gerow, M.A., & Kendall, P.C. (1997). Parent-focused and cognitive-behavioral treatments of antisocial youth. In D.M. Stoff, J. Breiling, & J.D. Maser (Eds.), *Handbook of antisocial behavior*. New York: John Wiley & Sons.

Steinberg, Z., & Knitzer, J. (1992). Classrooms for emotionally and behaviorally disturbed students: Facing the challenge. *Behavioral Disorders*, 17, 145-156.

Tarr, S.J., & Pyfer, J.L. (1996). Physical and motor development of neonates/infants prenatally exposed to drugs in utero: A meta-analysis. *Adapted Physical Activity Quarterly*, 13, 269-287.

Van Dyke, D.O., & Fox, A.A. (1990). Fetal drug exposure and its possible implications for learning in the preschool and school-age population. *Journal of Learning Disabilities*, 23(3), 160-163.

Walker, H.H. (1995). *The acting-out child: Coping with classroom disruption*. Longmont, CO: Sopris West.

Weber, R.C., & Thorpe, J. (1992). Teaching children with autism through task variation in physical education. *Exceptional Children*, 59, 77-86.

Wood, M.M., & Long, N.J. (1991). *Life space intervention: Talking with children and youth in crisis*. Austin, TX: Pro-Ed.

Yell, M.L. (1988). The effects of jogging on the rates of selected target behaviors of behaviorally disordered students. *Behavioral Disorders*, 13, 273-279.

Resources

Written

Dunn, J.M., Morehouse, J.W., & Fredericks, H.D.B. (1986). *Physical education for the severely handicapped: A systematic approach to a data based gymnasium*. Monmouth, OR: Teaching Research. This text is a "nuts and bolts" approach to working with students who have severe disabilities in the physical education setting. It combines detailed information on learning theory with a variety of examples to illustrate the principles that are advocated for use in the gymnasium.

Goldstein, A.P., Sprafkin, R.P., Gershaw, N.J., & Klein, P. (1980). *Skillstreaming the adolescent: A structured learning approach to teaching prosocial skills*. Champaign, IL: Research Press. This innovative program is designed to help adolescents develop competence in dealing with interpersonal conflicts, in increasing self-esteem, and in contributing to a positive classroom atmosphere.

Wood, M.M., & Long, N.J. (1991). *Life space intervention: Talking with children and youth in crisis*. Austin, TX: Pro-Ed. This is an updated version of the pioneering work of Fritz Redl with emphasis on the "intervention," since crisis implies verbal intervention. "Talking strategies" are presented and applied to particular types of problems.

Audiovisual

Autism: Breaking Through, videotape. (1988). Films for the Humanities & Sciences, Box 2053, Princeton, NJ 08543-2053. This film highlights the Daily Life Therapy program at Boston's Higaski School. The school's curriculum is based on the work of Dr. Kiyo Kitahara, who founded the first Higaski School in Tokyo, which serves 1,800 students, 25 percent of whom are autistic. The program emphasized physical exercise as a means of tension release and as a way to enhance self-concept.

Autism: The Child Who Couldn't Play, videotape. (1996). Films for the Humanities & Sciences, Box 2053, Princeton, NJ 08543-2053. This comprehensive overview of autism explores the frontiers of our understanding of autism and visits the Princeton Child Development Institute to see the results of the Institute's highly successful science-based approach to autism.

Conflict Resolution, interactive videodisc. (1993). University of South Florida, I'm Special Production Network, Tampa, FL 33620-8600. This Level III interactive application (external computer that uses a videodisc player as a peripheral device) teaches students how to send appropriate "I" messages, use active listening, and conduct the conflict resolution process. It is applicable to undergraduate students as well as to career teachers who will teach students with behavioral disabilities.

Electronic

Autism Society of America. Web site: **http://www.autism-society.org**. The home site of the Autism Society of America, providing information about membership, research interests of the Autism Society of America Foundation, and legislative and governmental news.

Center of the Study of Autism. Web site: **http://www.autism.org**. Provides autism and autism-related resources on the Internet, including information on nutrition, educational interventions, and research.

Division TEACCH. Web site: **http://www.unc.edu/depts/teacch/**. This is the site of the TEACCH program at the University of North Carolina. Provides information about TEACCH specifically and autism generally. Focuses on educational and communication approaches, supported employment, and research and training opportunities.

National Information Center for Children and Youth With Disabilities. Web site: **http://www.nichcy.org**. Home site for this organization, it provides news, publications, FAQs, training packages, and state resource sheets. It also provides fact sheets for specific disabilities and "briefing papers" on select issues such as PDD.

CHAPTER 11
References

Arnhold, R.W., & McGrain, P. (1985). Selected kinematics patterns of visually impaired youth in sprint running. *Adapted Physical Activity Quarterly*, 2, 206-213.

Bessler, H. (1990). The deaf sprinter: An analysis of starting techniques. *Palaestra*, 6(4), 32-37.

Block, M.E., Lieberman, L.J., & Conner-Kuntz, F. (1998). Authentic assessment in adapted physical education. *Journal of Physical Education, Recreation and Dance*, 69(3), 48-55.

Buell, C.E. (1966). *Physical education for blind children*. Springfield, IL: Charles C Thomas.

Butterfield, S.A. (1988). Deaf children in physical education. *Palaestra*, 6(4) 28-30, 52.

Dolnick, E. (1993). Deafness as culture. *The Atlantic*, 272(3) 37-53.

Dunn, J., Morehouse, J., & Fredericks, H.D. (1986). *Physical education for the severely handicapped: A systematic approach to data based gymnasium*. Austin, TX: Pro-Ed.

Dunn, J.M., & Ponticelli, J. (1988). *The effect of two different communication modes on motor performance test scores of hearing impaired children*. Abstracts of Research Papers, Reston, VA: American Alliance for Health, Physical Education, Recreation and Dance.

Ferrell, K. (1984). *Parenting preschoolers: Suggestions for raising young blind and visually impaired children*. New York: American Foundation for the Blind Press.

Graybill, P., & Cokely, D. (1993). *Introduction to the Deaf community*, videotape. Burtonsville, MD: Sign Media.

Graziadei, A. (1998). *Learning outcomes of deaf and hard of hearing students in mainstreamed physical education classes*. Unpublished doctoral dissertation, University of Maryland, College Park.

Individuals with Disabilities Education Act. (1997). *Individuals with disabilities education act amendments of 1997*. Washington, DC: U.S. Government Printing Office.

Kottke, F., & Lehmann, J. (1990). *Krusen's handbook of physical medicine and rehabilitation* (4th ed.). Philadelphia: W.B. Saunders.

Lieberman, L.J., & Carron, M. (1998). Health related fitness status of children with visual impairments. Poster session, AAHPERD national conference: Reno, Nevada.

Lieberman, L., & Cowart, J. (1996). *Games for people with sensory impairments: Strategies for including individuals of all ages*. Champaign, IL: Human Kinetics.

Lieberman, L.J., Dunn, J.M., van der Mars, H., & McCubbin, J.A. (2000). Peer tutors' effects on activity levels of deaf students in inclusive elementary physical education. *Adapted Physical Activity Quarterly*, 17(1), 20-39.

Lieberman, L.J., & Lepore, M. (1998). Camp Abilities: A developmental sports camp for children who are blind and deafblind. *Palaestra*, 14(1), 28-31.

Luckner, J. (1993). Developing independent and responsible behaviors in students who are deaf or hard of hearing. *Teaching Exceptional Children*, 26(2), 13-25.

Myklebust, H.R. (1946). Significance of etiology in motor performance of deaf children with special reference to meningitis. *American Journal of Psychology*, 59, 249-258.

Nowell, R., & Innes, J. (1997). *Educating children who are deaf or hard of hearing: Inclusion*. Reston, VA: ERIC Clearinghouse on Disabilities and Gifted Education. ERIC Digest # E557.

Paciorek, M.J., & Jones, J.A. (1994). *Sports and recreation for the disabled*. Carmel, IN: Cooper.

Schmidt, S. (1985). Hearing impaired students in physical education. *Adapted Physical Activity Quarterly*, 2, 300-306.

Stewart, D.A. (1991). *Deaf sport: The impact of sports within the deaf community*. Washington, DC: Gallaudet University.

Stewart, D., Dummer, G., & Haubenstricker, J. (1990). Review of administration procedures used to assess the motor skills of deaf children and youth. *Adapted Physical Activity Quarterly*, 7, 231-239.

Strassler, B. (1994). *Gallaudet University Football Centennial*. Washington, DC: Gallaudet University Department of Athletics.

USABA. (1998). *Athletes' profile*. Colorado Springs, CO: Author.

Winnick, J., & Short, F. (1999). *The Brockport physical fitness test manual*. Champaign, IL: Human Kinetics.

Winnick, J.P., & Short, F.X. (1986). Physical fitness of adolescents with auditory impairments. *Adapted Physical Activity Quarterly*, 3, 58-66.

Wyatt, L., & Ng, G.Y. (1997). The effect of visual impairment on the strength of children's hip and knee extensors. *Journal of Visual Impairment and Blindness*, 91(1), 40-46.

Resources on Visual Impairments

Written

Kuusisto, S. (1998). *Planet of the blind: A memoir*. New York: Dial Press. This acclaimed writer has been legally blind since birth. Readers gain insights into his experiences with blindness through his excellent prose.

Lieberman, L., & Cowart, J. (1996) *Games for people with sensory impairments: Strategies for including individuals of all ages*. Champaign, IL: Human Kinetics. This practical reference provides teachers and recreation specialists with 70 games that people with sensory impairments—both visual and hearing—can play.

Lieberman, L.J. & McHugh, B.E. (in press). Health related fitness status of children and youth with visual impairments. *Journal of Visual Impairment and Blindness*.

McInnes, J.M., & Treffry, J.A. (1982). *Deaf-blind infants and children: A developmental guide*. Toronto: University of Toronto Press.

Audiovisual

Portraits of Possibility, videotape. (1996). Insight Media, 2162 Broadway, P.O. Box 621, New York, NY 10024-0621; fax 212-799-5309. This 20-minute videotape explores issues raised by participation in sports for individuals who are blind. Training methods are included.

Electronic

American Foundation for the Blind (AFB). Web site: **http://www.afb.org**. Provides information about advocacy, resources, programs, and publications by AFB, a major organization serving persons who are blind. For more information: 11 Penn Plaza, New York, NY 10001; 800-232-5463.

International Blind Sport and Recreation Association (IBSA). Web site: **http://www.ibsa.es/**. Contains information about IBSA. For further information contact Enrique Sanz, IBSA President, c/o Quevedo, 1-1 28014, Madrid, Spain; phone (3491) 589-45-33/34/36; fax (3491) 589-45-37; e-mail ibsa@ibsa.es.

National Beep Baseball Association (NBBA). Web site: **http://www.nbba.org**. Offers information about NBBA competitions and how to play Beep Baseball. For more information: 9623 Spencer Highway, La Porte, TX 77571; phone 713-476-1592.

United States Association for Blind Athletes (USABA). Web site: **http://www.usaba.org**. Provides information about USABA, its events, competition schedules, newsletter, sponsors, and related sites. For further information: 33 North Institute Street, Colorado Springs, CO 80903; phone 719-630-0422; fax 719-630-0616; e-mail: usaba@usa.net

Resources on Deafness and Hard of Hearing

Audiovisual

Introduction to the Deaf Community, videotape. Graybill, P., & Cokely, D. (1993). Burtonsville, MD: Sign Media, 4020 Blackburn Lane, Burtonsville, MD 20866; 301-421-0268. Basic information is provided to assist hearing persons interact comfortably with Deaf persons. The running time is approximately 30 minutes.

Electronic

Comite International des Sports des Sords (CISS). Web site: **http://www.ciss.org**. Provides information on the CISS member organizations and committees, regional federations, calendar of sports events, CISS bulletin, and photo gallery. For more information contact John M. Lovett, President, 31 Clive Street, East Brighton, Victoria 3187, Australia.

Deaf Sports discussion list. E-mail address: deaf sport.juno.com. An electronic site for discussion of issues related to Deaf sports.

USA Deaf Sport Federation (USADSF). Web site: **http://www.usadsf.org**. Provides information about USADSF competition, events calendar, history of the organization, and news. For more information contact Dr. Bobbie Beth Scoggins, President, 911 Tierra Linda Drive, Frankfort, KY 40601-4633; TTY 801-393-7916; fax 801-393-2263.

Resources on Deafblindness

Written

Lieberman, L.J. (1996). Adapting games, sports, and recreation for children and adults who are deaf-blind. *Deaf-Blind Perspectives*, 3(3) 1-5. Available through Deaf Blind Link. This book presents information on how to adapt games and sports for persons who are deafblind.

Lieberman, L.J., & Downs, S.B. (1995). Physical education for students who are deaf-blind: A tutorial. *Brazilian International Journal of Adapted Physical Education*, 2, 125-143. This article shares principles and philosophies supporting the physical education of students who are deafblind.

Lieberman, L.J., & Taule, J. (1998). Including physical fitness into the lives of individuals who are deafblind. *Deafblind Perspectives*, 5(2), 6-10. This article presents how to include individuals who are deafblind in fitness activities such as running, biking, swimming, aerobics, and weight training.

Smith, T. (1994). *Guidelines: Practical tips for working and socializing with deafblind people*. Burtonsville, MD: Sign Media. This excellent resource offers a step-by-step guide to working and socializing with deafblind persons.

Electronic

American Association for the Deafblind. Web site: **http://www.tr.wou.edu/dblink/aadb.htm;** e-mail aadb@rols.com. For more information: 814 Thayer Avenue, Suite 302, Silver Springs, MD 20910-4500; TTY 301-588-6545; relay 800-735-2258; fax 301-588-8705.

CHAPTER 12

References

American Heart Association. (1998). *Heart and stroke guide*. Dallas, TX: Author.

Appleton, R.E. (1998). Epidemiology: Incidence, causes, and severity. In R.E. Appleton & T. Baldwin (Eds.), *Management of brain-injured children*. New York: Oxford University Press.

Baxter, K.F., & Lockette, K.F. (1995). Resistance training with stretch bands: Modifying for disability. In P.D. Miller (Ed.), *Fitness programming and physical disability*. Champaign, IL: Human Kinetics.

Brain Injury Association. (1997). *Basic questions about brain injury and disability*. Alexandria, VA: Author.

DiRocco, P.J. (1999). Muscular strength and endurance. In P. J. Winnick & F.X. Short (Eds.), *The Brockport physical fitness training guide*. Champaign, IL: Human Kinetics.

Ferrara, M., & Laskin, J. (1997). Cerebral palsy. In J.L. Durstine (Ed.), *ACSM's exercise management for persons with chronic diseases and disabilities*. Champaign, IL: Human Kinetics.

Finnie, N.R. (1997). *Handling the young child with cerebral palsy* (3rd ed.). Boston: Butterworth-Heinemann.

Levitt, S. (1995). *Treatment of cerebral palsy and motor delay* (3rd ed.). Cambridge, MA: Blackwell Scientific.

Lockette, K.F., & Keyes, A.M. (1994). *Conditioning with physical disabilities*. Champaign, IL: Human Kinetics.

Love, B.B., Orencia, A.J., & Billen, J. (1994). Stroke in children and young adults: Overview, risk factors and prognosis. In J. Billen, K. Mathews, & B.B. Love (Eds.), *Stroke in children and young adults*. Boston: Butterworth-Heinemann.

Mushett, C.A., Wyeth, D.O., & Richter, K.J. (1995). Cerebral palsy. In B. Golderberg (Ed.), *Sports and exercise for children with chronic health conditions*. Champaign, IL: Human Kinetics.

National Head Injury Foundation. (1989). *Basic questions about head injury and disability*. Washington, DC: Author.

Palmer-McLean, K., & Wilberger, J.E. (1997). Stroke and head injury. In J.L. Durstine (Ed.), *ACSM's exercise management for persons with chronic diseases and disabilities*. Champaign, IL: Human Kinetics.

Smith, S.M., & Tyler, J.S. (1997). Successful transitional planning and services for students with ABI. In A. Glang, G.H. Singer, & B. Todis (Eds.), *Students with acquired brain injury: The school's response*. Baltimore, MD: Paul H. Brookes.

Surburg, P.R. (1999). Flexibility/range of motion. In J.P. Winnick & F.X. Short (Eds.), *The Brockport physical fitness training guide*. Champaign, IL: Human Kinetics.

United Cerebral Palsy Associations, Inc. (1998). *Cerebral palsy—Facts and figures*. Washington, DC: Author.

United States Cerebral Palsy Athletic Association. (1997). *Classification and sports rules manual* (5th ed.). Newport, RI: Author.

Walker, B.R. (1997). Creating effective educational programs through parent-professional partnerships. In A. Glang, G.H. Singer, & B. Todis (Eds.), *Students with acquired brain injury: The school's response*. Baltimore, MD: Paul H. Brookes.

Winnick, J.P., & Short, F.X. (1999a). *The Brockport physical fitness test manual*. Champaign, IL: Human Kinetics.

Winnick, J.P., & Short, F.X. (Eds.). (1999b). *The Brockport physical fitness training guide*. Champaign, IL: Human Kinetics.

Ylvisaker, M., & Feeney, T.J. (1998). School reentry after traumatic brain injury. In M. Ylvisaker (Ed.), *Traumatic brain injury rehabilitation: Children and adolescents* (2nd ed.). Boston, MA: Butterworth-Heinemann.

Ylvisaker, M., Hartwick, P., & Stevens, M. (1991). School reentry following head injury: Managing the transition from hospital to school. *Journal of Head Trauma Rehabilitation*, 6(1), 10-22.

Resources

Written

American Heart Association. *Stroke connection*. Dallas, TX: Author. A bimonthly magazine published and written by stroke survivors and caregivers. Provides a wealth of post-stroke information. Reinforces health care provider messages about the importance of stroke prevention, early treatment, and rehabilitation. Can be obtained by contacting the American Heart Association, 7272 Greenville Ave., Dallas, TX 75231.

Berquist, W., McLean, R., & Kobylinski, B. (1994). *Stroke survivors*. San Rafael, CA: Ableforce. Firsthand account of the experience of having a stroke and the process of recovery. For purchase write to Ableforce, 84 Pilgrim Way, San Rafael, CA 94903.

Fitness Canada. (1994). *Moving to inclusion: Cerebral palsy*. Gloucester, Ontario: Active Living Alliance for Canadians with a Disability. One in a series of nine manuals providing physical education teachers with practical ways to integrate students with disabilities into regular physical education classes. Includes information about programming for inclusive physical education, manipulation skills, sport skills, fitness, outdoor education, intramural and interschool programs, and professional support services. Can be obtained by writing to Active Living Alliance for Canadians with a Disability, 1600 James Naismith Dr., Gloucester, Ontario, Canada, K1B 5N4.

Heartland Educational Agency II. (1993). *School reentry of students experiencing traumatic brain injury: Physical education considerations*. Johnston, IA: Heartland AEA 11. This paper provides current information regarding traumatic brain injury as it relates to physical education. Provides an overview of TBI and information relative to assessment, teaching, programming strategies, and sport-related issues. Can be acquired from Heartland Educational Agency II, 6500 Corporate Drive, Johnston, IA 50131-1603.

Audiovisual

Advancements in Traumatic Brain Injury, videotape. Films for the Humanities & Sciences, P.O. 2053, Princeton, NJ 08543-2053. This 19-minute video is narrated by Dr. George Zitnay, president of the Brain Injury Association and other experts who discuss innovative rehabilitation techniques. Several people with brain injury reveal how the injuries have impacted their lives.

Aerobics for Cerebral Palsy, videotape. National Handicapped Sports/Videotapes, 451 Hungerford Drive, Suite 100, Rockville, MD 20850. Illustrates vigorous exercise for ambulatory and nonambulatory individuals with muscular coordination difficulties. Program features prolonged warmup, followed by exercise using easy-to-follow upper-body movements. Low-impact rhythmic full-body actions are demonstrated.

A Stroke Survivor's Workout, videotape. American Heart Association, 7272 Greenville Avenue, Dallas, TX 75231. A 28-minute video showing flexibility and strength exercises. Some exercises are done from a standing position, using the support of a chair if needed, while other exercises are done while seated in a chair or wheelchair.

Electronic

The fitness challenge: Software for the Brockport Physical Fitness Test. Written by J. Winnick and F. Short, this software provides technical information on the health-related physical fitness test. It can be purchased from Human Kinetics, P.O. Box 5076, Champaign, IL 61825-5076.

National Resource Center for Traumatic Brain Injury. Web site: **http://www.neuro.pmr.vcu.edu**. Provides a wealth of information on brain injury such as a directory of experts in TBI, educational materials, assessment tools, audiovisuals, intervention kits, FAQs, comprehensive reference lists, announcements, and news.

National Stroke Association. Web site: **http://www.stroke.org**. Provides information about the association's mission, selected statistics pertaining to public understanding of stroke, stroke prevention guidelines, and other pertinent information.

United States Cerebral Palsy Athletic Association (USCPAA). Web site: **http://www.uscpaa.org**. Contains a description of the organization, information on the new national center in Newport, Rhode Island; sport news; the Paralympics; and products and links to other sites pertaining to sport for persons with disabilities.

CHAPTER 13

References

Adams, R. (1994). Physical activity and exercise guidelines. In H.C. Glauser (Ed.), *Living with osteogenesis imperfecta: A guidebook for families*. Tampa, FL: Osteogenesis Imperfecta Foundation.

Adams, R.C., & McCubbin, J.A. (1991). *Games, sports and exercises for the physically handicapped* (4th ed.). Philadelphia: Lea & Febiger.

Arthritis Foundation. (1996). *When your student has arthritis*. Atlanta, GA: Arthritis Foundation.

Bar-Or, O. (1997). Muscular dystrophy. In J.L. Durstine (Ed.), *ACSM's exercise management for persons with chronic diseases and disabilities*. Champaign, IL: Human Kinetics.

Cassidy, J.T., & Petty, R.E. (1995). *Textbook of pediatric rheumatology*. Philadelphia: Saunders.

Crandall, R., & Crosson, T. (Eds.). (1994). *Dwarfism: The family and professional guide*. Irvine, CA: Short Stature Foundation and Information Center, Inc.

DiRocco, P. (1999). Muscular strength and endurance. In J.P. Winnick & F.X. Short (Eds.), *The Brockport physical fitness training guide*. Champaign, IL: Human Kinetics.

Goodkin, D.E., & Rudick, R.A. (Eds.). (1996). *Multiple sclerosis: Advances in clinical trial design, treatment and future perspectives*. London: Springer.

Lockette, K.F., & Keyes, A.M. (1994). *Conditioning with physical disabilities*. Champaign, IL: Human Kinetics.

May, B.J. (1996). *Amputations and prosthetics*. Philadelphia, PA: F.A. Davis.

National Multiple Sclerosis Society. (1997). *What everyone should know about multiple sclerosis*. New York, NY: National Multiple Sclerosis Society.

Parry, G.J. (1993). *Guillain-Barre syndrome*. New York, NY: Theime Medical.

Scull, S.A., & Athreya, B.H. (1995). Childhood arthritis. In B. Goldberger (Ed.), *Sports and exercise for children with chronic health conditions*. Champaign, IL: Human Kinetics.

Short, F.X., McCubbin, J., & Frey, G. (1999). Cardiorespiratory endurance and body composition. In J.P. Winnick & F.X. Short (Eds.), *The Brockport physical fitness training guide*. Champaign, IL: Human Kinetics.

Surburg, P. (1999). Flexibility/range of motion. In J.P. Winnick, & F.X. Short (Eds.), *The Brockport physical fitness test training guide*. Champaign, IL: Human Kinetics.

Watson, D.F., & Lisak, R.P. (1994). Myasthenia gravis: An overview. In R.P. Lisak (Ed.), *Handbook of myasthenia gravis and myasthenic syndromes*. New York, NY: Dekker.

Winnick, J.P., & Short, F.X. (1999). *The Brockport physical fitness test manual*. Champaign, IL: Human Kinetics.

Resources

Written

Arthritis Foundation. (1996). *When your student has arthritis*. Atlanta, GA: Author. Brochure containing information about treatment, keeping up at school, physical education, and educational rights. To obtain, write the Arthritis Foundation, PO Box 7669, Atlanta, GA 30357-0669 or contact your local Arthritis Foundation chapter.

Fitness Canada. (1993). *Moving to inclusion: Amputation*. Gloucester, Ontario: Active Living Alliance for Canadians with a Disability. One in a series of nine manuals designed to provide physical education teachers with concrete examples in which to integrate students with disabilities into regular physical education. Includes informative ways of getting and keeping fit, adapting specific sports, and a number of resource materials. Obtain by writing to Active Living Alliance for Canadians with a Disability, 1600 James Naismith Dr., Gloucester, Ontario, Canada K1B 5N4.

Kimberg, I. (1997). *Moving with multiple sclerosis: An exercise manual for people with multiple sclerosis*. New York, NY: National Multiple Sclerosis Society. Manual provides information on passive range of motion and stretching exercises, active exercises, and coordination and balance exercises. Obtain from the National Multiple Sclerosis Society, 733 Third Ave., New York, NY, 10017-3288 or contact your local MS chapter.

Porter, P., Hall, C., and Williams, F. (1994). *A teacher's guide to Duchenne muscular dystrophy*. Tucson, AZ: Muscular Dystrophy Association. Manual provides information that includes developmental issues, school behaviors and attendance, and physical education participation. Obtain by writing the Muscular Dystrophy Association (MDA), 3300 East Sunrise Dr., Tucson, AZ, 85718-3208 or contacting your local MDA chapter.

Audiovisual

Aerobics for Amputees, videotape. Disabled Sports, USA, 451 Hungerford Dr., Suite 100, Rockville, MD 20850. This video depicts vigorous exercise for ambulatory individuals who have impaired balance and/or coordination—people who can exercise standing up but cannot do fancy footwork that may upset balance.

Disabled Sports, USA, introductory videotape. Disabled Sports, USA, 451 Hungerford Dr., Suite 100, Rockville, MD 20850. This video features competition footage and interviews with DS/USA athletes and staff. It can be purchased for $19.95.

Dwarfism: Born to Be Small, videotape. Films for the Humanities and Sciences, P.O. Box 2053, Princeton, NJ 08543-2053. This 51-minute video documents the lives of four families impacted by dwarfism. It can be purchased or rented.

MS Wheelchair Workout, videotape. New York City Chapter, National Multiple Sclerosis Society, 30 West 26th St., 9th Floor, New York, NY 10010-2094. This 30-minute video illustrates a variety of exercises designed to improve balance, coordination, flexibility, and strength, not only for persons with MS, but for others with extremity weaknesses as well.

Electronic

Disabled Sports, USA. Web site: **http://www.dsusa.org**. Home of Disabled Sports, USA, the largest multisport, multidisability organization in the country. It sponsors competition for amputees, and the site contains information on classifications, competition schedules, events, and more. The organization conducts year-round programs in 11 Olympic sports.

Flex-Foot, Inc. Web site: **http://www.flexfoot.com**. Provides information on the latest high-tech prosthetic devices available. Persons who have been successful in both sport and leisure by using the Flex-Foot devices are also highlighted.

CHAPTER 14

References

American College of Sports Medicine. (1997). *ACSM's exercise management of persons with chronic diseases and disabilities*. Champaign, IL: Human Kinetics.

American Medical Association. (1990). *Handbook of first aid and emergency care*. New York: Random House.

Centers for Disease Control. (1995). Progress toward global poliomyelitis eradication, 1985-1994. *Morbidity and Mortality Weekly Report*, 44(14), 273-275, 281.

Ferrara, M.S., & Davis, R.W. (1990). Injuries to wheelchair athletes. *Paraplegia*, 28, 335-341.

Figoni, S.F. (1997). Spinal cord injury. In *ACSM's exercise management for persons with chronic diseases and disabilities*. Champaign, IL: Human Kinetics.

Gayle, G.W., & Muir, J.L. (1992). Role of sports medicine and the spinal cord injured: A multidisciplinary relationship. *Palaestra*, 8(3), 51-56.

Goldberg, B. (Ed.). (1995). *Sports and exercise for children with chronic health conditions*. Champaign, IL: Human Kinetics.

Lockette, K., & Keyes, A.M. (1994). *Conditioning with physical disabilities*. Champaign, IL: Human Kinetics.

Miller, P.D. (Ed.). (1995). *Fitness programming and physical disability*. Champaign, IL: Human Kinetics.

National Spinal Cord Injury Association. (1999). *Spinal cord injury statistics*. Retrieved July 26, 1999 from the World Wide Web: **http://www.spinalcord.org/**.

National Wheelchair Basketball Association. (1999). *National Wheelchair Basketball Association official rules and case book 1998-1999*. Retrieved July 26, 1999 from the World Wide Web: **nwba.org/rules.html**.

Rimmer, J. (1994). *Fitness and rehabilitation programs for special populations*. Madison, WI: WCB Brown & Benchmark.

Spina Bifida Association of America (SBAA). (1999). *Facts about spina bifida*. Retrieved July 28, 1999 from the World Wide Web: **http://www.sbaa.org/html/sbaa_facts.html**.

Wells, C.L., & Hooker, S.P. (1990). The spinal injured athlete. *Adapted Physical Activity Quarterly*, 7, 265-285.

Winnick, J.P., & Short, F.X. (1999). *The Brockport physical fitness test manual*. Champaign, IL: Human Kinetics.

Winnick, J.P., & Short, F.X. (Eds.). (1999). *The Brockport physical fitness training guide*. Champaign, IL: Human Kinetics.

Winnick, J.P., & Short, F.X. (1985). *Physical fitness testing of the disabled*. Champaign, IL: Human Kinetics.

Winnick, J.P., & Short, F.X. (1984). The physical fitness of youngsters with spinal neuromuscular conditions. *Adapted Physical Activity Quarterly*, 1, 37-51.

Resources

Written

American College of Sports Medicine. (1997). *ACSM's exercise management of persons with chronic diseases and disabilities*. Champaign, IL: Human Kinetics. This book provides guidelines for exercise testing and programming for individuals with spinal cord injuries, polio, and post-polio syndrome. Excellent recommendations are provided regarding safety and precautions that should be taken to reduce risks during exercise.

Goldberg, B. (Ed.). (1995). *Sports and exercise for children with chronic health conditions*. Champaign, IL: Human Kinetics. This book provides guidelines for sport participation and exercise for children with spina bifida.

Lockette, K., & Keyes, A.M. (1994). *Conditioning with physical disabilities*. Champaign, IL: Human Kinetics. This book provides step-by-step descriptions on a variety of exercises for all the major components of physical fitness: flexibility, strength training, and aerobic conditioning. Specific guidelines are provided for working with individuals with spinal cord injuries, spina bifida, and polio.

Miller, P.D. (Ed.). (1995). *Fitness programming and physical disability*. Champaign, IL: Human Kinetics. This book contains a variety of "how-to" fitness training techniques for individuals with physical disabilities. Guidelines are provided on how to modify and adapt traditional exercises to include individuals with physical disabilities into existing fitness programs and routines.

Winnick, J.P., & Short, F.X. (1999). *The Brockport physical fitness test manual*. Champaign, IL: Human Kinetics. Manual containing specific health-related criterion-referenced test items in the areas of aerobic functioning, body composition, and musculoskeletal function and performance standards for individuals ages 10 to 17 with a variety of disabilities including spinal cord injuries.

Winnick, J.P., & Short, F.X. (Eds.). (1999). *The Brockport physical fitness training guide*. Champaign, IL: Human Kinetics. This guide complements the Brockport Physical Fitness Test and provides many helpful recommendations on how to develop and implement safe physical fitness programs for individuals. Specific recommendations are provided for individuals with spinal cord injuries.

Audiovisual

Aerobics for Quadriplegia, Aerobics for Cerebral Palsy, Aerobics for Paraplegia, Aerobics for Amputees, videotapes. (1995). Disabled Sports, USA, 451 Hungerford Drive, Suite 100, Rockville, MD 20850. These four videotapes provide aerobic routines for individuals with varying physical disabilities that progress from easy to demanding workouts. The instructors on the tapes provide guidelines to ensure the exercises are performed correctly and safely, and the exercises are demonstrated by youngsters and adults in wheelchairs.

Strength and Flexibility for Individuals With All Types of Physical Disabilities, videotape. (1995). Disabled Sports, USA, 451 Hungerford Drive, Suite 100, Rockville, MD 20850. Videotape providing a strength and flexibility routine using rubber tubes for resistance. Each exercise is presented by the instructor and modeled by a person with a disability.

Electronic

Assistive Technology. Web site: **www.nccn.net/~freed/at.html**. Contains a wealth of information on assistive technology that can be used to increase the independence of individuals with disabilities.

Association for Spina Bifida and Hydrocephalus. Web site: **www.asbah.demon.co.uk**. Contains the latest information on the treatment of spina bifida and shunts used to treat hydrocephalus.

Cure Paralysis Now Links. Web site: **www.cureparalysis.org/related/index.html**. Provides an extensive number of links to research projects investigating innovative procedures to cure spinal cord injuries.

Disabled Sports, USA. Web site: **www.dsusa.org**. Provides the latest information on events and workshops sponsored by DS/USA, as well as event results and extensive links to other related sites.

Orthotics and Prosthetics. Web site: **www.oandp.com**. Provides pictures and information on the latest developments in orthotics for individuals with disabilities.

Wheelchair Sports, USA. Web site: **www.wsusa.org**. Contains the latest information on WS/USA including rules, events, qualifying standards, event results, event calendar, publications, and links to other related sites.

CHAPTER 15

References

Adams, E. (1988). The multiple deficits of prenatal drug abuse. *Science Focus, 2,* 112-118.

Bennett, D.R. (1989). Epilepsy and the athlete. In B.C. Jordon (Ed.), *Sports neurology.* Gaithersburg: Aspen.

Browne, R.S. (1996). Evaluating and treating active patients for anemia. *The Physician and Sportsmedicine, 24,* 79-83.

Buettner, L.L., & Gavron, S.J. (1981). Personality changes and physiological effects of a personalized fitness enrichment program for cancer patients. Paper presented at the Third International Symposium on Adapted Physical Activity, New Orleans.

Centers for Disease Control and Prevention. (1999). *1999 HIV/AIDS surveillance report.* Available at **http://www.cdc.gov/nchstp/hiv_aids/stats/hasr1101.pdf**.

Chasnoff, I.J., Landeress, H.J., & Barrett, M.E. (1990). The prevalence of illicit drug or alcohol use during pregnancy and discrepancies in mandatory reporting in Pinnella County, Florida. *New England Journal of Medicine, 26,* 1202-1206.

Church, M.W., Eldis, F., Blakley, B.W., & Banule, E. (1997). Hearing, language, speech, vestibular, and dentofacial disorders in fetal alcohol syndrome. *Alcoholism, Clinical and Experimental Research, 21,* 227-237.

Coles, C.D., Platzman, K.A., Raskind, C.L., Hood, C.L., et al. (1997). A comparison of children affected by prenatal alcohol exposure and attention deficit, hyperactivity disorder. *Alcoholism, Clinical and Experimental Research, 21,* 150-161.

Conti, S.F., & Chaytor, E.R. (1995). Foot care for active patients who have diabetes. *The Physician and Sportsmedicine, 23,* 53-68.

Cratty, B.J. (1990). Motor development of infants subject to maternal drug use: Current evidence and future research strategies. *Adapted Physical Activity Quarterly, 2,* 110-125.

Disabella, V., & Sherman, C. (1998). Exercise for asthma patients. *The Physician and Sportsmedicine, 26,* 75-85.

Duda, M. (1985). The role of exercise in managing diabetes. *Physician and Sportsmedicine, 14,* 183-191.

Eichner, E.R. (1993). Sickle cell trait, heroic exercise, and fatal collapse. *The Physician and Sportsmedicine, 21,* 51-64.

Gaskin, A.D. (1994). Altered asthma treatment. *The Physician and Sportsmedicine, 22,* 38.

Gianta, L., & Steissguth, A.P. (1988). Parents and fetal alcohol syndrome and their caretakers. *Social Casework: The Journal of Contemporary Social Work,* sect. 453-459.

Helmrich, S.P., Ragland, V.R., Leung, R.W., et al. (1991). Physical activity and reduced occurrence of non-insulin dependent diabetes mellitus. *New England Journal of Medicine, 325,* 147-152.

Howze, K., & Howze, W.M. (1989). *Children of cocaine: Treatment and childcare.* Fact sheets presented at the Annual Conference of the National Association for the Education of Young Children (Atlanta, GA, November 2-5, 1989), 1-19.

Jacobson, J.L., Jacobson, S.W., & Sokol, R.J. (1996). Increased vulnerability to alcohol-related birth defects in the offspring of mothers over 30. *Alcoholism, Clinical and Experimental Research, 20,* 359-363.

Jimenez, C.C. (1997). Diabetes and exercise: The role of the athletic trainer. *Journal of Athletic Training, 32,* 339-343.

Kriska, A. (1997). Physical activity and the prevention of type II (non-insulin dependent) diabetes. *Research Digest, 10,* 1-6.

Kronstadt, D. (1991). Complex developmental issues of prenatal drug exposure. In R.E. Behrman (Ed.), *The future of children: Drug exposed infants,* 1(1), 36-49.

Lang, D. (1996). Seizure disorders and physical activity. *The Physician and Sportsmedicine, 10,* 24e-24k.

LaPerriere, A., Ironson, G., Antoni, M.H., et al. (1994). Exercise and psychoneuroimmendogy. *Medicine and Science in Sports and Exercise, 26,* 182-190.

Lee, F. (1995). Exercise and physical health: Cancer and immune function. *Research Quarterly for Exercise and Sport, 66,* 286-281.

Mahler, D.A. (1993). Exercise-induced asthma. *Medicine and Science in Sports and Exercise, 25,* 554-561.

Mitchell, J.H., Haskill, W.L., & Raven, P.B. (1994). Classification of sports. *Journal of the American College of Cardiology*, 24, 845-899.

Modlin, J., & Saah, H. (1991). Health and clinical aspects of HIV infection in women and children in the United States. In R. Faden, G. Geller, & M. Power (Eds.), *AIDS: Women and generation*. New York: Oxford University Press.

Niebyl, J.R. (1988). *Drug use in pregnancy*. Philadelphia: Lea & Febiger.

Osborn, J.A., Harris, S.R., & Weinberg, J. (1993). Fetal alcohol syndrome: A review of the literature with implications for physical therapists. *Physical Therapy*, 73, 599-607.

Roding, C., Beckwith, L., & Howard, T. (1990). Attachment in play in prenatal drug exposure. *Developmental and Psychopathology*, 1, 277-289.

Rupp, N.T. (1996). Diagnosis and management of exercise-induced asthma. *The Physician and Sportsmedicine*, 24, 77-87.

Sampson, P.D., Streissguth, A.P., Bookstein, F.L., Little, R.E., Clarren, S.K., Dehaene, P., Hanson, J.W., & Grahan, J.M. (1997). Incidence of fetal alcohol syndrome and prevalence of alcohol-related neurodevelopmental disorders. *Teratology*, 56, 317-326.

Spence, D.W., Galantino, M., Mossbert, K.H., & Zimmerman, S.O. (1990). Progressive resistance exercise: Effect on muscle function and anthropometry of a select AIDS population. *Archives of Physical Medicine and Rehabilitation*, 71, 644-648.

Steissguth, A.P., Aase, J.M., Clarren, S.K., Randels, D.P., LaDue, K.A., & Smith, D.F. (1991). Fetal alcohol syndrome in adolescents and adults. *Journal of the American Medical Association*, 265, 1961-1967.

Sternberg, S. (1997). Exercise helps some cancer, heart patients. *Science News*, 151, 269.

Stringer, W., Berezovskaya, M., O'Brien, W.H., Beck, K.C., & Casaburi, R. (1998). The effect of exercise training on aerobic fitness, immune indices, and quality of life in HIV+ patients. *Medicine and Science in Sports and Exercise*, 30, 11-16.

Taunton, J.E., & McCargar, L. (1995). Managing activity in patients who have diabetes. *The Physician and Sportsmedicine*, 23, 41-52.

Wagner, G., Rabkin, J., & Rabkin, R. (1998). Exercise as a mediator of psychological and nutritional effects of testosterone therapy in HIV+ men. *Medicine and Science in Sports and Exercise*, 30, 811-817.

WHO revises global estimates of HIV: Has more deaths of women, children. (1990). *Infectious Disease in Children*, 10, 3.

Winningham, M.L., & MacVicar, M.G. (1985). Response of cancer patients on chemotherapy to a supervised exercise program. *Medicine and Science in Sports and Exercise*, 17, 292.

Resources

Written

Adams, R.C., & McCubbin, J.A. (1991). *Games, sports, and exercises for the physically disabled*. Philadelphia: Lea & Febiger. Teachers will find this book to be an excellent guide to understanding various disabling conditions. An extensive portion of the book describes various ways to modify physical education activities.

American College of Sports Medicine. (1997). *ACSM's exercise management for persons with chronic diseases and disabilities*. Champaign, IL: Human Kinetics. This book covers 42 chronic diseases and disabilities. Each condition is briefly described, and suggestions are provided regarding exercise testing and programming.

Barnard, R.J., Jung, T., & Inkeles, S.D. (1994). Diet and exercise in the treatment of NIDDM: The need for early emphasis. *Diabetes Care*, 17, 1469-1472. Article addresses the strategy of using an intense exercise and diet program to manage non-insulin-dependent diabetes mellitus (NIDDM). Describes the use of this approach with 652 persons. The early introduction of this program proved to be effective in controlling this condition.

Chasnoff, I.J. (1986). Perinatal addiction: Consequences of intrauterine exposure to opiate and nonopiate drugs. In I.J. Chasnoff (Ed.), *Drug use in pregnancy: Mother and child*. Boston: MTD Press Ltd.

Goldberg, B. (Ed.). (1995). *Sports and exercise for children with chronic health conditions: Guidelines for participation from leading authorities*. Champaign, IL: Human Kinetics. An in-depth discussion of various chronic conditions is included in this book, along with recommended physical activities for each condition.

Kyle, J.M., Hershberger, S.L., Learman, J.R., & Melrose, J.K. (1995). Exercise bronchospasm in recreational athletes. *Your Patient and Fitness*, 9, 30e-30p. Diagnostic evaluation, optimal testing conditions, and treatment guidelines are presented in this article. Emphasis is placed on education and nonpharmacologic treatment of exercise-induced bronchospasm.

Osborn, J.A., Harris, S.R., & Weinberg, J. (1993). Fetal alcohol syndrome: Review of the literature with implications for physical therapists. *Physical Therapy*, 73, 599-607. An overview of the clinical and behavioral characteristics of children with fetal alcohol syndrome. Special attention paid to neuromotor differences and orthopedic abnormalities.

Palmer, J.K., Palmer, L.K., Michiels, K., & Thigpen, B. (1995). Effects of type of exercise on depression in recovering substance abusers. *Perceptual and Motor Skills*, 80, 523-530. Three types of exercise programs were examined to determine the effect on depressive symptoms. The weight-training program produced significant decrease in depressive symptoms.

Skinner, J.S. (1993). *Exercise testing and exercise prescription for special cases*. Baltimore, MD: Williams & Wilkins. A major emphasis is placed on exercise testing for different ages and gender. Three of the topics covered in this chapter are dealt with extensively in this book.

Thornton, J. (1997). Overcoming "protected child syndrome." *Physician and Sportsmedicine*, 25, 97-101. Article dealing with children with chronic illnesses such as asthma, diabetes, and cardiac disorders. Emphasis placed on shedding the shackles of overprotection, particularly as related to parents. Discussion regarding the benefits of exercise for these children.

26th Bethesda Conference. (1994). Recommendations for determining eligibility for competition in athletes with cardiovascular abnormalities. *Journal of the American College of Cardiology*, 24, 845-899.

Audiovisual

Diabetic Fitness: Oh, What a Feeling, videotape. (n.d.) Coronet/MTI Film and Video, 108 Wilmot Road, Deerfield, IL 60015. Presents physical activities that may enhance the overall fitness of the person with diabetes. The routines could be incorporated into the curriculum of a student with this condition.

Epilepsy, videotape. (n.d.) Third annual Ed-MED Conference, Institute for the Study of Developmental Disabilities, Indiana University, Bloomington, IN 47408. A presentation for teachers that addresses types of seizures and strategies for dealing with students having seizures in the school setting.

Fetal Alcohol Syndrome (FAS), videotape. (n.d.) Third annual Ed-MED Conference, Institute for the Study of Developmental Disabilities, Indiana University, Bloomington, IN 47408. This video reproduces a program on FAS that covered the following topics: history, characteristics, cause and effect, and examination of case studies.

Mentally-Handicapped and Epileptic: They Don't Make a Fuss, videotape. (n.d.) Films for the Humanities and Sciences, P.O. Box 2053, Princeton, NJ 08543-2053. This tape explores a person with mental retardation and epilepsy. Actual seizures are depicted, and psychosocial implications are discussed.

CHAPTER 16

References

Backburn, G.L., Duzer, J., Flonelers, W.D., et al. (1994). Report of the American Institute of Nutrition steering committee on healthy weight. *Journal of Nutrition*, 124, 2240-2243.

Centers for Disease Control and Prevention. (1994). Update: Prevalence of overweight among children, adolescents and adults—United States, 1988-1994. *Mobility and Mortality Weekly Report*, 46, 199-202.

Clark, N. (1995). Battling the bulge: Diet and exercise are key. *The Physician and Sportsmedicine*, 23, 21-23, 24.

Fitzgerald, G.K. (1997). Open versus closed kinetic chain exercise: Issues in rehabilitation after anterior cruciate ligament reconstructive surgery. *Physical Therapy*, 77, 1747-1754.

Joy, E., Clark, N., Ireland, M.L., Martire, J., et al. (1997). Team management of the female athlete triad. *The Physician and Sportsmedicine*, 25, 95-100.

Knight, K. (1979). Rehabilitating chondromalacia patellae. *The Physician and Sportsmedicine*, 7, 147-148.

Kraus, B.L., Williams, J.P., & Catterall, A. (1990). Natural history of Osgood-Schlatter disease. *Journal of Pediatric Orthopedics*, 10, 65-68.

Meisterling, R.C., Wall, E.J., & Meisterling, M.R. (1998). Coping with Osgood-Schlatter disease. *The Physician and Sportsmedicine*, 26, 39-40.

Oppliger, R.A., Harms, R.D., Herman, D.E., et al. (1993). Grappling with weight control. *The Physician and Sportsmedicine*, 23, 69-78.

Parr, R.B. (1996). Exercising when you're overweight: Getting in shape and shedding pounds. *The Physician and Sportsmedicine*, 24, 81-82.

Rencken, M.L., Chestnut, C.H., & Drinkwater, B.L. (1996). Bone density at multiple skeletal sites in amenorrheic athletes. *Journal of the American Medical Association*, 276, 238-240.

Sallis, R.E. (1998). Four diet tips for teens. *The Physician and Sportsmedicine*, 26, 33.

Saltin, B., & Rowell, L.B. (1980). Functional adaptation to physical activity and inactivity. *Federation Proceeding*, 39, 1506-1513.

Sherrill, C. (1998). *Adapted physical activity, recreation, and sport: Crossdisciplinary and lifespan* (5th ed.). Dubuque, IA: William. C Brown.

Short, F., McCubbin, J., & Frey, G. (1999). Cardiorespiratory endurance and body composition. In J. Winnick & F. Short (Eds.), *The Brockport Physical Fitness Training Guide*. Champaign, IL: Human Kinetics.

Smith, R.L., & Bronolli, J. (1989). Shoulder kinesthesia after anterior glenohumeral joint dislocation. *Physical Therapy*, 69, 106-112.

Surburg, P. (1999). Flexibility/range of motion. In J. Winnick & F. Short (Eds.), *The Brockport physical fitness training guide*. Champaign, IL: Human Kinetics.

Voss, D., Knott, M., & Iona, B. (1986). *Proprioceptive neuromuscular facilitation*. Philadelphia: Harper & Row.

Wall, E.J. (1998). Osgood-Schlatter disease. *The Physician and Sportsmedicine*, 26, 29-34.

Wichmann, S., & Martin, D.R. (1993). Eating disorder of athletes. *The Physician and Sportsmedicine*, 21, 126-135.

Winnick, J., & Short, F. (1999). *The Brockport Physical Fitness Test Manual*. Champaign, IL: Human Kinetics.

Resources

Written

Aronen, J.G., Chronister, R., Regan, K., & Hensien, M.A. (1993). Practical conservative management of iliotibial band syndrome. *The Physician and Sportsmedicine*, 21, 59-69. These authors discuss the means to treat the initial symptoms of what often is called runner's knee. Suggestions that focus on returning to complete activity are provided.

Fitzgerald, G.K. (1997). Open versus closed kinetic chain exercise: Issues in rehabilitation after anterior cruciate ligament reconstructive surgery. *Physical Therapy*, 77, 1747-1754. Provides an understanding of the difference between these types of exercises. Addresses some of the assumptions pertaining to closed kinetic chain exercises and certain misconceptions about this type of exercise.

Laskowski, E.R., Newcomer-Aney, K., & Smith, J. (1997). Refining rehabilitation with proprioception training. *The Physician and Sportsmedicine*, 25, 89-102. Overview of proprioception as part of a therapeutic protocol is followed by specific suggestions to develop this sense with ankle, low back, and shoulder. Both open and closed kinetic chain exercises are used to improve proprioception.

Surburg, P.R., & Schroder, J.W. (1997). Proprioceptive neuromuscular facilitation techniques in sports medicine: A reassessment. *Journal of Athletic Training*, 32, 34-39. Article explains the basis of proprioceptive neuromuscular facilitation (PNF), the different types of techniques, and the use of these techniques by athletic trainers. Compares the use of PNF techniques between these dates and a previous study and provides ideas regarding practical applications.

Winnick, J., & Short, F. (1999). *The Brockport physical fitness training guide*. Champaign, IL: Human Kinetics. Training manual provides information regarding the development of health-related physical fitness of children and adolescents with disabilities. Has an introductory chapter regarding health-related physical fitness concepts and three subsequent chapters on cardiorespiratory endurance and body composition, muscular strength and endurance, and flexibility/range of motion.

Audiovisual

Lose Weight the Safe Way, videotape. Films for the Humanities and Services, P.O. Box 2053, Princeton, NJ 08543-2053. Insights regarding possible problems with fad diets and the correct way to approach weight control. Running time is approximately 19 minutes.

The Silent Hunger: Anorexia and Bulimia, videotape. Films for the Humanities and Sciences, P.O. Box 2053, Princeton, NJ 08543-2053. Two basic questions are addressed: What are eating disorders? and What causes eating disorders? Running time is approximately 46 minutes.

CHAPTER 17

References

Caldwell, G.E., & Clark, J.E. (1990). The measurement and evaluation of skill within the dynamical system perspective. In J.E. Clark & J.H. Humphrey (Eds.), *Advances in motor development research.* New York: AMS Press.

Gallahue, D.L. (1996). *Development physical education for today's children.* Boston: McGraw-Hill.

Gallahue, D.L., & Ozmun, J. (1998). *Understanding motor development: Infants, children, adolescents, adults.* Boston: McGraw-Hill.

Gesell, A. (1954). The ontogenesis of infant behavior. In L. Carmichael (Ed.), *Manual of child psychology.* New York: Wiley.

Haubenstricker, J.L., & Seefeldt, V.D. (1986). Acquisition of motor skills during childhood. In V.D. Seefeldt (Ed.), *Physical activity and well being.* Reston, VA: AAHPERD.

Kamm, K., Thelen, E., & Jensen, J.L. (1990). A dynamical systems approach to motor development. *Physical Therapy,* 70(12), 763-774.

Magill, R. (1998). *Motor learning.* Dubuque, IA: WCB/McGraw-Hill.

McClenaghan, B.A., & Gallahue, D.L. (1978). *Fundamental movement: A developmental and remedial approach.* Philadelphia: Saunders.

McGraw, M. (1939). Later development of children specially trained during infancy. *Child Development,* 10, 1-19.

Roberton, M.A., & Halverson, L.E. (1984). *Developing children-their changing movement: A guide for teachers.* Philadelphia: Lea & Febiger.

Seefeldt, V. (1975). *Critical learning periods and programs of early intervention.* Paper presented to the National Convention of the American Alliance for Health, Physical Education and Recreation, Atlantic City.

Thelen, E. (1989). Dynamical approaches to the development of behavior. In J.A.S. Kelso, A.J. Mandell, & M.E. Schelsinger (Eds.), *Dynamic patterns in complex systems.* Singapore: World Scientific.

Thelen, E. (1995). Motor development. A new synthesis. *American Psychologist,* 50(2), 79-95.

Thelen, E., & Smith, L.B. (1994). *A dynamic systems approach to the development of cognition and action.* Cambridge, MA: MIT Press.

Thelen, E., & Smith, L.B. (Eds.). (1993*). A dynamic systems approach to development: Applications.* Cambridge, MA: MIT Press.

Resources

Written

Gallahue, D. (1996). *Developmental physical education for today's children.* Dubuque, IA: WCB/McGraw-Hill. This text contains an expanded and updated version of the Fundamental Movement Pattern Assessment Instrument originally developed by McClenaghan and Gallahue in 1978. Assessment guidelines for conducting both total-body configuration and segmental analysis are provided for 23 fundamental movement skills. Corresponding videotapes, devel-

oped by Arlene Ignico, Ball State University, are also available.

Gallahue, D.L., & Ozmun, J.S. (1998). *Understanding motor development: Infants, children, adolescents, adults.* Boston: WCB/McGraw-Hill. Contains a wealth of information on 23 fundamental movement skills. Line drawings depict initial, elementary, and mature stages of each.

Johnson-Martin, N.M., Attermeier, S.M., & Hacker, B. (1990). *The Carolina curriculum for preschoolers with special needs.* Baltimore, MD: Paul H. Brookes. Excellent resource for use with preschoolers in special education programs.

Johnson-Martin, N.M., Jens, K.G., Attermeier, S.M., & Hacker, B.J. (1991). *The Carolina curriculum for infants and toddlers with special needs.* Baltimore, MD: Paul H. Brookes. Innovating, developmentally appropriate activities for the very young with special needs.

Wessel, J. (1998). *I CAN: Pre-primary motor and play skills.* Northbrook, IL: Hubbard Scientific. Excellent curricular materials for preschool children in the regular or adapted physical education program.

Wessel, J.A., & Zittel, L.L. (1995). *Smart start preschool movement curriculum designed for children of all abilities.* Austin, TX: Pro-Ed. Excellent curricular materials for young children in the regular and adapted physical education program.

CHAPTER 18

References

Burton, A.W. (1990). Assessing the perceptual-motor interaction in developmentally disabled and handicapped children. *Adapted Physical Activity Quarterly,* 7, 325-337.

Burton, A.W. (1987). Confronting the interaction between perception and movement in adapted physical education. *Adapted Physical Education Quarterly,* 4, 257-267.

Cratty, B.J. (1986a). *Perceptual motor development in infants and children.* Englewood Cliffs, NJ: Prentice Hall.

Cratty, B.J. (1986b). *Adapted physical education in the mainstream.* Denver, CO: Love Publishing Company.

Gibson, J.J. (1977). The theory of affordance. In R. Shaw & J. Bransford (Eds.), *Perceiving, acting, and knowing: Toward an ecological psychology.* Hillsdale, NJ: Erlbaum.

Gibson, J.J. (1979). *The ecological approach to visual perception.* Boston: Houghton Mifflin.

Sherrill, C. (1998). *Adapted physical activity, recreation, and sport.* Boston: WCB/McGraw-Hill.

Winnick, J.P. (1979). *Early movement experiences and development: Habilitation and remediation.* Philadelphia: Saunders.

Resources

Written

Capon, J. (1994). *Perceptual-Motor development series.* Byron, CA: Front Row Experience. Perceptual-motor series of five books covering basic movement activities; ball, rope, and hoop activities; balance activities; beanbag, rhythm, and stick activities; and tire and parachute activities.

Cowden, J.E., Sayers, L.K., & Torrey, C.C. (1998). *Pediatric adapted motor development and exercise.* Springfield, IL: Charles C Thomas. Book providing exercises for increasing muscle tone and strength, decreasing muscle tone and reflex integration, and increasing sensory motor development (postural reactions and vestibular stimulation, visual-motor control, auditory discrimination, tactile stimulation, and kinesthetic and spatial awareness). Emphasis on organizing and conducting infant and toddler movement intervention programs.

Cratty, B.J. (1986). *Perceptual motor development in infants and children*. Englewood Cliffs, NJ: Prentice Hall. This comprehensive book provides a theoretical overview of perceptual-motor development and practical suggestions for programs.

Cratty, B.J. (1986). *Adapted physical education in the mainstream*. Denver, CO: Love. Chapter 5 in this book reviews and summarizes theories and models (perceptual-motor theories, recapitulation theories, cognitive approaches) related to movement abilities. Chapter 25 provides information and guidelines for in/out stimulation, including sensory stimulation.

Davis, W.E. (1983). An ecological approach to perceptual-motor learning. In R.L. Eason, T.L. Smith, & F. Caron (Eds.), *Adapted physical activity: From theory to application*. Champaign, IL: Human Kinetics.

Fink, B.E. (1989). *Sensory-motor integration activities*. Tucson, AZ: Therapy Skill Builders. Provides a collection of sensory-motor integration activities designed for elementary-age children. Many of the activities may be used to obtain goals associated with physical education.

Gibson, J.J. (1966). *The senses considered as perceptual systems*. Boston: Houghton Mifflin.

Johnson-Martin, N.M., Attermeier, S.M., & Hacker, B.J. (1995). *The Carolina curriculum for preschoolers with special needs*. Baltimore, MD: Paul H. Brookes. Presents curriculum sequences encompassing object manipulation, locomotion (walking, galloping/skipping, running, and hopping), stair activities, jumping, balancing, and ball throwing and catching.

Johnson-Martin, N.M., Jens, K.G., Attermeier, S.M., & Hacker, B.J. (1994). *The Carolina curriculum for infants and toddlers with special needs*. Baltimore, MD: Paul H. Brookes. This text includes a chapter on sensorimotor development and curriculum sequences related to visual pursuit and object permanence; auditory localization and object permanence; and understanding space, visual perception, and gross motor skills, including posture and locomotion, stair activity, jumping, and balance.

Sherrill, C. (1998). *Adapted physical activity, recreation, and sport*. Boston, MA: William C. Brown/McGraw-Hall. Chapter 10 presents information related to sensorimotor integration with particular attention to severe disability. Chapter 12 presents information related to perceptual-motor learning, perceptual-motor disorders, perceptual-motor theory, and implications for teaching.

Audiovisual

Cassettes, records, CDs, filmstrips, videos, multimedia, software, and books related to sensorimotor stimulation and perceptual-motor development may be purchased from Kimbo Educational, P.O. Box 477, Long Branch, NJ 07740-0477; Educational Activities, P.O. Box 87, Baldwin, NY 11510; Communication/Therapy Skill Builders, 555 Academic Court, San Antonio, TX 78204.

CHAPTER 19

References

Bailey, D.B., & Wolery, M. (1989). *Assessing infants and preschoolers with handicaps*. New York, NY: Merrill.

Bennett, T., Lingerfelt, B.V., & Nelson, D.E. (1990). *Developing individualized family support plans: A training manual*. Cambridge, MA: Brookline Books.

Bredekamp, S., & Copple, C. (1997). *Developmentally appropriate practice in early childhood programs—Revised*. Washington, DC: National Association for the Education of Young Children.

Brigance, A. (1999). *Brigance diagnostic inventory of early development*. Billerica, MA: Curriculum Associates.

Cowden, J.E., Sayers, L.K., & Torrey, C.C. (1998). *Pediatric adapted motor development and exercise: An innovative, multisystem approach for professionals and families*. Springfield, IL: Charles C Thomas.

Early Childhood Intervention Council of Monroe County. (1994). *Guidelines for screening and evaluation in the determination of eligibility for early intervention services for infants and toddlers (birth through two years)*. Rochester, NY: Author.

Eason, B.L. (1995). Infants and toddlers. In J.P. Winnick (Ed.), *Adapted physical education and sport*. Champaign, IL: Human Kinetics.

Fiorini, J., Stanton, K., & Reid, G. (1996). Understanding parents and families of children with disabilities: Considerations for adapted physical activity. *Palaestra*, 12(2), 16-23.

Folio, M.R., & Fewell, R. (1983). *Peabody developmental motor scales and activity cards*. Chicago, IL: Riverside.

Frankenburg, W.K., Dodds, J., Archer, P., Bresnick, B., Maschka, P., Edelman, N., & Shapiro, H. (1992). *Denver II training manual* (2nd ed.). Denver, CO: Denver Developmental Materials.

Furuno, S., O'Reilly, K.A., Hosaka, C.M., Inatsuka, T.T., Zeisloft-Falbey, B., & Allman, T.L. (1988). *HELP checklist: Ages birth to three years*. Palo Alto, CA: VORT.

Gallahue, D.L., & Ozmun, J.C. (1995). *Understanding motor development: Infants, children, adolescents, adults*. Madison, WI: Brown & Benchmark.

Garwood, S., & Sheehan, R. (1989). *Designing a comprehensive early intervention system: The challenge of Public Law 99-457*. Austin, TX: Pro-Ed.

Individuals with disabilities education act amendments of 1997. 20 U.S.C. 1400 (1997). Washington DC: Author.

Johnson-Martin, N., Jens, K.G., Attermeier, S.M., & Hacker, B.J. (1991). *The Carolina curriculum for infants and toddlers with special needs*. Baltimore, MD: Paul H. Brookes.

Linder, T. (1993). *Transdisciplinary play-based assessment*. Baltimore, MD: Paul H. Brookes.

McCall, R., & Craft, D.H. (2000). *Moving with a purpose: Developing programs for preschoolers of all abilities*. Champaign, IL: Human Kinetics.

McGonigel, M.J. (1991). Philosophy and conceptual framework. In M.J. McGonigel, R.K. Kaufman, & B.H. Johnson (Eds.), *Guidelines and recommendations for the Individualized Family Service Plan*. Bethesda, MD: Association for the Care of Children's Health.

Milani-Comparetti Motor Development Screening Test manual. (1987). Available from Meyer Children's Rehabilitation Institute, University of Nebraska Medical Center, 444 South 44th Street, Omaha, Nebraska 68131-3795.

Parks, S. (Ed.). (1992). *HELP strands: Curriculum-based developmental assessment birth to three years*. Palo Alto, CA: VORT.

Regulations of the commissioner of health state of New York—Subchapter H, Part 69-4.1 Definitions. (n.d.) Albany, NY: Author.

Resources

Written

Blanche, E.I., Botticelli, T.M., & Hallway, M.K. (1995). *Combining neuro-development treatment and sensory integration principles: An approach to pediatric therapy.* Tucson, AZ: Therapy Skill Builders. Explains both NDT and SI principles and offers suggestions for combining these two forms of treatment to enhance the motor development of infants and toddlers with special needs. Provides activities and worksheets to assist teachers in programming.

Bly, L. (1994). *Motor skills acquisition in the first year: An illustrated guide to normal development.* Tucson, AZ: Therapy Skill Builders. Illustrations of infant development during the first years are depicted. Allows teachers to note normal development and deviations in development. Provides activities to enhance development.

Bricker, D., Pretti-Frontczak, K., & McComas, N. (1998). *An activity-based approach to early intervention.* Baltimore, MD: Paul H. Brookes. Details how and why the activity-based approach can benefit any child ages birth to five.

Frankenburg, W., Dodds, J., Archer, P., Bresnick, B., & Shapiro, H. (1990). *Denver II training manual* (2nd ed.). Denver, CO: Denver Developmental Materials.

Johnson, L.J., LaMontagne, M.J., Elgas, P.M., & Bauer, A.M. (1998). *Early childhood education: Blending theory, blending practice.* Baltimore, MD: Paul H. Brookes. Combines best practices in early childhood education and early childhood special education to create environments in which all children can achieve.

Jordan, J.B., Gallagher, J.J., Hutinger, P.L., & Karnes, M.B. (1990). *Early childhood special education: Birth to three.* Reston, VA: Council for Exceptional Children—Early Childhood Division. Provides readers with updated information on early intervention policies and procedures.

Prudden, B. (1986). *How to keep your child fit from birth to six.* New York, NY: Ballantine Books. Provides an array of activities to stimulate the development of young children using both land and aquatic environments.

Rappaport M.L., & Schulz, L. (1989). *Creative play activities for children with disabilities: A resource book for teachers and parents.* Champaign, IL: Human Kinetics. Provides more than 250 play activities. Each activity includes simple directions, equipment needed, benefits of the activity, and adaptations for children with disabilities.

Electronic

National Association for the Education of Young Children. Web site: **www.naeyc.org**. Primary site for resources and best practices for the education of young children ages birth to eight years.

CHAPTER 20

References

Avery, M., Boos, S., Chepko, S., Gabbard, C., & Sanders, S. (1994). *Developmentally appropriate practice in movement programs for young children ages 3-5.* Reston, VA: Council on Physical Education for Children.

Bailey, D.B., & Wolery, M. (1989). *Assessing infants and preschoolers with handicaps.* New York, NY: Merrill.

Bredekamp, S., & Copple, C. (1997). *Developmentally appropriate practice in early childhood programs—Revised.* Washington, DC: National Association for the Education of Young Children.

Bredekamp, S., & Rosegrant, T. (Eds.). (1992). *Reaching potentials: Appropriate curriculum and assessment for young children* (Volume 1). Washington, DC: National Association for the Education of Young Children.

Bricker, D., & Pretti-Frontczak, K. (1996). *AEPS measurement for three to six years.* Baltimore, MD: Paul H. Brookes.

Bricker, D., & Woods Cripe, J.J. (1992). *An activity-based approach to early intervention.* Baltimore, MD: Paul H. Brookes.

Brigance, A. (1991). *Brigance diagnostic inventory of early development.* Billerica, MA: Curriculum Associates.

Clements, R. (1995). *My neighborhood movement challenges.* Oxon Hill, MD: AAHPERD.

Coker, C.A. (1996). Accommodating students' learning styles in physical education. *Journal of Physical Education, Recreation and Dance,* (67), 66-68.

Early Childhood Intervention Council of Monroe County. (1994a). *Guidelines for screening and evaluation in the determination of eligibility for early intervention services for infants and toddlers (birth through two years).* Rochester, NY: Author.

Early Childhood Intervention Council of Monroe County. (1994b). *Eligibility criteria: Preschool student with a disability.* Rochester, NY: Author.

Federal Register. (1999). *Assistance to states for the education of children with disabilities and the early intervention program for infants and toddlers with disabilities.* 34 CFR Parts 300-303. Vol. 64, No. 48. Washington, DC: Department of Education.

Folio, M.R., & Fewell, R. (2000). *Peabody developmental motor scales* (2nd ed.). Austin, TX: Pro-Ed.

Hills, T. (1992). Reaching potentials through appropriate assessment. In S. Bredekamp & T. Rosegrant (Eds.), *Reaching potentials: Appropriate assessment for young children* (Volume 1). Washington, DC: National Association for the Education of Young Children.

Individuals with disabilities education act amendments of 1997. 20 U.S.C. 1400 (1997). Washington, DC: Author.

Johnson-Martin, N., Attermeier, M.A., & Hacker, B.J. (1990). *The Carolina curriculum for preschoolers with special needs.* Baltimore, MD: Paul H. Brookes.

Linder, T. (1993). *Transdisciplinary play-based assessment.* Baltimore, MD: Paul H. Brookes.

Pangrazzi, R.P. (1998). *Dynamic physical education for elementary school children.* Needham Heights, MA: Allyn & Bacon.

Piaget, J. (1952). *The origins of intelligence in children.* New York: International Universities Press.

Regulations of the commissioner of education state of New York—Subchapter P, Part 200 Students with Disabilities. Albany, NY: Author.

Sherrill, C. (1998). *Adapted physical activity, recreation and sport: Crossdisciplinary and lifespan.* Boston, MA: WCB/McGraw-Hill.

Ulrich, D.A. (2000). *Test of motor development* (2nd ed.). Austin, TX: Pro-Ed.

Vygotsky, L. (1978). *Mind in society: The development of higher psychological processes.* Cambridge, MA: Harvard University Press.

Wessel, J.A., & Zittel, L.L. (1998). *I CAN primary skills: K-3.* Austin, TX: Pro-Ed.

Wessel, J.A., & Zittel, L.L. (1995). *SMART START: Preschool movement curriculum designed for children of all abilities.* Austin, TX: Pro-Ed.

Zittel, L.L. (1994). Gross motor assessment of preschool children with special needs: Instrument selection considerations. *Adapted Physical Activity Quarterly*, 11, 245-260.

Resources

Written

Avery, M. (1994). Preschool physical education: A practical approach. *Journal of Physical Education, Recreation and Dance*, 65(6), 37-39. Overviews the basic premises surrounding developmentally appropriate motor programs for young children and provides factors to consider when setting up a movement area.

Block, M. (1994). Including preschool children with disabilities. *Journal of Physical Education, Recreation and Dance*, 65(6), 45-49. Provides a rationale and strategies for including children with special needs in regular physical education settings.

Block, M.E., & Davis, T.D. (1996). An activity-based approach to physical education for preschool children with disabilities. *Adapted Physical Activity Quarterly*, 13(3), 230-246. Introduces activity-based intervention in physical education for young children with examples of how this intervention can be implemented in motor programs for youngsters with delays or disabilities.

Bricker, D., & Waddell, M. (1996). *AEPS curriculum for three to six years.* Baltimore, MD: Paul H. Brookes. A sequential curriculum that offers a flexible set of considerations, strategies, and activities designed to meet the needs of young children.

Clements, R.L. (1995). *My neighborhood movement challenges.* Oxon Hill, MD: AAHPERD. Contains a comprehensive overview of movement narratives and provides sample narratives for commonly used themes to teach motor skills and concepts.

Dummer, G.M., Connor-Kuntz, F.J, & Goodway, J.D. (1995). A physical education curriculum for all preschool students. *Teaching Exceptional Children*, 27, 28-34. Explains the process of curriculum development in eight easy steps. Identifies appropriate program goals and objectives in early childhood physical education.

Gilroy, P.J. (1985). *Kids in motion: An early childhood movement education program.* Tucson, AZ: Communication Skill Builders. Provides month-by-month objectives and activities to enhance the motor skills of young children.

Hammett, C.T. (1992). *Movement activities for early childhood.* Champaign, IL: Human Kinetics. Provides more than 100 child-tested movement activities that make learning new skills fun for preschoolers.

Ignico, A. (1994). Early childhood physical education: Providing the foundation. *Journal of Physical Education, Recreation and Dance*, 65(6), 28-30. Overviews the basic components (i.e., motor skills and concepts) of motor programming for young children. Identifies the need for health-related fitness activities to be embedded in games and activities.

Pica, R. (1991). *Special themes for moving and learning.* Champaign, IL: Human Kinetics. Presents various themes that can be incorporated into a regular physical education curriculum. Ideal for both preschool- and primary-age students.

Sanders, S. (1992). *Designing preschool movement programs.* Champaign, IL: Human Kinetics. Contains information on developing a preschool movement curriculum along with sample learning activities.

Satchwell, L. (1994). Preschool physical education class structure. *Journal of Physical Education, Recreation and Dance*, 65(6), 34-36. Offers an appropriate physical education class structure that centers around circle time, rhythms, fundamental movements, and closure. Provides suggested activities and equipment needed to conduct the lesson.

Sawyers, J.K. (1994). The preschool playground: Developing skills through outdoor play. *Journal of Physical Education, Recreation and Dance*, 65(6), 31-33. Discusses the importance to playground and factors to consider when selecting sites, equipment, and materials.

Torbert, M., & Schneider, L.B. (1993). *Follow me too: A handbook of movement activities for three to five year olds.* Menlo Park, CA: Addison-Wesley. Contains 49 child-tested active movement games that provide positive play experiences for young children.

Zittel, L.L. (1994). Gross motor assessment of preschool children with special needs: Instrument selection considerations. *Adapted Physical Activity Quarterly*, 11, 245-260. Provides an overview of criteria for selecting appropriate motor assessments and rates various assessments on their abilities to meet selected criteria.

Zittel, L.L., & McCubbin, J.A. (1996). Effect of an integrated physical education setting on motor performance of preschool children with developmental delays. *Adapted Physical Activity Quarterly*, 13, 316-333. Demonstrated that children with developmental delays can maintain their level of gross-motor skill and independence in an integrated physical education setting.

Audiovisual

Pica, R., & Gardzina, R. (1991). *Early elementary children moving and learning.* Champaign, IL: Human Kinetics. Contains 5 audiocassettes of original music that supplement the 40 developmentally appropriate lesson plans. Geared toward primary-age students.

Pica, R., & Gardzina, R. (1990). *More music for moving and learning.* Champaign, IL: Human Kinetics. Contains 6 audiocassettes with 62 songs that focus on typical early childhood themes. Great for both the preschool and primary years.

Pica, R., & Gardzina, R. (1990). *Preschoolers moving and learning.* Champaign, IL: Human Kinetics. Contains 5 audiocassettes and 40 lessons with 200 activities. Includes lessons that focus on locomotor and nonlocomotor and orientation skills.

Electronic

AAHPERD. Web site: **http://www.aahperd.org**. Primary link to all associations within the fields of health, physical education, recreation, and dance.

Council for Exceptional Children (CEC). Web site: **http://www.cec.sped.org**. Primary link to associations related to children with specific disabilities.

National Association for the Education of Young Children. Web site: **http://www.naeyc.org**. Primary site for resources and best practices for the education of young children.

PE Central. Web site: **http://www.pe.central.vt.ed**. Provides teaching resources for physical educators such as lesson ideas, instant activities, assessment ideas, and links to other interesting sites.

USPE. To subscribe, e-mail USPE at listserv@listserv.vt.edu. In body of message type: Subscribe USPE-L <your name>. Voluntary group of physical education teachers who talk about concerns, interests, and ideas on e-mail.

CHAPTER 21

References

Arnold, R., Ng, N., & Pechar, G. (1992). Relationship of rated perceived exertion to heart rate and workload in mentally retarded young adults. *Adapted Physical Activity Quarterly, 9,* 47-53.

Borg, G.A. (1998). *Borg's perceived exertion and pain scales.* Champaign, IL: Human Kinetics.

Bouchard, C., & Shephard, R.J. (1994). Physical activity, fitness, and health: The model and key concepts. In C. Bouchard, R.J. Shephard, & T. Stephens (Eds.), *Physical activity, fitness, and health: International proceedings and consensus statement.* Champaign, IL: Human Kinetics.

Caspersen, C.J., Powell, K.E., & Christenson, G.M. (1985). Physical activity, exercise, and physical fitness: Definitions and distinctions for health-related research. *Public Health Reports, 100,* 126-131.

Pate, R.R., (1988). The evolving definitions of fitness. *Quest, 40,* 174-178.

Surburg, P. (1999). Flexibility/range of motion. In J.P. Winnick & F.X. Short (Eds.), *The Brockport physical fitness training guide.* Champaign, IL: Human Kinetics.

Winnick, J.P., & Short, F.X. (1999). *The Brockport physical fitness test manual.* Champaign, IL: Human Kinetics.

Resources

Written

Adams, R.C., & McCubbin, J.A. (1991). *Games, sports, and exercises for the physically disabled.* Philadelphia, PA: Lea & Febiger. Provides a wide variety of games, sports, and exercises for persons with physical disabilities. They may be used as a means of improving strength, flexibility, endurance, aerobic capacity, and body composition.

American College of Sports Medicine. (1997). *Exercise in management for persons with chronic diseases and disabilities.* Champaign, IL: Human Kinetics. How to effectively manage exercise for someone with chronic disease or disability. Includes 40 chapters categorized by disabilities, diseases, or conditions written by persons with clinical and research experience in exercise programming. Prominent in the book are suggestions for exercise testing and exercise programming.

American Fitness Alliance. (2000). *2000 youth fitness resource catalog.* Champaign, IL: Human Kinetics. Outlines a number of available fitness-related materials including *The Brockport*

Physical Fitness Test Kit, FITNESSGRAM Test Kit, Physical Best Activity Guides, and *FitSmart,* the national health-related physical fitness knowledge test.

Council for Physical Education for Children (COPEC). (n.d.). *Physical activity for children: A statement of guidelines. National Association for Sport and Physical Education.* Reston, VA: NASPE.

Goldberg, B. (Ed.). (1995). *Sports and exercise for children with chronic health conditions.* Champaign, IL: Human Kinetics. Presents information on general issues related to sports, exercise, and chronic health conditions; and provides specific information on these topics in chapters organized according to 20 disabilities and conditions.

Lockette, K.F., & Keyes, A.M. (1994). *Conditioning with physical disabilities.* Champaign, IL: Human Kinetics. Covers general principles of physical conditioning including strength, aerobic, and flexibility training. Applies these principles to a number of physical disabilities including cerebral palsy, stroke, head injury, spinal cord injury, spina bifida, poliomyelitis, amputations, visual impairments, multiple sclerosis, and others. Includes separate chapters for upper-extremity, abdominal and trunk, and lower-extremity exercises.

Miller, P. (Ed.). (1995). *Fitness programming and physical disability.* Champaign, IL: Human Kinetics. An excellent book providing information on principles of conditioning for the development of health-related physical fitness; guidelines for developing resistance-training programs; modifications for using stretch bands or tubing as the mode of exercise; ways of maintaining and developing flexibility; exercises for the neck, shoulder, elbow, wrist, trunk, hip, knee, and ankle; intervention and emerging procedures associated with grand mal and petit mal seizures, ketoacidosis, hypoglycemia, autonomic dysreflexia, heat exhaustion, heat strokes, and postural and exercise hypotension; and information relating to wheelchairs, exercising in a wheelchair, and transfers (e.g., from wheelchair to floor or another chair).

Pate, R.R., Pratt, M., Blair, S.N., Haskell, W.L., Macera, C.A., Bouchard, C., et.al (1995). Physical activity and public health. *Journal of the American Medical Association, 273*(5), 402-407.

Rimmer, J. (1994). *Fitness and rehabilitation programs for special populations.* Madison, WI: Brown & Benchmark. Presents exercise prescriptions and guidelines for a variety of special populations. Focuses on fitness and rehabilitation programs for persons with the following diseases/conditions: aging, arthritis, obesity, diabetes, asthma and other pulmonary diseases, spinal cord injury, and mental retardation.

United States Department of Health and Human Services. (1996). *Physical activity and health: A report of the surgeon general.* Atlanta, GA: United States Department of Health and Human Services, Centers for Disease Control and Prevention, National Center for Chronic Disease Prevention and Health Promotion.

Winnick, J.P., & Short, F.X. (Eds.). *The Brockport physical fitness training guide.* Champaign, IL: Human Kinetics. Designed to be used in conjunction with *The Brockport Physical Fitness Test Manual.* Provides principles for fitness development in the areas of cardiorespiratory endurance, body composition, muscular strength and endurance, and flexibility/range of motion. Recent CDC/ACSM physical activity guidelines are incorporated. Focuses on youngsters with mental retardation, visual impairments, cerebral palsy, spinal cord injury, and amputations.

Audiovisual

Aerobics for Quadriplegia, Aerobics for Cerebral Palsy, Aerobics for Paraplegia, Aerobics for Amputees, videotapes. Disabled Sports, USA, 451 Hungerford Drive, Suite 100, Rockville, MD 20850. These four videotapes are a part of the "Fitness for Everyone" series developed by the National Handicapped Sports and Recreation Association (NHSRA). Instructors accompanied by individuals exercising in a wheelchair demonstrate easy to progressively more demanding aerobic activities. Helpful hints for safe and effective exercising also provided. Running time is 33 minutes for each videotape.

Strength and Flexibility for Individuals with All Types of Physical Disabilities, Tape #2, videotape. National Handicapped Sports and Recreation Association, 451 Hungerford Drive, Suite 100, Rockville, MD 20850. Presents dynamic isotonic strength exercises demonstrated by an instructor and followed by an individual with a disability. Exercises focus on the use of tubing for resistance. Running time is approximately 50 minutes.

Electronic

Fitness Challenge Software. (1999). American Fitness Alliance, Youth Fitness Resource Center, P.O. Box 5076, Champaign, IL 61825-5076. This computer software program supports the Brockport Physical Fitness Test (BPFT). It prints goals, results, and fitness plans for individual students; and separate reports can be generated for instructors or parents. Also includes the technical manual for the BPFT, which provides validity and reliability information.

FITNESSGRAM 6.0 Software. (1999). American Fitness Alliance, Youth Fitness Resource Center, P.O. Box 5076, Champaign, IL 61825-5076. Provides a sophisticated computer system for administering FITNESSGRAM. Has both teacher and student components. The student component allows students to enter their own test scores and track their progress. It also provides a system for monitoring their levels of physical activity through ACTIVITYGRAM, a subcomponent of the software.

CHAPTER 22

References

Adams, R.C., & McCubbin, J.A. (1991). *Games, sports, and exercises for the physically disabled* (4th ed.). Philadelphia: Lea & Febiger.

Cratty, B.J. (1971). *Movement and spatial awareness in blind children and youth.* Springfield, IL: Charles C Thomas.

New York State Education Department. (1966). *New York State physical fitness test for boys and girls grades 4-12.* Albany, NY: Author.

Resources

Written

American College of Sports Medicine. (1997). *ACSM's exercise management of persons with chronic diseases and disabilities.* Champaign, IL: Human Kinetics. This book provides recommendations for exercise testing and programming for a variety of disabilities that frequently have secondary postural problems.

Goldberg, B. (Ed.). (1995). *Sports and exercise for children with chronic health conditions.* Champaign, IL: Human Kinetics. This book

provides guidelines for sport participation and exercise for children with orthopedic and health impairments that can be applied to children with body mechanic and postural problems.

Lockette, K., & Keyes, A.M. (1994). *Conditioning with physical disabilities.* Champaign, IL: Human Kinetics. This book provides step-by-step descriptions on how to do a variety of exercises that can be easily adapted to develop exercise programs for students who need to remediate nonstructural postural deviations.

Miller, P.D. (Ed.). (1995). *Fitness programming and physical disability.* Champaign, IL: Human Kinetics. Contains a variety of "how to" fitness-training techniques that can be applied to students with severe postural problems such as scoliosis.

New York Posture Rating Test. (1966). In *New York State physical fitness test for boys and girls grades 4-12.* Albany, NY: New York State Education Department. An easy to use, comprehensive screening test for upper- and lower-back postural deviations that are shown in figure 22.1.

Posture Grid. (1991). In R.C. Adams & J.A. McCubbin, *Games, sports and exercises for the physically disabled* (4th ed.). Philadelphia: Lea & Febiger. This document describes how to make your own posture grid, which can then be used to evaluate postural deviations of the head, spine, legs, and feet.

Audiovisual

Growing Straighter and Stronger, videotape. (1983). The National Scoliosis Foundation. Phone: 781-341-6333; fax: 781-341-8333; e-mail: scoliosis@aol.com; or mail: 5 Cabot Place, Stoughton, MA 02072. The National Scoliosis Foundation sells and rents this outstanding videotape and teacher's guide that explains the purpose and value of scoliosis screening. Designed for students in grades five through seven. Running time approximately 15 minutes.

The Scoliosis Association, Inc. Phone: 800-800-0669 or 561-994-0669; fax: 561-994-2455; e-mail: normlipin@aol.com; Web site: **www.spine-surgery.com/Assoc/scoliosis.htm;** or mail: The Scoliosis Association, Inc., P.O. Box 811705, Boca Raton, FL 33481-1705. An excellent source for videotapes on the history, nature, and treatment of scoliosis.

Electronic

Kids Health. Web site: **http://kidshealth.org/index2.html**. Shows pictures and reviews a variety of braces commonly used to treat scoliosis in children.

Latest research on the treatment of scoliosis—the health resources directory site on scoliosis. Web site: **http://www.stayhealthy.com/**. Scoliosis Research Society. Web site: **www.srs.org/**. These sites contain extensive reference lists and reviews pertaining to research on the treatment of scoliosis.

Milwaukee Brace Information. Web site: **http://home2.swipnet.se/~w-27587/Milwaukee/Brace.html**. Provides pictures of various forms of the Milwaukee brace as well as a wealth of information on applications of the brace.

CHAPTER 23

References

Auxter, D. Pyfer, J., & Heuttig. (1997). *Principles and methods of adapted physical education and recreation* (8th ed.). Guilford, CT: Brown & Benchmark.

Ayres, J. (1972). *Sensory integration and learning disorders.* Los Angeles: Western Psychological Services.

Bannon, V. (1994). *Dance/movement therapy with emotionally disturbed adolescents*. Paper presented at the Safe Schools, Safe Students: A Collaborative Approach to Achieving Safe, Disciplined and Drug-Free Schools Conducive to Learning, Washington, DC.

Bornell, D. (1989). Movement discovery linking the impossible to the possible. In S. Grosse, C. Cooper, S. Gavron, & J. Stein (Eds.), *The best of practical pointers*. Reston, VA: AAHPERD, 120-138.

Boswell, B. (1993). Effects of movement sequences and creative dance on balance of children with mental retardation. *Perceptual & Motor Skills*, 77, 1290.

Boswell, B. (1989). Dance as creative expression for the disabled. *Palaestra*, 6(1), 28-30.

Caler, C., & Cronk, K. (1997). *Swing, rock, and roll dancing for the physically challenged*. Presented at the Eastern District Association of the American Alliance for Health, Physical Education, Recreation and Dance, Burlington, VT.

Harris, C. (1989). Dance for students with orthopedic conditions—Popular, square/folk, modern/ballet. In S. Grosse, C. Cooper, S. Gavron, & J. Stein (Eds.), *The best of practical pointers*. Reston, VA: AAHPERD, 155-172.

Hottendorf, D. (1989). Mainstreaming deaf and hearing children in dance classes. *Journal of Physical Education, Recreation, and Dance*, 60(9), 54-55.

Jay, D. (1991). Effect of a dance program on the creativity of preschool handicapped children. *Adapted Physical Activity Quarterly*, 8, 305-316.

Joyce, M. (1980). *First steps in teaching creative dance to children* (2nd ed.). Mountain View, CA: Mayfield.

Kindel, M. (1986). Wheelchair dancer. *A positive approach: A national magazine for the physically challenged*, Premier Issue, 41-43.

Kowalski, E., & Sherrill, C. (1992). Motor sequencing of learning disabled boys: Modeling and verbal rehearsal strategies. *Adapted Physical Activity Quarterly*, 9, 261-272.

Krebs, P. (1990). Rhythms and dance. In J. Winnick (Ed.). *Adapted physical education and sport*. Champaign, IL: Human Kinetics.

Krombholz, G. (1997). *Wheelchairdance*. Paper presented at the Achieving a Balance Conference. Western Illinois University.

Laban, R. (1963). *Modern education dance* (2nd ed.). Revised by L. Ullman. New York: Frederick A. Praeger.

Lazarus, J. (1990). Factors underlying inefficient movement in learning disabled children. In G. Reid (Ed.), *Problems in movement control*. New York: Elsevier.

Logsdon, B., Barrett, K., Broer, M., McGee, R., Ammons, M., Halverson, L., & Roberton, M. (1984). *Physical education for children: A focus on the teaching process* (2nd ed.). Philadelphia: Lea & Febiger.

Marsh, H., & Shavelson, R. (1985). Self-concept: Its multifaceted, hierarchical structure. *Educational Psychologist*, 20, 107-125.

National Association for Sport and Physical Education (1993). *NASPE standards for appropriate physical education practice*. Reston: VA: AAHPERD.

National Consortium for Physical Education and Recreation for Individuals with Disabilities (1995). *Adapted physical education national standards*. Champaign, IL: Human Kinetics.

New York State Education Department. (1995). *NYS learning standards*. Albany, NY: Author.

Pesetsky, S., & Burack, S. (1984). *Teaching dance for the handicapped: A curriculum guide*. Michigan Dance Association.

Preston, V. (1963). *A handbook for modern educational dance*. London: McDonald and Evans.

Riordan, A. (1989). Sunrise wheels. *Journal of Physical Education, Recreation and Dance*, 60(9), 62-64.

Roswal, P., Sherrill, C., & Roswal, G. (1988). A comparison of data-based and creative dance pedagogies in teaching mentally retarded youth. *Adapted Physical Activity Quarterly*, 5, 212-222.

Schmitz, N. (1989). Children with learning disabilities and the dance movement class. *Journal of Physical Education, Recreation and Dance*, 60(9), 59-61.

Schwartz, V. (1989). A dance for all people. *Journal of Physical Education, Recreation and Dance*, 60(9), 49.

Sherrill, C. (1998). *Adapted physical activity, recreation, and sport* (5th ed.). Boston: WCB/McGraw-Hill.

Sherrill, C., & Delaney, W. (1986). Dance therapy and adapted dance. In C. Sherrill (Ed.), *Adapted physical education and recreation* (3rd ed.). Dubuque, IA: Brown & Benchmark.

Silk, G. (1989). Creative movement for people who are developmentally disabled. *Journal of Physical Education, Recreation and Dance*, 60(9), 56-58.

Walden, P. (1999). What is Dance? *Spotlight on Dance*, 25(2), 8, 11-13.

Weikart, P. (1989). *Teaching movement and dance: A sequential approach*. Ypsilanti, MI: High/Scope Press.

Weiss, M., & Klint, K. (1987). "Show and tell" in the gymnasium: An investigation of developmental differences in modeling and verbal rehearsal of motor skills. *Research Quarterly for Exercise and Sports* 54, 190-197.

Wolf, G., & Launi, B. (1996). *The magic of learning through innovative dance therapy program*. Paper presented at the International Convention of the Council for Exceptional Children, Orlando, FL.

Woodard, R., Lewis, C., Koceja, D., & Surburg, P. (1996). The effects of an aerobic dance program on the postural control of adults with mental retardation. *Research Quarterly for Exercise and Sport* (supplement), A-124.

Resources

Written

Bennett, J., & Reimer, P. (1995). *Rhythmic activities and dance*. Champaign, IL: Human Kinetics. Although geared for teaching students without disabilities, this book provides rhythm activities and dances that can be easily modified.

Boswell, B. (1995). Enriching creative dance to facilitate balance skills of children with mental retardation. *Palaestra*, 11(2), 9.

Caler, C., & Cronk, K. (1997). *Swing, rock, and roll dancing for the physically challenged*. Provides great ideas and several dances already modified for students using crutches or a wheelchair. Contact Carol Caler and Karla Cronk in the Department of Health and Physical Education, Edinboro University, Edinboro, PA 16444.

Crain, C. (1981). *Movement and rhythmical activities for the mentally retarded*. Springfield, IL: Charles C Thomas. Provides basic knowledge and information as well as activities for teaching a movement/rhythmic activity program to individuals with mental retardation.

Fitt, S., & Riordan, A. (Eds.). (1980). *Dance for the handicapped—focus on dance IX.* Reston, VA: AAHPERD. A useful resource for those teaching older children and adolescents, this book shares different approaches, techniques, activities, and ideas for teaching dance to individuals with disabilities.

Gander, F. (1985). *Father gander's nursery rhymes.* Santa Barbara, CA: Advocacy Press. Excellent for use with young children, this book provides a wealth of rhymes from which teachers can easily develop action songs for children to perform basic movements.

Hill, K. (1976). *Dance for physically disabled persons: A manual for teaching ballroom, square and folk dances to users of wheelchairs and crutches.* Reston, VA: AAHPERD. Appropriate for those teaching adolescents and adults, this resource provides basic information helpful to teaching dance to individuals with physical challenges.

McGowan, H. (1999). Adapted aerobics: Sweat, reps and videotape. *New Mobility,* 22-24. This article provides an excellent review of several current exercise videos for persons with physical challenges.

Weikart, P. (1989, 1990). *Movement plus music; Movement to a steady beat.* High/Scope Press, 600 N. River St., Ypsilanti, MI 48198; phone: 313-485-2000. Appropriate for three- to seven-year-olds as well as older children, these resources provide useful activities for developing rhythmic competency.

Weikart, P. (1988). *Movement plus rhymes, songs, and singing games.* High/Scope Press, 600 N. River St., Ypsilanti, MI 48198; phone: 313-485-2000. This booklet provides lesson plans that combine key movement experiences with familiar folk activities. Most appropriate for three- to seven-year-olds, but may be adapted for older children.

Audiovisual

A World of Parachute Play, cassette/CD. (1999). Set to folk songs from different countries, children take imaginary trips around the world. Kimbo Educational Catalog, Department P, P.O. Box 477, Long Branch, NJ 07740-0477; 800-631-2187.

Chair Dancing and *Chair Dancing Around the World,* videotapes. (1999). Provides a fun way to improve muscle tone and develop aerobic fitness. Good for all ages and abilities. Each tape is 45 minutes. Kimbo Educational Catalog, Department P, P.O. Box 477, Long Branch, NJ 07740-0477.

Kidding Around With Greg and Steve, cassette/CD. (1986). Good for young and older children this recording helps stimulate positive interactions through movement and song. Kimbo Educational Catalog, Department P, P.O. Box 477, Long Branch, NJ 07740-0477; phone 800-631-2187.

In the Schoolyard by Sharon, Lois, & Bram, cassette/CD. (1980). Includes sing-a-longs and silly rhymes from which teachers can easily create action songs. Good for older children. Kimbo Educational Catalog, Department P, P.O. Box 477, Long Branch, NJ 07740-0477; phone 800-631-2187.

Movement and Dance: A Sequential Approach Instructors Video, Folk Dances I, II, III, and IV, videotape. (1988, 1989, 1991). Video by Phyllis Weikart is useful for older children, adolescents, and adults; it demonstrates 55 dances with the Say & Do process with the music. High/Scope Press, 600 N. River St., Ypsilanti, MI 48198; phone 313-485-2000.

Preschool Playtime Band by G. Stewart, cassette/CD. (1987). Excellent for young and older children, this recording includes ragtime to rock melodies with activities to encourage learning about rhythm. Kimbo Educational Catalog, Department P, P.O. Box 477, Long Branch, NJ 07740-0477; phone: 800-631-2187.

Preschoolers Moving & Learning and *Early Elementary Children Moving & Learning.* (1991). Champaign, IL: Human Kinetics. Great for children of all ages, *Moving & Learning* is a comprehensive program involving skills and activities arranged in a developmental progression, focusing on locomotor skills, body awareness, spatial concepts, and more. Complete with lesson plans and cassettes of music/songs designed for each lesson.

Rhythmically Moving by Phyllis Weikart, record series. (n.d.) Contains music designed to complement activity booklets. Appropriate for young and older children. High/Scope Press, 600 N. River St., Ypsilanti, MI 48198; phone 313-485-2000.

Rhythms on Parade by Hap Palmer, cassette/CD. (1989). Cassette tape useful for teaching children of all ages, this recording provides a variety of music to move rhythmically to. Kimbo Educational Catalog, Department P, P.O. Box 477, Long Branch, NJ 07740-0477; phone 800-631-2187.

Seat-A-Robics by Daria Allinovi, videotape. (n.d.) Provides an easy-to-follow aerobics and endurance workout segment. Includes tips for handling various wheelchair issues and specific instructions for quadriplegia. Contact Clarice Burns at SAR, P.O. Box 1253, Jamestown, NC 27282; phone 910-454-4610.

Seat Works, cassette/CD. (1999). Action songs of simple body movements designed to encourage participants to move their muscles. Kimbo Educational Catalog, Department P, P.O. Box 477, Long Branch, NJ 07740-0477; phone 800-631-2187.

Sing, Dance, 'N Sign by GAIA, cassette/CD. (1999). Introduces signs for learning concepts, numbers, and colors set to music. Kimbo Educational Catalog, Department P, P.O. Box 477, Long Branch, NJ 07740-0477; phone 800-631-2187.

Teaching Creative Dance to Children With Disabilities, videotape. (1987). Boswell, B., Physical Education Department, East Carolina University, Greenville, NC 27834; phone 919-757-4632. This introductory video is designed to assist educators in teaching movement elements of force, flow, and time to children with mental retardation.

Visions Exercise Video, Vol. III by Steele Fitness, videotape. (1999). Led by a certified fitness instructor, this is a challenging and fun video that provides warm-up, stretch, aerobic, strength, and cool-down segments. 30 minutes. Access to Recreation, 8 Sandra Court, Newbury Park, CA 91320-4302; phone 800-634-4351.

Electronic

Disabled Sports USA., Web site: **http://www.dsusa.org**. Provides information on the latest adapted fitness workshops and seminars.

VSAarts. Web site: **http://www.vsarts.org**. Provides updated information on the philosophical foundations of dance and disability as well as dance content and methodologies necessary to facilitate the evolution of inclusive dance.

CHAPTER 24

References

American Alliance for Health, Physical Education, Recreation and Dance—American Association for Active Lifestyles and Fitness. (1996). *Adapted aquatics: Position paper.* Reston, VA: Author.

American Red Cross. (1992). *Water safety instructor's manual.* St. Louis: Mosby Year Book.

American Red Cross. (1977). *Adapted aquatics.* Garden City, NY: Doubleday.

Anderson, L. (Ed.). (1992). *United States swimming handbook for adapted competitive swimming.* Colorado Springs: U.S. Swimming.

Block, M.E. (1994). *A teacher's guide to including students with disabilities in regular physical education.* Baltimore, MD: Paul H. Brookes.

Conatser, P. (1995). *Adapted aquatics swimming screening test.* Unpublished manuscript: Author. (Obtain from P. Conatser, Department of Physical Education, Slippery Rock University, Slippery Rock, PA 16057.)

Crawford, M. (1988). Adapted aquatics programming for persons with severe disabilities: An overview of current best practice standards. In P. Bishop (Ed.), *Adapted physical education: A comprehensive resource manual of definition, assessment, programming, and future predictions.* Kearney, NE: Educational Systems Associates.

Dunn, J., Morehouse, J., & Fredericks, B. (1986). *Physical education for the severely handicapped: A systematic approach to a data based gymnasium.* Austin, TX: Pro-Ed.

Dunn, K. (1981). PFDs for the handicapped: A question of responsibility. *The Physician and Sportsmedicine,* 9(8), 147-152.

Grosse, S.J., & Thompson, D. (Eds.). (1993). *Leisure opportunities for individuals with disabilities: Legal issues.* Reston, VA: AAHPERD.

Heckathorn, J. (1980). *Strokes and strokes.* Reston, VA: AAHPERD.

Hicks, L. (1988). Systematic desensitization of aquaphobic persons. *The National Aquatics Journal,* 4(1), 15-18.

Jankowski, L.W. (1995). *Teaching persons with disabilities scuba diving.* Montreal, Canada: Quebec Underwater Association.

Langendorfer, S.J., & Bruya, L. (1995). *Aquatic readiness.* Champaign, IL: Human Kinetics.

Lepore, M., Gayle, G.W., & Stevens, S.F. (1998). *Adapted aquatics programming: A professional guide.* Champaign, IL: Human Kinetics.

Lin, L.Y. (1987). Scuba divers with disabilities challenge medical protocols and ethics. *The Physician and Sportsmedicine* 15(6), 224-228, 233, 235.

Martin, K. (1983). Therapeutic pool activities for young children in a community facility. *Physical and Occupational Therapy in Pediatrics,* 3, 59-74.

Moran, J. (1961). Fear and aquatic instruction. In *DGWS Aquatics guide July 1961-1963.* Washington, DC: AAHPER.

National Center on Accessibility. (1996). *Swimming pool accessibility final report.* Martinsville, IN: Indiana University Department of Recreation and Park Administration.

Paciorek, M.J., & Jones, J.A. (1994). *Sports and recreation for the disabled: A resource manual.* Indianapolis, IN: Masters Press.

Petrofsky, J.S. (1995). Diving with spinal cord injury. Part I. *Palaestra,* 10(4), 36-41.

Priest, L. (1995). Aquatics. In J. Winnick (Ed.), *Adapted physical education and sport.* Champaign, IL: Human Kinetics.

Robinson, J., & Fox, A.D. (1987). *Diving with disabilities.* Champaign, IL: Human Kinetics.

Shurte, B. (1981). *Adapted aquatics bulletin #101.* Ann Arbor, MI: Danmar Products.

Skinner, A.T., & Thompson, A.M. (Eds.). (1983). *Duffield's exercises in water.* London: Bailliere Tindall.

Summerford, C.F. (1993). Apparatus used in teaching swimming to quadriplegic amputees. *Palaestra,* Spring, 54-57.

SwimAmerica. (n.d.) Organizational documents provided by SwimAmerica, 2101 N. Andrews Ave., Suite 107, Fort Lauderdale, FL 33311.

Tobin, R. (1990). The second decade: The progress of U.S. disabled rowing. *Row,* 1, 38.

Webel, G., & Goldberg, C. (1982). *Open boating: A handbook.* Oakland, CA: Office of Parks and Recreation.

Webre, A., & Zeller, J. (1990). *Canoeing and kayaking for persons with physical disabilities.* Newington, VA: American Canoe Association.

YMCA of the USA. (1999). *The youth and adult aquatics program manual.* Champaign, IL: Human Kinetics.

Resources

Written

Canadian Red Cross Society. (1989). *Adapted aquatics: Promoting aquatic opportunities for all.* Ottawa, Ontario: Author. This manual serves as a resource for the Canadian Red Cross adapted aquatic program.

Carter, M.J., Dolan, M.A., & LeConey, S.P. (1994). *Designing instructional swim programs for individuals with disabilities.* Reston, VA: AAHPERD. Easy-to-read resource that contains adaptations of aquatic skills, assessment and task-analysis forms, and general instructional tips for a variety of disabilities.

Jankowski, L.W. (1995). *Teaching persons with disabilities scuba diving.* Montreal: Quebec Underwater Association. A practical guide to introducing individuals with disabilities to the underwater world through scuba diving.

Lepore, M. (1999). *Aquatic activity achievement checklist.* Wilmington, DE: Author. This aquatic checklist is used in the West Chester University Adapted Aquatic Programs and at Camp Abilities at SUNY Brockport, NY.

Lepore, M., Gayle, G.W., & Stevens, S.F. (1998). *Adapted aquatics programming: A professional guide.* Champaign, IL: Human Kinetics. A comprehensive text including assessment forms, adaptations for a variety of unique needs, lift and transfer sequences, instructional techniques, and fitness information.

Osinski, A. (1993). Modifying public swimming pools to comply with provisions of the Americans with Disabilities Act. *Palaestra,* 9, 13-18. Very informative article that can help aquatic instructors and administrators meet the mandates of the Americans with Disabilities Act.

Paciorek, M.J., & Jones, J.A.. (1994). *Sports and recreation for the disabled: A resource manual.* IN: Masters Press. Important resource for all disabled sports and recreation activities. Includes sections on swimming, water skiing, and various boating activities.

Priest, E.L. (1996). *Adapted aquatics.* Kingwood, TX: Jeff Ellis & Associates. An addendum to the National Safety Council's Learn-To-Swim program materials, this manual provides information about developmentally appropriate practices and progressive skills that facilitate inclusion into aquatics.

YMCA of the USA. (1987). *Aquatics for special populations.* Champaign, IL: Human Kinetics. Although this is the manual used for the YMCA Leader in Aquatics for Special Populations, it is a concise resource for any pool operator or adapted aquatics specialist.

Audiovisual

Allen, J. (Director), and Priest, L. (1977). *Focus on Ability*, film. Available from select American Red Cross chapter offices.

Angelo, P. (1997). *Adapted Aquatics Teacher Training*, videotape. Available from the Department of Physical Education and Athletics, SUNY, Stony Brook, NY.

Bloomquist, L.E. (n.d.). *Adapted Aquatics I and II,* videotapes. Available from University of Rhode Island, Department of Physical Education and Exercise Science, Kingston, RI 02881.

Carter, M.J., Dolan, M.A., & LeConey, S.P. (1998). *Cincinnati Department of Parks and Recreation Adapted Aquatics*, videotape. Available from AAHPERD-AALR, Reston, VA. Designed to assist aquatic instructors providing appropriate sequences and progressions for aquatic skills. Accompanies the textbook by same authors and assessment checklists.

Courage Center of Minnesota. (1997). *Aquatics for Children With Disabilities*, videotape. Available from Courage Center of Minnesota, Golden Valley, MN. Video includes training on swimming for children with disabilities.

Macias, L., & Fagoaga, J. (n.d.) *Programming of Aquatic Activities for Children With Serious Motor Handicaps*, videotape. Available from TherEd Resource, P.O. Box 56153, Miami, FL 33256. Focus of this video is on therapeutic aquatic activities including treatment protocols, handling techniques, precautions, and specific methodology for a variety of diagnoses.

U.S. Swimming (n.d.). *Just Add Water*, videotape. Available from U.S. Swimming, 1 Olympic Plaza, Colorado Springs, CO 80909. The focus of this video is on adapted competitive swimming.

Electronic

American Canoe Association. Web site: **www.acanet.org/acanet.htm**. Site for information about canoe and kayak opportunities and the Disabled Paddlers Committee.

Aquajogger. Web site: **www.aquajogger.com**. Contains descriptions of aquatic exercise equipment and water fitness training information.

Aquatic Access. Web site: **www.aquatic-access.com**. Information related to aquatic lifts.

Aquatics International. Web site: **www.aquaticsintl.com**. Contains more than 100 links to a variety of aquatic programs.

Handicapped Scuba Association International. Web site: **http://ourworld.compuserve.com/homepages/hsahdq/**. Information on access to scuba programs for divers with disabilities.

National Center on Accessibility. Web site: **www.indiana.edu/~nca**. Contains information on all types of accessible recreation and on the swimming pool access project. In addition it contains an extensive bibliography on accessibility and pools.

Professional Association of Diving Instructors (PADI). Web site: **www.padi.com**. Information on general scuba diving instruction and certification.

Recreonics. Web site: **www.recreonics.com**. Dedicated to recreation facility structures and pool equipment including swim lifts.

Shake a Leg. Web site: **www.shakealeg.org**. Contains information about a world-renowned adapted sailing program.

Special Olympics International. Web site: **www.specialolympics.org**. Information about aquatics and other sports for individuals with mental retardation.

USA Swimming. Web site: **www.usswim.org**. Information on adapted competitive swimming highlighting an adapted swim discussion forum.

USRowing Association. Web site: **www.usrowing.org**. Information about rowing programs including the Adaptive Rowing committee is showcased.

Other Resources

Martin Marine Company, Kittery Point, ME 03905; phone 207-439-1507. Has developed adaptations to traditional rowing equipment for those individuals with physical disabilities.

National Association of Underwater Instructors (NAUI). 4650 Arrow Highway, Suite F-1, Montclair, CA 91763. This organization provides certification standards and instructor courses in scuba diving.

Special Sports Equipment, 11020 Solway School Road, Suite 103, Knoxville, TN 37931; phone 615-481-3557. Makes adapted rowing equipment for rowers with physical disabilities.

Wessel, J. (1976). *I CAN: Aquatic skills.* Northbrook, IL: Hubbard. (Now available from Pro-Ed, Austin, TX.) This resource is one of many tools within a larger curriculum called the I CAN instructional skills system. The aquatic information is self-contained in a box with task analysis of traditional swim skills and games and activities to reinforce these skills.

CHAPTER 25

References

National Wheelchair Basketball Association. (1998). *Official rules and case book 1998-1999.* Charlotte, NC: Author.

National Wheelchair Softball Association. (1996). *Wheelchair softball rules.* Hastings, MN: Author.

Play ball. (1997). *Sports 'N Spokes*, 23(2), 36-40.

Santa Barbara Parks and Recreation Department. (1993). *Rules for the blister bowl wheelchair football tournament.* Santa Barbara, CA: Author.

Special Olympics. (1999). *Official Special Olympics winter sports rules 1999-2001* (Rev. ed.). Washington, DC: Special Olympics International, Inc.

Special Olympics. (1996). *Official Special Olympics Summer Sports Rules 1996-1999* (Rev. ed.). Washington, DC: Special Olympics International, Inc.

United States Cerebral Palsy Athletic Association. (1997). *Classification and sports rules manual* (5th ed.). Newport, RI: Author.

World Organization Volleyball for Disabled (1997). *Official standing and sitting volleyball rules.* Haarlem, The Netherlands: De Vrieseborch.

Yilla, A.B., & Sherrill, C. (1998). Validating the beck battery of quad rugby skill tests. *Adapted Physical Activity Quarterly, 15,* 55-67.

Yilla, A.B., & Sherrill, C. (1994). Quad rugby illustrated. *Palaestra, 10*(4), 25-31.

Resources

Written

Bernotas, B. (1995). *Nothing to provide: The Jim Abbott story.* New York: Kodansha America, Inc. The biography of Jim Abbott who was born without a right hand. He overcame his disability and became a Little League and high school standout, a college All-American, an Olympian, and a major league baseball star.

Hedrick, B., Brynes, D., & Shaver, L. (1994). *Wheelchair basketball* (2nd ed.). Washington, DC: Paralyzed Veterans of America. Manual includes coaching philosophy, physiological and psychological training, practice planning and organization, and defensive and offensive play.

Special Olympics. (1985). *Sports skills program* (Rev. ed.). Washington, DC: Author. Separate manuals published for basketball, hockey, soccer, softball, and volleyball. Manuals include goals and objectives, skill assessment, task analyses, team tactics, modified games, and rules.

Audiovisual

Goal Ball: An Introductional Videotape, videotape. (n.d.) Winnipeg, Manitoba, Canada: Communication Systems Distribution Group, University of Manitoba. This tape examines the sport of goal ball. It can be purchased or rented.

Unified Sports/Basketball, videotape. (n.d.) Special Olympics, 1350 New York Ave., N.W., Suite 500, Washington, DC 20005. This five-minute tape features the Indianapolis Unified Basketball League, which is a pilot program. Teams are sponsored by local businesses and players, and coaches are interviewed.

Wheelchair Basketball Vols I & II, videotapes. (n.d.) Paralyzed Veterans of America, 801 18th St., NW, Washington, DC 20006. These 2 tapes run about 120 minutes each. Volume I contains a description of individual skills and drills. Volume II addresses team play and strategies, as well as information relative to nutrition and practice organization.

Electronic

International Blind Sports Association. Web site: **http://www.ibsa.es**. Contains information regarding the organization, sports rules and documents, IBSA news, and other information regarding sports for people with blindness.

International Wheelchair Basketball Federation. Web site: **http://www.iwbf.org/**. Contains information on classification, competition schedules, news, developments, and educational resources. The mailing address is IWBF, 1 Meadow Close, Shavington, Crewe, Cheshire, CW2 5BE England.

National Wheelchair Basketball Association. Web site: **http://www.nwba.org**. Provides information on current executive committee members, wheelchair basketball history, official rules, resources, and conference directory.

United States Quad Rugby Association. Web site: **http://www.quadrugby.com**. Provides information relative to the rules, classification, competition schedules, tournament results, news, helpful links, photo page, and a brief history of the game.

World Organization Volleyball for Disabled. Web site: **http://www.wovd.com**. Contains information regarding competition schedules, rules for both standing and sitting competition, as well as other general information about the game.

CHAPTER 26

References

Adams, R.C., Daniel, A.N., McCubbin, J.A., & Rullman, L. (1982). *Games, sports, and exercise for the physically handicapped* (3rd ed.). Philadelphia: Lea & Febiger.

Adams, R.C., & McCubbin, J.A. (1991). *Games, sports, and exercise for the physically handicapped* (4th ed.). Philadelphia: Lea & Febiger.

Cowart, J. (1989). Golf chirper for the blind. *Palaestra, 5*(3), 34-35.

Dunn, J.M., & Fait, H. (1997). *Special physical education: Adapted, individualized, developmental* (7th ed.). Dubuque, IA: Brown and Benchmark.

Dwarf Athletic Association of America. (1998). *Athletic handbook.* Lewisville, TX: Author.

Hattenback, R.T. (1979). Integrating persons with handicapping conditions in archery activities. In J.P. Winnick & J. Hurwitz (Eds.), *The preparation of regular physical educators for mainstreaming.* Brockport: State University of New York, College at Brockport. (ERIC Document Reproduction Service No. ED 222 028).

Havens, M.D. (1992). *Bridges to accessibility: A primer for including persons with disabilities in adventure curricula.* Hamilton, MA: Project Adventure.

Huber, J.H. (1991). An historic accomplishment: The first blind person to hike the Appalachian Trail. *Palaestra, 7*(4), 18-23.

International Blind Sports Federation. (1997). *IBSA technical rulebook.* Madrid, Spain: Author.

International Paralympic Committee. (1992-1993). *IPC handbook.* Loughborough, England: Author.

Kegel, B., & Peterson, J. (1989). Summer splash: A water sports symposium for the physically challenged. *Palaestra, 6*(1), 17-19.

Longo, P. (1989). Chair golf. *Sports'N Spokes, 15*(2), 35-38.

McCowan, L.L. (1972). *It is ability that counts: A training manual on therapeutic riding for the handicapped.* Olivet, MI: Olivet College Press.

Orr, R.E., & Sheffield, J. (1981). Adapted épée fencing. *Journal of Physical Education, Recreation and Dance, 52*(6), 42, 71.

Parks, B.A. (1997). *Tennis in a wheelchair.* White Plains, NY: United States Tennis Association.

Roos, M. (1991). Pass the adrenaline, please. *Palaestra, 8*(1), 44-46.

Roswal, G.M., & Daugherty, N. (1991). Whitewater rafting: An outdoor adventure activity for individuals with mental retardation. *Palaestra, 7*(4), 24-25.

Special Olympics. (1998). *Official Special Olympics summer sports rules, 1996-1999* (Rev. ed.). Washington, DC: Author.

Special Olympics. (n.d.) Gymnastics. Washington, DC: Author.

Spink, J. (1993). *Developmental riding therapy: A team approach to assessment and treatment*. Tucson, AZ: Therapy Skill Builders.

United States Cerebral Palsy Athletic Association. (1997). *Sports rules manual* (5th ed.). Newport, RI: Author.

Wachtel, L.J. (1987). Thoughts on a wilderness canoe trip. *Palaestra*, 3(4), 33-40.

Weber, R.C. (1991). Using Velcro to assist badminton players who are disabled or elderly. *Palaestra*, 7(3), 10-11.

Wheelchair Sports, USA. (1996). *Official rulebook of Wheelchair Sports, USA*. Colorado Springs: Author.

Winnick, J.P., & Short, F.X. (1985). *Physical fitness testing of the disabled*. Champaign, IL: Human Kinetics.

Wiseman, D.C. (1982). *A practical approach to adapted physical education*. Reading, MA: Addison-Wesley.

Resources

Written

Adams, R.C., & McCubbin, J.A. (1991). *Games, sports, and exercise for the physically handicapped* (4th ed.). Philadelphia: Lea & Febiger. Excellent resource providing suggestions for traditional and nontraditional games and sports suitable for participation by individuals with disabilities. Includes suggestions for modifying existing sport or game structures, rules, and procedures.

Grosse, S. (Ed.). (1991). *Sport instruction for individuals with disabilities: The best of practical pointers*. Reston, VA: AAHPERD. Book contains previously published *Pointers* and new articles on increasing opportunities for those with disabilities to participate in instructional sport programs. Adaptations are presented for students with crutches and with unilateral and bilateral upper-arm amputations in badminton, golf, archery, bowling, tennis, and table tennis.

Paciorek, M.J., & Jones, J.A. (1994). *Sports and recreation for the disabled*. Carmel, IN: Cooper. Perhaps the most comprehensive resource currently available, this manual uses a cross-categorical approach in discussing sports and recreation for people with disabilities. Information is provided on sport governing bodies for both able-bodied and disabled athletes. An overview of each sport, along with adapted equipment suppliers and manufacturers, accompanies each description.

Parks, B.A. (1997). *Tennis in a wheelchair*. White Plains, NY: United States Tennis Association. The best nuts and bolts manual on playing tennis in a wheelchair provides information not only on the history and rules/regulations of the game but also on drills, specialty shots, and wheelchair mobility as well as singles and doubles strategy.

Special Olympics. (1998). *Official Special Olympics summer sports rules, 1996-1999* (Rev. ed.). Washington, DC: Author. This manual contains information that organizers will need to conduct equitable competitions, including how to conduct official sporting events, how to place athletes in appropriate ability groups, and how coaches should prepare athletes for competition.

Audiovisual

Beyond the Barriers, videotape. (1998). Aquarius Health Care Videos, 5 Powerhouse Lane, P.O. Box 1159, Sherborn, MA 01770. This video features Mark Wellman, the first paraplegic to climb El Capitan in Yosemite National Park, and other individuals with disabilities engaging in such activities as sailing, body boarding, scuba diving, and hang gliding. It not only demonstrates the breadth of activities in which individuals with disabilities can engage but also provides an inspirational message for individuals of all ability levels.

How to Race, videotape. (n.d.). K. Carnes, 291 Comfort Drive, Henderson, NV 89014. This 30-minute video details chair setup and maintenance. Provides information on pushing techniques, sitting positions, use of compensators, transferring to and from chairs, and preliminary training techniques.

Making of a Champion: Track and Field, videotape. (n.d.) Special Olympics International, 1325 G Street, N.W., Suite 500, Washington, DC 20005. This video can be used as a training aid for athletes. Provides instruction on various techniques used to prepare athletes for competition.

Summer Sports Officials/Coaches Training Video, videotape. (n.d.) United States Association of Blind Athletes, 33 N. Institute Street, Colorado Springs, CO 80903. This video provides information about fundamentals and basic rules and their modifications from companion sports for able-bodied athletes. It addresses special equipment and adaptations. Sports include track and field, gymnastics, powerlifting, and wrestling.

Electronic

Ability Plus Inc. Web site: **http://www.abilityplus.org**. Home of a nonprofit organization providing year-round sports and recreation opportunities for people with disabilities in the northeastern United States and consultation services throughout the United States and abroad.

EQUEST. Web site: **www.equest.org**. Home of the nationally recognized and accredited equine-assisted therapy and rehabilitation program for children and adults with mental, emotional, and learning disabilities, located in Dallas, Texas.

National Sports Center for the Disabled. Web site: **www.nscd.org**. Home of the innovative, nonprofit organization in Winter Park, Colorado, that serves the year-round recreational needs of children and adults with disabilities.

Success Oriented Achievement Realized. Web site: **www.soarnc.org**. Home of SOAR, which sponsors success-oriented, high-adventure programs, such as wilderness backpacking, rock climbing, white-water rafting and canoeing, mountaineering, and wildlife studies for LD and ADD children and adults.

CHAPTER 27

References

Adams, R.C., & McCubbin, J.A. (1991). *Games, sports, and exercises for the physically disabled* (4th ed.). Philadelphia: Lea & Febiger.

Disabled Sports, USA. (1997). *Alpine competition rules for athletes with mobility related disabilities*. Rockville, MD: Author.

Leonard, E., & Pitzer, N.L. (1988). Special problems of handicapped skiers: An overview. *Physician and Sportsmedicine*, 16(3), 77-82.

Special Olympics. (1999). *Official Special Olympics winter sports rules 1999-2001* (Rev. ed.). Washington, DC: Special Olympics, Inc.

United States Association for Blind Athletes. (1999). *Alpine and nordic skiing*. Colorado Springs, CO: Author. Retrieved July 21, 1999 from the World Wide Web: **http://www.usaba.org/ Pages/description_pgs/alpinedesc.html**

Resources

Written

O'Leary, H. (1994). *Bold tracks: Teaching adaptive skiing*. Bolder, CO: Johnson Books. This book is a must for anyone who is going to be teaching skiing to individuals with disabilities.

Special Olympics. (1997). *The Special Olympics alpine skiing sports skills program guide*. Washington, DC: Special Olympics, Inc. This manual provides a "how-to" approach for teaching the basic skills involved in alpine skiing, including teaching suggestions, sample drills, and activities.

Special Olympics. (1995). *The Special Olympics floor hockey sports skills program guide*. Washington, DC: Special Olympics, Inc. This manual provides a "how-to" approach for teaching the basic skills involved in floor hockey. Teaching suggestions, as well as sample drills and activities, are provided for each skill.

Audiovisual

Adaptive Ski Teaching Methods, videotape. (n.d.) Available from DS/USA National Headquarters, 451 Hungerford Drive, Suite 100, Rockville, MD 20850; 301-217-0960. Instructional video providing basic techniques for working with three-track, four-track, and sit-skiers.

Advanced Three-Track and Four-Track Methods, videotape. (n.d.) Available from DS/USA National Headquarters, 451 Hungerford Drive, Suite 100, Rockville, MD 20850; 301-217-0960. This instructional video demonstrates how to teach advanced three- and four-track skiing techniques.

No Simple Road, videotape. (n.d.) Available from The Children's Hospital of Denver, Sports Program Office B385, 1056 East 19th, Denver, CO 90218; 303-861-6590. This is an inspirational videotape showing children with various disabilities skiing.

Electronic

Abledata Database of Assistive Technology for Winter Sports. Web site: **http://www.abledata.com/text2winter.htm**. Provides information and links to the latest assistive devices that can be used to assist individuals with disabilities to participate in winter sports.

British Sledge Hockey Association. Web site: **http://info.lut.ac. uk/research/paad/wheelpower/sledge.htm**. Provides information on sledge hockey rules, classifications and competitions.

Sit-skiing. **Web site: http://www.sitski.com/**. Provides information on the latest advances in sit-skiing equipment.

Special Olympics. **Web site: http://www.specialolympics.org**. This site provides links to all sport competitions and their corresponding rules and policies that are sponsored by Special Olympics, including alpine and nordic ski competitions.

United States Association for Blind Athletes (USABA). Web site: **http://www.usaba.org**. Contains the latest information on skiing and the annual national ski championships for skiers with visual impairments.

World Curling Foundation. Web site: **http://www.worldsport. com/ws/allsports/home/0,2217,0_0_0_1_19,00.html** Provides information on curling history, rules and strategies.

CHAPTER 28
References

Brasile, F. (1986). Do you measure up? *Sports 'N Spokes, 12*(4), 42-47.

Brasile, F., & Hedrick, B.N. (1996). The relationship of skills of elite wheelchair basketball competitors to the International Functional Classification System. *Therapeutic Recreation Journal, 30*(2), 114-127.

Burnham, R., Wheeler, G., Bhambhani, Y., Cumming, D., Maclean, I., Sloley, B.D., Belanger, M., Eriksson, P., & Steadward, R. (1993). *Performance enhancement in elite quadriplegic wheelchair racers through self-induced autonomic dysreflexia*. Vista '93 Conference, May 14-20, Jasper, AB.

Coutts, K.D., & Schutz, R.W. (1988). Analysis of wheelchair track performance. *Medicine & Science in Sport and Exercise, 20*, 188-194.

Curtis, K.A. (1981a). Wheelchair sportsmedicine: Part 3—Stretching routines. *Sports 'N Spokes, 7*(3), 16-18.

Curtis, K.A. (1981b). Wheelchair sportsmedicine: Part 2—Training. *Sports 'N Spokes, 7*(2), 16-19.

Gallahue, D.L., & Ozmun, J.C. (1998). *Understanding motor development* (4th ed.). Dubuque, IA: McGraw-Hill.

Goosey, V.L., & Campbell, I.G. (1998). Pushing economy and propulsion techniques of wheelchair racers at three speeds. *Adapted Physical Activity Quarterly, 15*(1), 36-50.

Hedrick, B., Byrnes, D., & Shaver, L. (1989). *Wheelchair basketball*. Washington, DC: Paralyzed Veterans of America.

Hedrick, B., Wang, Y.T., Moeinzadeh, M., & Adrian, M. (1990). Aerodynamic positioning and performance in wheelchair racing. *Adapted Physical Activity Quarterly, 7*(l), 41-51.

Higgs, C. (1993). *The rolling resistance of racing wheelchairs: The effect of rolling surface, rear-wheel camber and tire pressure*. Final report, Applied Sport Science Program, Fitness and Amateur Sport, Government of Canada.

Higgs, C. (1992a). Racing wheelchairs: A comparison of three- and four-wheeled designs. *Palaestra, 80*, 28-36.

Higgs, C. (1992b). Wheeling the wind: The effect of wind velocity and direction on the aerodynamic drag of wheelchairs. *Adapted Physical Activity Quarterly, 9*(l), 74-87.

Higgs, C. (1985). Propulsion of racing wheelchairs. In M. Ellis and D. Tripps (Eds.), *Proceedings of the 1984 Olympic Scientific Congress, Eugene, Oregon*. Champaign, IL: Human Kinetics.

Higgs, C. (1983). An analysis of racing wheelchairs used at the 1980 Olympic Games for the Disabled. *Research Quarterly for Exercise and Sport, 54*(3), 229-233.

LaMere, T.J., & Labanowich, S. (1984a). The history of sport wheelchairs—Part I: The development of the basketball wheelchair. *Sports 'N Spokes, 9*(6), 6-8; 10-11.

LaMere, T.J., & Labanowich, S. (1984b). The history of sport wheelchairs—Part II: The racing wheelchair 1956-1975. *Sports 'N Spokes, 10*(1), 12-15.

LaMere, T.J., & Labanowich, S. (1984c). The history of sport wheelchairs—Part III: The racing wheelchair 1976-1983. *Sports 'N Spokes, 10*(2), 12-16.

Meijs, P.J.M., Van Oers, C.A.J.M., Van de Woude, L.H.V., & Veeger, H.E.J. (1989). The effect of seat height on the physiological response and propulsion technique in wheelchair ambulation. *Journal of Rehabilitation Science, 2,* 104-107.

Moore, B., & Snow, R. (1994). *Wheelchair tennis: Myth to reality.* Dubuque, IA: Kendall Hunt.

O'Connor, T.J., Robertson, R.N., & Cooper, R.A. (1998). Three-dimensional kinematic analysis and physiological assessment of racing wheelchair propulsion. *Adapted Physical Activity Quarterly, 15*(1), 1-14.

O'Reagan, J.R., Thacker, J.G., KauzIarich, J.J., Mochel, E., Carmine, D., & Bryant, M. (1981). Wheelchair dynamics. In W.G. Stamp & C.A. McLaurin (Eds.), *Wheelchair Mobility 1976-1981.* Charlottesville, VA: Rehabilitation Engineering Center, University of Virginia.

Pugh, L.G.C.E. (1971). The influence of wind resistance in running and walking and the mechanical efficiency of work against horizontal and vertical forces. *Journal of Physiology* (London), 213, 795-808.

Shephard, R.J. (1990). *Fitness in special populations.* Champaign, IL: Human Kinetics.

Traut, L. (1989). Gestaltung ergonoisch relevanter Konstruktionsparameter am Antriebssystem des Greifreifenrollstuhls-Teil 1. *Orthopedaedie Technik, 7,* 394-398.

Van de Woude, L.H.V., Veeger, H.E.J., & Rozendal, R.H. (1990). Seat height in hand rim wheelchair propulsion: A follow up study. *Journal of Rehabilitation Science, 3,* 79-83.

Weisman, M., & Godfrey, J. (1976). *So get on with it: A celebration of wheelchair sports.* Toronto, Canada: Doubleday.

Wells, C.L., & Hooker, S.P. (1990). The spinal injured athlete. *Adapted Physical Activity Quarterly, 7,* 265-285.

Yilla, A.B. (1997). Express yourself. *Sports 'N Spokes* 23, 59-63.

Yilla, A.B., La Bar, R.H., & Dangelmaier, B.S. (1998). Setting up a wheelchair for basketball. *Sports 'N Spokes* 24(2), 63-65.

Yilla, A.B., & Sherrill, C. (1998). Validating the beck battery of quad rugby skill tests. *Adapted Physical Activity Quarterly, 15*(2), 155-167.

Resources

Written

Clark, R. (1986). *Wheelchair sports: Techniques and training in athletics.* Cambridge, England: Woodhead-Faulkner. This book, translated into English by Kristina Ehrenstrale, is a very basic introduction to wheelchair sports. It is well suited for use by young athletes, those just starting in the sport, and those seeking a nontechnical, easily read introduction to the topic.

Cooper, R.A. (1998). *Wheelchair selection and configuration.* New York: Demos. This book provides an in-depth, though somewhat technical, insight into all aspects of wheelchair design. While it is brief in the area of sport-specific wheelchair design, it belongs on the shelf of any professional with an interest in physical activity for wheelchair users.

LaMere, T.J., & Labanowich, S. (1984). The history of sport wheelchairs—Part I: The development of the basketball wheelchair. *Sports 'N Spokes, 9*(6), 6-8; 10-11.

LaMere, T.J., & Labanowich, S. (1984). The history of sport wheelchairs—Part II: The racing wheelchair 1956-1975. *Sports 'N Spokes, 10*(1), 12-15.

LaMere, T.J., & Labanowich, S. (1984). The history of sport wheelchairs—Part III: The racing wheelchair 1976-1983. *Sports 'N Spokes, 10*(2), 12-16. These articles contain a brief history of the development of wheelchairs until 1984. Most of the landmark developments are covered, and the reader is left with a vivid impression of the relentless drive of the athletes to improve their equipment.

Audiovisual

You Feel the Need for Speed, videotape. (1992). K.C. Racing, 291 Comfort Drive, Henderson, NV 89014. This instructional video, available in both North American and European VHS formats, covers all the basics of wheelchair racing. Adjustment and operation of a new chair, propulsion techniques, maintenance, and many tricks of the trade are covered. The narrator and producer of the video is Kenny Carnes, one of the top racers in the United States.

Wheelchair Basketball, text and videotape. (1989). Paralyzed Veterans of America, 801 Eighteenth Street NW, Washington, DC 20006. The authors (Hedrick, Byrnes, and Shaver) have all achieved considerable success as coaches in international wheelchair basketball competition. Their level of expertise is reflected in this comprehensive instructional text on wheelchair basketball and the accompanying series of videotapes.

Author Index

The italicized *f* and *t* following page numbers refer to figures and tables respectively.

Subject Index

The italicized *f* and *t* following page numbers refer to figures and tables respectively.

About the Authors

Diane H. Craft received a doctorate in adapted physical education from New York University and is a professor of physical education at the State University of New York, College at Cortland. She teaches undergraduate and graduate courses in adapted physical education. Her writing and consulting focuses on preparing teachers and early childhood educators to instruct all children, including those with disabilities, in inclusive physical education classes.

David L. Gallahue is a professor of kinesiology at Indiana University in Bloomington. He holds degrees from Indiana University (BS), Purdue University (MS), and Temple University (EdD). Dr. Gallahue is active in the study of motor development, sport, and fitness education of children. He is the author of several textbooks, numerous journal articles, and edited book chapters. Dr. Gallahue is a past president of the National Association for Sport and Physical Education (NASPE) and former chair of the Motor Development Academy and the Council of Physical Education for Children (COPEC). He is a recognized leader in children's motor development and developmental physical activity. His research focuses on environmental influences on the motor performance of young children across cultures.

Cathy Houston-Wilson received her doctorate from Oregon State University. She is an associate professor at the State University of New York, College at Brockport where she serves as the Coordinator of Teacher Certification. She also teaches undergraduate and graduate classes in adapted physical education with a concentration on early childhood education. Her research and writing focus on peer tutors and effective teaching methods in both general and adapted physical education.

Luke E. Kelly is a professor of kinesiology at the University of Virginia where he directs the masters and doctoral programs in adapted physical education. He received his doctorate from Texas Woman's University and his bachelors and masters degrees from the State University of New York, College at Brockport. Dr. Kelly works extensively with public schools on developing functional physical education curricula based on the Achievement-Based Curriculum model so that they accommodate the needs of all students. Dr. Kelly is a fellow in the American Academy of Kinesiology and Physical Education, a past president of the National Consortium for Physical Education and Recreation for Individuals with Disabilities, and director of the Adapted Physical Education National Standards Project.

Ellen M. Kowalski earned her bachelor's degree from SUNY, College at Geneseo, her certification in physical education from SUNY, College at Brockport, her master's from the University of Connecticut, and her doctorate in adapted physical education from Texas Woman's University. She is an associate professor at Adelphi University in New York where she prepares teachers in the undergraduate and graduate programs. Her research focuses on the infusion of disabilities knowledge throughout the entire teacher preparation curriculum to improve student competency to teach in an inclusive environment.

Patricia L. Krebs is the President and CEO of Special Olympics Maryland (SOMD), which serves more than 9,000 athletes. Prior to this appointment, she served as Director of Education for Special Olympics Inc. She earned her doctorate from the University of Maryland in 1979 and previously directed the undergraduate and graduate adapted physical education preparation programs at Adelphi University. In 1997 she was named one of Maryland's Top 100 Women and in 1999 led SOMD to receive the Maryland Association of Nonprofit Organizations Standards of Excellence Award, one of only seven nonprofit organizations in the state to receive this distinction.

Monica Lepore is the lead author of *Adapted Aquatics Programming: A Professional Guide* (1998), the director of two community adapted physical education/aquatics programs, and is the newly appointed chair of the Adapted Aquatics Specialty Committee of the AAHPERD-AAALF Aquatic Council. She teaches adapted physical education professional preparation classes at West Chester University of Pennsylvania and is a member of the Pennsylvania State AHPERD Adapted Activities Committee. Dr. Lepore is the assistant director and aquatic director of Camp Abilities at SUNY, College at Brockport. She lives in Wilmington, Delaware with her daughter Maria.

Lauren J. Lieberman received her PhD from Oregon State University in the Movement Studies Disabilities Program. She is an associate professor of adapted physical education at SUNY, College at Brockport and is the coordinator of the undergraduate concentration. Dr. Lieberman's specialization area is with individuals with sensory impairments, and she runs a developmental sports camp for children who are blind and deafblind at SUNY, College at Brockport each summer.

E. Michael Loovis, the 1999 AAALF President's Award recipient, earned his doctorate at The Ohio State University. He is currently a professor at Cleveland State University. He coordinates activities for the Northeast Regional Professional Development Center, which consists of a consortium of 31 school districts. His research interests include the development of "grounded" theory as it relates to instructional delivery in adapted physical education, including the use of ethnography.

Michael J. Paciorek received his PhD from Peabody College of Vanderbilt University. He is a professor and physical education coordinator at Eastern Michigan University where he teaches classes in adapted physical education and sport and motor development. Dr. Paciorek is the coauthor of the popular text *Sports and Recreation for the Disabled.* He has served on the board of directors for Special Olympics Michigan for 12 years and has many experiences in disability sport. He was a staff member on the United States Disabled Sports Team that competed at the 1992 Paralympic Games in Barcelona, Spain.

David L. Porretta received his bachelor's degree from Niagara University, his master's degree from Ithaca College, and his doctoral degree from Temple University. He is a professor at The Ohio State University and is responsible for graduate study in adapted physical activity. Dr. Porretta is involved in research pertaining to the motor performance of individuals with disabilities and is the recipient of the Hollis Fait Scholarly Contribution Award.

Sarah M. Rich earned degrees at the University of Connecticut and Texas Woman's University under the tutelage of Dr. Hollis Fait and Dr. Claudine Sherrill. In 1997, she retired from Ithaca College (New York) where she taught adapted physical education and therapeutic recreation for 25 years. Recently, Dr. Rich was inducted into the Ithaca College Sports Hall of Fame for her contributions to the Ithaca Women's Volleyball team and sports programs for athletes with disabilities.

Francis X. Short is an associate professor of physical education and sport and department chairman at the State University of New York, College at Brockport. He teaches courses in adapted physical education and motor development. He has authored and co-authored a number of publications related to the physical fitness of individuals with disabilities, including *The Brockport Physical Fitness Test.* He received the 2000 Professional Recognition Award from the Adapted Physical Activity Council of AAHPERD. Dr. Short holds degrees from Springfield College and Indiana University.

Paul R. Surburg earned a BS at Concordia Teachers College and a PhD from the University of Iowa. He is a professor at Indiana University and director of the adapted physical education program. He is a licensed physical therapist and has published research and pedagogical articles in sports medicine. His other research area is the motor functioning of individuals with mental retardation, with a focus on factors influencing information processing.

Joseph P. Winnick, EdD, is a distinguished service professor of physical education and sport at State University of New York, College at Brockport, where he has taught adapted physical education for more than 30 years. Renowned for his research in adapted physical education, he is the author of *The Brockport Physical Fitness Test Manual* and related resources, which present the best physical fitness test available for youths with disabilities. Dr. Winnick developed and implemented America's first master's degree professional preparation program in adapted physical education at Brockport in 1968 and since that time has secured funds from the U.S. Department of Education to support the program. He has received the Professional Recognition Award from the Adapted Physical Activity Council of AAHPERD, the G. Lawrence Rarick Research Award, and the Hollis Fait Scholarly Contribution Award. He earned master's and doctoral degrees from Temple University.

Abu B. Yilla received his master's and doctorate degrees from the Texas Woman's University and two bachelor's degrees, one from the University of Nottingham, England and one from the University of Texas at Arlington. He is an assistant professor at the University of Texas at Arlington. His research focus is the explication of elite disability sport, in particular, wheelchair basketball. He is a Paralympic medallist and has won 12 national championships in wheelchair basketball (8 in Great Britain, and 4 in the United States).

Lauriece L. Zittel received her doctorate from Oregon State University in Movement Studies in Disability. She is an associate professor at Northern Illinois University where she directs graduate programs in adapted physical activity. Her research focuses on environmental variables influencing the motor performance of young children with disabilities and those at risk for developmental delay.